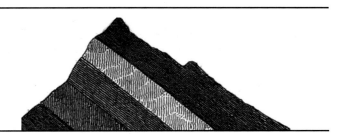

IGNEOUS PETROLOGY

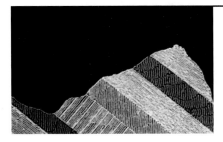

IGNEOUS PETROLOGY

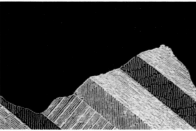

Myron G. Best
Eric H. Christiansen

Department of Geology
Brigham Young University

Blackwell
Science

Editorial Offices:

Commerce Place, 350 Main Street, Malden, Massachusetts 02148, USA
Osney Mead, Oxford OX2 0EL, England
25 John Street, London WC1N 2BL, England
23 Ainslie Place, Edinburgh EH3 6AJ, Scotland
54 University Street, Carlton, Victoria 3053, Australia
Other Editorial Offices:
Blackwell Wissenschafts-Verlag GmbH, Kurfürstendamm 57, 10707 Berlin, Germany
Blackwell Science KK, MG Kodenmacho Building, 7-10 Kodenmacho Nihombashi, Chuo-ku, Tokyo 104, Japan

Distributors:

USA
    Blackwell Science, Inc.
    Commerce Place
    350 Main Street
    Malden, Massachusetts 02148
    (Telephone orders: 800-215-1000 or 781-388-8250; fax orders: 781-388-8270)

Canada
    Login Brothers Book Company
    324 Saulteaux Crescent
    Winnipeg, Manitoba, R3J 3T2
    (Telephone orders: 204-837-2987)

Australia
    Blackwell Science Pty, Ltd.
    54 University Street
    Carlton, Victoria 3053
    (Telephone orders: 03-9347-0300; fax orders: 03-9349-3016)

Outside North America and Australia
    Blackwell Science, Ltd.
    c/o Marston Book Services, Ltd.
    P.O. Box 269
    Abingdon, Oxon OX14 4YN, England
    (Telephone orders: 44-01235-465500; fax orders: 44-01235-465555)

Acquisitions: Nancy Anastasi Duffy   Cover and interior design by Leslie Haimes
Development: Jill Connor   Typeset by The PRD Group, Inc.
Production: Irene Herlihy   Printed and bound by Sheridan Books/Ann Arbor
Manufacturing: Lisa Flanagan
Marketing Manager: Carla Daves

On the cover: Sections through the upper mantle of the Earth represented in magnified views of peridotite in cross-polarized light.

Printed in the United States of America
00 01 02 03 5 4 3 2 1

Library of Congress Cataloging-in-Publication Data
Best, Myron G.
    Igneous petrology / by Myron G. Best and Eric H. Christiansen.
        p. cm.
    Includes index.
    ISBN 0-86542-541-8
    1. Rocks, Igneous.   I. Christiansen, Eric II.   II. Title.

QE461 .B54 2001
552'.1–dc21   00-056480

CONTENTS

CONTENTS

THERMODYNAMICS AND KINETICS: AN INTRODUCTION

CHAPTER 4

SILICATE MELTS AND VOLATILE FLUIDS IN MAGMA SYSTEMS

CHAPTER 11

GENERATION OF MAGMA

CHAPTER 12

DIFFERENTIATION OF MAGMAS

CHAPTER 13

PETROTECTONIC ASSOCIATIONS

PREFACE

Igneous petrology in the last decades of the 20th century has exploded into a broad, multifaceted, increasingly quantitative science. Advances in geochemistry and physical and field petrology have forever changed our thinking about the origin and evolution of magmas, their dynamic behavior, and the way in which they are intruded and explosively extruded. Geophysics and mineral physics have provided new insights into the nature of the convecting mantle and its role as a giant heat engine driving magmatic and other global geologic processes. New tools of all kinds allow new ways of gathering petrologic data, while phenomenal developments in computers and computer software permit data to be stored, processed, and modeled in ways unimaginable as recently as a decade ago.

It has been a very daunting challenge to create within one book of reasonable length a balanced, comprehensive coverage of igneous petrology for the undergraduate student that embodies the classical as well as the new advances. We hope that this textbook will provide a foundation for future geologists that not only informs but nurtures the intellectual mindset, enabling them to pursue higher levels of professional endeavor. We have attempted to emphasize controlling petrologic processes in the formation of magmatic rocks while not sacrificing basic descriptive information about them, on which interpretations of their origin must be firmly based. Rather than the traditional chapter-by-chapter treatment of rock suites, our chapter organization is mostly by process—whether it be crystal melt equilibration, chemical dynamics of melts and crystals, physical dynamics of magma bodies, explosive magma eruption, or global tectonic controls on magmatism. The overarching themes of this textbook are the dynamic interactions between matter and energy and the ways in which transfers and transformations of gravitational and thermal energy drive changes in rock-forming systems.

We have designed this textbook as a balanced instructional tool for the college sophomore or junior. We assume the student is acquainted with basic chemistry, physics, mineralogy, and physical and historical geology. A background in optical mineralogy is desirable. As for mathematical background, a course in calculus will be helpful but not essential. The mathematical inclination and capability of geology students vary widely and so as to avoid intimidating some at the outset, we have generally limited the more quantitative material to certain chapters and setaside boxes.

Problem sets appear at the end of each chapter. We do not intend to minimize the growing importance of the quantitative facets of petrology. Some of the problems are amenable to attack by computers and spreadsheets. We assume that instructors have their own favorite computer-based teaching exercises. "Fundamental Questions Considered in the Chapter" provide a brief preview of each chapter. End-of-chapter "Critical Thinking Questions" provide an incentive for the student to think about the chapter contents. A comprehensive glossary is included at the end of the textbook together with a list of references cited. Space limitations permit citation of only the most crucial references or recent lucid summaries and select early classical works.

Because of space limitations, a chapter on Precambrian magmatic rock associations could not be included, but will appear in a forthcoming *Igneous and Metamorphic Petrology.*

This textbook obviously contains more than can be covered in one semester. It is expected that different classes will use different parts. For example, a class with limited background in physics and mathematics and desiring a more descriptive approach might use Chapters 1, 4, 5, 9–12, and selected parts of 13; Chapter 7 on fabrics and the first two-thirds of Chapter 2 could be used in an accompanying laboratory. Another class desiring a less descriptive, more quantitative approach might cover Chapters 1, 2 (last third), 3–6, and 11–13.

The senior author created the line drawings and diagrams and did the primary writing. The junior author thoroughly critiqued the entire work—greatly improving its clarity and accuracy—and provided many hours of beneficial discussion and computer assistance. Stephen T. Nelson offered valuable help. We gratefully acknowledge the constructive reviews and encouragement of Katherine Cashman (University of Oregon), Calvin Miller (Vanderbilt University), Suki Smaglik (Metro State College of Denver), Raj Sharma (Western Michigan), Charles Lesher (UC Davis), Michael Garcia (University of Hawaii), and Mihai Ducea (Cal Tech) without which this textbook would have been much less than it is. Simon Rallison first contacted us regarding doing a textbook with Blackwell Science and Jane Humphreys followed up. Nancy Duffy graciously allowed us the indulgence of doubling the initial contract period; her positive forebearance and advice at crucial times will forever be appreciated. Jill Connor was always cheerfully available for all kinds of help and information. Irene Herlihy patiently responded to endless queries and tactfully coordinated the illustrations and manuscript through to final publication. We also thank Lisa Flanagan, Manufacturing Manager, Nancy Whilton, Publisher/Science Books, and The PRD Group, who labeled the illustrations and typeset the text.

Finally, the senior author acknowledges the support of a very patient wife, Viv.

Myron G. Best   *May, 2000*
Eric H. Christiansen   *Provo, Utah*

Overview of Fundamental Concepts

FUNDAMENTAL QUESTIONS CONSIDERED IN THIS CHAPTER

1. What role is played by energy in its various forms to create magmatic rocks?

2. What is the source of internal thermal energy in the Earth; and how does it function as a giant heat engine to drive rock-forming processes?

3. What is the role of the mantle of the Earth in rock-forming processes?

4. In what way does mantle convection focus rock-forming processes in specific tectonic settings?

5. What do changes in geologic systems have to do with the formation of rocks?

6. What are the most significant properties of rocks, and what specific information does each property provide about rock-forming processes?

7. How does a petrologist study rocks to determine their nature and origin?

INTRODUCTION

This book is about **igneous rocks,** also called **magmatic rocks,** which form by cooling and solidification of magma. **Magma** is mobile molten rock material whose temperature is generally in the range of 700–1200°C (about 1300–2200°F) near the surface of the Earth. Magmatic rocks have textures, structures, and mineral and glass constituents indicative of that high-temperature magmatic ancestry. Temperatures sufficient to create magma only prevail beneath the surface, except in the very rare instances of impact of large asteroids on the surface of the Earth. When sampled and studied by geologists, magmatic rocks not only have (often) cooled, but in many cases have been brought by geologic processes to the surface from some considerable depth in the crust or mantle. Obviously, the origin of these once-hot rocks, and their exposure at the surface, involves flow of heat as well as movement of rock mass in the gravitational field of the Earth. Thus, interactions between heat and gravity are involved in the creation of magmatic rocks and their exposure at the surface of the Earth. To understand magmatic rocks and the related interactions between energy and matter is to understand, in a major way, how the planet Earth works.

With the development of concepts of plate tectonics in the 1960s all concepts of a static Earth became obsolete. Plate motion—a basic facet of the way the Earth works—manifests the interaction between gravity and outward flow of heat from the hot interior of a cooling, dynamic Earth. Oceanic lithosphere that is cooler and therefore denser than the underlying asthenosphere sinks at subduction zones. Plumes of hot mantle rock rising from near the core-mantle boundary and upwelling mantle beneath oceanic spreading junctures constitute the return circuit in the global mantle convection system. Seafloor spreading from the oceanic junctures maintains a constant global surface area, compensating for subduction.

Most of the magmatic activity in our planet occurs along the two linear tectonic regimes of plate convergence and divergence because that is where most interactions between energy and matter take place, generating magma from initially solid rock. Active volcanism occurs over less than 0.6% of the surface of the Earth, assuming a modest width of 100 km along the boundaries of converging and diverging plates (Plate I). Hidden from view beneath the sea, about three-fourths of global volcanism is estimated to occur along the world-encircling system of oceanic spreading ridges (Figure 1.1). On average, six to seven times greater volumes of magma are trapped within the crust as intrusions, compared to that extruded from volcanoes.

Localized and volumetrically minor magmatism far removed from plate boundaries is commonly related to mantle plumes ascending through the mantle. Such intraplate activity is manifest, for example, as volcanism in the Hawaiian Islands.

✳1.1 ENERGY AND THE MANTLE HEAT ENGINE

Without a critical amount of thermal energy within a planetary body there can be no movement of lithospheric plates or rise of mantle plumes and hence no magmatism, metamorphism, or tectonism. The geologically dead Moon, for example, has been too cold for billions of years for any such geologic activity. However, throughout its approximately 4.5-billion-year existence, the Earth has acted as a giant heat engine, powering all kinds of geologic processes. In this engine, the mantle of the Earth reigns supreme as the major source of driving energy. It is by far the most voluminous part (84%) of the planet, has the most mass (68%), and stores the most thermal energy. Ultimately, in one way or another, most magmatic rocks and magmas trace their ancestry to the mantle.

To understand how the Earth works as a heat engine driving rock-forming processes it is important to understand the various forms of energy, the ways they are transferred and converted into other forms, and the sources of thermal energy within the Earth.

1.1.1 Forms of Energy

Energy exists in various forms and is manifest in terms of motion, or potential for motion, and by the temperature of matter. An asteroid approaching the Earth, a high mountain from which boulders can be rolled downhill, an exploding volcano, and a hot lava flow all have energy, but in different forms. (Some forms of energy, such as magnetic energy, are important in man-made machines, but the geomagnetic field of the Earth is too weak to cause geologically significant movement of matter.)

Energy is commonly defined as the capacity for doing work. **Work,** w, is defined as the product of force, F, times a displacement over a distance, d, in the direction of the force

$$w = Fd \qquad 1.1$$

Thus, for example, energy is required to perform the work of shoving a thrust plate in an actively growing compressional mountain system or of throwing a fragment of rock from an explosive volcanic vent. An im-

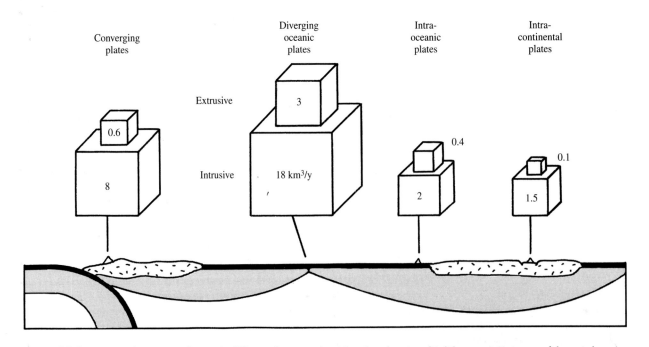

1.1   Global inventory of magma production in different plate tectonic settings (numbers in cubic kilometers). Estimates of the ratio between erupted magma and magma lodged as intrusions in the crust vary, depending on geologic factors and considerable uncertainties in the interpretations of the geologist. Production of basaltic magma predominantly in oceanic settings and mostly along ocean ridges far exceeds that of any other magma composition in any tectonic regime. (From Schmincke H-U. Vulkanismus, 2. Darmstadt, Wissenschaftliche Buchgesellschaft, 2000. With permission of Wissenschaftliche Buchgesellschaft, Copyright © 2000.)

Worked Problem Box 1.1 How much energy is required to lift this textbook 1 meter above the table?

Assume the book weighs 1 kg and the acceleration of gravity is 9.8 m/s². The increase in gravitational potential energy equivalent to the work, w, is $E_P = w = mgz = 1 \text{ kg} \times 9.8 \text{ m/s}^2 \times 1 \text{ m} = 9.8 \text{ J}$. (See the inside cover for units and conversions between them.) For comparison, one beat of the human heart consumes about 1 J and a small cup of water, 3.7 cm × 3.7 cm × 3.7 cm = 50 cm³ in volume, heated by 1000 J = 1 kJ of thermal energy raises its temperature by 5°C.

---

portant type of work in geologic systems is called **PV work,** where P is the pressure, such as possessed by a volcanic gas, and V is the volume of the gas. Expansion of pressurized gas does work in displacing magma out of a volcanic vent, creating an explosive eruption. Because **pressure** is defined as a force divided by the area over which it acts, $P = F/area$, and because volume, $V = area \times d$, then $PV = Fd = w$.

Kinetic energy is associated with the motion of a body. A body of mass, m, moving with a velocity, v, has kinetic energy

$$F_K = \tfrac{1}{2}mv^2 \qquad\qquad 1.2$$

A moving lava flow, ejecta thrown from an exploding volcano, and agitating molecules in a gas all have kinetic energy.

Potential energy is energy of position; it is potential in the sense that it can be converted, or transformed, into kinetic energy. A boulder cascading down a hill slope gains velocity and, therefore, kinetic energy as it loses potential energy. Potential energy can be equated with the amount of work required to move a body from one position to another in a potential field, in this instance, the gravitational field of the Earth. In lifting a boulder of mass m through a vertical distance z in the gravitational field of the Earth, whose acceleration is g, the amount of work equivalent to the **gravitational potential energy** is

$$E_P = mgz \qquad\qquad 1.3$$

The distance z is measured outward from the Earth above some reference level. Thermal energy within the Earth is expended to do the work of uplifting a mountain range, which imparts increased gravitational potential energy to the mountain mass.

Operating a bicycle tire pump demonstrates that mechanical work can be converted, or transformed, into thermal energy. As the pump handle is repeatedly depressed, the pump piston's rubbing on the inside of the cylinder produces frictional heating of the pump cylin-der; in addition, the work of compressing the air in the cylinder heats the air. The increased temperature of the tire pump is a manifestation of an increase in the thermal energy internally within the metal parts of the pump. The **thermal energy** of a body resides in the motions—kinetic energy—and the attractions—potential energy—of the atomic particles within it. An increase in the internal thermal energy of a solid is associated with greater kinetic energy via faster motion of the atoms and is manifest in a greater temperature, T. This motion can become sufficiently vigorous to break atomic bonds momentarily so that the solid becomes a flowing liquid, or, if bonds are fully broken, a gas. The term *heat* is sometimes used synonymously with *thermal energy*, but, strictly speaking, **heat** is transferred thermal energy caused by a difference in temperature between bodies. For example, the thermal energy of a magmatic intrusion is reduced as heat moves into the surrounding cooler wall rocks, heating them to a higher T.

The *joule*, J, is the fundamental unit of energy (see the inside cover for units used throughout this textbook).

1.1.2 Flow and Transformation of Energy

In nature, energy moves, is transferred, or flows from place to place. Energy is also exchanged, converted, or transformed, from one form into another. Thus, decay of an unstable radioactive U nucleus emits high-speed smaller particles whose kinetic energy is transformed into thermal energy that heats the mineral hosting the U atom. As rocks adjacent to a magmatic intrusion are heated, they expand and exert an increased pressure on adjacent rocks, displacing them outward and doing PV work on them. Thermal energy and work are, therefore, interconvertible. And work can be converted into thermal energy—such as in a tire pump. PV work is a transfer of energy due to a difference in pressure; heat is a transfer of thermal energy due to a difference in temperature, T. In all such flows and transformations of energy the total amount is rigorously and quantitatively conserved in agreement with the law of conservation of energy, also called the first law of thermodynamics.

This law claims that the total amount of energy and mass in the universe is constant. The total amount of energy is not added to or subtracted; it only moves about and is converted to other, perhaps less obvious, forms. In all such flows and transformations we are concerned with *changes* in the amount of energy. In contrast, the total, or absolute, amount of energy residing in a system is difficult to evaluate and generally is unimportant.

1.1.3 Heat Flow in the Earth

Within Earth systems the transfer of thermal energy, or flow of heat, is especially important and is therefore considered further here. Movement of thermal energy is obviously involved in magmatic rock-forming processes, such as heating solid rock so it melts, forming magma. On a larger scale, cooling oceanic lithosphere becomes denser and sinks as subducting slabs into the hotter, less dense upper mantle. Without heat, the Earth would be geologically dead.

An increment of heat, Δq, transferrred into a body produces a proportional incremental rise in its temperature, ΔT, given by

$$\Delta q = C_P \,\Delta T \qquad\qquad 1.4$$

where the proportionality constant, C_P, is the **molar heat capacity,** with units of J/(mol degree). The subscript in C_P indicates the heat capacity is for a condition of constant pressure, a common geologic situation, as for example, when rock is being heated by a nearby magmatic intrusion at a particular pressure in the Earth. The heat capacity based on mass is the **specific heat,** with units of J/(g degree). Because 1 calorie of heat (1 cal = 4.184 J) raises the T of 1 g of water 1 degree (Celsius or Kelvin), the specific heat of water is 1 cal/g degree. However, rocks generally have specific heats of 0.25–0.3 cal/g degree, which means that a given amount of heat, Δq, can raise the T of a mass of rock three to four times more than it can an equal mass of water. In other words, water absorbs relatively large amounts of heat per unit of mass for a particular T increase, ΔT; it is an effective thermal transfer agent and moderator. Use of water in building heating systems and in automobile radiators is serendipitous because water is also inexpensive and readily available. In geologic systems, water absorbs considerable heat from nearby magmatic intrusions and as it moves through cracks can effectively transport this heat to distant rock, changing its T.

Heat can be transferred in four different ways: radiation, advection, conduction, and convection. Commonly, two or three of these act in unison, as in the cooling of the lava flow in Figure 1.2. **Radiation** involves emission of electromagnetic energy from the surface of a hot body into transparent cooler surroundings, such as the Sun into surrounding space or a hot lava into the atmosphere. In a vacuum, this energy moves at 277,800 km/s, the speed of light. Radiation is insignificant in cool rocks because they are opaque, but the effectiveness of radiative transfer increases exponentially with T as rocks become more transparent above about 1200°C. **Advection** involves flow of a liquid through openings in a rock whose T is different from that of the liquid. Because all rocks near the surface of the Earth are fractured on some scale and because these fractures are, at least partly, filled with water, advection is a significant heat transfer process. For example, hot water heated by a nearby magmatic intrusion advects through cracks in cooler rock, heating it while moderating the T of the water. The greater heat

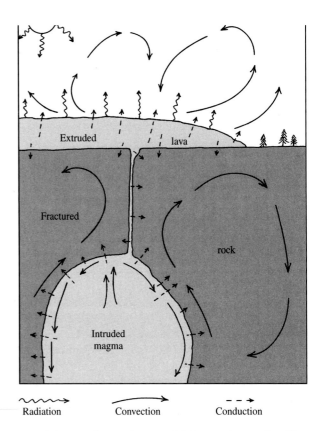

1.2   Schematic diagram (not to scale) showing four modes of heat transfer. Heat from an intrusive body of magma, in which convection may occur, conducts into the cooler wall rock, where heat is further transferred away by advective flow of heated groundwater through interconnected cracks. Heat is mainly dissipated from the top of the lava by conductive and radiative transfer into the overlying air, which expands and buoyantly convects upward so that cooler air descends, is heated, expands, and ascends.

capacity of water than of rock makes advective heat transport more effective. Advective heat transfer is also important where magma penetrates cooler rock.

Conduction. Transfer of kinetic energy by vibrating atoms in any material is called **conduction** of heat. Heat cannot be conducted through a perfect vacuum because of the absence of atoms. Imagine a box filled with rigid balls (representing atoms) all interconnected by springs (representing atomic bonds). If a ball in one corner of the box is set into motion (i.e., is given kinetic or internal thermal energy) all of the balls in the box eventually will be set into motion and given kinetic energy, but the motion and energy of any individual ball are less than those for the ball in the corner because the initial energy input is dissipated throughout the box. Internal kinetic energy moves throughout the box, manifest in heat conduction. Heat always flows from a hotter region, where atomic motion is greater, to a cooler region, where motion is less. A cool metal pan on a hot stove becomes hot as a result of conduction through the metal. Heat from a magmatic intrusion conducts into the enclosing cooler rocks, which become hotter, while the magma cools. (In this instance, conduction acts in concert with advection of water moving through cracks in the wall rocks.) For a given volume, a hot body conducts heat away faster if its enclosing surface area is larger; this is why air-cooled engines have attached fins to dissipate the heat faster.

The difference in T between adjacent hotter and cooler masses, called the **thermal gradient,** is reduced and may eventually be eliminated over sufficient time, provided heat is not restored to the hotter mass. The rate at which heat is conducted over time from a unit surface area, called the **heat flux** or **heat flow,** is the product of the thermal gradient and the thermal conductivity, or

heat flow = thermal conductivity
× thermal gradient   1.5

Because of their extremely low **thermal conductivity,** compared with that of familiar metals, rocks are considered to be thermal insulators. All other factors being equal, copper conducts heat nearly 200 times faster than rock. Because of the low thermal conductivity of rock and the large dimension of the Earth (radius about 6370 km), little heat has been conducted from the deep interior over the lifetime (4.5 Ga) of the Earth (Verhoogen, 1980).

Thousands of measurements all over the planet since the 1950s reveal that the surface heat flow from the hotter interior averages about 0.09 watt/meter^2 (W/m^2). If one recalls the wattage of a common incandescent light bulb, say 60–100 W, this is an extremely small quantity of heat! Significant variations in the heat flow and corresponding geothermal gradient depend on the plate tectonic setting. The **geothermal gradient,** or **geotherm,** expressed as the change in temperature divided by the depth interval over which it occurs, or $\Delta T/\Delta z$, has been found to vary from hundreds of degrees per kilometer beneath oceanic spreading ridges to about 20–30°C/km in active orogenic belts alongside convergent plate junctures to as low as 7°C/km in the nearby deep-sea trench. These variations in gradient might reflect lateral variations around the globe in thermal conductivity of rocks, in their radiative transparency, or in heat transferred by another mechanism. As the first two possibilites involving lateral variations in rock properties are unreasonable, the possibility of another mechanism of heat transfer should be considered.

Another argument illustrates that global heat flow may not be solely by conduction. If a modest geotherm of 20°C/km is extrapolated to a depth of, say, 200 km, the temperature there would be 200 km × 20°C/km = 4000°C. This is an impossible T because it exceeds the

---

Special Interest Box 1.2 Experimental petrology of the deep interior of the earth.

Laboratory devices that can create the *P-T* conditions prevailing to the center of the Earth provide important information on which minerals might constitute the deep interior. All of the devices produce high pressures in basically the same manner by squeezing a sample between opposing pistons or anvils to which force is applied. Various gasket materials are used to contain the sample between the anvils. Bench-top piston-cylinder devices can create pressures as much as 40 kbar (4 GPa) equivalent to depths of roughly 120 km. These consist of two opposing hard-metal pistons, one driven against the other by a hydraulic jack, and the sample is constrained within a cylinder into which the pistons move. The sample is heated by a furnace that surrounds the cylinder and can be opened at the end of the experiment so a blast of cool air can be directed at the sample assembly to "quench" the high *P-T* run products. Multianvil presses can achieve pressures to as much as 35 GPa (roughly 850-km depth), weigh several tons, and consist of four or six hydraulic pistons that move from their rigidly supported cylinders and converge symmetrically upon a tetrahedral or cubic sample assembly. This can be a block of soapstone that serves as an extrudable gasket between the carbide anvils at the ends of the pistons. Inside the block is the sample surrounded by a small electrical resistance furnace that is destroyed during the experiment. The third and most intriguing device consists of two small faceted diamonds held between the jaws of a hand-size "nutcracker" device (Jayaraman, 1984). The mechanical leverage exerted by the nutcracker arms pulled together by a large screw and the small diameter (about 100 micrometers = 100 microns) of the faces of the opposing gem diamonds can create pressures exceeding those at the center of the Earth. Not only are the diamond anvils very strong, they are also transparent to a laser beam for heating the sample, to light used for directly observing the sample, and to X rays to do diffraction analysis of the sample while under high *P* and *T*. Yet another high-*P-T* technique employs shock waves created by firing a projectile at a fixed target.

Information provided by high-*P-T* devices has revolutionized understanding of the mineralogical composition of the interior of the Earth and has answered questions posed by seismic data. Although much remains to be learned, the upper mantle appears to consist of peridotite—a rock made mostly of Mg-rich olivine and subordinate monoclinic and orthorhombic pyroxenes, and Mg-rich garnet, or at shallow depths a complex Cr-Fe-Mg-Al spinel. At a depth of 410 km, orthorhombic olivine begins to transform into a denser cubic Mg-Fe silicate and other complex dense phases; pyroxenes transform into denser cubic garnetlike minerals. Below 670 km, the lower mantle consists of still denser cubic Mg-Fe-Ca-Al silicates, whose atomic structure is like that of the mineral perovskite, $CaTiO_3$, plus cubic magnesiowüstite, $(Mg,Fe)O$, whose structure is like that of halite, $NaCl$.

melting T of mantle rock at that depth (Figure 1.3; as measured in the laboratory) and because seismic shear waves, which cannot pass through a liquid, are propagated throughout the mantle. Obviously, the measured near-surface geotherm cannot be extrapolated far into the interior of the Earth to obtain the T: That is, the geothermal gradient is not constant with respect to depth. There are at least two possible reasons for a substantially reduced geotherm at depth in the Earth so wholesale melting does not occur. One is that another mechanism of more efficent heat transfer prevails in the deep mantle, and the other is the presence of a concentration of heat producing rock nearer the surface. Both turn out to be true.

Convection. Movement of material having contrasting temperatures from one place to another is **convection**. Movement is caused by significant differences in density of different parts of the material so that, under the influence of gravity, less dense expanded material rises and more dense sinks. For example, soup in a pan on a hot stove convects as it warms and expands at the bot-

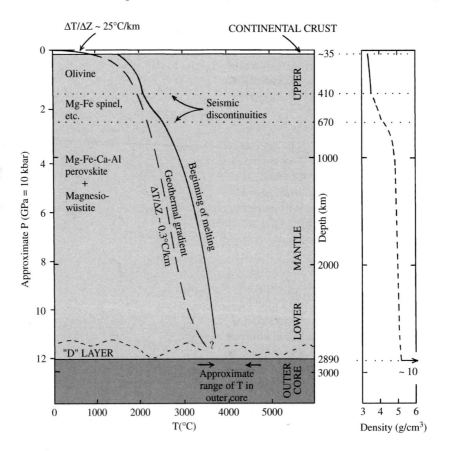

1.3   Relations among pressure, temperature, mineral composition, density, and melting conditions with respect to depth in the mantle and outer core of the Earth. Beginning-of-melting temperatures of mantle silicate rock and core Fe alloy have been determined experimentally in the laboratory (Special Interest Box 1.2). The geothermal gradient, or geotherm (dashed line) must lie below melting temperatures in the *solid* mantle and also pass appropriately through the 410- and 670-km phase transitions of olivine to spinel and spinel to Mg-Fe-Ca-Al perovskite plus wüstite, which cause discontinuities in seismic velocity. Note that the geotherm has a more or less constant slope through the convecting mantle of only about 0.3°C/km. The geotherm in the D″ layer and lithosphere is much greater because of less efficient conductive heat transfer in these lower and upper thermal boundary layers, respectively. Note the exaggerated thickness of the continental crust, which averages about 35 km, at the top of diagram.

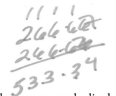

tom, becomes less dense, and buoys upward, displacing cooler, denser soup at the top of the pan, which sinks toward the bottom to complete the circuit. Density contrasts driving convection can also be related to contrasts in composition. For example, surface evaporation of water in saline lakes in hot arid regions increases the surface salt concentration, making the water more dense and causing it to sink, even though it may be warmer than underlying less saline water. It should be emphasized that, unlike heat transfer by radiation and conduction, convection depends upon gravity. Without gravity, there is no buoyancy force to act on density contrasts that move matter.

Because convection is generally associated with fluid bodies, such as soup in a pan and bodies of water but also the gaseous atmosphere, it may seem surprising that this mode of heat transfer is possible in the solid rock mantle. Indeed, the reality of mantle convection was not a part of geologists' thinking until the acceptance of the lateral motion of plates and continental drift in the 1960s. The paradox, on the one hand, of a solid mantle that transmits seismic shear waves and, on the other hand, of a fluid mantle capable of convection is resolved by a consideration of the factor of time in viscous bodies. **Viscosity,** a measure of the resistence to flow, is illustrated by tar (asphalt). On an average 24°C day a body of viscous tar, like a brittle solid, can be broken into sharp-edged fragments by a hammer blow. But over a period of several hours this same body of tar flows under its own weight into a flat blob. Tar at 24°C is more viscous than honey, which is more viscous than water. Because hot mantle rock is about a billion billion times—10^{18}, or 18 **orders of magnitude**—more viscous than 24°C tar, the rate of convective flow in the mantle is measured not in centimeters/hour, as for tar, but at most in centimeters/year—the speed at which lithospheric plates move. One way of defining the difference between a fluid and a solid is the time scale of their measurable flow.

Slabs of cooler oceanic lithosphere, mostly mantle rock, too viscous to convect within themselves, sink into the underlying hotter and relatively less dense underlying mantle. Computed tomography of the Earth using seismic waves (instead of X rays, used in scanning a person's body) has shown that at least some subducting lithospheric slabs sink all the way through the mantle and come to rest on top of the dense metallic core (Figure 1.4). Convectively upwelling hotter mantle at oceanic ridges and associated seafloor spreading complement subduction of lithosphere.

Even though the rate at which the mantle convects, on the order of a few centimeters/year, seems minuscule, it is far greater than the rate of heat transfer

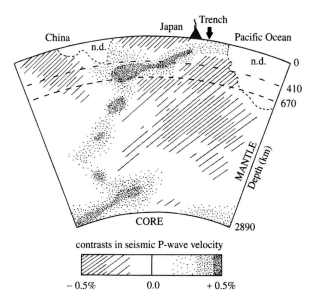

contrasts in seismic P-wave velocity

−0.5%   0.0   +0.5%

1.4 Seismic tomography cross section of the mantle beneath the Japan subduction zone. The most densely stippled regions have seismic primary-wave velocities as much as 0.5% greater than average mantle and delineate the cooler, denser lithospheric slab, which has been subducting for most of the Cenozoic and has not conductively heated up to the T of the surrounding mantle. Note the segmented nature of the slab at midmantle depths and crumpled deeper slab that rests atop the core. Diagonally ruled regions represent hotter mantle where velocities are as much as 0.5% less than average. n.d., regions of no data; dashed lines, seismic discontinuities at depths of 410 and 670 km (Figure 1.3). (From a color diagram created by Rob van der Hilst of the Massachusetts Institute of Technology in Levi [1997; see also Grand et al., 1997]. Reproduced here as a modified black and white version with his kind permission.)

by conduction. The fact that lithospheric slabs can sink convectively over tens of millions of years about 2800 km through the hotter mantle but still maintain a recognizable cooler T throughout their approximately 100-km thickness demonstrates how slow they are conductively heated. Conversely, if heat conduction were more rapid than convection, the subducting slabs would absorb heat and lose their density contrast and identity before sinking very far.

A second style of convection in the mantle, which can apparently operate independently of plate motion, consists of columns of relatively hotter mantle a few hundred kilometers in diameter that are rising vertically toward the base of the lithosphere. First proposed by Morgan (1971), **mantle plumes** had been confirmed to exist by the end of the century through seismic tomography imaging (see, for example, articles in the March 19, 1999, and May 14, 1999, issues of *Science*). The source of these plumes is apparently at the base of the mantle in a so-called D″ layer (Figure 1.5). Perhaps 10–20% of the heat driving mantle

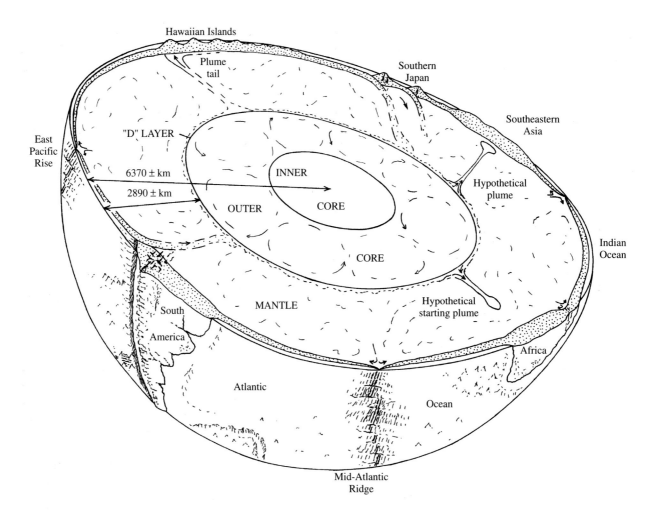

1.5   Northern part of the Earth sliced off to reveal the convecting interior. Thickness of lithosphere (stippled pattern) is exaggerated to show details. Note descending oceanic lithospheric slabs in subduction zones and mantle plumes, two of which are hypothetical, rising from the bumpy D″ layer at the base of the mantle.

convection comes from cooling of the core and perhaps all of this drives plumes. Seismological investigations reveal the D″ layer has considerable relief (Jeanloz and Romanowicz, 1997), and from time to time and place to place, a thick bulge becomes sufficiently buoyant to move upward as a hotter, lower viscosity, but still solid-rock plume, drawing nearby D″ layer with it. The decorative "lava lamp" in some homes is a colorful model. Many geologists believe the head of the plume partially melts in the shallow mantle, producing massive outpourings of basaltic magma onto the crust, forming huge continental and oceanic basalt plateaus. The plume tail, which can persist for tens of millions of years, is believed to be responsible for volcanic island chains, such as the Hawaiian.

In conclusion, heat transfer mainly in the upper mantle is manifest in movement of lithospheric plates, whereas plumes transfer heat from the core and lowermost mantle.

1.1.4 Implications of Mantle Convection

No apologies need be made for an extended discussion of mantle convection in this introductory chapter. Mantle convection is of critical importance in understanding how the Earth works as a gigantic heat engine driving geologic processes that create magmatic rocks. Descending cooler, denser lithospheric slabs and complementary upwelling of hotter mantle at spreading ridges and ascending deep hot mantle plumes constitute a whole-mantle convective system, one consequence of which is a small geothermal gradient of only a few tenths of a degree per kilometer throughout most of the mantle (Figure 1.3). Another consequence is substantial lateral variations in surface heat flow, which would not be expected if heat flow were wholly governed by conduction. Rock-forming processes, especially the creation of magmas and magmatic rocks, are strongly focused near the surface of the Earth by mantle convection. Figure 1.5 (see also Plate I) shows that magmatism is localized in relatively narrow belts along convergent and divergent plate boundaries and in so-called hot spots above mantle plumes, virtually to the exclusion of any other surface area of the Earth.

Associated with focused magmatic activity is the concept of **petrotectonic associations:** that specific types of rocks are found together in specific tectonic regimes. Although basaltic rocks composed mostly of plagioclase and pyroxene are created in most of the tectonic settings diagrammed in Figure 1.5, there are significant differences from one tectonic regime to another in their chemical compositions, particularly in so-called trace elements such as Sr, Ba, Ta, and Nb. Also, the types of associated rocks are different. For example, andesites are common in convergent plate subduction zones but are essentially absent at oceanic spreading ridges. Rhyolites and their plutonic granitic rock counterparts are widespread along continental margin subduction zones but rare where two oceanic lithospheric plates converge, as in the island arcs of the western Pacific. These petrotectonic associations are discussed further in Chapter 13.

1.1.5 Energy Budget of the Earth

With all of this heat within the Earth one cannot but wonder, What is its origin? Is the thermal energy in the Earth the result of a one-time investiture, or is it being replenished somewhere as it is being expended elsewhere? Are **energy sources** being exhausted, or are they still operative to compensate for **energy sinks?** Countless volcanic eruptions mainly from oceanic ridges over eons of geologic time have dissipated heat from the interior of the Earth into the oceans and atmosphere, from which it is radiated into outer space, the ultimate heat sink. So why are volcanic eruptions still occurring?

The largest source of energy driving terrestrial processes, roughly 50,000 times all other sources, is radiant thermal energy from the Sun. The 70% trapped in the atmosphere drives the global hydrologic system of moving masses of air, water, and sediment. Radiant solar energy does not conduct very far into the ground, perhaps only a few meters in sunny areas. Although the surface heat flow from the interior of the Earth is minute compared to the solar influx, it is perhaps 20 times greater than all of the energy dissipated in volcanism and tectonic activity.

A major source of internal heat within the Earth is the radioactive decay of the long-lived isotopes ^{238}U, ^{235}U, ^{232}Th, ^{40}K, and ^{87}Rb, which have half-lives of billions of years. Most investigators (e.g., Stacey, 1992; Verhoogen, 1980) calculate that this heat source is probably at least half and possibly approaching 100% of the total for the Earth. The uncertainty stems from the fact that concentrations of these isotopes are highly variable in different types of rock and where and in

what quantity these isotopes occur are poorly known. Overall, concentrations are greatest in the continental crust in granites, lower in basalt, in minute but uncertain amounts in the much more voluminous peridotitic mantle, and probably nonexistent in the core. Because of radioactive decay over eons of Earth history, the thermal energy produced when Archean rocks were created, 2.5–4.0 Ga, was roughly three times that of today; at 4.5 Ga, when the Earth was born, the rate was six times greater. Additionally, in that youthful Earth, decay of short-lived radioactive elements, such as ^{26}Al (half-life of 0.7 My), may have been significant.

Other important sources of internal thermal energy in the Earth (Verhoogen, 1980) are due to tides and to "original" heat. Tidal deformation of the solid Earth and oceans due to the gravitational pull of the Sun and Moon is dissipated as thermal energy, but this contribution is estimated to be an order of magnitude less than that of radioactive decay. In addition to current heat production by radioactive decay and dissipation from tides, some original heat inherited from the formation of the Earth at 4.5 Ga remains. Formation of the Earth is now generally believed to involve accretion of solid particles from a hot but cooling solar nebula of condensing gas and dust. As these particles and larger bodies (planetismals), themselves formed by collection of dust in the nebula, accreted into a proto-Earth, their gravitational potential energy was transformed into kinetic and then into thermal energy. Compression of these particles by additional accretion of more solids on top added more thermal energy. Compression does work on rock in the interior of the Earth, which is transformed into thermal energy, raising the rock T. Once the rock is compressed, no more thermal energy is created because no more work is done. Continuing capture of Sun-orbiting debris and impact of these fragments as asteroids onto the Earth for about 600 million years raised T further. The total energy in this growth process is estimated to have been sufficient to raise the T of the Earth tens of thousands of degrees. But the actual T increase was less, by some unknown amount, because heat was radiating and convecting away in the primitive atmosphere during accretion.

If, as is generally believed, the accreted Earth was initially chemically homogeneous, a large amount of thermal energy was generated during formation of the core as dense iron particles segregated from the molten silicate mantle by gravity settling. The calculated thermal energy gained from the loss of gravitational potential energy in core segregation is more than sufficient to produce the current surface heat flow, throughout the history of the Earth. Yet another source of heat related to the core is the ongoing solidification of the liquid outer core, releasing latent heat of crystallization. In other words, the core is currently heating the mantle.

Obviously, there are ample sources of internal thermal energy to drive the mantle heat engine.

It should not be forgotten that most geologic processes depend not only on thermal energy but also on gravity. Without gravity, matter would be dispersed indefinitely by the thermal processes of expansion, melting, and even vaporization. But on the other hand, gravity pulls matter together, compressing it. Interacting thermal energy and gravity constitute a push-pull in global geologic processes.

❈1.2 GRAVITY, PRESSURE, AND GEOBARIC GRADIENT

Thermal energy is manifest in temperature, T; the rate at which T increases into the interior of the Earth is the geothermal gradient, or $\Delta T/\Delta z$. In parallel fashion, the force of gravity acts on mass to produce pressure, P, in the interior of the Earth; the rate at which P increases into the interior of the Earth is the **geobaric gradient**, or $\Delta P/\Delta z$.

As previously defined, pressure is the force acting over a particular area. Within the Earth, pressure caused by the weight of overlying rock is called load, lithostatic, or **confining pressure**. It is denoted throughout this text by P; the context will indicate whether confining pressure or the chemical element phosphorus, P, which is not italicized, is being discussed. There are many units of pressure (listed inside the cover). In the International System (SI) of units, the unit of pressure is the *pascal,* denoted Pa, but the most commonly used unit in petrologic work is the *bar:* 1 bar $= 10^5$ Pa (0.1 MPa) and 1000 bar $= 1$ *kbar* $= 10^8$ Pa (0.1 GPa). Also, 1 bar $= 0.9869$ atmospheres (atm) $= 14.504$ pounds per square inch, the mean atmospheric pressure at sea level.

The magnitude of P increases with depth in the Earth in a predictable manner. To evaluate this increase, we note that bodies of rock more than several kilometers beneath the surface are hot and, over long periods of geologic time, behave as viscous fluids. Like water, which seeks its own level because it has no intrinsic strength, hot rocks have low strengths and flow readily, particularly over long periods of geologic time. This concept is implied in mantle convection and isostasy. The confining pressure, P, at the base of a vertical column of rock, considered as fluid, with cross-sectional area, A, equals

$$P = \frac{F}{A} = \frac{mg}{A} \qquad 1.6$$

from Newton's first law, $F = mg$, where m is mass and g the acceleration of gravity. In terms of the **density,** ρ = mass/volume of the rock, equation 1.6 becomes

$$P = \frac{\rho V}{A} = pgz \qquad 1.7$$

where z is the height of the rock column, considered positive downward here. In the Earth, the geobaric gradient is

$$\frac{\Delta P}{\Delta z} = \rho g \qquad 1.8$$

ρ being the mean density of the column of rock. In the continental crust, where the mean density $\rho \sim 2.7$ g/cm^3, $\Delta P/\Delta z = 270$ bar/km (~ 27 MPa/km). Actual densities in the continental crust range between 2.2 and 3.0 g/cm^3. (The less familiar SI unit of density is kg/m$^3 = 10^{-3}$ g/cm^3.) In the basaltic oceanic crust, $\rho \sim 3.0$ g/cm^3. In the upper mantle, where $r \sim 3.3$ g/cm^3, $\Delta P/\Delta z = 330$ bar/km. These values of the geobaric gradient are useful in converting P into an approximate equivalent depth in the crust or mantle, or vice versa. For example, at a depth of 30 km in the crust, $P = 30$ km $\times 270$ bar/km $= 8100$ bar $= 8.1$ kbar.

❈1.3 ROCK-FORMING PROCESSES AS CHANGING STATES OF GEOLOGIC SYSTEMS

All geologic processes in the dynamic Earth, including rock-forming processes, involve energy changes and interaction between energy and mass. Some geologic processes are driven wholly, or in large part, by changes in thermal energy and involve heat flow or transformation of heat into other forms of energy, or the reverse; cooling of an intrusive magmatic dike is a thermal process. Other geologic processes involve work, which can be considered as a mechanical or physical process in which, for example, rock is crushed into smaller pieces or magma is expanded by internal gas pressure into a greater volume. Still other geologic processes involve chemical reactions and movement of atoms, such as their organization into a well-ordered crystal of feldspar as a silicate melt cools. Most geologic processes are a combination of changes in several forms of energy. However, changes in thermal and gravitational potential energies dominate on a global scale. Hence, T and P are important characterizing variables in changing geologic systems.

All natural changes in a system move it toward a state of lowest possible energy, which is the most stable of all possible states. An example, familiar to any mountain hiker, illustrates these fundamental concepts. In Figure 1.6 are three hypothetical identical boulders; two are positioned on a hill slope and the third is in a lower valley. Each boulder has a different vertical position above some reference elevation and hence has a different gravitational potential energy. If boulders A and B are dislodged from their positions of rest, gravity causes them to fall down into the valley alongside boulder C, where their gravitational potential energy is the lowest possible. The potential energy given up is

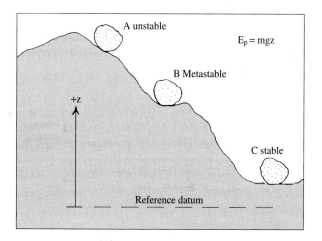

1.6 Concept of stability. Three boulders possess different amounts of gravitational potential energy, $E_p = mgz$. The tendency of changing natural systems to move toward a state of minimum energy may be referred to as the principle of parsimony, or laziness!

transformed into kinetic energy and ultimately into heat and mechanical work of breaking rock.

In this example, the three boulders initially possessed three different gravitational potential energies representing three possible energy states of a geologic system. A **system** is simply a part of the universe that is set aside in one's mind for the purpose of study or discussion. All else, the remainder of the universe, is the **surroundings.** As the hillside boulders were dislodged, their **state**—the particular conditions defining their properties or energy—changed. Once boulders A and B join C in the valley, all three are at the same lowest possible state of gravitational potential energy.

Equilibrium is a state that has no tendency to change spontaneously. The net result of forces acting on an equilibrium system is zero. Atoms may move about in a chemical system but at equilibrium nothing happens over time. Any slight disturbance will not result in any permanent change, as the system will return to its original condition. Boulder A satisfies some of the equilibrium conditions but not all. It is **unstable,** because disturbed even slightly it will cascade downhill to the lowest energy level alongside boulder C, which lies in a state of **stable equilibrium.** Boulder C in the valley may be rocked from side to side, say, by an earthquake or by a hiker but will readily return to its original, lowest-energy, position. Boulder B represents a state of **metastable equilibrium** whose energy is more than the lowest state but is prevented from moving to that more stable, lower-energy state by an energy barrier, called the **activation energy,** represented by an elevated lip on the terrace on which the boulder rests. A metastable state can persist indefinitely; if subject to only a small disturbance—represented, for example, by the hiker's slightly rocking boulder B on the terrace on the hill slope—the metastable state will return to its original

configuration. Only a more forceful shove by the hiker will send the boulder over the lip of the terrace downhill to stable state C.

Glass and virtually all high-P and -T magmatic minerals are metastable under atmospheric conditions. Their thermodynamic energy (Chapter 3) is not the lowest possible under atmospheric conditions. Glass is a solid aggregate of atoms more or less randomly arrayed as if it were liquid: that is, it is amorphous. Whether of human or natural volcanic origin, glass eventually crystallizes into an aggregate of stable crystals under atmospheric conditions. Most volcanic glass is of late Cenozoic age, little glass is early Cenozoic, and so on. Diamond, formed at depths of at least 150 km in the mantle, has a much larger activation energy barrier to overcome than glass, and this characteristic prevents it from converting into stable graphite of minimal energy for hundreds of millions of years while lying in an African kimberlite pipe in the shallow crust. However, if a diamond is heated to several hundred degrees in the oxygen-rich atmosphere, the provided thermal energy supplies the necessary activation energy so that the metastable diamond decomposes into stable CO_2.

Without metastability there would be no magmatic rocks exposed at the surface of the Earth, only minerals such as quartz, calcite, gypsum, and clays that are stable at atmospheric conditions.

These concepts apply to chemical equilibria where chemical reactions involving movement of atoms are occurring, to thermal equilibria where heat is being transferred to parts of a system at different temperatures, and to other types of equilibria. In evaluating the stability and equilibrium of any system, one must carefully examine the nature of all possible changes in state. Thus, a particular system, such as boulder C, would be chemically unstable if it were limestone subject to acid rain. When asking the question, What state is most stable?, the next question should always be, Under what conditions?

In the dynamic Earth, rock-forming geologic systems are dislodged from states of higher energy and move naturally to a more stable state of lowest possible energy. For example, lava extruded at the summit vent of a volcano loses gravitational potential energy and gives up heat to its cooler surroundings as it flows down slope and solidifies into solid rock. Both forms of energy change to a new lower-energy level. An energy gradient is available through which to move. The direction of transfer is always from a higher to a lower energy level, in the one case elevation and in the other T.

❈1.4 ROCK PROPERTIES AND THEIR SIGNIFICANCE

Thus far, our approach to understanding rock-forming geologic processes has been deductive, starting with general principles and illustrating them by specific examples. In practice, however, the geologist is usually faced with the opposite, inductive problem. Given a particular mass of rock and its observable properties, the geologist asks, What was the state of the past geologic system in which it was created? What geologic processes of energy transfer and transformation and movement of matter were involved? What caused the changes in state of the rock-forming system, perturbing a previous state of equilibrium and producing a new state of stable equilibrium? From the only tangible record of the system—the rock itself—the geologist must work backward, trying to comprehend the rock-forming system in which it was created. So, what properties of the rock are most significant? And what specifically do they tell us?

Among many rock properties—such as aesthetic, electrical, magnetic, and mechanical—the rock properties of most significance for the geologist are composition, fabric, and field relations.

1.4.1 Composition

Rocks consist of minerals and locally occurring glass (actually, an amorphous solid, not a mineral) that are made of atoms of the chemical elements. Three basic compositional properties can be recognized in any rock: the concentrations of chemical elements in the bulk or whole rock, the character of the minerals and glass in which they reside, and the amounts of the different minerals and glass (Figure 1.7).

Whole-Rock Chemical Composition. An analysis of a rock for its chemical elements, irrespective of mineral

WHOLE ROCK CHEMICAL COMPOSITION	MINERALOGICAL COMPOSITION OF ROCK			MODAL COMPOSITION OF ROCK
Concentrations of chemical elements or oxides in the entire rock	Types of minerals present and their individual chemical compositions			Proportions of the different types of minerals constituting the rock

	Vol.%
A	13.7
B	15.9
C	70.4
Total	100.0

	Wt.%		A	B	C			Wt.%
				Wt.%				
SiO_2	65.71		37.17	99.82	64.50		A	16.0
TiO_2	0.51		3.14	0.05	0.00		B	15.8
Al_2O_3	16.16		14.60	0.06	20.25		C	68.2
Fe_2O_3	0.92		3.75	0.03	0.47			
FeO	4.30		26.85	0.01	0.00		Total	100.0
MgO	0.68		4.23	0.00	0.00			
CaO	0.36		0.17	0.00	0.48			
Na_2O	3.25		0.15	0.00	4.72			
K_2O	7.87		8.25	0.01	9.60			
H_2O	0.41		1.35	0.02	0.28			
Total	100.17		99.66	100.00	100.30			

1.7   The three compositional aspects of a rock. The modal composition is expressed on a volume percent volume basis, and from this, using mineral densities for biotite (A), quartz (B), and alkali feldspar (C), the mode in weight percentage (wt.%) can be calculated.

constituents, yields its **whole-rock chemical composition,** also referred to as its **bulk chemical composition,** or sometimes simply as its chemistry. The bulk chemical composition is expressed in terms of weight concentrations on a percentage basis of chemical species such as SiO_2, Al_2O_3, H_2O, and CaO. Thus, in a 100-gram sample of the hypothetical rock (Figure 1.7), 65.71 grams, or 65.71 wt.%, would be SiO_2; 16.16 g, or 16.16 wt.%, would be Al_2O_3; and so on. The SiO_2 and Al_2O_3 actually occur as Si, Al, and O atoms in the minerals making up the rock.

Magmatic rocks contain virtually all of the approximately one hundred chemical elements, but in concentrations that vary continuously over a restricted range. Why such variations occur and how they can be used to elucidate its origin are major topics explored in this textbook. Although complex, the chemical composition of a magmatic rock reflects the conditions of generation of the magma in a particular global tectonic setting and subsequent evolution of the magma, creating a possibly diverse petrotectonic association.

Mineralogical Composition. The types of minerals constituting the rock *and* their chemical compositions are the **mineralogical composition** of the rock. The types of minerals—such as biotite, plagioclase, or olivine—and perhaps glass, in the case of a volcanic rock, can be identified by naked eye or a magnifying lens in hand samples, a petrographic microscope using thin sections, and by X-ray diffraction and other laboratory techniques. Chemical compositions of minerals can be roughly approximated by optical techniques using a petrographic microscope but are most accurately determined by an electron microprobe analyzer.

Of the thousands of known mineral species, only seven major minerals or mineral groups make up most magmatic rocks. These are quartz and solid solutions of feldspars, micas, amphiboles, pyroxenes, olivine, and Fe-Ti oxides.

The minerals that make up a magmatic rock depend on the *P* and *T* of the last equilibrium state of the magma system and its chemical composition. In most rocks, this final state is "frozen into," or quenched in, the rock and persists metastably and indefinitely. For example, whether cristobalite or quartz or other polymorphs of silica (SiO_2) crystallize in a magma depends on the prevailing *P* and *T* of the system. But some magmas cannot crystallize any silica polymorph at all; because of an insufficient concentration of SiO_2, that which occurs in the magma is consumed in formation of olivines, pyroxenes, and so on.

Modal Composition. The volumetric proportions, on a percentage basis, of the different minerals in a rock are referred to as the **modal composition,** or simply the **mode.** It can be roughly estimated through examination by eye or in thin section under a microscope.

Further Comments. Modal, chemical, and mineralogical compositions of a rock are correlated, but not always in an obvious manner. For example, in two rocks of the same bulk composition containing, say, 63 wt.% SiO_2, one may contain quartz but the other not, depending on what other minerals are present. A granite containing 75 wt.% SiO_2 in its bulk chemical composition must not only have quartz as a constituent mineral, but in substantial amounts in the mode because no other common rock-forming silicate mineral has more than about 66 wt.% SiO_2. Rocks with abundant biotite in the mode will have relatively large concentrations of K_2O in the bulk rock chemical composition. Some minor or trace chemical constituents in the whole rock chemical composition, such as Sr, will be found in solid solution in major minerals, in this case feldspar.

A chemist investigating a rock would be expected to go no further than an analysis of composition and perhaps theoretical interpretations based on thermodynamics and kinetics (Chapter 3). However, rock-forming processes reflect changing states of geologic systems that operate in a context of *space and time*. This context is especially manifest in the field relations and fabric of the rock body. The task of the geologist interpreting rocks, therefore, goes far beyond simple chemical study.

1.4.2 Field Relations

Field relations are the larger-scale features of a mass of rock discerned in exposures (outcrops) in hills and mountains and man-made road cuts, quarries, and mines and seen in drill core and on geologic maps. The study of field relations in layered volcanic rock bodies is called **stratigraphy.** Field relations include, but are not restricted to, such characteristics as

1. the nature of contacts of a rock mass with neighboring bodies, whether sharp or gradational, conformable or cross-cutting;

2. the relative chronology to neighboring bodies, whether older or younger;

3. the spatial aspects of the fabric and composition, whether they are uniform over the extent of the mass or vary in some way, particularly near contacts or borders;

4. the dimensions of the mass.

Field relations provide insights into the causes of changes in state of the rock-forming system, or why transfer or transformation of energy occurred to drive the rock-forming geologic process. Field relations of magmatic rocks reflect the physical dynamics of the body of magma from which it was created: for example, whether it was extruded onto the seafloor or on top of dry ground, or intruded deep in the crust. Understanding the evolution of long-lived extrusive magma systems depends on a thorough investigation of the stratigraphic features of the volcano.

1.4.3 Fabric

In this textbook, the term **fabric** is used for all of the noncompositional properties of a mass of rock discernible on scales of observation from the outcrop and hand sample to the microscopic. Fabric encompasses both texture (sometimes called *microstructure*) and structure. **Texture** refers to grain characteristics—including grain size and shape, intergrain relations, and amount of glass—generally seen at the scale of a hand sample or smaller. **Structure** refers to features seen at a scale of a hand sample or larger, such as bedding in a pyroclastic deposit from an explosive volcano, columnar joints in a lava flow, or chunks of foreign wall rock within a magmatic intrusion. Tectonic and structural geologists also refer to folds and faults as structures. A sharp distinction between texture and structure is not always possible to make, hence the utility of the all-inclusive term **fabric.** Also, there is no sharp distinction between some structures and field relations.

Fabric chiefly reflects the time-dependent path of change in state of the rock-forming system. The rate of cooling—fast versus slow—of the magma largely dictates the grain size of a magmatic rock. How fast magma rises in the crust and boils can determine whether it erupts placidly as vesicular lava or violently explodes to create far-flung pyroclastic particles. Successive changes in state are commonly recorded in overprinted fabrics. In detail, the variety of time-dependent paths and the resulting fabrics are virtually infinite.

✳1.5 HOW PETROLOGISTS STUDY ROCKS

As a subdiscipline of geology, **petrology** seeks to understand the nature of rocks and their origin. Past rock-forming *processes* in dynamic geologic systems are recorded in the observable rock *product*—its field relations, fabric, and composition. Sometimes, petrologists approach the cause-effect, process-product correspondence in the opposite way by creating models or doing experiments simulating rock-forming, or petrologic, systems and then making comparisons with properties of real rocks.

In their attempts to understand the origin of rocks and explain how the mostly inaccessible Earth has worked through billions of years of geologic history recorded in rocks, petrologists theorize, using paper-and-pencil and computer models, and experiment, using simple model magmatic systems in the laboratory. In his presidential address to the Geochemical Society in 1969, James B. Thompson, Jr. (1970) had this to say regarding petrology:

> We who are concerned with such [petrologic] matters fall mainly in one or more of three categories: (1) those who collect rocks and study them; (2) those who try to duplicate rocks in the laboratory; and (3) those who worry

about rocks. All three facets, the observational, the experimental, and the theoretical should clearly be integrated in a healthy science. Description as an end in itself is sterile, experiments that answer unasked questions are irrelevant, and theory unchecked by fact is useless. Theory, at its best, is in many ways a link between the other two. Good theory can thus be used to interpret or reinterpret the results of observation in order to ask more significant questions of experiment and vice versa.

Petrologists study rocks as interpretive historians who try to "solve" or explain geologic problems and puzzles. They ponder and try to answer fundamental questions, such as the following:

1. Where and how did a particular magma originate?
2. What was the nature of the source of the magma and the changing system in which it was generated?
3. How was the magma transported and emplaced, that is, moved from its source to where the magmatic rock is now found?
4. What dynamic physical and chemical processes affected the magma as it solidified into rock?
5. How does the particular petrologic process relate to its global tectonic setting?
6. How can modern petrotectonic associations be used to infer tectonic processes and regimes in ancient rocks?
7. How did the planet Earth originate and evolve?
8. What is the effect of petrologic processes on our society and life in general?

The intent of this textbook is to provide a background sufficient to *begin to answer* these petrologic questions, and many others of a more specific nature. Interpretations of the origin of rocks, or **petrogenesis,** must be based on accurate **petrographic** data (modal composition and fabric) as well as field, chemical, isotopic, experimental, and theoretical data.

Petrology has become a global science that is finding increasing integration into other fields of Earth science. An understanding of petrology leads to a better understanding of how the Earth works.

In his presidential address to the Mineralogical Society of America, Peter Robinson (1991, p. 1781) said:

> In petrologic studies . . . the observations that are possible are conditioned by the experience and training of the observer. Petrologic research is an attempt to define the sequence of events in the formation of the rocks, the conditions under which the events happened, and why they happened. Petrology can never be totally quantitative, i.e., "getting numbers out of rocks" . . . observational and descriptive petrography from the atomic through the microscopic and macroscopic to the petrologic map scale [are essential in] providing the framework for interpretations.

Careful observation and innovative ways of thinking about rocks can lead to new discoveries and overturn established dogma. That an interpretation is espoused by esteemed experts is no guarantee of its validity. Continual reassessment and well-documented criticism of established truths make for better science.

Petrology at the dawn of the 21st century is a vibrant, dynamic discipline in which unsolved problems and controversies abound. Numerous puzzles remain to be explained and resolved. Many have not even been discovered!

Summary

The central themes of this textbook developed in more detail in subsequent chapters are these: On a global scale, rock-forming geologic processes are driven by interactions between matter and energy, chiefly thermal and gravitational. These interactions are strongly focused along convergent and divergent plate boundaries and above ascending mantle plumes, the two expressions of a convecting mantle, which is the major heat engine driving geologic processes. Without thermal energy in the Earth it would be geologically dead. Major sources of thermal energy in the interior of the Earth are radioactive decay and "original" heat generated at the time of its creation. Rock-forming processes focused by mantle convection create characterisitc petrotectonic associations of magmatic rocks that are more or less unique to each tectonic setting in which they occur.

In the dynamic Earth, transfers and transformations of different forms of energy and movement of matter cause changes in states of geologic systems. The direction of change tends toward the lowest possible energy state of stable equilibrium. However (by good fortune!), metastable states can persist indefinitely, allowing the petrologist to study magmatic mineral assemblages stabilized at high P and T and thus gain insight into rock-forming processes. Other clues to petrogenesis are found in the chemical, isotopic, and modal composition of the rock body and its fabric and field relations. Aspects of the time-dependent path taken during changing states of the rock-forming system are recorded in the rock fabric, such as the rate of cooling of the magma. Field relations provide insights into the flow of matter and energy in the system, for example, whether the magma was extruded explosively from a volcano or intruded into the deep crust.

Petrology seeks to understand the nature and origin of rocks and the way the whole Earth works by making critical observations, asking relevant questions, and integrating all available information into a self-consistent interpretive petrogenetic model.

CRITICAL THINKING QUESTIONS

These questions may not have definite, specific answers. More than one answer may be valid.

1.1   Describe examples of interaction between energy and matter that create magmatic rocks.

1.2   Where do rock-forming processes occur in relation to global tectonics?

1.3   Describe and give examples of the different forms of energy.

1.4   Cite an example of how energy is transferred, transformed, and conserved in a geologic system.

1.5   Characterize four modes of heat transfer and indicate their significance in different geologic systems and for the whole Earth. Why is less heat transferred into the substrate beneath a lava flow than out of the top of the flow?

1.6   In what way is the Earth a heat engine capable of doing work?

1.7   Why is the mantle important in creation of magmatic rocks?

1.8   Describe two modes of convection in the mantle and indicate one example that provides specific evidence for each.

1.9   Contrast the heat capacity of water and that of rock. Why is this contrast important?

1.10 Account for the inflections in the geotherm from the surface to the mantle-core boundary.

1.11 Account for the paradox of fluidlike convection in the solid rock mantle.

1.12 What is a petrotectonic association?

1.13 What are sources of internal heat in the Earth?

1.14 Why does T increase with depth in the Earth? P?

1.15 Characterize a geologic system that is in an unstable state. In a state of stable equilibrium. Of metastable equilibrium. How can these states be changed?

1.16 Briefly describe the significant properties of rock masses and tell how each property provides particular information about the rock-forming process.

1.17 Contrast the three different compositional properties of a rock.

1.18 Discuss the goals of petrology and the way a petrologist accomplishes them.

PROBLEM

1.1   What is the confining pressure, P, at the base of the oceanic crust 4 km below the seafloor? What is P 30 km below the base of this crust, assuming the density of peridotite is 3.3 g/cm^3? Assume that the ocean is 5 km deep and the density of seawater is 1 g/cm^3. Express your answers in bars, kilobars, megapascals, and gigapascals. (*Answers:* 1700 bars = 1.7 kbar = 170 MPa = 0.17 GPa; 11,600 bars = 11.6 kbar = 1160 MPa = 1.16 GPa)

Composition and Classification of Magmatic Rocks

FUNDAMENTAL QUESTIONS CONSIDERED IN THIS CHAPTER

1. How are rocks sampled in the field and analyzed in the laboratory to determine their chemical, modal, and mineralogical composition?

2. What do these analyses disclose regarding the composition of magmatic rocks?

3. How can analytical data be presented to reveal petrogenetically significant similarities, patterns, and contrasts?

4. How are magmatic rocks classified so as to convey meaningful petrogenetic information concerning the origin and evolution of the magmas from which they solidified?

5. What special petrogenetic information do the trace element and isotopic compositions of magmatic rocks provide?

INTRODUCTION

Magmatic rocks possess a seemingly endless variety of chemical, modal, and mineralogical compositions from which the petrologist must extract information about how rocks form. Samples can be collected, laboratory analyses made, similarities and contrasts in composition sought through graphed data and classification, compositional patterns found, and interpretative models hypothesized.

Compositions of magmatic rocks, together with their fabrics and field relations (Chapters 7, 9, and 10), are the essential "nuts and bolts" from which interpretive models (Chapters 11–13) can be constructed. After a

brief introduction to sampling and analytical techniques for obtaining compositional data on rocks the chapter focuses on three compositional attributes of rocks:

1. Mineralogical, modal, and major element compositions that are the bases for different classifications of magmatic rocks

2. Trace element compositions

3. Isotope ratios

All of these compositions provide crucial information regarding the sources and evolutionary paths of magmas.

❉2.1 ANALYTICAL PROCEDURES

Modern petrology relies heavily on quantitative, or numerical, data on rock and mineral compositions obtained by a variety of analytical devices and techniques. Data can be stored and processed and interpretive models created by using computers.

2.1.1 Sampling

The validity and usefulness of any compositional data are crucially dependent upon the quality of the sample collected in the field. "Garbage" samples collected in the field are liable to yield "garbage" analytical results. The importance of this crucial first step cannot be ignored or minimized. If the analysis of the sample is to be truly representative of the body of rock, the sample should be as free as possible of overprinting effects of alteration and weathering. This is especially important for chemical and isotopic analyses. Virtually all primary, high-temperature (high-T) minerals and glasses are metastable near-atmospheric conditions. Except for quartz, they are subject to replacement by secondary minerals more stable at near atmospheric temperatures such as carbonates, clays, and ferric oxides. This **weath-**

ering is especially common in warm, wet climates and in subaqueous environments. The chemical composition of the rock also changes during these secondary replacement processes, including but not restricted to the addition of water and oxidation of ferrous iron. It is difficult to know to what extent mobile chemical elements in the original rock have experienced gains or losses in their concentration. Hence, the sample must be as fresh and unweathered as possible. Silicate minerals should have vitreous luster with distinct grain boundaries and well-defined cleavage or fracture surfaces. The rock sample should be free of patchy discoloration due to films of secondary manganese or iron oxide, or a whitish clouding of feldspar caused by conversion to clay minerals. Secondary carbonate and silica (e.g., chalcedony) should be absent. Pervasively weathered samples crumble from the outcrop, whereas unweathered rock breaks with difficulty under the hammer and yields sharp-edged pieces. At an outcrop, a hammer may be unable to break far enough into the rock through a weathered rind to obtain a fresh sample. A portable diamond core drill may be required. As a final test for the quality of a sample, a thin section should be examined with a petrographic microscope.

Many rocks have also experienced **alteration** by hot gases around volcanic vents and by hot aqueous (hydrothermal) solutions farther beneath the surface. Alteration can create many of the same secondary minerals as does weathering but also includes conversion of primary magmatic minerals to somewhat higher T zeolites, chlorite, serpentine, epidote, and others.

If a fresh sample can be obtained, the petrologist must next determine how large it should be and where it should be taken from a heterogeneous rock body. Different sampling plans must be adopted to solve different problems. In most plans, the size of the sample should be representative of the outcrop; therefore, it must be many times larger than the dimensions of the coarsest grains. Obviously, a representative sample of a coarse phaneritic granite containing phenocrysts of alkali feldspar 3 cm in length must be considerably larger than a representative sample of an aphanitic, nonporphyritic basalt.

For chemical analyses, samples should be pulverized to at least-200 mesh to ensure homogeneity of the powder. Rock powders are contaminated by pulverizing machines. Alloy steel pulverizers add contaminating Fe as well as possible Cr, Co, Ni, and Mn. Ceramic pulverizers add Al. Agate adds Si. Corrections may be applied to analytical results for such contaminants.

2.1.2 Analyses

Accuracy and Precision. Any measured value should be accompanied by a statement of accuracy and precision indicating the reliability of the value; otherwise, it has little meaning.

Precision, or reproducibility, is a number that indicates how much statistical variation from the average or mean value occurs in replicate determinations (see, for example, Le Maitre, 1982). The greater the number of determinations of a particular quantity in a sample the smaller is the precision and the more reliable is the average value. Suppose, for example, the precision of analysis of, say, CaO in basalt samples is ±0.25 wt.%. If analysis of one sample yields an average value of 8.45 wt.% and of another yields 8.75 wt.% it might be supposed that these two values are significantly different. However, if the ± uncertainties are taken into consideration the first analysis actually lies between 8.45 + 0.25 = 8.70 and 8.45 − 0.25 = 8.20 wt.% and the second lies between 8.75 + 0.25 = 9.00 and 8.75 − 0.25 = 8.50 wt.% Hence, the two analyses actually overlap in the range of 8.50 to 8.70 wt.% and it is possible the two samples have the same CaO concentration in that range. Additional replicating analyses to reduce the precision would be necessary to resolve the question whether or not the two samples have the same CaO value.

Accuracy is less easily determined; it is an indication of how close the measurement is to the "true" value. But what is the true value? For chemical analyses, accuracy is an expression of how the result for a standard sample analyzed in a petrologist's laboratory compares with the "accepted" value (Govindaraju, 1989) for the standard sample.

Modal Analyses. Determination of the volumetric proportions of the minerals that make up a rock—its modal composition or mode—can be done by various techniques, yielding different degrees of precision and accuracy (van der Plas and Tobi, 1965). The quickest, but least accurate technique, adequate for preliminary work, is a visual estimate of mineral proportions in a hand sample or thin section (Figure 2.1). Greater accuracy and precision can be obtained on sawed, polished slabs on which a transparent overlying grid is placed. The proportion (or percentage) of grid intersections falling on a particular mineral indicates its proportion in the rock. The same concept underlies commercially available electromechanical point counters used on thin sections of rocks. Sometimes troublesome distinctions between alkali feldspar, plagioclase, and quartz can be overcome by selective staining (e.g., Bailey and Stevens, 1960).

Because the volumetric proportions of minerals in a rock are based on their proportions on a surface area, the modal composition of a rock having a preferred orientation of inequant mineral grains that is based only on one surface generally will be inaccurate. For example, the mode of a rock in which biotite flakes are strongly oriented in planar fashion would yield, on a surface parallel to that plane, an apparent overabun-

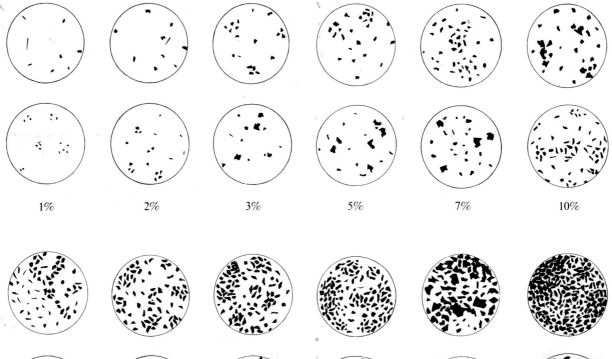

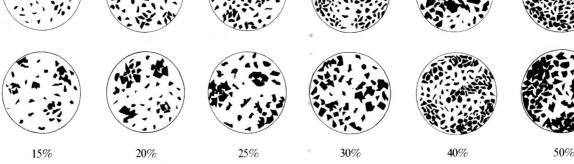

2.1   Charts to aid the visual estimation of modal proportions of minerals in rocks. (From Terry RD, Chilingar GV.)

dance of biotite at the expense of the other minerals. In rocks that contain oriented tablets of plagioclase, an examination in hand sample of a rock surface parallel to their {010} direction would reveal little if any polysynthetic twinning, thus possibly leading to an erroneous modal determination for plagioclase. (This twinning is the surest criterion for recognition of plagioclase in hand sample.) To overcome biasing the mode in anisotropic rocks, it is necessary to average the analysis on three mutually perpendicular surfaces, generally oriented parallel and perpendicular to the planar and/or linear fabric elements.

Modal analyses can also be performed by analyses of digital images on a computer. Digital images of rocks can be produced from backscattered electron intensities that portray the average atomic number of minerals or by element mapping under the beam of an electron microprobe (discussed later). The Rietveld X-ray diffraction method provides modes of wholly crystalline (nonglassy) rocks. However, because the sample is pulverized, this method ignores textures of mineral grains, which might be important in the case of secondary mineral overgrowths and replacements.

Chemical Analyses. Since the 1960s, rapid instrumental methods of chemical analyses of rocks and minerals have taken the place of tedious wet methods requiring skilled chemical analysts. All of these instrumental methods are comparative, based upon a comparison of the intensity of some measured quantity in an unknown sample with the intensity in a standard sample of known composition. Some instrumental methods have very low detection limits that are measured in parts per million (**ppm**) or even in parts per billion (**ppb**) by weight. The detection limit, specific for an element, is what the method can measure in a sample above the background of instrumental "noise." Although *element* concentrations are determined in all of these methods, major elements are nonetheless reported as *oxides*, following the traditional presentation of the results of wet methods of analysis.

Some methods, such as atomic absorption spectrophotometry (AA), flame photometry, emission spectroscopy, and inductively coupled plasma spectrometry (ICP), depend upon detection of shifts in outer electrons in the atom as the sample is heated to extreme temperatures. In X-ray fluorescence spectrometry (XRF), atoms in the sample are bombarded with X rays of sufficient energy to eject an inner orbital electron. As an outer electron drops into its place, energy is released as fluorescent Xrays whose wavelength is characteristic of the atom, whether Ca, Si, Al, etc. The intensity of the wavelength-specific X ray is proportional to the concentration of the chemical element in the sample. Other methods utilize the radioactivity of atomic nuclei. For example, in gamma-ray spectrometry, the intensities of gamma rays of different energies are a measure of the amount of naturally decaying K, Th, and U isotopes. In neutron activation analysis (NAA), artificial radioactive isotopes are induced by reaction with neutrons in a reactor and the energies of decaying radiation monitored. Ratios of isotopes of the elements are determined by a mass spectrometer, an instrument that distinguishes among nuclei of differing masses accelerated through a magnetic field.

Chemical analysis of minerals has been revolutionized by the electron microprobe, which is essentially an electron microscope fitted with XRF spectrometers that has the capability of analyzing a 1- to 10-micrometer-diameter volume of a mineral in situ in a polished thin section of a rock. Thus, spatial variations in composition within a mineral grain can be determined.

An example of an unusually complete whole-rock chemical analysis is shown in Table 2.1. Oxides > 0.1 wt.% (weight percent; in a 100-gram sample) are said to be **major elements,** whereas **trace elements** contain <0.1 wt.% of the element, or more conventionally, <1000 ppm. This limit is rather arbitrary. Some elements are not consistently major or trace elements. For example, Ti is a major element in basalts because of abundant Fe-Ti oxides and Ti-bearing pyroxenes, but it is a trace element in dunites, which are made of more than 90 modal % olivine that contains virtually no Ti. About 99% of most rocks are made of 11 major-element oxides: SiO_2, TiO_2, Al_2O_3, Fe_2O_3, FeO, MnO, MgO, CaO, Na_2O, K_2O, and P_2O_5. Iron typically occurs in two oxidation states (ferric, Fe_2O_3, and ferrous, FeO) in rocks and minerals, but most instrumental methods cannot distinguish between them and so total Fe is expressed as either Fe_2O_3t or as FeOt.

Most silicate rocks contain **volatiles** such as water, carbon dioxide, sulfur, fluorine, and chlorine. H_2O^+ is structural or combined water in the form of the hydroxyl ion, (OH^-) in amphiboles and micas and in secondary limonite and clays and as molecular H_2O in glasses. H_2O^- is "dampness," or water absorbed on grain surfaces and in pore spaces that can be driven off by heating to 110°C. Significant amounts of CO_2 are present if secondary carbonate, usually calcite, has been introduced during weathering or alteration. Total volatiles in a rock can be determined by weighing an aliquot of rock powder, heating it to 1000°C, and weighing again to determine the **loss on ignition (LOI).** To facilitate comparisons, many rock analyses are recalculated to a volatile-free basis totaling 100.00 wt.%.

Because analyses of rocks and minerals always involve human and instrumental errors and because not all elements are analyzed, the total of major and trace constituents is never exactly 100.00 wt.%. Generally speaking, if an analysis lists all major oxides, including H_2O, and the total lies between 98.8 wt.% and 100.8 wt.% it is considered to be acceptable.

Chemical analyses of average common rock types are listed in Table 2.2.

Table 2.1   Whole-Rock Chemical Composition of Basalt from the Columbia River Plateau, Sample BCR-1[a]

SiO_2	54.06	Ag	27*	Er	3.63	Nd	28.8	Tb	1.05
TiO_2	2.24	As	650	Eu	1.95	Ni	(13)	Te	(4.9*)
Al_2O_3	13.64	Au	(0.66*)	F	490	Pb	(13.6)	Th	5.98
Fe_2O_3	3.59	Ba	681	Ga	22	Pr	6.8	Tl	0.3
FeO	8.88	Be	(1.6)	Gd	6.68	Rb	47.2	Tm	0.56
MnO	0.18	Bi	47*	Ge	1.5	Re	0.84	U	1.75
MgO	3.48	Br	(72*)	Hf	4.95	Rh	(0.23*)	V	407
CaO	6.95	Cd	130*	Hg	(7.9*)	S	410	W	(0.44)
Na_2O	3.27	Ce	53.7	Ho	1.26	Sb	0.62	Y	38
K_2O	1.69	Cl	59	In	92*	Sc	32.6	Yb	3.38
P_2O_5	0.36	Co	37	La	24.9	Se	(88*)	Zn	129.5
H_2O^+	0.75	Cr	(16)	Li	12.9	Sm	6.59	Zr	190
H_2O^-	0.81	Cs	0.96	Lu	0.51	Sn	(2.7)		
CO_2	0.03	Cu	(19)	Mo	(1.6)	Sr	330		
LOI	1.67	Dy	6.34	Nb	(14)	Ta	0.81		
Total	99.93								

[a]Major element oxides in wt.%. Less certain values in parentheses. *, Trace element concentration in parts per billion (ppb); all other trace elements in parts per million (ppm).
Data from Govindaraju (1989).

Table 2.2 Average Chemical Compositions of Some Common Rock Types (Recalculated Volatile-Free to Total 100%) and Their Normative Compositions[a]

	PHONOLITE	SYENITE	TRACHYTE	GRANITE	RHYOLITE	GRANODIORITE	DACITE	DIORITE	ANDESITE
n	340	517	534	2485	670	885	651	872	2600
SiO_2	57.43	59.63	62.31	71.84	73.95	66.91	65.98	58.34	58.70
TiO_2	0.63	0.86	0.71	0.31	0.28	0.55	0.59	0.96	0.88
Al_2O_3	19.46	16.94	17.27	14.43	13.48	15.92	16.15	16.92	17.24
Fe_2O_3	2.85	3.09	3.04	1.22	1.50	1.40	2.47	2.54	3.31
FeO	2.07	3.18	2.33	1.65	1.13	2.76	2.33	4.99	4.09
MnO	0.17	0.13	0.15	0.05	0.06	0.08	0.09	0.12	0.14
MgO	1.09	1.90	0.94	0.72	0.40	1.76	1.81	3.77	3.37
CaO	2.78	3.59	2.38	1.85	1.16	3.88	4.38	6.68	6.88
Na_2O	7.96	5.33	5.57	3.71	3.61	3.80	3.85	3.59	3.53
K_2O	5.36	5.04	5.07	4.10	4.37	2.76	2.20	1.79	1.64
P_2O_5	0.18	0.30	0.21	0.12	0.07	0.18	0.15	0.29	0.21
Q		0.83	5.00	29.06	32.87	22.36	22.73	10.28	12.37
C				0.92	1.02	0.26			
Or	30.96	29.29	29.41	24.50	25.44	16.11	12.82	10.42	9.60
Ab	35.48	44.34	46.26	31.13	30.07	31.73	32.07	29.96	29.44
An	1.50	7.24	7.05	8.04	4.76	17.34	20.01	24.40	26.02
Lc									
Ne	16.50								
Di	6.89	5.35	2.14				0.11	4.67	4.84
Wo	0.73								
Hy		4.16	2.06	3.37	1.34	7.40	5.73	12.56	9.49
Ol									
Mt	4.05	4.41	4.33	1.75	2.14	2.00	3.53	3.63	4.74
Il	1.18	1.60	1.34	0.58	0.54	1.03	1.09	1.80	1.65
Ap	0.41	0.70	0.49	0.28	0.17	0.42	0.34	0.68	0.50

	TRACHYANDESITE	TRACHYBASALT	BASALT	BASANITE	NEPHELINITE	ANORTHOSITE	LHERZOLITE	HARZBURGITE	DUNITE
n	232	161	3594	165	176	104	179	206	93
SiO_2	59.30	49.99	49.97	45.16	41.81	51.12	45.43	43.73	41.04
TiO_2	1.10	2.44	1.87	2.56	2.74	0.65	0.45	0.28	0.10
Al_2O_3	17.03	16.89	15.99	14.99	14.76	26.29	4.39	2.57	1.95
Fe_2O_3	3.32	3.75	3.85	4.02	5.64	0.98	5.15	6.00	3.85
FeO	3.27	6.28	7.24	7.65	6.35	2.10	7.44	7.09	10.05
MnO	0.16	0.16	0.20	0.16	0.27	0.05	0.17	0.16	0.76
MgO	2.62	5.25	6.84	8.71	6.58	2.16	30.31	36.34	40.66
CaO	5.06	8.03	9.62	10.39	12.25	12.69	5.68	3.18	1.08
Na_2O	4.44	4.02	2.96	3.62	4.93	3.20	0.59	0.34	0.21
K_2O	3.27	2.59	1.12	2.00	3.56	0.66	0.27	0.15	0.09
P_2O_5	0.42	0.60	0.35	0.75	1.10	0.09	0.12	0.14	0.21
Q	7.80								
C									0.80
Or	19.00	15.06	6.52	11.61	3.16	3.86	1.50	0.83	0.47
Ab	36.80	29.39	24.66	12.42		23.16	4.66	2.60	1.69
An	16.58	20.10	26.62	18.38	7.39	49.71	7.99	4.17	1.17
Lc					13.57				
Ne		2.23		9.55	21.95	1.89			
Di	3.95	11.85	14.02	21.03	32.36	8.61	13.54	6.93	
Wo									
Hy	6.06		15.20				21.48	21.13	14.48
Ol		8.28	1.50	12.38	2.32	2.01	36.31	46.22	67.38
Mt	4.73	5.36	5.49	5.72	7.95	1.40	7.00	7.94	5.20
Il	2.07	4.55	3.49	4.77	5.05	1.22	0.79	0.50	0.18
Ap	0.97	1.38	0.82	1.74	2.51	0.21	0.26	0.30	0.47

[a]The number of analyses averaged is represented by n. The rock-type names are those used by the author of the report in which the analyses were published; that is, the names are not based on the IUGS classification. In most instances there is little discrepancy between the original report writer's name and the IUGS name. Data from Le Maitre (1976).

☀2.2 MINERAL COMPOSITION OF MAGMATIC ROCKS

The minerals that crystallize from most magmas include only a very small number of the thousands of known mineral species. Major **rock-forming minerals** include only olivine, pyroxene, amphibole, mica, feldspars, quartz, feldspathoids, and Fe-Ti oxides (chiefly magnetite and ilmenite). All of these are solid solutions, except quartz, although it, too, contains variable concentrations in the ppm range of $Al^{3+} + Li^{+}$ substituting for Si^{4+} and $(OH)^{-}$ for O^{2-}. Very rare magmatic carbonatite rocks are composed chiefly of carbonate minerals.

Because of the limited number of major minerals, their identification is relatively straightforward in hand sample with the aid of a hand lens and in thin section with a petrographic polarizing microscope. X-ray diffraction analysis can be useful for mineral identification in very fine-grained rocks, but the electron microprobe is the ultimate tool. In addition to the standard mineral properties such as cleavage, hardness, and habit, the petrologist can use **mineral associations** as a means of mineral identification in hand samples. For example, Figure 2.2 shows highly generalized mineral associations in common subalkaline rocks. (In less common alkaline rocks mineral associations are more complex.) Most magmatic rocks contain feldspar; of the two feldspar solid-solution series—alkali feldspar and plagioclase—the latter is more widespread. One of the most easily identified minerals in rocks is biotite; once recognized, its presence suggests the probable coexistence of quartz, alkali feldspar, plagioclase, and amphibole but not magnesian olivine and perhaps not pyroxene. More reliably, quartz cannot coexist with pale green magnesian olivine, so once one mineral is recognized for certain the other cannot be present, at least under equilibrium conditions. Other "forbidden" associations are feldspathoids (leucite and nepheline) with quartz or orthopyroxene.

Appendix A tabulates chemical compositions of select major rock-forming solid solutions occurring in magmatic rocks. Such tabulations are generally more useful for the petrologist than mineral formulas because they are readily compared to whole-rock chemical compositions presented in the same tabular format.

The standard references for rock-forming minerals are the five-volume work by Deer, Howie, and Zussman (1962) and their abridged, single-volume paperback (1997).

Minerals that occur in small modal proportions of no more than a few percent and do not influence the naming of a rock are sometimes referred to as **accessory minerals.** Some crystalline phases occur only as accessory minerals, such as apatite and zircon, rarely as

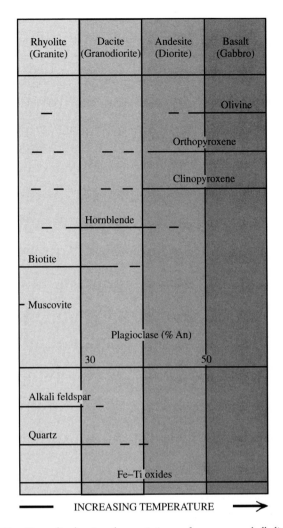

2.2 Generalized **mineral associations** of common subalkaline rocks plotted according to relative temperature of equilibration. Rock types listed across top (rhyolite, dacite, etc.) are aphanitic or glassy, usually volcanic. Corresponding phaneritic, usually plutonic, rock types (granite, diorite, etc.) are in parentheses. An andesite (diorite), for example, commonly contains intermediate composition plagioclase (An_{40-50}, or andesine), clinopyroxene, orthopyroxene, hornblende, and Fe-Ti oxides; possible minor constituents are biotite, quartz, and olivine. Greater Mg/Fe ratio in mafic solid-solution minerals and greater Ca/Na and K/Na ratios in plagioclase and alkali feldspar solid solutions, respectively, correlate with increasing temperature.

major rock-forming minerals. Although seemingly insignificant, these minerals can be petrologically important because they contain large concentrations of trace elements (discussed later) and some can be used for age determinations. Common accessory minerals are listed in Table 2.3.

2.2.1 Glass

Glass, not a mineral, originates from magmas that lose heat so rapidly that atoms in the silicate melt have insufficient opportunity to organize into the regular geometric arrays of crystals. Instead, the melt solidifies into

Table 2.3   Generally Compatible Trace Elements and the Minerals in Which They Occur

MAJOR MINERAL	SIMPLE FORMULA	COMPATIBLE TRACE ELEMENTS
Olivine	$(Mg, Fe)_2SiO_4$	Ni, Cr, Co
Orthopyroxene	$(Mg, Fe)SiO_3$	Ni, Cr, Co
Clinopyroxene	$(Ca, Mg, Fe)_2(Si, Al)_2O_6$	Ni, Cr, Co, Sc
Hornblende	$(Ca, Na)_{2\text{-}3}(Mg, Fe, Al)_5$ $(Si, Al)_8O_{22}(OH, F)_2$	Ni, Cr, Co, Sc
Biotite	$K_2(Mg, Fe, Al, Ti)_6$ $(Si, Al)_8O_{20}(OH, F)_4$	Ni, Cr, Co, Sc, Ba, Rb
Muscovite	$K_2Al_4(Si, Al)_8O_{20}(OH, F)_4$	Rb,Ba
Plagioclase	$(Na, Ca)(Si, Al)_4O_8$	Sr, Eu
K-feldspar	$KAlSi_3O_8$	Ba, Sr, Eu
ACCESSORY MINERALS[a]		
Magnetite	Fe_3O_4	V, Sc
Ilmenite	$FeTiO_3$	V, Sc
Sulfides		Cu, Au, Ag, Ni, PGE[b]
Zircon	$ZrSiO_4$	Hf, U, Th, heavy REEs
Apatite	$Ca_5(PO)_3(OH, F, Cl)$	U, middle REEs
Allanite	$Ca_2(Fe, Ti, Al)_3(O, OH)$ $(Si_2O_7)(SiO_4)$	Light REEs, Y, U, Th
Xenotime	YPO_4	Heavy REEs
Monazite	$(Ce, La, Th)PO_4$	Y, light REEs
Titanite (sphene)	$CaTiSiO_5$	U, Th, Nb, Ta, middle REEs

[a]Accessory minerals constitute only a small fraction of rock but their very high partition coefficients create a disproportionate influence on bulk distribution coefficients.
[b]**Platinum group elements:** Ru, Rh, Pd, Os, Ir, Pt.

a very viscous amorphous glass—a supercooled liquid solution of O, Si, Al, Ca, K, and so on. It is, therefore, common in extruded lavas but is also found along margins of thin dikes emplaced in the shallow cool crust. Rarely, glass is produced locally by frictional processes in fault zones (creating pseudotachylite), by impact of large meteorites, by some lightning strikes (creating "fulgurite"), and by burning of underground coal. No sharp transition appears between the liquid solution and amorphous solid upon heating or cooling (Bouška, 1993). Rhyolitic glass is particularly widespread because the relatively high viscosity of the silicate melt from which it forms hinders crystallization during rapid cooling, particularly of extruded magmas. Rhyolitic glass is colorless in thin section but gray, black, or dark red-brown in hand samples of massive obsidian because of minute crystals of dark colored minerals. Increasing concentrations of Fe, generally accompanied by increasing Ca, Mg, and decreasing Si and K, produce increasingly darker brown colors so that basaltic glass in thin section is honey- to cinnamon-brown and jet black in hand sample.

All glass is metastable at near-surface conditions and is susceptible to replacement by more stable minerals (Section 7.1.1). Geologically older glasses are more rare; most are Cenozoic. Glasses are also susceptible to loss of relatively mobile Na and K.

✳2.3 CHEMICAL COMPOSITION OF MAGMATIC ROCKS

2.3.1 Variation Diagrams

Chemical compositions of rocks and minerals are conventionally presented by petrologists in two formats: tables of oxide and/or element concentrations—as in Tables 2.1 and 2.2 and Appendix A—and graphs where points represent the concentrations of chemical constituents. These graphs, called **variation diagrams,** show trends or patterns in the chemical data. Modal data can also be presented in variation diagrams. Three common types of diagrams are used by petrologists:

1. Cartesian graph of two variables (x and y)
2. Triangular diagram
3. Normalized diagrams (see Section 2.5)

The Cartesian diagram (Figure 2.3) best portrays the absolute concentrations of any compositional parameter and is amenable to quantitative interpretations; however, in order to represent all n constituents in a rock, $(n - 1)$ plots are required. The triangular plot presents one more constituent than can be represented in one Cartesian graph but cannot show absolute concentrations of the three variables, only their ratios; it is most useful in portraying trends in variation in suites of rocks but cannot be used to extract any quantitative in-

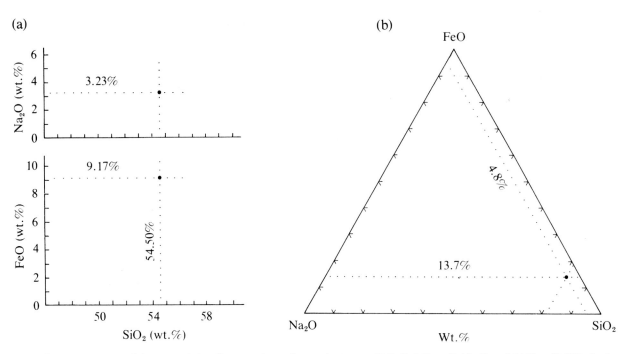

2.3 Plotting compositional data on **variation diagrams.** A sample contains 9.17 wt.% FeO, 3.23 wt.% Na$_2$O, and 54.50 wt.% SiO$_2$. In the Cartesian diagram (a), FeO and Na$_2$O are plotted against SiO$_2$ (all in wt.%). To plot SiO$_2$, Na$_2$O, and FeO on the triangular diagram (b), *they must be recalculated to total 100.00.* First, they are summed, 54.50 + 9.17 + 3.23 = 66.90 and a recalculation multiplier found, 100.00/66.90 = 1.495. Second, the wt.% of each constituent is recalculated; Si$_2$O is 54.50 × 1.495 = 81.48, FeO = 13.71, and Na$_2$O = 4.83. The total of the three recalculated oxides is now 81.5 + 13.7 + 4.8 = 100.0 These recalculated values can then be plotted so that each apex represents 100 wt.% of a constituent and the leg of the triangle opposite the apex is the locus of points representing 0 wt.% of that constituent. A line parallel to the leg of the triangle opposite the FeO apex and 13.7% of the way toward that apex is the locus of points representing 13.7 wt.% FeO. Similarly, the line labeled 4.8 % is the locus of points representing 4.8 wt.% Na$_2$O. The intersection of these two lines is a point that represents the relative SiO$_2$, FeO, and Na$_2$O wt.% in the sample. Note that it is only necessary to draw lines for any two of the three constituents represented in the diagram because the third variable is the difference from 100% of the other two.

formation. In either type of diagram, two or more constituents may be combined into one variable or ratios of elements may be represented by a variable.

2.3.2 Continuous Spectrum of Rock Compositions

One of the most frequently used Cartesian diagrams, especially for chemical classification of volcanic rocks (discussed later), is the total alkalies (Na$_2$O + K$_2$O) versus silica (SiO$_2$) diagram. Figure 2.4 is such a TAS diagram, in which are plotted over 41,000 published whole-rock analyses of silicate magmatic rocks of all compositions of all ages from all over the world. (Rare carbonatites are excluded.) This plot has two especially significant attributes:

1. Magmatic rocks constitute a continuous spectrum of compositions, lacking natural breaks or discontinuities. Variation diagrams plotting any other combination of elements show the same continuity. This continuous spectrum introduces an arbitrariness into chemical classification of rocks, as discussed later. More importantly, the spectrum poses some of the most fundamental questions in igneous petrology: How is such a large and continuous compositional

range of rocks created? Is there a corresponding range in magma compositions? If so, what processes of generation of magmas from solid source rock could yield such a range? Or, are magmas generated from solid rock more restricted in composition but subsequently diversified in some manner?

2. Variations in silica (about 30–80 wt.%) and total alkalies (0–20 wt.%) occupy only a part of a possible range of 100% in each variable. This attribute of the worldwide compositional spectrum begs the question, What petrologic factors dictate this restricted range? Why, for example, are there no magmatic rocks that contain 95 wt.% SiO$_2$, or 50 wt.% total alkalies?

These fundamental questions are addressed in later chapters, especially Chapters 11–13.

It is interesting to note that the major rock-forming minerals that make up magmatic rocks form a polygonal envelope surrounding the global rock spectrum (Figure 2.5). Variations in the modal proportions of major minerals can produce any rock within the envelope, whose corners are represented by Fa (Fe-olivine

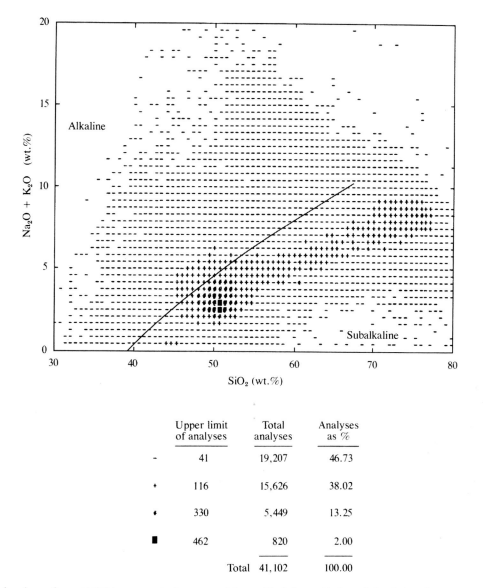

	Upper limit of analyses	Total analyses	Analyses as %
-	41	19,207	46.73
✦	116	15,626	38.02
✦	330	5,449	13.25
■	462	820	2.00
		Total 41,102	100.00

2.4   Chemical analyses of over 41,000 igneous rocks from around the world of all ages. Each symbolized plotted point represents a particular number of analyses falling within the indicated range (0–41, 42–116, etc.). Nearly half (46.7%) of all igneous rocks are widely scattered over the diagram (dash symbol), whereas slightly more than half (53.3%) are tightly clustered in a central band. Note the still higher concentration of analyses near 2.5 wt.% ($Na_2O + K_2O$) and 50 wt.% SiO_2, corresponding roughly to basalt, the dominant magmatic rock type on Earth. (Compiled by and furnished courtesy of R. W. Le Maitre, University of Melbourne, Australia.)

end member); Ne, Lct (two of the most common feldspathoids); Kfs; and Qtz. Some accessory minerals, such as magnetite and ilmenite, plot well outside this envelope but, because of their small concentrations, do not disperse whole-rock chemical compositions very far toward them in the diagram.

Variation diagrams of modal compositions of rocks are also continuous spectra, such as Figure 2.6.

✳2.4 CLASSIFICATION OF MAGMATIC ROCKS

Scientists have traditionally sought regularity, order, and predictability in their investigations of the natural world.

However, for the petrologist, the continuity of rock compositions revealed in Figures 2.4 and 2.6, the seemingly endless variety of fabric, and the wide range of geologic environments in which rocks form pose formidable obstacles to erecting a well ordered, simple, single rock classification. Unlike in the plant and animal kingdoms, which have discrete species, no such natural divisions exist in rocks. Rocks are more like complex, highly variable biological ecosystems; minerals constituting a rock are like the plant and animal species constituting an ecosystem.

Despite the obstacles, a consistent classification of rocks is essential for communication with other petrologists, who should all speak the same language of clas-

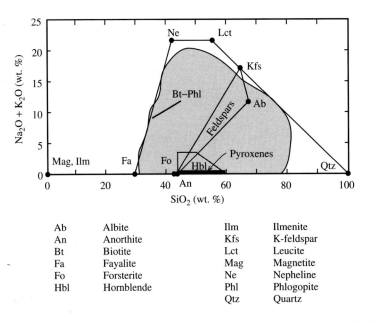

Ab	Albite	Ilm	Ilmenite
An	Anorthite	Kfs	K-feldspar
Bt	Biotite	Lct	Leucite
Fa	Fayalite	Mag	Magnetite
Fo	Forsterite	Ne	Nepheline
Hbl	Hornblende	Phl	Phlogopite
		Qtz	Quartz

2.5 End-member compositions of major magmatic rock-forming minerals compared with the field of worldwide magmatic rocks (shaded) from Figure 2.4. Triangular field of feldspar solid solution is outlined by the K-feldspar-albite-anorthite end members. Trapezohedral field is hornblende solid solutions.

sification; a particular rock name should convey the same meaning to every petrologist, regardless of his or her native tongue. In addition, classification serves as an important means of systematizing information. Through appropriate and relevant classification, meaningful patterns in composition, fabric, field relations, and, therefore, origin can be perceived. As all classifications of rocks are the fruits of the human mind attempting to erect discrete subdivisions where none exists in the natural, uninterrupted continuum of rock properties, every classification is, to some degree, arbitrary and imperfect.

There are many different criteria for classification; consequently, many different labels exist for the very

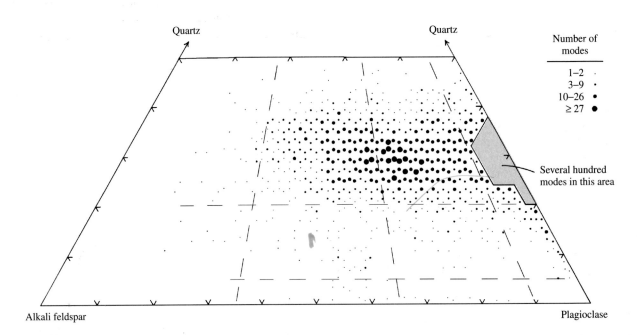

2.6 Modal composition of 4368 samples from 102 Late Triassic to Late Cretaceous plutons in the central part of the Sierra Nevada batholith, California. The number of sample modes plotting within a triangular area 2% on a side is coded by the indicated symbol; the greatest number of modes in any such triangle is 35. Note the continuity of modal variation across dashed-line boundaries between rock-type compartments from Figure 2.8. (Data from Bateman, 1992.)

same rock. Each has its own benefit and use; none can combine the merits of all. "A rock may be given one name on the ground of field occurrence and from hand lens examination, only to require another when it is studied in thin section, and perhaps a third when it is chemically analyzed. . . . Different schemes have different objects in view" (Williams et al., 1982, p. 68; see the following pages in this classic text for an extended discussion of igneous rock classification).

2.4.1 Classification Based on Fabric

We review here only the most fundamental rock terminology based on fabric as it is generally introduced in a beginning geology course. Magmatic fabric (Chapter 7) is essentially governed by time-dependent (kinetic) processes in the solidifying magma, such as its rate of heat loss, or cooling.

Four principal types of fabric occur in magmatic rocks: phaneritic, aphanitic, glassy, and volcaniclastic. The first two refer to the dominant crystal grain size, which ranges over several orders of magnitude, from $<10^{-6}$ to 10 m.

Phaneritic applies to rocks that have mineral grains sufficiently large to be identifiable by eye (minute accessory minerals excepted). This texture is typical of rocks crystallized from slowly cooled intrusions of magma. **Aphanitic** rocks have mineral grains too small to be identifiable by eye and require a microscope or some other laboratory device for accurate identification. Aphanitic texture is most common in rapidly solidified extruded magma but can also be found in marginal parts of magma intrusions emplaced in the cool shallow crust. Some magmatic rocks contain essentially two grain-size populations and few of intermediate size; such texture is said to be **porphyritic.** The larger grains are **phenocrysts,** and the smaller constitute the **groundmass,** or **matrix.** Porphyritic aphanitic rocks are far more common than porphyritic phaneritic rocks. **Glassy,** or vitric, rocks contain variable proportions of glass, in contrast to **holocrystalline** rocks made entirely of crystals. A **vitrophyre** is a porphyritic rock that contains scattered phenocrysts in a glassy matrix.

The fragmental fabric of **volcaniclastic** rocks is produced by volcanic processes, chiefly explosive, producing broken fragments of glass, rocks, and/or minerals. Classification of volcaniclasts parallels that of sedimentary clasts according to their particle size, as follows:

	< 2 mm	2–64 mm	> 64 mm
volcaniclasts	**ash**	**lapilli**	**block, bomb**
sedimentary clasts	clay, silt, sand	granule, pebble	cobble, boulder

Consolidation of volcaniclasts produces volcaniclastic rock types that are classified according to their particle size (Figure 2.7).

2.4.2 Classification Based on Field Relations

The location where magma was emplaced provides a basis for rock classification. Some petrologists recognize three categories for rocks solidified from magmas emplaced onto the surface of the Earth (volcanic or extrusive), into the shallow crust (intrusive hypabyssal), and into the deep crust (intrusive plutonic). The first and the last categories are readily distinguished on the basis of their field relations but less directly on the basis of their grain size, **degree of crystallinity** (proportion of crystals to glass), and mineralogical composition.

Magmas emplaced onto the surface of the Earth as coherent lava flows or as fragmental deposits form **extrusive, or volcanic rocks.** These rocks are typically aphanitic and glassy. Many are porphyritic. Some have fragmental (volcaniclastic) fabric. High-T disordered feldspars are common, so that alkali feldspar, where present, is a clear sanidine. Other minerals that occur only at high-T and low-P in volcanic environments—including leucite, tridymite, and cristobalite—are found in some volcanic rocks. Amphiboles and biotite, especially where they occur as phenocrysts, are commonly partially altered to fine-grained anhydrous aggregates of Fe oxides, pyroxenes, and feldspars. Phenocrysts of feldspar and quartz commonly contain inclusions of glass.

Intrusive, or **plutonic, rocks** form where magma was intruded into preexisting rock beneath the surface of the Earth as **intrusions, or plutons.** Plutonic rocks are typically phaneritic. Monomineralic rocks composed

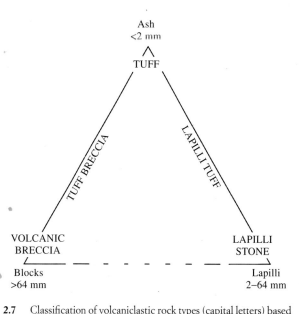

2.7   Classification of volcaniclastic rock types (capital letters) based on size of clasts.

only of plagioclase, or olivine, or pyroxene are well known but rare. Amphiboles and biotite are commonly partially altered, usually to chlorite. Some granites contain muscovite, which is exceedingly rare in volcanic rocks. Perthite—an intergrowth of sodic and potassic feldspar—is widespread and reflects slow cooling and exsolution in initially homogeneous alkali feldspar.

Characteristics of intermediate-depth **hypabyssal rocks** are not clearly distinct from those of volcanic and plutonic rocks. Many occur in shallow crustal dikes, sills, and plugs that represent feeding conduits for surface extrusions of magma. But dikes and sills are also intruded deep in the crust. Hypabyssal rocks can have fabric similar to that of plutonic and volcanic rocks. Because of these ambiguities, many petrologists tend to categorize magmatic rocks in the field simply as plutonic or volcanic.

2.4.3 Classification Based on Mineralogical and Modal Composition

Mineralogical Mnemonics. **Felsic** is a mnemonic adjective derived from the words *feldspar* and *silica*. It is a useful appellation for rocks that contain large proportions of feldspar with or without quartz and/or its polymorphs, tridymite and cristobalite. Granite and rhyolite made mostly of feldspar and quartz are examples of felsic rocks. The term *felsic* also applies to rocks containing abundant feldspathoids, such as nepheline, and to these rock-forming minerals as well. **Mafic** is a mnemonic adjective derived from the words *magnesium* and *ferrous/ferric*. *Mafic* is a less cumbersome term than the synonymous *ferromagnesian*. It refers to major rock-forming biotite, amphibole, pyroxene, olivine, and Fe-Ti oxide solid solutions as well as rocks that contain large proportions of them, such as basalt. **Ultramafic** rocks are especially rich in Mg and Fe and generally have little or no feldspar; an example is the olivine-pyroxene rock called peridotite. **Silicic** rocks contain large concentrations of silica, manifested by an abundance of alkali feldspar, quartz, or glass rich in SiO_2. Examples are rhyolite and granite. The term **sialic** is used less frequently for rocks rich in Si and Al that contain abundant feldspar and is used especially with reference to the continental crust.

Color is usually the first rock property noticed by the novice. However, a particular rock type can possess a wide range of colors; granites, as just one example, can be nearly white, shades of gray, green, red, and brown. These widely ranging colors reflect equally widely variable colors of the dominant rock-forming feldspars, whose pigmentation is a complex function of minute mineral inclusions, exsolution, and small concentrations of elements such as Fe in solid solution; none of these factors may be petrologically very significant and in any case may be difficult to determine.

Color is not a valid basis of rock classification and can, in fact, be highly misleading. **Color index** has been defined as the modal proportion of dark-colored minerals in a rock. But, in view of the fact that dominant rock-forming feldspars can be light- to dark-colored, a more accurate index should be defined on the basis of the proportion of mafic minerals. **Leucocratic** and **melanocratic** rocks can be defined as having 0–30% and 60–100% modal mafic minerals, respectively.

Rock Types. The classification of magmatic rocks most familiar to the beginning geology student is that of rock types. In contrast to the broadly defined compositional labels just described, a **rock type** has a narrowly defined composition and a particular fabric. Familiar rock types include rhyolite, andesite, and basalt (all aphanitic) and granite and diorite (both phaneritic). Many rock-type labels have a long and obscure history stemming from miners' jargon (Mitchell, 1985); many are coined from geographic locales, such as andesite from the Andes Mountains of western South America. About 800 igneous rock-type names are listed in the classic four-volume work of Johannsen (1931–1938), written toward the end of an era when petrology was mostly descriptive petrography and the coining of new rock names was in vogue. Today, most of these names have, fortunately, been abandoned and petrologists need have only a working knowledge of a few dozen major igneous rock-type names (Le Maitre, 1989).

Regrettably, however, few of these major names have had consistent usage among petrologists. One petrologist's andesite has been another's basalt. Personal biases and backgrounds have been strong factors in schemes of classification. If rock compositions were clustered into isolated clumps on any variation diagram it would be a simple matter to draw a line around each cluster and append a rock-type name to it. However, as Figures 2.4 and 2.6 show, compositions are not clustered but consist of a continuum. There are at least two approaches to nomenclature within this continuum:

1. Flexible, loosely defined limits could be defined, leaving the details to the individual petrologist guided by the circumstances and need at hand. However, this approach has over the decades resulted in considerable confusion in the geologic literature.

2. The continuous spectrum could be subdivided along specific, well-defined limits that follow as closely as possible a usage agreed upon by as many petrologists as possible. This is the approach of the International Union of Geological Sciences Subcommission on the Systematics of Igneous Rocks, hereafter referred to as the IUGS. The IUGS system of classification (Le Maitre, 1989) is a *universal standard* that can eliminate individual biases and contradictions among petrologists.

The IUGS rock-type classification for phaneritic (generally plutonic) rocks, which consist mostly (>10 modal % but usually more) of felsic minerals is set out in Figure 2.8. Quartz-rich felsic rocks are also classified in Figure 2.9; these are collectively referred to as **granitic rocks** or **granitoids.** Three special fabric categories of granitic rocks are not shown in these figures. A porphyritic aphanitic to finely phaneritic rock having abundant phenocrysts and occurring in a pluton (in-trusion) is called **porphyry;** depending on its modal composition it may be a granite porphyry, granodiorite porphyry, or other. Uniformly fine-grained phaneritic, very leucocratic granites composed almost entirely of feldspar and quartz that typically occur in thin dikes within a coarser-grained, somewhat more mafic granitic pluton are **aplite.** Commonly associated with aplite are equally leucocratic rocks called **pegmatite;** these are phaneritic rocks of highly variable grain size in which

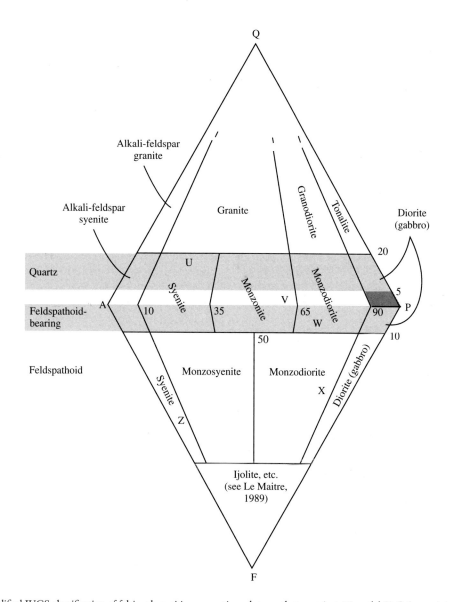

2.8 Slightly modified IUGS classification of felsic, phaneritic, magmatic rock types that contain >10 modal % Q (quartz) + A (alkali feldspar) + P (plagioclase) + F (feldspathoids). Coordinates of critical field corners along the A-P join (10, 35, 65, 90) refer to modal percentages of P. Numbers on right side of diagram are modal percentages of Q (5, 20) and of F (10). Mineralogical prefixes on left side of diagram are modifiers to be appended to rock-type names in associated lightly shaded or unshaded fields. For example, a rock whose mode in terms of QAPF plots at U is a quartz syenite, V is a monzonite, W is a feldspathoid-bearing monzodiorite, X is a feldspathoid monzodiorite, and Z is a feldspathoid syenite. Rocks plotting in the small darkly shaded parallelogram near the P apex are **anorthosite** if containing <10% mafic minerals, or, if containing >10% mafic minerals, are gabbro or diorite. **Gabbro** contains plagioclase more calcic than An_{50} and is further classified according to Figure 2.10a. **Diorite** has plagioclase less calcic than An_{50}. The composition of plagioclase can be approximated in a thin section by optical techniques. In the lower part of the diagram, **ijolite** is a rock composed of nepheline and clinopyroxene; it is essentially a phaneritic equivalent of aphanitic to glassy nephelinite (Figure 2.12). For a supplementary classification of phaneritic felsic rocks (granitoids) containing abundant plagioclase and >20% quartz see Figure 2.9. (Redrawn from Le Maitre, 1989, Figure B.4.)

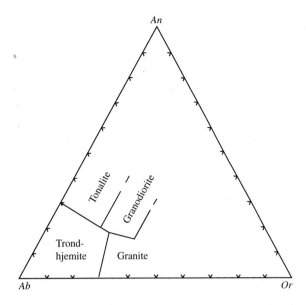

2.9   Classification of phaneritic felsic rock types (granitoids) containing abundant plagioclase and >20% quartz based on proportions of normative feldspars. (Redrawn from Barker, 1979.)

individual crystals are several centimeters to several meters.

Gabbros—phaneritic rocks made of plagioclase, pyroxene, and olivine—are classified in Figure 2.10a and phaneritic ultramafic rocks that contain <10 modal % felsic minerals in Figure 2.10b.

2.4.4 Classification Based on Whole-Rock Chemical Composition

There are many advantages of a numerical chemical classification. Insights are provided regarding the na-

ture, origin, and evolution of magmas. Rigorous comparisons can be made between members of suites of rocks and petrotectonic associations. The advantages of chemical classifications are obvious for very fine-grained rocks, whose mineralogical compositions may be difficult to determine, and certainly for glassy rocks (but beware of loss of Na and other possibly mobile elements). Aphanitic and glassy volcanic rocks can correspond more closely to the composition of the magma from which they formed than do porphyritic and phaneritic rocks, which may have been derived from magmas that experienced crystal accumulation during their evolution. Magmatic rocks whose characterizing minerals have been obliterated by alteration or metamorphism can be analyzed to reveal their original nature, provided diagnostic chemical elements have not been significantly mobilized during recrystallization.

However, an inherent weakness of purely chemical classifications is they have little or nothing to say regarding the effects of geologic processes on fabric and of different $P-T$ conditions that govern mineralogical composition.

<u>Aphanitic and Glassy Rock Types</u>.   A rigorously quantitative chemical classification of aphanitic and glassy, usually volcanic, rocks must be tempered by the fact that most rock-type names were established decades, and in some instances centuries, ago, when few if any chemical analyses were available and names were based upon mineralogical and modal compositions. Linking chemically based names to mineralogical/modal names was accomplished as follows.

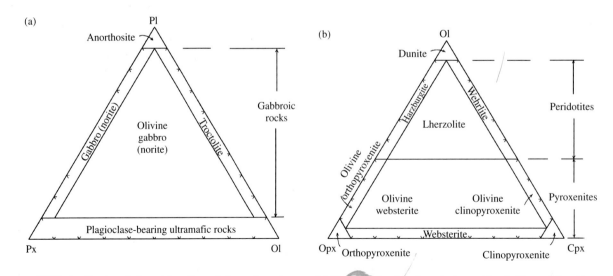

2.10   IUGS classification of phaneritic mafic and ultramafic rock types. (a) Classification of mafic rocks (see gabbro in Figure 2.8 and its caption) based on the proportions of plagioclase, pyroxene, and olivine. **Norite** has more orthopyroxene than clinopyroxene (see Le Maitre, 1989, p. 17, for details). (b) Classification of **ultramafic** rocks that are composed of orthopyroxene, clinopyroxene, and olivine. For hornblende-bearing ultramafic rocks see Le Maitre, 1989, Figure B.8. (Redrawn from Le Maitre, 1989.)

Le Maitre (1976) compiled 26,373 published rock analyses given a rock-type name by the original author(s). All analyses were sorted as to rock-type label, such as "andesite" and "dacite," irrespective of the classification scheme used. These compiled analyses are shown on a total alkalies-silica diagram in Figure 2.11. Overlap between the two fields of andesite and dacite reflects inherent variability in their composition—an attribute of all rock types no matter how defined. Nonetheless, averages of these two rock types are quite different. Average compositions of common magmatic rock types, which represent the opinions of thousands of petrologists over many decades, are listed in Table 2.2. All of the common volcanic rock-type names were so examined and bounding lines drawn on a total alkalies-silica diagram in such a way as to recognize a "consensus" composition (Le Bas et al., 1992). The resulting IUGS rock-type classification of aphanitic and glassy volcanic rocks is shown in Figure 2.12. Rock samples to be classified should be as fresh as possible (unweathered and unaltered). Analyses must be recalculated to 100% volatile-free before plotting.

A rock of basaltic composition in which the grain size is marginally phaneritic and transitional into gabbro is **diabase** (alternatively called **dolerite** by United Kingdom geologists). Diabase commonly occurs in dikes and sills but also constitutes local lava flows. An olivine-rich basalt or picrobasalt having MgO >18 wt.% is called **picrite** if $(Na_2O + K_2O) = 1-3$ wt.% and komatiite if $(Na_2O + K_2O) < 1$ wt.% and TiO_2 is low, generally <1 wt.%. Komatiites are commonly ultramafic and composed essentially of olivine and pyroxene so that they are chemically a peridotite, but their glassy to aphanitic texture precludes use of this phaneritic name.

The chemical classification can be appended to fabric **heteromorphs** that solidified from chemically similar magmas but have different fabrics. For example, chemically defined rhyolite can be, depending on fabric, rhyolite tuff, rhyolite breccia, rhyolite obsidian (wholly glass), rhyolite vitrophyre, and rhyolite pumice (vesicular glass).

A preliminary IUGS classification for volcanic rocks based upon modal proportions of phenocrysts (Figure 2.13) may be used in the field and before chemical analyses are available. This classification should never be final because the groundmass of porphyritic aphanitic or glassy rocks will always be poorer in plagioclase than the assemblage of phenocrysts because of the way magmas crystallize (Chapter 5). A rock containing sparse plagioclase as the dominant or sole phenocryst could be a dacite, rhyolite, or trachyte in terms of Figure 2.12.

<u>Absolute Concentration of Silica.</u> Except for the very rare carbonatites, silica (SiO_2) is the principal oxide constituent of magmatic rocks and serves as a basis for broadly defined classifications.

Some petrologists use a classification based on silica concentration in the rock analysis, as follows (Figure 2.12):

Silica concentration (wt.%)	Name
>66	**acid**
52 to 66	**intermediate**
45 to 52	**basic**
45 or less	**ultrabasic**

As defined here, acid and basic have no reference whatsoever to hydrogen ion content, or pH, as used in chemistry. (Long ago it was erroneously believed that SiO_2 occurred as silicic acid and metallic oxide components, such as CaO and FeO, as bases in magmas.) These four categories have no direct correlation with modal quantity of quartz in the rock, although as a general rule, acid rocks do contain quartz and ultrabasic ones do not. Two rocks having identical concentrations

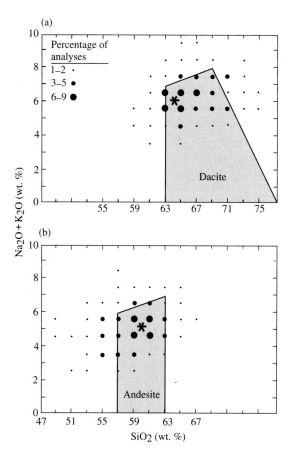

2.11 Total alkalies versus silica plots of analyzed rocks. In a worldwide database 727 analyses are designated as dacite (a) and 2864 analyses as andesite (b). Symbols indicate percentage of analyses falling within cells whose sides are 2% SiO_2 by 1% $(Na_2O + K_2O)$. Shaded fields of dacite and andesite from Figure 2.12. Asterisk is average of analyses. (Redrawn from Le Bas et al., 1992.)

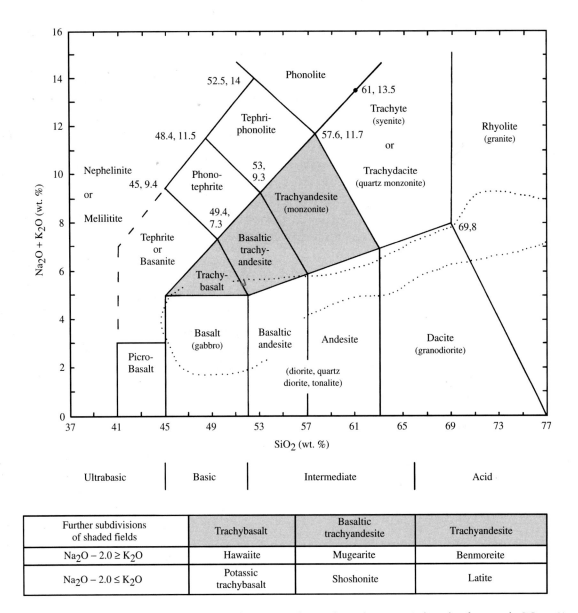

2.12 IUGS classification of aphanitic and glassy volcanic rock types. Coordinates of critical points are indicated as, for example, SiO₂ wt.% = 69 and (Na₂O + K₂O) wt.% = 8 at the common corner of the fields of trachyte, rhyolite, and dacite. Rocks plotting in the shaded area may be further subdivided into sodic and potassic rock types as shown in the box below the main part of the diagram. Figure 2.18 shows an alternate classification based on K₂O versus SiO₂. The distinction between trachyte ($Q < 20\%$) and trachydacite ($Q > 20\%$) is based on the amount of normative quartz, Q, from a recalculation in which $Q + An + Ab + Or = 100$. The amount of normative olivine, Ol, in the rock distinguishes tephrite (<10%) from basanite (> 10%). Rock-type names for more or less corresponding common phaneritic rocks are indicated in parentheses. Dotted line encloses 53% of the rocks plotted in Figure 2.4. (Redrawn from Le Maitre, 1989.)

of silica may have widely different quantities of quartz, and two rocks of similar quartz content may have different silica concentrations, depending upon the composition and quantity of other minerals in the rock. Roughly speaking, acid rocks are silicic, basic are mafic, and ultrabasic are ultramafic.

The CIPW Normative Composition. Near the beginning of the 20th century, three petrologists (W. Cross, J. P. Iddings, and L. V. Pirsson) and a geochemist (H. S. Washington) devised an elegant procedure

(from whose surnames the acronym CIPW is formed) for calculating the chemical composition of a rock into a *hypothetical* assemblage of water-free, standard minerals. These standard **normative minerals** (Appendix B) are designated in italics, such as Q, An, Ol, to distinguish them from the actual rock-forming minerals in the rock. Normative minerals are some of the simple end members of the complex solid solutions the actual minerals in the rock comprise. A complex solid solution, such as hornblende, is represented by several simpler normative minerals. The assemblage of normative

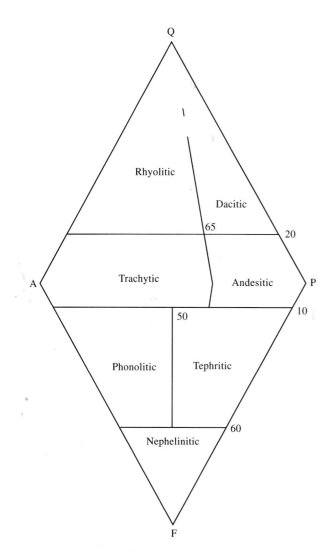

2.13 *Preliminary* classification of aphanitic and glassy rock types for use in cases in which an accurate chemical analysis is not available. This is a useful classification of rocks in the field or in thin section based upon the proportions of phenocrysts. Some rocks that are chemically rhyolite or dacite (Figure 2.12) contain quartz as the only phenocryst. (Redrawn from Le Maitre, 1989.)

minerals constitutes the **CIPW normative composition,** or simply, the **norm,** of the rock, which can be calculated according to the rules in Appendix B.

What are the benefits of the normative calculation? Because of extensive solid solution in the major rock-forming minerals, substantial variations in whole-rock chemical composition may not be evident in any obvious variations in mineralogical or modal composition. Basaltic rocks are an example. Much the same assemblage of plagioclase, clinopyroxene, olivine, and Fe-Ti oxides can constitute basalt, trachybasalt, and basanite (Figure 2.12). The norm facilitates comparisons between these basaltic rocks as well as others in which solid-solution minerals conceal whole-rock chemical variations. Aphanitic and, especially, glassy rocks are

readily compared. Mica- and amphibole-bearing rocks that crystallized from hydrous magmas can be compared with rocks lacking hydrous minerals that crystallized from dry magmas of otherwise similar chemical composition. Moreover, rock compositions cast as norms can be easier to relate to the results of experimental laboratory studies of simplified, or model, rock systems (Chapter 5).

Silica Saturation. In its allocation of silica first to normative feldspars and then to pyroxenes and finally to quartz, the normative calculation (Appendix B) emphasizes the concentration of SiO_2 relative to oxides of K, Na, Ca, Mg, and Fe in the rock. The relative amounts of these oxides are compared on a molecular, rather than weight, basis. If there is insufficient silica in the rock to make normative pyroxenes from the amounts of these other oxides, then some FeO and MgO is instead allocated to normative olivine, which requires relatively less silica than Fe-Mg pyroxene; the silica deficiency is thus compensated. This chemical balance may be seen in the reaction

$$(Mg,Fe)_2SiO_4 + SiO_2 = 2(Mg,Fe) SiO_3$$
$$\text{olivine} \qquad\qquad \text{orthopyroxene}$$

Note that there are equal molar proportions (1:1) of SiO_2 and (Mg,Fe)O in orthopyroxene, but half as much SiO_2 as (Mg,Fe)O in olivine, or SiO_2:(Mg,Fe)O = 1:2. (In the norm, orthopyroxene is represented by the normative mineral hypersthene, *Hy*.) In rocks that still have a deficiency of silica after eliminating all of the orthopyroxene, some silica must be reassigned from albite to nepheline, a silica-poor mineral. This chemical balance may be seen in the reaction

$$NaAlSiO_4 + 2SiO_2 = NaAlSi_3O_8$$
$$\text{nepheline} \qquad\qquad \text{albite}$$

Once again, note the difference in relative molar proportion of SiO_2:Na_2O = 6:1 in albite and 2:1 in nepheline. Creating one mole of nepheline from one mole of albite liberates more silica than does conversion of one mole of orthopyroxene to one mole of olivine. Hence, modest silica deficiencies in rocks are manifest by olivine in lieu of orthopyroxene, whereas greater deficiencies are manifest by nepheline in lieu of sodic plagioclase.

The normative calculation serves as a model for a crystallizing magma and illustrates the concept of the **degree of silica saturation.** Consider a simple hypothetical magma consisting only of O, Si, Al, and Na (Figure 2.14). If there is an excess of molar SiO_2 relative to that needed to make albite from Na_2O, that is, $SiO_2/Na_2O > 6$, then the magma can crystallize quartz in addition to albite. (In a natural magma, the albite would be in solid solution in plagioclase and/or alkali feldspar.) This magma and the corresponding rock are

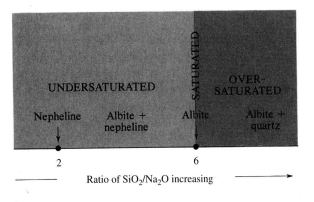

2.14   Degree of silica saturation in a model system consisting only of Si_2O, Al_2O_3, and Na_2O.

silica-oversaturated. If the magma contains SiO_2 and Na_2O in the exact ratio of 6, then these two constituents can only combine into albite; the magma and rock are silica-saturated. If the molar ratio $SiO_2/Na_2O < 6$ but > 2 in the magma, then there is insufficient SiO_2 to combine with all of the Na_2O into albite and some nepheline is created instead; the magma and rock are silica-undersaturated. If the molar ratio $SiO_2/Na_2O = 2$ in the magma, then there is insufficient SiO_2 to combine with the Na_2O to create any albite at all and only nepheline can be produced; the magma and rock still qualify as silica-undersaturated.

In real magmas and corresponding rocks that contain Mg, Fe, Ca, K, Ti, and so on, in addition to O, Si, Na, and Al, the concept of silica saturation still applies. In the classification that follows, the degree of satura-

tion is manifested in normative minerals (shown in italic letters) and with less accuracy by real minerals (in parentheses).

1. **Silica-oversaturated** rocks contain Q (quartz or its polymorphs—cristobalite and tridymite), such as granite.
2. **Silica-saturated** rocks contain Hy, but no Q, Ne, or Ol (no quartz, feldspathoids, or olivine), such as diorite and andesite.
3. **Silica-undersaturated** rocks contain Ol and possibly Ne (Mg-olivine and possibly feldspathoids, analcime, perovskite, melanite garnet, and melilite), such as nepheline syenite.

Alumina Saturation. Al_2O_3 is the second most abundant constituent in most magmatic rocks and provides another means of classification, especially for felsic rocks, such as granitic ones (Figure 2.15). The **alumina saturation index** is defined as the *molecular* ratio $Al_2O_3/(K_2O + Na_2O + CaO)$, which equals 1 in feldspars and feldspathoids. In magmas crystallizing feldspars and/or feldspathoids, any excess (ratio > 1) or deficiency (<1), respectively, of Al_2O_3 must be accommodated in mafic or accessory minerals. In alumina-oversaturated, or **peraluminous**, rocks, excess alumina is accommodated in micas, especially muscovite, in addition to Al-rich biotite, and in aluminous accessory minerals such as cordierite, sillimanite, or andalusite, corundum, tourmaline (requires boron), topaz (fluorine), and almandine-spessartine garnet. (But beware: The latter three minerals also occur as vapor-phase precipitates in some metaluminous rocks.) After alloca-

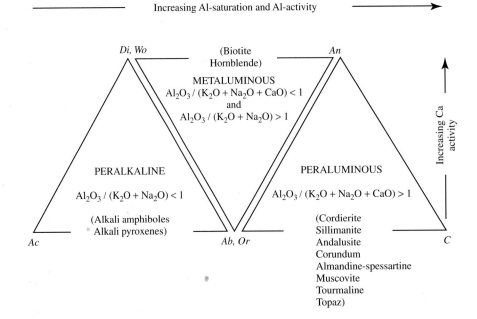

2.15   Classification of felsic rocks on the basis of degree of aluminum saturation. Ratios are molar. Apices of triangles are normative minerals. Diagnostic real minerals are listed in parentheses (muscovite, cordierite, etc.) for each of the three saturation categories.

tion of CaO for apatite, peraluminous rocks contain normative corundum, *C.* In alumina-undersaturated, or **metaluminous,** rocks, deficiency in alumina is accommodated in hornblende, Al-poor biotite, and titanite (but its stability also depends on other compositional properties of the magma including oxidation state). After allocation of CaO for apatite, metaluminous rocks contain normative anorthite, *An,* and diopside, *Di* (or wollastonite, *Wo*). A further constraint on metaluminous rocks is that they have $Al_2O_3/(K_2O + Na_2O) > 1$, whereas **peralkaline** rocks have $Al_2O_3/(K_2O + Na_2O) < 1$. In peralkaline rhyolites and granites the alumina deficiency (alkali excess) is accommodated in alkali mafic minerals such as aegirine end-member pyroxene ($NaFe^{3+}Si_2O_6$) and the alkali amphiboles riebeckite richterite (Appendix A), and aenigmatite in which Fe_2O_3 and TiO_2 substitute for Al_2O_3. Peralkaline rocks contain normative acmite or sodium metasilicate (*Ac* or *Ns*) and lack normative *An.* Real feldspars in peralkaline rocks contain little of the anorthite end member. Peralkaline rhyolites can be further subdivided into **comendites** in which $Al_2O_3 > 1.33$ FeO + 4.4 (on a wt.% basis), and **pantellerites,** in which $Al_2O_3 < 1.33$ FeO + 4.4. Peralkaline rocks can be silica-oversaturated, -saturated, or undersaturated, as in, for example, comenditic and pantelleritic trachytes.

An inherent weakness of classifications depending on the ratios of alumina or silica to alkalies is that Na and K can be mobilized and transferred out of a magma by a separate fluid phase. For example, escaping steam from cooling hot lava flows carries dissolved Si, Na, and K. However, Al tends to be less mobile. Initially metaluminous magma can, therefore, become peraluminous after alkali loss. Glasses can also lose alkalies relative to Al during high-*T* alteration or during weathering. A clue to preferential alkali loss is the presence of metaluminous minerals as phenocrysts, formed prior to extrusion, in a glassy matrix.

2.4.5 Rock Suites

Each of the chemical categories just described may embrace several rock types that share a common chemical attribute. Thus, silica-undersaturated rocks include the phonolite, tephrite, basanite, nephelinite, and melilitite rock types in Figure 2.12 and metaluminous rocks include the more common rhyolite, dacite, andesite, and their phaneritic plutonic counterparts. Peralkaline rocks include rhyolites and trachytes. These and other compositionally related or kindred groups of rock types are called **rock suites.**

Since the beginning of the 20th century, petrologists have recognized that suites of kindred magmatic rock types occur in particular geographic areas. Thus, the volcanic rocks in islands of the Atlantic Ocean were found to be more highly concentrated in alkalies relative to silica than rock types around the margin of the Pacific Ocean. This simple twofold division of magmatic rocks into alkaline (Atlantic) and subalkaline (Pacific) rock suites persists today, though it is now realized to be an overly simplistic characterization of these two oceanic regions. For example, the major part of the huge Hawaiian shield volcanoes in the Pacific are made of subalkaline basalt, but alkaline rocks form late capping lava sequences. Both alkaline and subalkaline lavas occur in western Mexico near the Pacific rim.

Despite their relatively small volume worldwide, **alkaline** rocks account for most of the hundreds of rock-type names in the geologic literature because of unusually great variation in chemical, mineralogical, and modal composition. Regrettably, there is no consensus among petrologists as to the precise definition of alkaline rocks, and their classification continues to be challenging (Mitchell, 1996). Alkaline rocks have a relative excess of alkalies over silica (Figure 2.16) but the exact ratio of these constituents has not been established. Most are silica-undersaturated and contain normative nepheline and real feldspathoids (nepheline, leucite). Alkaline rocks commonly include one or more of analcime, alkali feldspar, alkali-rich amphiboles; Na-Ti-Al-rich clinopyroxenes, biotite-phlogopite solid solutions; olivine; and no orthopyroxene or quartz. However, some very rare rocks known as lamproites contain significant modal proportions of leucite yet are quartz normative by virtue of their very low concentration of Al_2O_3. Alumina deficient peralkaline rocks are also sometimes con-sidered to be alkaline, even though they may be silica-oversaturated. Because Na and K are relatively abundant in alkaline rocks, a twofold subdivision into sodic and potassic series is used (bottom of Figure 2.12).

More common **subalkaline** rocks are usually silica-saturated or silica-oversaturated and lack normative nepheline. Real minerals include combinations of feldspars, hornblende, augite clinopyroxene, orthopyroxene, biotite, quartz in more silica-rich rocks, and olivine in less silica-rich rocks. Subalkaline rocks have been subdivided into the **tholeiitic** and **calc-alkaline** suites. (These two terms emerged from a tangled history spanning many decades. The calc-alkaline label originated in a now virtually abandoned classification scheme of M. A. Peacock proposed in 1931. Tholeiitic originated in the mid-1800s when it was applied to basalts from near Tholey, Saarland, western Germany.) As the term is used today, tholeiitic rocks show stronger enrichment in Fe relative to Mg than do calc-alkaline rocks and generally have less variation in silica, whereas the calc-alkaline suite shows enrichment in silica and alkalies (Figure 2.17; see also Miyashiro, 1974). Tholeiitic and calc-alkaline rocks typify subduction zones, where their composition correlates in a general way with the nature of the crust in the overriding plate. The tholeiitic suite of relatively Fe-enriched basalt, andesite, and dacite develops chiefly in island arcs where two oceanic plates converge. The calc-alkaline suite of

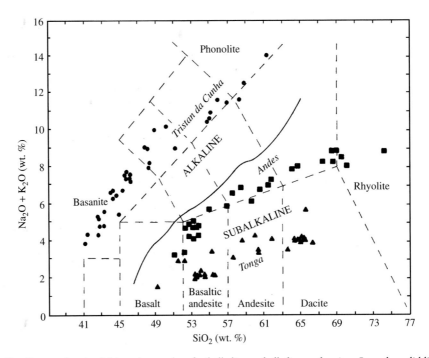

2.16 Total alkalies-silica diagram showing fields and examples of **subalkaline** and **alkaline rock suites.** Irregular solid line separates the field of nepheline-normative rocks from rocks having no normative nepheline in the 15,164-sample database of Le Bas et al., (1992). Light dashed lines delineate the IUGS volcanic rock-type classification from Figure 2.12. Note that a single rock type, such as basalt, can be either alkaline (*Ne*-normative) or subalkaline (*Hy*-normative). The alkaline volcanic suite of basanite, phonotephrite, tephriphonolite, and phonolite (filled circles) is from Tristan da Cunha, a volcanic oceanic island near the intersection of the Mid-Atlantic and Walvis Ridges in the South Atlantic Ocean (Le Roex et al., 1990). The subalkaline volcanic suite from the oceanic island arc of Tonga (filled triangles) is mostly basaltic andesite, andesite, and dacite (Cole, 1982). Subalkaline-suite rocks from Volcan Descabezado Grande and Cerro Azul in the southern volcanic zone of the Andes in central Chile (filled squares) are mostly basaltic andesite, trachyandesite, trachydacite, and rhyolite (Hildreth and Moorbath, 1988).

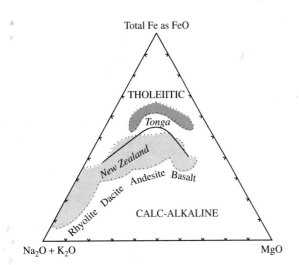

2.17 Subalkaline rocks can be subdivided into **tholeiitic and calc-alkaline rock suites.** **AFM diagram** in terms of alkalies ($Na_2O + K_2O$), total Fe as FeO, and MgO. Solid line separates fields of tholeiitic rocks, exemplified by volcanic rocks from the Tonga island arc in the Pacific Ocean (Figure 2.16), from calc-alkaline rocks, exemplified by most of the volcanic rocks from the North Island of New Zealand. Approximate range of rhyolite, dacite, andesite, and basalt rock types in New Zealand is indicated. Data from Cole (1982).

less Fe-enriched basalt, andesite, dacite, and rhyolite develops on thicker continental crust above subducting lithosphere. A plot of K_2O versus SiO_2 (Figure 2.18) reveals that subalkaline rocks of increasing concentration of K_2O correspond to increasing thickness of the continental crust. **Low-K rocks,** which are essentially equivalent to the tholeiitic suite, occur where the crust is oceanic and relatively mafic or where the continental crust is thin. **Medium-K** and **high-K** calc-alkaline rocks develop on increasingly thicker continental crust. The **absarokite-shoshonite-banakite** association is found in some subduction zones overlain by thick crust but also in some island arcs.

A Brief Note on Petrotectonic Associations. One reason petrologists plot whole-rock chemical compositions on diagrams such as Figures 2.16–2.18 is to focus attention on the petrogenesis of different rock suites in contrasting global tectonic settings, that is, petrotectonic associations. Such variation diagrams prompt many questions: How do the thickness and nature of the crust influence magma composition? Can one parental magma, such as of basaltic composition created in the mantle, evolve into different daughter compositions depending on the

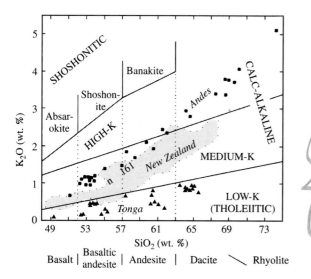

2.18 Subdivision of subalkaline rocks according to K_2O versus SiO_2. (Redrawn from Ewart, 1982.) Volcanic rocks from the oceanic island arc of Tonga (filled triangles; see also Figures 2.16 and 2.17), where the crust is only about 12 km thick, define the **low-K,** or **tholeiitic,** series consisting of basalt, basaltic andesite, andesite, and dacite rock types. These rock-type names are from Figures 2.12 and 2.16. The same rock types in the North Island of New Zealand (continental sialic crust about 35 km thick; 161 rock analyses in shaded area) are mostly of the **medium-K** series. (Analyses for Tongan and New Zealand rocks from Cole, 1982.) Basalt, basaltic andesite, andesite, dacite, and rhyolite (filled squares) from Volcan Descabezado Grande and Cerro Azul in the southern volcanic zone of the Andes in central Chile (Hildreth and Moorbath, 1988), where the continental crust is about 45 km thick, belong to the medium- to **high-K** series. Note that all three of these rock suites diverge from basalt in the lower left corner of the diagram. The **shoshonitic** series of still more K-enriched rocks is found in a few subduction zones in thick continental crust and some island arcs.

crust through which it ascends to the surface? Or do magmas moving up from a mantle source differ in composition? Does the oxidation state of Fe somehow influence the evolution of contrasting magma suites? Why should magmas forming the Atlantic Ocean island of Tristan da Cunha differ so distinctly from magmas forming the Tongan island arc in the Pacific (Figure 2.16)? These and many other questions concerning petrotectonic associations are considered further in Chapter 13.

2.4.6 Classification of Basalt

Because basalts are by far the most abundant rock type on Earth (and possibly in the inner planets) and are found in virtually all global tectonic settings, their classification deserves special consideration. Basalts, like all magmatic rocks, define a continuous compositional spectrum (Figures 2.12 and 2.19b). Any classification must artificially divide this continuum.

Yoder and Tilley (1962) used the normative tetrahedron (Figure 2.19a) to portray the wide range of basalt compositions. Because of the difficulty of plotting and visualizing composition points within a three-dimensional tetrahedron, data points can be projected onto the triangular base of the tetrahedron in Figure 2.19b. Three *basalt rock types* can be recognized according to their degree of silica saturation:

1. Quartz-hypersthene normative ($Q + Hy$) **quartz tholeiite**

2. Olivine-hypersthene normative ($Ol + Hy$) **olivine tholeiite**

3. Nepheline-normative (Ne) **alkaline basalt**

Tholeiitic basalts make up the oceanic crust and, on continents, large flood basalt plateaus and some large intrusions. Alkaline basalt is the most common rock type in the alkaline rock suite and occurs in oceanic islands, such as Hawaii, and in some continental settings.

Different parental basalt magmas evolve into contrasting daughter suites of less mafic, more silica-rich rocks in different global tectonic settings. This is one facet of petrotectonic associations. In Figure 2.16, some basalts are silica-undersaturated and are associated with more felsic rocks of the alkaline suite, whereas others are silica-saturated and are associated with more felsic rocks of the subalkaline suite. In Figure 2.18, subalkaline basalt is either tholeiitic (low-K) or calc-alkaline (medium- and high-K).

2.5 TRACE ELEMENTS

Previous sections of this chapter have emphasized the great, but limited diversity in major element and mineralogical compositions of magmatic rocks as well as systematic patterns in these parameters. The remainder of this chapter is an introduction to the behavior of trace elements and isotopes in magmas, showing how they serve as powerful **petrogenetic indicators** of magmatic processes. As just one example, the trace element and isotopic composition of basalts is at least as variable as their major element concentrations and provide significant petrogenetic information on the origin and evolution of basalt magmas in different tectonic settings.

About 90 of the known chemical elements occur in rocks and minerals in trace concentrations, arbitrarily set at <0.01 wt.% = 1000 ppm. Concentrations as low as 1 ppb can be detected for some elements. These low concentrations are insufficient to stabilize any major rock-forming mineral but in many cases do stabilize accessory minerals such as zircon. Unlike major elements (Si, Al, Ca, etc.), whose variations are mostly limited to a factor of <100, trace element concentrations can vary by as much as a factor of 1000 (three orders of magnitude). This fact, together with the way trace elements

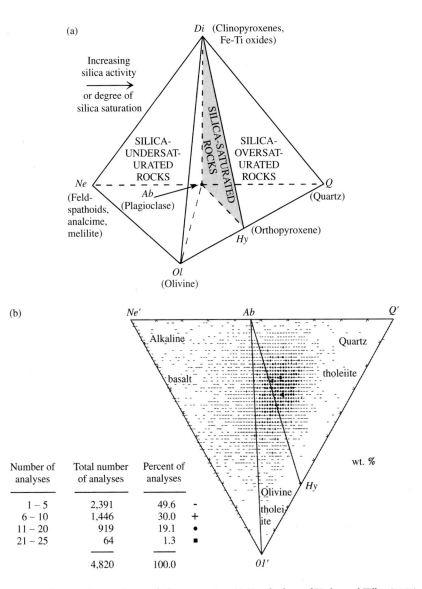

2.19 Classification of basaltic rocks according to degree of silica saturation. (a) Tetrahedron of Yoder and Tilley (1962) showing variable degree of silica saturation in basaltic rocks. Italicized normative minerals define the degree of saturation. Real minerals (in parentheses) can be used as guides in the absence of normative data. The shaded plane represents **silica-saturated** basaltic rocks separating the volume on the left of **silica-undersaturated** basaltic rocks that contain normative olivine (*Ol*) and possibly normative nepheline (*Ne*), and modal feldspathoids from **silica-oversaturated** basaltic rocks on the right that contain normative quartz (*Q*). (b) Base of tetrahedron. Compositions of over 4800 basalts lying within the tetrahedron have been projected onto its base from the *Di* apex and are thus *Di*-bearing. Apices of triangle are adjusted normative minerals: *Ol′* = *Ol* + [0.714 − (Fe/(Fe + Mg))0.067]*Hy; Ne′* = Ne + 0.542 *Ab; Q′* = Q + 0.4 *Ab* + 0.25 *Hy.* (Data compiled and plotted through the courtesy of Roger W. Le Maitre, University of Melbourne, Australia.)

are distributed between coexisting minerals and liquids in geologic systems, qualify them as highly significant indicators of petrologic processes.

2.5.1 Partition Coefficients and Trace Element Compatibility

Generation of magma from solid rock in the Earth involves only *partial* melting. Where upper mantle peridotite is partially melted, the resulting magma consists of crystals of pyroxenes and olivines in equilibrium with a liquid solution of ions of O, Si, Al, Mg, Na, and so on, called a **melt.** In this magma, ions of **incompati-** **ble trace elements** prefer to be dispersed in the loosely structured (on the atomic scale) melt and are *excluded* from the more restrictive, less tolerant crystalline structure of the coexisting pyroxenes and olivine. On the other hand, ions of **compatible trace elements** are tolerated and largely remain *included* in the crystalline phases. The contrast between these two categories of trace elements is formalized by a simple concentration ratio called the **partition coefficient,** *D.*

$$2.1 \quad D^{crystal}_{melt} = \frac{(Concentration\ in\ mineral)}{(Concentration\ in\ melt)}$$

Thus, compatible trace elements have $D > 1$. For example, Sr, Ba, and Eu are compatible elements that partition strongly into feldspars in silicic magmas. Cr, Ni, and Co are compatible in olivine and orthopyroxenes in basaltic magmas. On the other hand, incompatible elements, such as Rb, Li, Nb, and rare earth elements, have $D \ll 1$ and partition only weakly into the major minerals found in most magmas.

Incompatible trace elements cannot readily substitute for major elements in crystalline phases because of dissimilar ionic charge and/or radius (Figure 2.20; Table 2.4). Thus, Be^{2+} is typically incompatible because its small size (0.45 Å in sixfold coordination) pre-

Table 2.4 Trace Elements Substituting for Major Elements of Similar Ionic Size and Charge (see Figure 2.20)

MAJOR ELEMENT	SUBSTITUTING TRACE ELEMENT(S)
Si	Ge, P
Ti	V
Al	Ga
Fe	Cr, Co, Ni
Mg	Cr, Co, Ni
Ca	Sr, Eu, REEs
Na	Eu
K	Rb, Ba, Sr, Eu

cludes substitution for divalent ions and its low charge precludes substitution for similarly sized Si^{4+}. Another typically incompatible element in most major minerals is U^{4+} because it has both a large ionic charge and a large radius (1.0 Å in eightfold coordination). However, compatibility depends on what minerals exist in the magma. Hence, in a silicic magma in which zircon ($ZrSiO_4$) is crystallizing, U^{4+} is compatible because it substitutes for Zr^{4+}, whose radius is 0.85 Å. (This substitution, incidentally, makes possible isotopic dating of zircon.)

No single partition coefficient describes the behavior of a particular trace element in all magmas. The composition of the magma and that of the mineral both affect the value of D. Coefficients for the same element in the same mineral generally increase as the magma becomes more silicic; variations of a factor of 10 are common (Figure 2.21). Decreasing magma temperature (T) also corresponds with increasing coefficients. Cooler, more silicic melts are more tightly structured, causing trace elements to be rejected and forced into coexisting crystals. The effect of pressure (P) on partition coefficients is apparently small and in the opposite direction of T; thus, the effect on coefficients of increasing P and T with depth in the Earth might more or less cancel out. The oxidation state of the magma affects the partition coefficient of europium, Eu. In reduced magmas, europium exists mostly as Eu^{2+}, rather than the usual Eu^{3+} of other rare earth elements, and is a compatible element in plagioclase, as is Sr. Obviously, selection of a partition coefficient depends on many factors (e.g., Rollinson, 1993); only a few representative coefficients are presented in Table 2.5. The behavior of important trace elements in magma systems is summarized in Table 2.6.

Because a particular trace element has different affinities for different minerals in a magma, a **bulk partition coefficient**, D_{bulk}, for the behavior of a particular element in the whole magma must be formulated, as follows:

$$2.2 \quad D_{bulk} = X_1 D_1 + X_2 D_2 + X_3 D_3 \ldots$$

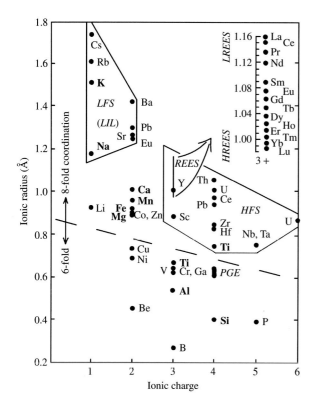

2.20 Radii and classification of positively charged ions of major (bold letters) and trace elements. Radii based on eightfold coordination in upper part of diagram and on sixfold in lower part. Rare earth elements (**REEs**) in center of diagram are plotted on an expanded scale in upper right. On the basis of **ionic potential** (charge/radius), most elements can be subdivided into two categories surrounded by polygons, namely, (1) Low field strength (LFS) elements, more commonly called **large-ion lithophile (LIL) elements**, in upper left; (2) **high-field-strength (HFS) elements** in right center. The lithophile designation arises from an affinity for silicate rocks, as contrasted with elements having an affinity for metallic phases (siderophile) containing Fe, Co, Ni, and so on, as in the core of the Earth, or for sulfide phases (chalcophile) containing S, Cu, Zn, and so on. Ionic potential also serves as a rough index of the **mobility** of cations of the elements, that is, their solubility in aqueous solutions; elements with low (<3) and high (>12) potential tend to be more soluble and mobile than elements in midrange. PGE, **platinum group elements** (Ru, Rh, Pd, Os, Ir, Pt). (Data from Shannon, 1976.)

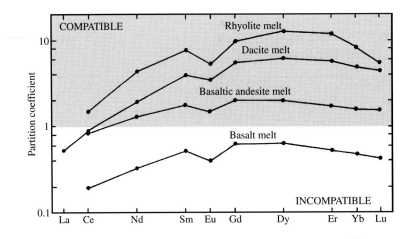

2.21 Partition coefficients for REEs between amphibole and indicated melts. REEs are more compatible in more silicic and lower-*T* melts. (Redrawn from Rollinson, 1993.)

where X_1, X_2, and so on, represent the weight fraction, expressed in decimal format (e.g., 0.25) of each mineral, and D_1, D_2, are their respective partition coefficients for the particular element.

Simple mathematical equations can be used to construct models of trace element behavior in petrologic processes (Haskins, 1983; Hanson, 1978). These theoretical models are based on idealized assumptions that approximate a particular process in a natural geologic system. Because of the uncertainties in what partition coefficient actually applies to a particular magma system, the petrologist seeks *patterns in trace element behavior* rather than specific details. Trace element models are most appropriately used to evaluate a hypothesized process conceived from other information, such as the fabric, major element, and modal compositional variations and field relations. Applications of the models are discussed in Chapters 11–13.

2.5.2 Rare Earth Elements

Rare earth elements (REEs) are a mostly coherent group of elements that can be especially useful in testing petrogenetic hypotheses. REEs comprise atomic numbers 57 (La) to 71 (Lu) (Figure 2.20). Promethium (61) is not found naturally. Yttrium (Y, 39) has an ionic charge of 3 +, like that of REEs, and a radius (1.019 Å) similar to that of holmium (1.015 Å) and is, therefore, sometimes included as a REE. The existence of a divalent ion of Eu in magmas has been noted; also, Ce^{4+} may exist in some very oxidized magmas. Two properties of REEs make them especially useful as petrogenetic indicators:

1. They are generally insoluble in aqueous fluids; hence, they are useful in altered or weathered rocks.

2. Trivalent ions of REEs have decreasing radii with respect to increasing atomic number, from La (1.160 Å) to Lu (0.977 Å) (Figure 2.20). This small

but systematic variation from light REEs to heavy REEs causes significant differences in their behavior and partition coefficients (Figure 2.22). Because of their slightly larger sizes, light REEs are generally more incompatible in common silicate minerals than are the heavy REEs. Plagioclase, because of the similarity of Eu^{2+} to Ca^{2+}, will accommodate much more of this trace element than immediately adjacent lighter and heavier trivalent REEs, creating a positive **Eu anomaly** (Figure 2.22). The behavior of REEs in garnet-bearing basaltic magmas is striking because of a 1000-fold difference in partition coefficients between La and Lu.

To smooth out the otherwise sawtoothlike absolute abundances of odd and even atomic numbers (the Oddo-Harkins effect), the concentration of a REE in a rock is divided by the concentration of the same element in average chondritic meteorites (Table 2.7). This sample/chondrite ratio is then plotted on a logarithmic scale (Figure 2.23). **Chondrites** are used as the basis of comparison because they are thought to have accreted to form the inner planets in the solar system and thus have a chemical composition like that of the entire primitive Earth. If all REEs had the same partition coefficient in all minerals in a magma system, the chondrite-normalized pattern would be flat—a horizontal line. However, few magmatic rocks have such a pattern.

The sloping, arcuate, and even spiky patterns of rocks provide important information on the sources and processes in the origin of the magmas (Figure 2.23). For example, lunar basalt (Taylor, 1982) has a pronounced negative europium anomaly that provides an amazing insight into the early history of the Moon. It is believed that lunar basalt magmas were generated by partial melting from a Eu-depleted lunar mantle formed as a gravitative accumulation in the bottom of the primordial "magma ocean" in which crystallizing and floating plagioclase took up most of the compatible

Table 2.5 Partition Coefficients for Some Trace Elements

	U	R$_B$	K	B$_A$	S$_R$	Y$_B$	Y	N$_B$	E$_U$	L$_A$	C$_E$	Z$_R$	T$_I$	V	C$_R$	N$_I$
BASALT MAGMA																
Plagioclase	0.01	0.07	0.17	0.23	1.83	0.067	0.03	0.01	0.34	0.19	0.1	0.048	0.04			
Clinopyroxene	0.04	0.031	0.038	0.026	0.06	0.62	0.9	0.005	0.51	0.056	0.09	0.1	0.4	1.35	34	1.5–14
Orthopyroxene		0.022	0.014	0.013	0.04	0.34	0.18	0.15	0.05		0.02	0.18	0.1	0.6	10	5
Olivine	0.002	0.01	0.007	0.01	0.014	0.014	0.01	0.01	0.007	0.007	0.006	0.012	0.02	0.06	0.7	6–29
Magnetite						1.5	0.2	0.4	1.0	2.0	2.0	0.1	7.5	26	153	29
Garnet		0.042	0.015	0.023	0.012	11.5	9	0.02	0.49	0.01	0.03	0.65	0.3		2	
RHYOLITE MAGMA																
Plagioclase	0.093	0.041	0.1	0.31	4.4	0.09	0.1	0.06	2.1	0.38	0.27	0.1	0.05			
K-feldspar	0.02	0.5		4.3	3.76	0.0015	1		2.6	0.07	0.04	0.03				
Quartz	0.025	0.04	0.013	0.022		0.017			0.056	0.015	0.014		0.038			
Biotite	0.167	4.2		5.4	0.5	0.54			0.87	3.18	0.3					
Hornblende		0.014	0.08	0.044	0.022	8.38	6	4	5.14		1.5	4	7		5.2	
Zircon	340					527			16	17	17					
Apatite						24	40	0.1	30	14.5	35				190	
Allanite	15.5					31			111	2595	2279	0.1	0.1	15.5	380	
Titanite								6.3		4						

Data from Rollinson, 1993.

Table 2.6   Trace Element Characteristics Useful in Evaluating Petrogenesis of Rocks

ELEMENT	CHARACTERISTICS AND INTERPRETATIONS
Ni, Co, Cr	Typically highly compatible elements. High concentrations (e.g., Ni = 250–300 ppm, Cr = 500–600 ppm) of these elements indicate derivation of parental magmas from a peridotite mantle source. Declining concentrations of Ni and to a lesser extent Co in a rock series suggest olivine fractionation. Decrease in Cr suggests spinel or clinopyroxene fractionation.
PGE, Cu, Au, Ag	Strongly partitioned into immiscible sulfide melts. A series of mafic magmas that lack sulfides may show increases in these elements. In most other magma series, these are compatible elements that decline with increasing silica.
V, Ti	Typically compatible elements in ilmenite and titanomagnetite, although Ti can become enriched in some mafic magmas that lack these oxide minerals.
Nb	Incompatible element in most magmas. However, because it substitutes somewhat for Ti, residual titanates (such as rutile) may cause depletions of Nb in subduction-zone magma sources. Nb has a lower solubility in aqueous fluids than other equally incompatible elements.
Zr, Hf	Characteristically incompatible in mafic magmas and not readily substituting in mantle phases. In zircon-saturated (silicic) magmas both may behave as compatible elements.
P	Characteristically incompatible in mafic magmas but becomes a compatible element in intermediate and silicic magmas where apatite is a stable phase.
Ba	Substitutes for K in micas, K-feldspar, and to a lesser extent amphibole. A change from incompatible to compatible behavior in a magma series may indicate an increasing role for one of these phases.
Rb	Incompatible element in most magma, but it substitutes for K in micas and K-feldspar in silicic magmas, though not as strongly as Ba.
Sr, Eu	Substitute readily for Ca in plagioclase and K in K-feldspar. Declining Sr concentrations indicates feldspar removal from a series of related magmas. Sr is more incompatible under mantle conditions because of the absence of feldspar.
REEs	Generally, the *trivalent* rare earth elements are incompatible in basaltic magmas. Garnet more readily accommodates heavy REEs than light REEs and a steep REE pattern may indicate garnet remained in a mantle residue. Titanite prefers the middle REEs. Apatite, monazite, and allanite have very high partition coefficients for light REEs; consequently, light REEs are commonly compatible elements in rhyolitic magmas that have these minerals. Zircon and xenotime prefer heavy REEs but their abundance in natural magmas is rarely sufficient to make the heavy REEs behave as compatible elements.
Y	Generally behaves incompatibly, as do middle to heavy REEs. It has a high partition coefficient in garnet and to a lesser extent in amphibole. Its behavior is strongly affected by REE-rich accessory minerals such as apatite and especially xenotime.

Data from Green (1989).

Eu; the plagioclase rock today forms the light-colored lunar highlands which contrast with the dark, basaltic mare lowlands. In contrast, the most widespread basalt on Earth, mid–ocean ridge tholeiitic basalt (MORB; Wilson, 1989), reflects extensive partial melting of mantle peridotite from which light REEs had been previously extracted. Yet another contrast is found in the negatively sloping pattern for **adakite** (Drummond and Defant, 1990), a type of dacite found in some subduction zones. Because its pattern is virtually a mirror image of the garnet coefficient pattern in Figure 2.22, petrologists believe that adakite magmas are generated by partial melting of the oceanic crust under high pressure where garnet is left behind after melting. Most of the incompatible light REE partition into the adakite partial melt and most of the compatible heavy REE stay behind in the garnet.

2.5.3 Other Normalized Trace Element Diagrams

The utility of the normalized REE diagram led to the development of similar **normalized trace element diagrams** that involve a wider variety of generally incompatible trace elements. One common approach is to normalize the trace element abundances in a rock sample with respect to **primitive mantle** concentrations (Table 2.7). These are average chondritic meteorite values but multiplied by a factor of 2.9. Adjustments must be made in K, Rb, and P because the two alkalies are volatile and may not have chondritic abundances in the Earth. Phosphorus may have been extracted from the mantle during formation of the core of the Earth. Other normalizations are with respect to chondrite or mid–ocean ridge basalts. In a primitive-mantle-normalized diagram (see Figure 13.2) a smooth, positively sloping pattern is obtained for mantle-derived mid–ocean ridge basalts (MORB). Generally, elements with the lowest partition coefficients are listed on the left and increasingly more compatible elements are found toward the right. However, it must be remembered that the partition coefficient for a particular element is not the same in every magma. A wide variety of similar diagrams have been used in the literature, so it is wise to pay careful attention to the exact scheme of normalization used for any trace element diagram.

Detailed discussion of the meaning of these trace element patterns is deferred to Chapters 11–13. Here, it is important to realize that there are significant differences between the trace element patterns of magmatic rocks—even in just one rock type, such as basalt—found in different tectonic settings and belonging to different petrotectonic associations. Thus, mid–ocean ridge basalts are markedly impoverished in the most in-

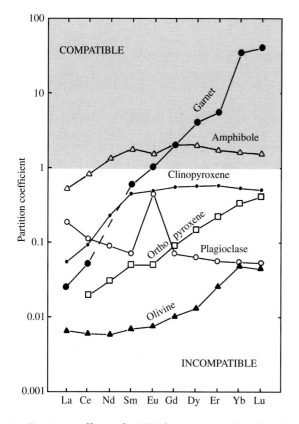

2.22 Partition coefficients for REEs between minerals and basaltic melt. Garnet and plagioclase are the principal aluminous phases in basaltic magmas, but the former is stable at pressures above which plagioclase is stable. Note the striking contrast in their pattern of coefficients. Amphibole is another aluminous phase stable in hydrous systems. (Redrawn from Rollinson, 1993.)

Table 2.7 Element Concentration (ppm) in Chondrite, Primitive Mantle, and Normal Mid–Ocean Ridge Basalt (N-MORB)

	Chondrite	Primitive Mantle	N-MORB
Rb	2.32	0.635	0.56
Ba	2.41	6.989	6.3
Th	0.029	0.085	0.12
U	0.008	0.021	0.047
Nb	0.246	0.713	2.33
Ta	0.014	0.041	0.132
K	545	250	600
La	0.237	0.687	2.5
Ce	0.612	1.775	7.5
Sr	7.26	21.1	90
P	1220	95	510
Nd	0.467	1.354	7.3
Sm	0.153	0.444	2.63
Zr	3.87	11.2	74
Hf	0.1066	0.309	2.05
Eu	0.058	0.168	1.02
Ti	445	1300	7600
Gd	0.2055	0.596	3.68
Tb	0.0374	0.108	0.67
Dy	0.254	0.737	4.55
Y	1.57	4.55	28
Ho	0.0566	0.164	1.01
Er	0.1655	0.48	2.97
Tm	0.0255	0.074	0.456
Yb	0.17	0.493	3.05
Lu	0.0254	0.074	0.455

Data from Sun and McDonough (1989).

compatible elements, including Ba, Rb, and Th; otherwise their trace element pattern is quite smooth and nearly flat. MORB magmas are believed to be generated by extensive partial melting of mantle peridotite, but this would not create a depletion in the less compatible elements on the left. And neither would any crystallization effects. It is, therefore, believed that their mantle source must have experienced a previous partial melting event, or events, that preferentially extracted the most incompatible elements. This positively sloping pattern characterizes magmas derived from a **depleted source**—a magma-generating region that is depleted in the most incompatible elements. The depleted nature of the worldwide MORB source was one of the first fruits of studies of trace elements in rocks.

Calc-alkaline basalt magmas erupted in island arcs have dramatically different trace element patterns (see Figure 11.19). These basalts are **enriched** in the most incompatible elements, especially Ba, Rb, and K, but are strongly depleted in elements with high field strengths—Nb and Ta—relative to adjacent elements on the diagram. These depletions yield a spiked, irreg-

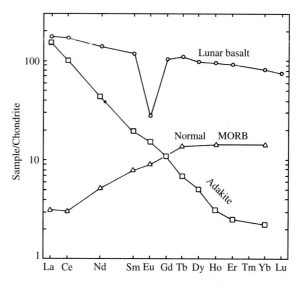

2.23 Very different **chondrite-normalized REE patterns** in three rocks. Compare patterns of partition coefficients in Figure 2.22, especially the mirror image of lunar basalt and plagioclase and of adakite and garnet.

ular pattern that, nonetheless, has an overall negative slope. This pattern is believed to reflect magma generation involving hydrous fluids in the mantle source overlying the downgoing lithospheric slab; the negative anomaly in Nb and Ta reflects a lower solubility of these elements in the migrating fluids.

❋2.6 ISOTOPES

Isotopes of an element are atoms whose nuclei contain the same number of protons but a different number of neutrons. Different isotopes are denoted with their atomic weights (protons + neutrons) as a superscript. Thus, all hydrogen atoms contain one proton in their nucleus and one electron that determines its chemical behavior. However, there are three H isotopes: hydrogen (^1H), deuterium (^2H), and tritium (^3H), which have zero, one, and two neutrons, respectively.

Isotopes are introduced in beginning geology courses because of their importance in determinations of the absolute age of rocks and minerals. In addition, like trace elements, isotopes serve as useful petrogenetic indicators of

1. processes of magma generation and evolution
2. T of crystallization
3. thermal history
4. other geologic processes, such as advective migration of aqueous fluids around hot magmatic intrusions

Investigations of trace elements and isotopes often go hand in hand because many of the petrologically important isotopes, except O and H, are of trace elements, namely, Rb, Sr, Pb, U, Th, Sm, and Nd. Geochemical investigations of trace elements and isotopes in magmas derived more or less directly from the mantle have provided significant constraints on the nature of this remote region of the interior of the Earth—once the exclusive and sole domain of geophysicists.

The isotopic proportions of elements in geologic materials are complex functions of their history and depend on whether the isotopes are unstable (radioactive) or stable and whether the isotopic system has remained closed.

2.6.1 Stable Isotopes

Stable isotopes do not decay. The stable isotope ratios of a particular element can only change by various physical and chemical **isotopic fractionation** phenomena in which one isotope is preferentially incorporated into one phase over another coexisting phase. For example, isotopes of ^{18}O are heavier than ^{16}O, and consequently molecules of $H_2^{18}O$ are heavier than $H_2^{16}O$ molecules. Because the vapor pressure, or escaping tendency, of a vibrating molecule is inversely proportional to its mass, evaporation of seawater enriches the overlying atmospheric vapor in the lighter molecules that contain ^{16}O

and ^1H isotopes, compared to the liquid seawater. Meteoric (rain) water derived from this vapor is also enriched in the lighter molecules and isotopes. Isotopic fractionation can occur between the isotopes of any element, but is greater for lighter elements, where there are larger relative differences in the masses of two isotopes. Hence, the mass difference between ^1H and ^2H is 100% and fractionation is considerable. Smaller fractionation occurs for heavier ^{16}O and ^{18}O, where the difference in mass is only about 12 percent. Isotopes of still heavier elements have even smaller relative mass differences; isotopic fractionation is too small to be detected for elements heavier than about Ca.

As a consequence of fractionation processes, stable isotopes can be used to trace the materials and processes involved in the evolution of many petrologic systems and the T at which they form. Because natural waters contain H, O, C, and S, the isotopes of these elements furnish valuable information regarding fluid-rock interactions in geologic systems. The importance of O exceeds that of the other isotopes because of its abundance in all kinds of rocks as well as natural fluids; it is the only stable isotope considered further here. See Faure (1986) for discussion of other stable isotopic systems.

<u>Oxygen Isotopes</u>. Stable isotope fractionation occurs among the three isotopes of oxygen, ^{16}O, ^{17}O, and ^{18}O, whose *average* percentages in natural materials are 99.76%, 0.04%, and 0.20%, respectively; these proportions yield the atomic weight given in periodic tables of the elements. Geochemists define the isotope fractionation factor, $\alpha_{A-B} = R_A/R_B$ where $R_A = {}^{18}O/{}^{16}O$ in phase A and $R_B = {}^{18}O/{}^{16}O$ in phase B. Because these factors vary only in the thousandths place (generally between 1.000 and 1.004), stable isotope compositions are cited in the delta notation, wherein the isotope ratio *in a particular phase* is expressed as a fractional deviation from a standard of accepted composition

$$2.3 \quad \delta^{18}O\text{‰} = \left[\frac{(R_{sample} - R_{standard})}{R_{standard}} \right] \times 1000$$

Thus, the measured $\delta^{18}O$ value is expressed in parts per thousand, or per mil ‰. The standard used for comparison is usually Vienna Standard Mean Ocean Water (VSMOW), which has a composition similar to that of average ocean water. Positive $\delta^{18}O$ values are enriched in the heavy ^{18}O isotope, negative in light ^{16}O.

Fractionation of stable isotopes can occur during crystallization of minerals from liquids for much the same reason it does during evaporation. In crystals containing small light ions, such as Si^{4+}, the higher vibrational component of the internal energy can be reduced by bonding with heavier isotopes. Thus, in a rock in which the two phases quartz and magnetite crystallized together under equilibrium conditions, the quartz will be more enriched in ^{18}O and the magnetite in ^{16}O. Fractionation is T-dependent but rather insensitive to P. A pair of minerals formed in nature at equilibrium can thus be used as a **geothermometer.** The basis for the O-isotope geothermometer is the T-dependent **isotope exchange reaction** between two minerals, such as quartz and magnetite

$$2Si^{16}O_2 + Fe_3^{18}O4 = 2Si^{18}O_2 + Fe_3^{16}O_4$$

Exchange reactions of this sort commonly occur in the presence of some kind of liquid in which the isotopes can move about freely. Oxygen isotope ratios between many mineral pairs, including quartz-magnetite, plagioclase-magnetite, plagioclase-pyroxene, quartz-plagioclase, and quartz-muscovite have been measured experimentally over a wide range of temperatures. Consequently, the T at which coexisting minerals crystallized can be deduced from their isotopic ratios. Underlying assumptions in this geothermometer are that the coexisting minerals reached isotopic equilibrium with one another at some T and with the liquid agent, such as a melt or metamorphic aqueous fluid, and that the isotopic compositions of the minerals have not changed since equilibration. However, the second assumption is commonly not valid because many geologic systems are open to migrating fluids that perturb the isotopic ratios. In slowly cooled plutonic rocks, some reequilibration of isotopes may occur, leading to T estimates lower than those of initial crystallization. Moreover, some minerals reequilibrate more readily than others. For example, quartz is relatively resistant to isotopic reequilibration compared to biotite or magnetite.

Separation of crystals from coexisting melt in crystallizing magmas (fractional crystallization) results in only minor isotopic fractionation of most stable isotope ratios, oxygen included. In order of their tendency to concentrate ^{18}O from the melt, quartz is highest, then alkali feldspar, plagioclase, muscovite, pyroxene, hornblende, olivine, biotite, and ilmenite; magnetite is the lowest. Extensive separation of minerals from a mafic magma may yield a change of only 1 per mil or so in the residual melt. The same magnitude of changes is possible as a result of varying degrees of partial melting of source rocks.

It is therefore not surprising that the oxygen isotope ratio, $\delta^{18}O$, of most magmatic rocks has a limited range of only about $+5\%$ to $+13\%$ (Figure 2.24). Mantle peridotite and mantle-derived mafic magmas are about $+6\%$. In contrast, sedimentary rocks are much higher, shales $+15$ to $+20\%$ and carbonate rocks to as high as $+33\%$. These high ratios result from two factors: Fractionation factors for clay minerals and calcite in equilibrium with water are large and

2.24 Oxygen isotope composition of rocks and natural waters. (Redrawn from Rollinson, 1993, and Taylor and Sheppard, 1986.)

these minerals form in sedimentary environments at low temperatures where isotope fractionation is a maximum. It appears that the mantle is the source of most magmas but that some of the more silicic magmas having higher $\delta^{18}O$ values were contaminated by sedimentary rock.

Isotopic exchange reactions between magmatic rocks and hot aqueous fluids (hydrothermal solutions) advecting through them result in lowering of $\delta^{18}O$ values. Meteoric water ($+6$ to -40%) advecting around shallow cooling magmatic intrusions (Figure 1.2; see also Figure 4.12) substantially lowers the $\delta^{18}O$ of the altered rock (Figure 4.14). Maps of $\delta^{18}O$ values can reveal fluid pathways.

2.6.2 Radiogenic Isotopes

Unstable, or radioactive, isotopes decay by nuclear processes into daughter **radiogenic isotopes,** which may be of the same or commonly of a different element as the parent. For example, radioactive ^{238}U decays into ^{206}Pb at a rate such that one-half of the parent ^{238}U is transformed into the daughter ^{206}Pb in about 4.5 Gy. This and other radioactive isotopic systems provide information on the absolute age of a mineral from that time the parent was initially lodged in it. (An absolute **age** measured backward from the present is expressed in anna; in International System [SI] units, 10^6 years = **Ma** [mega anna] and 10^9 years = **Ga** [giga anna]. An interval of time, such as the half-life just cited, is denoted in elapsed years, $4.5 \times 10^9 y$ or 4.5 Gy.) As time passes the ratio of radioactive parent and daughter isotopes changes.

Because of their potentially different chemical behavior and mobility, parent and daughter isotopes might be susceptible to differential separation in an open isotopic system. For example, the decay of radioactive ^{40}K in biotite crystals yields daughter ^{40}Ar, an inert, noble gas. Heating the biotite to modest temperatures (>300°C) can promote the release, by diffusion, of the unbonded Ar from the crystal on geologic time scales. Comparisons of different minerals and isotopic systems having different **closure temperatures** provide insights into the thermal history of rocks (see, for example, Cliff, 1985).

Radioactive isotopes and their daughters behave differently as do other trace elements in geologic systems, making them valuable petrogenetic tracers. The most important difference lies in their contrasting compatibility in mantle-basaltic systems; thus, in the following, the degree of compatibility *increases to the right* and individual parent-daughter isotopic pairs are listed on the same line:

Rb>   Sr

    Th>   Pb

        U>   Pb>

                Nd > Sm

                Hf > Lu

Hence, because Rb is the most incompatible, it is strongly concentrated in partial melts of the mantle that rise and solidify as crustal rock. In contrast, Sr, Sm, and Lu are least concentrated in the crust relative to the Rb-depleted mantle. Nd and Sm isotopes, on the other hand, are hardly fractionated from one another during partial melting and crystallization because of their very similar ionic radii. However, both are quite immobile and Nd-Sm systems, therefore, remain closed in many geologic environments where hydrothermal solutions and melting cause opening of the Rb-Sr system, mobilizing Rb but not Sr. The Th-Pb and U-Pb isotope systems are complex and the three elements have differing mobilities in addition to contrasting compatibilities; they are not discussed further in this textbook.

Rubidium-Strontium Systematics.

Rubidium occurs in nature as the isotopes ^{86}Rb and ^{87}Rb; the latter is radioactive and decays by beta emission to ^{87}Sr with a half-life of 48.8 Gy. The present relative abundance of these isotopes—72.17% ^{86}Rb and 27.83% ^{87}Rb—is the same in all rocks and minerals, regardless of age. Apparently, these heavy isotopes were thoroughly mixed in the primeval Earth and have not experienced fractionation since then regardless of the geologic processes that have acted upon them.

The same ionic charge of Rb^+ and K^+ and similar ionic radii (1.61 Å and 1.51 Å, respectively, based on eightfold coordination, Figure 2.20) means that Rb readily substitutes for K in micas and K-feldspar. Rocks and minerals that have high concentrations of K also tend to have relatively high Rb, although the K/Rb ra-

tio is not uniform in all materials, ranging over more than four orders of magnitude. The crystal chemical characteristics of Sr are a little more complicated than those of Rb, but essentially follow Ca, because of identical ionic charge and similar ionic radii (Sr^{2+} 1.26 Å; Ca^{2+} 1.12 Å). Consequently, Sr is relatively concentrated in calcic minerals such as plagioclase, apatite, and calcite; however, Ca^{2+} sites in calcic pyroxenes are too small for the slightly larger Sr^{2+} ions.

Strontium has four stable isotopes, ^{88}Sr, ^{87}Sr, ^{86}Sr, and ^{84}Sr, whose relative abundance is 82.5%, 7.0%, 9.9%, and 0.6%, respectively. But because ^{87}Sr is a decay product of ^{87}Rb, its exact abundance in a rock or mineral depends not only upon the amount of ^{87}Sr present when the material formed, but also upon the concentration of Rb and the age. Materials rich in Rb, such as micas and alkali feldspars, will obviously contain considerable ^{87}Sr, especially if they are old. As isotopic *ratios* are more accurately measured by mass spectrometers than the absolute amount of a single isotope, the abundance of ^{87}Sr is conventionally expressed as the ratio $^{87}Sr/^{86}Sr$. The number of atoms of ^{86}Sr in a mineral is constant, because it is a stable isotope not formed as a decay product of any other naturally occurring radioactive isotope. The relationships among the present day measurable $^{87}Sr/^{86}Sr$ ratio; the initial ratio $(^{87}Sr/^{86}Sr)_0$ when the rock or mineral formed at time zero; its present day, measureable $^{87}Rb/^{86}Sr$ ratio; the age in t years since the formation of the rock or mineral at time zero; and the decay constant λ $(= 1.42 \times 10^{-11}\ y^{-1})$ for ^{87}Rb, is expressed by the equation

$$2.4 \quad {}^{87}Sr/^{86}Sr = ({}^{87}Sr/^{86}Sr)_0 + ({}^{87}Rb/^{86}Sr)(e^{\lambda t} - 1)$$

This is a linear equation of the form $y = b + mx$, where $b = ({}^{87}Sr/^{86}Sr)_0$ and $m = (\lambda t - 1)$. A plot (Figure 2.25) of $x = {}^{87}Rb/^{86}Sr$ and $y = {}^{87}Sr/^{86}Sr$ measured on separated minerals from one igneous rock, or on a group of genetically related whole rocks from a single igneous or metamorphic body that has behaved as a closed system since $t = 0$, yields a straight line called an **isochron.** The intercept (b) of the isochron on the y axis is the initial ratio $({}^{87}Sr/^{86}Sr)_0$. From the slope of the line m = $(\lambda t - 1)$, the age of the rock from the time of crystallization can be calculated. Because of the long half-life of ^{87}Rb, the present-day $^{87}Sr/^{86}Sr$ ratio measured on a mass spectrometer for samples only a few million years old is essentially the same as the initial ratio.

The initial ratio $({}^{87}Sr/^{86}Sr)_0$ is an especially valuable petrogenetic tracer because it is a record of the Rb/Sr ratio of the magma source. Magmas derived by partial melting of source rocks with high Rb/Sr ratios, or contaminated by such material, such as old continental crust, inherit this geochemical property in a high initial ratio. Sources in the peridotitic mantle, where Rb/Sr ratios are very low, yield magmas with low initial ratios.

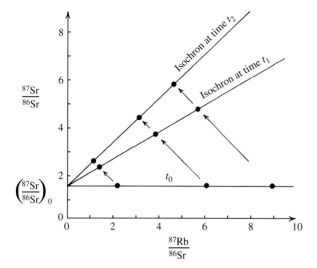

2.25 Schematic Rb-Sr **isochron** diagram for comagmatic rocks or minerals decaying through time. Three samples of rocks or minerals (solid circles) having identical initial $(^{87}Sr/^{86}Sr)_0$ ratios crystallized at the same time, t_0, from a magma or related magma, after which the isotopic system remained closed. The initial $^{87}Rb/^{86}Sr$ ratios of the three samples differed because of different Rb/Sr ratios; muscovite, for example, would have a greater ratio than plagioclase. Subsequent to time t_0, individual rocks or minerals track along straight lines having a slope of -1 due to decay of ^{87}Rb to ^{87}Sr. At any one time, such as t_1, and so on, points representing the analyzed samples define a straight line isochron whose positive slope is dictated by the age of the system since crystallization occurred to close it.

Although discussions in later chapters will explore specific implications of Rb-Sr isotopic compositions to magma genesis, it is pertinent here to present generalized models of Sr-isotope evolution in the Earth. In a simple model that assumes an initial meteoritic composition and no fractionation of Rb and Sr during the entire history of the Earth, all rocks would have the same $^{87}Sr/^{86}Sr$ ratio today after its 4.5-Gy history. This uniformity in $^{87}Sr/^{86}Sr$ ratio is decidedly incorrect. Instead, a single-stage model can be assumed (Figure 2.26a) in which a partial melting event occurred in the mantle at, say, 1.5 Ga that generated magma of greater Rb/Sr ratio than the mantle source because of the differing incompatibility of these two elements. Solidification of this magma created crust whose $^{87}Sr/^{86}Sr$ ratio evolved at a greater rate along a steeper path because of the higher Rb/Sr ratio than in mantle peridotite. However, this single-stage model is still not valid because crustal rocks have a *range* of $^{87}Sr/^{86}Sr$ ratios, to as high as about 0.74, and mantle rocks and magmas derived from them have ratios of about 0.702–0.711. A multistage model must, therefore, apply in which Rb-enriched magmas have been repeatedly extracted from the mantle throughout Earth history (Figure 2.26b). The granitic continental crust has been derived, probably indirectly, from the mantle over most of the history of the Earth; older crustal segments now have the highest $^{87}Sr/^{86}Sr$ ratios.

Samarium-Neodymium Systematics. There are many isotopes of these two light REEs, but the one of most relevance to geochronology and petrology is ^{147}Sm, which decays by alpha emission to ^{143}Nd with a half-life of 106 Gy. The abundance of the radiogenic daughter product is referred to by a ratio with the stable ^{144}Nd isotope, that is, $^{143}Nd/^{144}Nd$. The Sm-Nd isotopic system is similar to the Rb-Sr system just described and is handled in much the same manner. However, one important difference between the two isotopic systems is the much greater half-life for ^{147}Sm decay, which limits the usefulness in dating to rocks more than 1 Ga or so in age.

The initial ratio $(^{143}Nd/^{144}Nd)_0$ serves as a valuable petrogenetic tracer. Rocks with high Sm/Nd ratios (i.e., light REE–depleted) develop higher isotope ratios with the passage of time, whereas rocks with low Sm/Nd ratios (light REE–enriched) develop lower isotope ratios (Figure 2.27). This behavior means that enriched sources like the continental crust have *lower* $(^{143}Nd/^{144}Nd)_0$ but *higher* $(^{87}Sr/^{86}Sr)_0$. Another important contrast with Sr isotopes is the very small difference in the $^{143}Nd/^{144}Nd$ ratios (0.510 to 0.514, or <1%), which stems from the small variations in Sm/Nd ratios of 0.1 to 2.0 in rocks, in contrast to Rb/Sr ratios, which vary from 0.005 to over 100. The small differences in Sm and Nd are related to their similar crystal chemical behavior. Because of this similarity, there are no reservoirs on Earth especially rich in Sm or Nd like the Sr-rich oceanic waters that produce Sr-rich marine carbonates and the Rb-rich continental crust.

Sm-Nd isotopes offer significant advantages over other systems such as Rb-Sr and U-Pb because Sm and Nd occur in major minerals such as pyroxenes and plagioclase and are not easily mobilized. Comparison of Nd and Sr isotopic ratios can constrain models of magmatic evolution, for example, mantle partial melting and crustal contamination, to a better degree than use of Sr isotopes alone.

As a final note regarding Sr and Nd isotopic ratios, $^{143}Nd/^{144}Nd$ and $^{87}Sr/^{86}Sr$ in the remainder of this textbook will always refer to the initial ratio.

2.6.3 Cosmogenic Isotopes: Beryllium

Cosmogenic isotopes are produced when high-energy cosmic rays interact with nuclei of atoms in the atmosphere or on the surface of the Earth to produce new isotopes. Many of these cosmogenic isotopes are unstable and decay to other isotopes, including ^{14}C—which is the basis of a dating method for material less than about 40,000 years old. Isotopes of H (tritium), Al, Cl, and Be are also formed; only the latter is discussed here.

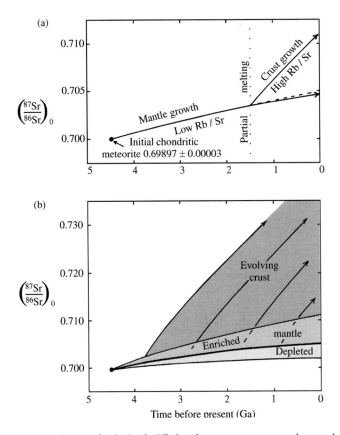

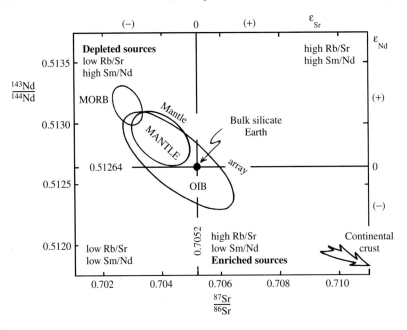

2.26 Schematic strontium isotope evolution diagrams for the Earth. Filled circle represents an assumed parental chondritic meteorite whose initial $(^{87}Sr/^{86}Sr)_0$ ratio at 4.5 Ga, when the Earth accreted from primordial material, was 0.69897 ± 0.00003. Note that growth lines are slightly curved because the amount of ^{87}Rb decaying to ^{87}Sr decreases through time. (a) Single-stage model with hypothetical partial melting event at 1.5 Ga. Thereafter, mantle growth is at a slightly lower rate because of extracted Rb from a depleted mantle. (b) Multistage model in which the mantle has been partially melted many times throughout its history to create a growing, evolving crust. Metasomatic processes create an incompatible-element-*enriched* mantle (Section 11.2.2); other parts of mantle are relatively *depleted* through partial melting processes.

2.27 Sr and Nd isotope ratio correlations. All terrestrial rocks are derived from a **primordial bulk silicate Earth.** The mantle array is defined by relatively depleted, mantle-derived, mid-ocean ridge basalt (MORB) and more enriched ocean island basalt (OIB), as well as fragments (xenoliths) of the suboceanic mantle (labeled MANTLE) brought up in erupted OIBs. Continental rocks plot well off the diagram to the lower right. The ε_{Sr} and ε_{Nd} notation refers to the difference between the measured $^{87}Sr/^{86}Sr$ and $^{143}Nd/^{144}Nd$ ratios in rock samples and the reference bulk Earth ratio (for further details see Rollinson, 1993). (Redrawn from Rollinson, 1993.)

Of the three isotopes of beryllium, only ^9B is stable, whereas ^7Be and ^{10}Be are radioactive isotopes constantly produced as cosmic rays fragment stable isotopes of oxygen and nitrogen in the atmosphere. ^7Be decays to stable ^7Li with a half-life of only 53 days. The more geologically useful ^{10}Be transforms by beta decay to stable ^{10}B with a half-life of 1.5 million years. After production, the cosmogenic Be isotopes are rapidly removed from the atmosphere by precipitation of rain and snow. Because Be is relatively insoluble in water, it is then absorbed or otherwise incorporated into organic and inorganic sediment particles that settle to the bottom of the oceans. As the oceanic crust is subducted and partially melted, the ^{10}Be isotopes are partitioned into ascending and erupting magmas, providing unequivocal evidence that oceanic sediment supplied at least some of the mass of subduction zone magma. Moreover, if detected, ^{10}Be isotopes provide insights into the time involved in subduction, magma generation, and ascent. If no sediment were subducted, no ^{10}Be would be found in any associated volcanic rocks, but, in fact, several investigators have found traces of ^{10}Be in rocks erupted above subduction zones. Of course, the absence of ^{10}Be does not prove that sediment was not involved because ^{10}Be has a very short half-life.

SUMMARY

Magmatic rocks have a broad continuous spectrum of chemical, mineralogical, and modal compositions that are produced by a virtually infinite variety of conditions in the magma source and subsequent evolutionary processes. Variation diagrams facilitate presentation of compositional data so that meaningful differences or similarities and evolutionary patterns can be discerned.

Rock classifications attempt to systematize these compositional continua, as well as recognizing a wide range of fabrics and field relations, in order to understand better the origin of rocks. One rock can be given several labels depending on the intent of the applied scheme of classification.

Magmatic rocks are subdivided on the basis of field relations into plutonic and volcanic, whose fabric is chiefly phaneritic, on the one hand, and aphanitic, glassy, or volcaniclastic, on the other. Mineralogical mnemonics are used to convey modal attributes. Specific rock-type names in plutonic/phaneritic rocks are based on the modal proportion of alkali feldspar, plagioclase, quartz, and feldspathoids for felsic rocks and the proportion of plagioclase, clinopyroxene, orthopyroxene, and olivine for mafic and ultramafic rocks. Other major rock-forming minerals nonessential to

their classification include amphibole, biotite, and Fe-Ti oxides. Rock-type names for volcaniclastic rocks depend on clast size. Names for volcanic/aphanitic (glassy) rock-types depend mostly on the relative amounts of $(Na_2O + K_2O)$ and SiO_2.

The CIPW normative composition of magmatic rocks, together with actual mineral constituents, facilitates their classification according to degrees of silica and alumina saturation. Other coherent comagmatic kindreds and trends that can be discerned on variation diagrams include the alkaline rock suite, whose member rock types are enriched in alkalies relative to silica, and the subalkaline suite. This widespread suite of rocks, in which alkali concentrations are modest so that rocks are usually silica saturated to oversaturated, is further subdivided into the tholeiitic subsuite showing relative Fe enrichment and the calc-alkaline subsuite in which evolved rocks are less Fe-enriched and more felsic. Still other comagmatic kinships can be discerned with respect to variations in SiO_2 and K_2O. A particular rock type such as basalt can occur in different rock suites and can be found in different global tectonic settings. However, specific rock suites tend to be associated with specific tectonic settings, forming petrotectonic associations. Thus, tholeiitic basalt makes up most of the oceanic crust produced at spreading ridges; tholeiitic basalt and tholeiitic andesite (and locally dacite) compose island arcs at convergent ocean-ocean plate junctures; calc-alkaline basalt, andesite, dacite, and rhyolite (and their phaneritic equivalents) constitute medium- to high-K magmatic rocks at continental margin subduction zones; and midplate oceanic islands are of tholeiitic and alkaline basaltic rocks. Highly alkaline rocks occur in relatively stable continental cratons and in continental rifts where basalt and rhyolite also form a bimodal association.

Contrasting sources, global tectonic settings, and evolutionary processes affecting magmas are recorded in trace element and isotopic-tracers in rocks. In contrast to major elements, whose distribution between crystals and melt in a magma is controlled by phase equilibria, trace elements have more widely ranging concentrations, which reflect partitioning according to compatibility constraints between different major and accessory minerals and coexisting silicate melt. Trace element partitioning patterns can indicate specific minerals with which partial melts equilibrated at their source and which minerals might have been subsequently fractionated from the magma. Radiogenic isotope ratios can be used to establish the chronology of magmatic processes and place constraints on magma sources and interactions between a magma and its surroundings. The stable isotopes of oxygen serve as petrogenetic tracers as well as geothermometers of mineral growth.

CRITICAL THINKING QUESTIONS

2.1   What are the hindrances, rationale, and justification for classification of magmatic rocks? Is classification just meaningless busywork?

2.2   Briefly indicate the basis for each of the several classifications of magmatic rocks described in this chapter.

2.3   Contrast rock types with rock suites.

2.4   What are three fabric heteromorphs of granite?

2.5   From Figures 2.8 and 2.12, describe systematic variations in mode and silica among granite, granodiorite, and diorite. From Table 2.2, describe and explain systematic changes in major oxides.

2.6   Explain in detail how it is possible for a quartz-normative basalt to contain no real quartz in its mode, only plagioclase, pyroxene, olivine, and Fe-Ti oxides. A nepheline normative basalt of the same mode containing no actual nepheline. What is the role of pyroxene?

2.7   Discuss silica saturation in a simple model magma consisting of O, Si, Al, and K, which can potentially crystallize quartz, K-feldspar, leucite ($KAlSi_2O_6$), and kalsilite ($KAlSiO_4$). Create a diagram for this hypothetical system like Figure 2.14.

2.8   Discuss differences in the major and trace element compositions of basalt and relate these differences to global tectonic setting.

2.9   What is the relation between crustal thickness and K_2O at a given SiO_2 content for magmatic rocks in subduction zones?

2.10   What are the bases for distinguishing between compatible and incompatible trace elements? Give examples of each in rhyolitic and in basaltic magmas.

2.11   What are the rationale and the goal of investigations of trace elements in magmatic rocks? Of isotopes?

2.12   What factors control variations in the ratios of stable isotopes? Radiogenic isotopes?

2.13   Contrast the behavior and implications of Sm-Nd and Rb-Sr isotope systems.

PROBLEMS

2.1   Classify the garnet and spinel peridotites from Table 11.1 using Figure 2.10b.

2.2   Plot average granite and granodiorite in Table 2.2 in an enlarged photocopy of Figure 2.8 using the amounts of normative *Q, Or,* and *Ab* in lieu of modal Q, A, and P.

2.3   Calculate the normative composition of a magmatic biotite and hornblende from Appendix A. Classify these minerals, *as if they were rocks,* as to degree of silica saturation. Discuss how crystallization of these minerals in magmas controls the availability of silica for potential crystallization of quartz.

2.4   From problem 2.3 classify the selected biotite and amphibole as to degree of alumina saturation.

2.5   Calculate the normative compositions of the pantellerite in Table 13.10, the phlogopite-rich leucite lamproite in Table 13.11, and the shoshonite in Table 13.6. Classify these rocks in the total alkali-silica diagram (Figure 2.12). Discuss whether these rocks belong to the alkaline rock suite.

2.6   Calculate the bulk partition coefficients for Ni and for Rb in the garnet peridotite in Table 11.1 using partition coefficients in Table 2.5.

2.7   Construct a chondrite-normalized diagram for the spinel peridotite in Table 11.1. Describe this diagram. What rock in Figure 2.23 does the pattern of the spinel peridotite most closely resemble?

Thermodynamics and Kinetics: An Introduction

FUNDAMENTAL QUESTIONS CONSIDERED IN THIS CHAPTER

1. What are the basic concepts of thermodynamics and why is thermodynamics important in petrology?

2. How can the flows of matter and energy in changing rock-forming systems be predicted and interpreted using thermodynamic models?

3. How do phase diagrams reflect the thermodynamic stability of different states of matter?

4. What is kinetics, why is it important in petrology, and in what ways do kinetic factors limit the application of thermodynamics to rock-forming systems?

INTRODUCTION

Unlike sedimentary systems, in which many flows of energy and matter are readily observed and studied in modern environments, natural magmatic systems are usually not amenable to direct investigation. Only a few nonexplosively extruded magmas can be examined at close range. Accordingly, igneous petrologists have turned to controlled laboratory experiments and theoretical models to enhance their understanding of the behavior of magma systems, supplementing inferences from rocks. Theory and experiment can often demonstrate what is impossible and what is possible in a geologic system. At a minimum, they provide fruitful insights.

However, despite their advantages, such investigations of natural geologic systems have inherent weaknesses. Laboratory studies can never duplicate the large dimensions and long time scales of geologic systems—as large as the entire Earth and commonly millions of years. Natural systems are always more complex than the experiment or model used to simulate them, necessitating simplifying assumptions that may deviate from the real world to varying degrees. The ultimate test of the validity of any petrologic interpretation lies in the fabric, composition, and field relations of real rocks. Nature does not lie, but our interpretations of it can.

With these caveats in mind, we consider in this chapter one of the foundation stones of petrogenetic theory—thermodynamics—and the kinetic factors that limit its application to real petrologic systems.

✳3.1 WHY IS THERMODYNAMICS IMPORTANT?

In Section 1.3, the concepts of stability and equilibrium were introduced, showing that rock-forming processes in changing geologic systems tend toward a state of stable equilibrium in which the energy is the lowest possible. Metastable states having less than the minimal possible energy can persist indefinitely because of an activation energy barrier impeding the change. Stability and equilibrium were illustrated by hypothetical examples of three boulders that have differing amounts of gravitational potential energy because of their different position on hilly terrain and of hot lava flowing downhill and losing heat, thus reducing its potential *and* thermal energy.

The melting of rock to generate magma suffices to illustrate the inadequacy of minimization of thermal and gravitational potential energies as a universal indicator of how all systems spontaneously change. During the spontaneous melting of mantle rock in an ascending plume beneath the island of Hawaii, the plume system has moved to a state of *higher* gravitational potential energy while simultaneously losing thermal energy by adiabatic cooling (discussed later). Is the "rule" for attainment of stable equilibrium cited in Section 1.3 in error, or is there another form of energy that is minimized in these melting systems, and in natural processes in general?

It turns out that a state of stable equilibrium in rock-forming systems does indeed have the lowest possible energy, but of a new form. This new form of energy, or energy function, incorporates potential, kinetic, thermal, chemical, and mechanical (work) energies in such a way as to be generally applicable to all natural systems. Evaluation of this new energy function allows one to predict the direction of changes that naturally occur in rock systems as their *T*, *P*, and chemical composition change. J. W. Gibbs formulated this new energy function about a century ago. Called the Gibbs free energy, this new energy function can be considered as a thermodynamic potential energy by analogy with gravitational potential energy in systems, such as the hypothetical boulders in Figure 1.6, whose stability is governed by gravity.

To begin to appreciate this new Gibbs free energy and how it can be applied to a rock-forming system, some elementary concepts of thermodynamics must be introduced. **Thermodynamics** is a set of mathematical models and concepts that describe the way changes in *T*, *P*, and chemical composition affect states of stable equilibrium in rock-forming systems. Thermodynamics is a powerful petrologic tool if an appropriate **model** that accurately mimics the real rock system can be constructed to enhance understanding of the system's working. The goals of thermodynamics in petrology are essentially twofold:

1. Predict how rock-forming systems respond to changes in *P*, *T*, and chemical composition
2. Interpret the nature of *P*, *T*, and chemical composition in ancient rock-forming systems from the chemical compositions of the constituent minerals and glass in the magmatic rock

The plan of the following pages is, first, to present some "nuts and bolts" of thermodynamic models and concepts, including types of simple ideal systems, properties of state that characterize systems, ideal processes of change, and energy changes; and, second, to introduce the concept of entropy as a necessary ingredient in formulating the Gibbs free energy.

✳3.2 ELEMENTARY CONCEPTS OF THERMODYNAMICS

3.2.1 Thermodynamic States, Processes, and State Variables

Geologic **systems**—parts of the universe set aside in our mind for investigation—are commonly more complex than, for example, the systems of a laboratory chemist. Geologic systems can be as large as the entire Earth and may endure for millions of years; they tend to be poorly definable and ever changing. Ideal end-member systems (Figure 3.1) are as follows:

1. **Isolated system:** No matter or energy can be transferred across the boundary of the system and no work can be done on or by the system. Considering the span of time over which most identifiable geologic systems operate, no part of the Earth, nor even all of it, can be considered perfectly isolated because transfers of energy and movement of matter typify the dynamic Earth.

2. **Open system:** The opposite of an isolated system. Matter and energy can flow across the boundary and work can be done on and by the system. Most geologic systems are open, at least in the context of their long lifetimes.

3. **Closed system:** Energy, such as heat, can flow across the boundary of the system, but matter cannot. The composition of the system remains exactly constant. Because movement of matter is slow across boundaries of, for example, a rapidly cooling thin dike, it can be considered to be virtually closed. On the other hand, a large intrusion of magma that remains molten for tens of thousands

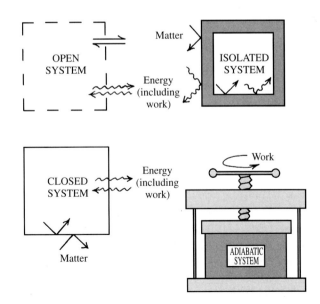

3.1 End-member thermodynamic systems. Classification is according to flows of matter and energy across their boundary with the surroundings.

of years while various fluids move across the wall-rock contact is not a closed system.

4. **Adiabatic system:** A special category of an isolated system is one in which no heat can be exchanged between the system and the surroundings. It is thermally insulated, but energy can be transferred across the system boundary through work done on or by it. Thus, an ascending, decompressing mantle plume or magma body cools as it expands against and does PV work on the surroundings, into which little heat is conducted because the rate of conduction is so slow.

Any **thermodynamic state** of a system can be characterized by its **state properties,** or **state variables.** Some state properties are **extensive** in that they depend on the amount of material, such as the mass and volume of some chemicals constituting a system. Other state properties are **intensive** and are independent of the amount of material present; they have a definite value at each point within the system, such as T, P, and concentration of a particular chemical species. **Density** (mass/volume) is also an intensive property. For example, at $T = 25°C$ and $P = 1$ atm, the density of stable graphite anywhere within a crystal is 2.23 g/cm³ whereas that of diamond is 3.51 g/cm³.

P and T are extremely important intensive variables whose change in rock-forming systems is responsible for changes in their state; for example, increasing T causes solid rock to melt. Values of the geobaric and geothermal gradients in the Earth can be used to approximate P and T at a particular depth, assuming these gradients to be uniform (Sections 1.1.3 and 1.2). Otherwise uniform gradients can locally be perturbed, such as the geothermal gradient around a shallow crustal magma intrusion.

Rock-forming geologic processes can be thought of as **thermodynamic processes:** those that affect a thermodynamic system as it changes from one state to another. What happens in transit—the time-dependent kinetic path—between the initial and final states is not important at this point, but the nature of kinetic paths will be considered later in Section 3.6. In a thermodynamic process there could be all kinds of work done on or by the system, flows of heat in or out, chemical reactions and movement of matter, tortuous paths taken, and so forth. However, two ideal, end-member thermodynamic processes can be recognized: irreversible and reversible. In an **irreversible thermodynamic process** the initial state is metastable, and the spontaneous change in the system leads to a more stable, lower-energy final state. The conversion of metastable volcanic glass into more stable crystals—a process called devitrification—under near-atmospheric conditions is one example of an irreversible process. The energy trough in which metastable glass is stuck has to be surmounted, overcoming the activation energy, before

it can move to a lower energy state. The atomic bonds in the glass must be broken or reformed into more stable crystalline structures. Devitrification of glass occurs spontaneously in the direction of diminishing energy, never in reverse (Figure 3.2).

In a **reversible process,** both initial and final states are stable equilibrium states and the path between them is a continuous sequence of equilibrium states. A reversible process is never actually realized in nature; it is a hypothetical concept that is used to make the mathematical models of thermodynamics work, and that's all that can be said without an extended discussion.

3.2.2 First Law of Thermodynamics

Suppose that a change in the internal energy, dE_i of a rock system, such as a mineral grain, is produced by adding some amount of heat to it, dq. As a result of absorbing heat, the mineral grain expands by an increment in volume, dV, doing an increment of PV work, $dw = PdV$, on the surrounding mineral grains. According to the **first law of thermodynamics,** or **law of conservation of energy,** the increase in internal energy due to heat absorbed is diminished by the amount of work done on the surroundings, or

$$3.1 \quad dE_i = dq - dw = dq - PdV$$

In the "scientific" convention, heat added to (or work done on) a system, as in this formulation, is positive and work done by (or heat withdrawn from) the system on its surroundings is negative.

3.2.3 Enthalpy

A new state property, the **enthalpy,** H, is *defined* here as

$$3.2 \quad H = E_i + PV$$

Upon differentiation, this equation becomes $dH = dE_i + PdV + VdP$. Combining with equation 3.1 gives

$$3.3 \quad dH = dq + VdP$$

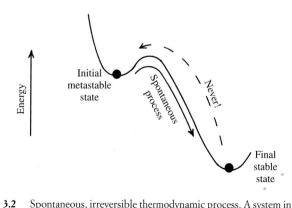

3.2 Spontaneous, irreversible thermodynamic process. A system in an initial higher-energy metastable state moves spontaneously to a lower-energy, stable state. The reverse never happens spontaneously.

For **isobaric** (constant-pressure) changes, $dP = 0$, equation 3.3 becomes

3.4 $dH_p = dq_p$

Although H may appear to be simply another and unnecessary label for q, the enthalpy is significant in providing a measure at constant P of how much heat is gained or lost from a system. Three possible changes in a system are accompanied by a gain or loss of heat, as follows:

1. Chemical reactions, such as the reaction of quartz plus calcite creating wollastonite plus CO_2.

2. A change in state, such as crystals melting to liquid, that occurs at a fixed T once that T is reached in heating a system.

3. A change in T of the system where no change in state occurs, such as simply heating crystals below their melting T. Combining equations 1.4 (in differential form) and 3.4 gives $dq_p = C_p\, dT = dH_p$ or

3.5 $C_p = (dH/dT)_p$

The last two changes can be illustrated by a plot of enthalpy versus T for the diopside $CaMgSi_2O_6$ system as it is heated at constant P (Figure 3.3). As diopside crystals absorb heat up to their melting point, (3., above), T increases proportionally with the heat capacity, C_p. The slope of this line is $(dH/dT)_p = C_p$. At the melting T absorbed heat does not increase T, but is consumed in breaking the atomic bonds of the crystalline structure to produce the more random liquid array. This relatively large amount of absorbed heat at the *constant* T of melting, (2., above), is the **latent heat of melting,** or **enthalpy of melting,** ΔH_m. Once melting is complete, additional input of heat into the system raises the T of the melt proportional to its heat capacity. The slopes of the T-H lines above and below the melting point are the heat capacities of the melt and the crystals, respectively.

Note in Figure 3.3 that the latent heat involved in changing the state of the system from liquid to crystal, or vice versa, is similar to the heat absorbed in changing the T of the crystals or liquid by hundreds of degrees. Thus, melting of rock in the Earth to generate magma absorbs a vast amount of thermal energy, which moderates changes in T in the system.

Chemical reactions, (1., above), either release or absorb heat: They are either **exothermic** or **endothermic,** respectively. If catalyzed by a spark, the reaction between hydrogen and oxygen releases a burst of heat—a rapid exothermic reaction, or explosion. Solidification of magma exothermically releases the latent heat of crystallization. In contrast, dissolution of potassium nitrate in water endothermically absorbs heat so that the container becomes cold. Thus, whether heat is released or absorbed in a chemical reaction provides no *consistent* clue as to the direction a reaction moves spontaneously.

3.2.4 Entropy and the Second and Third Laws of Thermodynamics

Another way to look at spontaneity is in changes in the distribution or concentration of energy. Spontaneous thermal processes lead to a more even concentration of heat. A bowl of hot soup on the table eventually reaches the same T as the room. Heat flows spontaneously from a hot body to a cold, eliminating the difference in T. Without an uneven concentration of thermal energy, or the opportunity for heat flow, no work can be done. As heat flows from an intrusion of hotter magma into the cooler wall rocks, PV work of volumetric expansion is performed on them. The heat also drives endothermic chemical reactions of wall rock metamorphism. Water in an enclosed lake on a high plateau has gravitational potential energy relative to sea level, but so long as it is isolated from sea level and cannot flow, the concentration of energy is uniform and no work can be done. However, in the natural course of events, a river drains the lake into the sea, forming a process path along the potential energy gradient between the high- and low-potential-energy levels. Work can then be done, driving turbines to generate electricity, eroding the river channel, transporting sediment, and so on.

One statement of the **second law of thermodynamics** is that spontaneous natural processes tend to even out the concentration of some form of energy, smoothing the energy gradient. A hot lava flow extruded from a lofty volcano cools to atmospheric T as it descends down slope, thereby reducing differences in thermal

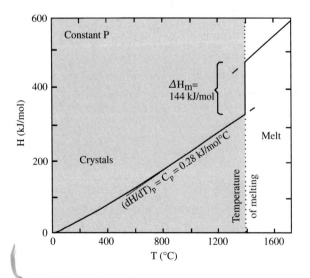

3.3 Enthalpy-T relations for $CaMgSi_2O_6$ at 1 atm. (From Weill DF, Hon R, and Navrotsky A, The igneous system $CaMgSi_2O_6$-$CaAl_2Si_2O_8$-$NaAlSi_3O_8$: Variations on a classic theme by Bowen. In: Hargraves RB, ed., Physics of Magmatic Processes. Copyright © 1980 by Princeton University Press. Reprinted with permission of Princeton University Press.)

and gravitational potential energy between initial and final states in accordance with the second law.

Eventually, billions of years from now, all of the thermal energy in the Earth will be consumed in tectonism, volcanism, and other processes and dispersed into outer space. No mountains or volcanoes will be erected and erosion in the solar-powered hydrologic system will wear everything down to some common level (assuming the Sun does not run out of nuclear energy!). Without differences in the concentration of thermal and gravitational potential energy no geologic work can be accomplished and the planet will be geologically dead!

The measure of the uniformity in concentration of energy in a system is called the **entropy, S.** The more uniform the concentration of some form of energy, the greater the entropy. The geologically dead planet will have maximal entropy.

Another, more useful, way to define entropy is to relate it to the internal disorder in the system. This provides an alternate statement of the second law: In any spontaneous process in an isolated system there is an increase in entropy, that is, an increase in disorder. The law in this form is illustrated by Figure 3.4, where

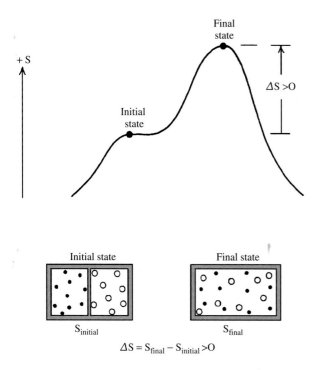

$$\Delta S = S_{final} - S_{initial} > 0$$

3.4 Entropy increases in a spontaneous, irreversible process in an isolated system. Bottom left, a hypothetical isolated system—a box filled with atoms of two gases (black and white balls) separated by an impermeable wall. Bottom right, the wall has been removed in the box, and the atoms of the two gases have mixed spontaneously and irreversibly as a result of their motion (kinetic energy). An increase in disorder or randomness of the atoms in the system and an increase in entropy, $\Delta S > 0$. No change in energy has occurred. (Redrawn from Anderson, 1996.)

white and black balls in the boxes represent molecules of two different gases. The spontaneous mixing of the two gas molecules results in an increase in "mixed-upness," disorder, randomness, or entropy. Note that there is no accompanying change in energy in this mixing process. Thus, another driving "force" for a spontaneous process is an increase in entropy, even though there may be no change in the energy.

At decreasing T, crystals become increasingly ordered, less atomic substitution is possible, and their entropy decreases. The **third law of thermodynamics** states that at absolute zero, where the Kelvin temperature is zero ($0K = -273.15°C$), crystals are perfectly ordered and all atoms are fixed in space so that the entropy is zero.

A convenient way to think of relative entropies is that a gas made of high-speed molecules in random trajectories has a greater entropy than the compositionally equivalent liquid array, which, though still somewhat disordered, has linked atoms. The compositionally equivalent crystalline solid has still lower entropy, because its atoms form an ordered array. As an example, for water,

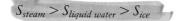

$$S_{steam} > S_{liquid\ water} > S_{ice}$$

3.2.5 Gibbs Free Energy

The boulder-on-the-hill example (Figure 1.6) has two major flaws as an analogy for the way natural systems, in general, change spontaneously from a higher to a lower energy state. First, it isn't always gravitational potential energy that is minimized. Second, the analogy does not take into account the fact that in an isolated system a process can proceed spontaneously without any change in energy, but it does proceed with increasing entropy (Figure 3.4). To overcome these two flaws, a new extensive property of a system is *defined* in such a way as to serve as a universal directionality pointer for spontaneous reactions. This new property is called the **Gibbs free energy,** G, and is defined by the expression $G = H - TS$. Combining with equation 3.2

$$3.6 \quad G = E_i + PV - TS$$

In differential form this becomes

$$3.7 \quad dG = dE_i + PdV + VdP - TdS - SdT$$

Remembering the work-pressure-volume relation ($dW = PdV$), we can write a parallel expression for the heat-temperature-entropy relation

$$3.8 \quad dq = TdS$$

Equation 3.7 can be simplified by substituting this equality and also by making a substitution from equation 3.2 to obtain

$$3.9 \quad dG = VdP - SdT$$

Equation 3.9 is a useful thermodynamic expression that allows us to make powerful statements regarding the direction of changes in geologic systems as the independent intensive variables of state, T and P, change. The extensive entropy and volume properties of the system are also relevant factors.

If P and T remain the same through any spontaneous change in state, that is, $dP = dT = 0$, then, from equation 3.9

$$3.10 \quad dG_{P,T} = 0$$

This is simply the condition for a minimum (or maximum) in G in P-T space where the slope of the tangent to the energy function is zero, or horizontal. Figure 3.5 shows a system that has moved to a state of minimal energy and stable equilibrium from a higher-energy, metastable state at constant P and T in a closed (constant composition) system. Note that the energy change, $\Delta G_{P,T}$, between the initial metastable state and the final stable state is negative: $\Delta G_{P,T} < 0$.

Without further justifying details (see Anderson, 1996) a summary statement can be made as follows: In some spontaneously changing systems, increasing entropy is the dominant factor, whereas in others, decreasing energy is the dominant factor. The *Gibbs free energy* is formulated in such a way that it *always decreases in a spontaneous change in the state of a system*. In other words, the Gibbs free energy of the final stable state is lower than that of the initial metastable state (Figure 3.5). The analogy with the boulders in the valley (Figure 1.6) is now complete. The Gibbs free energy is a **thermodynamic potential energy** that, like gravitational potential energy for the hypothetical boulders, is the lowest possible in a state of stable equilibrium for a changing system.

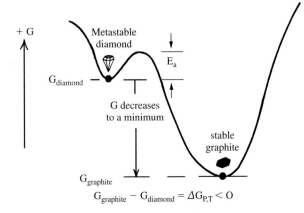

3.5 Gibbs free energy decreases in a spontaneous change in a closed system where the initial and final states are at the same P and T. In this example, the energy of diamond is greater than that of graphite at the same P and T, or $G_{diamond} > G_{graphite}$, so the change in energy, $\Delta G_{P,T}$, in the spontaneous process is negative, or $G_{diamond} - G_{graphite} = \Delta G_{P,T} < 0$. Note the activation energy barrier, E_a, that must be surmounted in order for the change to occur.

✳3.3 STABILITY (PHASE) DIAGRAMS

The foregoing concepts of stability couched in equation 3.9 can be portrayed graphically. For a system of constant chemical composition, the two intensive variables that govern the state of the system, whether it is stable or unstable, are P and T. On a diagram where P is usually represented on the y-axis and T on the x-axis, it would be expected that over a particular range of P and T one state of the system is more stable and over another range in P and T another state is more stable, with some sort of boundary between these two states. Across the boundary, the difference in the free energy of the two states, ΔG, changes sign, as the G of the more stable state in its stability field is exceeded by the G of the less stable state. A **stability diagram** shows which state is more stable as a function of P, T, or other variables.

In rock-forming systems, different states are manifested by the existence of different phases or assemblages of phases. A **phase diagram** shows which phase, or assemblage of two or more phases, is more stable as a function of P, T, or other variables. A **phase** is defined as a part of a system that is chemically and physically homogeneous, bounded by a distinct interface with adjacent phases, and physically separable from other phases. Phases may be gaseous, liquid, or solid. *Minerals in rocks are solid crystalline phases*, whereas glass is a solid amorphous phase that has no ordered atomic structure. Crystalline grains of olivine and plagioclase in a basalt are two separate and distinct solid phases. Even though made of albite ($NaAlSi_3O_8$) and anorthite ($CaAl_2Si_2O_8$) end members, a homogeneous plagioclase solid solution crystal is a single phase because the end members cannot be physically separated.

Figure 3.6 is a phase diagram that portrays stability relations of a hypothetical crystalline phase (crystals) and its compositionally equivalent melted liquid phase separated by the melting (freezing or crystallization) curve. Melting and crystallization are fundamental changes in state that are caused by changes in P and T. Obviously, such changes are the very core of dynamic magma systems and deserve special consideration. Other phase diagrams that portray stability relations of only crystalline phases are of no less importance in petrology.

Over the range of P and T where crystals are more stable in Figure 3.6, the Gibbs free energy of crystals is less than that of liquid, $G_c < G_l$. On the other hand, in the P-T region where liquid is more stable, the reverse is true, or $G_c > G_l$. For any value of P and T exactly on the melting (crystallization) curve that is the boundary line the two regions, crystals and liquid coexist together in equilibrium and their free energies are equal, $G_c = G_l$.

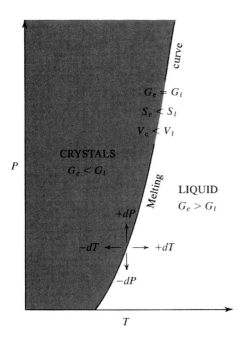

3.6 Schematic phase diagram for a crystalline solid and the liquid that forms by melting it. The positively sloping melting curve indicates that increasing P causes the melting T to increase and that the stability field of denser, smaller-molar-volume crystals is expanded at increasing P relative to the liquid phase.

The significance of any P-T phase diagram, such as Figure 3.6, can be better appreciated if G is made a variable in the third dimension, plotting along an axis perpendicular to the P and T axes (Figure 3.7). In this diagram, the functional relationship among G, P, and T for crystals is represented by a curved surface, and that for the liquid state by a second curved surface. The free-energy surfaces for these two phases differ because their entropies and molar volumes differ at different P and T. The two surfaces intersect in a line that is the locus of points in G-P-T space for which $G_c = G_l$, or $\Delta G = 0$, where crystals and liquid are in equilibrium. This line is, of course, the melting (crystallization) curve and its projection onto the two-dimensional P-T plane is the melting curve in Figure 3.6.

Thus, any two-dimensional P-T phase diagram portraying phase stabilities implies that the Gibbs free energy is minimal for the indicated stable phase(s), even though it is not explicitly shown.

3.3.1 Slope of the Melting Curve

The slope of a melting curve on a P-T phase diagram happens to be quite significant and warrants further consideration here. Imagine a system in which crystals and liquid coexist together in equilibrium. The **molar volume** (inverse of the density, or volume/mass) and entropy of the two phases are not equal:

$$V_c < V_l$$
$$S_c < S_l$$

Qualitatively, we can predict how the equilibrium will shift if P and/or T changes by applying **LeChatelier's principle**: if a change occurs in the state of a system it will respond in such a way as to minimize or moderate the effects of the change. States of smaller molar volume are more stable at higher P because if P increases on any phase, its volume will diminish (its density increases), moderating the effect of the increased P. Hence, in Figure 3.6, if P increases by a positive increment, $+dP$, on the system where crystals and liquid are initially in equilibrium, the inequality in molar volumes of the two phases dictates that the state of lesser volume—crystals—is stabilized. In contrast, more disordered states of higher entropy are more stable at higher T. Hence, a positive perturbation in T, $+dT$, stabilizes the higher-entropy liquid at higher temperatures. Because crystals are stabilized at higher P but liquid at higher T, the melting curve can only have a positive slope in P-T space. In other words, higher P shifts the melting T to higher values.

Manipulation of equation 3.9 leads to the same predictions. For a small change in only T, dT, holding P constant so that $dP = 0$, this equation reduces to

$$3.11 \quad dG_p = -S dT$$

This expression indicates that the small positive increment in dT causes the incremental change in free energy to be more negative for the higher-entropy

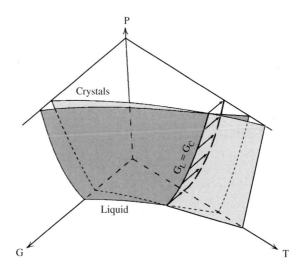

3.7 Perspective view of free energy surfaces in three-dimensional G-P-T space for a hypothetical crystalline solid and the liquid produced by melting it. The free energy surfaces intersect in a curved line where crystals and liquid coexist in equilibrium, and that is the locus of points for which $G_{crystals} = G_{liquid}$. Projection of this line parallel to the G axis (arrows) onto the P-T plane is the melting curve (thick line) shown in Figure 3.6. Note that the free energy surface for liquid lies at higher G at lower T than the equilibrium line so that crystals are more stable in this region. Conversely, the free energy surface for liquid lies at lower G at greater T than the equilibrium line so that liquid is more stable in this region.

liquid than for crystals. Hence, increasing T in a system in which crystals and liquid are initially in equilibrium, $G_c = G_l$, on the melting curve shifts the system into the stability field of liquid. A negative dT shifts the system into the stability field of crystals. If T is constant and P changes, equation 3.9 reduces to

3.12 $dG_p = VdP$

Therefore, from an initial state in which crystals and liquid are in equilibrium, an increase in P ($+dP$) stabilizes the state of lesser volume, or crystals, as this minimizes dG.

Such melting curves are generally positively sloping for minerals in the Earth. Some melting curves determined experimentally in the laboratory are shown in Figure 3.8. Only very rarely does a melting curve have a negative slope; the familiar water-ice system is a disconcerting example.

An important consequence of the positively sloping melting curve is that hot crystals at P and T near the melting curve can melt by **decompression,** reducing P on the system ($-dP$). This can be accomplished **isothermally,** without changing T, and even by simultaneously decreasing ($-dT$) somewhat. This is obviously an important phenomenon where hot mantle rock decompresses and melts during ascent in plumes and in upwellings beneath oceanic spreading ridges.

Clapeyron Equation.   From equation 3.9, a quantitative expression can be obtained for the slope of the equilibrium boundary between two phases in a P-T phase diagram. Again, note that for crystals in equilibrium with liquid, $G_c = G_l$. In a new state at a different P and

T but still on the melting curve so that crystals and liquid coexist in equilibrium, the change in free energies for the two phases must be equal

$$dG_c = dG_l$$

or from equation 3.9

$$V_c dP - S_c dT = V_l dP - S_l dT$$

Collecting terms

3.13 $(V_c - V_l)dP = (S_c - S_l)dT$

$$\frac{dP}{dT} = \frac{(S_c - S_l)}{(V_c - V_l)} = \frac{\Delta S}{\Delta V}$$

From this **clapeyron equation,** the slope of the melting curve of a mineral on a P-T diagram can be calculated if the appropriate values of molar volume and entropy for the phases are known. More generally, this equation gives the slope of the boundary line separating stability fields of any two phases, or two assemblages of phases, of identical composition in a closed system.

3.3.2 Determination of Phase Diagrams

Most diagrams of petrologic relevance have been determined by painstaking experiments in the laboratory using various types of devices that simulate the elevated P-T conditions that prevail in the crust and mantle of the Earth (see Special Interest Box 1.2). Mixtures of chemical compounds from the chemical stockroom, synthetic reactive glasses, or finely pulverized natural minerals and rocks appropriate to the particular investigation are maintained at a desired P and T in the device for a sufficient period to attain a state of stable equilibrium. The products of the experiment are then examined to determine what equilibrium phase(s) existed at the desired P and T. Many experiments at different P and T are required to "map out" the stability fields in P-T space, and their boundaries.

In contrast to the method of direct synthesis of new phases from other reactant phases, a phase diagram can also be determined from thermodynamic data on possible phases that may be stable in a particular compositional system. Changes in Gibbs free energies can be calculated as a function of changes in P and T to establish which phase(s) is (are) the more stable. Relevant thermodynamic data include molar volumes of phases measured by various techniques and entropies measured in a laboratory calorimeter, a device that reveals the heat produced or absorbed in a chemical reaction. Direct synthesis methods also yield thermodynamic data for calculations. Wherever possible, it is prudent to compare results from independent methods of determination of a phase diagram as a possible check on accuracy.

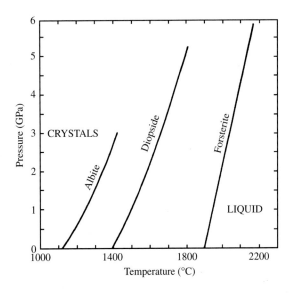

3.8   Melting (freezing) curves for some pure end-member minerals. (Redrawn from Boyd and England, 1961; Davis and England, 1963.)

✳3.4 THERMODYNAMICS OF SOLUTIONS: SOME BASIC CONCEPTS

The thermodynamic models developed thus far only apply to systems of pure minerals and their corresponding liquids created by melting. In such systems of constant chemical composition, equilibrium states and directions of change in state are governed solely by P and T. However, a pure phase in any form in geologic systems is rare. Even "pure" natural quartz in the clearest of crystals may contain small amounts, in the parts per million (ppm) range, of Al, Li, Fe, and Mn in solid solution. All rock-forming minerals are solid solutions, including the common feldspars, micas, amphiboles, pyroxenes, olivines, and Fe-Ti oxides. **Silicate liquids,** also referred to as **melts,** which form by melting of minerals and crystallize to form minerals, are ionic solutions of O^{2-}, Si^{4+}, Al^{3+}, Ca^{2+}, Na^{1+}, and so on. All natural gaseous phases are mixtures of H_2O, CO_2, N_2, and so on. Nature is full of phases that are solutions. Better understanding of these solid, liquid, and gaseous solutions and the ways in which they react together and equilibrate requires additional thermodynamic tools and models.

3.4.1 Components and Mole Fractions

A **solution** is a homogeneous mixture of two or more chemical components in which their concentrations may be freely varied within certain limits. The components of a system are the chemical constituents needed to make the phases in it. So that our list of constituents is not too long and unwieldy, we add the provision that the **components** of a system are the *smallest number* of chemical entities required to describe the composition of every phase that exists in the system. Hence, the phases constituting a system must be known in order to decide what the components are. For example, consider a hypothetical magma system composed of three phases: leucite, $KAlSi_2O_6$; K-feldspar, $KAlSi_3O_8$; and a melt that is approximately $KAlSi_4O_{10}$. What are the components? Possible sets of chemical entities by which the composition of all three phases can be expressed include the following:

1. O, Si, Al, K
2. SiO_2, Al_2O_3, K_2O
3. $KAlSi_2O_6$, $KAlSi_3O_8$, $KAlSi_4O_{10}$
4. $KAlSi_2O_6$, SiO_2

Of these possibilities, the least number of components is possibility (4) and therefore $KAlSi_2O_6$ and SiO_2 constitute the correct components. $KAlSi_2O_6$ plus SiO_2 expresses the composition of K-feldspar while $KAlSi_2O_6$ plus 2 SiO_2 expresses the composition of the melt. Note that the components in a system are not necessarily or simply the chemical formulas of the phases in it.

To express the proportions of a component A in a phase or a system the **mole fraction,** X_A, can be defined as

$$3.14 \quad X_A = \frac{n_A}{\Sigma n} = \frac{n_A}{(n_A + n_B + n_C + \cdots)}$$

where n_A is the number of moles of component A in the phase or system and n_B, n_C, . . ., are the number of moles of other components B, C, and so on. For example, the composition of a particular plagioclase in the plagioclase solid-solution series between the end-member components $NaAlSi_3O_8$ (albite, or Ab) and $CaAl_2Si_2O_8$ (anorthite, or An) is commonly expressed in terms of the mole fraction of the calcic component as, say, An_{37}. The mole fraction is actually 0.37, whereas 37 is the percentage of the end-member component ($100 \times$ mole fraction). Note that for a binary system of two components A and B, $X_A + X_B = 1$, or $X_A = 1 - X_B$. Thus, in the example just cited, the mole fraction (in percentage notation) of the albite end-member is Ab_{63}.

3.4.2 Partial Molar Volume

Suppose white and black sand grains are physically mixed to produce an intermediate speckled "solution." If the volume of white sand alone is V_W and of black sand alone is V_B, the molar volume, V_m, of the mixture is

$$3.15 \quad V_m = X_W V_W + X_B V_B$$

In this simple additive relationship, the molar volume of intermediate physical mixtures plotted against mole fraction is simply a straight mixing line (Figure 3.9a). The molar volume of any intermediate mixture depends upon the mole fraction of the two end-member components.

But chemical solutions are unlike this simple physical mixture of inert sand grains. At the atomic level significant interactions occur in the form of mutual attraction and repulsion of charged ions. For the usual case of attraction between cations and anions in a melt solution, the volume of intermediate mixtures of two components is no longer a straight line, but a convex downward loop as shown in Figure 3.9b. The molar volume of some intermediate solution, V_m, in the chemical mixture of the two pure end-member components is less by an amount ΔV_{mix} because of the interionic attractions.

The molar volume of any intermediate solution between its end-members can be determined by measuring its density and calculating its reciprocal. But how can the molar volume of *one component in the solution* be determined? If the molar volumes of intermediate solutions of the two components are known (represented as the convex downward curve in Figure 3.9b),

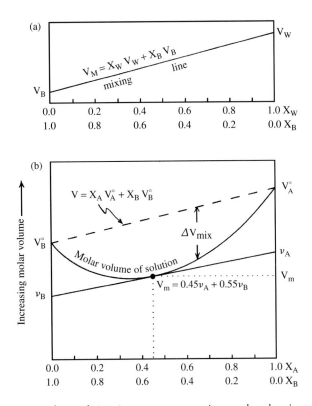

3.9 Volume relations in two-component mixtures plotted against mole fraction of the two end-member components. (a) Physical mixtures of white and black sand grains have molar volumes, V_M, lying along a linear mixing line. (b) Molar volumes of a chemical solution, in which there is attraction between constituent atoms, lie along a convex downward curve. The greater the magnitude of the attraction the greater is the magnitude of ΔV_{mix}. The molar volumes of the pure end-member components in their standard states are $V°_A$ and $V°_B$. Molar volume of the solution at $X_A = 0.45$ is V_m. Partial molar volumes of components A and B in the solution at $X_A = 0.45$ are v_A and v_B.

a simple geometric construction gives the desired **partial molar volume** of each component in the solution, for any mole fraction of the components. One simply draws a straight tangent line to the molar volume curve in Figure 3.9b at the desired mole fraction; the intercepts of this tangent line on the two vertical axes give the desired partial molar volumes, v_A and v_B, of each component in the solution at that particular composition.

The meaning of a partial molar volume can be illustrated as follows: The volume of pure water at 0°C and 1 bar is 18.0 cm³/mol, and at 950°C and 1 kbar it is 99.3 cm³/mol (Burnham et al., 1969). If, to a vast body of hydrous $NaAlSi_3O_8$ melt at 950°C and 1 kbar, is added 1 mole of water, the increase in volume is only 22.3 cm³ (Ochs and Lange, 1997). The increase in volume under these conditions is less than one-fourth that of adding 1 mole of water to a reservoir of pure water because the Na, Al, and Si ions in the $NaAlSi_3O_8$ melt chemically bond to the additional

H and O ions in water. The volume increase that results from the addition of an infinitesimally small amount of one component—water, in this case—to a solution of that component plus $NaAlSi_3O_8$ is the partial molar volume of water in the melt. The partial molar volume of water in a $NaAlSi_3O_8$ melt is compared to the molar volume of pure water at 3.5 kbar in Figure 3.10.

It may be noted that the molar volume of the solution, V_m, at some particular composition is

$$3.16 \quad V_m = X_A v_A = X_B v_B$$

3.4.3 Partial Molar Gibbs Free Energy: The Chemical Potential

Introduction of the concept of the partial molar volume has its own intrinsic merits but, in addition, serves as an analogy for another important thermodynamic quantity, the partial molar Gibbs free energy.

In Figure 3.11, the Gibbs free energy of plagioclase solid solutions that are mixtures of two components—Ab and An—is a convex downward curve. It must lie below a straight line connecting the free energies of the two pure components Ab and An in their *standard states* because dissolution of one in the other in a solid or a liquid solution of plagioclase is a spontaneous process; hence $\Delta G_{mix} < 0$.

Because of its special importance in thermodynamic models, the **partial molar Gibbs free energy,**

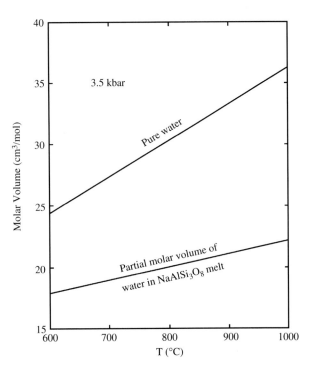

3.10 Comparison of molar volume of pure water and partial molar volume of water in a $NaAlSi_3O_8$ melt at the same P and T. (Data from Burnham et al., 1969; Ochs and Lange, 1997.)

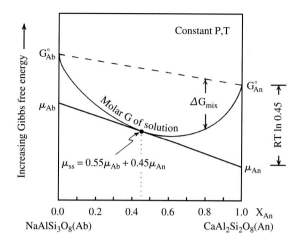

3.11 Schematic Gibbs free energy relations in a two-component plagioclase solid solution plotted against mole fraction of the two pure end-member components $NaAlSi_3O_8$ (Ab) and $CaAl_2Si_2O_8$ (An) for fixed P and T.

is usually designated by a special label, μ, and is called the **chemical potential.** Formally, it is defined by

$$3.17 \quad \mu_A = \left(\frac{\partial G_A}{\partial X_A}\right)_{P,T,X_B,X_C,X_D}$$

These symbols mean that the chemical potential of component A in a solution equals the infinitesimally small change in G that accompanies the addition of an infinitesimally small amount of component A to a large reservoir of solution at constant P, T, and mole fractions of all other components (X_B, X_C, X_D ...).

At this point, a brief digression is necessary to explain what is meant by "standard state." A **standard state** is a carefully defined reference state for a phase or component that is necessitated by the lack of absolute values for the thermodynamic parameters E_i, H, and G in their defining equations cited earlier. Because only differences in Gibbs free energies and in enthalpies between two states can be evaluated, one of these must be some reference value in order to find the value of the second real state of interest to the petrologist. Standard state values of thermodynamic quantities, usually denoted by the superscript °, for example $G°$, are commonly tabulated for pure phases at atmospheric conditions of 25°C (298.15K) and 1 bar (e.g., Robie and Waldbaum, 1968; Anderson, 1996).

The total change in the Gibbs free energy for the most general petrologic system is expressed in the **master equation of chemical thermodynamics**

$$3.18 \quad dG = VdP - SdT + \Sigma_i \mu_i dX_i$$

where the term $\Sigma_i \mu_i dX_i = \mu_A dX_A + \mu_B dX_B + \mu_c dX_c + \ldots$. This important equation shows that changes in the Gibbs free energy are dependent, through (1) the chemical potential, on changes in the

concentrations (X_A, X_B, etc.) of the components in the solution; (2) molar volume, on changes in P; and (3) molar entropy, on changes in T. In other words, stability in geologic systems of variable chemical composition, or where compositionally variable solutions are present, depends upon X, P, and T. We now have a general thermodynamic tool that is applicable to all systems, not just those whose phases are of fixed composition, to which the restricted equation 3.9 applies.

The chemical potential in chemical systems is analogous to gravitational potential energy in gravitational systems (Figure 1.6). The most stable state is the one of lowest potential. If there is a difference in the chemical potential of a component in adjacent phases, then some of the component in the phase of greater potential will spontaneously move into the lower-potential phase. For example, if $\mu_{water}^{melt} > \mu_{water}^{gas}$, then some water will move into the gas phase from the melt. At equilibrium in chemical systems, chemical potentials of the same component in different phases must be the same. (All boulders at the same elevation in the valley have equal gravitational potential energy and are in equilibrium with one another.) Hence, in a hydrous magma system at equilibrium the chemical potential of water in the silicate melt solution must equal the chemical potential of water in any associated gas phase, which must equal the chemical potential of water in biotite solid solution crystals suspended in the melt, and so on:

$$3.19 \quad \mu_{water}^{melt} = \mu_{water}^{gas} = \mu_{water}^{biotite} = \cdots$$

Similarly, for another component, such as CaO

$$3.20 \quad \mu_{CaO}^{melt} = \mu_{CaO}^{gas} = \mu_{CaO}^{biotite} = \cdots$$

3.4.4 P-T-X Phase Diagram

The state of any homogeneous system at equilibrium is uniquely determined by the intensive state properties T, P, and X, where X represents the concentration—specifically the mole fraction—of all of the chemical components the system comprises. A change in any one or all of these intensive variables may modify the equilibrium in a system. For example, it may be seen in Figure 3.12 that a liquid system can be made to crystallize by changes in any of the three intensive variables. Because of the limitations of a two-dimensional page of paper, phase diagrams generally portray only two variables, either P-T, P-X, or T-X. Additional variables can be represented in special projections.

❈3.5 APPLICATION OF THERMODYNAMICS TO SOLUTIONS

3.5.1 Fugacity and Activity

In Figure 3.11, because of the spontaneous mutual dissolution of components A and B in mixtures of the

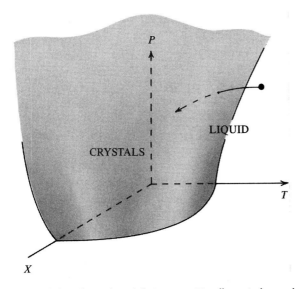

3.12 Stability of crystals and their compositionally equivalent melt depends on the three intensive variables P, T, and X. In this perspective view of a three-dimensional phase diagram, the shaded surface represents the values of P, T, and X where crystals coexist in equilibrium with liquid and $G_{crystals} = G_{liquid}$. Behind this melting or crystallization surface into the page crystals are more stable whereas outside the enclosed space liquid is more stable. Compare with Figure 3.6, which shows only the two-dimensional P-T plane at constant X through this diagram. From an initial liquid state, any general change in P, T, and/or X can move the system across the equilibrium (freezing) surface into the stability field of crystals.

two, the chemical potential of each component is less than the corresponding standard state value of the free energy per mole of the pure component, or $\mu_A < G_A^\circ$ and $\mu_B < G_B^\circ$. The amount of this difference at fixed P and T for component A is

$$3.21 \quad \mu_A - G_A^\circ = RT\ln X_A$$

where the molar gas constant $R = 8.3145$ J/K mole and T is in degrees kelvin. (It is convenient in calculations to use the conversion from natural to base 10 logarithms: $\ln X = 2.303\log X$.) The text by Anderson (1996) or others on thermodynamics may be consulted for a rigorous derivation of this equation. Here, we simply note that in a system at fixed T by combining $dG_T = VdP$ (Equation 3.12) with the perfect gas law for 1 mole

$$3.22 \quad V = \frac{RT}{P}$$

we have

$$3.23 \quad dG_T = VdP = \left(\frac{RT}{P}\right)dP = RT\ln dP$$

Because most natural gases are mixtures of different species such as H_2O, CO_2, and F, we note that the **partial pressure** of one gas component, A, is related to the total pressure of the gaseous solution, P_{total}, by Dalton's law

$$3.24 \quad P_A = X_A P_{total}$$

The foregoing equations apply to perfect or ideal gases at low pressures and high temperatures. For real gases under general conditions we can *define* a function so that the form of equation 3.21 is preserved, as follows

$$3.25 \quad \mu_A - G_A^\circ = RT\ln f_A$$

where f_A is called the **fugacity** of component A. The fugacity—Latin for "fleetness"—may be considered to be an equivalent or effective partial pressure that is used to calculate free energy differences for real gases. The ratio f_A/P_A is called the **fugacity coefficient** and is a measure of the departure of a real gas from a perfect one. As f_A approaches P_A where the density of the gas is very small, the coefficient approaches unity. The fugacity can also be viewed as an "escaping tendency" in situations in which a vapor is in equilibrium with a liquid, such as steam over a pot of boiling water, or a vapor is in equilibrium with a solid.

For real liquid and solid solutions an analogous equation can be written as

$$3.26 \quad \mu_A - G_A^\circ = RT\ln a_A$$

where the **activity** of component A, a_A, like the fugacity, may be thought of as an "effective concentration" or "effective availability" of A because the activity proxies for mole fraction in real liquid and solid solutions. Through their relation to chemical potential, the activity and fugacity "drive" reactions that distribute components between coexisting phases in a system. Thus, for example, from equation 3.19, if $\mu_{water}^{melt} > \mu_{water}^{gas}$, some water should migrate from the melt into the gas phase to maintain equilibrium between the melt and gas phases. Equations 3.25 and 3.26 are models that express the deviation of a real geologic solution, represented in the chemical potential, μ, from the ideal, pure substance whose standard free energy of formation from the elements, G°, can be found in a table of thermodynamic quantities (e.g., Robie and Waldbaum, 1968).

3.5.2 Equilibrium Constants

Chemical reactions proceed until one of two things happens: (1) One of the reactants is used up or (2) the reaction reaches an equilibrium state in which all reactants and products are still present in some particular proportions. The movement of reactant atoms making product phases is balanced by the equal and opposite movement of product atoms making reactant phases. The following sections deal with this type of chemical reaction. The following simple reaction at a fixed P and T models the equilibrium between olivine and orthopyroxene solid solutions in a natural, basaltic silicate melt solution (Carmichael et al., 1974)

3.27 $\quad Mg_2SiO_4 \quad + \quad SiO_2 \quad = \quad 2MgSiO_3$

$\qquad$ in olivine $\qquad$ in melt $\qquad$ in orthopyroxene

At equilibrium

3.28 $\quad \Delta G = 2\mu_{MgSiO_3}^{Opx} - \mu_{Mg_2SiO_4}^{Ol} - \mu_{SiO_2}^{melt} = 0$

where the superscripts on the chemical potentials indicate solid and liquid solutions in which the subscripted components are dissolved. The chemical potential of silica in the melt is given by an expression like equation 3.26

3.29 $\quad \mu_{SiO_2}^{melt} = G_{SiO_2}^{\circ\, gl} + RT\ln a_{SiO_2}^{melt}$

where $G_{SiO_2}^{\circ\, gl}$ is the free energy of formation of silica in a standard state of glass at $P = 1$ atm and the T of interest. Two additional equations can be written for the chemical potentials of Mg_2SiO_4 and $MgSiO_3$ in olivine and orthopyroxene solid solutions, respectively. Inserting these three equations into 3.28 and collecting like terms we find that at *equilibrium*

3.30 $\quad 2G_{Mg_2SiO_3}^{\circ\, Opx} - G_{Mg_2SiO_4}^{\circ\, Ol} - G_{SiO_2}^{\circ\, melt} =$

$\qquad \dfrac{-RT\ln\left(a_{MgSiO_3}^{Opx}\right)^2}{\left(a_{Mg_2SiO_4}^{Ol} a_{SiO_2}^{melt}\right)}$

or

3.31 $\quad \Delta_r G^\circ = RT\ln K$

where

3.32 $\quad K = \dfrac{\left(a_{MgSiO_3}^{Opx}\right)^2}{\left(a_{Mg_2SiO_4}^{Ol} a_{SiO_2}^{melt}\right)}$

is the **equilibrium constant,** a fixed value, at any particular P and T, which depends upon tabulated standard state free energies of formation, $_rG^\circ$. In general, if K is large, the reaction will mainly create products; if it is small, the reaction will mainly create reactants.

3.5.3 Silica Activity, Silica Buffers, and Silica Saturation

From equation 3.30, values of $\log a_{SiO_2}^{melt}$ as a function of T can be calculated for the equilibria in the model system forsterite-enstatite-melt at 1 atm (Worked Problem Box 3.1). These values can be plotted to yield the line labeled *Enstatite-Forsterite* in Figure 3.13. This line represents equation 3.27, which is an example of a magmatic **buffer reaction.** In this buffer reaction, represented also in the equilibrium constant, equation 3.32, the equilibrium between the melt and the two pure solids (having unit activity) fixes, or buffers, the activity of the silica component in the melt at a fixed

---

Worked Problem Box 3.1

Calculate the activity of silica as $\log a_{SiO_2}^{melt}$ for a melt in equilibrium with pure forsterite and pure enstatite at 1000K (= 727°C) and 1 atm.

In the case of pure solids of forsterite olivine Mg_2SiO_4 and enstatite orthopyroxene $MgSiO_3$, their activities are unity. Their effective concentrations are 1. Making these substitutions, equation 3.30 becomes

$\Delta_r G^\circ = 2_r G_{MgSiO_3}^{\circ\, En} - _r G_{Mg_2SiO_4}^{\circ\, Fo} - _r G_{SiO_2}^{\circ\, melt}$

$\qquad = -RT\ln\left(\dfrac{1}{a_{SiO_2}^{melt}}\right) = -2.303RT\log\left(\dfrac{1}{a_{SiO_2}^{melt}}\right)$

The standard state free energies, $_rG^\circ$, in gram formula weights or moles are tabulated in Robie and Waldbaum (1968). The standard state of silica glass can be used for SiO_2 and of clinoenstatite for $MgSiO_3$. At 1000K (= 727°C)

$\Delta_r G^\circ = 2(-300.565) - (-425.117) - (-173.943)$

$\qquad = -2.070$ kcal/mole

-2.070 kcal/mole (4186 J/kcal)

$\qquad = -8665.020$ J/mole

-8665.020 J/mole $= -2.303 RT\log\left(\dfrac{1}{a_{SiO_2}^{melt}}\right)$

$\qquad = -2.303\ (8.3145\ \text{J/deg mole})$

$\qquad (1000\text{K})\left(\log \dfrac{1}{a_{SiO_2}^{melt}}\right)$

Solving

$\left(\log \dfrac{1}{a_{SiO_2}^{melt}}\right) = -\log a_{SiO_2}^{melt} = 0.45$

T and P. Suppose, for example, the magma system is perturbed by addition of silica, perhaps by grains of quartz from the wall rock of an intrusion that are dissolved in the melt, momentarily increasing the silica activity. To restore equilibrium, some forsterite must react with the melt to form enstatite, restoring or buffering the activity of silica to its former value at the fixed T and P. Provided both solid phases and melt coexist at constant T and P, the activity of silica is buffered (fixed) at a constant value. (A system of liquid and solid water [ice] at 0°C and 1 atm is a thermal buffer in that the T of the system is fixed at 0°C as long as both phases coexist.) If T increases in the model magma system containing melt in equilibrium with pure forsterite and enstatite crystals, then the activity of silica must increase (to lesser negative log values) because the system is constrained to lie on the Enstatite-Forsterite buffer curve.

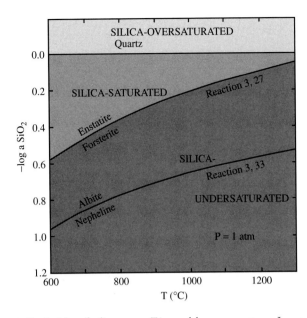

3.13 Activity of silica versus *T* in model magma systems. Lower activities of silica correspond to lower degrees of silica saturation.

For a magma not constrained to the buffer curve, such as one whose silica activity and *T* place it above the Enstatite-Forsterite buffer curve in Figure 3.13, only enstatite is stable; all of the potential forsterite in the magma has been consumed into enstatite as a result of the relatively greater silica activity. Silica activity and *T* are *not* mutually fixed, or buffered, *off* the buffer curve.

The upper limit of silica activity where $a = 1$ and log $a_{SiO_2} = 0$ occurs in magmas in which the melt is in equilibrium with quartz or one of its polymorphs. This equilibrium is represented by a straight line labeled *Quartz* at the top of Figure 3.13.

A model magma system having a lower silica activity than represented by the Enstatite-Forsterite buffer curve in Figure 3.13 is one that contains coexisting albite and nepheline. This pair of pure solids in equilibrium with melt constitute another buffer reaction

$$3.33 \quad NaAlSiO_4 + SiO_2 = NaAlSi_3O_8$$
$$\text{nepheline} \quad \text{in melt} \quad \text{albite}$$

This buffer reaction serves to constrain lesser activities of silica to a unique value at fixed *T* and *P* provided nepheline, albite, and melt coexist.

Note that in Figure 3.13 the range in silica activity is only about $10^{0.1}$ between a model basaltic magma that contains forsterite and enstatite at a reasonable $T = 1200°C$ and a rhyolite magma that contains quartz at 800°C, even though the bulk silica concentrations in these two magmas are about 45 and 75 wt.%. This is a striking demonstration of the fact that silicate melts are not ideal solutions. If they were ideal, the silica activities would correspond to the mole fractions of silica.

Ghiorso and Carmichael (1987) show that the activity of silica ranges over only about two orders of magnitude, about 10^{-2} to $10^{0.0}$, through the entire range of natural silicate magma compositions.

Figure 3.13 for model magma systems involving pure end-member solid phases is readily extended to naturally occurring, more chemically complex magmas. It is obvious that the concept of silica activity is simply a thermodynamic expression of the century-old concept of silica saturation observed in rocks as described in Section 2.4.4 and illustrated in Figures 2.14 and 2.19a. Indeed, the quartz (or silica glass), Enstatite-Forsterite, and Albite-Nepheline curves in the silica activity diagram of Figure 3.13 correspond to boundaries between fields of contrasting saturation of silica constructed by petrographers over the past century in Figure 2.19. Although not explicit in Figure 2.16, more alkaline rocks of lesser silica/(K_2O + Na_2O) ratio are less silica-saturated and their corresponding magmas have lower silica activity, compared with subalkaline rocks. In other words, decreasing silica saturation corresponds to decreasing silica activity.

A similar concept is that of alumina saturation, discussed in Section 2.4.4 and illustrated in Figure 2.15.

3.5.4 Oxygen Buffers

Fe, Si, and O make up a large proportion of magmas and rocks and are involved in important reduction-oxidation reactions, or **redox reactions,** in rock-forming systems. Two model redox reactions in magmatic systems that involve oxygen in equilibrium with the silicate melt are the following:

$$3.34 \quad 4Fe_3O_4 + O_2 = 6Fe_2O_3$$
$$\text{magnetite} \qquad \text{hematite}$$

$$3.35 \quad 3Fe_2SiO_4 + O_2 = 2Fe_3O_4 + 3SiO_2$$
$$\text{fayalite} \qquad \text{magnetite} \quad \text{quartz}$$

In these reactions, the relatively reduced solid reactant on the left is oxidized by combination with oxygen to form a more oxidized solid or solids product on the right. The equilibrium constants for these reactions at fixed *P* and *T* are

$$3.36 \quad K_{P,T} = \frac{\left(a_{Fe_2O_3}^{Hem}\right)^6}{\left(a_{Fe_3O_4}^{Mag}\right)^4 f_{O_2}}$$

$$3.37 \quad K_{P,T} = \frac{\left(a_{Fe_3O_4}^{Mag}\right)^2 \left(a_{SiO_2}^{Qtz}\right)^3}{\left(a_{Fe_2SiO_4}^{Fa}\right)^3 f_{O_2}}$$

With the simplification that the activities of the pure crystalline solids are unity, oxygen fugacities may be calculated at chosen temperatures in the same way that the silica activity was calculated in Worked Problem Box 3.1 using tabulated standard free energies of formation in Robie and Waldbaum (1968). Such calculations are plotted in Figure 3.14 as the Hematite-Magnetite (HM) and Quartz-Fayalite-Magnetite (QFM) buffer curves. These curves and the redox reactions they represent are magmatic oxygen buffers in the same sense as the magmatic silica buffers discussed previously. For example, if the amount of oxygen is perturbed in a hypothetical magma system at fixed P and T in which magnetite, quartz, and fayalite are in equilibrium, the system adjusts through the redox reaction, equation 3.35, so that equilibrium is restored and the oxygen fugacity remains at its initial buffered (fixed) value.

Five other calculated solid-oxide buffer curves are also shown in Figure 3.14. (The Nickel-Nickel Oxide, or NNO, buffer reaction does not occur in natural magmas but is commonly used for reference.) These seven buffer curves are used extensively to characterize the oxidation state of magmas. They show the strong dependence of the stability of Fe-bearing phases on f_{O_2} and T.

The values of the oxygen fugacities in Figure 3.14, as small as 10^{-30} bar, are unimaginably minute and have little meaning connected with a particular partial pressure of oxygen. It should be realized that fugacity is only a thermodynamic tool that serves as a means of expressing the equivalence of the chemical potential of a particular component, in this case oxygen, in all coexisting phases—solid, liquid, and gaseous—in a system at equilibrium. There is not even any necessity that a separate gas phase actually exists in the system; the oxygen in the system may only exist as dissolved ions in the melt and in solid-solution minerals. The oxygen fugacity is simply an index of the redox state in a magma as represented by states of crystal-melt equilibrium. Moreover, the value of the fugacity is not meaningful unless T is also specified. For example, the same oxygen fugacity of 10^{-14} bar corresponds to the highly reduced Quartz-Fayalite-Iron (QFI) equilibrium at 1100°C but highly oxidized Hematite-Magnetite (HM) equilibrium at 600°C.

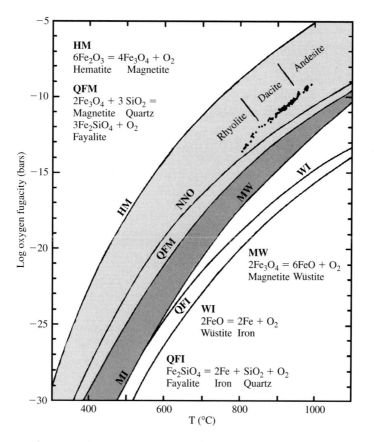

3.14 Buffer curves representing reduction-oxidation mineral reactions in the system O-Si-Fe. (Redrawn from Eugster and Wones, 1962.) In every buffer reaction, the right side is stable *below* the corresponding curve. Light shading indicates stability field of magnetite in systems containing excess quartz, that is, in which silica activity is unity. Dark shading indicates stability field of magnetite in silica-free systems that extends to lower oxygen fugacities, to the MW buffer curve. Dots represent oxygen fugacity and T at which coexisting Fe-Ti oxides crystallized in a suite of 65 rhyolite, dacite, and andesite magmas erupted in 1912 from Novarupta caldera at the head of the Valley of Ten Thousand Smokes, Katmai National Park, Alaska. (Redrawn from Hildreth, 1983.) The continuity and coherence of these data and their trend parallel to the HM and QFM buffers indicate the eruptions sampled a compositionally variable magma system that had been buffered during evolution by other redox reactions involving chemically complex solid and liquid solutions.

The Earth as a whole has a wide range of redox conditions, from the metallic iron (Fe)-nickel core, through the ferrous (Fe^{2+}) silicate mantle, to the hydrosphere and oxygen-rich atmosphere, where ferric (Fe^{3+}) minerals such as hematite and goethite are stable. At sea level, $P_{total} = 1$ bar approximately for air and, because the mole fraction of oxygen in it is 0.21, from Dalton's law (equation 3.24) the partial pressure of oxygen $P_{O_2} = 0.21$ bar. At equilibrium, this P_{O_2} must be equal to the fugacity of oxygen in a melt at the surface of the Earth, thus setting an upper limit of $f_{O_2} = 0.21$ bar ($10^{-0.68}$ bar) on the oxygen fugacity in natural melts. Only in lunar basalts and in very rare terrestrial basalts that formed from magma intruded into and reduced by interaction with coal does iron occur in the metallic state, or as wüstite (FeO). Most magmas have oxygen fugacities such that magnetite is the stable phase, that is, the shaded region of Figure 3.14. As ascending magmas contact near-surface meteoric ground waters or the oxygen-rich atmosphere, Fe^{2+} in the magma can be at least partly oxidized to Fe^{3+}, depending on kinetic factors. At highest oxygen fugacities hematite is stable. Hence, iron-rich basaltic lavas extruded onto the surface are locally red because of the presence of very finely divided particles of hematite. Provided magnetite persists in equilibrium with hematite in an HM buffer, decreasing T results in the system's experiencing decreasing oxygen fugacity as oxygen is consumed in production of hematite. Once all of the magnetite is gone, the system leaves the buffer curve and enters the field of stable hematite, and T and f_{O_2} are no longer interdependent variables.

How do other factors, such as confining pressure (P) and P_{H_2O}, in magmatic systems influence redox equilibria? As for P, Kress and Carmichael (1991) show that, provided the magma system remains closed, any change in P less than 30 kbar (i.e., depths above the upper mantle) changes the oxygen fugacity by less than one-half of a log unit. Therefore, the oxidation state of a closed magma system rising from a source in the upper mantle will essentially reflect the oxidation state of that source. For example, if a closed magma system originates on the QFM buffer, it virtually stays on it during ascent. As for dissolved water in melts, it has commonly been assumed that more water-rich melts would be more oxidized because of the thermal dissociation of water into oxygen and hydrogen ions and the loss of the more easily diffusing hydrogen. However, there is increasing evidence (Moore et al., 1995) that dissolved water by itself has *no* effect on the oxidation state of iron in natural melts.

3.5.5 Fe-Ti Oxide Buffers: Oxygen Geobarometers and Geothermometers

Buffer reactions like the ones just described provide the means of evaluating the intensive variables in magmatic systems at which minerals crystallized under equilibrium conditions using their chemical compositions and thermodynamic models. Equilibria involving oxygen and minerals are the basis for oxygen **geobarometers** (indicating f_{O_2} of crystallization) and **geothermometers** (T).

Since the pioneering and now classic publication of Buddington and Lindsley (1964), the chemical composition of coexisting oxide phases in the system FeO-Fe_2O_3-TiO_2 has been an important source of information on the f_{O_2} and T of the magma in which they have crystallized. Two major solid-solution series occur in this system: cubic, or isometric, ulvöspinel-magnetite solid solutions essentially between Fe_2TiO_4 and Fe_3O_4 and rhombohedral ilmenite-hematite solid solutions essentially between $FeTiO_3$ and Fe_2O_3. The compositions of the two solid solutions are strongly dependent on f_{O_2} and T. Hence, determinations of the compositions of an equilibrium pair of cubic and rhombohedral Fe-Ti oxides in a rock, usually by electron microprobe analysis, give the values of the f_{O_2} and T at which they equilibrated. This is the method used by Hildreth (1983) to determine f_{O_2} and T in the rhyolite-dacite-andesite magma sequence plotted in Figure 3.14.

Ghiorso and Sack (1991) provide detailed instructions for recalculating the compositions of natural oxides, which always contain Mn, Al, V, and so on, in solid solution, before plotting in terms of Fe and Ti, and for evaluating whether the coexisting compositions represent equilibrium. Carmichael et al. (1974, p. 87) demonstrate that confining pressure, P, has little effect on the equilibrium compositions; an increase in P from 1 to 5000 bars on the solid phases at 1000°C only increases log f_{O_2} by 0.073.

✳3.6 KINETICS

The basic questions with regard to changes in the states of petrologic systems are: In what direction does a spontaneous process proceed, and how fast will it proceed? The answer to the first question is that systems move toward a more stable state of lower Gibbs free energy or chemical potential. With regard to the second question, there is usually some inertia or resistance to the changing of a state. A metastable state seldom has unrestricted access to a lower energy more stable state because of an activation energy barrier or hump between the initial and final states (Figure 3.5). High-T minerals, such as diamond and olivine, are mostly metastable at the cool, wet, oxygen-rich surface of the Earth. Yet these minerals do not convert into stable minerals, such as clays, carbonates, gypsum, and hematite, while sitting in museum drawers in a human lifetime, and they had not done so for millennia before they were collected from exposures at the surface of the Earth.

A limitation of classical thermodynamics is that it cannot predict the time required for a change in state to occur—whether 10 minutes, 10 years, or 10 million years. The thermodynamic driving "force" for changes in state, namely, the difference in free energies or chemical potentials between the initial and final states, is independent of the factor of time. The **kinetic path** between these states is not considered at all. **Kinetics** deals with the time-dependent dynamics of systems—the time rates of movement of matter and flow of energy—that allow them to move from one state to another, such as during crystallization of a melt.

Basically, there are two kinetic factors of concern in changing systems (Putnis and McConnell, 1980):

1. An activation energy barrier between the initial and final states that causes the initial metastable one to persist although there is another more stable state of lower energy.

2. The formation of a new metastable phase rather than a stable phase.

3.6.1 Activation Energy

The **activation energy barrier** can be large or small and depends upon the nature of the system undergoing change (Figure 3.5). For example, the transformation of metastable diamond to more stable graphite is impeded by a very large activation energy barrier in the form of strong covalent carbon bonds that prevent reorganization into more stable graphite (or carbon dioxide) in which carbon atoms are weakly held together with van der Waals bonds. This energy can be supplied by high T in a laboratory oven, converting diamond into carbon dioxide. The purpose of catalysts in the chemical industry is to lower the activation energy barrier in some way.

For any thermally activated process, such as melting of rock and recrystallization of diamond to graphite, the probability of a particular atom's having sufficient energy to get over the energy barrier is $e^{-E_a/RT}$ where E_a is the activation energy, R ($= 8.3145$ J/K mol) is the gas constant, and T is in degrees Kelvin. The kinetic rate of some thermally activated process is

$$3.38 \quad kinetic\ rate = Ae^{-E_a/RT}$$

sometimes written as rate $= A \exp(-E_a/RT)$. In this **Arrhenius equation,** A is called the *frequency factor* or *preexponential constant.* It is more or less independent of T and indicates the frequency with which an atom

The logarithmic forms of the **Arrhenius equation** (3.38) are

$$\ln (rate) = \ln A + (-E_a/RT)$$
$$\log (rate) = \log A + (-E_a/2.303RT)$$

using natural base $e = 2.718$ logarithms (ln) and base 10 logarithms (log). In either logarithmic expression, the equation has the form $y = a + bx$, which is a straight line on a plot of log (*rate*) for the y axis and $1/T$ for the x axis. The slope, $dy/dx = -E_a/2.303\ R$, and the intercept, b, on the log (*rate*) axis at $1/T = 0$ is log A (Figure 3.15). Kinetic processes represented by a straight line on such a plot therefore obey the Arrhenius equation and are thermally activated. The slope of the line gives the activation energy; more steeply sloping lines correspond to greater activation energies. For example, the activation energy for interdiffusion of Fe and Mg in Mg-rich olivine from 1000 to 1100°C can be calculated from Figure 3.15 as $E_a = 2.303 \times R \times$ slope $= 2.303 \times 8.3145$ J/K mole $\times 0.49$K/$0.000080 = 117$ kJ/mole where the rise in slope is 0.49 log unit for a run of 0.000080 unit of $1/T$.

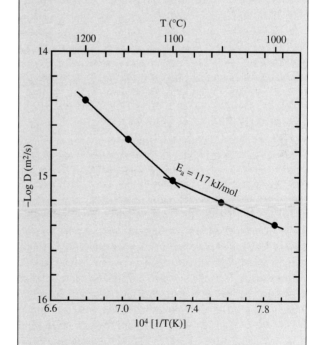

3.15 Kinetics of interdiffusion of Fe^{2+} and Mg^{2+} in olivine (Fo_{90}) along its *c*-axis. The experimentally measured data (filled circles) are plotted in terms of the negative logarithm to the base 10 of the diffusion coefficient, D, on the vertical axis against reciprocal degrees Kelvin times 10^4 on the lower horizontal axis and degrees Celsius on the upper horizontal axis. The two straight but nonparallel lines indicate that the diffusion is a thermally activated kinetic process obeying the Arrhenius equation (3.38) but with different activation energies above and below 1100°C. (Redrawn from Buening and Buseck, 1973.)

Worked Problem Box 3.2

Calculation of the activation energy for interdiffusion of Fe and Mg in olivine.

moves toward the activated state. This is a transient, or temporary, higher energy intermediate state on the hump between the metastable and stable states (see Figure 3.5). The drastic influence of T on kinetic processes that obey the Arrhenius equation can be illustrated with a simple example. The segregation (ordering) of K and Na ions that occurs in slowly cooling alkali feldspars to form perthitic intergrowths has an activation energy of 230 kJ/mol. The relative rates of ordering at 1000K (727°C) and at 300K (27°C, i.e., near-atmospheric T) are

$$e^{-(230,000J/mol)/(8.314J/K\ mol\ \times\ 1000K)}$$

$$= e^{-27.66} \sim 10^{-12} \text{ and}$$

$$e^{-(230.000J/mol)/(8.314J/K\ mol\ \times\ 300K)}$$

$$= e^{-92.21} \sim 10^{-40}$$

(The constant A, being the same at both temperatures, is ignored in this example.) Thus, the rate is 10^{28} times faster (10^{-40} versus 10^{-12}) at the higher T. As a rule of thumb, a rise in T of 10 degrees doubles a reaction rate.

Because many rock-forming processes, including crystallization, are thermally activated, mineral assemblages in higher-T magmatic systems tend to reach states of stable thermodynamic equilibrium more readily than do lower-T metamorphic systems. *Tend* in the previous sentence does not mean always! The fact that magmatic processes occur at high T does not universally guarantee attainment of states of stable equilibrium.

3.6.2 Overstepping and Metastable Persistence and Growth

Changing states of a system involve **overstepping.** This means that changing conditions must go beyond the actual equilibrium value before the system responds and actually changes its state. Freezing rain in some climates is an example. Drops of water cooled metastably below the freezing T (= 0°C) do not crystallize until hitting the ground where ice crystals nucleate and grow. This important kinetic phenomenon is illustrated by Figure 3.16, which shows hypothetical free energy versus T curves for two equivalent composition states, A and B. These may be two polymorphs, crystals and their melt, or two mineral assemblages of the same bulk chemical composition. As T decreases for stable state A, a critical $T = T_e$ is reached, where the free energies of the two states are equal so that the two coexist in equilibrium. At $T < T_e$ state B has lesser free energy and is more stable than A. However, because of sluggish mobility of atoms, the transition from A to B does not occur instantaneously at T_e; instead a certain amount of overstepping below T_e is required to effect the change. In this case, the

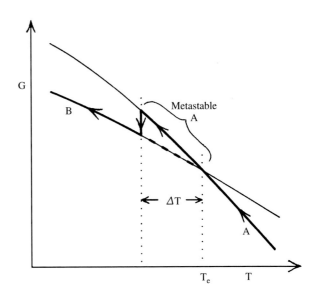

3.16 Free energy-T relations (thinner lines) for two hypothetical compositionally equivalent states A and B. These states could be two mineral polymorphs, mineral assemblages, or crystals and corresponding melt. State A is stable for $T > T_e$ because of lesser free energy than state B. For $T < T_e$ the opposite is true. The two states are in equilibrium, or coexist stably together, at $T = T_e$. With decreasing T, state A can persist metastably below T_e for some amount of undercooling, ΔT, before finally transforming into B. This irreversible path is shown by the heavy solid line marked by arrows. A less likely path corresponding to an instantaneous change at $T = T_e$ is shown by the heavy dashed line segment.

overstepping is an undercooling, ΔT, that may be a few to several tens of degrees. In many instances, state A may persist metastably, and indefinitely, for hundreds of degrees below its stable T. This, in fact, is the fate of virtually all minerals formed at high T in magmatic rocks. We might anticipate that the increasing ΔG at decreasing T would provide an increasing driving force for the transformation; however, the exponentially decreasing mobility of atoms at decreasing T overrides this force and consequently the reaction may not happen or may be incomplete. In contrast, in a system that is becoming hotter, transformations tend to occur with less overstepping because the increasing driving force is enhanced by increasing particle mobility.

Tend in the previous sentence emphasizes that there are exceptions. Sometimes the exceptions have significant geologic consequences. For example, subducting slabs of oceanic lithosphere conductively heat very slowly as they sink into hotter asthenosphere (Figures 1.4 and 1.5). Consequently, the transformation of olivine to the more dense spinel structure, which begins at about 410-km depth beneath plate interiors, is delayed to greater depths because of the refrigerated slab (Kirby et al., 1996). Instead of the transformation occurring incrementally under equilibrium conditions, a large mass of metastable olivine on which *PV* work has been done might catastrophically transform to the smaller volume spinel, releasing this stored work energy as seismic energy. This overstepped transformation could account for the origin of deep focus earthquakes, which has puzzled seismologists since they were first discovered beneath Japan in 1928 by K. Wadati.

Some metastable crystals actually *nucleate and grow* outside their stability field (Putnis and McConnell, 1980, p. 97). Formation of the high-*P* polymorph of $CaCO_3$, aragonite, in hot springs at the surface of the Earth and in shallow marine environments far outside its high-*P* stability field is an example. Another is the occurrence of cristobalite and tridymite in devitrified silicic glass and as precipitates from the vapor entrapped in vesicles in rhyolite; both of these silica polymorphs nucleate and grow below 867°C at atmospheric *P* within the stability field of quartz (see Figure 5.1). The driving "force" for the crystallization, which is expressed by the free energy difference between the initial phase (silicic glass) and the silica polymorphs, $\Delta G_{G \rightarrow Q}$, is considerable (Figure 3.17). In such cases, formation of the phase with the lowest activation energy barrier is favored, even if it is not the most stable; such a phase will commonly have the closest similarity in atomic structure to the parent phase and have the least ΔG. This is an example of the **Ostwald step rule:** that in a change of state the kinetically most favored phase may form at an intermediate energy step, not necessarily the step of least possible free energy. One or more intermediate kinetically favored phases may form stepwise (Figure 3.17) until the truly stable state is attained.

Metastable persistence and nucleation and growth of new metastable minerals outside their stability field are notorious phenomena that have plagued the experimental determination of phase diagrams by synthesis techniques (Section 3.3.2). Countless experimental results have been published that portray phase assemblages which do not represent states of equilibium, causing confusion and doubts as to what phases are truly stable at a particular *P* and *T*. Reversed experiments that proceed both forward and backward across a *P-T* phase boundary may bracket its position and can be more reliable than unreversed experiments that begin with the phases in one stability field and move across the boundary in only one direction into the adjacent field. Another experimental procedure that overcomes kinetic difficulties involves placing together all possible phases ("seeds") that might be stable at the desired *P* and *T* conditions in the device and then examining the products after the experiment to see which stable phases grew and which unstable ones were destroyed.

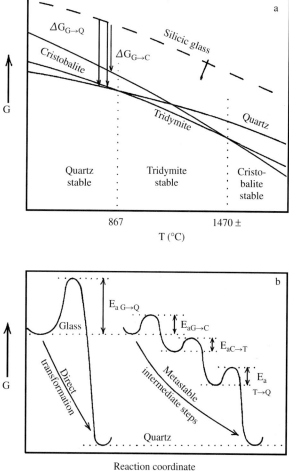

3.17 Schematic free energy relations for some silica phases. (a) Silica glass (dashed line) that occurs in obsidian and other rhyolite rock types is metastable at all *T* in the diagram because of its highest free energy. Devitrification of glass by crystallization to any of three crystalline polymorphs (plus alkali feldspar) has a similar change in free energy at about 800°C. But the largest free energy change, $\Delta G_{G \rightarrow Q}$, is to the most stable polymorph, quartz. However, the kinetics of crystallization to metastable cristobalite or tridymite appears to be favored in many rhyolites. (b) The reason why silicic glass does not always devitrify directly to quartz (and alkali feldspar) lies in the large activation energy barrier between glass and quartz. More commonly, glass devitrifies to metastable cristobalite and/or tridymite, which may eventually transform to stable quartz; this stepwise process is an example of **Ostwald's step rule.**

Summary

Thermodynamics provides models and tools for the prediction and interpretation of changing states in rock-forming systems. Because of the manner in which the Gibbs free energy is formulated, changes in *P, T,* and/or *X* spontaneously lead to states of lowest possible energy. States of lesser molar volume (greater density) are thermodynamically more stable at higher *P,* and states of greater entropy are more stable at higher T. Phase (stability) diagrams are a convenient graphical device to portray stability relations among different states of a particular system as a function of P, T, and X.

Most rock-forming minerals and, of course, all melts and gases are solutions or mixtures of two or more chemical components whose mole fractions X_A, X_B, X_C, . . . can be varied within certain limits, producing smoothly varying properties of these phases without destroying their integrity. The partial molar Gibbs free energy (chemical potential) is minimized in spontaneous changes in a system. At equilibrium, the chemical potential of the same component must be the same in whichever phases it occurs in the system.

The fugacity and activity can be considered as effective or equivalent concentrations of components in real solutions in geologic systems that deviate from ideal solutions. If a magmatic system is buffered by a state of equilibrium between coexisting solid phases, the activity of some component, such as silica, is fixed at constant P and T. Other solid redox equilibria buffer the oxygen fugacity in magma systems. Compositions of Fe-Ti oxides are especially useful in providing values of the oxygen fugacity and T at which they crystallized at equilibrium.

In all natural systems, there is a contest between changing intensive variables which prompt a change in the state of a system, and time, which controls through kinetic phenomena whether a state of equilibrium is actually attained. Despite the driving "force" provided by differences in free energy, higher-energy metastable states do not instantaneously move to lower-energy more stable states because of activation energy barriers. Hence, metastable states persist in changing systems in the form of minerals existing well outside their P-T-X stability field. P-T-X variables must overstep field boundaries by a certain amount in order to create the most stable phase(s). New metastable phases may actually nucleate and grow outside their stability field.

CRITICAL THINKING QUESTIONS

3.1 How does an understanding of the basic concepts of thermodynamics provide insights into the behavior of rock-forming systems?

3.2 What types of flows of matter and energy occur in end-member thermodynamic systems?

3.3 Why is the Earth and its atmosphere not an isolated system?

3.4 Contrast intensive and extensive state properties. Changes in what properties are petrologically most important?

3.5 State the first, second, and third laws of thermodynamics and discuss their application to petrology.

3.6 Contrast heat, molar heat capacity, T, and enthalpy and explain how they are interrelated (review Section 1.1.3).

3.7 Explain the concept of entropy and give examples of entropy changes in natural processes.

3.8 In what way is the Gibbs free energy a thermodynamic potential energy?

3.9 What is a phase diagram, and how can it be read to provide information concerning the stability of different states of a system?

3.10 In a P-T phase diagram, why does the *slope* of the melting curve of a pure mineral such as albite (Figure 3.8; see also Figure 3.6) increase slightly with increasing P? (*Hint:* What is the effect of increasing P on the difference in molar volume, ΔV, between liquid $NaAlSi_3O_8$ and crystalline albite? Does one of these two phases compress more than the other so that ΔV changes with increasing P?)

3.11 Explain the meaning of the master equation of chemical thermodynamics (3.18) and of the chemical potential.

3.12 What thermodynamic quantities are used to describe solutions? Give examples.

3.13 What is meant by the activity of a component? The fugacity?

3.14 What is a buffer reaction? Give an example.

3.15 Describe how the activity of silica in a melt can be related to the degree of silica saturation in magmas and rocks.

3.16 In Figure 2.15, how does increasing Ca activity stabilize calcic pyroxene, hornblende, and anorthite?

3.17 Describe redox buffer reactions applicable to magma systems and what they tell about their oxygen fugacity.

3.18 On photocopies of Figure 3.14, color shade the stability fields of the following mineral equilibria in the system O-Si-Fe: (a) fayalite as the sole phase; (b) fayalite plus magnetite; (c) metallic iron plus fayalite.

3.19 Describe the Fe-Ti oxide geothermobarometer.

3.20 What is kinetics?

3.21 Discuss the kinetic obstacles to attainment of stable equilibrium in changing states of thermodynamic systems.

PROBLEMS

3.1 The amount of heat required to melt 1 mole of diopside crystals at atmospheric P is the molar enthalpy of melting, $\Delta H_m = 144$ kJ/mol. This amount of heat would raise the T of 1 mole of diopside crystals just below the melting temperature by how many degrees Celsius? (*Answer:* 514°C)

3.2   Using densities from a mineralogy text, determine the relative molar volumes of diamond and graphite. What does the positive slope of the diamond-graphite equilibrium boundary line in *P-T* space tell about the relative entropies of graphite and diamond? Predict the *relative* compressibility, that is, how much the molar volume changes with increasing *P,* of graphite and diamond. Then show in a *P-T* diagram how the *slope* of the equilibrium curve changes at increasing *P.* Justify your answer, indicating any assumptions made.

3.3   Sketch the melting curve in *P-T* space for the water system. This will require you to decide on the relative entropies of liquid and solid water and their relative molar volumes. Isothermally, what change in *P* (increasing or decreasing) causes ice to melt? What possible implication might this have for the interior of a planetary satellite, such as Europa, that has an ice crust?

3.4   In *T-G* space the free energy lines for crystals and liquid have negative slopes and the slope of the liquid line is greater than that of crystals (Figure 3.7). Why? Why is the opposite true of the lines in *P-G* space?

3.5   At the center of the Earth, *P* is probably $>3 \times 10^6$ bar and *T* is possibly about 5000°C. Make an approximate calculation for the relative effects of *P* and *T* on the change in free energy ($\Delta G_{center} - \Delta G_{surface}$) of a mineral reaction for which $\Delta V = 1$ cm^3/mol and $\Delta S = 2$ J/K mol in going from the surface to the center of the Earth (Atkins, 1978, p.157). (*Hint:* Base the approximation on equation 3.9 that $\Delta G_{center} - \Delta G_{surface} = \Delta V[P_{center} - P_{surface}] - \Delta S[T_{center} - T_{surface}]$.) Discuss your result, especially in the light of the mineralogical changes within the mantle shown in Figure 1.3.

3.6   In a photocopy of Figure 3.9 show the partial molar volumes of the two components A and B in a solution whose composition is $X_A = 0.2$.

3.7   From Figure 3.11, express the chemical potential of component Ab in a solution whose composition is $X_{Ab} = 0.7$ in terms of a Gibbs standard state free energy of formation and an (RT ln) term. Show the value of this chemical potential on the appropriate axis of the diagram. Also show and give an algebraic expression for the chemical potential of the solution at $X_{Ab} = 0.7$.

3.8   From tabulated free energies of formation from the elements in Robie and Waldbaum (1968) calculate the activity of silica at 600, 900, and 1200°C in the reaction nepheline + silica = (high) albite to verify the corresponding buffer curve in Figure 3.13.

3.9   From tabulated free energies of formation from the elements in Robie and Waldbaum (1968) calculate the fugacity of oxygen at 600, 900, and 1200°C in the reaction 6 hematite = 4 magnetite + O$_2$ to verify the corresponding buffer curve in Figure 3.14.

3.10   Calculate the activation energy for interdiffusion of Fe and Mg in Mg-rich olivine from 1100 to 1200°C in Figure 3.15. (*Answer:* 239 kJ/mol)

Silicate Melts and Volatile Fluids in Magma Systems

FUNDAMENTAL QUESTIONS CONSIDERED IN THIS CHAPTER

1. What is the nature of magma and its essential melt fraction?

2. What volatiles exist in magma systems and how are they dissolved in the melt?

3. How does the release of dissolved volatile from melts impact magmatic behavior and that of geologic systems in general?

INTRODUCTION

Everyone is familiar with glowing extrusions of incandescent magma and explosions of gas-charged magma from volcanoes. Hidden from direct observation are intrusions of magma beneath the surface of the Earth. Nothing is more fundamental to the behavior of these magma systems and the origin of magmatic rocks than the nature and behavior of magma. The key ingredient in this mobile molten material is a liquid silicate solution, or melt. In this chapter, the basic atomic structure of melts is introduced as a means to understand the dynamic behavior of bodies of magma discussed in more detail in Chapter 8. Especially important in this dynamic behavior is the role of volatiles that are dissolved in melts and can be released to cause explosive volcanism and many other important geologic phenomena.

✳4.1 NATURE OF MAGMA

Magma is a term first introduced into geologic literature in 1825 by Scope, who referred to it as a "compound liquid" consisting of solid particles suspended in a liquid, like mud. Measurements on extruded magma (lava), together with evaluations of mineral geothermometers in magmatic rocks and experimental determinations of their melting relations, indicate that temperatures of magmas near the surface of the Earth generally range from about 1200°C to 700°C; the higher values pertain to mafic compositions, the lower to silicic. Very rare alkali carbonatitic lavas that contain almost no silica have eruptive temperatures of about 600°C. Extruded magmas are rarely free of crystals, indicating that they rarely are superheated above temperatures of crystallization. Densities of magmas range from about 2.2 to 3.0 g/cm^3 and are generally about 90% of that of the equivalent crystalline rock.

Magma in general consists of a mobile mixture of solid, liquid, and gaseous phases. The number and nature of the phases constituting a magma depend, under stable equilibrium conditions, on the three intensive variables—P, T, and X (concentrations of chemical components in the magma). At sufficiently high T, any rock melts completely to form a homogeneous liquid solution, or **melt.** Except for carbonatite magmas, melts consist mostly of ions of O and Si—hence the alternate appellation **silicate liquid**—but always contain in addition significant amounts of Al, Ca, H, Na, and so on.

Examples of different types of magmas are shown schematically in Figure 4.1. Only in some unusually hot systems will a magma consist wholly of melt and no other phases. In most instances, a melt is only part of the whole magma, but is always present and gives it mobility. Hence, *melt and magma are generally not the same*. To a significant extent, the properties of the melt largely govern the overall dynamic behavior of the whole magma. Rare magma systems consist at equilib-

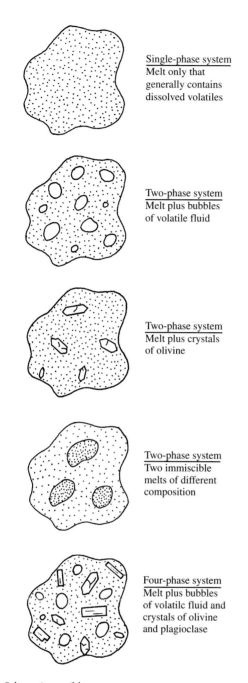

Single-phase system
Melt only that
generally contains
dissolved volatiles

Two-phase system
Melt plus bubbles
of volatile fluid

Two-phase system
Melt plus crystals
of olivine

Two-phase system
Two immiscible
melts of different
composition

Four-phase system
Melt plus bubbles
of volatile fluid and
crystals of olivine
and plagioclase

4.1   Schematic possible magmas.

rium of two physically distinct melts—one essentially of carbonate and the other of silicate, or both are silicate but one is silicic and the other very rich in Fe. Each of these **immiscible melts** has distinctive properties, such as density. Oil and water are familiar immiscible liquids.

4.1.1 Atomic Structure of Melts

The configuration of ions in a melt—its **atomic structure**—largely dictates many of its significant properties.

In pictorial representations of crystalline, liquid, and gaseous states, individual atoms have to be drawn

as fixed in position relative to one another, but these are only their average, or instantaneous, positions. Even in crystals above absolute zero (0K), individual ions have motion. In glasses that are supercooled very viscous melts, ions experience vibrational motion: small periodic displacements about an average position. But at temperatures above a glass-melt transition, approximately two-thirds to three-quarters the melting T in degrees Kelvin, ions in the melt have more mobility and can break their bonds with neighboring ions and wander about, forming new configurations. In a flowing melt, bonds are broken and bond angles and distances are distorted, but after deformation ceases, the ionic array may have sufficient time to reform into a "relaxed" equilibrium structure.

Many studies of melts in the laboratory using nuclear magnetic resonance, vibrational spectroscopy, and X-ray analyses reveal a lack of long-range (on the scale of more than a few atomic bond lengths) **structural order** and symmetry that characterize crystals. However, melts possess a short-range structural order in which tetrahedrally coordinated Si and Al cations are surrounded by four O anions and octahedrally bonded cations such as Ca and Fe^{2+} surrounded by six O anions roughly resemble those in crystals. Because silica is the most abundant constituent in most natural melts, the fundamental structural unit is the $(SiO_4)^{4-}$ tetrahedron, as it is in silicate minerals. Conceptual models of the atomic structure of silicate liquids can be constructed on the basis of these observations. Figure 4.2 depicts these models for liquid silica (SiO_2) and $CaMgSi_2O_6$; the latter in crystalline form is diopside pyroxene.

Because the entropy of melting of crystalline silica (i.e., the change in entropy from the crystalline to the liquid state) is relatively small, there can be little change in the degree of order in the atomic structure of the melt relative to the crystalline state. Thus, a model for liquid silica is a three-dimensional network of somewhat distorted Si-O tetrahedra, not unlike the corresponding structure of crystalline silica. Short-range order is roughly similar to that in the crystalline state, but long-range order, as would be evident in a symmetrical crystal lattice, is absent. The silica melt can be viewed as a three-dimensional network of interlinked chains, or **polymers,** of Si-O tetrahedra.

On the other hand, in the model of the $CaMgSi_2O_6$ melt, these stringlike polymers are shorter, less intricately linked, and interspersed among octahedrally coordinated cations of Ca and Mg. This melt is not as polymerized as liquid silica.

Four different types of ions can be recognized in these models (Figure 4.2) on the basis of their relation to the polymers: (1) **Network-forming cations** of Si^{4+} within the interconnected tetrahedra of the polymers are strongly linked by (2) **bridging oxygens.**

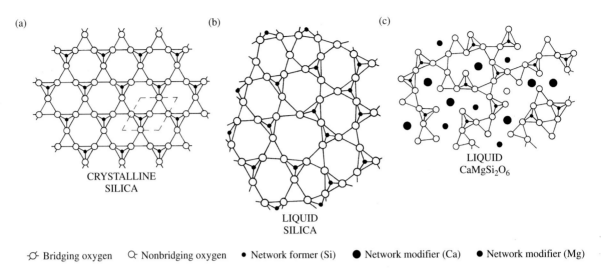

4.2   Conceptual models of atomic structures of silicate melts compared with the symmetric lattice of a crystalline solid. (a) Crystalline silica (high tridymite). Layers of hexagonal rings of Si-O tetrahedra with alternating apices pointing up and down are stacked on top of one another, creating a three-dimensional structure in which each oxygen is shared by two silicons. Tetrahedra with apices pointing up have the upper apical oxygen left out of the drawing so as to reveal underlying silicon. Dashed line indicates outline of one unit cell in the lattice. (b) Model of liquid silica. Si-O tetrahedra are slightly distorted relative to the crystalline lattice. Long-range order is absent. Structure is highly polymerized because all tetrahedra are interconnected by bridging oxygen anions. (c) Model of liquid $CaMgSi_2O_6$ showing less polymerization than that of liquid silica. Note presence of network-modifying cations (Ca and Mg) and nonbridging oxygen, neither of which occurs in the silica melt. (Redrawn from Carmichael et al., 1974, p. 133.)

(3) **Network-modifying cations** of Ca and Mg are more weakly bonded to (4) **nonbridging oxygens** in nontetrahedral bonding arrangements. The ratio of nonbridging oxygens to network-forming, tetrahedrally coordinated cations—chiefly Si and Al—is a measure of the **degree of polymerization** in a melt; small ratios correspond to high degrees of polymerization. In completely polymerized liquid silica, the ratio = 0. In partially polymerized liquid $CaMgSi_2O_6$ it is = 2/1 = 2.

The atomic structure of naturally occurring melts is more complex than these simple models. Despite considerable research, many details are not understood. Other ions of different size, charge, and electronegativity, such as Al^{3+}, Ti^{4+}, Fe^{3+}, P^{5+}, H^-, or F^- make natural melts more complex. In this milieu, mobile cations compete for available anions, principally oxygen, in order to satisfy bonding requirements and to minimize the free energy of the melt (Hess, 1995). This is not quite the same situation as in crystals, where cations have more or less fixed sites of a particular coordination in the ordered lattice. In addition to the widespread $(SiO_4)^{4-}$ tetrahedra in melts, there are less abundant neighboring tetrahedra of more negatively charged $(Al^{3+}O_4)^{5-}$ and $(Fe^{3+}O_4)^{5-}$. The ionic charge and size of network-modifying cations, which generally form weaker bonds with nonbridging oxygens, can play an important role in melt structure. Network modifiers most commonly include monovalent K and Na; divalent Ca, Mg, Fe, and Mn, and more highly charged, but less abundant high-field-strength cations including P^{5+}, Ti^{4+}, and the still less abundant trace elements.

The most important dynamic property of a melt—its viscosity—depends strongly on its atomic structure. Because **viscosity** is a measure of the ease of flow of a melt and the mobility of ions, it should be intuitively obvious that more highly polymerized melts are more viscous. Alternatively, it can be said that, because nonbridging oxygen anions are less strongly bonded to neighboring cations than bridging oxygens to Si and Al, viscosity correlates with the ratio of nonbridging to bridging oxygens. Increased concentrations of some components can depolymerize melts and reduce viscosity. Even small weight proportions of dissolved water or fluorine can depolymerize silicate melts, making them much less viscous. Also, high-field-strength, network-modifying cations whose charge is generally >3+ have a strong affinity for oxygen anions and may successfully compete against network-forming Si^{4+}, Al^{3+}, and Fe^{3+}, thus depolymerizing the melt. The role of Fe in melt structures is especially significant because it occurs in two oxidation states. Fe^{2+} appears to be exclusively a network modifier, whereas Fe^{3+} can be either a network modifier or a network former. Changes in the oxidation state can therefore affect the degree of polymerization of a melt.

Increasing pressure appears to reduce the degree of polymerization somewhat. Because octahedral coordination of Si and Al is favored in crystalline structures at high P over tetrahedral coordination, similar coordination changes might occur in melts at high P. Some experiments suggest that Al more readily shifts toward octahedral coordination with increasing P than does Si.

In conclusion, water-free ("dry") rhyolite melts have virtually no nonbridging oxygens and are nearly completely polymerized and highly viscous. In andesite melts the ratio of nonbridging oxygens to network-forming, tetrahedrally coordinated cations is about 0.2, and in basalt melts it is 0.4–1.2 (Mysen, 1988). Consequently, mafic, silica-poor melts are significantly less polymerized and less viscous than dry silicic melts.

※4.2 VOLATILE FLUIDS IN MELTS

Evidence for the participation of volatiles generally in magmatic systems includes widespread hydrous minerals such as micas and amphiboles and explosive volcanic eruptions. Nonetheless, so little was understood about volatiles in magma systems into the early decades of the 20th century that it was common to trivialize their significance or to blame a magmatic "mystery" on devious "fugitive elements." But, beginning with pioneering experiments by R. W. Goranson in the 1930s on the solubility of water in silicate melts, the mysteries began to disappear with the light of understanding. It is now clear that even modest amounts of volatiles, most commonly water, have a profound, and, for Earth at least, virtually universal influence on magmatic behavior.

4.2.1 Nature of Volatiles

In magmas at equilibrium, a particular ion resides in the melt, in any coexisting crystals, and in a possible separate gas phase. Some ions, such as Ca, Mg, Al, Ti, and Si, are more concentrated under equilibrium conditions in crystalline and melt phases in the magma and constitute condensed constituents. In contrast, volatile constituents, or **volatiles,** are chemical species that at near-atmospheric P but high T of magma systems exist as a gas or vapor, including H_2O (steam), CO_2, H_2, HCl, N_2, HF, F, Cl, SO_2, H_2S, CO, CH_4, O_2, NH_3, S_2, and noble gases such as He and Ar. Most volatiles consist of only six low-atomic-weight elements—H, C, O, S, Cl, and F. At equilibrium, small concentrations of volatiles are dissolved in the coexisting melt and any crystalline phases that may be present. Oxygen, the most abundant ion in magma, occurs in significant amounts in all three possible coexisting phases—solid, liquid, and volatile. Volatiles in most magmas are dominated by water and generally to a lesser extent by carbon dioxide.

As confining pressure, P, increases, initially dispersed molecules in a gas are forced closer together, increasing its density and altering other properties such as its capacity to carry other chemical elements in solution: Si, Fe, Hg, and so on. Above the **critical point,** gaseous (vapor) and liquid states are no longer distinguishable; there is no abrupt density change in the two phases above the critical point. For pure H_2O, the critical point lies at 218 bars and 371°C; for pure CO_2, at 73 bars and 31°C. Thus, at depths of more than a kilo-

meter there is no longer any familiar distinction between liquid and gaseous water or liquid and gaseous carbon dioxide, and each is one fluid. For this reason, in this text, we will refer to a liquid phase that consists chiefly of volatiles and has a density generally $<2 g/cm^3$ as a **volatile fluid,** or simply a **fluid.** Because the **specific volume** is the reciprocal of density, volumes of geologic fluids are $>0.5 \ cm^3/g$.

At depths of more than a few kilometers and over a wide range of P and T water has a density near that of surface waters in lakes and streams ($1 \ g/cm^3$; Figure 4.3). In erupting, low-P volcanic systems, the term *gas* may be used in lieu of *fluid* because at depths of less than a few kilometers and magmatic temperatures $>700°C$ the density of water (Figures 4.3 and 4.4) is only 0.1–0.0001 g/cm^3 (specific volume = 10–10,000 cm^3/g).

A volatile fluid should not be confused with a silicate liquid, or melt, made mostly of condensed constituents and whose density is generally $>2.2 \ g/cm^3$. Melts generally contain dissolved volatiles, and many melts are in equilibrium with a separate volatile fluid phase.

The **fluid pressure,** P_f, of a separate fluid phase in a magma system is the sum of the partial pressures (Section 3.5.1) of the different volatile constituents, $P_f = P_{H_2O} + P_{CO_2} + P_{SO_2} \ldots$. If water is the only volatile in the separate fluid phase, then $P_f = P_{H_2O}$. The confining pressure, P, is conceptually different from fluid pressure and partial pressures (Figure 4.5). P and P_f can vary independently in geologic systems. P depends essentially upon the depth of burial of the system. P_f

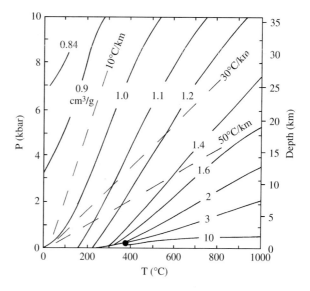

4.3   Specific volume (cubic centimeters per gram) of pure water as a function of P and T. For reference, geothermal gradients of 10°C, 30°C, and 50°C/km are shown as dashed lines. Filled circle is the critical point of pure water. (Data from Burnham et al., 1969.)

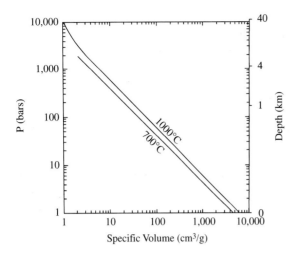

4.4   Specific volume of pure water as a function of *P* (and depth) for two temperatures.

depends upon *P*, *T*, the volatile solubility in the melt, and the amount of the volatile in the magma system, which is related to its origin and evolution. Generally, $P > P_f$, but in fluid saturated systems they are equal. For $P < P_f$, the magma is oversaturated in volatiles and the rocks surrounding the magma system, like a pressure cooker, must be strong enough to contain the excess fluid pressure, or else the rocks rupture and an explosion occurs.

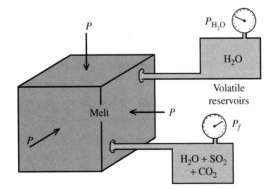

4.5   Schematic diagram illustrating the contrasts in meaning of confining pressure, *P*; fluid pressure, P_f; and partial pressure of individual volatile species, such as P_{H_2O} and P_{CO_2}. A hypothetical container on the left holds a high-*T* melt, with or without suspended crystals, which contains dissolved volatiles H_2O, CO_2, and SO_2. A confining pressure *P* is exerted on the melt with equal magnitude on all sides as in a plutonic environment. On the side of the container are two small membranes. To one of these membranes, permeable only to H_2O molecules, is attached a reservoir of pure H_2O at the same *P* and *T* as the melt and in equilibrium with it. A pressure gauge on the reservoir shows the partial pressure, P_{H_2O}, in the melt. A second reservoir containing all three volatiles is attached to the melt container at the other membrane, which is permeable to all three volatiles. A pressure gauge here shows $P_f = P_{H_2O} + P_{CO_2} + P_{SO_2}$.

4.2.2 Solubilities of Volatiles in Silicate Melts

The concentration of a particular volatile that can be uniformly dissolved in a melt—its **solubility**—depends on *P*, *T*, and chemical composition of the magma. Volatile solubilities are fundamental to several facets of magmatic behavior (Johnson et al., 1994).

Rock fabrics and explosive volcanism suggest that the solubility of volatiles in a melt decreases with decreasing *P* at lesser depth in the Earth. Solidified lava flows are commonly vesicular, whereas deep plutonic rocks are not, suggesting that volatile bubbles are released from melts whose volatile solubility decreases at lower *P*. Pillows of basaltic lava formed in deeper ocean depths have fewer, smaller vesicles, or none at all, compared to pillows formed in shallow ocean depths, even though the overall total volatile content of the magma that formed the pillows is similar in both.

Measurements of the amount of volatiles dissolved in a melt from their concentration in the solidified rock would generally be inaccurate. Cooling bodies of magma lose water to the surroundings during solidification. However, rapidly quenched glasses in fresh oceanic basalt pillows appear to retain pristine volatile contents of ~0.5 wt.% water. Minute volumes of melt can be entrapped inside growing crystals during cooling of the magma (see Figure 6.18). Some of these **melt inclusions** that are now glass appear to preserve pristine volatile concentration of the melt. For example, Wallace et al. (1995) measured 2–6 wt.% dissolved water in melt inclusions encased within quartz phenocrysts in the rhyolite Bishop Tuff of eastern California; higher water contents were found in the earliest erupted material apparently derived from the *top* of the magma chamber. Sobolev and Chaussidon (1996) measured water concentrations of melt inclusions hosted in olivine phenocrysts in more than 100 samples of basalts that they considered to have had a source in the upper mantle. Basaltic lavas erupted along oceanic ridges were found to contain <0.5 wt.% water; those in subduction zones, 1–3 wt.%

Crystallization of micas and amphiboles in magmas indicates water concentrations of at least 3–5 wt.% (Johnson et al., 1994).

The solubility of volatiles in melts as a function of *P* and *T* can be determined quantitatively in the laboratory (Special Interest Box 4.1).

The fundamental nature of volatile solubilities stems from the fact that the partial molar volume of a volatile in a silicate melt is significantly less than its volume in a separate pure phase at a corresponding *P* and *T* (for example, Figure 3.10).

In the equilibrium

4.1   *volatile-rich melt = volatile-poor melt*
+ volatile fluid

Special Interest Box 4.1 Experimental determination of water solubility in a melt as a function of P and T

Rock powder or other chemical constituents of the desired composition together with water in known weight proportions are placed in an inert metal (Pt, Pd, or Au) foil capsule, which is welded shut and placed in a hydrothermal pressure vessel (Figure 4.6). This vessel or "bomb" consists of a hollow cylinder of high-strength alloy steel into which an end plug is screwed. The plug is fitted with steel tubing to carry a fluid (usually inert gas such as Ar) under pressure from a pump. The bomb is placed in a furnace, at the desired T, then the fluid pressure is raised to the desired value. Inside the bomb, the fluid bears against the flexible walls of the impervious foil capsule to yield the desired confining pressure, P, on the material inside. After sufficient time has passed for a state of equilibrium to be attained within the capsule, it is very rapidly cooled, or quenched, to room T in less than a minute or so by dropping the bomb into a bucket of water or by directing a blast of cold compressed air at it. Any melt present in the capsule is quenched to glass. If the concentration of water in the capsule system exceeds the solubility in the melt at the particular P and T of the experiment, so that the system is oversaturated, a separate water phase exists in the melt. Upon rapid quenching, bubbles of fluid water are frozen into a glass as vesicles. If the concentration is less than the solubility, so that the system is water-undersaturated, no vesicles are present. By making a number of experimental runs at different P, T, and X_{water}, the solubility can be mapped out as a function of P and T to yield a diagram like Figure 4.7.

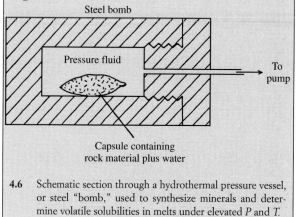

Steel bomb

Pressure fluid

To pump

Capsule containing rock material plus water

4.6 Schematic section through a hydrothermal pressure vessel, or steel "bomb," used to synthesize minerals and determine volatile solubilities in melts under elevated P and T.

Increasing P drives the reaction toward the left, the state of lesser volume. Therefore, greater concentrations of volatiles can be forced into a melt at greater P. On the other hand, increased T favors an expanded state of greater volume; hence, the equilibrium shifts to the right and the concentration of dissolved volatiles in the melt is less at greater T. Thus, changes in P and T have opposite effects on volatile solubility, but the effect of P dominates.

Water. Experimental data in Figure 4.7 show the substantial increase in water solubility with respect to increase in P. On a weight percentage basis, the solubility of water in silicic melts is slightly greater than in mafic melts. The solubility curves in Figure 4.7 indicate the *maximal* amount of water that can be contained within the indicated melt at a particular P and T. The curves can also be considered as saturation curves in that a melt containing a lesser concentration of dissolved water than the solubility value at a particular P is water-undersaturated. On the other hand, a melt containing an excess concentration is water-saturated and the magma system consists of this saturated melt plus a separate phase of fluid water. The analogy with the concept of silica saturation (Section 2.4.4) should be apparent.

Many naturally occurring melts, particularly at their site of generation from solid source rock, are probably undersaturated in water.

Early investigations indicated that the solubility of water in a melt is approximately proportional to $(P_{H_2O})^{0.5}$, suggesting that H_2O is dissolved in a melt according to the reaction

4.3 $H_2O + O^{2-} = 2(OH)^-$

in melt   in melt

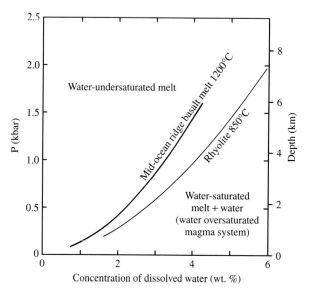

4.7 Solubility of water in silicate melts. (Redrawn from Moore et al., 1998.)

the volume of the single-phase homogeneous melt containing dissolved fluid is less than that of a two-phase system of melt plus fluid, or

4.2 $V_{volatile\text{-}rich\ melt} < (V_{volatile\text{-}poor\ melt} + V_{volatile\ fluid})$

This solution process breaks silicate polymers by substituting the two hydroxyl ions for one bridging oxygen in the melt (Figure 4.8). Even small amounts of water by weight drastically depolymerize a silicate melt. This effect follows from the large difference between the weight per mole of water and the weight of condensed constituents (Si, Ca, Fe, etc.) of silicate melts.

Continuing experimental investigations confirm that water solubility is more complicated than suggested by equation 4.3 (McMillan, 1994). Thus, infrared spectroscopic analyses indicate that an increasing *proportion* of water at increasing P is dissolved in rhyolitic melts as molecular H_2O rather than $(OH)^-$ (Figure 4.9). The maximal concentration of dissolved $(OH)^-$ is about 2 wt.%.

Other Volatiles. Other volatile species are generally less abundant in magma systems than water but can, nonetheless, be significant (Carroll and Webster, 1994).

Carbon dioxide is generally less concentrated than water in melts, except in kimberlite, carbonatite melts, and possibly mid–ocean ridge basalt. Like that of water, carbon dioxide solubility in silicate melts increases with increasing P and decreasing T but is about an order of magnitude less than that of water at comparable P (Figure 4.10). Solubility also generally increases with decreasing silica concentration and, in some melts at high P, with increasing water content. Infrared spectroscopy suggests that in silica-poor melts, carbon dioxide is dissolved as $(CO_3)^{2-}$ ions, whereas in silicic melts it is dissolved as CO_2 molecules. Apparently, the dissolution mechanism depends upon the degree of polymerization of the melt (Holloway and Blank, 1994). Carbon dioxide cannot combine with bridging oxygens in a polymerized silicic melt as water does. In CO_2-saturated water-bearing melts, the separate fluid is a mixture of CO_2 and H_2O, even though the melt may not be water-saturated.

Precipitates of native sulfur and noxious sulfurous fumes, such as "rotten-egg" gas, around volcanic vents testify to ubiquitous sulfur in magmas. In most cases, its concentration, generally less than a few thousand parts per million, is exceeded by that of water and only barely by that of carbon dioxide in some magmas. Nonetheless, some calc-alkaline magmas contain significant concentrations. Sulfur solubility depends strongly on the composition and oxygen fugacity of the magma, in addition to P and T. Sulfur dissolves as the reduced sulfide ion, S^{2-}, in generally water-poor ultramafic and mafic magmas but as the oxidized sulfate ion, $(SO_4)^{2-}$, in generally more water-rich intermediate to silicic magmas. Fluid species in reduced and oxidized systems are H_2S and SO_2, respectively. In the reaction

$$4.4 \quad S^{2-} + 2O_2 = (SO_4)^{2-}$$

the activities of sulfide and sulfate species dissolved in a melt are related to the square of the oxygen fugacity. If this fugacity is below (less than) the quartz-fayalite-magnetite (QFM) buffer (Figure 3.14), the equilibrium shifts to the left, reducing the sulfur. Above the buffer, it shifts to the right, oxidizing it. Another complication is that in reduced mafic magmas an immiscible sulfide-iron-copper melt, which depresses the concentration of S in the melt, can form. In some silicic magmas that are more oxidized, sulfate ions can combine with Ca, if its activity is sufficient, to form stable anhydrite ($CaSO_4$), which also depresses the sulfur content of the coexisting melt. Removal of the immiscible sulfide melt or the anhydrite crystals from the magma will diminish the remaining amount of sulfur in it.

Although not generally abundant, concentrations as much as several weight percentages of fluorine have been found in some silicic glasses. Experimentally measured solubilities in melts can be as much as 10 wt.%; the higher solubilities occur in hydrous melts. The exact way in which fluorine is dissolved in silicate melts is uncertain and probably complex, but, by whatever means, F causes substantial depolymerization. Although the effect of P on F solubility is poorly known, reduced P does not reduce the F solubility to the extent it does that of water (Carroll and Webster, 1994, p. 262). Whereas at 1 atm granitic melts contain only about 0.1 wt.% water, some simple model Na-Ca-Al-Si-O melts contain as much as 10 wt.% F.

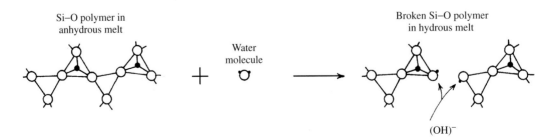

4.8   Some dissolved water in silicate melts forms hydroxyl ions, $(OH)^-$, which break O-Si-O polymers, reducing the degree of polymerization.

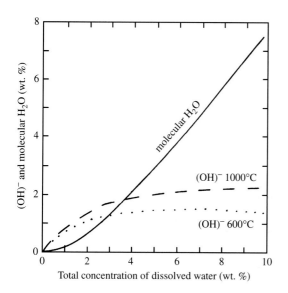

4.9 Dissolved water in rhyolitic melts exists in ionic, $(OH)^-$, and molecular, H_2O form. The concentration of $(OH)^-$ is dependent upon T, but molecular H_2O concentration is independent of T. For example, in a melt at 1000°C that contains 4 wt.% total water, about half is dissolved as $(OH)^-$ ions and half as molecular H_2O. For a 1000°C melt containing 2 wt.% total water, about 1.5 wt.% is dissolved as $(OH)^-$ and 0.5 wt.% as molecular H_2O. (From Silver et al., 1990.)

The maximal solubility of Cl in silicate melts appears to be less than about 2 wt.%. Cl probably bonds with network-modifying cations, especially Na, K, and Fe. If the magma is water-saturated so that a separate aqueous fluid coexists with the melt, Cl partitions strongly into the aqueous fluid phase, in which the presence of Cl^- ions enhance metal solubilities.

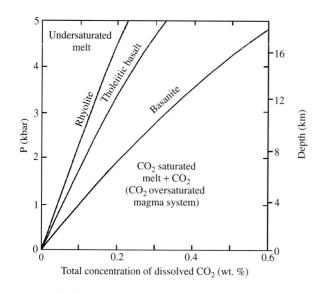

4.10 Solubility of carbon dioxide in some silicate melts. Note that more CO_2 can dissolve in less polymerized mafic and especially silica-undersaturated melts. (From Holloway and Blank, 1994.)

Still other volatiles have solubilities in melts generally less than hundreds of parts per million and are not considered further here, except to note that dissolved phosphorus and boron appear to enhance water solubility.

4.2.3 Exsolution of Volatiles from a Melt

As magmas leave their upper mantle or lower crustal source, rise to shallower depths and cool, their melts can become saturated. Once saturated, the excess dissolved volatiles are released from the melt and separate into a distinct coexisting fluid phase in the process called **exsolution** or **boiling**.

Although the exsolution of volatile fluids from melts is a complex phenomenon, two ideal end-member processes that encompass the spectrum of real phenomena can be recognized. Water is used as the sole volatile to illustrate these two processes, as follows:

1. An ascending, decompressing, initially volatile undersaturated melt can become saturated as P decreases, exsolving fluid, generally in shallow crustal, volcanic environments. For example, in Figure 4.11, decompression of an initially water-undersaturated rhyolite melt at about 850°C that contains 4 wt.% dissolved water becomes water saturated at 1 kbar at a depth of about 4 km.

2. A stagnant, isobaric (P constant) magma that is initially volatile, undersaturated but losing heat to the surroundings, and cooling can become water-saturated by crystallization of minerals such as olivine, pyroxene, feldspar, and quartz. As these anhydrous minerals crystallize, the water concentration in the residual melt increases, in some cases sufficiently to lead to saturation, overriding the effect of increasing solubility of water in the cooling melt. This phenomenon has been called **retrograde, resurgent,** or **second boiling,** because it occurs by decreasing T, the reverse of the familiar cause of boiling water. Because all magmas eventually cool, and most consequently crystallize, more or less isobaric exsolution induced by crystallization can potentially affect all magma systems if volatile concentrations and other factors are appropriate.

Exsolution of volatiles from melts is an exothermic process that causes them to cool in the same way that evaporation of vapor from a body of liquid water causes cooling. In addition to this cooling by exsolution, ascending and decompressing magmas also cool as a result of adiabatic expansion of the melt and fluid phases. The total refrigeration of a magma system might be as much as hundreds of degrees in some systems. An example is afforded by kimberlite magmas which are unusually rich in CO_2 and H_2O that exsolve relatively deep in the crust. Field relations indicate that pipelike bodies of kimberlite breccia were apparently emplaced in the shallow crust at no more than 200°C

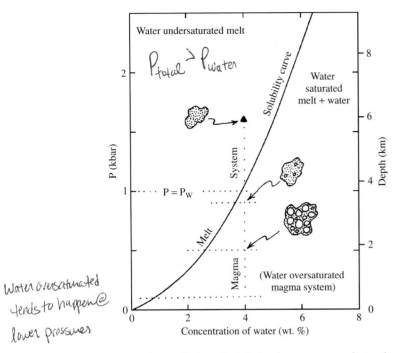

Handwritten annotations on figure: "P_total ⊃ P_water" ; "Water oversaturated tends to happen @ lower pressures"

4.11 Evolution of a hypothetical closed magma system during decompression from an initially water-undersaturated state. The initial magma is a crystal- and bubble-free melt at 1.6 kbar, corresponding to a depth of 6 km, and contains 4 wt.% dissolved water and no other volatiles.

or so because of minimal thermal effects on wall rock and the absence of thermal conversion of rare inclusions of coal into coke.

After solidification, only a part of the volatiles initially dissolved in the melt is preserved in the magmatic body as Cl and F in accessory apatite and S in accessory sulfides, together with H_2O in micas, amphiboles, and apatite. Volatiles can be preserved in glass formed by quenching of melt.

Interactions of Volatile Species: Composition of Exsolved Fluids. In most magmas, several volatile species are dissolved in the melt. These species do not necessarily exsolve individually in pure form in some sort of sequence depending on their contrasting concentrations and solubilities. Rather, interactions between species are widespread and, because of substantial mutual solubilities, exsolved fluids are generally mixtures of two or more species.

Because of its order-of-magnitude lower solubility, CO_2 tends to exsolve from melts at greater depths and at lesser degrees of crystallization than does water. Even though a melt is undersaturated in water, small concentrations are dissolved in any exsolved CO_2. These mixed CO_2-H_2O volatile bubbles can rise through the body of magma and collect at the top of the chamber. Or the mixed fluid may escape entirely from the magma chamber. In either case, the water content of the initial magma is reduced.

Significant concentrations of sulfur are partitioned into an aqueous fluid exsolved from silicic melts (Keppler, 1999). S partitioning is governed mostly by oxygen fugacity, much less by P and T. Under reducing conditions where H_2S is stable, an order of magnitude more sulfur is sequestered in the fluid compared to the fluid in equilibrium with oxidized magmas in which SO_2 is stable. During the June 1991 explosive activity of Mount Pinatubo in the Philippines, the 17×10^9 kg (17 megatons) of SO_2 released into the atmosphere was one to two orders of magnitude more than the total amount of S that could have been degassed out of the *erupted* volume of magma. The outgassed amount of S is based on analyses of the concentration of S (60–90 ppm) in melt inclusions entrapped in growing phenocrysts in the preeruption magma that was subsequently blown out of the volcano. Calculations (Problem 4.5) show that the 17×10^9 kg could have reasonably been derived by scavenging S from throughout a 40–90 km^3 volume of magma beneath the volcano (indicated by seismic data), rather than just the 5–10 km^3 erupted. If the calculations are creditable, it is also necessary to assume that long-term migration of aqueous fluid bubbles containing the scavenged S had occurred and this fluid had accumulated in the upper part of the magma chamber, which furnished most of the erupted material.

✳4.3 CONSEQUENCES OF FLUID EXSOLUTION FROM MELTS

Exsolution of volatiles in magma systems plays a surprisingly varied role in geologic phenomena (Figure 4.12). Some of these occur in the shallow crustal volcanic environment, where the magma system can vent to the surface and interact with atmospheric gases. Other phenomena occur in confined plutonic magma systems in the deeper crust, where interactions with meteoric groundwaters are possible.

4.3.1 Explosive Volcanism

Exsolving and expanding volatiles provide the driving force for explosive volcanic eruptions.

Hypothetical Model. This fundamental concept can be illustrated by a hypothetical magma system that is rising adiabatically through the crust (Figure 4.11; see also Problem 4.6). The model system, which contains 4 wt.% dissolved water in the melt as the sole volatile, is assumed to be closed, whereas few real systems are truly closed. Equilibrium is assumed to prevail between all phases, whereas in real systems, especially highly viscous silicic magmas, sluggish kinetic properties retard attainment of equilibrium (discussed later). As the magma decompresses during ascent, the initially water-undersaturated melt becomes water-saturated and begins to exsolve water at $P = P_{water} = 1$ kbar at a depth

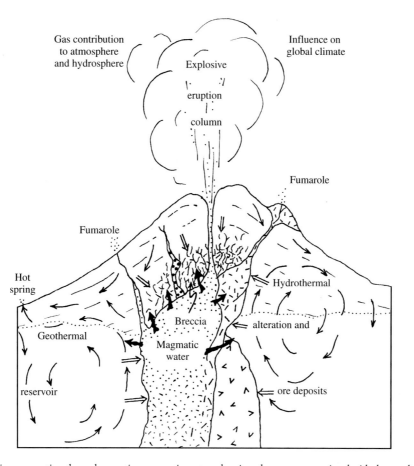

4.12 Schematic geologic cross section through an active magmatic system showing phenomena associated with the exsolution of volatiles from magma. An explosive eruption column of gas and pyroclastic material rises above the vent of a volcano built on an irregular erosion surface (dotted line) carved into older rock. **Fumaroles** vent gases on the flanks of the volcano. Volcanic gases contribute to the atmosphere and hydrosphere of the Earth and influence its global climate. Roof rocks overlying the multiple intrusive bodies have been brecciated. Large cells of advecting water (long lines with arrow barbs) move through fractures and other open channels in the wall rocks around the cooling intrusions, which provide heat to drive the advection. Meteoric groundwater is heated near the intrusions, expands, and rises as cooler water moves inward to take its place, is heated, and so on. Juvenile, or magmatically derived, water is expelled from the cooling magma bodies (heavy arrows). Heated meteoric and/or juvenile water constitutes a resource potentially capable of providing geothermal power. Also, because the heated water carries dissolved Si, Ca, Fe, Cu, Au, Pb, and other material, it is a hydrothermal solution that can precipitate potentially valuable ore minerals as well as quartz and other uneconomic minerals in veins and alters the rocks through which it percolates (double-line arrows).

near 4 km. During continued ascent to lower P, the melt and the magma system follow increasingly divergent paths in the diagram. The *closed water-oversaturated magma system* (magma plus separate water fluid phase) tracks straight down the diagram along the constant 4 wt.% water line. The *water-saturated melt,* on the other hand, moves down the solubility (saturation) curve. The concentration of water that can be held in solution in the melt decreases at lower pressures, so water exsolves, forming bubbles of compressed steam. At about 0.9 kbar, the melt contains about 3.6 wt.% water; the remaining 0.4 wt.% in the closed magma system has exsolved into the separate fluid phase. As P continues to decrease, more water exsolves and the mass of water in the separate fluid phase increases relative to that of dissolved water in the melt. At 0.5 kbar, about 2.6 wt.%

water is still dissolved in the melt and 1.4 wt.% resides in many, larger bubbles. At 0.1 kbar, <1 wt.% water is in the melt and the remaining is in bubbles.

Not only does the mass of exsolved water increase during adiabatic decompression in this model closed magma system, but each mass unit also experiences a tremendous increase in volume. Two factors are responsible for this volumetric expansion:

1. The volume of the exsolved water is much greater than the partial molar volume of the same mass of dissolved water in the melt: 99.3 versus 22.3 cm^3/mole for water in a $NaAlSi_3O_8$-H_2O system at 950°C and 1 kbar.

2. The exsolved water expands at lower pressure according to the perfect gas law, $PV = nRT$. In our model, the mass of the exsolved water has in-

creased during decompression; that is, the number of moles of water, n, has increased. Because RT remains essentially constant and P has decreased, V increases accordingly, hundreds of times (Figure 4.4). It is not surprising that relatively small volumes—on the order of 1–10 km^3—of erupting hydrous magma can generate the gigantic cauliflower clouds of ash-laden steam rising tens of kilometers above a volcano that are familiar hallmarks of countless climactic explosive eruptions, such as those of Mount Saint Helens and Mount Pinatubo.

This compounding of factors—continued exsolution and expansion of the exsolved water—reduces the density of the magma, promoting, at least, extrusion of buoyant, bubble-bearing magma as a lava flow, but, in many instances, leading to explosive eruption.

Real Magma Systems. Of course, in real, generally open, magma systems, exsolution is a complex interplay of many factors, including decreasing P and T, initial volatile concentration in the melt, types of volatile species, changing solubilities, and interactions with the atmosphere and surrounding wall rocks. These factors modify the details of explosive eruption, even though the tremendous volumetric expansion of the magma still occurs.

Most magmas cool and crystallize en route to the surface, augmenting exsolution due to decompression. Most magmas do not behave as perfectly closed systems. Every volcano vents gas before and after explosive events. Not all of this gas is **juvenile**, that which resides in the melt from its place of origin in the deep crust or upper mantle. Some vented gas may be heated meteoric groundwater derived from atmospheric precipitation, and some may be atmospheric gases. Mixtures of all these fluids are typical.

Ample evidence indicates that many magma bodies, especially silicic ones, have higher concentrations of volatiles, especially water, in their upper part than in their lower. Hence, eruptions are initially highly explosive as the uppermost, volatile-rich part of the magma chamber is tapped, but as eruptions continue, they tend to be less explosive as less volatile-rich magma is erupted.

Additional, often very significant factors in volcanic eruptions are kinetic. The most important of these kinetic factors are the viscosity of the melt and the rate of ascent of the magma body. These two factors can conspire to cause different dynamic scenarios. On the one hand in a slowly ascending and decompressing low-viscosity mafic magma, exsolving volatile fluid can readily segregate into bubbles that may be able to escape from the magma into openings in the surrounding wall rocks or escape relatively harmlessly out of a vent. Such magma might extrude from a volcanic vent as a coher-

ent, nonexplosive lava flow. **Vesicles** in the solidified lava, which define **vesicular fabric,** are remnants of the gas bubbles. On the other hand, in a highly viscous silicic magma cooling and crystallizing in a shallow crustal chamber, and perhaps decompressing, slow release of dissolved volatiles from the highly polymerized melt could lead to a state of disequilibrium in which the volatile pressure in the magma system exceeds the confining pressure on it so that it is an **overpressured** system. In other words, water exsolution lags behind that dictated by decreasing P (Figure 4.11). Slow release of volatiles is exacerbated by the fact that as water exsolves the melt becomes more polymerized and more viscous so that the release is further decelerated. Overpressured systems can rupture the overlying roof rocks, as a lid on a pressure cooker can fail. Or the system might, for another reason, suddenly be unroofed, as at Mount Saint Helens in 1981, when a moderate earthquake shook the oversteepened volcano summit, causing it to slide off the top of the bulging magma chamber. Whether the overlying load of roof rock is removed or ruptures, the magma is suddenly decompressed, as is a can of soda pop from which the lid is removed. Overpressured bubbles of volatile fluid rupture their intervening melt walls, producing fragments of melt that quench to form vitroclasts, plus possible phenocrysts and phenocryst fragments, or phenoclasts (Figure 4.13). All these bits and pieces of the former coherent magma, together with possible fragments of rock torn from the explosive conduit and vent, are collectively called **pyroclasts;** a deposit made of them has **pyroclastic fabric.**

Such overpressured volcanic systems illustrate that the pressure of a magma system cannot be assumed equal to the confining pressure, P, evaluated from the geobaric gradient (Section 1.2). In this case, the confining pressure due to the load of overlying rock, P, is less than fluid pressure, P_f, in the magma system. Burnham (1985) showed that exsolution and expansion of >2 wt.% water from a crystallizing silicic magma at depths of no more than a few kilometers have the capacity to do PV work, rupturing the roof rocks overlying the magma body. In other words, $P_f > P + $ *strength of roof rocks.* As rocks fracture, openings are created, decompressing the magma system, leading to further exsolution and, in some instances, explosive venting of the gas-charged magma. Fracturing creates **breccia**—rock fragments ranging widely in size but commonly several centimeters in diameter—and **brecciated fabric.** Void spaces between fragments serve as channels for advective heat transfer and migration of hydrothermal solutions and provide openings for deposition of metals from them (Figure 4.12).

4.3.2 Global Atmosphere and Climate

There is wide consensus among geologists that the atmosphere and hydrosphere of the Earth were pro-

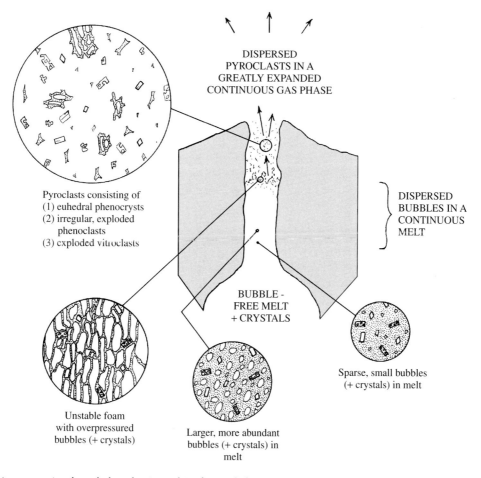

DISPERSED
PYROCLASTS IN A
GREATLY EXPANDED
CONTINUOUS GAS PHASE

Pyroclasts consisting of
(1) euhedral phenocrysts
(2) irregular, exploded
phenoclasts
(3) exploded vitroclasts

DISPERSED
BUBBLES IN A
CONTINUOUS
MELT

BUBBLE -
FREE MELT
+ CRYSTALS

Sparse, small bubbles
(+ crystals) in melt

Unstable foam
with overpressured
bubbles (+ crystals)

Larger, more abundant
bubbles (+ crystals) in
melt

4.13 Schematic cross section through the volcanic conduit of an exploding magma system. Circular diagrams are "snapshots" of the state of the expanding magma as a function of depth and P in the conduit. Because of limitations of the diagram area, the hundreds-fold expansion of the volatile fluid phase cannot be accurately represented.

duced over the 4.5 Gy of its evolution by degassing of the interior through generation of volatile-bearing magmas and varying degrees of exsolution of these volatiles during magma ascent and cooling.

There is a correlation between some types of volcanism and global climate, as first suggested by Benjamin Franklin, the early American statesman and scientist. After the eruption of 12.3 km^3 of basalt lava and about 0.3 km^3 of ash at Laki (Lakagígar), Iceland, beginning June 1783 (Francis, 1993; Rampino et al., 1988), Franklin noted a persistent "dry" fog and faint sun during the severe 1783–84 winter in Europe and eastern North America. In contrast to this predominantly lava eruption, the Tambora, Indonesia, eruption in April 1815 was probably the largest ash-producing eruption in the past 10,000 years. Its eruption column of gas and pyroclastic material may have reached a height of 50 km from an estimated 50 km^3 of pyroclastic material, causing darkness for as much as 2 days 600 km away. The following year, 1816, was the notorious "year without a summer" when mean temperatures in Europe were about 1°C cooler and repeated summertime frosts in New England caused devastating crop failures.

The contrasts between the Laki and Tambora eruptions suggest that the amount of ash inserted into the atmosphere cannot be responsible for global climate change. The real culprit—sulfur—was only identified after predominantly ash eruptions at Mount Saint Helens, Washington, in 1980 and at El Chichón, Mexico, in 1982. Although of similar, small volume, about 0.35 km^3 each, the global cooling effect in the northern hemisphere from Mount Saint Helens was nil, but that of El Chichón was about 0.5°C. The erupted El Chichón magma was unusually rich in sulfur; contained as much as 2.5 wt.% total SO_2; included phenocrysts of pyrrhotite ($Fe_{1-x}S$) and anhydrite ($CaSO_4$); and injected about 10^{13} g of H_2SO_4 into the stratosphere, about one hundred times more than Mount Saint Helens. The magma in the 1991 eruption of Pinatubo in the Philippines was similarly sulfur-rich, contained anhydrite phenocrysts, and produced about twice as much stratospheric H_2SO_4 as El Chichón, but an order of magnitude less than that estimated for Tambora and Laki.

Therefore, more important than dust-sized ash particles, which fall to the ground in a few months, are acid **aerosols**. In the stratosphere, about 25 km above the ground, SO_2 reacts with $(OH)^-$, created by photodissociation of water vapor, to form micrometer-size droplets of sulfuric acid, H_2SO_4. This aerosol, which continues to form for several years after an eruption, is not washed out by atmospheric precipitation to form acid rain, as is the 10–20 times greater amount of man-made SO_2 injected into the lower atmosphere (Symonds et al., 1994). The H_2SO_4 aerosol absorbs

and backscatters incoming solar radiation, heating the stratosphere but restricting normal solar heating of the atmosphere.

4.3.3 Fumaroles, Hydrothermal Solutions, Ore Deposits, and Geothermal Reservoirs (Figure 4.12)

Exsolved volcanic gases can vent to the atmosphere in large volumes at a rapid rate in explosions, or much more slowly but in at least comparable volumes over many years from **fumaroles** (called *solfataras* if sulfurous). Fumaroles can be located at a summit crater or on the flank of an active volcano and in recently emplaced extrusions of lava and pyroclastic material. Cooling intrusive magma heats adjacent wall rock and any meteoric groundwater included (Figure 4.12). Hot fluids exsolved from the magma can mix with meteoric water, both surface water and groundwater, to create hot springs, such as the famous Yellowstone National Park in the United States and the Wairakei area on the North Island of New Zealand. Because of extensive contamination with near-surface meteoric water, seawater, and atmospheric gases, which are highly oxidizing, the exact nature of the exsolved gas species and their solubilites are difficult to determine. However, thermodynamic modeling (e.g., Symonds et al., 1994) indicates many elements are transported at 800–900°C as chloride complexes ($NaCl$, KCl, $FeCl_2$, $ZnCl_2$, $PbCl_2$, $CuCl$, $SbCl_3$, $MnCl_2$, $NiCl_2$, MoO_2Cl_2), and some elements as sulfide, fluoride, and carbonate complexes.

Beneath the surface of the Earth at elevated P hot aqueous fluids called **hydrothermal solutions** (Barnes, 1979; Henley et al., 1984; Brimhall and Crerar, 1987) carry many of the same elements as fumarolic gases. These solutions are essentially brines whose total dissolved solids range to as much as 50 wt.%. The dissolved species are mostly Na, K, Ca, and Cl and lesser amounts of Mg, Br, SO_4, H_2S, CO_2, and possibly NH_3 but include concentrations as high as 1000 ppm of elements such as Au, Ag, Cu, Zn, Pb, and Mo. Fluids exsolved from decompressing and cooling magmas are more or less neutral in pH, but as their T decreases, Cl- and S-bearing species ionize and the hydrothermal solutions become carriers for metals as well as developing acid potential for wall-rock interaction. Partition of metals such as Cu from the silicate melt into the aqueous fluid can be nearly complete: The fluid/melt partition coefficient is nearly infinite. Juvenile magmatic fluids, as well as heated meteoric water or groundwater and by-product fluids derived from metamorphic reactions, are sources of hydrothermal solutions. The oxygen isotopic composition of rocks and minerals (Section 2.6.1) demonstrates that ^{18}O-enriched meteoric water can advect through rock openings for tens of kilometers around intrusive magma bodies (Figure 4.14). Thermal energy released from the cooling in-

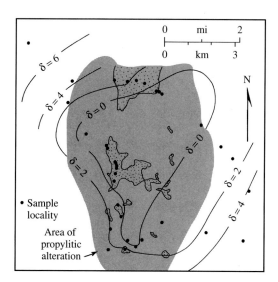

4.14 Effects of meteoric water-rock interactions around cooling magmatic intrusions in the Bohemia mining district, Oregon. Volcanic country rocks (unpatterned) are enriched in ^{16}O toward dioritic intrusions (stippled) because of exchange reactions between the country rocks ($\delta = +6$) and heated meteoric groundwater ($\delta = -9$). Shaded area delineates propylitically altered rocks. $\delta = \delta^{18}O/^{16}O$ (see Section 2.6.1). (Redrawn from Taylor, 1974.)

trusions supplies heat that powers these circulating water systems. Hot springs (see Figure 11.17) on the seafloor along oceanic spreading ridges are obvious examples of advective water systems driven by magmatic heat. Because of their solvent capabilities at high T, hydrothermal solutions advectively migrating through large volumes of rock over tens of thousands of years can leach significant amounts of valuable metal ions from the wall rocks through which they percolate.

Whether derived from wall rock or partitioned into exsolving fluids from the crystallizing melt, or both, dissolved metals can be precipitated from hydrothermal solutions to form economically significant hydrothermal **ore deposits** (e.g., Guilbert and Park, 1986; Brimhall and Crerar, 1987; Clarke, 1992). Commonly associated with these hydrothermal deposits are tabular or sheetlike veins of various minerals, such as quartz. These owe their origin largely to the strongly T-dependent solubilities of compounds in water. Also associated with the deposits are surrounding, more extensive and obvious halos of hydrothermally altered wall rock produced by thermal and chemical interaction of the migrating acid fluids with their wall rock (Rose and Burt, 1979). These altered rocks serve to indicate the presence and character of the smaller, more focused ore deposit. Widespread replacement of magmatically crystallized alumino-silicate minerals—especially feldspars—by clay minerals and other low-T sheet silicates creates **argillically altered rock.** Replacement of feldspars, micas, amphiboles, and pyroxenes

by epidote, actinolite, chlorite, albite, and calcite create **propylitically altered rocks.**

Another economically significant resource related to magma systems is the **geothermal reservoir;** this consists of a large volume of underground supercritical water, largely if not entirely of meteoric origin, lodged in open spaces within the rock, which can be tapped and used to drive electric power turbines. Geothermal resources have been developed in regions of young—generally <1 Ma—volcanism because associated underground bodies of unerupted magma that are invariably present have not cooled sufficiently for the thermal energy to be dissipated.

SUMMARY

Magma is high-T mobile rock material that generally includes crystals and always includes melt, a silicate liquid solution that contains dissolved volatiles, chiefly H_2O and many other compounds of H, C, O, and S. The kinetic properties of a magma are largely determined by the atomic structure of the melt, which resembles crystalline silicates but lacks their symmetric long-range order. Network-forming Si and Al cations are linked to bridging oxygen anions to form polymers. Silicic melts are polymerized to varying degrees; silicic melts in which the ratio of nonbridging oxygens to network-forming ions is small are most polymerized and consist of three-dimensional ionic networks that possess high viscosity. In less silicic, more mafic melts, there are more nonbridging oxygens, the degree of polymerization is less, and melts are less viscous. Dissolved water and fluorine depolymerize melts, reducing their viscosity.

Volatiles such as H_2O, CO_2, and SO_2 are gases at magmatic temperatures and near-atmospheric pressures but become compressed into denser fluids more than a kilometer below the surface of the Earth, where there is no physical distinction between a supercritical gas and fluid. Because dissolved volatiles have smaller partial molar volumes in melts relative to a separate pure phase at corresponding P and T, their solubilities in melts increase substantially with increasing P and less markedly at decreasing T. Concentrations of several weight percentages of dissolved water, as both molecular H_2O and hydroxyl ion, $(OH)^-$, can occur in natural melts at high P and depolymerize them because bridging oxygens are replaced by $(OH)^-$. Modest weight concentrations of water translate into larger molar concentrations because of the contrast in the molar weight of silicate and water. Depolymerizing fluorine, in concentrations of several weight percentages, also occurs in some melts, but, unlike water, these concentrations can be retained in the melt even at atmospheric P. Carbon dioxide, sulfur, and chlorine solubilities are about an order of magnitude less than those of water and fluorine.

Most melts are probably initially undersaturated in volatiles where magmas are generated in the crust or upper mantle. However, as magmas ascend, decompress, cool, and crystallize, melts become volatile saturated and a separate volatile fluid phase exsolves from the melt. Ascending melts saturate in CO_2, S, F, and Cl at greater depth than they do in H_2O because of their lesser solubilty.

Exsolution of volatile fluids from magmas is associated with several significant geologic phenomena. Explosive volcanism is driven by volatile exsolution and drastic volumetric expansion of the exsolved volatile fluid. These processes create pyroclastic deposits and fabrics, have contributed to the formation of the atmosphere and hydrosphere of the Earth throughout its long history, and can influence global climate, chiefly because of SO_2 ejected high into the stratosphere, where it forms H_2SO_4 aerosols that block solar heating of the surface of the Earth. Volatile exsolution in shallow magma bodies can generate sufficient PV energy to fracture overlying roof rocks, producing a mass of brecciated rock or at least channelways for emplacement of veins of quartz and ore minerals. Hydrothermal solutions alter preexisting rocks and form ore deposits.

CRITICAL THINKING QUESTIONS

4.1   Describe the nature of magmas and the combinations of solid, liquid, and gaseous phases that are possible.

4.2   Describe the atomic structure of silicate melts and the role played by different ions in governing the degree of polymerization.

4.3   Describe the polymerization of Mg_2SiO_4 melt. (You may need to consult a mineralogy text to refresh your memory of the atomic structure of forsterite olivine.)

4.4   Characterize volatile fluids and indicate how they are distinguished from silicate liquids (melts).

4.5   Describe the volatile species that are commonly dissolved in silicate melts, their concentrations, the factors that control their solubilities, and their effects on the atomic structure of the melt.

4.6   Discuss how volatiles exsolve from melts and variations in the concentrations of H_2O, CO_2, F, Cl, and S in mixed volatile fluid phases exsolved from melts.

4.7   Describe factors governing explosive volcanism.

4.8   Describe fabrics that are a direct consequence of volatile exsolution.

4.9   How does volatile exsolution from magmas influence global climate?

4.10   What economic benefits accrue from volatile fluids?

PROBLEMS

4.1   Make a schematic sketch of three different five-phase magma systems after the style of Figure 4.1. Label the phases represented.

4.2   Verify that 5 wt.% water in a $NaAlSi_3O_8$ melt is equivalent to 15 wt.% on a molar basis. (*Hint:* Convert the chemical analyses in weight percentage of albite [Appendix A] plus 5 wt.% water to mole percentage by dividing each oxide weight by its formula weight. Recalculate the total molar weight to 100.00%.)

4.3   Show from equation 4.3 why the solubility of water, as $(OH)^-$, should be proportional to $(P_{H_2O})^{0.5}$.

4.4   In a plot of P versus concentration of dissolved volatiles in melt draw schematic solubility curves for CO_2 and H_2O. If an initially volatile-undersaturated melt decompresses at more or less constant T, which volatile, CO_2 or H_2O, exsolves first?

4.5   In the erupting 1991 Mount Pinatubo magma assume that the partition coefficient of SO_2 is $D = C_{fluid}/C_{melt} = 47$, appropriate to a relatively oxidized state (Keppler, 1999). From the maximal concentration of S in the melt (90 ppm), what was the weight percentage SO_2 in the coexisting aqueous fluid phase? If it is assumed that 1 wt.% water was released from an all-melt magma body whose volume was 90 km^3, how many kilograms of SO_2 was released? (Use 2.2 g/cm^3 for the melt density.) Discuss your results.

4.6   A body of silicic melt that contains 5 wt.% water and has a density of 2.2 g/cm^3 occupies a volume of 1 km^3 at 1000°C in a crustal magma chamber. What is the volume of melt plus exsolved water at the surface (1 atm), assuming all 5 wt.% water exsolves and T remains 1000°C? What is the ratio of the volume of steam to melt at the surface? What is the ratio if $T = 700°C$ upon eruption? What is the ratio if only 4 wt.% water has exsolved in a nonequilibrium system at 1 atm and 700°C? Discuss your answers in terms of the gigantic "cauliflower" clouds of ash and steam that typically develop over exploding volcanoes.

Crystal-Melt Equilibria in Magmatic Systems

FUNDAMENTAL QUESTIONS CONSIDERED IN THIS CHAPTER

1. How are crystal-melt equilibria portrayed in phase diagrams?

2. What do phase diagrams tell about crystallization and melting in magma-rock systems as intensive variables change?

3. How are chemical variations created in evolving melts and in associated crystalline products during crystallization of magmas?

4. How do stabilities of major rock-forming minerals depend on intensive variables in magmas?

INTRODUCTION

Knowledge of the nature of chemical interactions between coexisting melt and crystals in magmas as intensive parameters change is essential to an understanding of how magmas are generated from solid rock at their sources, how magmas evolve and crystallize during ascent and cooling, and how the composition of an igneous rock can provide information on these parameters. This chapter deals with *equilibrium* relations between major rock-forming minerals and their associated melts in simple model and more complex multicomponent systems under controlled conditions. Such equilibria indicate what is possible and impossible in real magmas and serve as guides for their behavior.

Crystal-melt equilibria are represented in phase diagrams that have been determined in the laboratory for hundreds of systems of a wide range of composition at crustal and upper mantle pressures and temperatures.

Useful summaries and discussions of phase diagrams and crystal-melt equilibria can be found in Ernst (1976), Morse (1980), and Hess (1989).

✳5.1 PHASE DIAGRAMS

Phase diagrams were introduced in Section 3.3 as graphical models that portray states of stable equilibrium in terms of the intensive variables P, T, and X. Though not explicitly shown, any phase diagram implies the following (Figure 3.7):

1. A stable state represented in a stability field has the lowest possible Gibbs free energy or chemical potential for the values of the intensive variables delineating the field.

2. A boundary line between stability fields of different phases is the locus of points representing intensive parameters for which the phases are in equilibrium.

Because of experimental constraints and the limitations of a two-dimensional page of paper, phase diagrams generally portray only two variables, either P-T, P-X, or T-X. Three or more variables can be represented in special projections.

5.1.1 Phase Rule

A useful formulation that aids in the interpretation of phase diagrams is the **phase rule,** which provides an inventory of the number of phases, components, and degrees of freedom in a system at equilibrium. For definitions and examples of phases refer to Sections 3.3 and Figure 4.1 and for components Section 3.4.1. The **degrees of freedom,** or the **variance,** of a system is the number of independent intensive variables that must be specified to characterize the state of equilibrium fully.

A phase rule can easily be formulated for a system of only one component, $C = 1$, such as the silica system shown in *P-T* space in Figure 5.1. In this one-component system, as in most two-dimensional phase diagrams, the areas, lines, and points have special significance and meaning, as follows:

1. **Stability fields** are areas over which a phase or assemblage of phases is stable.

2. **Equilibrium boundary lines** lie between stability fields and represent values of intensive variables where phases in adjacent stability fields coexist stably.

3. **Triple points** are where equilibrium boundary lines meet. All phases in adjacent stability fields coexist in equilibrium.

Note the descending dimensionality—2, 1, and 0—of these three geometric elements. In Figure 5.1, one crystalline phase, such as α-quartz, has the lowest possible Gibbs free energy of any polymorph for the range of P and T within its labeled field. In this one-phase field of α-quartz, both P and T can be varied independently over a range of values without affecting the state of equilibrium or the stability of α-quartz. Therefore, in this one-phase stability field the state of equilibrium is **divariant,** or has two degrees of freedom, $F = 2$. Although stable throughout this field, the properties of the α-quartz, such as its exact density, vary slightly with respect to P and T. Along any boundary line between their individual stability fields, two phases, say, liquid and coesite, are in equilibrium. Their free energies are equal. Only one variable, P or T, may be independently changed along a boundary line representing the equilibrium between the two phases. In selecting an arbitrary value for P, say, 60 kbar, there is no freedom in what T may be: It is 2650°C. For the free choice of the

value of one intensive variable, P, then, T is uniquely fixed, or vice versa. This equilibrium is **univariant,** or the degrees of freedom, $F = 1$. Where three phases coexist in equilibrium at a triple point, say, at just over 2400°C and about 44 kbar, where liquid, coesite, and β-quartz coexist, the equilibrium is **invariant:** there are no degrees of freedom $F = 0$; there is no freedom in either P or T; they are unique for the three-phase equilibrium of liquid, coesite, and β-quartz. In summary, it appears the phase rule for a one-component system is $F = 3 - \Phi$ where Φ is the number of phases present; for each additional phase coexisting in equilibrium, the variance of the system decreases by 1.

In formulating a phase rule for multicomponent systems at equilibrium, we first take an inventory of the *number of possible variables*. These are generally P, T, and X. If other factors influence states of equilibrium, they must be added to the list. The composition of each equilibrium phase is denoted by the mole fractions of its components, X_i, X_j, X_k . . . (Section 3.4.1). But as the sum of these mole fractions is 1, we actually only have $(C - 1)$ independent mole fractions as variables for each phase, where C is the number of components. So, if the total number of phases is Φ, then the total number of compositional variables is $\Phi(C - 1)$. Therefore, the overall total number of possible variables in the system at equilibrium is $2 + \Phi(C - 1)$, where the 2 represents P and T.

We next consider *how many relations or equations* exist to evaluate the $2 + \Phi(C - 1)$ variables in the system. At equilibrium, the chemical potential of a particular component, i, must be the same in every phase, a, b, c, . . . in the system in which the component occurs, or

$$\mu_i^a = \mu_i^b = \mu_i^c = \cdots$$

and similarly for component j

$$\mu_j^a = \mu_j^b = \mu_j^c = \cdots$$

and so on, for each component. Hence, for each component there are $\Phi - 1$ equations, and for the entire system, $C(\Phi - 1)$ equations. These $C(\Phi - 1)$ equations reduce the number of total variables so that the variance of the system, that is, the difference between the number of possible variables and the equations defining them, is

5.1 $F = [2 + \Phi(C - 1)] - [C(\Phi - 1)]$ or

   $F = 2 + C - \Phi$

This is the Gibbs **phase rule.** For a one-component system the phase rule becomes $F = 2 + 1 - \Phi = 3 - \Phi$, as earlier. Remember that the number 2 in the phase rule refers to the variables P and T. If P (or T) is fixed, or constant, then the phase rule becomes $F = 1 + C - \Phi$, and if both P and T are fixed, then $F = C - \Phi$.

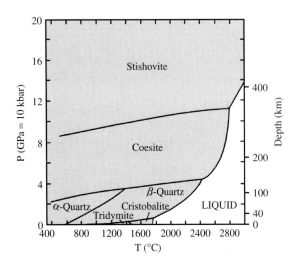

5.1   One-component silica phase diagram. Stability fields of crystalline polymorphs are shaded. (Redrawn from Swamy et al., 1994.)

The phase rule tells us the maximum number of intensive variables, F, that may independently vary without changing the number of phases, Φ, in a system of C components at equilibrium. A knowledge of the phases present at equilibrium is necessary to determine C and to apply the rule, from which the variance can then be determined. For example, as a result of the metastable persistence of some phase into the stability field of another phase or phase assemblage, or of the new growth of a metastable phase where it shouldn't form, the phase rule might be $F < 0$. This negative variance reflects an excess number of phases in the system and indicates a state of disequilibrium.

✱5.2 MELTING OF A PURE MINERAL AND POLYMORPHISM

5.2.1 Volatile-Free Equilibria

Additional useful concepts can be learned from the one-component silica phase diagram (Figure 5.1). The melting curve is equivalent to the freezing or crystallization curve if T is decreasing; this curve lies between the two fields where the liquid and crystals are in a state of dynamic equilibrium. This means, that for any combination of P and T along the curve, atoms are organizing from the liquid into crystals at the same rate that other atoms from dissolving crystals are being added to liquid. On this curve, there can be *any proportion* of the two phases—a bit of liquid and lots of coesite, or vice versa; provided they coexist in any proportion, equilibrium prevails. This suggests that the melting line is like the buffer curves in Figures 3.13 and 3.14. Provided phases coexist as indicated in the reaction represented by the line, that is, crystals = melt in Figure 5.1, or hematite = magnetite + O_2 in Figure 3.14, the equilibrium is univariant. There is only one degree of freedom among the intensive variables; choose some T and the other variable along the other axis of the diagram is uniquely determined.

Qualitatively, we can predict how a state of equilibrium will shift if P or T changes by applying Le Chatelier's principle and noting the inequalities in molar volumes and entropies, as explained in Section 3.3.1. States of smaller molar volume are more stable at higher P and the more disordered states of higher entropy are more stable at higher T. Slopes of equilibrium boundary lines in P-T space can be determined from the Clapeyron equation (3.13).

Silica polymorphs are widespread in the crust of the Earth and provide insight into pressure of crystallization. Tridymite and cristobalite are restricted to low-P, high-T volcanic environments. Only rarely is the high-P silica polymorph coesite found in crustal rocks, but it, as well as stishovite, can occur in silica-rich meteorite-impact rocks where transitory high P prevailed.

P in the mantle is adequate to stabilize coesite and stishovite, but the low silica activity in most mantle systems precludes crystallization of any silica polymorph.

5.2.2 Melting of a Pure Mineral in the Presence of Volatiles

Volatiles, especially water, are virtually always present in magmatic systems and can have a profound impact on their phase relations, including how they melt.

$NaAlSi_3O_8$-H_2O System. The system has served as a model of granitic magmas since the 1930s. Without water, dry albite melts along a positively sloping line in P-T space (Figures 5.2 and 3.8). Melting shifts to higher T at higher P as predicted from Le Chatelier's principle and the Clapeyron equation (3.13).

Addition of water to the system so it is water-saturated drastically lowers the T of melting by as much as 300°C at 2 kbar and 500°C at 10 kbar, compared to that of the dry system (Figure 5.2). Addition of water to the melt lowers the activity of the albite

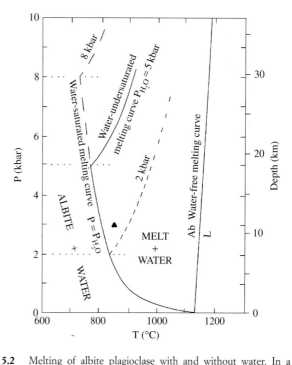

5.2   Melting of albite plagioclase with and without water. In a dry, water-free system the virtually straight melting line (line labeled Ab/L) has a steep positive slope. In contrast, the melting curve in a hydrous $NaAlSi_3O_8$ system in which $P_{H_2O} = 5$ kbar has a negative slope where water-saturated up to 5 kbar and then turns to a positive slope in the water-undersaturated region above 5 kbar where $P_{H_2O} < P$. For a $NaAlSi_3O_8$-H_2O system in which $P_{H_2O} = 8$ kbar the negatively sloping water-saturated melting curve (long-dashed line) extends up to $P = 8$ kbar at which pressure it turns positive. The water-undersaturated part of the melting curve in a hydrous system in which $P_{H_2O} = 2$ kbar is shown by the short-dashed line. (Redrawn from Burnham and Davis, 1974; Boettcher et al., 1982.)

component, but this reduction can be compensated by reducing the T until finally the activity becomes unity and albite crystallizes. A similiar depression of the freezing T of water occurs when ethylene glycol, or antifreeze, is added to it, a phenomenon familiar to any owner of an automobile who wants to prevent water in the radiator from freezing in cold climates. Another example is the depression of the freezing T of seawater, relative to pure water, because of the dissolved salt. But dilution cannot be the complete explanation, as can be appreciated from the following data. At 2 kbar, a water-saturated $NaAlSi_3O_8$ melt contains about 6.4 wt.% H_2O, which is equivalent to about 20 mole % H_2O because of the large difference in the weight per mole of water and albite. In contrast, addition of 6.4 wt.% SiO_2 to $NaAlSi_3O_8$ lowers the melting T only 20°C at 2 kbar (Burnham, 1979). Therefore, the more substantial freezing point depression due to addition of water is related to the way it dissolves in a melt, breaking Si-O polymers.

The depression of the water-saturated melting curve at increasing P is a general phenomenon in silicate mineral systems.

Figure 5.2 shows the effects of decompression on an ascending water-saturated melt. For example, a water-saturated melt (filled triangle) at 3 kbar and 850°C will track along an adiabatic decompression curve, experiencing only a slight drop in T. The system will become crystalline at $P < 2$ kbar.

Suppose there is only sufficient water to saturate the melt at, say, 5 kbar. In this system, the water pressure, $P_{H_2O} = 5$ kbar, and so for $P > 5$ kbar the system is water-undersaturated and the melting curve will track off on a positive slope in Figure 5.2.

The depression of the melting T resulting from dissolved CO_2 is substantially less than that of water at the same P because of its lower solubility in the melt. A mixture of H_2O and CO_2, however, depresses the fluid-saturated melt curve significantly. Dissolved fluorine depresses the melting T more than CO_2 because of its greater solubility in the melt.

❋5.3 PHASE RELATIONS IN BINARY SYSTEMS

In a one-component phase diagram, two relevant intensive variables—P and T—can be conveniently represented on a two-dimensional sheet of paper. The compositions of all phases are fixed; there is no variation in their composition. In two-component, or **binary** (and more complex), systems a choice must be made as to which of the intensive variables to hold constant in representing equilibria in two dimensions. Virtually all phases in rock-forming systems have variable composition so most binary diagrams portray the proportions of two components, either on a weight or a mole basis,

along the horizontal axis. The vertical axis can be P, with T fixed in an isothermal P-X diagram, or it can be T, with P fixed in an isobaric T-X diagram. For a complete understanding of the system, more than one of these diagrams has to be examined or special projection devices employed.

5.3.1 Basic Concepts: $CaMgSi_2O_6$ (Di)-$CaAl_2Si_2O_8$ (An) System at P = 1 atm

The binary $CaMgSi_2O_6$-$CaAl_2Si_2O_8$ system was one of the first elucidated, in 1915, by N. L. Bowen. It serves a twofold purpose: first, as an introduction to reading binary phase diagrams; second, as a demonstration of some basic concepts of crystal-melt equilibrium in a simple model "basalt" magma in which plagioclase and pyroxene crystallize. For brevity, the $CaMgSi_2O_6$ component is sometimes designated as Di and the $CaAl_2Si_2O_8$ component as An. In this binary system, the chemical formulae of the two components are the same as the chemical formulae of crystalline phases. However, *in many systems, formulae of phases are different from formulae of the components.*

<u>Crystallization and Liquidus Relations.</u> This binary isobaric T-X diagram can be introduced by means of the concept of freezing (melting) point depression as another component is added. In Figure 5.3, T_f is the

---

Special Interest Box 5.1 N.L. Bowen, pioneer experimental petrologist

It has been said that "just as modern biology would be unthinkable without the overarching genius of Darwin, modern igneous petrology would be unthinkable without the overarching genius of Norman Levi Bowen" (Young, 1998, p. 253). Bowen not only attacked many of the core problems in petrology but did so decades ahead of anyone else and with such insight that subsequent investigations have only built upon his pioneering work. Added to his experimental acumen was a gift for fluency in writing that enabled him to broadcast the results of a new discipline in a way that almost any geologist could understand. His 1928 book, *The evolution of the igneous rocks*, served as the fundamental textbook for half a century and is still widely used. His determinations of fundamental phase equilibria in the plagioclase, pseudobasalt, and granite systems are described in this chapter. Other contributions and the debates they generated are referred to in subsequent chapters. The biography by Young (1998) is especially interesting, not only about Bowen, but about igneous petrology in general for Bowen occupied a central position in this discipline for much of the first half of the 20th century.

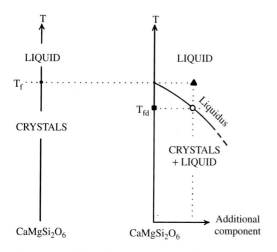

5.3   Depression of the melting T of diopside due to addition of another component to the system.

freezing or crystallization T of a pure $CaMgSi_2O_6$ melt (or melting T of pure crystals of diopside) in a one-component system. Suppose at T_f, where melt (liquid) and diopside crystals coexist in equilibrium, another component, which does not form a solid solution with diopside, such as $CaAl_2Si_2O_8$, is added to and dissolved in the melt. This dilutes the melt in the original single component, $CaMgSi_2O_6$, and depresses the freezing T of the now-two-component liquid to, for example, T_{fd}. With the T of the system still at T_f the diopside crystals have dissolved into the melt and the system consists only of the two-component melt (represented by the filled triangle in Figure 5.3). At T_{fd} the melt (represented by an open circle), which is in equilibrium with crystals of pure diopside (filled square), contains all of the added component, $CaAl_2Si_2O_8$, plus some additional $CaMgSi_2O_6$. Adding still more of the second component depresses the freezing T still lower as the composition of the melt shifts farther to the right along the downward inclined dashed line. This line representing the locus of depressed freezing (melting) T points as more of the second component is added to the melt is called the **liquidus.**

Figure 5.3 is a part of the left-hand side of Figure 5.4, which is the complete T-X phase diagram of the $CaMgSi_2O_6$-$CaAl_2Si_2O_8$ system at a fixed $P = 1$ atm. On the vertical T axis on the left side of Figure 5.4 is the freezing T of a pure $CaMgSi_2O_6$ melt at 1392°C and on the right T axis is the freezing T of pure $CaAl_2Si_2O_8$ melt at 1553°C. Between these two pure end members, melts of mixtures of the two components freeze at declining temperatures, as indicated by the two oppositely sloping liquidus line segments that merge at point E, called the **eutectic point.** Mixtures of the two components are completely liquid (a crystal-free melt) at temperatures above the **liquidus;** this is a second meaning of this term.

To understand more of the vast amount of information in this binary diagram, especially in the two-phase fields of liquid + diopside and liquid + anorthite, it is necessary to get acquainted with some special lines and another rule. An **isopleth** is a line of constant composition in terms of the components in the diagram. In Figure 5.4, it is a vertical line drawn, for example, at An_{90}, that represents 90 wt.% of $CaAl_2Si_2O_8$ and 10 wt.% of $CaMgSi_2O_6$. An **isothermal line,** or **isotherm,** is a line of constant T; it is a horizontal line drawn, for example, at 1400°C. The intersection of these two lines is a point (filled triangle) representing the T (1400°C) and *bulk* chemical composition (An_{90}) of a system that consists of two phases—liquid (melt) and pure anorthite crystals. But what is the composition of the liquid, and what is the modal proportion of liquid and crystals? To answer these questions, note that the isotherm intersects the liquidus line at a point, designated L and shown by an open circle, and also intersects an isopleth through pure An at a point, designated S (for solid) and shown by a filled square. The line segment of the isotherm connecting points L and S is called a **tie line;** it connects two stably coexisting phases at points representing their compositions in terms of the system components. Point S obviously is equivalent to the composition of anorthite crystals, $CaAl_2Si_2O_8$. Point L represents the composition of the coexisting melt at 1400°C, found by drawing a vertical isopleth through point L to the base of the diagram and reading An_{62}. Thus, the **liquidus** is the locus of points representing the composition of liquid coexisting with a solid phase (or phases) at a particular T; this is a third meaning of the term. The compositions of crystals and liquid just determined are analogous to the mineralogical composition of a rock.

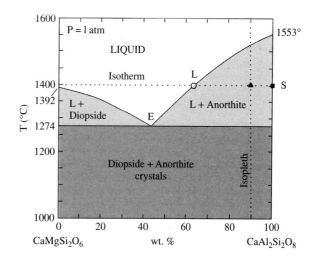

5.4   Binary system $CaMgSi_2O_6$(Di)-$CaAl_2Si_2O_8$(An) at 1 atm. Diagram is simplified from Yoder (1976), who discusses complications due to the fact that the pyroxene is not pure $CaMgSi_2O_6$ but contains Al in solid solution.

What about the modal composition, or the proportion of the two phases, in the system posed as the second question? This is solved by using the lever rule, which can be formulated as follows: The isothermal tie line passing through the bulk composition point (filled triangle) has end points S and L, representing coexisting solid and liquid phases at 1400°C, respectively. Imagine this tie line to be a mechanical lever carrying masses S and L and resting on a fulcrum, the filled triangle, in Figure 5.5. For equilibrium (as on a balanced "teeter-totter" in a children's playground), the weight fraction of the solids, S, multiplied by their lever arm, y, must equal the weight fraction of liquid, L, multiplied by its lever arm, x, or $Sy = Lx$. But as $S + L = 1$, $Sy = (1 - S)x = x - Sx$ or

$$5.2 \quad S = \frac{x}{(y + x)}$$

which is the **lever rule.** Thus, we measure on the diagram the distances x and y, say, in millimeters, and calculate the ratio, in this case 0.72, which is the weight fraction or proportion of solids, anorthite crystals, in the system at 1400°C whose bulk chemical composition is An_{90}. This fraction multiplied by 100 is the weight percentage (wt.%) crystals, 72%; the remaining 28 wt.% is melt. To prevent possible confusion, note that in this example the bulk composition point lies closer to the solids (crystals) point, S, than it does to the liquid point, L; accordingly, there is a greater proportion of crystals than liquid.

An instructive exercise is to track the crystallization of a melt, say An_{90}, from above the liquidus. As the melt cools, the first crystals, of pure anorthite, appear at 1520°C. As T decreases, and equilibrium prevails, more anorthite precipitates, accordingly driving the coexisting remaining **residual liquid** down the liquidus toward more Di-rich compositions. These increasingly more Di-rich residual liquid compositions together with changing modal proportions of liquid and anorthite crystals can be tracked by drawing a series of isothermal tie lines, dropping isopleths to the composition axis from the liquid composition end point of the tie line that lies on the liquidus, and applying the lever rule to determine the modes. At 1274°C, as more heat is withdrawn form the system, its T remains fixed at this **eutectic temperature** T_e, until all of the liquid of **eutectic composition,** E, has crystallized to pure anorthite and diopside in 90/10 weight ratio.

Figure 5.6 shows thermal relations in the cooling system above and below T_e. The T in the system drops continuously as heat is lost where $T > T_e$. But, at $T_e =$ 1274°C, no change in T occurs during further loss of heat from the system because of the compensating contribution of the latent heat, or enthalpy, of crystallization of diopside (ΔH_m^{Di}, Figure 3.3) and of anorthite, ΔH_m^{An}. Only after all of the melt has been crystallized into the diopside and anorthite and their latent heats dissipated can T once again fall.

The eutectic point, E, is the only invariant point in the isobaric binary diagram. For fixed P, from the phase rule, $F = 1 + C - \Phi = 1 + 2 - 3 = 0$. This means that every intensive variable in the system is fixed at this point: P (given as 1 atm), T (1274°C), and the composition of all three coexisting phases, melt $X_L = An_{42}$ (Di_{58}) and pure crystals of anorthite and diopside. This unique eutectic point lies at the juncture of four stability fields:

1. Liquid

2. Diopside + anorthite, below an isothermal horizontal line at 1274°C

3. Liquid + diopside, a crudely triangular field on the Di-rich side of the diagram

4. Liquid + anorthite, another crudely triangular field on the An-rich side of the diagram

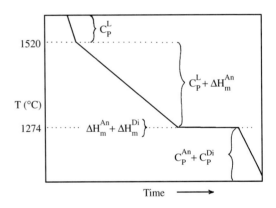

5.6 Hypothetical cooling history (*T*-time relations) for a model magma An_{90} in the binary system $CaMgSi_2O_6$-$CaAl_2Si_2O_8$ losing heat to its surroundings. Compare Figure 3.3. The time rate of cooling (loss of thermal energy) of the magma is assumed to be constant, that is, heat loss is proportional to time. The crystal-free melt falls relatively rapidly in temperature above the liquidus ($T = 1520°C$) because only its specific heat, C_P^L, must be dissipated. During its excursion through the two-phase region ($T = 1520 - 1274°C$), the crystallizing magma cools more slowly because the latent heat of crystallization (the enthalpy of melting, ΔH_m^{An}) must also flow out of the system. At the eutectic T, 1274°C, the invariant system remains fixed in T for a period as the latent heat of crystallization of both anorthite and diopside is dissipated into the surroundings. In well-insulated, intrusive natural magma systems, the latent heat is dissipated very slowly and therefore crystallization takes a long time.

5.5 Lever rule. See text for explanation.

In the isobaric one-phase field consisting only of melt above the liquidus, divariant equilibria prevails, $F = 1 + C - \Phi = 1 + 2 - 1 = 2$. Two degrees of freedom means that T and X must both be specified in order to know all about the system: that is, to fix a point in the field by an intersecting isopleth and isotherm. In the isobaric two-phase fields of the binary system, univariant equilibrium prevails, $F = 1 + C - \Phi = 1 + 2 - 2 = 1$. For example, if anorthite + liquid are stable in a system, specification of one variable, say T, uniquely fixes the composition of the liquid, the only other intensive variable. *Modal proportions of phases do not constrain intensive variables in equilibria, but chemical compositions of coexisting phases do.* This important principle is the basis of mineral geothermometers and geobarometers discussed in Sections 3.5.5 and 5.8.

The fourth and final significance of the **liquidus** is that it is a saturation line. At a T of, say, 1350°C in Figure 5.4, melts whose compositions lie between An_{20} and An_{55} are unsaturated in any crystalline phase: No crystals coexist with these melts because their concentrations have not exceeded their solubility (activities <1). However, melts more enriched in $CaAl_2Si_2O_8$ than An_{55} are oversaturated in this component, and accordingly crystals of anorthite coexist at equilibrium with these melts.

The $CaMgSi_2O_6$-$CaAl_2Si_2O_8$ system serves as a very simplified model of how basaltic magmas crystallize. Magmas with relatively large concentrations of $CaAl_2Si_2O_8$ precipitate a calcic plagioclase as the **liquidus phase,** the first crystalline phase to appear at the liquidus with decreasing T from a wholly liquid state. After a calcic plagioclase has grown over a range of T, a diopside-rich pyroxene then co-precipitates at the eutectic. Such basaltic magmas would be expected to contain high-T phenocrysts of plagioclase in a finer matrix that includes pyroxene. Basaltic magmas that contain more $CaMgSi_2O_6$ have pyroxene phenocrysts and plagioclase only occurs in the matrix.

<u>Melting and Solidus Relations</u>. Melting accompanying increasing T is the reverse of crystallization, provided equilibrium is maintained.

A perhaps unexpected result of heating an aggregate of diopside and anorthite crystals, *in any modal proportion,* is that melting occurs at the same $T = 1274$°C, the eutectic temperature, and yields the same unique composition melt, An_{42} (Di_{58}), the eutectic composition. This unique melting at the eutectic point for any anorthite-diopside "rock" can be visualized by drawing a series of horizontal isotherms at increasing T in the anorthite + diopside stability field. No melting occurs until the $T = T_e = 1274$°C. In this binary system, this 1274°C isotherm is a **solidus,** at all temperatures below which any mixture of the components consists only of crystalline solids, provided equilibrium prevails.

Although any proportion of diopside and anorthite crystals begins to melt at 1274°C, and the composition of the first "drop" of melt is An_{42}; continued melting at higher T follows one of two paths. Model "rocks" that contain more than 42 wt.% anorthite crystals cannot rise in T above 1274°C until all of the diopside in the rock has melted. Input heat is absorbed in the latent heat of melting of diopside, ΔH_m^{Di}, at constant T (compare Figure 5.6). Slightly above 1274°C only crystals of anorthite remain unmelted, and these are in equilibrium with a liquid that is just slightly more enriched in $CaAl_2Si_2O_8$ than the eutectic composition. After the diopside is completely melted, more heat added to the system simply results in an increase in T proportional to the heat capacity of the system of melt plus anorthite. As T increases, and if equilibrium prevails, an increasing amount of anorthite dissolves in the melt, increasing the concentration of $CaAl_2Si_2O_8$ in the liquid. These changing liquid compositions and modal proportions of melt and anorthite crystals can be tracked by drawing a series of isothermal tie lines, dropping isopleths to the composition axis from the liquid composition end point of the tie line that lies on the liquidus, and applying the lever rule to determine the modes. Finally, at 1520°C, the An_{90} bulk-composition isopleth intersects the liquidus, the proportion of solids in the system by the lever rule is now zero, or melting is complete. The system is entirely liquid at $T > 1520$°C.

5.3.2 Mg_2SiO_4-SiO_2 System at 1 atm

The phase diagram (Figure 5.8) of this simple, yet significant system, elucidated by Bowen and Anderson (1914), has a eutectic point, E, similar to that in the $CaMgSi_2O_6$-$CaAl_2Si_2O_8$ system. Melts lying between about 61 and 70 wt.% silica crystallize in similar manner to melts in that system. The Mg_2SiO_4-SiO_2 system provides valuable insight about magma generation and evolution and is especially important in demonstrating phase incompatibility, reaction relation, contrasts between equilibrium and fractional crystallization, incongruent melting, and liquid immiscibility.

<u>Phase Incompatibility</u>. Countless observations of magmatic rocks indicate that certain minerals never occur together, except in rare accidental circumstances. There can be two reasons for this. Because of the way natural magmas originate and evolve, minerals that typically precipitate at highest temperatures from the least-evolved systems are seldom, if ever, found to coexist with minerals that precipitate at lowest temperatures from highly evolved magmas. For this reason, high-T forsterite and low-T albite are unlikely associates, even though they can coexist stably. In contrast, nepheline-quartz and forsterite-quartz are pairs of

Advanced Topic Box 5.2 G-T-X sections for the binary system CaMgSi$_2$O$_6$ (Di)-CaAl$_2$Si$_2$O$_8$ (An) at 1 atm

It may be recalled from Section 3.3 that a phase diagram portraying stable equilibria in terms of T and X or any other combination of intensive variables implies a minimum of the Gibbs free energy, G, or chemical potential for the stable phases. Thus, the P-T diagram in Figure 3.6 indicates which phases are stable as a function of P and T from the G-P-T diagram in Figure 3.7. Similarly, the binary T-X diagram for the CaMgSi$_2$O$_6$-CaAl$_2$Si$_2$O$_8$ system in Figure 5.4 indicates which phases are stable as a function of T and X from the sequence of G-X diagrams at different temperatures in Figure 5.7. This figure is presented here to emphasize that stable phases and phase assemblages have minimal free energies or chemical potentials, even though they are not explicitly shown in a phase diagram plotted in terms of intensive variables. The presentation here follows Anderson (1996).

In Figure 5.7, the G of a physical mixture of diopside and anorthite crystals with no mutual solid solubility is simply a straight line between G_{Di} and G_{An}. On the other hand, G for any solution, whether liquid silicate (melt) or solid crystals, is a convex-downward curve, or loop (Figure 3.11). Because $dG/dT = -S$, the G straight line representing the mix of crystals and the G loop representing liquid both move upward to greater G with decreasing T, but the G loop moves more, for a given increment in T, because the entropy of a liquid is greater. At T_1 above the liquidus, liquid of any composition between the two end-member components CaMgSi$_2$O$_6$ (Di) and CaAl$_2$Si$_2$O$_8$ (An) has a lesser G than a mixture of diopside and anorthite crystals; hence, the liquid is more stable and the crystal mix is metastable. At T_5 below the solidus, the opposite is true. At T_2, G for both liquid and crystals has increased, but more for the liquid, so that the G of liquid CaAl$_2$Si$_2$O$_8$ has become greater than that of anorthite crystals, but the G of liquid CaMgSi$_2$O$_6$ remains less than that of diopside crystals. From Figure 3.11, the tangent to the free energy curve for a solution intercepts the An vertical axes at the chemical potential of the An component in the solution, $\mu_{An}^{liquid} = \mu_{An}^{crystals}$. This equality means that the melt on the liquidus at $T = T_2$ in the T-X phase diagram is in equilibrium with crystals of pure anorthite. As T decreases below T_2, that tangent point slides along the loop toward greater Di concentrations because the loop is rising relative to the $G_{An}^{crystals}$, and at some $T < T_2$ but $> T_3$ the left end of the loop coincides

with $G_{Di}^{crystals}$, that is, at the melting T of pure diopside crystals. At T_3 the tangent relation described holds for both components. As T decreases below T_3 the two tangent points converge and finally at $T = T_e$ there is only one tangent because μ_{An} is the same in all three phases and a different value of μ_{Di} is the same in all three phases.

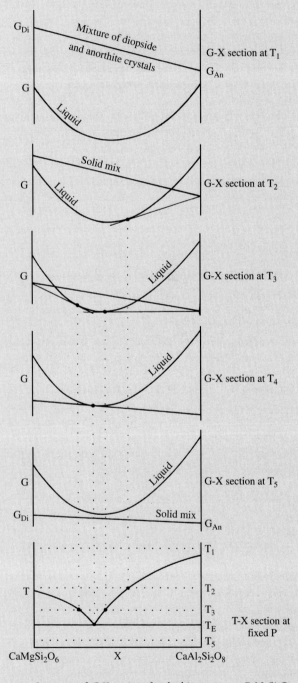

5.7 Sequence of G-X sections for the binary system CaMgSi$_2$O$_6$ (Di)-CaAl$_2$Si$_2$O$_8$ (An) at 1 atm. Top five frames are G-X sections for five temperatures. Lowest frame is the T-X section (Figure 5.4). (From Anderson, 1996.)

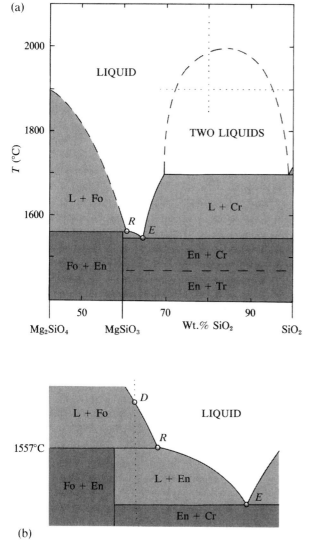

5.8 Binary system Mg_2SiO_4-SiO_2 at 1 atm. Stability fields that contain melt (silicate liquid) plus a crystalline phase are lightly shaded; stability fields that contain two crystalline phases are shaded dark. Proportions of the silica component along the bottom of the diagram are with respect to the binary system MgO-SiO_2, of which this is a part. Fo, forsterite; En, enstatite; Cr, cristobalite; Tr, tridymite; L, liquid. Pure forsterite contains 42.7 wt.% SiO_2, enstatite 59.85, point R about 61, and point E about 65. Lower part of the figure is an enlarged and slightly distorted region of the phase diagram in the vicinity of the peritectic reaction point, R. (Redrawn from Bowen and Anderson, 1914.)

minerals that are thermodynamically unstable. Rare magmatic rocks might contain both magnesian olivine and quartz, but one or both will likely be corroded, or partially resorbed, indicating a state of disequilibrium that was frozen into the rock. The Mg_2SiO_4-SiO_2 system confirms that forsterite and quartz cannot coexist in equilibrium at 1 atm (or at any P, for that matter) but react to form more stable, intermediate composition enstatite.

Reaction Relation. Consider the crystallization behavior of a melt cooling from 1900°C and containing 50 wt.% silica, a concentration found in basalts. The 50 wt.% isopleth intersects the liquidus at about 1815°C, at which T forsterite crystals begin to precipitate. After further cooling to 1558°C, the system consists, at equlibrium, of about 60 wt.% forsterite crystals and 40 wt.% melt, from the lever rule. At 1556°C, the system is below the solidus and consists of about 58 wt.% forsterite crystals and 42 wt.% enstatite crystals. At 1557°C, there are stably coexisting forsterite, enstatite, and a liquid, R, slightly more silica-rich (about 61 wt.%) than enstatite crystals (59.85 wt.%). To explain the disappearance of 2 wt.% forsterite between 1558°C and 1556°C, these crystals must be **resorbed,** or dissolved into the melt, R, at 1557°C according to the reaction

5.3 $Mg_2SiO_4 + SiO_2 = 2MgSiO_3$ + latent heat

forsterite   in melt   enstatite

This is an example of a **reaction relation** in which crystals of one composition react with melt of another composition yielding crystals of a third composition.

At 1557°C, the isobaric equilibrium in this binary system of the three phases—forsterite, enstatite, and liquid R—is invariant. Point R on the liquidus is called a **peritectic.**

Crystallization of a melt which contains 59.85 wt.% silica and 40.15 wt.% MgO—the composition of enstatite—further illustrates what happens in crystal-melt equilibria at the peritectic reaction point, R. As this unique melt cools, forsterite begins to crystallize at about 1600°C and continues to precipitate until at 1557°C it is resorbed into the melt R, yielding at 1558°C only enstatite as the sole final phase in the cooling system. The system is "stuck" isothermally at 1557°C until all of the forsterite and melt are consumed in the reaction to produce enstatite before the T of the system can decrease as heat is continually withdrawn from it. Note the thermal similarity with eutectic behavior (Figure 5.6).

Reaction relations of the sort just described are universal in natural magmas and are a factor contributing to the great compositional diversity of magmatic rocks (Figure 2.4). The following paragraphs begin to explain how and why this is so.

Equilibrium versus Fractional Crystallization. Magmas crystallize in some manner between two ideal end members:

1. **Perfect equilibrium crystallization,** in which crystals continually react and reequilibrate completely with melt as _P-T-X_ conditions change. Crystal-melt reaction relations are reversible at any point in the process.

2. **Perfect fractional crystallization,** in which crystals are immediately isolated, removed, or fractionated from the melt as soon as they form so that no crystal-melt reaction relation occurs as *P-T-X* conditions change. Reactions are irreversible.

Perfect equilibrium crystallization obviously has stringent requirements. At any particular set of values of intensive variables, each phase must be homogeneous and uniform in composition; the melt must everywhere be exactly the same composition. Rates of changes in intensive variables must be slower than the slowest kinetic process so the magma can keep up with the changes. Changes in intensive variables can be reversed at any time and a former state of equilibrium involving unique melt and crystal compositions can be restored or recovered. There can be no physical separation of phases. The total composition of the system cannot change; the initial composition of the system must be maintained, so it is strictly a closed system. Every part communicates with every other part. This ideal reversible process of equilibrium crystallization may only be approached, rarely, in deep plutonic systems. The examples given previously of crystallization along the 50 and 59.85 wt.% silica isopleths described equilibrium crystallization.

Fractional crystallization is quite different. It is realized to varying degrees in virtually all magmas because reaction relations between crystals and melt are incomplete as a result of sluggish kinetics. In the extreme, ideal case, no reactions whatsoever occur between liquid and crystals, once they have precipitated, during changing states of the system. To prevent reaction there must be a separation, isolation, or **fractionation** of melt from crystals. This fractionation can occur in one of three ways, or by combinations thereof:

1. It can occur by separating whole crystals and liquid as a result of differences in their densities or other dynamic processes in magmas.

2. Crystals and melt simply do not react because the kinetics are too slow. In the ever-present contest between changing intensive variables in magma systems and kinetically controlled equilibration between phases, the latter loses.

3. Fractionation can occur by growing an armoring or protective layer of another composition on the initial crystal, thereby effectively separating it from the liquid and precluding any reaction between them. This is a common and widespread means of fractionating crystals from coexisting melt in natural magmas. In a fractionating basaltic magma, modeled by the Mg_2SiO_4-SiO_2 system, stable enstatite might form at the interface between melt and forsterite, creating a **reaction rim** on the metastable forsterite at the peritectic. This overgrowth can develop during relatively rapid changes in intensive

variables so the magma system "tries to catch up" by precipitating the new stable phase before the metastable crystals have been eliminated by reaction with the melt.

Fractional crystallization can be illustrated by a thought experiment using the Mg_2SiO_4-SiO_2 system. Consider again a melt that contains 50 wt.% silica (Figure 5.8). At 1815°C, forsterite crystals begin to precipitate, but are immediately isolated or fractionated—by either 1. or 2.—from the liquid. In effect, the bulk composition of the system is all melt more silica-rich than 50 wt.% as the system has no "memory" of its initial composition. Further cooling results in further precipiation of forsterite, which is immediately isolated from the enclosing melt. Eventually, as fractionation continues, the residual melt—which may be considered to be the whole system that has "forgotten" about its previously precipitated crystals—reaches the peritectic, *R,* at 1557°C. Here, fractionation might be accomplished by forming reaction rims of stable enstatite around previously precipitated forsterite crystals as *T* decreases. Continued cooling results in further precipitation of enstatite, which is fractionated from the system. Ultimately, the residual melt reaches the eutectic, *E,* at 1543°C, where enstatite and cristobalite (the silica polymorph that is stable at this *T*) coprecipitate. Crystallization is complete.

Important contrasts between equilibrium and fractional crystallization are immediately apparent, as follows:

1. For this particular composition (50 wt.% silica), the residual fractionated melt was not all consumed at the peritectic as it was in equilibrium crystallization.

2. The range of temperatures over which fractionation occurred is greater than that of equilibrium crystallization.

3. The compositional path of evolving residual melts portrayed on a variation diagram, the **liquid line of descent,** has a more extended range in a fractionating system than does the path defined by successive melts during equilibrium crystallization (Figure 5.9). However, by either mode of crystallization, **residual melts** after some crystallization has occurred in the Mg_2SiO_4-SiO_2 system are enriched in SiO_2 and depleted in MgO relative to the initial melts. This same liquid line of descent—depletion in MgO and enrichment in SiO_2—typifies many natural crystallizing multicomponent magmas.

4. The crystalline products of fractional crystallization are also more extended in their range of compositions than those that develop by equilibrium crystallization (Figure 5.9). The final crystalline product created by closed system equilibrium crystallization—forsterite plus enstatite—occurs in a

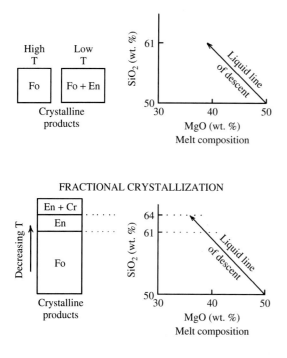

FRACTIONAL CRYSTALLIZATION

5.9 Crystals and liquids produced during equilibrium crystallization (top) and fractional crystallization (bottom) of an initial melt that contains 50 wt.% silica in the binary system Mg$_2$SiO$_4$-SiO$_2$ at 1 atm. In the equilibrium process, forsterite crystals appear in the initial stage of cooling and the final crystalline products are forsterite plus enstatite in a proportion whose bulk composition is exactly 50 wt.% silica. During fractional crystallization, the crystalline products, which are shown here as a hypothetical gravity accumulation, are forsterite, then enstatite, and finally enstatite plus cristobalite with decreasing T. The **liquid line of descent** on the MgO-SiO$_2$ variation diagram resulting from fractional crystallization is more extended in composition than from equilibrium crystallization. The increase in silica and decrease in MgO of the residual melt during either mode of crystallization displayed here are common in many natural magmas.

proportion such that the bulk composition, 50 wt.% silica, is exactly that of the initial melt. Fractional crystallization, in contrast, has yielded an additional crystalline phase, cristobalite. This final cristobalite + enstatite "rock" is silica-oversaturated, whereas the first crystals that precipitated—forsterite—manifest an undersaturated system.

In the model Mg$_2$SiO$_4$-SiO$_2$ system, forsterite and enstatite constitute a **reaction pair** that forms across the peritectic. In real fractionating magmas containing additional components, a reaction pair of olivine and orthopyroxene is manifested as a reaction rim of the latter on the former in a single grain. On a larger scale, this reaction pair might be manifested in layers of olivine-bearing rock overlain by orthopyroxene-bearing rock in a layered mafic intrusion (compare lower-left diagram in Figure 5.9). In evolving intermediate composition calc-alkaline magmas, a reaction re-

lation between clinopyroxene and melt produces hornblende. Incomplete reactions of this sort create hornblende reaction rims around anhedral, unstable clinopyroxene in dioritic rocks (Plate II). At lower T, hornblende may react with a more evolved melt to yield biotite. Two or more reaction pairs might develop in a fractionating multicomponent magma and constitute a **discontinuous reaction series.** Many different reaction series are possible in magmas, depending on their bulk chemical composition and intensive variables.

Incongruent Melting. Crystallization of a melt that contains 59.85 wt.% silica and 40.15 wt.% MgO—the composition of enstatite—was discussed to illustrate the principle of the reaction relation. In the reverse sense, if crystals of pure enstatite are heated at 1 atm, they are found to melt in an unexpected manner. In contrast to the familiar congruent melting of ice to liquid water of the same composition, **incongruent melting** of enstatite yields a liquid that is slightly more silica-rich *plus* forsterite crystals at its melting point of 1557°C (Figure 5.8 and reaction 5.3). As more heat is absorbed into the 59.85 SiO$_2$-40.15 MgO system, forsterite dissolves in the silicate liquid, and eventually, at about 1600°C, melting is complete, finally yielding a melt of the same composition as that of enstatite. Rather than having a unique melting T point at some particular P, as do albite, forsterite, diopside (Figure 3.8), and some other pure end-member minerals, pure enstatite has a melting T *range* through which the silicate liquid, which coexists with other crystals, has a different composition from the initial solid. As shown later, all solid solution minerals also melt incongruently.

It may now be realized that any "rock" mixture of diopside and anorthite crystals also melts incongruently (except for the one unique eutectic proportion) in Figure 5.4. The initial melt has a eutectic composition regardless of the proportion of diopside and anorthite crystals. The incongruent melting behavior of mineral solid solutions and rocks has profound implications not only for magma systems but for the whole Earth. Incongruent melting has produced the global-scale "differentiation" of the planet into a mantle and more silica-rich, Mg-poor crust over the course of its 4.5-Gy evolution. Small degrees of partial melting of the peridotitic mantle of basically olivine and pyroxene has generated more Si-rich, Mg-poorer partial melts forming the crust.

Liquid Immiscibility. In the Mg$_2$SiO$_4$-SiO$_2$ system a homogeneous high-T melt lying between 70 and 100 wt.% silica will split into two stable **immiscible liquids** upon cooling below the convex upward **solvus,** shown by a dashed line in Figure 5.8. For example, an initial liquid at 2000°C that contains 80 wt.% silica splits into two melts at 1900°C; one contains 72 wt.% silica and the other 96 wt.%. The less abundant, more silica-rich immiscible melt will form as drops in the mass of less siliceous liquid. Because of differing densities, the two melts could eventually segregate into contrasting horizontal layers in the gravity field. Subsequently, as the two immiscible melts cool below 1700°C, cristobalite precipitates from each, but in greatly differing amounts, found from the lever rule. It is a property of immiscible melts that the crystalline phases in equilibrium with each are the same, but in different modal proportions. Liquid immiscibility appears to be rare in natural multicomponent magmas.

*5.4 CRYSTAL-MELT EQUILIBRIA IN REAL BASALT MAGMAS

The two binary systems examined so far provide valuable insights into crystal-melt equilibria in simple, model mafic systems. The effect of additional chemical constituents, such as Al, Fe, and Na on the behavior of real mafic magmas has only been hinted. Other binary diagrams could be explored, as could three- and four-component (ternary and quaternary) diagrams, which are discussed at length in Morse (1980). Instead, at this point as a "reality check" we consider crystal-melt equilibria in two real basalt compositions. Although these two basalts are real rocks composed of many components, they are only two points in the wide compositional spectrum of all rocks (Figure 2.4). What has been gained over the simple binary model systems by an examination of real multicomponent systems is limited by their unique compositions.

5.4.1 Makaopuhi Basalt

Nature provided an especially instructive experiment for petrologists in the 1965 eruption of Kilauea, a volcano on the island of Hawaii, when lava partly filled the small preexisting Makaopuhi crater to form a lava lake. Soon after a hard crust developed on the lake, U.S. Geological Survey petrologists drilled into its still molten interior. Samples of magma were collected at various depths in the lake and temperatures were measured with a thermocouple. Photomicrographs of thin sections made of the quenched samples (Wright and Okamura, 1977) clearly show the sequence of precipitation of crystalline phases in this basaltic magma (Plate III, Figure 5.10, and Table 5.1). Each sample is presumed to represent an equilibrium assemblage (except as noted later) of crystals and melt at the indicated *T* near atmospheric pressure in a closed magma system.

The liquidus of the Makaopuhi basalt lies just above 1200°C and the solidus is at about 965°C. Throughout most of this 235°C range of crystallization between the

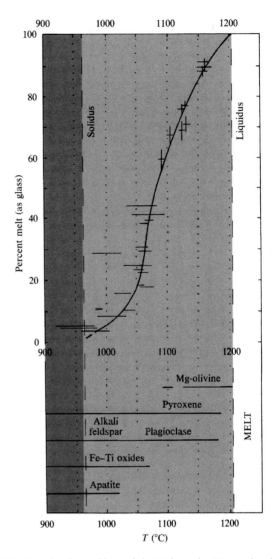

5.10 Crystal-melt equilibria of the Makaopuhi, Hawaii, basalt at about 1 atm. See also color photomicrographs, Plate III. Temperatures, measured in drill hole, and amounts of melt, determined in thin sections as glass, are indicated by horizontal line segments and crosses; length of line segments indicates uncertainty in measured values. Heavy curve is best-fit line to data showing how the amount of melt varies with *T*. Range of *T* over which each mineral precipitated from the melt is shown at the bottom. Approximate range of solid solution compositions from high to low *T* is Fo 82 to 76 in olivine, Mg/Fe 1.2 to 0.36 in clinopyroxene, and An 71 to 30 in plagioclase. (Redrawn from Wright and Okamura, 1977.)

liquidus and solidus, pyroxene and plagioclase solid solutions coprecipitate. In simple binary systems only one pure crystalline phase precipitates from a melt until it reaches a eutectic or peritectic, but with more components in a natural magma, two or more crystalline solid solutions can coprecipitate. The crystalline silicate phase at the liquidus (the liquidus phase) is a Mg-rich olivine that occurs as relatively large euhedral crystals (Plate IIIa). At about 1180°C, much smaller, somewhat

Table 5.1 Bulk Chemical Compositions (wt.%) and Modal Composition (vol.%) of Basalts Whose Melting Relations Are Shown in Figures 5.10 and 5.11 and Plate III (Makaopuhi). Data from Wright and Okamura (1977) and Green (1982).

	MAKAOPUHI	HIGH-AL OLIVINE THOLEIITE
SiO_2	50.24	49.93
TiO_2	2.65	1.34
Al_2O_3	13.32	16.75
Fe_2O_3	1.41	
FeO	9.85	11.40t
MnO	0.17	0.18
MgO	8.39	7.59
CaO	10.84	9.33
Na_2O	2.32	2.92
K_2O	0.54	0.37
P_2O_5	0.27	0.19
Total	100.00	100.00

MODE		
Olivine	5	
Pyroxene	51	
Plagioclase	30	
Fe-Ti oxides	9	
Glass	5	
Alkali feldspar	trace	

darker clinopyroxene and colorless plagioclase crystals begin to coprecipitate with olivine. With decreasing T, increasing amounts of these three crystalline phases coprecipitate. Although not visible in the photomicrographs, plagioclases become more sodic and pyroxenes and olivines become more enriched in Fe relative to Mg: That is, the Fe/(Fe + Mg) ratio increases, as the T of crystallization decreases. Olivines below 1100°C are corroded and embayed into anhedral shapes and therefore were unstable in the melt. Had there been more time during cooling of that level of the lava lake, the resorption of olivine would have been complete. Resorption of olivine into the melt with falling T is like that seen for some compositions in the simple binary system Mg_2SiO_4-SiO_2. This confirms that the reaction relation 5.3 persists in natural mafic magmas, but only in those of appropriate silica-saturated composition, that is, tholeiitic basalts.

It is obvious from the changes in the color of the glass in Plate III that the melt also changed composition as the T changed. From 1170°C to 1075°C the glass becomes darker brown as a result of enrichment of Fe and Ti in the residual melt. This enrichment was created because the concentrations of these elements are less in the crystallizing phases than in the bulk

magma. An accompanying enrichment of incompatible Na, K, and P in the melt also occurred because the partition coefficient of these elements in the coprecipitating minerals is <1 (Section 2.5.1). Ultimately, at about 1070°C, the activity of ilmenite in the melt reached 1; that is, the melt became saturated with respect to ilmenite and it precipitated, together with magnetite. Consequently, Fe and Ti were removed from the melt, thus causing the color of the glass to change from redbrown to pale gray. At about 1030°C, the melt became saturated with respect to the phosphate mineral, apatite, which then precipitated. As the residual melt became sufficiently enriched in Na and K near the solidus temperature, alkali feldspar precipitated as thin rims on earlier-formed calcic plagioclases.

5.4.2 Basalt Magmas at High Pressures and High Water Concentrations

Crystal-melt equilibria of naturally occurring rocks can be determined in the laboratory by using equipment like that used for determination of simple phase diagrams. However, instead of mixtures of pure reagent compounds, finely pulverized rock is the starting material.

Figure 5.11 presents phase relations as a function of P and T for a tholeiitic basalt, such as the one listed in Table 5.1. Obviously, crystal-melt equilibria are strikingly different for dry (water-free) and water-saturated conditions. Relative to dry conditions, the water-saturated liquidus and solidus are depressed almost 600°C at 10 kbar (near the base of the average continental crust). Melting and crystallization occur over a much broader temperature range, about 400°C, in the water-saturated system, where amphibole is stable over a wide range of P and T. The stability of this hydrous, aluminous phase suppresses the stability of plagioclase, which is only stable near the solidus at relatively low pressures in the water-saturated system.

Figure 5.11 also reveals that mineral stability in basaltic systems is strongly dependent on P. At $P >$ 20 kbar, corresponding to depths greater than about 70 km in the upper mantle, any basalt system—wet or dry—is dominated by clinopyroxene and garnet solid solutions. The garnet contains a substantial proportion of the pyrope end member ($Mg_3Al_2Si_3O_{12}$) but also Fe and Ca. Elements normally sequestered in plagioclase at low P, such as Ca, Na, K, and Al, occur in the high-P clinopyroxene solid solution known as *omphacite*, which contains a substantial amount of the jadeite end member ($NaAlSi_2O_6$). Increasing P increases the activity of sixfold coordinated Al^{VI} at the expense of fourfold Al^{IV} in aluminosilicate melts because the sixfold coordination is a more compact, or smaller-volume, entity. This stabilizes Al^{VI}-coordinated crystalline phases such as garnets and aluminous jadeitic pyroxenes in

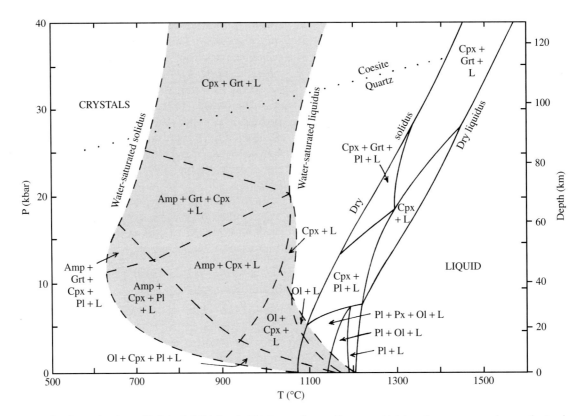

5.11 Generalized crystal-melt equilibria in tholeiitic basalt. This is actually two diagrams, which have been superposed to emphasize the striking influence of water on liquidus and solidus temperatures and on crystalline phase assemblages. Equilibria in the water-free "dry" system are shown by solid lines. Equilibria in the water-saturated system indicated by dashed lines and shaded area. Quartz-coesite polymorphic transition is shown by dotted line. (Redrawn from Green, 1982.)

lieu of Al^{IV}-coordinated feldspars. With increasing P, calcic, more aluminous plagioclases destabilize before sodic, less aluminous plagioclases. These phase relations demonstrate why feldspars are crustal phases and are not normally stable in the sub-continental mantle.

The dense (3.3–3.4 g/cm^3) high-P rock made essentially of red pyropic garnet and green omphacite clinopyroxene (Appendix A) that is of basaltic bulk chemical composition is called **eclogite.**

❈5.5 FELDSPAR-MELT EQUILIBRIA

Because nearly all magmatic rocks contain feldspar, it is imperative that their fundamental phase relations be understood. We begin with a discussion of the three binary systems (Kf-An, Ab-An, Kf-Ab) and then assemble these into the **feldspar ternary** of the three-component Ab-An-Kf.

In the two model binary systems already considered, crystalline solids have fixed compositions with no solid solution between end-member components. On the other hand, the binary plagioclase system (Ab-An) exhibits complete solid solution between end-member components $NaAlSi_3O_8$ (Ab) and $CaAl_2Si_2O_8$ (An) whereas the other two binary systems exhibit only par-

tial mutual solubility. Feldspars serve as models for other major rock-forming minerals that are also solid solutions.

5.5.1 $KAlSi_3O_8$ (Kf)-$CaAl_2Si_2O_8$ (An) Binary System: Limited Solid Solution

This system (Figure 5.12) resembles the $CaMgSi_2O_6$-$CaAl_2Si_2O_8$ system except there is a limited mutual solubility between the K- and Ca-feldspar components of just a small weight percentage. A slender, wedge-shaped one-phase stability field labeled Kf_{ss} (K-feldspar solid solutions) lies along the left-hand T axis and a similar stability field labeled An_{ss} (anorthite solid solutions) lies along the right-hand T axis. Above an isotherm through the eutectic, the boundary line between these wholly crystalline fields and the two-phase fields of liquid + Kf_{ss} and liquid + An_{ss} is the solidus line. The solidus line converges with the T-axes at the melting points of pure anorthite and K-feldspar.

For a magma system $An_{70}Kf_{30}$ at 950°C, an isothermal tie line shows that crystals of $An_{96}Kf_4$ are in equilibrium with liquid $An_{21}Kf_{79}$. The lever rule indicates that 64 wt.% of this system is crystals, 36 wt.% liquid. With increasing T, the solubility of the $KAlSi_3O_8$ component in anorthite crystals decreases, from about 4 wt.% at 850°C to about 2 wt.% at 1100°C.

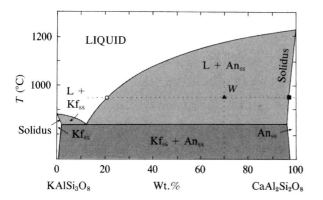

5.12 The water-saturated system $KAlSi_3O_8$ (Kf)-$CaAl_2Si_2O_8$ (An) at 5 kbar. The presence of a separate water-rich phase makes this system ternary, but for our purposes the H_2O component can be ignored. At 1 atm this binary system is complicated by a large stability field of leucite (Problem 5.7), which is eliminated at high-P and water-saturated conditions as shown here. The symbol Or is commonly used to denote the $KAlSi_3O_8$ component, but our use of Kf is a reminder that different atomic structural forms of potassium feldspar exist; orthoclase is only one of these. The subscript ss on the crystalline phases An and Kf denotes solid solutions. (Redrawn from Yoder et al., 1957.)

5.5.2 $NaAlSi_3O_8$ (Ab)-$CaAl_2Si_2O_8$ (An) Binary Plagioclase System: Complete Solid Solution

At 1 atm, the plagioclase phase diagram for mixtures of the $NaAlSi_3O_8$ (Ab) and $CaAl_2Si_2O_8$ (An) end-member components (Figure 5.13) consists of a convex upward liquidus and convex downward solidus between the melting points of pure anorthite and pure albite. Within the univariant two-phase region, $L + Pl_{ss}$ (liquid + plagioclase solid solutions), specification of a T uniquely fixes the equilibrium compositions of both liquid and crystals, given by the ends of an isothermal tie line where it intersects the liquidus and solidus loops, respectively. Alternatively, specification of the composition of one phase fixes the composition of the other, as well as T. The modal proportion of crystals to liquid at any particular T depends on the bulk composition of the univariant system and can be determined from the lever rule.

At any T within the univariant two-phase region, $L + Pl_{ss}$, plagioclase solid solutions are always more calcic (anorthitic) than is the coexisting liquid, which is more sodic (albitic). For example, the first crystals precipitating from a liquid An_{40} have a composition of An_{78}. The progress of crystallization can be followed by drawing a series of isothermal tie lines at decreasing T. As T decreases, more plagioclase precipitates, and it, as well as the liquid, become more albitic. The liquid line of descent in the plagioclase system yields more Na-Si-rich residual melts.

Any plagioclase of intermediate composition between pure end member Ab and pure An melts incon-

gruently with increasing T, yielding a liquid more albitic and crystals more anorthitic than the original. For example, initial melting of crystals An_{40} yields liquid whose composition is An_8. As T increases, the melting progresses, creating a larger proportion of liquid to crystals, both of which become more enriched in $CaAl_2Si_2O_8$, as may be seen by drawing a series of isothermal tie lines at successively higher temperatures. In a sense, the process involves an infinite number of incongruent melting steps. At 1425°C the last crystal to be consumed, An_{78}, is in equilibrium with a melt of the original bulk composition, An_{40}.

It may seem paradoxical that in a closed system progressive crystallization makes *both* crystals and liquid more albitic and both more anorthitic for progressive melting. This happens because the modal proportions of solid and liquid solutions change sympathetically and continuously with the concomitant continuous changes in their compositions as T changes. This phenomenon can be shown in the following hypothetical reaction, which is an example of a **mass-balance equation,** for a system whose bulk composition is An_{40}:

5.4 72.9 wt.% liquid $An_{29.2}$

  + 27.1 wt.% crystals $An_{70.3}$ at 1387°C

  = 80.6 wt.% liquid $An_{32.3}$

  + 19.4 wt.% crystals $An_{73.6}$ at 1400°C

<u>Continuous Reaction Relations.</u> In order for continuous sympathetic changes in compositions and in ratios of coexisting solid and liquid solutions to occur, **continuous reaction relations** must take place between them as intensive variables (T in this case) continuously change. Continuous reaction relations in which reactants and products coexist *over a range of* T are to be contrasted with the discontinuous (peritectic) reaction relation at a unique, single T in the system Mg_2SiO_4-SiO_2. To maintain a constant state of equilibrium be-

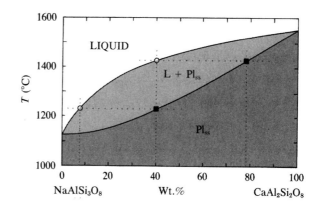

5.13 The binary system $NaAlSi_3O_8$ (Ab)-$CaAl_2Si_2O_8$ (An) at 1 atm. (Redrawn from Bowen, 1928.)

tween crystals and melt in the $NaAlSi_3O_8$-$CaAl_2Si_2O_8$ system there is an exchange reaction between coupled ions

5.5 $Na^+Si^{4+} = Ca^{2+}Al^{3+}$

Exchange must occur by migration (diffusion) of these ions across the interface between the melt and already formed crystals as intensive variables change. Obviously, the larger the crystals to be modified in composition; or the more viscous the melt, which makes ions less mobile; or the faster the change in intensive variables; or combinations of these conditions, the less chance there is for equilibrium to be maintained in the system.

Therefore, we would expect that perfect, reversible equilibrium crystallization in the plagioclase system would only occur under exceptional circumstances. Indeed, this expectation is borne out by the rocks themselves; perfectly homogeneous plagioclases of uniform composition throughout are rare in magmatic rocks. Instead, compositionally inhomogeneous, or **zoned,** plagioclases are far more common and result from incomplete reaction relations during fractional crystallization as intensive variables change. Before complete reaction with the melt can occur by diffusional processes additional crystalline material of different composition precipitates. Then, before the melt can react with that newly accreted crystalline material, changing conditions inhibit further reaction. As the process continues, the melt never has a chance to equilibrate fully with the whole crystal, which becomes zoned as a result. Bowen (1928) referred to such zoned solid solution crystals as well as accumulated crystals in a plutonic mass as a **continuous reaction series** (Figure 5.14).

During perfect fractional crystallization of a liquid, such as An_{40} (Figure 5.14), each liquid fraction, isolated from all previously precipitated crystals, is effectively a new system with no knowledge of its prior history. Because of the lack of reaction, Na^+Si^{4+} ions are conserved and $Ca^{2+}Al^{3+}$ ions depleted in a relative sense in the liquid. Carried to completion, perfect fractional crystallization theoretically creates a residual melt that is ultimately $NaAlSi_3O_8$, at which composition it precipitates pure albite (An_0). Hence, the crystalline products in this hypothetical example of a continuous reaction series range from An_{78} to An_0. Fractionating, evolving liquids and related crystals progress toward more sodic and silicic and less calcic compositions. It is worth emphasizing once again here that fractional crystallization significantly extends the range of T over which crystals precipitate, as well as extending the range of compositions of liquid and solid solutions, relative to equilibrium crystallization (Figure 5.15).

Continuous reaction relations occurred in the Makaopuhi magma system as the melt reacted only partially with previously precipitated olivine, pyroxene, and plagioclase solid solutions. Though inconspicuous to the naked eye in Plate III, continuous changes in the chemical composition of these solid solutions are revealed by microprobe analyses of the minerals in the quenched lava lake samples. The two mafic silicates become more Fe-rich at the expense of Mg and plagioclases become more NaSi-rich with decreasing T.

Influence of Other Components on the Plagioclase System. Increasing P increases liquidus and solidus temperatures in the plagioclase system by only several degrees Celsius per kilobar. In contrast, addition of other chemical components, especially water, to the system depresses the liquidus and solidus by hundreds of degrees (Figure 5.16). Addition of $CaMgSi_2O_6$ (Di) depresses anorthitic compositions but not albitic compositions, so that small changes in T yield large changes in the equilibrium compositions of the coexisting melt and crystals. Dissolved water not only depresses the liquidus and solidus but in multicomponent real magmas stabilizes more calcic plagioclase. For example, in subduction zone basalt magmas that are typically more water rich, crystallizing plagioclase is more anorthitic, to An_{90-95}, than is plagioclase in relatively dry mid–ocean ridge basalt magma (Johnson et al., 1994).

5.5.3 $NaAlSi_3O_8$ (Ab)-$KAlSi_3O_8$ (Kf) Binary Alkali Feldspar System

Dry, or with only small concentrations of water at relatively low P, this system is complicated by a large stability field of leucite ($KAlSi_2O_6$), whose composition cannot be expressed in terms of the components Ab and Kf. However, at between 2 and 3 kbars under water-saturated conditions the stability field of leucite disappears and the system becomes truly binary (Figure 5.17). Solid solution is complete between the $NaAlSi_3O_8$ and $KAlSi_3O_8$ components; the solidus and liquidus form loops on each side of a minimum-melting composition. This minimum resembles a eutectic, in that evolved residual liquids move to it and there precipitate an alkali feldspar solid solution, about Kf_{30}. At $P_{H_2O} = 3$ kbar (water-saturated conditions at 3 kbar) any feldspar precipitated from a melt is a homogeneous alkali feldspar solid solution. As any feldspar cools below the solidus, its isopleth eventually intersects the convex-upward **solvus,** below which the single feldspar unmixes, or exsolves, under equilibrium conditions, into two stable alkali feldspar solid solutions. For example, at 600°C an initially homogeneous feldspar Kf_{60} exsolves into a K-rich feldspar solid solution Kf_{68} and a Na-rich feldspar solid solution Kf_{21}. The lever rule indicates that their proportions are about 83 wt.% and

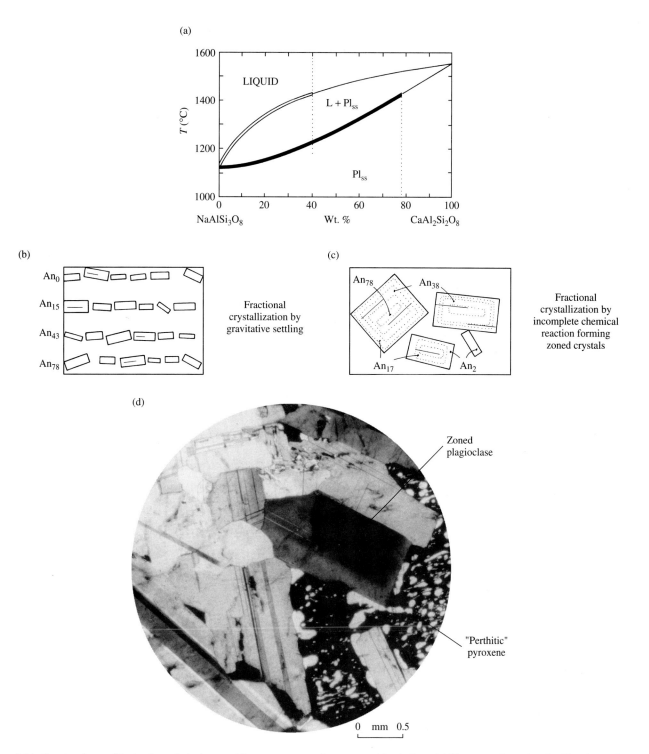

5.14 Fractional crystallization in a plagioclase model magma system whose composition is An_{40}. (a) The range of transient liquid and crystal compositions during cooling of the magma is indicated by the double and thick black lines, respectively, along the liquidus and solidus. (b) Stacks of crystals (a continuous reaction series) that are more albitic toward the top representing schematically the product of fractional crystallization by gravitative segregation of crystals in a less dense melt. The same sort of sequence, but rotated 90°, could occur by side-wall crystallization along the steep border of an intrusion, as more albitic plagioclases would precipitate into the intrusion. (c) Schematic product of fractional crystallization resulting from incomplete reaction relations; zoned crystals are more albitic toward their margins. All An values are arbitrary except for initial crystals, which are An_{78}. (d) Normal compositional zoning in plagioclase under cross-polarized light in gabbro, Skaergaard intrusion, Greenland. Twinned grain in center of view has been carefully oriented to show lighter gray interference color in slightly more albitic rim. Very slow interdiffusion of NaSi and CaAl ions in plagioclase has not taken place to erase its zoning. Adjacent pyroxene grain (black in extinction orientation) has experienced subsolidus exsolution by means of more rapid diffusion of Ca-rich and Ca-poor phases, forming a "blebby perthitic" intergrowth (white) within the original homogeneous crystal.

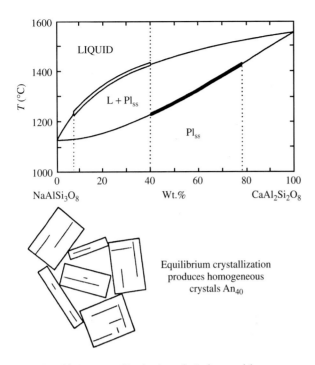

5.15 Equilibrium crystallization in a plagioclase model magma system whose composition is An₄₀. The final products of equilibrium crystallization are homogeneous crystals An₄₀.

17 wt.%, respectively. These two feldspars produced by the exsolution process usually segregate within the original crystal as thin subparallel lamellae, forming the intergrowth known as **perthite.** At lower temperatures the mutual solubility decreases, that is, the **miscibility gap** widens, as the feldspar structure tightens, so that at 500°C, for example, Kf_{75} and Kf_8 coexist at equilibrium. Perthitic intergrowths in volcanic rocks are generally not visible even with a microscope, because the magmas cool so quickly that the diffusion-controlled exsolution does not create visible lamellae; these exceedingly fine, cryptoperthitic intergrowths can, however, be discerned by X-ray diffraction analysis. Perthite is commonly visible in more slowly cooled plutonic rocks, even with the naked eye. Sluggish rates of diffusion within the crystal that preserve metastable compositions prevent the use of perthites as a geothermometer, but they do furnish information on cooling rates.

At higher pressures, such as 5 kbar, in water-saturated systems the solidus loop is depressed so far that it intersects the solvus, forming an isothermal boundary line between the two-feldspar and liquid + feldspar fields (Figure 5.17b). In this case, the minimum in the liquidus is a eutectic. The phase diagram now resembles the Kf-An diagram (Figure 5.12), except that the extent of solid solution between the two end members is greater in the alkali feldspars. Starting liquids that lie between about Kf_{19} and Kf_{52} and crystallize in the equilibrium manner, as well as liquids of

any composition undergoing extreme fractional crystallization, ultimately yield two feldspars, Kf_{19} and Kf_{52}. Each of these phases may subsequently experience slight exsolution upon cooling below the solvus.

The contrasting phase relations depicted in Figure 5.17 prompted Tuttle and Bowen (1958) to classify granites into two textural categories, hypersolvus and subsolvus (Figure 5.18). **Hypersolvus granites,** and some syenites, crystallize from relatively dry magmas whose phase relations are governed as are those in Figure 5.17a where the chief or sole feldspar is perthite; accompanying mafic minerals are commonly anhydrous. **Subsolvus granites** and other felsic rocks have two distinct feldspars that crystallize directly from the melt (Figure 5.17b); accompanying mafic minerals are commonly hydrous amphiboles and biotite.

5.5.4 KAlSi₃O₈ (Kf)-NaAlSi₃O₈ (Ab)-CaAl₂Si₂O₈ (An) Ternary Feldspar System

The ternary feldspar system is of paramount importance in petrology because most rocks contain ternary feldspar solid solutions. It also serves as an introduction to "reading" ternary phase diagrams. For realism and for hastening of reaction rates, laboratory studies of feldspars have involved water, making the system quaternary. However, to keep the discussion simple, water pressure is assumed constant and can be ignored in an isobaric diagram.

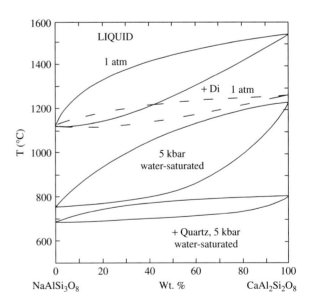

5.16 Comparison of liquidus and solidus temperatures in the plagioclase system at 1 atm with projected liquidi and solidi curves in systems with additional components. (Redrawn from Johannes, 1978.) Dashed curves labeled Di are for a system at 1 atm in which diopside coprecipitates. (Redrawn from Morse, 1980.)

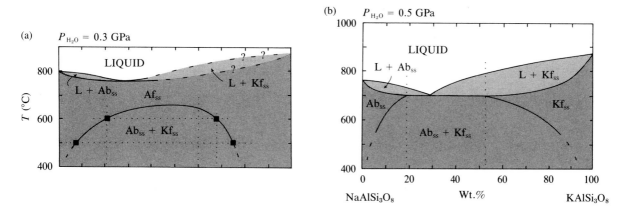

5.17 The $NaAlSi_3O_8$ (Ab)-$KAlSi_3O_8$ (Kf) system under water-saturated conditions at 3 and 5 kbar (0.3 and 0.5 GPa). Granite magmas crystallizing under conditions like those in (a) yield hypersolvus textures, whereas in (b) they yield subsolvus textures (Figure 5.18 and Plate IV). Dashed lines in (a) are inferred liquidus and solidus where experimental data are lacking. Subscript ss refers to solid solutions. (Redrawn from Yoder et al., 1957; Morse, 1970.)

In ternary phase diagrams, T varies along an axis perpendicular to an equilateral triangle on which are represented the proportions of the three components (review Figure 2.3b), in this case Kf, Ab, and An. Thus, the four intensive variables T, X_{Kf}, X_{Ab}, and X_{An} form a triangular prism whose three side faces are the three binary feldspar systems in Figures 5.12, 5.13, 5.17, and 5.19a. The upper bounding surface of the prismatic volume (Figure 5.19b) is the **liquidus surface,** above which, at higher temperatures, any mixture of components is liquid. This three-dimensional liquidus surface is analogous to the familiar topographic surface of the Earth in that it has hills and valleys whose exact configuration can be represented by isothermal contour lines drawn on the surface. A contour line on the liquidus surface is the intersection of an isothermal plane

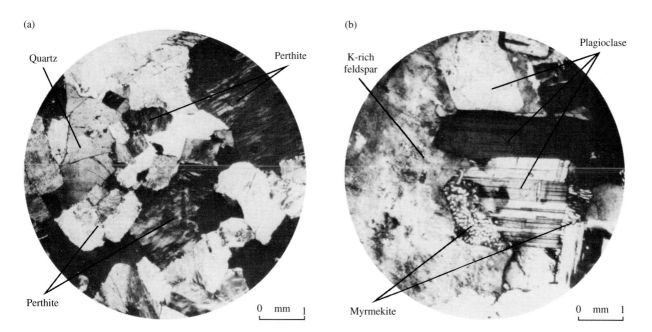

5.18 Textural types of granites formed at contrasting water pressures during crystallization. Photomicrographs of thin sections under cross-polarized light. (a) **Hypersolvus** granite formed under conditions like that in Figure 5.17a where only one initially homogeneous alkali feldspar precipitated from the melt. This single phase subsequently unmixed at temperatures below the solvus from perthite, an intergrowth of Na-rich and K-rich alkali feldspars. Associated mafic minerals are commonly anhydrous: Fe-rich pyroxenes and olivines. (b) **Subsolvus** granite formed under conditions like that in Figure 5.17b, where two discrete feldspars—a sodic plagioclase and a potassic alkali feldspar—coprecipitated from the melt; each may experience slight unmixing at lower temperatures, but this may not be obvious under the microscope. Associated mafic minerals are typically hydrous: biotite and amphibole. Turbid appearance of alkali feldspar is caused by clay alteration and possible minute fluid inclusions. **Myrmekite** is a vermicular intergrowth of quartz and sodic plagioclase.

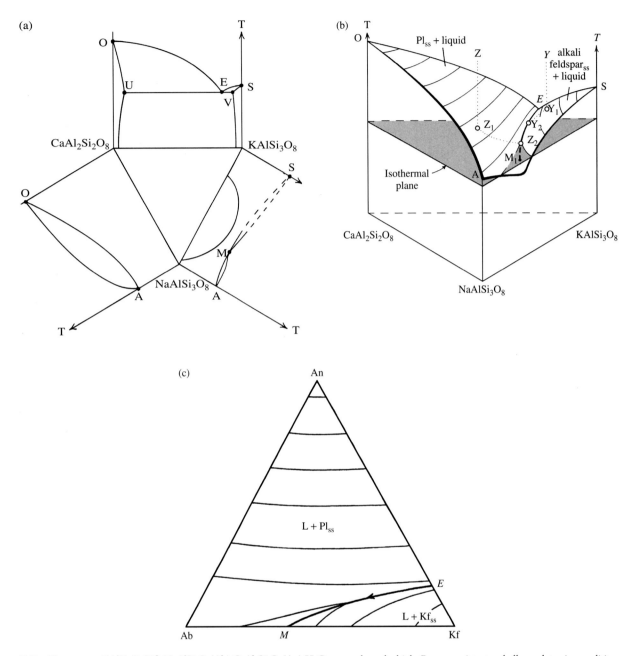

5.19 The system $KAlSi_3O_8(Kf)$-$NaAlSi_3O_8(Ab)$-$CaAl_2Si_2O_8(An)$-H_2O at moderately high P appropriate to shallow plutonic conditions. (a) Three binary systems shown in previous figures are linked around a compositional equilateral triangle in which proportions of the three components $CaAl_2Si_2O_8$, $KAlSi_3O_8$, and $NaAlSi_3O_8$ are represented. The binary Kf-An system has been slightly distorted for clarity. Points *A, O, S, V,* and so on, refer to points in subsequent figures of this ternary system. (b) Perspective view of the three binary systems folded up to form a three-dimensional triangular prism whose axis represents *T.* The underlying solidus and solvus surfaces (Figure 5.20) have been omitted for clarity. The upper surface of the prism is the liquidus surface in three-dimensional *T-X* space on which curved isothermal contour lines are drawn parallel to the compositional base of the prism. One isothermal plane (shaded) parallel to the base is shown cutting the liquidus surface along isothermal contour lines on each side of the two-feldspar-liquid boundary line, EM_1. This line lies in the thermal valley of the liquidus surface in which falling *T* is toward M_1. The liquidus phases to the left of the boundary line are plagioclase solid solutions; these coexist stably with any melt on this part of the liquidus surface. The liquidus phases to the right of the boundary are alkali feldspar solid solutions. Melts lying on the boundary are in equilibrium with both feldspar solid solutions. The end point of the boundary line does *not* lie on the $NaAlSi_3O_8$-$KAlSi_3O_8$ join, but within the ternary with a small amount of dissolved $CaAl_2Si_2O_8$. (See Nekvasil and Lindsley, 1990; Brown, 1993, for a discussion of the complexities at the termination of the boundary line.) (c) Projection of isothermal contour lines and boundary line EM_1 from the triangular prism in (b) onto its base.

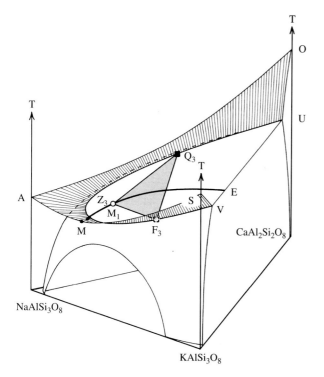

5.20 Perspective view of the three-dimensional solvus and solidus surfaces in the system $KAlSi_3O_8$-$NaAlSi_3O_8$-$CaAl_2Si_2O_8$-H_2O at a moderately high P and T appropriate to shallow plutonic conditions. The T-X prism has been rotated clockwise about 120° relative to the view in Figure 5.19b, and the liquidus surface in it has been omitted for clarity. Thus, these two figures complement each other. The igloo-shaped solvus surface intersects the overlying spoon-shaped solidus surface (ruled thin lines) along the line VF_3Q_3U. (It may help in visualizing this three-dimensional diagram by thinking of a triangular blanket—the solidus surface—held by three people at points O, A, and S in a rainstorm so that the water-soaked blanket is stuck to an underlying, flat-topped boulder—the solvus.) The line of intersection of the solvus and solidus is the locus of compositions of ternary feldspar solid solutions at a particular set of intensive variables shown in Figure 5.21. The curved boundary line EM_1 lies on the common solidus-solvus surfaces as well as the liquidus surface and is projected onto the compositional base in Figures 5.19c and 5.21. Lines Q_3F_3, F_3Z_3, and Z_3Q_3 define the sides of an isothermal, three-phase triangle (shaded) parallel to the composition base of the prism that represents a melt, Z_3, coexisting stably with plagioclase, Q_3, and K-rich alkali feldspar, F_3.

with the curved liquidus surface. A two-dimensional "topographic map" of the three-dimensional configuration of the liquidus surface can be created by projecting the contour lines parallel to the T axis of the prism onto the triangular compositional base (Figure 5.19c).

A perspective view of the solidus and solvus surfaces in the ternary is shown in Figure 5.20. The intersection of these two surfaces is the arcuate line UQ_3F_3V. This line is the locus of points representing the compositions of ternary feldspar solid solutions, such as the stable pair Q_3 and F_3, at a particular set of intensive

variables. Isothermal tie lines that connect these composition points are projected onto the ternary "map" in Figure 5.21.

Consider the crystallization behavior of an initially all-liquid system such as Z in Figure 5.19b. Upon cooling down the isopleth (dotted line parallel to the T axis), the system reaches the liquidus surface at Z_1, where plagioclase Q_1 begins to precipitate (Figure 5.21). As the system cools further, the melt follows a curved path Z_1Z_2 down the liquidus surface, precipitating more plagioclase, which is progressively more enriched in Na and to a much lesser extent, K. The curved path on the liquidus results from the fact that a ternary solid solution feldspar is being withdrawn from the bulk system; at any particular T, the liquid moves directly away from the equilibrium composition of the plagioclase. Ultimately, the liquid reaches the two-feldspar-liquid boundary line EM_1 at Z_2 where an alkali feldspar solid solution rich in K, such as F_2, coprecipitates with plagioclase Q_2 from the melt. This coprecipitation must occur, regardless of whether crystallization occurs through the equilibrium or fractional process, because the amount of the $KAlSi_3O_8$ component in the initial, bulk composition of the system exceeds the amount that can be dissolved in any plagioclase. In other words, there is a miscibility gap between plagioclases, which contain only limited Kf in solid solution, and alkali feldspars, which contain only limited An in solid solution. As two feldspars, both increasingly sodic at lower T, precipitate, the melt also becomes more sodic, until at Z_3 plagioclase Q_3 and alkali feldspar F_3 coexist in equilibrium (Figures 5.20

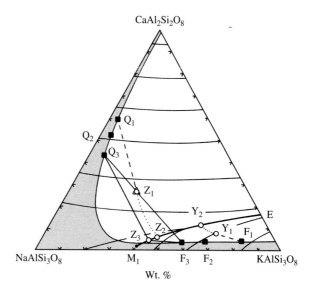

5.21 Projection of the system $KAlSi_3O_8$-$NaAlSi_3O_8$-$CaAl_2Si_2O_8$-H_2O at moderately high P and T. Same as Figure 5.19c but includes projected compositional points and the arcuate line of intersection of the solidus and solvus VF_3Q_3U from Figure 5.20. (Redrawn from Yoder et al., 1957; Morse, 1970.)

and 5.21). These three compositions (melt Z_3 and feldspars Q_3 and F_3) are significant because under conditions of equilibrium crystallization the tie line Q_3F_3 connecting the two ternary feldspars passes exactly through the bulk composition of the system Z_1. This means that, except for one drop of remaining melt Z_3, the system is crystalline and composed of feldspars Q_3 and F_3, in proportions given by the lever rule. Note that the higher T assemblage of liquid Z_2 and feldpars F_2 and Q_2 form a three-phase, isothermal triangle, which encloses the bulk composition point Z_1. At that higher T, the modal proportions of the three coexisting phases could be found by the technique described in Figure 2.3b, but adapted for a scalene triangle.

A wholly liquid system Y in Figure 5.19b upon cooling first crystallizes K-rich alkali feldspar F_1 as it impinges upon the liquidus at Y_1. With continued precipitation of K-rich feldspar during cooling, the melt moves along a curved path to Y_2 on the two-feldspar-liquid boundary line EM_1 (Figure 5.21), where a rather calcic plagioclase coprecipitates. Only in exceptionally K-rich magmas does an alkali feldspar crystallize before plagioclase. In most magmas, plagioclase precipitates first and continues to do so over a broad range of T before alkali feldspar coprecipitates, if at all.

The orientation of isothermal, two-feldspar tie lines, such as Q_3F_3 in Figure 5.21, as well as the compositions of the coexisting feldspars represented by their end points, depend upon P and T (Figure 5.22). If P can be independently evaluated, then an equilibrium pair of coexisting feldspars can serve as a geothermometer for the T of crystallization (e.g., Fuhrman and Lindsley, 1988).

Fractional crystallization in the ternary feldspar-melt system produces plagioclases and alkali feldspars that have more albitic compositions at decreasing T. For the common case of normally zoned feldspars, rims are more albitic than cores, which are more calcic in plagioclase and more potassic in alkali feldspars (Figure 5.22).

Fractional crystallization with decreasing T yields melts that evolve along curved fractionation paths on the ternary liquidus surface (Figure 5.23). Ultimately, the most evolved melt may reach a composition near M_1. One set of liquid lines of descent (among an infinite number of possible residual melt paths) corresponding to the highlighted fractionation path is shown in the variation diagram in Figure 5.23b. Liquid lines of descent are not straight lines but instead show a smoothly varying curvature where changing plagioclase solid solutions are not the only fractionating crystalline phase but a sharp inflection (in this case at about 66.8 wt.% SiO_2) where alkali feldspar solid solutions begin to cofractionate. Residual melts commonly display a decreasing Ca/(Na+K) ratio in magmas frac-

tionating plagioclase. Curved liquid paths on variation diagrams are characteristic of ternary and more complex multicomponent systems that fractionate compositionally variable solid solutions.

5.5.5 KAlSi$_3$O$_8$ (Kf)-NaAlSi$_3$O$_8$ (Ab)-SiO$_2$ (silica)-H$_2$O: The Granite System

Pertinent relations in this water-saturated system at 2 kbar are projected onto the ternary diagram in Figure 5.24. A prominent valley, or thermal minimum, in the liquidus surface near the center of the diagram is defined by a steep liquidus surface descending from the silica apex and a more gently sloping liquidus surface from the $KAlSi_3O_8$ and $NaAlSi_3O_8$ apices. Melts that contain more than about 40 wt.% of the silica component crystallize quartz as the liquidus phase. For example, quartz precipitates from melt A at about 820°C; as T decreases, continued crystallization of quartz drives the residual melt down the liquidus directly away from the silica apex, because quartz has a fixed composition, along a (dotted) straight line in the diagram. Once the residual melt reaches the curved boundary line at

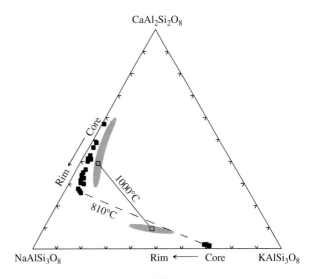

5.22 Compositions of ternary feldspars and the orientation of tie lines connecting equilibrium pairs of plagioclase and alkali feldspar solid solutions depend upon intensive variables. 1000°C tie line connects coexisting equilibrium feldspar compositions (open squares) in a trachybasalt whose approximate T of crystallization was determined from coexisting Fe-Ti oxides. (Data from Smith and Carmichael, 1969.) Other zoned feldspars in the trachybasalt plot in the shaded bands. 810°C tie line connects coexisting alkali feldspar and rim plagioclase (filled squares) in a block of rhyolite pumice in a tuff whose T of crystallization—at higher water pressures than the trachybasalt—was determined from the model of Fuhrman and Lindsley (1988). (Data from Best et al., 1995.) Zoned feldspars in rhyolite pumice shown by filled squares. Note crossing 810°C and 1000°C tie lines. Normally zoned alkali feldspars and plagioclases have rims that are enriched in NaAlSi$_3$O$_8$ relative to their cores. Note that the miscibility gap between plagioclase and alkali feldspar solid solutions is less at higher T and lower water pressure.

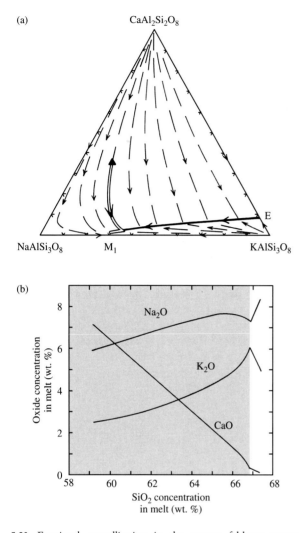

(a)

CaAl$_2$Si$_2$O$_8$

NaAlSi$_3$O$_8$ M$_1$ KAlSi$_3$O$_8$

(b)

Oxide concentration in melt (wt. %)

Na$_2$O

K$_2$O

CaO

SiO$_2$ concentration in melt (wt. %)

5.23 Fractional crystallization in the ternary feldspar system KAlSi$_3$O$_8$-NaAlSi$_3$O$_8$-CaAl$_2$Si$_2$O$_8$. (a) Schematic representative fractionation paths of residual melts on the liquidus surface. Curvature of paths results from removal of ternary feldspar solid solutions from the evolving melts. The filled triangle and double line leading away from it represent a hypothetical andesitic initial melt that fractionates to an evolved rhyolitic composition. (b) Liquid lines of descent of the hypothetical fractionating melt in (a). The shaded part indicates the fractionation of plagioclase only from the melt, the unshaded part where two coprecipitating feldspars fractionate from the melt as it tracks down T along the two-feldspar boundary EM_1. Note that fractionation yields residual evolved melts that are increasingly more enriched in Si and Na, but impoverished in Ca.

point A_1 between the fields of quartz + liquid and alkali feldspar solid solutions + liquid, potassic alkali feldspar, initially about Kf$_{81}$, begins to coprecipitate with quartz. Ultimately, as the residual melt almost reaches the lowest T in the thermal valley, represented by liquid at point M, it is entirely consumed into crystallizing quartz and feldspar. Under equilibrium crystallization and closed system behavior, the tie-line triangle representing the three-phase assemblage melt

M + quartz + Kf$_{57}$ no longer encloses the bulk composition point A for the crystallizing system. Less siliceous, more feldspathic systems initially precipitate alkali feldspar as the liquidus phase. As alkali feldspar solid solutions precipitate, residual melts track along a curved path on the liquidus surface to the boundary line where quartz coprecipitates.

Above 3.6 kbar in the water-saturated system the minimum in the thermal valley of the liquidus surface is replaced by a eutectic; the liquidus has been depressed in T sufficiently to intersect the solvus (compare Figure 5.17). Increasing P in water-saturated systems also shifts the position of the minimum and eutectic toward more Ab-rich compositions (Figure 5.25). Other modifications in intensive variables, such as addition of CaAl$_2$Si$_2$O$_8$, F, and CO$_2$ to the system, also change the position of the minimum or eutectic somewhat (Johannes and Holtz, 1996). The shift in the minimum away from the silica apex corresponds to an expansion of the stability field of quartz on the liquidus at higher P. This expanded stability field may account for the common occurrence of partially resorbed quartz phenocrysts in silicic volcanic rocks, such as that illustrated in Figure 6.19. For example, consider a

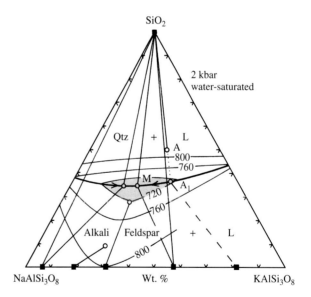

SiO$_2$

2 kbar water-saturated

Qtz   +   L

A

800

760

M

720

A$_1$

760

Alkali   Feldspar

+   L

800

NaAlSi$_3$O$_8$ Wt. %   KAlSi$_3$O$_8$

5.24 Liquidus relations and phase compositions at P_{H_2O} = 2 kbar (water-saturated conditions) in the **granite system** KAlSi$_3$O$_8$(Kf)-NaAlSi$_3$O$_8$ (Ab)-SiO$_2$-H$_2$O. As projected here onto the feldspar-silica composition plane, the system can be considered as ternary. Representative isothermal tie lines connect coexisting phases. Filled squares are crystal compositions; open circles are liquid compositions. Isothermal contour lines with temperatures in Celsius show form of the liquidus surface. Thermal valley in the liquidus surface where T < 720°C is shaded. (Reprinted by permission of the publisher from Tuttle OF and Bowen NL, Origin of granite in the light of experimental studies in the system NaAlSi$_3$O$_8$-KALSi$_3$O$_8$-SiO$_2$-H$_2$O, Boulder, Colorado, The Geological Society of America © 1958.)

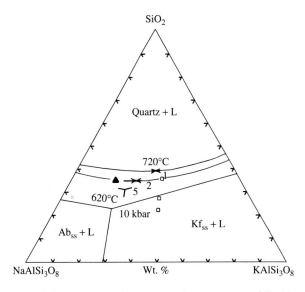

5.25 Shift in minima and eutectics on the water-saturated liquidus surface with respect to *P* in the system KAlSi$_3$O$_8$(Kf)-NaAlSi$_3$O$_8$ (Ab)-SiO$_2$-H$_2$O. Data projected onto the anhydrous triangular base. Minima (opposing arrows on boundary line) and eutectics (triple line intersection) are for 1, 2, 5, and 10 kbar. The minimum becomes a eutectic at about 3.6 kbar. The minimum *T* at 1 kbar is 720°C, and the eutectic *T* at 10 kbar is 620°C. Compare Figure 5.24. For comparison, the 2 kbar minimum and 5- and 10-kbar eutectics for water activity of 0.25 (undersaturated conditions) are shown as open squares. (Redrawn from Johannes and Holtz, 1996.)

magma system on the boundary line at 2 kbar that consists of quartz + alkali feldspar + melt and whose bulk composition is represented by the filled triangle in Figure 5.25. At lower *P* as the magma ascends to the surface, the system becomes stranded in the stability field of alkali feldspar + melt; consequently, the previously precipitated quartz is now unstable and dissolves into the melt, but only partly if insufficient time is available, as would occur during rapid ascent and eruption.

<u>Origin of Granite Magmas.</u>   The KAlSi$_3$O$_8$-NaAlSi$_3$O$_8$-SiO$_2$ system has been called the **granite system** because many granites, and their compositionally equivalent aphanitic and glassy counterparts, are predominantly alkali feldspar and quartz (Figure 5.26). Granites contain only a small amount of CaAl$_2$Si$_2$O$_8$, which appears as the anorthite end member of plagioclase. However, in granodiorites and associated dioritic rocks, this component is more substantial and its absence in the model system hinders direct application to such magmas. The presence of Fe, Ti, Mg, and so on, which stabilize various mafic minerals, again especially in dioritic compositions, is another difference between this model system and real magmas. Nonetheless, this system serves as a very useful model for crystal-melt equilibria in felsic magmas that form huge volumes of plutonic

and volcanic rock, particularly along convergent plate margins.

It was said in ancient times that "all roads lead to Rome." In the granite ternary, all fractionation paths lead liquids into the bottom of the thermal valley in the liquidus surface—a unique composition that contains more or less equal proportions of the three end-member components (Figure 5.27a). This convergence of residual melts during fractionation into the ternary minimum on the liquidus surface is not limited to initial compositions that lie within this ternary. Fractionation of some magmas that initially have only small concentrations of Na and K and modest Si can, nonetheless, yield residual melts that move into this ternary system and thence to the thermal valley as fractionation continues. Subequal proportions of alkali feldspar and quartz coexisting with melt at this low-*T* minimum has, for good reason, been referred to as **petrogeny's residua system.** The striking coincidence of petrogeny's residua system with the favored composition of granites portrayed in Figure 5.26 supports the hypothesis, asserted by Bowen (e.g., 1928) throughout the first half of the 20th century, that granite mag-

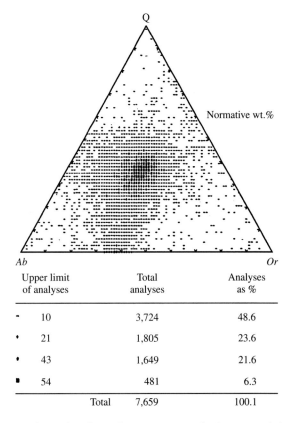

Upper limit of analyses	Total analyses	Analyses as %
10	3,724	48.6
21	1,805	23.6
43	1,649	21.6
54	481	6.3
Total	7,659	100.1

5.26 Chemical analyses of 7659 granitic rocks from around the world that have > 80 wt.% normative *Q* + *Ab* + *Or.* Note that about 50% of the analyses fall within only about 5% of the diagram. Compare with Figures 5.24, 5.25, and 5.27. (Analyses compiled and plotted courtesy of Roger W. Le Maitre of the University of Melbourne.)

mas originate by extreme fractional crystallization of basaltic parent magma. But Bowen's assertion was repeatedly criticized by more field-oriented petrologists who could not find any direct evidence for the approximately ninefold greater volume of more mafic fractionates that should be necessarily associated with the huge granitic batholiths of mountain belts. (The amount of granite components sequestered in a typical basalt magma is only about 10%.)

Another origin for granite magmas that answers the field-relation criticism is to partially melt sialic rocks in

instead appear to grade over some distance from granite into country rock, suiting a transformation front. Confusing the issue is the fact that magmatic granitic rocks have a variable and locally pervasive overprint of subsolidus recrystallization that makes them appear to have originated through metamorphism.

In some respects, the granite debate reflected the mutual distrust of protagonists. Field-oriented granitizers were highly skeptical of experimental studies that employed small Pt crucibles in a furnace allegedly mimicking huge plutons, and Bowen and associates lamented the lack of understanding of basic chemistry and physics of the granitizers. Since the 1950s, petrologists have adopted a more holistic integrated viewpoint that experiments and theory are both useful in constraining and testing hypotheses based on observations of real rocks, and vice versa.

Special Interest Box 5.3 Do granites originate by solid-state granitization or from magmas?

The long-standing and often heated debate (Pitcher, 1997, chapter 1) between magmatists and granitizers peaked near the middle of the 20th century and was finally resolved by the classic investigation of Tuttle and Bowen (1958). They showed that the restricted composition of granitic rocks (Figure 5.26) corresponds closely with a thermal valley in the liquidus surface in the model granite system (Figures 5.24, 5.25, 5.27) where a melt coexists in equilibrium with alkali feldspar and quartz. Thus, granitoids are easily concluded to be the product of crystallization of a corresponding magma that either is a residual melt produced by fractional crystallization of a parental magma of more mafic composition (basalt) or is generated by partial melting of sialic continental rock, or is both. Besides this rational explanation in terms of crystal-melt equilibria, geologists began finding, in the decades after 1958, vast silicic pyroclastic deposits of mostly glass fragments. Single eruptions of thousands of cubic kilometers of pyroclasts could only mean that no lesser volume of magma existed in a subterranean preeruption magma chamber.

The granitization viewpoint was based on field relations of granite plutons (Read, 1957). To avoid the **"room problem"**—how space was provided for a magmatic intrusion in the crust—granitizers claimed that the granitic rock was not intruded as magma but was simply transformed, or granitized, preexisting rock, a sort of subsolidus, or solid-state, metamorphism. Chemical diffusion, or at least mobilization and transport of chemical constituents in migrating fluids, on a very grand and pervasive scale would be required. But chemical diffusion was first shown by Bowen (1921) to be much too slow and other means of element transport lacked a rational thermodynamic basis. But in support of the granitization argument is the fact that some plutons lack sharply defined, intrusive contacts and

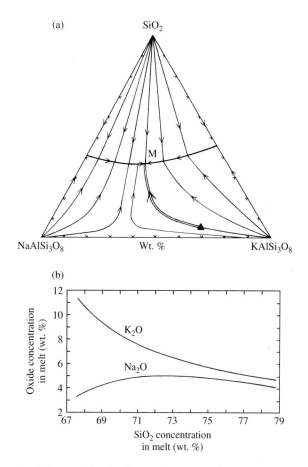

5.27 Effects of fractional crystallization in the granite system $KAlSi_3O_8$(Kf)-$NaAlSi_3O_8$(Ab)-SiO_2-H_2O. (a) Fractionation paths of residual evolving melts at 1-kbar water-saturated conditions. (b) Liquid lines of descent for fractionation of the melt represented by the filled triangle and the double line path in (a). (Redrawn from Tuttle and Bowen, 1958.)

the continental crust. The initial partial melt of a rock that contains substantial amounts of the components $KAlSi_3O_8$, $NaAlSi_3O_8$, and SiO_2 in any proportion is a melt of the ternary minimum T composition. Thus, some shales and feldspathic sandstones and preexisting granitic rocks, and their metamorphosed equivalents, are potential source rocks for generation of granite magma by partial melting.

Whether derived by fractionation of more mafic parent magma or by partial melting of sialic crustal rock, the restricted composition of granites (Figure 5.26) is governed by crystal-melt equilibria in the $KAlSi_3O_8$-$NaAlSi_3O_8$-SiO_2 system. These fundamental phase relations, elucidated by Tuttle and Bowen (1958) in a classic study, laid to rest the then-popular argument that granites were created by "granitization," the transformation of solid rocks *without* involvement of a silicate melt. Granitization cannot explain why the composition of granites should be so restricted.

✳5.6 CRYSTAL-MELT EQUILIBRIA INVOLVING ANHYDROUS MAFIC MINERALS: OLIVINE AND PYROXENE

Like plagioclases, olivines constitute a complete solid-solution series, and their phase equilibria are similar. The more complex pyroxene solid solutions are considered next.

Compositions of common pyroxenes, mostly in subalkaline rocks such as basalt are conveniently represented in the pyroxene quadrilateral, which is a part of the ternary system $CaSiO_3$-$MgSiO_3$-$FeSiO_3$. Low-P liquidus relations (Figure 5.28a) are complex because of the coexistence of solids whose compositions lie well outside the ternary and its three defining components. For highly magnesian compositions, an Mg-rich olivine is the liquidus phase, as it is in the Mg_2SiO_4-SiO_2 system (Figure 5.8). Hence, incongruent melting of enstatite still prevails with addition of Fe and Ca in

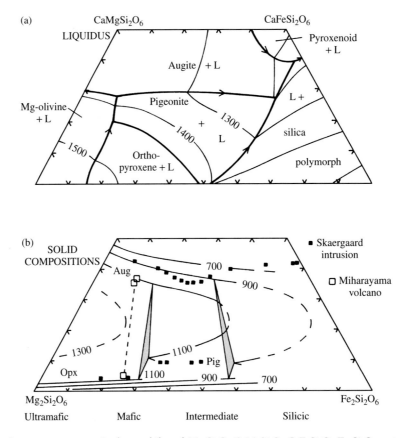

5.28 Phase relations of common pyroxenes in the quadrilateral $Mg_2Si_2O_6$-$CaMgSi_2O_6$-$CaFeSi_2O_6$-$Fe_2Si_2O_6$ at 1 atm. (a) Liquidus surface with 100°C contours and boundary lines. (Redrawn from Huebner and Turnock, 1980.) (b) T-dependent pyroxene compositions. (Redrawn from Lindsley, 1983.) Augites (Aug) occur in a band across the top of the quadrilateral, orthopyroxenes (Opx) in a narrower band across the bottom, and pigeonites (Pig) just above them. Representative partial contour lines on the solvus are labeled in degrees Celsius for geothermometry after suitable corrections (Lindsley, 1983) have been made for other components in solid solution. Isothermal tie lines at 1100°C and 900°C connect coexisting equilibrium compositions of three pyroxenes (shaded triangles). Filled squares are pyroxenes from the Skaergaard intrusion, Greenland. Open squares and connecting dashed tie-line are phenocrysts from the 1950–51 basalt lava flow of the Mihara-yama volcano, Japan. Approximate correlation between rock suites and pyroxene compositions is shown below quadrilateral.

solid solution. For highly Fe-rich compositions, a large field of Fe-rich melts coexist with a silica polymorph. This occurs because, unlike magnesian compositions, an Fe-rich olivine + quartz (or one of its polymorphs) is more stable than ferrosilite pyroxene ($FeSiO_3$) under most magmatic conditions. Instead of hedenbergite ($CaFeSi_2O_6$), a complex pyroxenelike solid, or "pyroxenoid," is the liquidus phase in the upper-right apex of the quadrilateral. The remaining liquidus phases are pyroxenes. Augite is a Ca-rich clinopyroxene that crystallizes in a wide range of natural magmas. In alkaline magmas that have low silica activity (Section 3.5.3), it is the sole pyroxene and contains substantial amounts of Al, Ti, Fe^{3+} (all three partly substituting for Si in tetrahedral sites), and Na (for Ca) in solid solution. Uncommon pigeonite is a Ca-poor clinopyroxene crystallizing at low P in subalkaline basalt and some andesite magmas. Orthopyroxenes contain no more than 5 mole % of the calcic component and crystallize only in subalkaline magmas, usually with augite. Fractionating melts migrate down T toward more Fe-rich compositions and ultimately can reach near $CaFeSi_2O_6$. Thus, Fe/(Fe + Mg) ratios increase in more evolved, lower T-melts and in their crystalline products.

Phase relations below the liquidus (Figure 5.28b) are complicated by polymorphism in enstatites, metastability in subcalcic augites and pigeonites, and subsolidus exsolution, none of which is discussed here. Equilibrium compositions of two- and three-pyroxene assemblages that can stably coprecipitate from a melt are indicated by isothermal tie lines and tie-line triangles in Figure 5.28b. Equilibria are strongly dependent upon T and less upon P. Such equilibrium pyroxene pairs can therefore serve as geothermometers for crystallization T after their compositions have been appropriately projected from multicomponent space onto the $Mg_2Si_2O_6$-$CaMgSi_2O_6$-$CaFeSi_2O_6$-$Fe_2Si_2O_6$ quadrilateral and plotted on the diagram of nearest P (for details see Lindsley, 1983). For example, coexisting augite and orthopyroxene phenocrysts from a basalt lava flow extruded in 1950–51 from Mihara-yama volcano, Japan, indicate crystallization at slightly more than 1100°C; for comparison, the highest T of lava measured during the eruption was 1125°C.

Pyroxenes and olivines in the highly fractionated Skaergaard intrusion in northeast Greenland show the degree to which Fe enrichment can be produced by a strongly fractionating tholeiitic basalt magma. The relatively reducing conditions (Frost and Lindsley, 1992) in the fractionating Skaergaard magma body, to as much as 2 log units below the QFM buffer (Figure 3.14), undoubtedly played a significant role in stabilizing Fe-rich clinopyroxenes and pure fayalite olivine. Relatively oxidizing magmas in which the oxygen fugacity is higher have more Fe^{3+} and consequently less Fe^{2+} to stabilize Fe-rich pyroxenes and fayalite.

5.7 CRYSTAL-MELT EQUILIBRIA IN HYDROUS MAGMA SYSTEMS

Volatiles are generally present in terrestrial magma systems, can profoundly depress liquidi and solidi, and stabilize volatile-bearing phases. In this section, some aspects of crystal-melt equilibria in hydrous magma systems are examined.

5.7.1 Equilibria in the Granodiorite-Water System

Granodiorite is a plagioclase-rich granitic rock (Figure 2.8) that contains more mafic minerals, generally biotite and hornblende, than granite. Granodiorite and its aphanitic to glassy volcanic counterpart, dacite, are widespread in convergent plate margin settings.

To determine melt equilibria with two feldspars and quartz in the proportions in which they occur in a typical granodiorite, Whitney (1988) studied a synthetic rock free of mafic minerals that has weight percentages of components $CaAl_2Si_2O_8$ 19.8, $NaAlSi_3O_8$ 37.3, $KAlSi_3O_8$ 19.8, and SiO_2 23.1. Several generalities can be gleaned from a T-X_{H_2O} diagram at constant P (Figure 5.29). Sequential and overlapping crystallization of solid phases with decreasing T is once again evident, as

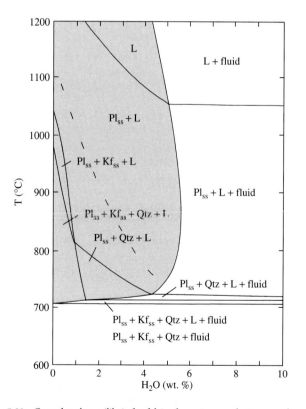

5.29 Crystal-melt equilibria for felsic phases in a synthetic granodiorite at 2 kbar. (Composition R5; redrawn from Whitney). Region of water-undersaturated melt is shaded. Dashed line indicates the boundary line for stable quartz (+ Pl_{ss} + L) at 8 kbar, showing the expansion of its stability field at increased P. L, silicate liquid (melt). (Redrawn from Whitney, 1975 cited in Whitney, 1988.)

in Figures 5.10 and 5.11. Plagioclase is the liquidus phase, continues to crystallize all the way to the solidus, is the sole precipitating solid for a large range of T, and is only joined by quartz and alkali feldspar (Kf_{ss}) near the solidus. For water contents > 3–5 wt.%, appropriate to stabilize biotite and amphibole in natural multi-component magmas, quartz and alkali feldspar only precipitate within <20°C of the solidus. Depending upon how much of the $NaAlSi_3O_8$ component is dissolved in plagioclase, at least one-half of the system does not crystallize until quartz and alkali feldspar precipitate. These relations are compatible with textural relations in many granodiorites in which plagioclases are robust euhedral grains mingled with generally more anhedral quartz and alkali feldspar grains that appear to have grown from the interstial melt lying between the plagioclases. Piwinskii (1968) and subsequently many others have shown that in real granodiorite magmas, Fe-Ti oxides, biotite, and hornblende along with common accessory phases zircon and apatite crystallize at high T together with plagioclase; the mafic crystals also tend to be more or less euhedral in granodiorites. The origin of melts forming widespread, late-stage dikes of leucocratic alkali feldspar-quartz aplite in granodiorites can also be appreciated (see Figure 9.3).

5.7.2 Equilibria Involving Melt and Micas and Amphiboles

All of the crystalline phases heretofore discussed do not contain volatiles as essential constituents. The stability of micas and amphiboles, on the other hand, depends on the existence of water in the system because $(OH)^-$ anions are essential in their crystalline structure. Water concentrations > 4–5 wt.% stabilize amphibole in andesitic-dacitic magmas (Merzbacher and Eggler, 1984); somewhat less water stabilizes biotite. Micas and amphiboles also contain generally lesser concentrations of F and Cl, which substitute for $(OH)^-$, and variable concentrations of Fe^{2+} and Fe^{3+}. Therefore, their equilibria in magma systems depend on the fugacities of H_2O, F, Cl, and O, as well as P and T.

<u>Stability of Mg-Fe Biotites.</u> Discussed in this section is how oxygen fugacity, bulk chemical composition of the magma, T, and P influence the Fe/(Fe + Mg) ratio and stability of biotite solid solutions (Figure 5.30). Considering first the pure Fe end-member biotite annite ($KFe_3^{2+}AlSi_3O_{10}(OH)_2$), it may be noted that this phase is not stable at oxygen fugacities above the HM buffer curve in Figure 3.14, where sanidine + hematite are more stable, nor below the WI buffer, where fayalite (pure Fe-olivine) and potassium aluminum silicates are more stable. The following reactions constrain the stability of annite

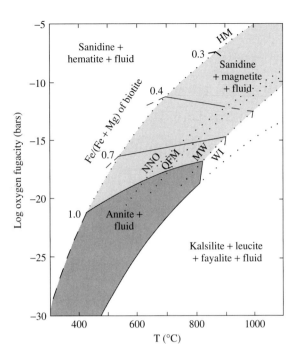

5.30 Stability of biotite as a function of oxygen fugacity and T at 2070 bars. Compare with Figure 3.14 for identification of buffer curves (dotted lines) in the Fe-Si-O system. Stability field of the pure iron (Fe/(Fe + Mg) = 100) end-member biotite annite is the dark shaded area. Breakdown assemblages from reactions 5.6 to 5.8 are indicated for annite. Stability limits for Fe-Mg biotite solid solutions (light shade) are indicated by subhorizontal line segments labeled along the HM buffer curve with their Fe/(Fe + Mg) values. (Redrawn from Wones and Eugster, 1965.)

5.6 $KFe_3^{2+}AlSi_3O_{10}(OH)_2 + 3O_2 = KAlSi_3O_8$

         annite   fluid   sanidine

      $+ \ 3Fe_2^{3+}O_3 + H_2O$

       hematite   fluid

5.7 $KFe_3^{2+}AlSi_3O_{10}(OH)_2 + 0.5O_2 = KAlSi_3O_8$

         annite   fluid   sanidine

      $+ \ Fe_2^{3+}Fe^{2+}O_4 + H_2O$

       magnetite   fluid

5.8 $2KFe_3^{2+}AlSi_3O_{10}(OH)_2 = KAlSiO_4$

         annite   kalsilite

      $+ \ KAlSi_2O_6 + 3Fe_2^{2+}SiO_4 + 2H_2$

       leucite   fayalite   fluid

These reactions may be viewed as **decomposition (breakdown) reactions** in which an assemblage of other phases takes the place of a single crystalline phase, annite. The three oxidation-reduction reactions are represented by the three heavy line segments bounding the $T\text{-}f_{O_2}$ stability field of annite (dark-shaded) in Figure 5.30 (see also the buffer curves in Figure 3.14).

More Mg-rich biotites are stable to higher T and higher oxygen fugacities. Thus, a biotite in which Fe/(Fe + Mg) = 0.4 is stable (light shaded field in Figure 5.30) to as much as 1000°C and f_{O_2} as much as 10^{-12} bar whereas annite is stable to only a little over 800°C and 10^{-17} bar. Decomposition reactions involving real Mg-Fe biotites, which also contain Ti, Mn, F, and Cl, are more complex than those listed for the annite end member.

Three aspects of biotite stability in magmatic rocks are explained by the phase relations in Figure 5.30:

1. In many volcanic rocks, such as common calc-alkaline andesites, dacites, and rhyolites, that have cooled in the oxygen-rich atmosphere from relatively more reduced extruded magmas, biotites are typically bronze- or copper-colored. These are commonly mistaken for phlogopite, the Mg analog of annite. However, phlogopite is never stable in such calc-alkaline magmas and occurs only rarely in some highly alkaline, ultramafic rocks. Instead, the bronze or copper color is caused by minute included particles of hematite, locally with other Fe-Ti oxides, in the biotite. Hematite is produced by partial subsolidus reequilibration of biotite to the high oxygen fugacity (0.21 = $10^{-0.68}$ bar) of the atmosphere via a reaction like reaction 5.6. There may be no sanidine from the decomposition reaction remaining because of its solubility in the high-T steam escaping from the cooling rock.

2. In a wide range of magmas, biotite is stable with magnetite and alkali feldspar, suggesting buffering by a reaction like reaction 5.7.

3. In high-T magmas, irrespective of oxygen fugacity, biotite is less stable than anhydrous silicates plus an aqueous fluid. The specific thermal breakdown reaction for the annite end-member is reaction 5.8.

Thermal Stability of Volatile Minerals. The general form of thermal breakdown (endothermic decomposition) reactions for volatile-bearing minerals is

5.9   *thermal energy + volatile-rich mineral*

      = volatile-free or volatile-poor mineral(s)

      + volatile fluid

Absorption of heat by the volatile-rich mineral liberates the volatiles from it: That is, hydrous minerals are "dried out" by absorption of heat.

The **upper thermal stability limit** of a volatile-bearing mineral lacking solid solution in P-T space is shown in Figure 5.31a and the breakdown band for a volatile-bearing solid solution crossing a magma solidus in Figure 5.31b. From the Clapeyron equation

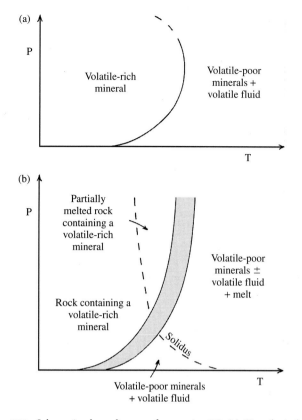

5.31 Schematic phase diagrams for reaction 5.9. (a) Hypothetical volatile-rich mineral that has no solid solution. At increasing P the breakdown curve steepens and eventually has a negative slope. (b) Hypothetical solid solution, such as amphibole, that decomposes over a range of P and T (shaded). Hypothetical solidus (dashed line) for a magma in which the solid solution crystallizes and intersects the decom-position band. Compare Figure 5.11 for a tholeiitic basalt system. Volatile-poor minerals may actually be volatile-free.

(3.13), because the high-T decomposition products have a greater volume and entropy than the volatile-bearing solid, the breakdown curve has a positive slope. Because the volatile fluid is more compressible than the other phases in the reaction, this slope steepens at increasing P and actually becomes negative at high P. Hence, Mg-rich micas and amphiboles are stable into the upper mantle, but not deeper (see, for example, amphibole stability in Figure 5.11).

It is important to realize the contrasting influence of volatiles on the stability of volatile-free minerals, such as feldspars, on the one hand and that on volatile-bearing minerals, such as micas, on the other hand. The upper thermal stability limit of volatile-free minerals represented in their melting curve is depressed to lower T at increasing volatile concentration at increasing P, partly because the mineral components in the melt are diluted by the dissolved volatiles. On the other hand, if volatiles are an essential component in a mineral, then increasing volatile fugacity stabilizes the phase to higher T.

<u>Case History from Mount Saint Helens</u>. The importance of these stability relations in common hydrous magma systems in which both hydrous and anhydrous minerals crystallize can be illustrated for the Mount Saint Helens, Washington, dacite magma episodically erupted during the 1980–1986 activity. The preeruption magma equilibrated at about 900°C and 2.2 kbar or a depth of 8 km. Magmas extruded after the catastrophic explosive eruption of May 18, 1980, contain irregularly shaped, partially resorbed amphibole phenocrysts enclosed within a reaction rim aggregate of orthopyroxene, clinopyroxene, plagioclase, and ilmenite (Figure 5.32); these same crystalline phases occur as phenocrysts and matrix constituents in the dacite. Two observations indicate that in the preeruption magma chamber the amphibole partially reacted with the melt to produce the more stable anhydrous mineral assemblage in the reaction rim:

1. The aggregate occurs where the amphibole is surrounded by melt (now glass) and not where it is in tight contact with another crystalline phase, such as plagioclase, or in the interior of the amphibole crystal.
2. The bulk chemical composition of the aggregate is not equivalent to that of amphibole.

The reaction, as calculated by Rutherford and Hill (1993), involved seven parts amphibole and three parts melt on a weight basis.

How was the magma system displaced out of the stability field of amphibole so as to produce the partial resorption? One possibility is an influx of new, higher-*T*, less hydrous magma into the chamber, but there is no independent evidence for this. Another explanation compatible with the observations of the eruptive activity and its products involves relatively slow decompression of an ascending magma, and possibly, but

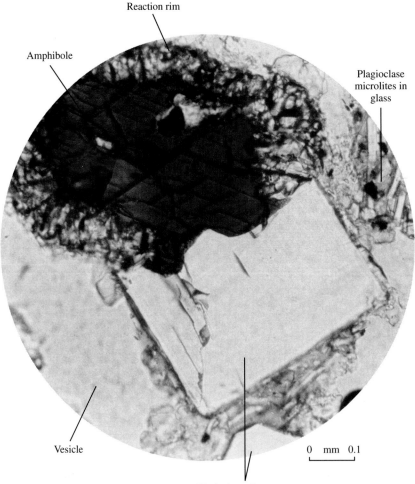

5.32 Reaction rim around a partially resorbed amphibole phenocryst. Photomicrograph under plane-polarized light of post–May 18, 1980, Mount Saint Helens dacite. Anhedral amphibole phenocryst (dark color with characteristic cleavage) is partly surrounded by a reaction rim aggregate of pyroxene + plagioclase + Fe-Ti oxides. Note that the rim occurs between the amphibole and glassy matrix, which was a reactive melt at the time the rim grew, but not adjacent to a contacting "armoring" plagioclase phenocryst. (Photomicrograph courtesy of M.J. Rutherford.)

not necessarily, loss of its dissolved water. Though many explosive pyroclastic eruptions followed the catastrophic May 18, 1980, event, they were of declining intensity and volume. Meanwhile, a viscous lava dome slowly extruded from the vent; the magma feeding this dome is believed to have followed a path indicated by one of the dotted lines in Figure 5.33.

Experimental studies of the rate of amphibole breakdown and rim production indicate ascent rates of 15–50 m/h for post–May 18 magma extrusions. These relatively slow rates would allow some conductive cooling to the wall rocks as well as cooling by the loss of water from the magma. In contrast, ascent rates > 66 m/h must have occurred for the explosively erupted May 18 magma in which no reaction rims developed on the amphibole phenocrysts.

✳5.8 GEOTHERMOMETERS AND GEOBAROMETERS

Determination of intensive parameters in long dead magma systems using equilibrium assemblages of coexisting minerals is a major endeavor of modern petrology. Many quantitative mineral geothermometers and geobarometers have been calibrated by experiment and thermodynamic models, but a detailed presentation is beyond the scope of this textbook; only a few have been briefly introduced. Oxygen fugacity and T can be determined from coexisting equilibrium mag-

netite and ilmenite solid solutions (Section 3.5.5). T can be determined from coexisting alkali feldspar and plagioclase and from clino- and orthopyroxene solid solutions (Sections 5.5.4 and 5.6). A geobarometer for determination of P during crystallization of some felsic magma systems relies on the equilibrium concentration of Al in hornblende; calibration is sensitive to T, oxygen fugacity, and the crystalline phases coexising with a water-saturated melt (Anderson and Smith, 1995). Finally, the water fugacity in a magma can be determined from an equilibrium assemblage of biotite + magnetite + K-feldspar (discussed earlier) in a rock if P, T, and oxygen fugacity can be independently determined (Bohlen et al., 1980).

5.8.1 Assessing States of Equilibrium in Rocks

Use of mineral geothermometers and geobarometers, as well as other interpretations of rocks, depends on a state of stable equilibrium prevailing in the magma system when the crystals formed. Mineral compositions must not have been modified since that equilibrium state and must be accurately known, and all relevant components in the system must be accounted for (Frost and Lindsley, 1992). Stable equilibrium is generally assumed in the case of euhedral crystals, which are bounded on all sides by rational crystal faces, surrounded by glass or aphanitic crystalline material. Such textures are only necessary, not sufficient, to establish a state of stable equilibrium. Various tests have been devised to ascertain whether the pair of minerals used in geothermobarometry are valid representations of equilibrium.

States of disequilibrium, or metastable phases frozen into the rock because of sluggish kinetics and incomplete reaction relations in dynamic magma systems, are more readily discerned and verified than equilibrium states. Disequilibrium states are suggested by the following:

1. Isolated anhedral (noneuhedral) grains that resulted from partial resorption or dissolution into the melt.

2. Reaction rims of one or more phases surrounding another anhedral crystalline phase. Phaneritic rocks may have reaction rims of amphibole jacketing pyroxene (Plate II) and biotite surrounding amphibole that are created by incomplete reaction relations in a fractionating magma.

3. Compositionally zoned grains also originate through incomplete reaction relations within a single solid-solution series, such as plagioclase. Zoning in plagioclase is obvious in thin sections in cross-polarized light but in other solid solution crystals is generally less obvious. In some cases, the outermost rims of an assemblage of zoned minerals represent equilibrium precipitates.

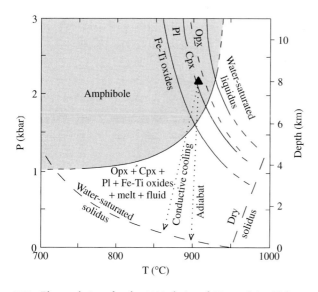

5.33 Phase relations for the 1980 dacite of Mount Saint Helens, Washington. Compare Figure 5.31. Boundary lines (dashed where approximate) are for $P = P_{fluid} = P_{H_2O}$ and oxygen fugacity near the NNO buffer. Stability field of the amphibole in the dacite is shaded. The filled triangle represents equilibration conditions in the magma just prior to its paroxysmal eruption of May 18, 1980. Dotted lines are hypothetical paths of ascending magmas erupted subsequently in the 1980–86 eruptions. (Redrawn from Rutherford and Hill, 1993.)

Advanced Topic Box 5.4 The MELTS computer software model

MELTS is a thermodynamically-based model by Ghiorso and coworkers (e.g., Ghiorso, 1997) of crystal-melt equilibria calibrated from thousands of experiments. The model is menu-driven and interactive and computes phase compositions and proportions by minimization of the chemical potential of magma systems. MELTS models real magmas crystallizing in either the fractional or equilibrium way under specified conditions: initial conditions and constraints under which the system evolves. For example, specification of P, oxygen fugacity, and system composition allows one to track as a function of T the progress of crystallization in a closed system under equilibrium conditions; evolving solid and liquid compositions, viscosities, and densities are output data.

Other indicators of disequilibrium in rocks include experimentally known unstable or unlikely (in an evolutionary sense) mineral assemblages: for example, a volcanic rock that contains magnesian olivine and quartz, or magnesian olivine and/or pyroxene with sanidine, or grains of two distinct compositions in a single solid-solution series having no miscibility gap or solvus. Such disequilibrium assemblages commonly originate through mixing of two dissimilar magmas.

❊5.9 A BRIEF COMMENT REGARDING SUBSOLIDUS REACTIONS IN MAGMATIC ROCKS

All magmatic rocks solidified from magma at high T must have cooled to near-atmospheric T before a petrologist studies them. Plutonic rocks must also have decompressed. At decreasing T, reaction rates are sluggish, causing metastable, magmatic phases to persist well outside their field of stability (Section 3.6.2). Subsolidus mineral reactions and recrystallization are commonly incomplete, leading to states of disequilibrium that can sometimes be detected by the phase rule. Secondary, subsolidus mineral grains are typically very small and not readily identified even with a microscope.

Cooling volcanic rocks emplaced on the surface of the Earth are also subjected to a significant increase in oxygen fugacity relative to pristine deep-seated magmas, which are equilibrated generally within a few log units of the QFM buffer (Figure 3.14). Consequently, the predominantly ferrous Fe in primary magmatic minerals and glass is at least partially oxidized to the ferric state, stabilizing hematite (or at least excess magnetite). The tiny grains of secondary hematite redden Fe-bearing phases, as explained for biotites. Amphiboles and biotites also suffer dehydration reactions because of the decrease in water fugacity that affects erupted magmas. These subsolidus decomposition effects may be difficult to distinguish from reaction relations involving melt above the solidus. However, grain

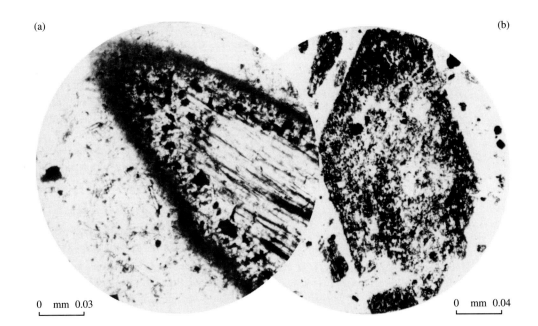

(a)   (b)

0   mm 0.03   0   mm 0.04

5.34 Subsolidus decomposed hornblende phenocrysts. (a) Partially decomposed hornblende with rim of fine-grained Fe-Ti oxides, pyroxenes, and feldspars. (b) Completely decomposed (pseudomorphed) amphibole in diorite porphyry. Fine-grained breakdown assemblage similar to that in (a) mimics shape of original mineral.

sizes tend to be larger in reaction rims, whereas sub-solidus breakdown products are generally very fine-grained anhydrous and oxidized secondary minerals jacketing or completely replacing both amphiboles and biotites (Figure 5.34).

In the plutonic environment, on the other hand, magmatic rocks generally cool to some extent below the solidus T before much decompression occurs, per-haps millions of years later by tectonic uplift and ero-sional denudation. Therefore, the P-T path followed is much different, and so is the nature of secondary min-eralogical changes. As the crystalline plutonic assem-blage cools, it passes into the stability fields of lower T, but elevated P, hydrous phases such as epidote miner-als and other hydrated Ca-Al silicates, chlorites, and, at lowest temperatures, clay minerals (Plate IV).

Summary

Crystal-melt equilibria in simple model systems indi-cate what is possible and impossible in chemically more complex, multicomponent natural magma systems. The phase rule can be used to assess states of equilib-rium and the degrees of freedom in magmatic systems. For example, both intensive variables may be freely varied without changing the state of equilibrium in di-variant stability fields in P-T or T-X space. Drawing isopleths, isotherms, and tie lines and applying the lever rule permit compositions and proportions of co-existing melt and crystals to be determined in equilib-rium systems.

Essentially isothermal decompression of initially sub-solidus water-free rocks can lead to melting, whereas decompression of water-saturated magmas can cause crystallization.

Solid solutions, some pure minerals such as ensta-tite, and polymineralic rocks melt incongruently to a liquid and crystals whose compositions differ from that of the starting material. Incongruent melting ac-counts for much of the diversity of igneous rocks and the global differentiation of the Earth into crust and mantle.

Denser crystalline phases are stabilized in magmas at high P. Under upper mantle conditions, a basalt magma can contain Mg-rich garnet and a complex Na-Al-Ca clinopyroxene instead of plagioclase, Ca-Mg-Fe pyroxene, and olivine.

In crystallizing multicomponent magmas and in melting source rocks, reaction relations at the ionic level between liquid and solid solutions continuously modify their compositions as intensive variables change. Compounded with kinetic factors in dynamic magma systems, reaction relations in crystallizing magmas al-low for a wide range of possible liquid lines of descent

taken by residual melts and reaction series taken by crystalline products. A significant kinetic factor is the degree to which a magma fractionates during crystal-lization because reaction relations between melt and crystals are precluded. Fractionation can occur by moving the crystal away from the parcel of melt where it precipitated, by creating a reaction rim about the metastable crystal, or by changing intensive parameters in the magma faster than the rate at which reaction occurs.

Fractionation extends the range of T over which crystallization takes place, extends the range of residual melt compositions, and creates more diverse crystalline products, or reaction series. Liquid lines of descent produced as solid solutions fractionate are curved on a variation diagram. More evolved residual melts are commonly enriched in Si and have greater (Na + K)/Ca and Fe/(Fe + Mg) ratios.

Volatiles, especially water because of its high solu-bility in melts, have a profound influence on crystal-melt equilibria. By depolymerization of the silicate melt, liquidus and solidus temperatures are depressed by as much as several hundreds of degrees. Volatile-bearing minerals, especially major rock-forming micas and amphiboles, are stabilized in volatile-rich magmas but decompose into volatile-free minerals at high T. Amphibole and biotite stabilization requires 3–5 wt.% dissolved water in andesite-dacite melts. Stability of Fe-bearing hydrous minerals such as biotite and amphi-bole depends on fugacities of H_2O, O_2, Cl, and F in addition to P and T. Nonvolatile mineral compositions can also be influenced by volatiles; thus, more calcic plagioclase is stabilized in water-rich magmas. Sub-solidus decomposition (alteration) of primary high-T magmatic precipitates yields fine-grained aggregates of ferric oxides and anhydrous minerals at low water and high oxygen fugacity in extruded magma and fine-grained, more hydrous phases in confined plutonic sys-tems to the extent that water is available.

Because the equilibrium chemical composition of crystallizing solid solutions in magma systems depend upon the prevailing values of intensive variables, the compositions of coexisting minerals in a rock serve as geothermometers and geobarometers.

CRITICAL THINKING QUESTIONS

5.1 Describe how the phase rule is derived, what it means, and how it is used in reading phase dia-grams and interpreting magmatic mineral assem-blages.

5.2 Contrast the effects of decompressing a mag-matic system from a P just above the solidus un-der water-free and water-saturated conditions.

5.3   Liquidus temperatures in simple binary systems (e.g., Figures 5.4 and 5.8) used to model multicomponent magma systems are hundreds of degrees higher than the temperatures at which liquidus phases precipitate in real magmas (e.g., Figures 5.10 and 5.11). Why?

5.4   A liquidus line in a binary phase diagram has four meanings. Describe them.

5.5   Contrast the state of a water-free and a water-saturated tholeiite system at 1100°C and 10 kbar (Figure 5.11).

5.6   Cite two scenarios that would allow a thermally insulated, water-rich intrusive magma to crystallize rapidly—developing a fine-grained texture—as if it were extruded onto the surface of the Earth into the cool atmosphere.

5.7   In a magmatic rock, what compositional aspect—modal or mineral—defines the intensive parameters prevailing at the time the magma crystallized? Explain.

5.8   Contrast crystallization in a magma system dominated by a eutectic versus a system in which continuous reaction relations take place over a range of *P-T* conditions. Which has the potential to develop the most diverse crystalline products? Which do you think prevails in real magmas? Discuss.

5.9   Describe all of the ways in which a crystallizing magma can fractionate crystals from melt.

5.10   Contrast equilibrium versus fractional crystallization with respect to operative range of *T,* range of residual melt compositions, and nature of crystalline products. Cite examples.

5.11   Describe incongruent melting and its significance.

5.12   Account for the contrasting mineralogical compositions of basalt and eclogite.

5.13   Why are liquid lines of descent on variation diagrams curved for fractionating systems involving solid solutions?

5.14   Partially resorbed anhedral phenocrysts in volcanic rocks are not uncommon. Describe possible origins in both open and closed magma systems.

5.15   Contrast the changes in the magmatic system that created the breakdown of amphibole in Figures 5.32 and 5.34.

5.16   How common are water-saturated granitic magmas? Most of the experimentally determined phase relations for granite systems are for water-saturated conditions. Does this necessarily mean real magmas are water-saturated? What textures can be found in water-saturated magmas? What is the implication of the widespread assemblage of biotite + alkali feldspar + magnetite in

granitic rocks? (Hint: Write a balanced mineralogical reaction involving these three minerals.)

5.17   Is it true that minerals having higher melting temperatures crystallize from magmas at relatively higher temperatures? Critically evaluate this by examining the relative melting temperatures of quartz (β-form in Figure 5.1), forsterite olivine, diopside pyroxene, and albite feldspar (Figure 3.8). With what mineral(s) does quartz commonly coprecipitate (Figure 2.2)? What really controls crystallization temperatures of minerals from melts?

5.18   On the basis of the concepts of crystal-melt equilibria set out in this chapter, how should Figure 3.12 be modified?

PROBLEMS

5.1   Describe the melting and crystallization of a bulk composition An$_{20}$ from above the liquidus and from below the solidus, respectively, in the CaMgSi$_2$O$_6$-CaAl$_2$Si$_2$O$_8$ system at 1 atm. What are the proportions of phases at 1300°C?

5.2   Thousands of basaltic deposits worldwide contain fragments of phaneritic rock composed of about 65% magnesian olivine, 25% enstatite, 8% diopside, and 2% Mg-Al-Cr spinel. Name this rock from Chapter 2, recalculate its mode so that olivine plus enstatite equals 100%, and draw an appropriate isopleth on Figure 5.8 for their recalculated proportions. (Because the densities of the two minerals are nearly the same, no correction is needed to convert modal volume into weight percentage.) Because these fragments are pieces of the upper mantle from which basalt magmas are derived by partial melting, describe the melting relations of this recalculated model mantle rock. How much melt is produced at the initial melting *T?* What is the composition of the first partial melt? How does it compare in composition with an average basalt (Table 2.2)?

5.3   Using MELTS (Advanced Topic Box 5.4) on the Makaopuhi basalt magma, (a) confirm that residual melts become more enriched in Fe and Ti until Fe-Ti oxides precipitate; (b) determine the chemical composition of the residual melt just above the solidus; (c) track the changing chemical compositions of major solid solutions during crystallization. Discuss your results.

5.4   Make photocopies of Figure 5.11. Color the stability field of amphibole. Is amphibole a stable phase in extruded basalt magmas? Discuss. In the dry system, contrast (by colors) the stability fields with plagioclase and with garnet. What controls these contrasting stabilities? What is the maximum depth to which olivine is stable in water-free or water-saturated tholeiitic basalt? How do you reconcile this limited stability of olivine in basalt relative to what it is in upper mantle peridotite (Figure 1.3)?

5.5 Prove that the mass balance equation 5.4 is valid by multiplying the weight percentages of liquid and crystalline phases times the weight fractions of the $NaAlSi_3O_8$ and $CaAl_2Si_2O_8$ components on each side of the equation.

5.6 On a photocopy of Figure 5.27a, draw the path of a fractionating melt whose initial composition is 5 wt.% SiO_2 and 25 wt.% $KAlSi_3O_8$. Construct lines of liquid descent for Na_2O and K_2O versus SiO_2 for this fractionating melt. Compare your diagram with Figure 5.27b.

5.7 Sketch the binary $KAlSi_2O_6$-SiO_2 phase diagram at 1 atm on the basis of the following information: Pure K-feldspar melts incongruently at 1150°C to equal amounts of liquid and leucite crystals; melting is complete at about 1560°C. Pure leucite ($KAlSi_2O_6$) melts at 1685°C and pure silica at 1713°C. A binary eutectic lies at about 990°C and 54 wt.% silica where a silica polymorph, K-feldspar, and melt are in equilibrium. Describe equilibrium and fractional crystallization of a system that contains 30 wt.% silica.

5.8 Sketch the phase diagram for the binary system Mg_2SiO_4 (forsterite)-Fe_2SiO_4 (fayalite) at 1 atm keeping in mind that solid solution is complete, as it is in the plagioclase system. The melting T of pure fayalite is 1205°C and of pure forsterite (Figure 3.8) about 1890°C. Describe the liquid and crystalline products of perfect fractional crystallization of a selected melt composition, including a statement about the Fe/(Fe + Mg) ratio.

5.9 On a P-T diagram, sketch approximate relative positions and slopes of liquidus and solidus curves for a water-saturated granitic magma. Superpose on this diagram the relative approximate positions of the breakdown curves for muscovite, biotite, and hornblende, using the following observations for guidelines: Phenocrysts of muscovite almost never occur in rhyolite, but fine-grained muscovite does form as a subsolidus alteration product. Primary muscovite occurs only in some peraluminous granites. Biotite and hornblende are common in granite and as phenocrysts in rhyolite and dacite. Amphibole is abundant in higher-T diorites.

5.10 Using Advanced Topic Box 5.2 and Figure 5.7 as guides, draw a sequence of G-X diagrams at different temperatures for the $NaAlSi_3O_8$-$CaAl_2Si_2O_8$ system.

Chemical Dynamics of Melts and Crystals

6

FUNDAMENTAL QUESTIONS CONSIDERED IN THIS CHAPTER

1. How do atomic-scale, time-dependent kinetic phenomena govern changing states of melts? How do melts flow as viscous liquids?

2. What is viscosity, what factors control the viscosity of melts, and how does viscosity impact the dynamic behavior of melts?

3. What is diffusion, what factors control rates of atomic and thermal diffusion, and how do these rates impact crystallization and vesiculation of melts?

4. How do crystals nucleate, grow, and modify their shape and size?

5. How do exsolved volatile bubbles nucleate, grow, and fragment melts?

INTRODUCTION

In dynamic magmatic systems, changing intensive parameters and energy conditions cause the state of the system to change, but only insofar as kinetic factors allow. Thus, the rate at which the T of a magma decreases is commonly faster than the rate at which kinetically controlled crystallization can progress, creating compositionally zoned crystals and reaction rims in the fractionating magma. Sluggish kinetic rates prevent homogenization of the zoned crystal and equilibration via reaction relations with the enclosing melt, even though temperatures may be relatively high. Instantaneous change from an initial metastable state to a new state of more stable, perfect equilibrium almost never takes place in magmatic systems because of slow kinetics.

In this chapter, we focus on kinetic factors that operate on the atomic level in controlling two basic dynamic aspects of melts—solidification and vesiculation-fragmentation. Whether the same granitic melt solidifies to a massive metastable glass; to an aphanitic or to a phaneritic aggregate of quartz, feldspar, and biotite crystals; to highly vesicular pumice; or to fine ash particles depends on the interplay between rates of changing environmental conditions, such as T, and kinetic (time-dependent) process rates. The nature of the kinetic history, or kinetic path, taken by the changing melt during solidification or vesiculation is recorded in the fabric of the resulting rock, discussed in the following chapter.

Section 3.6 introduced kinetic phenomena. A comprehensive treatment is Lasaga (1998).

The first three sections of this chapter deal with three so-called **transport phenomena** (Figure 6.1):

1. Transport of viscous material, sometimes referred to as a transfer of momentum (1 = mass × velocity)

2. Transfer of atoms, usually referred to as chemical diffusion

3. Transfer of heat by conduction (Section 1.1.3)

Each of these three transport processes has a driving force and a proportionality constant that is an intrinsic property of the material. Each depends in some way on the movement of atoms.

✳6.1 VISCOSITY OF MELTS

On an atomic scale in melts, viscosity is a measure of the mobility of atoms, how readily atomic bonds can be stretched, broken, and reformed with neighboring

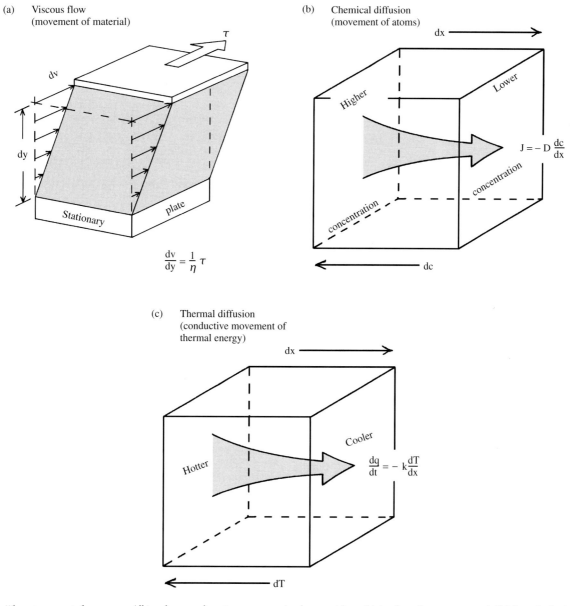

(a) Viscous flow
(movement of material)

τ

dv

dy

Stationary   plate

$$\frac{dv}{dy} = \frac{1}{\eta}\, \tau$$

(b) Chemical diffusion
(movement of atoms)

dx

Higher   Lower

$J = -\,D\dfrac{dc}{dx}$

concentration   concentration

dc

(c) Thermal diffusion
(conductive movement of
thermal energy)

dx

Hotter   Cooler

$$\frac{dq}{dt} = -\,k\frac{dT}{dx}$$

dT

6.1 Three **transport phenomena.** All involve a gradient in some quantity that provides a driving force for transport and all follow similar laws. (a) **Viscous flow** in a cube of material (dashed lines) is driven by a force applied tangentially to its upper surface area. This applied shear stress (force/area = stress), τ, produces a change in the shape of the body so that the cube becomes a parallelepiped. The change in shape, called strain, ε, becomes more extreme with time, that is, the applied shear stress creates a certain strain rate, $d\varepsilon/dt$, whose magnitude depends on the viscosity of the particular material, η. The strain rate is equivalent to a velocity gradient, dv/dy, which is shown by arrows of increasing length over the distance dy. (b) **Chemical diffusion** is driven by a change in concentration, dc, of some particular atoms between a region of higher concentration (left face of the cubical volume of material) and lower concentration (right face); these two faces are separated by a distance, dx. The concentration gradient, dc/dx, produces a flow of atoms, J, down the gradient. (c) **Thermal diffusion**, or **heat conduction**, is driven by a gradient in temperature over a distance, dT/dx, and produces a flow of heat over time, dq/dt. The proportionality constants for these three types of transport are intrinsic properties of the material. For viscous flow the constant is the coefficient of viscosity, or viscosity, η; for chemical diffusion it is the diffusion coefficient, or diffusivity, D; and for heat conduction it is the thermal conductivity, k.

atoms. Solidification of melts and exsolution of volatile components from them, which are the two principal themes of this chapter, involve the movement of atoms, or the lack thereof. The ease with which atoms can move about is expressed on a larger macroscopic scale in a body of material, such as a lava flow, in its resis-

tance to flow or deformation. On both scales, **viscosity** is a manifestation of mobility. Less mobility corresponds to greater viscosity. "Fluidity" and viscosity are inversely related (opposites).

Unquestionably, viscosity is the most important property of melts that controls the dynamic behavior of magmas. Segregation of partial melts in upper mantle and lower crustal sources, magma ascent to shallower depths, intrusion, extrusion as lavas or as explosive fragments, and crystallization all depend on the viscosity of the melt.

Viscosity is formally defined in Figure 6.1a. A force exerted parallel to the surface area of a viscous body produces a shear stress, τ (sress = force/area), that deforms the body by viscous flow in a time-dependent way. This flow deformation can be measured by a velocity gradient, dv/dy, or alternatively by a rate of change in shape, or strain rate, $d\varepsilon/dt$. The proportionality constant relating the shear stress (the "driving force") and the resultant flow is the coefficient of Newtonian viscosity, η

$$6.1 \quad \tau = \eta \frac{dv}{dy} = \eta \frac{d\varepsilon}{dt}$$

For brevity, the coefficient of viscosity can simply be called the *viscosity*.

A convenient way to measure viscosity is by determining how fast a sphere of contrasting density falls or rises through the material (see Stokes law in Section 8.3.3).

The unit of viscosity is the Pa s (Pascal × second) or the poise (10 poise = Pa s). Typical viscosities (in Pa s) include air 10^{-5}; water 10^{-3}; honey 10^1; cool asphalt (tar) 10^8. Silicate glasses have viscosities $> 10^{12}$. In terms of its atomic structure, glass is a liquid, but in terms of its behavior (atomic mobility and capacity to flow), it is a solid. One way to distinguish between a solid and a liquid is by the time scale for flow. Liquids, such as water and honey, can flow appreciably in minutes because of low viscosity, whereas solids "flow" negligibly, or not at all, over periods measured in years. For example, the solid mantle of the Earth, whose viscosity is on the order of 10^{20} Pa s, convectively flows at a rate of a few centimeters or less per year, the rate of plate motion.

The two principal factors governing viscosity of melts are their composition—as it controls the degree of polymerization—and their T. More polymerized, more silica-rich melts, which have smaller ratios of nonbridging oxygens to network-forming cations (Section 4.1.1), are more viscous. Ions cannot move about readily, especially those that make up polymers, because the strong Si = O and Al = O bonds must be stretched or broken. Less polymerized silica-poor melts and silica-rich melts that contain dissolved water and/or fluorine have lower viscosities. For any melt composition, higher T reduces the viscosity by "loosening" the melt structure through the increased kinetic energy of the atoms; ionic mobility is enhanced.

Figure 6.2 illustrates the strong dependence of melt viscosity on T and composition. For example, the dependence on T is shown by the fact that between 700°C and 1000°C the viscosity of a water-free rhyolite melt decreases by six orders of magnitude, from about 10^{15} to 10^9 Pa s. The dependence on composition is shown by the fact that there is a difference of about 10^5 Pa s in viscosities at 1200°C between a silica-poor (46 wt.%) alkali-olivine basalt melt and a silica-rich (77 wt.%) rhyolite melt, both water-free. Besides the silica concentration factor, dissolved water drastically lowers the viscosity of highly polymerized, silica-rich melts because the polymers are broken by hydroxyl ions. Addition of 6 wt.% water to a rhyolite melt with 77 wt.%

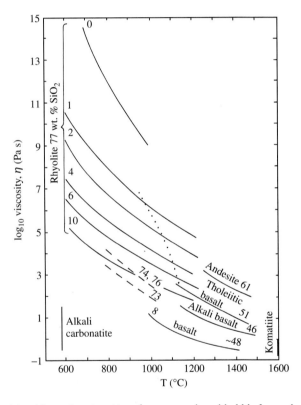

6.2   Newtonian viscosities of some crystal- and bubble-free melts as a function of T at 1 atm. P dependences are negligible. Some viscosities pertain to metastable melts at temperatures below their liquidus where they would contain crystals under equilibrium conditions. Concentrations of silica in weight percentage are indicated as well as of water in weight percentage at low-T end of curves. (Data from Shaw, 1965; Murase and McBirney, 1973; Webb and Dingwell, 1990; Dawson et al., 1990; Huppert and Sparks, 1985.) Viscosities for rhyolite melts that contain 1.5 wt.% fluorine and 6 wt.% water at 10 kbar are dashed lines (peraluminous, 74 wt.% silica; metaluminous, 76; peralkaline, 73). (Data from Baker and Vaillancourt, 1995.) The dotted line indicates the viscosity of residual melts resulting from fractional crystallization of Makaopuhi basalt melt (Table 5.1) with 0.5 wt.% water on the QFM buffer calculated from the MELTS program; the most evolved melt has 73 wt.% silica and 0.3 wt.% water.

silica at 800°C lowers the viscosity by seven orders of magnitude, from about 10^{12} to 10^5 Pa s. Most of the reduction in viscosity in hydrous silicic melts occurs in the first 2 wt.% or so of added dissolved water because that first 2% is mostly $(OH)^-$ (Section 4.2.2). Dissolved water has less effect on basalt melts because they are less polymerized regardless of volatile content. Dissolved fluorine also reduces polymerization and viscosity, but, unlike water, which has very limited solubility in melts at low P, fluorine can remain in solution in the melt and can therefore enhance the mobility of lavas extruded onto the surface of the Earth. Rhyolite melts having the lowest viscosity are those that contain both dissolved water and fluorine. The lowest viscosities of any known terrestrial melts are found in ultramafic high-T komatiites and essentially silica-free, alkali carbonatites extruded at about 585°C from Oldoinyo Lengai volcano in the East African Rift (Dawson et al., 1990). The 15 orders of magnitude difference in viscosity between carbonatite and dry rhyolite melts extruded near 600°C is illustrated by the contrasting morphological characteristics of their extruded lavas—the former having the appearance of muddy water and the latter forming steep bulbous domes that can be higher than their horizontal diameter. Widespread basaltic lavas that are no more than a couple of orders of magnitude more viscous than cooler carbonatitie melts form sheetlike flows, again in striking contrast with mushroom-shaped rhyolite extrusions.

Melt viscosity is only weakly dependent on P. An isothermal increase in P of 20 kbar reduces the viscosity of mafic melts by less than 50% and silicic melts by a slightly greater factor (Richet and Bottinga, 1995). This negative dependence on P follows because bridging oxygens have a greater molar volume than nonbridging, and, therefore, from Le Chatelier's principle, a positive increment in P increases the ratio of nonbridging to bridging oxygens that depolymerizes the melt, reducing the viscosity.

The viscosity of melts that carry suspended crystals (i.e., magma) and contain bubbles of volatile fluid, as well as those that have an intrinsic yield strength do not follow the form of Equation 6.1 (see Section 8.2.2 for a discussion of non-Newtonian magmas).

If a melt is deformed slowly, its distorted atomic structure has time to adjust or relax; it behaves as a liquid. However, for increasingly rapid deformation (increasing strain rate) at constant T, particularly in cooler more viscous polymerized melts, the melt crosses a threshold between liquid and solid behavior, called the *glass transition,* defined as the T at which $\eta = 10^{12}$ Pa s. If rapid deformation takes place at viscosities above this glass transition, the melt cannot adjust: its **relaxation "time"** is exceeded. It behaves as if it were a solid, breaking as any brittle solid does if the applied stress is excessive.

Advanced Topic Box 6.1   Calculation of melt viscosity

Shaw (1972) developed a model from which the viscosity of homogeneous melts can be calculated from their chemical composition. This model assumes the viscosity follows an Arrhenius relation $\eta = A \exp(-E_a/RT)$ where A is a constant, E_a is the activation energy of viscous flow, R is the gas constant, and T is in degrees Kelvin. Although many melts depart to varying degrees from Arrhenian behavior, Shaw's model has nonetheless proved useful. The full details cannot be dealt with here; only a "cookbook" presentation is given. The viscosity, η, is found from the equation

$$2.303 \log_{10}\eta = s(10^4/T) - 1.5s - 6.4$$

For example, calculation of the viscosity of a hydrous rhyolite (granite) melt at $T = 1073K$ (=800°C) is presented in Table 6.1, where determination of the value of s is described (minor constituents, such as MnO and P_2O_5, are ignored). Numbers in the Moles column are found by dividing the Wt.% values by the Formula wt. of the corresponding oxide. Note that the moles of Al_2O_3 and Fe_2O_3 are doubled. The mole fraction, X, is found by dividing the Moles value by the sum of Moles, or 1.908. In the next column, X_i, the mole fractions of Fe_2O_3, FeO, and MgO are added together (giving 0.009), as are CaO and TiO_2 (0.004), and Na_2O and K_2O (0.057). Shaw's values of the partial molar activation energies of SiO_2 in binary systems with categories of other oxides, s_i^o, are listed in the next column. These values multiplied by the mole fraction of SiO_2 in the particular melt are listed in the $s_i^o X_{SiO_2}$ column. Finally, at the bottom of Table 6.1, the sum of the $X_i(s_i^o X_{SiO_2})$ values listed in the last column is divided by $(1 - X_{SiO_2})$ to give s, which is inserted into the viscosity equation. The viscosity calculated from this equation is $10^{5.24}$ poise = $10^{4.24}$ Pa s. The viscosity measured by Shaw (reference in his 1972 paper) is $10^{4.63}$ Pa s.

Hess and Dingwell (1996) and Baker (1996) have devised models based on experimental data that account for the non-Arrhenian T dependence of viscosity for hydrous rhyolite (granite) melts.

In the Shaw, Hess-Dingwell, and Baker models no dependence on P is included, as this is negligible.

Ryan and Blevins (1987) have compiled experimentally determined viscosities of a wide range of melt compositions.

Table 6.1   Calculation of the viscosity of a hydrous rhyolite melt at 1073K as described in Advanced Topic Box 6.1.

	Wt.%	Formula Wt.	Moles	X	X_i	s_i°	$s_i^\circ X_{SiO_2}$	$X_i(s_i^\circ X_{SiO_2})$
SiO_2	71.90	60.085	1.195	0.627				
Al_2O_3	12.15	101.961	0.238	0.125	0.125	6.7	4.201	0.525
Fe_2O_3	0.57	159.692	0.007	0.004	0.009	3.4	2.132	0.019
FeO	0.52	71.846	0.007	0.004				
MgO	0.04	40.311	0.001	0.001				
CaO	0.27	56.079	0.005	0.003	0.004	4.5	2.82	0.011
TiO_2	0.09	79.899	0.001	0.001				
Na_2O	3.94	61.999	0.064	0.033	0.057	2.8	1.756	0.100
K_2O	4.32	94.203	0.046	0.024				
H_2O	6.20	18.015	0.344	0.180	0.180	2.0	1.254	0.226
SUM	100.00		1.908					0.881

$s = [\Sigma \, X_i \, (S_i^\circ \, X_{SiO_2})] \div (1 - X_{SiO_2}) = 0.881 \div 0.373 = 2.362.$ $2.303 \log_{10} \eta = s \, (10^4/T) - 1.5s - 6.4 = 2.362 \, (10^4/1073) - (1.5 \times 2.362) - 6.4 = 12.07.$ $\eta = 10^{5.24}$ poise $= 10^{4.24}$ Pa s.
Data from Shaw (1972).

✳6.2 CHEMICAL DIFFUSION

Diffusion, the second transport phenomenon (Figure 6.1b), is the movement of individual atoms or molecules through a group of atoms and driven by a concentration gradient. The material in which atoms are moving does not itself necessarily move. Atoms and molecules diffuse quickly through gaseous bodies where the atoms are widely separated and loosely associated; for example, aromatic molecules from brewing coffee disperse many meters in minutes into the surrounding air. Diffusion is much slower in a liquid and depends strongly on its viscosity and *T;* higher *T* expands the atomic structure and, together with lower viscosity, allows greater freedom of movement for diffusing atoms. Diffusion is still slower in solids where atoms are tightly bonded and have limited kinetic motion relative to those in liquids.

Bowen (1921) first drew attention to the significance of diffusion in magma systems. But it was not until a half-century later, beginning with Shaw (1974), that other experimental petrologists began to explore this important transport process.

Diffusion plays many roles in petrologic processes, some of which are growth of crystals from melts and other multicomponent solutions, exsolution of volatiles from a melt, reaction relations between melt and crystals to produce equilibrium compositions among major as well as trace elements, mixing and contamination of magmas, and isotopic exchange between crystals and fluids. Because the distance over which diffusion takes place depends on time and *T,* measurement of small-scale compositional inhomogeneities within rock-forming minerals and glasses, such as by an electron microprobe, can provide important constraints on their thermal history if the relevant diffusional properties are known. The **closure temperature** at which

T-dependent diffusion effectively ceases, or becomes nil in a geologic time frame, must be taken into account in interpreting isotopic ages, mineral geothermometers and geobarometers, and any other aspect of a "frozen-in" state of compositional equilibrium.

If rates of diffusion (and other kinetic processes) were infinitely fast, thermodynamic equilibrium would be accomplished instantaneously as changes in intensive variables occurred. However, diffusion is slow in melts and can be extremely slow in crystals, so that vestiges or relics of prior states may be preserved in magmatic rocks. Fortunately for the petrologist, sluggish rates of diffusion can provide insights into prior metastable states and make possible inferences about petrologic history.

6.2.1 Types of Diffusion

Three routes of diffusion are possible in an aggregate of mineral grains, a rock.

1. **Surface diffusion** occurs essentially over a two-dimensional surface area, but generally has a component of atomic movement in the third dimension. Thus, ions can move about over the free surface of a mineral grain through a static liquid in contact with it.

2. **Grain-boundary (intergranular) diffusion** occurs along mutual boundaries between contacting mineral grains. Because of the hindering effects of the adjacent grain, diffusion is slower than for surface diffusion, though both utilize the exterior of a grain. A very thin film of fluid along grain boundaries may facilitate diffusion.

3. **Volume diffusion** occurs within any single homogeneous phase, such as a crystal or a melt body. Volume diffusion through solids is generally the slow-

est, commonly by many orders of magnitude, of the three types of diffusion.

6.2.2 Theory and Measurement

In liquids, atoms have considerable mobility and one atom can collide billions of times a second with its neighbors. In solids, atoms vibrate thermally in a more or less fixed position but can, from time to time, randomly jump by pure chance to a new site. If the concentration of a particular atom varies in some direction, then the random jumps of those atoms will result in a net movement, a flow, or flux, toward the lower concentration region, down the concentration gradient, smoothing out the overall gradient over time (Figure 6.3a).

The mathematical expression, known as **Fick's first law of diffusion,** for the diffusive flow, J_i, of atoms i down their concentration gradient, dc/dx, in the x direction is

$$6.2 \quad J_i = -D_i\left(\frac{dc}{dx}\right)$$

The minus sign that precedes the right-hand expression accommodates the fact that atoms flow spontaneously toward a lower concentration: That is, dc_i is intrinsically negative (Figure 6.1b). The **diffusion coefficient,** or **diffusivity** D_i, has units of square meters per second (m^2/s) and a magnitude related to the frequency at which atoms jump and their jumping distance. Large or highly charged ions strongly bonded with their neighbors jumping through a highly viscous fluid or a crystalline solid have small diffusivities. The transfer rate, or diffusion flux, J_i, can be considered as a ratio of the driving force, dc_i, to the resistance, dx/D_i, for motion of the diffusing atom. In isotropic phases, such as melts, where properties are the same in all directions, the diffusion rate is independent of direction

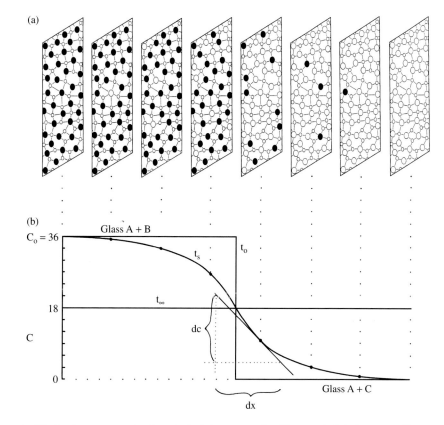

6.3 Schematic atomic diffusion in two contacting hypothetical glasses or melts. Glass on the left that is *initially* composed only of atoms A (small white balls) and B (large black balls) is placed in contact at time t_0 with a glass on the right *initially* composed only of atoms A and C (large white balls). Each glass is assumed to extend infinitely far from their mutual contact: Each is a semi-infinite medium. Because of the initial steplike concentration profile of B atoms, line labeled t_0 in (b), and of C atoms across the contact, B atoms diffuse into the B-free glass on the right and C atoms diffuse in coupled manner into the C-free glass on the left. After some time, t_s diffusion has smoothed the concentration gradient of B atoms, curve labeled t_s in (b), represented in the panel-like slices through the two glasses shown in (a). Note that the glass represented in the three panels on either side of the contact have experienced a change in concentration of atoms B and C whereas the end panels still have the initial concentration. After infinite time, t_∞, the concentration gradient of B atoms is eliminated, straight horizontal line labeled t_∞ in (b), and the two glasses are of the same uniform composition: Concentrations of atoms A, B, and C are the same throughout. In (b), the initial concentration of B atoms in the left glass is c_0 and the scale along the vertical concentration axis is c. The concentration gradient at any position x along the horizontal axis is dc/dx. The equation for the line labeled t_s is the mathematical solution (Jost, 1952, p. 20) for a common experimental setup of Fick's second law of diffusion. Experimentally measurable variables c, x, and t allow evaluation of the diffusivity D.

and only one D value applies. However, most crystals are anisotropic; hence values of the diffusion coefficient and diffusion rates are somewhat different in different crystallographic directions.

Fick's first law describes a steady state in which the concentration profile remains constant through time. But the extent of chemical diffusion in dynamic petrologic systems is time-dependent, expressed in Fick's second law of diffusion, a second-order differential equation (Jost, 1952; Crank, 1975). A graphical solution of this equation is shown in Figure 6.3b for two bodies of glass of contrasting composition juxtaposed along a planar boundary and extending infinitely far in either direction. Another setup might involve a crystal of fayalite (Fe_2SiO_4) in physical contact along a smooth planar interface with a crystal of forsterite (Mg_2SiO_4). At high T, Mg^{2+} ions measurably diffuse into the fayalite and Fe^{2+} ions migrate in the opposite direction into the forsterite. In this case, ionic motion is coupled if electrostatic balance and the appropriate stoichiometrical characteristics of olivine are to be preserved. If left in contact for a sufficient time at high T, the concentrations of Mg and Fe cations become uniform across the two crystals, and they become one homogeneous crystal (if their crystallographic orientations were initially identical).

6.2.3 Factors Governing Diffusivities

The diffusivity, D, of a particular atom in a melt or volatile fluid is inversely correlated with its radius, r, and the viscosity of the medium but directly proportional to T. This relationship is formalized in the **Stokes-Einstein equation**

$$6.3 \quad D = \frac{kT}{6\pi\eta r}$$

where k is the Boltzman constant, T is in degrees Kelvin, and η is the Newtonian viscosity. This model predicts that D is only weakly dependent on P because of the weak dependence of η on P.

The dependence of diffusion on T is described by an Arrhenius equation (equation 3.38 and Worked Problem Box 3.2)

$$6.4 \quad D = D_0 \exp(-E_a/RT)$$

Therefore, the three variables T, D_0, and E_a govern the diffusivity, D, in a solid or liquid medium. The Arrhenius equation indicates diffusion is always more rapid at higher T.

Diffusion in Crystals. In crystals, for relevant magmatic temperatures, the diffusivity, D, ranges over many orders of magnitude. Figure 6.4 shows a range from 10^{-23} m^2/s to 10^{-8} m^2/s. The fastest diffusion is for "trace" diffusion of H_2 along the a axis of Mg-rich

olivine and the slowest is for coupled CaAl-NaSi chemical diffusion in bytownite plagioclase, An$_{80}$. Note the large activation energy for diffusion in plagioclase, 516 kJ/mole, which is not surprising in view of the strong bonding of Al and Si to O. In other words, the slower diffusion of network-forming Si and Al is the limiting rate and not the faster diffusion of network modifying Na and Ca. CaAl-NaSi interdiffusion in plagioclase is more than 10 orders of magnitude slower than diffusion of Na and K in alkali feldspar at about 1000°C, and the difference increases as T decreases. This fact readily accounts for the ubiquity of zoned plagioclases in contrast to the less common, generally more subtle zoning seen in alkali feldspars and other rock-forming minerals. Oxygen isotope exchange in calcic plagioclase is orders of magnitude faster than

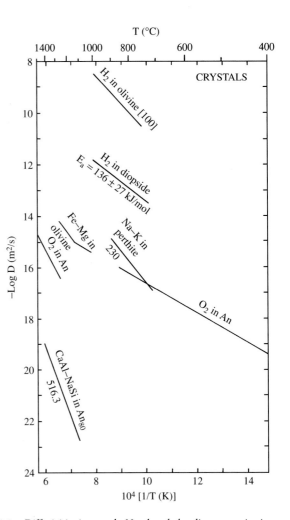

6.4 Diffusivities in crystals. Numbers below lines are activation energies in kilojoules per mole. Diffusion of H_2 in olivine along the a axis [100] and in diopside (in which diffusion appeared to be isotropic). (From Ingrin et al., 1995.) Homogenization of K and Na concentration in a perthite and Fe-Mg coupled diffusion along [001] in olivine. (From Freer, 1981.) Oxygen self-diffusion in anorthite (An) and coupled CaAl-NaSi diffusion in calcic plagioclase An$_{80}$ at 1 atm (homogenizing fine-scale oscillatory zoning). (From Grove et al., 1984.)

CaAl-NaSi homogenization. In the gabbroic Skaergaard intrusion, plagioclases are commonly zoned, yet the oxygen in some of them is from low-*T* (250–400°C) hydrothermal solutions.

Diffusion in crystals is accomplished by movement of atoms utilizing **point defects** (Figure 6.5). These imperfections in the atomic structure originate as the crystal grows and persist regardless of what subsequently happens to it. Vacancy concentrations are on the order of several per million lattice sites and attain some particular equilibrium value depending upon *T*.

Diffusion in Melts. Diffusivities are generally larger in melts than in crystals because of the more expanded, or "looser," atomic structure of melts. Figures 6.6 and 6.7 indicate diffusivities that range from 10^{-18} m²/s to 10^{-6} m²/s. Comparison of diffusion of univalent alkali metal and divalent alkaline earth ions in water-free rhyolite melt shows that smaller, less charged ions diffuse relatively faster.

Dissolved water reduces the degree of polymerization and the viscosity of rhyolite melts, consequently hastening rates of diffusion of many chemical species by as much as four orders of magnitude (Figure 6.7a). The diffusion of H_2 in hydrous silicic and perhaps basaltic melts is the fastest of any species known thus far. Diffusion of volatiles such as Cl, S, H_2, and CO_2 is enhanced in hydrous melts (Figure 6.7b). Infrared spectroscopy suggests that water diffuses as the H_2O molecule, whereas $(OH)^-$ is effectively immobile. Once H_2O molecules infiltrate dryer melts, they react with bridging oxygens, as in reaction 4.3 (Figure 4.8), to form bridging $(OH)^-$.

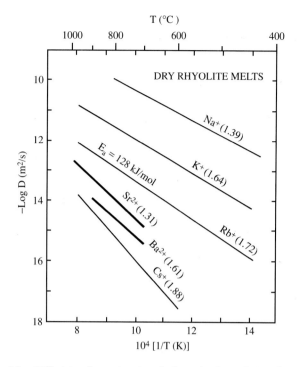

6.6 Diffusivity of some ions in anhydrous rhyolite melt. Numbers in parentheses are ionic radii in ångstroms. (Redrawn from Hofmann, 1980.)

6.2.4 Average Diffusion Distance

A useful relationship first formulated by Einstein in his classic investigation of Brownian movement is

$$6.5 \quad x = \sqrt{Dt} \quad \text{or} \quad t = \frac{x^2}{D}$$

where the diffusion time, *t*, increases as the square of the "average diffusion distance," or "penetration length," *x*, and inversely as the diffusivity, *D*, of the diffusing species. Thus, the time required to double the distance of diffusion for a given *D* is four times longer. A gas molecule in air at normal atmospheric conditions travels an average distance of 0.1 m in 1 second because $D \sim 10^{-2}$ m²/s. For silicate melts, a typical $D \sim 10^{-12}$ m²/s means that an average diffusion distance of about 6 m requires 1 million years (= 3.15×10^{13} s). In a crystal, a diffusivity of, say, 10^{-20} m²/s allows diffusion over an average distance of only 1 mm (10^{-3} m) in 3.2 million years.

6.2.5 Soret Diffusion

Strictly speaking, diffusion of a chemical species *i* is driven by a gradient in its chemical potential, $d\mu_i/dx$. For a phase to be in equilibrium, the chemical potential of all components must be uniform; all compositional gradients are smoothed out by diffusion. The chemical potential of *i* in a phase depends not only on its concentration but also on *P* and *T* (Section 3.4.3). Accord-

6.5 Schematic types of **point defects** in a crystal lattice. (a) Perfect lattice that never exists in real crystals. (b) Vacancy or Schottky defect is an atom missing from a normal site; an adjacent atom can move into the vacancy and another into its place and so on to accomplish diffusion. (c) and (d) Interstitial, impurity, and substitutional atoms can move through a lattice.

(a) Perfect lattice

(b) Vacancy

(c) Self interstitial and impurity interstitial

(d) Substitutional impurities

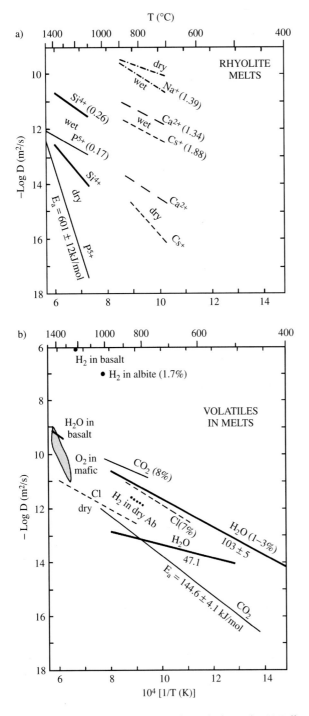

6.7 Diffusivities of some ions in rhyolite and other melts. (a) Diffusion in rhyolite melts that contain 6 wt.% water ("wet") and about 0.1 wt.% ("dry"). Numbers in parentheses are ionic radii in ångstroms. Rare earth elements have similar diffusivities to Si and P. (Redrawn from Watson, 1994.) (b) Diffusion of volatiles in rhyolite melts except as otherwise noted. Bottom three lines are for melts that contain 0.1 wt.% water ("dry"); upper three lines are for melts that contain the (indicated) amount of dissolved water. Shaded lens represents chemical diffusion of O_2 in mafic melts. (Redrawn from Wendlandt, 1991.)

ingly, ions can diffuse in a chemically homogeneous melt that has a thermal gradient. Discovered by C. Soret in 1879, this **Soret diffusion** has been the subject

of considerable debate among petrologists concerning its significance in creating diverse compositions from a uniform parent magma (magmatic differentiation). Experiments reveal that relatively smaller network modifying ions of lesser charge, such as Fe^{+2}, Mg^{+2}, and Ca^{+2}, migrate through a thermally nonuniform body of melt from the hotter to the cooler part. Network-forming Si^{+4} and larger network modifying ions, including K^+ and Na^+, remain in the hotter part. However, as chemical gradients so imposed tend to be readily erased by ordinary chemical diffusion and thermal gradients are smoothed even faster (see next section), Soret diffusion is believed to be of little consequence in magmatic systems (Lesher and Walker, 1991).

✳6.3 DIFFUSION OF HEAT

Heat conduction is the third transport phenomenon of relevance to petrology (Figure 6.1c). As conduction of heat requires transfer of kinetic energy through atomic networks (Figure 6.8), governing equations are similar to those of chemical diffusion. Consequently, heat conduction is often referred to as **thermal diffusion.**

For the variation in T in one dimension, z, Fourier's law for the time rate of heat transfer (compare equation 1.5) is

$$6.6 \quad dq/dt = -k\left(\frac{dT}{dz}\right)$$

where dq is the increment of heat transferred in time dt and k is the **thermal conductivity,** which for rocks is about 2–3 W/m degree. The negative sign is a reminder that heat flows toward the lower temperature. This equation is analogous to Fick's first law for chemical diffusion (Equation 6.2). Another expression for the intrinsic thermal property of a material is its **thermal diffusivity,** $\kappa = k/\rho C$, where C is the specific heat, and ρ is the density. κ is the ratio of the ability of a material to conduct heat relative to its accumulative capacity.

Because it has units of square meters per second, the thermal diffusivity can be compared with the chemical diffusivity, D. The thermal diffusivity is about 5×10^{-7} m^2/s for common dry rocks, but is somewhat less for rocks that contain water or air in pore spaces. Rocks having strongly anisotropic fabric, such as schists, have a slightly greater diffusivity parallel to the foliation than across it. Thermal diffusivities of melts are on the order of 10^{-6} to 10^{-7} m^2/s—as much as eleven orders of magnitude greater than chemical diffusivities (Figures 6.6 and 6.7)! Therefore, static melts solidify by conductive heat transfer well before significant diffusional transfer of ions can occur. Thermal gradients are conductively smoothed much faster than most chemical gradients. However, it will be shown later that more rapid transfer of chemical constituents

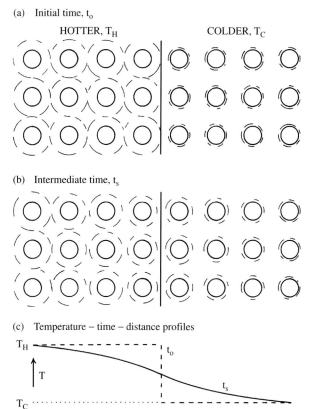

(a)   Initial time, t_0

HOTTER, T_H COLDER, T_C

(b)   Intermediate time, t_s

(c)   Temperature – time – distance profiles

6.8   Schematic atomic model of thermal diffusion or heat conduction. (a) Semi-infinite hotter and colder bodies initially at T_H and T_C, respectively, are brought into contact at time, t_0. Lighter line arcs represent vibrational thermal motion of atoms represented by heavier line circles. Note greater thermal energy in hotter body on left than in colder on right at t_0. (b) After some time, t_s, vibrational energy has been imparted from hotter to colder body in an exponentially decreasing amount from left to right. (c) Graphs of T in bodies at times t_0 and t_s. Eventually, after infinite time, both bodies will be at some uniform intermediate T. Note similarity of smoothed thermal gradient at time t_s to smoothed concentration gradient produced by atomic diffusion in Figure 6.3. In thermal diffusion, however, the kinetic energy of atomic vibration is transferred, whereas in chemical diffusion the atom itself is transferred, a more difficult and slower process.

can take place if the melt is convecting or if fluid bubbles are buoyantly migrating through the melt.

Like chemical diffusion (equation 6.5), a thermal transfer time, t, can be defined as

$$6.7 \quad t = \frac{z^2}{\kappa}$$

Thus, t increases as the square of the dimension, z, of the body. For a constant thermal diffusivity and all other parameters remaining the same, doubling the dimension of a body increases its conductive heating or cooling time by a factor of 4.

6.3.1 The Role of Body Shape on Conductive Cooling

The shape of a conductively cooling body also influences its cooling time. The overall rate of conductive

heat loss is a trade-off between the surface area of a body—from which heat is dissipated—and its volume—which holds the thermal energy. The surface area/volume ratio, therefore, controls the rate of conductive heat loss and thermal equilibration between a body and its surroundings, smoothing gradients in T. The most thermally retentive shape is a sphere, for which the ratio is least. However, smaller spheres cool faster than larger because their surface area/volume ratio is larger (see Problem 6.6); this is one reason why the small Moon is cold and the Earth is still hot and geologically active. The least thermally retentive shapes are long small-diameter rods and thin sheets because their surface area is large compared to their volume.

Heat is conducted at different rates from different parts of a nonspherical body. For example, in a cube, heat conducts away into the surroundings faster from corners than along planar sides because of the larger mass (volume) of the surroundings into which heat can sink.

❋6.4 INTERFACIAL ENERGY

A brief digression from our discussion of kinetic topics is made here to explore the significant role that the surface area/volume ratio plays in crystal-melt equilibria in magmatic systems. It will be seen that this ratio can have an important bearing on rock texture.

Picture a single cubical unit of halite with the Cl anions and Na cations at alternate corners. Half of the ionic bonds are unsatisfied, so it has a large energy and is, therefore, highly unstable. Within larger atomic arrays, Na and Cl ions have proportionately fewer overall unsatisfied bonds at corners and edges. Thus, larger crystals have a lower surface-related energy and are more stable.

But crystals do not exist in nature as isolated entities; they have some sort of neighboring phases in contact with them. The **interface** between a crystal and its neighboring crystals, melt, or fluid is a layer a few atom diameters thick that differs in structure and thermodynamic properties from the interior of the crystal. The intergrain layer between adjacent crystals is an incoherent mismatch of crystal lattices (Figure 6.9a) where constituent atoms have more energy than interior ones because they are more loosely bonded or have unbalanced bonds. The arrangement of atoms in an interphase layer between crystal and liquid is again different from the crystal interior and has aspects of both phases (Figure 6.9b).

The energy associated with a solid-solid or a solid-liquid interfacial layer is here referred to as the **surface free energy**, γ, with units of joules per square centimeter (J/cm^2) (since the energy is usually related to a particular area). It is defined as the change in energy of the system per unit area of interface generated at constant composition, T, and P. The creation of a surface re-

(a)

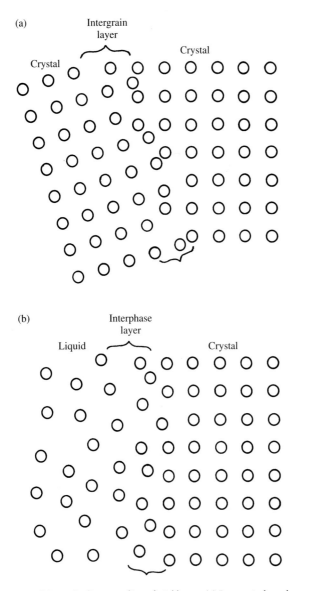

(b)

6.9   Schematic diagram of interfacial layers. (a) Intergrain layer between two crystals, either of different phases or of the same phase, showing misorientation of lattices across the interface. (b) Interphase layer between a liquid and a crystal.

quires work, an input of energy, as in cleaving a mineral grain, tearing bonds apart, and creating a free surface.

Because the total energy of a phase is the sum of its surface free energy, γ, and its Gibbs free energy, G, the total energy of progressively smaller volumes of a phase increases as the surface area/volume ratio increases. This means that many small particles are less stable (have greater energy) than one large particle of equivalent volume but lesser surface area (Figure 6.10).

In a liquid body bounded by a gas phase, unbalanced atomic bonds on the surface of the liquid tend to pull it inward, giving rise to a **surface tension.** This attractive force, which has units of energy/area (J/cm^2) like surface energy, makes isolated liquid droplets and soap bubbles spherical, that is, the shape with the smallest surface/volume ratio (lowest energy).

Unlike isolated spherical drops of liquid surrounded by a gas, crystals can possess significantly different surface energies on different crystallographic planes because the geometry of atomic bonding is different; this anisotropy can exist even in isometric crystals. It is the reason why crystals that grow freely in an unrestricted liquid environment are not spheres but have a characteristic crystal habit, such as tabular, platy, or columnar. These **euhedral** crystals are bounded by their characteristic crystal faces. Under equilibrium conditions, crystals growing without any restriction in liquids do so in such a way as to minimize their total energy. Planes of easiest cleavage in feldspars, micas, and amphiboles have relatively lowest energy, and, in these minerals at least, these same planes are developed during crystal growth, minimizing the overall surface energy of the crystal (Kretz, 1966). Hence, in euhedral platy mica, {001} has low energy and is typically prominent, as is {010} in euhedral tabular plagioclase and {110} in euhedral columnar amphibole.

An irregularly shaped, **anhedral** mineral grain bounded by nonrational faces has greater surface free energy than a euhedral crystal of the same phase and volume and is, therefore, less stable than the euhedral equilibrium shape (Figure 6.10).

Therefore, mineral grain sizes and shapes spontaneously adjust toward a state of lower surface free energy. Minimization of the surface energy results in coarser grain size as the surface area/volume ratio is reduced. Modification of grain shape can occur. The minimization principle is an important factor in the

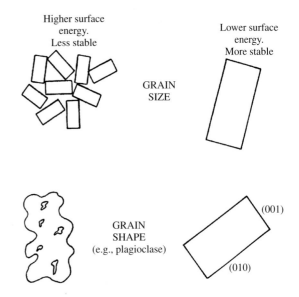

6.10   Influence of grain size and shape on the stability of a phase of *equivalent* volumes (shown schematically here as equivalent areas). One large particle is more stable than many small ones of the same shape because of their greater surface energy contribution. A euhedral crystal bounded by characteristic crystal growth faces is more stable than an irregularly shaped anhedral grain.

evolution of rock fabric, particularly where relatively high rates of diffusion at elevated temperatures are capable of modifying grain boundaries.

Thus, the thermodynamic stability of a phase volume depends on its size and shape, *in addition to P, T,* and concentrations of chemical components (Section 3.4.3).

❋6.5 CRYSTALLIZATION

The three kinetic processes of viscous mobility, transport of atoms, and transport of heat, as well as the tendency of grain systems to minimize their surface free energy, provide a foundation on which to consider crystallization of melts.

Creation of a new phase from any preexisting phase always involves two independent, consecutive kinetic processes—nucleation followed by growth. A growing crystal in a cooling melt must start from an embryonic cluster of ions, called a **nucleus,** probably tens to hundreds of ångstoms in diameter, that possesses all of the characteristics of the crystal. Because the symmetrical lattice of a crystal is usually quite different from the disordered array of ions in a melt, a substantial reorganization of ions is required to produce the crystal nucleus. Once viable, other kinetic factors come into play to allow the accretion of ions onto the nucleus; this is crystal growth. Nucleation phenomena exert a major control on the textures of magmatic rocks, particularly their grain size, as well as their crystallinity and vesicularity. Growth phenomena chiefly influence crystal shapes in magmatic rocks.

Many theories have been proposed for nucleation and crystal growth in melts (summarized by Dowty, 1980; Lofgren, 1980; Cashman, 1990). Most of these models apply to simple, one-component melts so their validity for multicomponent melts in natural magmatic systems is uncertain. In any case, insights from the simple models are useful. As always, the textures of real rocks provide the final test of how correct a theoretical model might be.

6.5.1 Why Is It Important to Study Nucleation and Crystallization?

The application of the material discussed in this chapter to real rocks may seem remote. Connections are mainly deferred to the following chapter on rock fabric. However, to put the discussion of kinetics in perspective it may be beneficial at this point to digress briefly and comment on one of the most fundamental of all rock properties. This property, recognized when a student first becomes acquainted with igneous rocks in the field or laboratory, is grain size.

Igneous rocks obviously possess a wide range of grain size, from submicroscopic (<0.001 mm for an optical microscope) grains to the giant crystals of pegmatites, which can be several meters. This is a range of seven orders of magnitude. Some magmatic rocks have

essentially no crystals at all and are instead composed of an amorphous glass. The range in grain size of most rocks is only two to three orders of magnitude. The most common phaneritic plutonic rock—granite—generally has grains 1–20 mm whereas the most widespread aphanitic volcanic rock—basalt—has grains 0.1–1.0 mm. What kinetic process(es) permits such a wide range of grain size but commonly favors a more restricted range? Rate of cooling does control grain size, as usually indicated in elementary geology texts, but is cooling rate the only factor?

If one were to examine thousands of all types of magmatic rocks around the world, it would soon become apparent that some minerals, such as magnetite and olivine, are invariably small, less than a few millimeters, regardless of the magma in which they form. Although phenocrysts of olivine, rarely to as much as 5 mm, occur in basalts, rocks having phenocrysts of magnetite visible to the naked eye (>1 mm) are virtually nonexistent. Upward of 10% Fe-Ti oxides are common in basalts and andesites, for example, but they are invariably small groundmass grains. Even in phaneritic rocks with centimeter-size felsic and mafic silicate minerals, Fe-Ti oxides are generally much smaller. Why is this? What factors allow plagioclases to form phenocrysts 1 cm or more across in many volcanic rocks, and alkali feldspars to form phenocrysts 5 cm across in some granites, and giant crystals meters across in pegmatites? Obviously, cooling rate alone cannot account for the difference in sizes of different crystals growing in the same magma.

Answers to these questions depend on the interplay between nucleation and growth rates for different mineral species in the melt as intensive parameters change in the solidifying magma system.

6.5.2 Nucleation

Countless experiments have amply confirmed W. Ostwald's discovery in 1897 that every phase transformation requires some degree of overstepping beyond equilibrium conditions (Section 3.6.2) to accomplish **nucleation** of a new phase. A second concept is that some phases typically nucleate more readily than others from melts. Kinetic barriers to nucleation are mineral-specific.

Two types of nucleation process can provide a "seed" on which ions in the melt subsequently can accrete during crystal growth: heterogeneous and homogeneous nucleation.

<u>Homogeneous Nucleation.</u> Homogeneous nucleation occurs as a consequence of spontaneous, random fluctuations in the disordered array of ions *within* a uniform body of melt. These transient fluctuations result in a momentary ordered array of clustered ions—a potential nucleus or embryo—that happen to form in the thermally agitating milieu of otherwise disorganized ions. One might imagine a flat tray on which lie

closely, but not tightly, packed marbles. If the tray is agitated, momentary clusters of marbles in arrays of six-fold symmetry appear from place to place on the tray and then immediately disappear. Whether similar transient clusters of organized ions in melts can serve as viable nuclei for further accretion of ions during crystal growth depends upon their size and the amount of overstepping—the driving "force" for nucleation.

A new crystalline phase is stabilized once its free energy becomes less than the melt, $G_{crystal} - G_{melt} = \Delta G < 0$. This thermodynamic driving force can be caused by a change in T, P, or concentration of some component, or combinations of these changes. Consequently, the melt becomes saturated with respect to the stabilized crystals. Because the most common and easily understood change in geologic systems involves a decrease in T, the following discussion focuses on **undercooling** as the means of overstepping.

Very small embryos have a substantial surface free energy, γ, relative to the volumetric $G_{crystal}$ that requires overstepping to become stable (Figure 6.11). At T_e, an embryo of any radius is unstable. For small undercooling, ΔT, small embryos have a relatively large surface energy that makes them unstable with respect to the melt. They might, however, become stable nuclei by growing larger beyond some critical radius, r_c (Figure 6.12). Larger clusters are more stable, but they are less likely to occur by random thermal fluctuations. For large ΔT, even small transient embryos can be stable nuclei because of the increasing difference between the free energies of the crystalline phase and melt, which

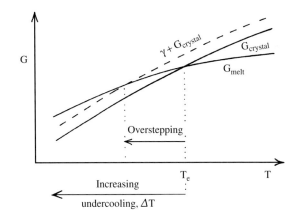

6.11 Schematic plot of free energies of melt and crystal in a one-component system as a function of T. Free energies of these two phases are equal and cross over at the equilibrium T_e. The surface free energy, γ, that must be added to the free energy of the crystal phase for very small crystalline nuclei, because of their large surface area/volume ratio, increases their total free energy to the dashed line. Consequently, the cross over of melt and nuclei free energies is shifted to some T below T_e, so that a nucleus can only be stable below some amount of **overstepping** below T_e. The **undercooling**, ΔT, increases to the left below T_e.

overrides the surface free energy contribution of the embryo (Figure 6.11). Therefore, the rate of formation of nuclei increases for increasing ΔT. But as T continues to fall below T_e, the probability of transient fluctuations in the atomic array in the melt or parent crystal must decrease because of the decreasing thermal motion of atoms in the increasingly more viscous melt; random fluctuations having the crystalline array are decreasingly likely to occur.

Several complicating factors in the nucleation of multicomponent melts may make this one-component model only a crude approximation to what actually happens. Different minerals in natural melts begin to crystallize at different temperatures and continue to do so over a range of T. Increasing cooling in such melts stabilizes solid solutions of changing composition in a melt. Moreover, while one mineral may be nucleating abundantly, another simultaneously stable mineral may not be nucleating or nucleating sluggishly.

Unraveling these sorts of complexities by laboratory studies has proved to be difficult and largely unsuccessful because homogeneous nucleation is a random phenomenon, thus making experimental results inconsistent. The statistically random fluctuations involved in homogeneous nucleation are more likely to occur in less polymerized and therefore less viscous melts; dissolved fluid content, especially water, is thus significant.

Despite the lack of quantitative data, some qualitative inferences have been made regarding relative rates of homogeneous nucleation of common rock-forming minerals. Experimental petrologists (e.g., Kirkpatrick, 1983) have long recognized that minerals with simpler

---

Advanced Topic Box 6.2 Theoretical model of homogeneous nucleation

The change in Gibbs free energy accompanying formation of a crystal embryo, assumed to be spherical, from a melt at $T < T_e$ is

$$\Delta G = G_c - G_l = \frac{4/3\pi r^3 (g_c - g_l)}{V} + 4\pi r^2 E_s$$

where r is the radius of the embryo, V is the volume of an atom in the embryo, and g_c and g_l are the free energies per atom in the liquid and crystalline states. The $4\pi r^2 E_a$ term is the surface energy contribution of the embryo. For it to be stable, $\Delta G < 0$. The $4/3\pi r^3(g_c - g_l)/V$ term is negative at $T < T_e$ and ΔG can only become negative when this term exceeds the surface energy term for some increasing r, called the critical embryo radius, r_c (Figure 6.12). For larger ΔT the nucleation driving force, $g_c - g_l$, is an increasingly larger negative number and the $4/3\pi r^3(gc - g_l)/V$ term exceeds the surface energy term for smaller r_c.

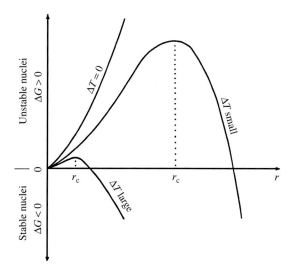

6.12   Relation between radii of homogeneously formed nuclei, r, and free energy in a one-component system at three different amounts of undercooling, ΔT. Nuclei are unstable when $\Delta G > 0$ and stable when $\Delta G < 0$.

atomic structures nucleate with greater ease than tectosilicate (framework) feldspars and quartz in melts of appropriate composition. Carmichael and others (1974, p. 164) have developed a theoretical model in which the nucleation rate is proportional to the square of the entropy change, ΔS_m, during melting of a particular mineral. Small entropy changes occur in melting of cristobalite, albite, and K-feldspar (0.3–0.7 entropy unit per atom in gram formula weight), more for anorthite (1.2), still more for pyroxenes and olivines (1.9–2.1), and greatest for Fe-Ti oxides (2.5–2.6). Minerals with larger entropy change should nucleate more readily, yielding a hierarchy of ease of nucleation, as follows: Fe-Ti oxides (easiest, yielding most abundant nuclei), olivine, pyroxene, plagioclase, and alkali feldspar (least). It may seem surprising that framework silicates would nucleate more slowly from a polymerized granitic melt than mafic minerals and especially Fe-Ti oxides, whose atomic structures are so different. However, greater differences in atomic structure between melt and nucleating mineral, as reflected in the larger entropies of melting, may provide a stronger driving force for nucleation at a particular undercooling.

If these conclusions are valid, they might provide an explanation for the local occurrence of large alkali feldspars and universally small Fe-Ti oxides in magmatic rocks. With only a few nuclei for feldspar-forming components to accrete onto, crystals would be large, and for Fe-Ti oxide components the reverse. (An analogy can be found in growing fruit. To create robust large apples, for example, the orchardist reduces the number of embryonic fruit in some way; in some growing seasons, nature does this by means of a late frost that kills many blossoms.) However, this explanation may be premature because the possible effect of heterogeneous nucleation has not been considered, nor has that of crystal growth rate, which must certainly influence crystal size.

Heterogeneous Nucleation. It is common knowledge that crystals readily nucleate on any existing surface in contact with a melt. This phenomenon is, in fact, a hindrance in experimental investigations of homogeneous nucleation in precious metal containers. The existence of an interface with any contrasting material against the melt can overcome the activation energy barrier so that hetereogeneous nucleation may occur more readily for small ΔT than homogeneous nucleation (Lofgren, 1983; Putnis and McConnell, 1980, p. 104). Existing surfaces can be the solid walls of the melt container or wall rock in the case of a natural magma body.

Existing "seed" crystals in the magma are especially significant in overcoming the difficult nucleation step in crystallization. Overgrowths on the seed are readily facilitated if that phase is stable in the system. Another mineral may also grow around the seed crystal; possible examples are common biotite overgrowths around zircons. Some existing crystals may be earlier-formed crystals. Others may be foreign crystals, or **xenocrysts,** which may have been removed by "erosion" of the wall rock during flow of the magma or introduced into it by mixing with a compositionally contrasting magma. Still other seeds may be **restite crystals** that are undissolved refractory remnants of the source rock from which the magma was generated by partial melting processes in the deep crust or mantle.

Minute crystalline entities, microscopically invisible, may serve as seeds for crystal formation. These might have survived an episode of brief melting above the liquidus and could be of restite or xenocryst derivation in magmas extracted rapidly from their source. Some melts seem to have a "memory" of their thermal history—such as how long they were heated at a particular T above the liquidus—that influences their crystallization behavior below the liquidus.

Other potential interfaces for heterogeneous nucleation are walls of volatile bubbles in the melt (Davis and Ihinger, 1998). In this case, there can be an interplay, even feedback, between exsolution of fluid from the melt and nucleation. Crystals nucleate and grow on bubble walls, causing more saturation of fluid in the melt, which leads to more exsolution, and so on.

Many petrologists believe that heterogeneous nucleation is common, if not dominant, in natural magmas. However, there is little conclusive data to confirm this belief.

6.5.3 Crystal Growth

Once nuclei are viable, growth of crystals can occur as additional ions become attached. Like nucleation, the rate of crystal growth is related to the degree of undercooling of the system, ΔT (Figure 6.13). Increasing

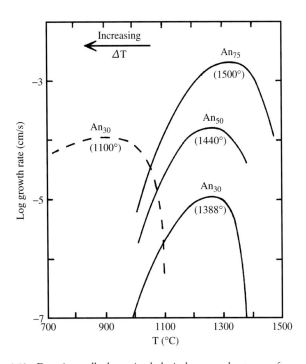

6.13 Experimentally determined plagioclase growth rates as a function of degree of undercooling, ΔT, in their equivalent melts. The composition of the crystals is indicated in mole percentage and the corresponding liquidus T is in parentheses below each curve. Solid lines are in dry melts at 1 atm; note decreasing peak growth rate at decreasing T. Dashed line is for growth in a water-saturated melt at 2 kbar; the dissolved water depolymerizes the melt and promotes faster peak growth, despite the lower T. (Redrawn from Lasaga, 1998; see also Fenn, 1977.)

undercooling provides a stronger driving force for growth, but with falling T the increasing melt viscosity retards ionic mobility. For this reason, the growth rate is a bell-shaped curve.

There are more experimental data on crystal growth than on nucleation in geologically relevant melts. Some experiments report data in terms of the degree of undercooling, ΔT, others in terms of the cooling rate, $\Delta T/\Delta t$. These parameters are obviously related, but in dynamic magma systems the latter is more meaningful than an apparent one-step drop in T implied in a ΔT value.

<u>Influence of Undercooling on Crystal Shape.</u> Many experiments on crystal growth (e.g., Lofgren, 1980; Swanson and Fenn, 1986) have demonstrated that as ΔT and $\Delta T/\Delta t$ increase, crystals increasingly depart from an equilibrium habit of characteristic crystal faces. This departure from euhedral shapes occurs because, with changing intensive parameters in the system, usually falling T, diffusion of atoms in the cooler, more viscous melt and conduction of latent heat away from the growing crystal are less able to keep up. Crystals become less compact with increasing ΔT and $\Delta T/\Delta t$.

Actual grain shapes vary with respect to the particular mineral, melt composition, and amount of undercooling. A general pattern seen in laboratory experiments can be illustrated with plagioclase as an example (Figures 6.14 and 6.15). For $\Delta T/\Delta t$ less than a few degrees/hour (ΔT less than tens of degrees), crystals have a euhedral tabular habit; these are typical of phaneritic plutonic rocks and slowly grown phenocrysts in volcanic rocks. For $\Delta T/\Delta t$ of tens of degrees/hour (ΔT on the order of 100°C) crystals are hollow, **skeletal,** and H-shaped forms; these are found in some glassy and aphanitic volcanic rocks. For greater ΔT and $\Delta T/\Delta t$, **dendritic,** branching, and **feathery** forms develop. For $\Delta T/\Delta t$ of hundreds of degrees/hour and very high effective ΔT, radiating, three dimensional sprays of fibrous to needlelike crystals called **spherulites** develop, probably after the melt drops below the glass transition ($\eta > 10^{12}$ Pa s). Rapidly cooled submarine basalt pillow lavas show dendritic and skeletal forms as well as radiating intergrown plagioclase and pyroxene.

These kinetically controlled crystal habits result from different factors during the growth process, including the following:

1. Phenomena at the crystal-melt interface as ions become attached to the surface of the growing crystal
2. Diffusion of ions through the melt to the growing surface
3. Removal of latent heat of crystallization from the crystal-melt interface
4. Viscous flow of melt past the crystal face

Whichever of these kinetic rates is slowest dictates the overall rate of growth. In a one-component system, such as ice crystals growing in water, diffusion and viscous flow are irrelevant because necessary ions in the proportions of the crystal are everywhere and always present, but ion attachment to and dissipation of latent heat from the growing crystal face are rate-controlling. In multicomponent melts, interfacial reactions, diffusion, and dissipation of latent heat appear to be significant, in that order, for increasing ΔT and $\Delta T/\Delta t$.

In the growth of robust, compact crystals having characteristic habits (tabular, prismatic, etc.) the controlling process is the attachment of ions to the growing crystal face (Dowty, 1980; Kirkpatrick, 1981); this is interface-controlled growth. Chemical diffusion is relatively rapid near liquidus temperatures and still faster dissipation of heat can keep pace with the slow cooling rate of the melt. However, even for slow growth in the plutonic environment, euhedral plagioclases generally have slight compositional zoning that indicates perfect equilibrium was not attained because, once a crystal forms, adjustment of the CaAl/NaSi ra-

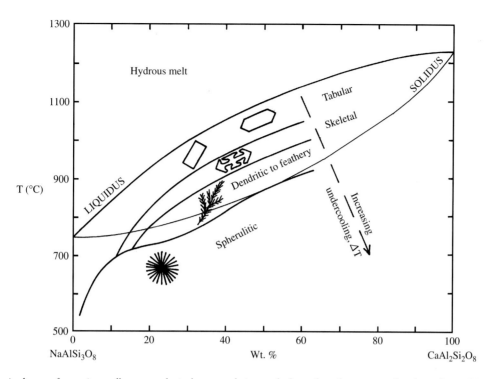

6.14 Schematic shapes of experimentally grown plagioclase crystals in a melt depend on the amount of undercooling, ΔT. Compare Figure 5.13. Heavier lines delineate regions of undercooled-controlled crystal morphology, schematic representations of which are shown. Compare photomicrographs of actual crystals in Figure 6.15. (Redrawn from Lofgren, 1980.)

tio in it by diffusion-controlled reaction relations is exceedingly slow.

Crystal forms become more open and skeletal and have an increasing number of reentrant angles at increasing ΔT and $\Delta T/\Delta t$. Dendritic and feathery crystals have yet larger (disequilibrium) surface area/volume ratios. Growth may result from the need to dissipate latent heat, as exemplified by wintertime formation of feathery ice crystals on windows. The exact way in which exponentially decreasing rates of chemical diffusivity with decreasing T enter into crystal growth is not clear, although many theoretical models have been proposed.

<u>Compositional Boundary Layers and Implications</u>. Compositional gradients defining a **compositional boundary layer** can develop in the melt beside a growing crystal (Figure 6.16). While incompatible ions become enriched in the layer compatible ions become depleted, slowing crystal growth. For growing solid solutions (e.g., plagioclase) the boundary layer becomes depleted in the high-T component ($CaAl_2Si_2O_8$) and enriched in the low-T component ($NaAlSi_3O_8$) relative to the melt outside the layer. In some way, which is not yet completely understood, kinetic phenomena in boundary layers adjacent to growing plagioclases in intermediate composition magmas create fine-scale oscillatory zoning in the crystal (Figure 6.17). In hydrous melts,

growing anhydrous crystals create a boundary layer enriched in water that hastens chemical diffusion, depresses the liquidus, and possibly locally saturates the melt in water.

Small volumes of melt, called **melt inclusions,** can be entrapped within a growing crystal (Figure 6.18).

If convective movement of the melt in the boundary layer beside a growing crystal is negligible, then incompatible components can become saturated in the boundary layer and a new phase can be stabilized (Bacon, 1989) including accessory apatite (incompatible P) and zircon (Zr).

6.5.4 Crystal Size in Magmatic Rocks

We return now to the question posed in Section 6.5.1, namely, What controls grain size in magmatic rocks?

The number of nuclei in a volume of melt, called the **nucleation density,** is the integrated effect of a varying rate of nuclei formation over some period. During this time, crystal growth is probably also taking place, consuming the mass of necessary chemical components in the melt for a particular mineral. If the growth rate is relatively slow, many nuclei might form before the mass is consumed; grain size would consequently be small, perhaps aphanitic. This occurs typically in volcanic environments where the rate of cooling of melts is relatively fast, or in some plutonic situations where a sud-

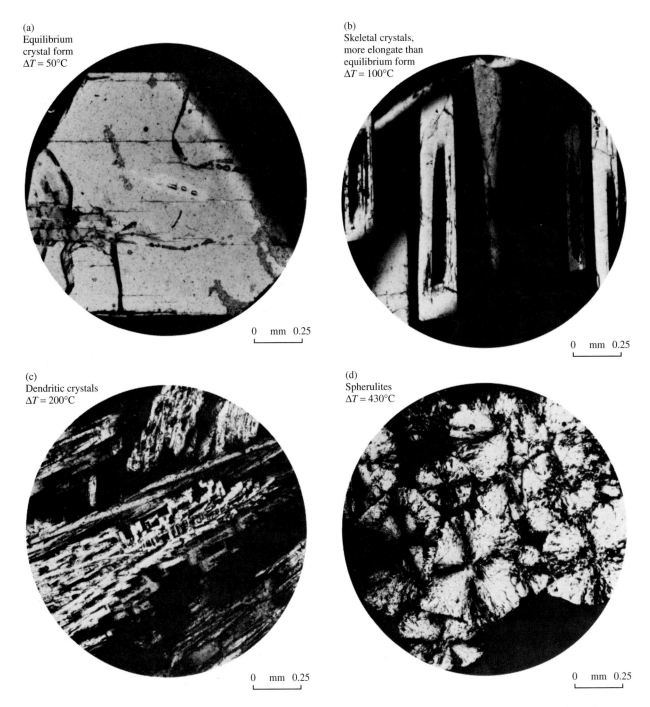

(a) Equilibrium crystal form $\Delta T = 50°C$

(b) Skeletal crystals, more elongate than equilibrium form $\Delta T = 100°C$

(c) Dendritic crystals $\Delta T = 200°C$

(d) Spherulites $\Delta T = 430°C$

0   mm 0.25

6.15 Shapes of plagioclase crystals grown experimentally from melts as a function of undercooling, ΔT. Photomicrographs under cross-polarized light. **Spherulites** are radiating spherical aggregates of fine needlelike crystals that are optically extinguished in the N-S and E-W directions, forming the black crosses. (Photographs furnished by Gary E. Lofgren and NASA. Reproduced by permission from Lofgren GE. An experimental study of plagioclase crystal morphology. Am. J. Sci. 1974; 274:243–273.)

den release of water occurs to create a pressure quench, forming a porphyry. Some melts, especially highly viscous ones, can thermally quench to form glass. If, on the other hand, growth rate is fast relative to nucleation rate, then fewer nuclei form by the time the mass is consumed; the grain size is larger, perhaps

phaneritic but possibly more variable. This occurs typically in deep confined plutonic settings where heat loss is slow.

Shaw (1965; see also Brandeis and Jaupart, 1987) proposed that grain size is proportional to the ratio of crystal growth rate, $\mathcal{G}$, to nucleation rate, N,

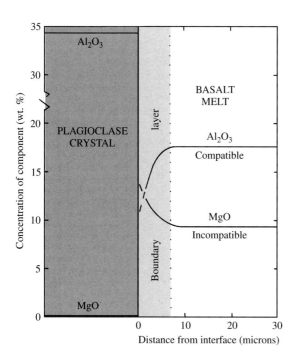

6.16 During relatively rapid growth of a crystal, concentration gradients form in the melt adjacent to the crystal and define a **compositional boundary layer.** Compatible elements, such as Al, diffusing to the growing crystal of plagioclase (about An_{86}) in the basalt melt define an exponentially decreasing concentration in the boundary layer adjacent to the crystal. Incompatible elements, such as Mg, mostly excluded from the growing crystal, must diffuse away from it and form an exponentially increasing concentration toward the crystal. Uniform concentrations of chemical elements on each side of the boundary layer in the basalt melt and plagioclase crystal are indicated by straight horizontal lines. Concentrations of oxides of the elements shown here were determined by electron microprobe analyses. (Modified from Bottinga et al., 1966.)

$$6.8 \quad \text{average crystal size } \alpha \left(\frac{\mathcal{G}}{N} \right)^{1/4}$$

Extrapolation of theoretical and experimental studies of nucleation and crystal growth in simple, usually one-component systems to multicomponent natural melts is a tenuous step and has not led to a thorough understanding of the kinetic factors governing such a fundamental rock property as grain size. Attempts to quantify kinetic factors from textures of real rocks in the simplest of thermal environments—such as a thin basaltic dike—have met with limited success. The fact that some granitoid plutons contain quite variable grain size on an outcrop scale whereas others have strikingly uniform grain size throughout tells us that overall cooling rate and magma composition are not the whole story. What role does the distribution of water or some sort of seeds in magmas play?

Investigations by chemical engineers into industrial crystallization have yielded insights into the kinetics of crystallization. Their concepts have been adapted to

rocks by Marsh (1988) and have been applied to basalts and silicic volcanic rocks by Cashman (e.g., 1990).

✳6.6 SECONDARY OVERPRINTING PROCESSES MODIFYING PRIMARY CRYSTAL SIZE AND SHAPE

Two additional factors that play roles in governing magmatic grain size and shape are crystal dissolution and secondary equilibration of grain margins after primary growth.

6.6.1 Crystal Dissolution

Resorbing reaction relations are common in magmas, consuming, to varying degrees, previously precipitated phases (e.g., olivine in Makaopuhi basalt magma, Figure 5.10, and quartz phenocrysts in decompressing rhyolite magma, Section 5.5.5). Dissolution of crystals occurs during mixing of dissimilar crystal-laden magmas that are striving to reach a state of internal equi-

6.17 **Oscillatory zoning** in plagioclase in an intrusive diorite porphyry, central Utah. Photomicrograph in cross-polarized light. Thin concentric zones have differing optical extinction orientations that manifest an alternating relative enrichment in albitic and anorthitic end members from one zone to the next. The zoning is usually not strictly oscillatory, as is a sine function; each peak increases in Ab content away from the crystal center and drops more sharply. Peaks are typically 5–15 mole % in amplitude and the wavelength is 10–100 micrometers. These oscillations are commonly superposed on a larger-scale normal zonation toward a more sodic rim. The cluster of grains defines **cumulophyric** texture. See Pearce and Kolisnik (1990) for further description of oscillatory zoning in plagioclase.

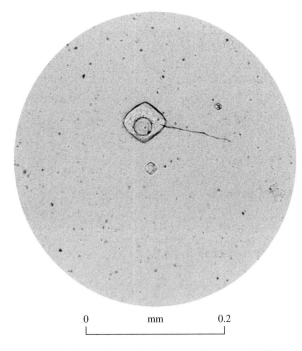

0   mm   0.2

6.18 **Melt inclusions** (now glass) in quartz phenocryst in silicic volcanic rock. Photomicrograph under plane-polarized light. The three inclusions quenched from melt entrapped in host quartz crystal as it was growing. Inclusions have a slightly smoothed rhombohedral "negative crystal" form that indicates the melt and crystal were in equilibrium at entrapment. Spherical fluid bubble in largest inclusion is of uncertain origin; it may not indicate volatile saturation at time of entrapment. Line is a crack in host crystal.

librium and during the evolution of magmas that contain unstable restite crystals and assimilated xenocrysts from foreign country rock.

Partially dissolved crystals of different origins are evident in many volcanic rocks and are preserved because the magma solidified more rapidly than the crystal could completely dissolve; such is generally not the case in more slowly cooled and more nearly equilibrated plutonic systems. Unstable crystals are readily apparent from their embayed and corroded forms and, in volcanic rocks, abundance of irregularly shaped melt inclusions (now glass; Figure 6.19). Some rocks preserve partially resorbed crystals rimmed by a later precipitated phase or assemblage of phases—a reaction rim between the dissolving crystal and the enclosing melt (Figure 6.20). For example, a quartz xenocryst plucked off the wall rock and incorporated into a basalt melt dissolves because the melt is undersaturated in silica (silica activity <1; Section 3.5.3). Dissolution of foreign material is called **assimilation;** it contaminates the magma and contributes to the wide compositional diversity of magmas and igneous rocks on Earth.

Relevant kinetic processes for dissolution might involve reactions at the crystal-melt interface, diffusion, absorption of heat by the crystal, and convective move-

ment of the melt beside the crystal—the same factors involved in crystal growth. Zhang et al. (1989) find that the melt immediately adjacent to a dissolving crystal saturates with its constituents almost instantaneously, so that convective processes are required for continued effective dissolution. The rate of dissolution is proportional to the square root of time. At low P and moderate undersaturation, felsic minerals and pyroxenes dissolve faster than olivine, which dissolves faster than accessory minerals. This is a trend of generally decreasing solubility and mirrors the trend of ease of nucleation cited previously. These solubility differences could play an important role in which minerals are most likely to be preserved as restite crystals in magmas ascending out of their source.

6.6.2 Textural Equilibration: Grain Boundary Modification

An underappreciated and little understood phenomenon that also affects magmatic grain size and shape involves secondary reequilibration of grain margins as an aggregate strives to minimize its surface free energy (Section 6.4). Textural equilibration is favored if temperatures remain high in the system. Concepts of textural equilibrium developed chiefly by metallurgists have been applied to magmatic textures by Maaløe (1985) and Hunter (1987), among others.

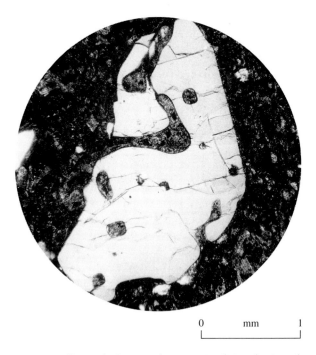

0   mm   1

6.19 **Partially resorbed** quartz phenocryst in silicic volcanic rock. Photomicrograph under plane-polarized light. Deep, irregularly shaped embayments indicate crystal was unstable in the melt prior to its solidification into an aphanitic groundmass. Irregularly shaped apparent inclusions embedded in quartz crystal may only be narrow embayments extending from third dimension.

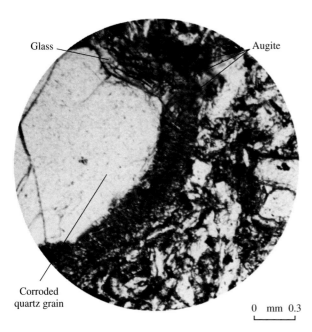

Glass

Augite

Corroded
quartz grain

0   mm 0.3

6.20 Partially resorbed quartz xenocryst in a basalt surrounded by a reaction rim of minute rod-shaped clinopyroxenes.

Modification of Grain Size. Crystal size can increase by the process of **Ostwald ripening** (Figure 6.21). Large and small crystals of the same phase dispersed in a communicating fluid or melt at fixed P and T are relatively more and less stable, respectively, because of their differences in surface free energy per volume (Figure 6.10). Consequently, smaller crystals tend to be consumed at the expense of larger, more stable grains. (This is the petrologic counterpart of the economic dictum that the rich get richer and the poor get poorer.) The result of this equilibrating "ripening" is an overall increase in average grain size of the system. Jurewicz and Watson (1985) experimentally confirmed that in a partially crystallized granite magma held at constant P and T for several days, average quartz grain diameter increased while number of grains decreased. Grain diameter adjustment was a response to minimization of surface energy, not progressive growth because the overall volume of quartz remained constant. Park and Hanson (1999) observed the same phenomenon for olivine in a model basalt system.

Modification of Grain Shape. In Ostwald ripening, grain shape may be modified as well as grain size, both according to the requirement to minimize surface free energy. Therefore, cooling rates initially control shapes of crystals growing freely in a melt (Figures 6.14 and 6.15), but subsequent modifications in shape may occur, if time and diffusion rates permit.

Consider three grains A, B, and C in mutual contact. A cross section drawn perpendicular to this common grain-boundary line defines a triple point between the grains cited (Figure 6.22). Interfacial energies between the grains, γ_{AB}, γ_{BC}, and γ_{AC} can be considered as surface tensions and represented as force vectors. At equilibrium, these must sum to zero around the triple point. For a particular set of crystalline phases whose interfacial energies are dictated by their compositions, T, and P, equilibrium is achieved by the grains adjusting their mutual angles, α, β, and δ, at the triple point so that

$$6.9 \quad \frac{\gamma_{AB}}{(\sin \alpha)} = \frac{\gamma_{BC}}{(\sin \beta)} = \frac{\gamma_{AC}}{(\sin \delta)}$$

If all three grains have the same interfacial energy, the three interfacial angles must be equal, or 120°, because they must sum to 360°. A tightly packed, void-free aggregate of polyhedrons that have uniform volume and minimal overall surface area and, therefore, surface energy is shown in Figure 6.23. In such an aggregate viewed in thin section, triple-junction interfacial angles approach 120°. This geometry is seen in very slowly cooled gabbroic and phaneritic ultramafic rocks (Figure 6.24). A remarkably similar geometry is found in highly vesicular glassy basalt (Figure 6.25), where gas-bubble shape is dictated by surface tension. The extent to which the angle is near 120° (but see Kretz [1966] for exceptions) in mineral aggregations means that the grains had similar interfacial energies, regardless of their crystallographic orientation. Interfacial angles were balanced and grain boundaries migrated during prolonged textural adjustment at elevated T by means of differential solution and growth via atomic diffusion.

Different crystallographic surfaces of a crystal actu-

(a)   (b)

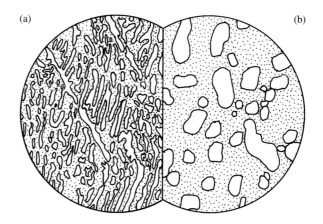

6.21 Ostwald ripening of ammonium thiocyanate dendrites. Grain coarsening in this low-T organic compound models what may happen in a high-T silicate system. (a) Immediately after cooling to 50°C the initial single-crystal dendrite has already segmented into discrete rods within the melt (stippled). (b) Same field of view as (a) after 9.8 hours. The initial dendritic crystal with a large surface area has been stabilized by formation of larger crystals with overall lesser surface area; crystal volume remained the same. Note that the isolated crystals in (b) *did not individually nucleate*. Field diameter is about 0.4 mm. (Drawn from a photomicrograph in Means and Park, 1994.)

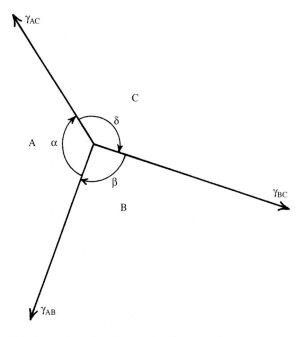

6.22 Intergrain surface free energies. These energies, γ_{AB}, γ_{AC}, γ_{BC}, are represented by vectors along boundaries of grains about a triple-grain boundary junction of three grains A, B, and C.

ally have different energies. The lowest-energy faces stabilize the crystal volume during growth. In plagioclases grown from melts the typical habit is tabular with largest faces parallel to {010}. This suggests preferred growth on higher-energy faces parallel to {100} or {001} so that the lowest-energy and most stable {010} faces dominate the enclosed volume.

An example of how this variable surface energy might influence crystal shapes of contacting grains in a magmatic rock is shown in Figure 6.26. Although the validity of extrapolating grain boundary modification in a low-T ammonium compound, as in this figure, to high-T silicate systems can be questioned, the obvious textural adjustment of the former should convince us to be cautious in the interpretation of the latter.

✳6.7 VESICULATION AND FRAGMENTATION OF MAGMA

Exsolution of volatiles from melts that produce fluid bubbles—the process of **vesiculation**—has many parallels to crystallization that produces crystals. Bubbles are preserved in solidified rock as **vesicles.** Both vesiculation and crystallization follow kinetic paths of nucleation and then growth. Both kinetic paths accompany changes in the state of the magma system caused by changes in intensive parameters, chiefly decreasing T and P. Crystallization during cooling and decompression during ascent induce exsolution of volatile fluids from initially volatile undersaturated melts (Section 4.2.3). The increasing mass of exsolved volatiles,

chiefly water in most systems, compounded by volumetric expansion of the volatile fluid at decreasing P, causes bubble growth and potential fragmentation of the magma during explosive volcanism. This and other consequences of volatile exsolution are discussed in Section 4.3. Kinetic processes of magma degassing near the surface of the Earth are the focus of this section.

6.7.1 Nucleation and Growth of Bubbles—Vesiculation

<u>Nucleation.</u> According to homogeneous nucleation theory, energy is required to create the interface of an embryonic cluster of volatile molecules that originate by random fluctuations in their concentration in the melt (Sparks et al., 1994). As for crystal nucleation, only clusters of a critical radius are stable with that radius determined by the Gibbs free energy difference between melt and the surface free energy of the cluster (compare Figures 6.11 and 6.12 for nucleation of crystals).

In experiments on bubble nucleation in a hydrous rhyolite magma, Hurwitz and Navon (1994) found no

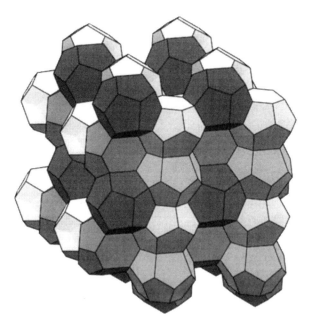

6.23 An array of polyhedrons that has minimal surface area. The equal-volume polyhedrons are of two types: 14-sided polyhedrons of 12 pentagonal and 2 hexagonal faces and 12-sided polyhedrons of 12 faces that are somewhat distorted pentagons (Weaire and Phelan, 1994). This minimal surface area geometry is found in aggregates of soap bubbles and vesicles in thread-lace scoria (Figure 6.25). It is also approximated in some crystalline aggregates (e.g., Figure 6.24) in which there has been textural adjustment over long time at high T. In two-dimensional slices, bubble or grain boundaries have ~120° triple junctions: $\alpha = \beta = \delta = 120°$ in Figure 6.22. (Image courtesy of Ken Brakke.)

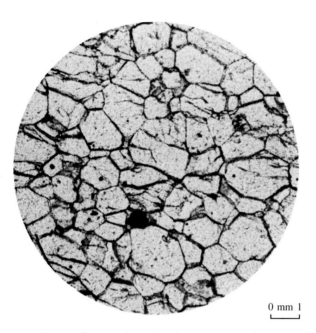

6.24 Dunite (olivine rock) under plane-polarized light showing ~120° grain boundary triple junctions. This texture reflects adjustment of grain boundaries to achieve minimal surface energy. Compare Figure 6.23. Black grain is a Cr-Al-Mg-Fe spinel.

clear evidence for homogeneous nucleation even for decompression-induced oversaturation, ΔP, as much as 130 MPa, equivalent to almost 6 km of ascent. However, they found that substantial *heterogeneous* nucleation occurred for ΔP as small as <1 MPa where minute Fe-Ti oxide grains served as bubble sites to overcome the kinetic barrier. Observed bubble densities were on the order of 10^6 bubble/cm^3 s. Bubble nucleation was slower on other minute "crystallites" of biotite, apatite, and zircon but required tens of megapascals of oversaturation. In apparently crystal-free melts, bubbles still appeared to be heterogeneously formed, perhaps on submicroscopic entities of unknown character.

Klug and Cashman (1994) found bubble "densities"—number of vesicles in a volume of natural silicic pumice—to be similar to those created heterogeneously in experiments and also noted greater vesicularity in glass that contains small crystals of crystals. Radial pipelike vesicles in basalt pillows and concentrations of vesicles in horizontal zones in the interiors of basalt flows may reflect bubble nucleation along crystallization fronts in these cooling magma bodies.

These experimental and empirical observations lend credence to heterogeneous bubble nucleation and indicate a link with crystallization.

Growth. A volatile fluid bubble in a static volume of melt adopts a spherical shape as a consequence of surface tension, which acts to minimize its surface area. Bubbles can grow to a diameter of as much as 10 m in low-viscosity basaltic magmas (Cashman and Mangan, 1994). However, most bubbles are much smaller, typically between a centimeter and a few micrometers, regardless of melt composition. Creation of a high density of bubbles in a melt produces, after solidification, a glassy silicic **pumice** or basaltic **scoria** in which the proportion of vesicle volume to whole rock is quite high. Corresponding bulk rock densities are low enough (several tenths of a gram per cubic centimeter) that pumice and scoria can float on water (density = 1 g/cm^3).

If bubbles are uniformly sized spheres, their theoretical maximum packing occupies 74% of the total volume; for nonuniform spheres the packing approaches 85%. Many natural pumices and scorias have vesicularities of 70–80%. In the very highly vesicular "thread-lace" basaltic scoria, also called *reticulite,* which has 95–98% voids (Figure 6.25), the vesicles are not perfect spheres but have a close-packed polyhedral shape (compare Figure 6.23) with only thin intervening filaments of glass. In silicic melts, the equilibrium bubble volume should be about 75% at depths of about 450–750 m and exceed 99% at 1 atm for common dissolved water contents of 4–6 wt.% (Klug and Cashman, 1996). Vesicularities upward of 93% are observed

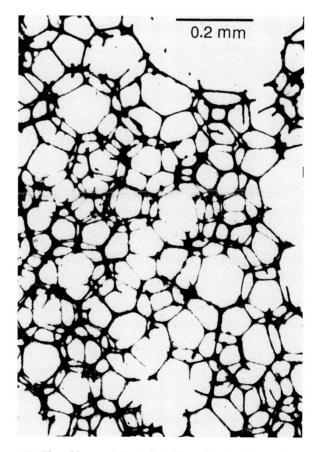

6.25 Thread-lace scoria (reticulite), Hawaii. Vesicle volume is about 98%. Compare Figure 6.23. (From Wentworth and Macdonald, 1953.)

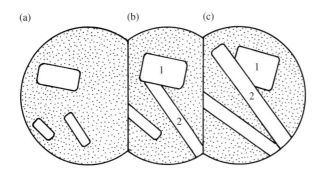

(a)   (b)   (c)

6.26 Grain-shape modification in coalesced grains. Tabular crystals of ammonium thiocyanate whose habit resembles that of plagioclase experienced modification of shape during cooling in a melt from 83°C–61°C over a little more than 3 hours. (a) Separately nucleated and growing crystals. (b) Growing crystal 2 impinges upon crystal 1. (c) The lower-energy, more stable large face of growing crystal 2 has encroached into what must be a higher energy and less stable face in crystal 1. The conventional textural interpretation of the common boundary of crystals 1 and 2 in (c) would be that crystal 1 grew against pre-existing crystal 2, which in this case is incorrect. Diameter of fields of view is about 0.3 mm. (Redrawn from a photomicrograph in Means and Park, 1994.)

in fibrous pumice (Figure 6.27) in which vesicles are rod-shaped, probably as a result of stretching flow of the bubbly melt just prior to solidification. The optimal packing limit for rods, rather than spheres, is about 93%.

Many investigators have noticed a bimodal size distribution in vesicles in pumice—some are significantly larger than others, not unlike porphyritic texture for crystal sizes (see Figure 7.28b). Such a distribution could mean that some larger bubbles began to form before a nucleation "cascade" occurred during high oversaturation accompanying eruption that created many smaller bubbles. Alternatively, it may reflect coalescence of smaller bubbles to form larger ones.

Common silicic pumice has an overall surface area as much as 0.5 m^2/g corresponding to a sheet of glass 1 m^2 in area and only 0.87 micrometer thick (Whitham and Sparks, 1986). The energy expended in creating this surface area from an equidimensional volume of viscous melt is obviously considerable. (One may think of the work of rolling out a layer of much less viscous pie dough not nearly as large in area or as thin.) But this expenditure is only part of the total energy capacity in an exploding magma system. Additional energy is expended in breaking the pumice into pyroclasts and blasting them tens of kilometers against gravity into the atmosphere.

1 mm

6.27 Fibrous, or woody, pumice lapilli in a pyroclastic deposit. Photomicrograph by scanning electron microscopy.

Growth of bubbles depends on several factors in addition to volumetric expansion of an existing volatile fluid in the bubble as the magma decompresses:

1. The melt viscosity that resists stretching of the bubble walls during volumetric expansion
2. Coalescence, or physical merging, of two or more bubbles into a larger one
3. Diffusion of volatiles through the melt walls to an existing bubble as intensive parameters continue to change, adding mass
4. Ostwald ripening, whereby small bubbles are consumed into larger ones by differential diffusion of volatiles through bubble walls (Mangan and Cashman, 1996)
5. Rate of magma ascent, or the rate of decompression and loss of heat
6. Volatile concentration and solubility

Which of these factors are most important in determining bubble growth in different magmas? Diffusivity depends strongly on viscosity, so this important parameter as well as concentrations of volatile species and solubilities must be considered. Rate of magma ascent also reflects magma supply rates and conduit attributes. Therefore, volatile concentration and solubility, ascent rate, and especially viscosity can be taken as the principal controlling factors in bubble growth.

One or more of these factors may be relevant in the contrast between nonexplosive extrusion of coherent magma, however bubbly it is, and explosive blasting of fragments of magma from a volcanic vent.

6.7.2 Melt Fragmentation and Explosive Volcanism

No one is really sure where, how, or why the transition takes place between a bubble-rich melt and an exploding magma. A common notion is that fragmentation occurs in a rising column of vesiculating magma (Figure 4.13) as the bubble volume exceeds a critical packing limit of 70–80%. However, the occurrence of unexploded pumice having higher vesicularity casts doubt on this notion (Gardner et al., 1996). A more fruitful line of inquiry may be to examine the three factors just enumerated, namely, volatile concentration and solubility and viscosity of the melt as well as magma ascent rate. Because viscosity may be the dominant factor, a comparison between the behavior of a low-viscosity basaltic magma and a highly viscous rhyolitic one might provide useful insights.

Explosive Basaltic Volcanism. Basaltic melts have not only low viscosity but high T so chemical diffusivities are large. Also, dissolved volatile concentrations, especially of water, tend to be lower than in silicic melts. A slowly ascending, decompressing, and cooling column of basaltic magma might experience near equilibrium bubble growth that keeps pace with changing intensive parameters. Larger bubbles might rise buoyantly through the magma fast enough to escape harmlessly from the top of the column. Exsolved volatiles may also be dissipated into openings in wall rock around the column. Variably degassed bubbly magma can erupt as coherent lava or in a mildly explosive manner.

More vigorous eruption occurs in basaltic **lava fountains,** which consist of molten blobs as large as bathtubs ejected hundreds of meters above the vent. Because of large bubble densities in fountain ejecta that imply large volatile oversaturation, Mangan and Cashman (1996) suggest that a rapidly ascending column of magma overshoots its saturation pressure and experiences a disequilibrium nucleation "runaway" and subsequent explosive degassing at less than about 100-m depth. The upward accelerating magma falls apart, much as water spray does in a high-speed fire hose.

Explosive Rhyolitic Volcanism. The dissolved water concentrations, commonly in the range of 3–6 wt.%, lower the viscosities of rhyolitic melts to only a few orders of magnitude more than basaltic. However, as a rhyolitic melt exsolves water into growing bubbles, the intervening melt walls between bubbles become drastically more viscous, impeding further bubble growth, both by restricting diffusion of more water into the bubble and by retarding viscous stretching of the bubble wall in response to volumetric expansion of the steam. Although increased viscosity would be expected to slow magma ascent, the increased volume of bubbles decreases the overall density of the magma, making it more buoyant and able to ascend faster. A high rate of magma ascent exacerbates the state of disequilibrium. Faster deformation of the magma during faster ascent in the volcanic conduit might cause the viscous melt to exceed its characteristic relaxation time (Section 6.1) so that bubble walls are, in effect, glass and the excess internal fluid pressure in the bubbles ruptures the walls. The result is explosive fragmentation of the magma.

A typical and important attribute of explosive rhyolitic deposits is the presence of a range of fragment sizes—ash, lapilli, and local blocks (Section 2.4.1). Ash is composed mostly, if not entirely, of glass shards that are largely ruptured bubble walls, whereas lapilli and blocks are composed mostly of unexploded pumice. Why is it that not every bubble in the exploding melt bursts? One possible reason for this heterogeneity in fragmentation (Gardner et al., 1996; Klug and Cashman, 1996) may be different degrees of bubble coalescence possibly resulting from uneven partitioning of strain in the melt during shearing flow accompanying rapid ascent and eruption from the volcanic conduit. Flow is indicated by widespread elongate, rod-shaped vesicles in pumice fragments (Figure 6.27).

Laboratory experiments have revealed fresh insights into explosive processes (Mader et al., 1994; Sugioka and Bursik, 1995). Test cells that contained CO_2 dissolved in HCl-K_2CO_3 solutions to mimic a volatile-rich silicate melt were monitored by high-speed photography. During decompression of the test-cell system, bubbles rapidly nucleated uniformly throughout the solution over a diffuse region, rather than at a well-defined, downward-propagating disruption surface. The expanding foam fragments heterogeneously where the bubble density and expansion rate conspire to rupture bubble walls.

Summary

A change from some initial metastable state to a final more stable state is never instantaneous because it depends on time-dependent kinetic factors. The kinetic path between initial and final states of a melt depends, ultimately, on the mobility of ions. Melt viscosity and diffusion of ions are the two most important ionic transport phenomena because their time-dependent rates range over many orders of magnitude.

Melt viscosity is a measure of the resistance to flow, which depends on the mobility of atoms in it. Viscosity is less at high T because of the loosened atomic structure and in less polymerized melts that have less silica and/or more dissolved water and fluorine. Viscous melts have large relaxation times so that rapid deformation may cause them to break as if they were solid.

Atomic diffusion is net migration down a concentration gradient via random jumps. Diffusion is enhanced at high T and is generally fastest in gases, less in melts, and slowest in crystals. Large and highly charged ions diffuse slowly through more viscous melts, on the order of a few meters in 1 million years. Diffusion through crystals depends on the existence of point defects. The slowest diffusion involves exchange of Na^+Si^{4+} for $Ca^{2+}Al^{3+}$ in plagioclases even at high magmatic temperatures, preserving fine-scale (micrometer) compositional zoning indefinitely.

Thermal diffusivities are generally several orders of magnitude faster than chemical diffusivities. Consequently, static, nonconvecting bodies of melt cool and solidify before atoms can migrate more than a meter or so.

Crystallization is a two-step kinetic process that begins by formation of a nucleus and follows by accretion of ions onto it (crystal growth). Homogeneous nucleation requires significant undercooling so that the increasing difference between the free energies of the crystal and melt overcomes the surface energy contribution. Heterogeneous nucleation on various sorts of preexisting surfaces—solid and fluid—in the melt largely bypasses the kinetic barrier and could be the rule rather than the exception in magmatic systems. Shapes of crystals growing freely in melts depend on the degree of undercooling (cooling rate). Euhedral crystals having characteristic faces grow just below their liquidus T, but with increasing undercooling, crystals become less compact and are skeletal, then dendritic and feathery, and finally spherulitic for hundreds of degrees undercooling. If crystals grow faster than rates of diffusion, a compositional boundary layer develops at the crystal-melt interface.

Grain size in magmatic rocks is a compound function of nucleation and crystal growth rates. For similar growth rates, high nucleation density (integrated nucleation rate) yields small grains and low nucleation density large grains. A hierarchy in ease of nucleation depends on mineral structure and solubility in melts; among rock-forming minerals, Fe-Ti oxides nucleate readily and form small grain-size populations, mafic silicates nucleate somewhat more slowly, and commonly sluggish nucleation of felsic minerals, especially alkali feldspar, creates larger crystals.

Primary grain sizes can be increased (Ostwald ripening) and grain shapes modified because of the tendency for minimization of surface free energy.

Like crystallization, vesiculation is a kinetic phenomenon controlled by nucleation and growth of volatile bubbles in melts. Bubble nucleation, at least in silicic melts, appears to occur heterogeneously on minute crystals (crystallites). Bubble growth depends mostly on volatile concentration and solubility, ascent rate of magma in the crust, and, especially, viscosity. Low-viscosity basalt magmas are usually extruded as somewhat bubbly lava flows because vesiculation keeps pace with the rate of ascent in the conduit. Mildly explosive lava fountaining occurs locally, apparently as a result of rapid ascent and delayed catastrophic nucleation and growth of bubbles. Greater departure from equilibrium occurs during vesiculation of rhyolitic melts, especially during rapid ascent, because of their much greater viscosity, which increases as volatiles exsolve into bubbles. Exactly how factors such as excessive internal gas pressure in bubbles, viscosity and strength of bubble walls, melt relaxation time, bubble concentration, and shear flow during eruption interact to produce explosive volcanism has yet to be fully elucidated.

CRITICAL THINKING QUESTIONS

6.1   Contrast and characterize the three time-dependent transport phenomena with regard to mechanisms at the atomic scale, governing equations, driving forces, and proportionality constants.

6.2   Characterize and account for the viscosity of melts in terms of their composition, T, P, and concentration of dissolved volatiles.

6.3   List and discuss factors that govern the rate of diffusion of ions through liquids and solids.

6.4 Describe the role of diffusion in petrologic processes and how a knowledge of diffusion rates can provide insight into rock-forming phenomena.

6.5 Account for the large contrast in rates of chemical and thermal diffusion.

6.6 Discuss the nature of a phase interface and the importance of surface free energy in the kinetic paths and textural equilibria that influence rock fabric.

6.7 Contrast homogeneous and heterogeneous nucleation.

6.8 Describe a homogeneous nucleation model for a one-component melt and indicate limitations in applying this model to natural muticomponent melts.

6.9 Comment on the apparent relative nucleation rates of olivine, pyroxene, and plagioclase in the Makaopuhi basalt (Plate III), assuming growth rates are uniform.

6.10 Describe factors that control the growth and dissolution of crystals in a melt.

6.11 Discuss factors that govern grain size in magmatic rocks; grain shape.

6.12 What similarities and contrasts exist between nucleation and growth of bubbles and crystals?

6.13 Describe possible kinetic controls on fragmentation of melts and explosive volcanism.

6.14 What kinetic factors might be responsible for the highly explosive eruptions of some silicic magmas?

PROBLEMS

6.1 From Figure 6.2: (a) Plot viscosities, extrapolated if necessary, at 1100°C versus weight percentage silica for dry silicate melts. Explain your plot relative to the chemical composition of the melts and their atomic structure. (b) Plot viscosities of rhyolite melts with 77 wt.% silica at 800°C versus dissolved water concentration. Explain the drastic decrease in viscosity in the first 2 wt.% or so of dissolved water, followed by a lesser effect at higher water concentrations. How does the speciation of dissolved water, as $(OH)^-$ or as molecular H_2O, enter into your explanation?

6.2 Calculate the viscosity of Makaopuhi basalt melt at 1200°C using the chemical composition listed in Table 5.1 and the Shaw equation in Advanced Topic Box 6.1. How does your answer compare with the viscosity of $10^{1.51}$ Pa s measured experimentally by Shaw (1969)?

6.3 Calculate the activation energy of diffusion of O_2 through anorthite crystals (Figure 6.4).

6.4 Compare the diffusivities of water, H_2, and O_2 in a basalt melt at 1200°C. Calculate the average diffusion distance for these three species over 1 day. Could this have any bearing on the common red color of basaltic lapilli around explosive vents? Explain.

6.5 Freer (1981) cites $D = 9.66 \times 10^{-16}$ m^2/s at 1373K (=1100°C) for interdiffusion of Fe and Mg in Mg-rich olivine. Calculate the order of time to homogenize a compositionally zoned olivine crystal whose radius is 2 mm.

6.6 Demonstrate graphically that the ratio of surface area to volume, $4\pi r^2/(4/3)\pi r^3$, for spheres increases rapidly as the radius, r, diminishes to submillimeter dimensions. (*Suggestion:* Plot select values of r from 0.0002 to 0.015 mm against corresponding values of the ratio.)

6.7 Draw hypothetical homogeneous nucleation and crystal growth rate curves for a single mineral as a function of undercooling, and use these to illustrate the contrasts in the kinetic paths of glassy, aphanitic, and phaneritic textures.

Kinetic Paths and Fabric of Magmatic Rocks

FUNDAMENTAL QUESTIONS CONSIDERED IN THIS CHAPTER

1. How do the time-dependent kinetic phenomena discussed in the previous chapter control the fabric of magmatic rocks?

2. How can the fabric of a magmatic rock be used to obtain insights and sometimes specific information regarding the kinetic path that the magma followed during creation of the rock?

INTRODUCTION

Like magmatic rock compositions, magmatic rock fabrics comprise a wide and continuous spectrum of attributes. Fabrics provide rich petrologic information regarding the time-dependent kinetic path of the transition between the magmatic state and the final solidified rock, as well as its subsolidus history after solidification. The kinetic history recorded in rock fabrics takes different paths (Figure 7.1). In human affairs, there are many possible "paths" for a journey (i.e., a change in a person's "state") between San Francisco and Paris, but an observer in Paris (at the final "state") can readily differentiate the relaxed, well-tanned traveler who made a week-long stopover on the beach of a Caribbean island from the haggard red-eyed overnight traveler. In similar manner, the appearance of a rock—its fabric—is predominantly a record of the kinetic path taken during the solidification of the magmatic system. Other factors can influence fabric besides kinetics, as in the contrast in water pressure in development of hypersolvus versus subsolvus texture in granite (Figure 5.18). Of course, contrasting kinetic paths can result from contrasting magma compositions, as in the

contrast between paths in highly viscous silicic versus much less viscous basaltic magma.

The importance of diverse kinetic paths in development of fabric is most obvious when comparing fabrics in one type of solidifying magma. For example, the same identical granitic (rhyolitic) magma can solidify into the following **fabric heteromorphs:** phaneritic, porphyritic phaneritic, aphanitic, porphyritic aphanitic, glassy, vitrophyric, pumiceous, or pyroclastic. As this chapter will show, additional distinct fabrics can be found in the granitic-rhyolitic composition type. Fabrics record kinetic paths in much the same way that mineral compositions record intensive variables in the final thermodynamic state of the system.

Some kinetic paths recorded in rocks are relatively simple, such as the creation of glassy fabric by drastic undercooling of a melt. Other paths involve more than one kinetic process operating simultaneously or in close sequence, such as rather rapid crystallization and volatile exsolution that result in an aphanitic vesicular fabric. A particular fabric attribute, such as grain size, can evolve along multiple paths, including crystallization, crystal dissolution, textural equilibration, and fragmentation processes.

Fabric encompasses noncompositional properties of a rock that comprise textures and generally larger-scale structures. There is no sharp distinction between these two. **Textures,** also called **microstructures,** are based on the proportions of glass relative to mineral grains and their sizes, shapes, and mutual arrangements that are observable on the scale of a hand sample or thin section under a microscope. **Structures** are larger-size features generally seen in an outcrop, such as bedding in a pyroclastic deposit or pillows in a submarine lava flow. Features related to exsolution of volatiles and fragmentation of magma can occur on a wide range of scales. One rock can have more than one texture and one or more structures. Altogether, textures and struc-

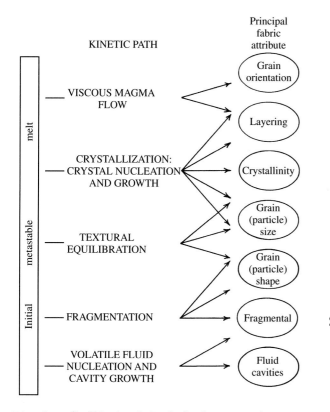

KINETIC PATH

Principal fabric attribute

VISCOUS MAGMA FLOW

CRYSTALLIZATION: CRYSTAL NUCLEATION AND GROWTH

TEXTURAL EQUILIBRATION

FRAGMENTATION

VOLATILE FLUID NUCLEATION AND CAVITY GROWTH

melt — metastable — Initial

Grain orientation

Layering

Crystallinity

Grain (particle) size

Grain (particle) shape

Fragmental

Fluid cavities

7.1   Generalized kinetic paths involved in the creation of magmatic fabric as a metastable melt transforms to a solid magmatic rock. Most fabric attributes evolve along multiple paths. For example, layering can develop by magma flow or by crystallization processes.

tures constitute the rock fabric, for which responsible multiple kinetic paths of formation can be interpreted.

There is also no sharp distinction between fabric and field relations. But the latter pertains to aspects of the whole rock body as a genetic entity, usually on a scale of observation no smaller than an outcrop, such as a lava flow from its source to terminus. Field relations of intrusive magmatic rocks are specifically addressed in Chapter 9 and of extrusive rocks in Chapter 10.

The links between major kinetic paths and resulting principal fabric attributes are shown schematically in Figure 7.1. Following is an outline of how this chapter is organized according to these kinetic paths. The outline lists specific fabrics to be described. Numbers in parentheses refer to figures and plates illustrating the specific fabric.

Section 7.1   Crystallization path → crystallinity and grain size

    7.1.1   Glassy (7.2–7.6)
        Vitrophyric (7.4, Plate IIIa)
        Perlitic (7.4, 7.5)
        Spherulitic (devitrification of glass) (7.6, 6.15d)

    7.1.2   Aphanitic (7.7–7.9, 7.26)
        Aphyric

        Cryptocrystalline
        Microcrystalline (7.7–7.9, 7.35)
        Felsitic (7.7, 7.26, 7.35)
        Felty (7.8b)
        Intergranular (7.9)
        Intersertal (Plate IIId)

    7.1.3   Phaneritic (7.10, 7.11, 7.13, 7.14, 7.16, 7.17, 7.20)
        Equigranular (7.17)
        Inequigranular (7.10, 7.11, 7.13, 7.14)
        Seriate (7.10)
        Pegmatitic (7.11)

    7.1.4   Porphyritic (7.4, 7.7–7.9, 7.13)
        Vitrophyric (7.4, Plate IIIa)
        Cumulophyric (6.17, 7.34)

    7.1.5   Ophitic and poikilitic (7.14, 7.15, 7.22)

Section 7.2   Crystallization path → crystal shape
        Euhedral (7.3b, 7.15, 7.16, Plate IIIa)
        Subhedral (7.10, 7.17)
        Anhedral (7.10, 7.16)
        Skeletal (7.3c, 6.15b)
        Dendritic (6.15c)
        Feathery (7.3b)
        Spherulitic (7.6, 6.15d)
        Glass (melt) inclusions (6.18)
        Hypidiomorphic-granular (7.16, 5.18b, Plate IV)
        Aplitic (7.17)
        Partial resorption (5.32, 6.19, Plate II)

Section 7.3   Crystallization path → inhomogeneous grains
        Normal zoning (5.14b)
        Oscillatory zoning (6.17)
        Reaction rim (5.32, 6.20, Plate II)
        Rapakivi (7.18)
        Decomposition (rim) (5.34)
        Exsolution (5.14b, 5.18a)

Section 7.4   Textural equilibration → secondary grain boundary modification
        Graphic (7.20)
        Granophyric (7.21)
        Myrmekitic (5.18b)

Section 7.6   Nonexplosive exsolution of volatiles → volatile-fluid cavities
        Vesicular (7.23, 7.24, 7.28b, c)
        Pumiceous (7.23)
        Scoriaceous
        Amygdaloidal (7.24)
        Vuggy (7.25)

Additional textural and structural features specific to restricted rock types and their formative processes are discussed in later chapters. For additional description and illustrations of textures the classic work of Williams, Turner, and Gilbert (1982) or the shorter treatment by Nockolds, Knox, and Chinner (1978) can be consulted.

This chapter is designed to be a companion and sequel to Chapter 6, which introduced some concepts of the kinetics or chemical dynamics of solidifying and vesiculating melts. The intent is to apply these concepts,

(a)

(b)

0 cm 2

0 mm 0.04

7.2   **Glassy** texture in high-silica rhyolite **obsidian.** (a) Hand sample shows characteristic conchoidal fracture of the broken amorphous glass. Note faint, alternating lighter and darker **flow layers** whose definition is apparent in the photomicrograph in plane polarized light (b), which shows minute, spiderlike **crystallites** that are too small to display any interference colors under cross-polarized light. Differing concentrations of these abundantly nucleated crystallites, perhaps related to differing amounts of shear during viscous flow, create the alternating lighter and darker layers seen in the hand sample.

however tentative they may be, to facilitate a better understanding of the origin of magmatic rock fabrics.

✳7.1 FABRICS RELATED TO CRYSTALLIZATION PATH: CRYSTALLINITY AND GRAIN SIZE

The modal percentage of mineral grains relative to glass—their crystallinity—ranges from 0% to 100% in magmatic rocks. Grain size ranges widely, from submicroscopic grains much less than about 0.001 mm, which is about the smallest that can be discerned with an optical microscope, to the giant crystals of pegmatites, which can be as much as several meters in dimension. Crystallinity and grain size depend upon rates of crystal nucleation and growth, as discussed in Section 6.5.4.

7.1.1 Glassy Texture

Glass is basically a highly viscous liquid, disordered on an atomic scale, formed from a polymerized silicate melt that was cooled too rapidly for crystallization to occur. Although natural, massive high-silica glass, or **obsidian** (Figure 7.2a), appears in hand samples to have zero crystallinity, few natural glasses are entirely lacking in crystals. High magnification (Figure 7.2b) reveals that obsidian contains abundantly nucleated submicrometer-size **crystallites** that experienced limited growth in the highly viscous glass. Though much less viscous, basaltic melt solidifies as a glass in drastically undercooled margins of submarine lava pillows extruded on the seafloor (Figure 7.3) and in thin pahoehoe lava flows on land. Thin streamers of basalt melt

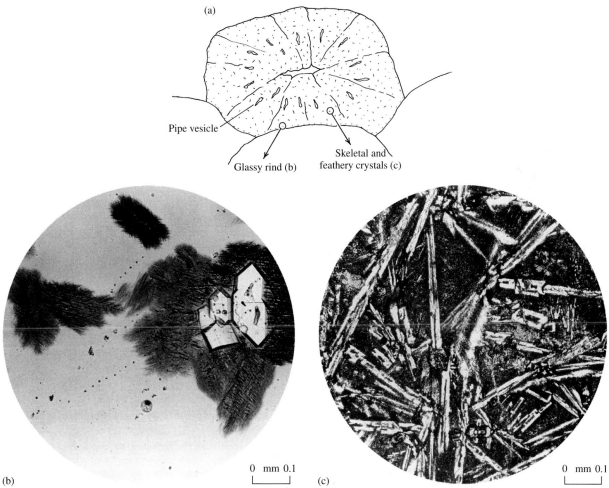

7.3   Fabric of a basalt **pillow.** (a) Idealized cross section showing concentric zonal variation from outermost glassy rind, underlying zone of skeletal crystals, and inner, more crystallized part with radial pipe vesicles and shrinkage cracks. **Pipe vesicles** may form normal to an inward advancing front of crystallization that promotes volatile exsolution (Philpotts and Lewis, 1987). (b) The glassy outer rind a few millimeters thick was produced by quenching of the hot magma against cold seawater and consists of basaltic glass in which are embedded sparse phenocrysts of **euhedral** olivine that crystallized before extrusion on the seafloor. **Feathery** crystals of clinopyroxene apparently nucleated heterogeneously on the olivine during the rapid quenching and substantial undercooling of the melt. (Photograph of a sample collected from the submersible *Alvin* along the Mid-Atlantic Ridge courtesy of A. E. Bence.) (c) Beneath the glassy rind is a mosaic of **skeletal** crystals of plagioclase, pyroxene, and olivine produced by slightly less, but still substantial, undercooling. The innermost, and most voluminous, zone of fabric development (not shown), reflecting the relatively slowest rate of heat loss in the core of a typical pillow, is a partly glassy to holocrystalline, intersertal to intergranular aggregate of plagioclase, olivine, pyroxene, and spinel, usually exhibiting some noncompact crystal shapes. (Photograph from a Deep Sea Drilling Project core sample courtesy of A. E. Bence.)

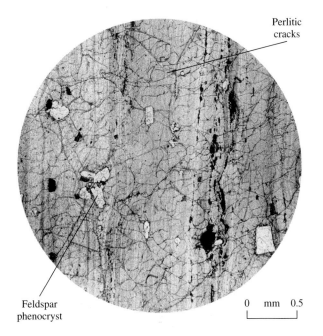

Perlitic cracks

Feldspar phenocryst

0   mm   0.5

7.4   A combination of vitric (glassy) and phyric (porphyritic) texture is **vitrophyric.** In this photomicrograph of a silicic **vitrophyre,** phenocrysts of feldspar and opaque (black) Fe-Ti oxides lie in a **flow-layered** glass that has hydrated to produce **perlitic** cracks (see also Figure 7.5).

ejected from lava fountains in Hawaiian-type eruptions also quench to glass (see Figure 7.29b). Thin dikes and margins of thicker dikes of a wide range of composition that are emplaced in the cool shallow crust can also be of glass.

0   cm   1

7.5   Nests of concentric curved fractures in glass constitute **perlitic** texture. Internal reflectance of light imparts a lustrous, pearly-gray color to hand samples. Small nodules of black, unhydrated, and uncracked obsidian (called *Apache tears*) may be located in the core of the nested cracks.

All glass is metastable and therefore susceptible to secondary hydration, devitrification, and other types of alteration (Bouška, 1993) that progress over time to achieve a more stable state. Hydration and devitrification are kinetic processes that depend on diffusion of water molecules or $(OH)^-$ and H^- ions and on nucleation and growth of crystals, respectively, in the glass. Hydration is faster than devitrification in silicic glasses.

Some massive glass having a waxy luster and dark color in hand sample into which 6–16 wt.% water has been absorbed is called **pitchstone.** More common hydrated silicic glass, occurring in subaerial environments, is called **perlite** because it typically has a pearl-gray color, but some is green, red, or brown. As much as 6 wt.% of perlite is H_2O^- absorbed at near-atmospheric T and can be liberated by heating the glass to only 110°C; the isotopic composition of this water indicates that it is meteoric, not magmatic, in origin.

0   cm   2

7.6   **Spherulitic** texture in high-silica rhyolite obsidian. **Spherulites** are spherical to ellipsoidal clusters of radiating fibrous alkali feldspar and a polymorph of SiO_2, here in a black glassy matrix. Faint, relict flow layering that extends from lower left to upper right is accentuated in the spherulites in this hand sample, indicating their growth after active flow of the magma. Individual spherulites in volcanic rocks can range in diameter from less than 1 mm to 1 m or so. A phenocryst may be located at the center of the spherulite, where, in the original glass or drastically undercooled melt, it allowed hetereogeneous nucleation of crystals to occur. Spherulites are secondary devitrification features, *not phenocrysts.*

On the other hand, obsidian generally has <1 wt.% H_2O^+ of a magmatic isotopic composition that was dissolved in the high-silica melt prior to solidification. It is part of the atomic structure of the glass, is held tightly in the glass, and can only be liberated from the glass by heating to hundreds of degrees. **Perlitic** texture (Figures 7.4 and 7.5) develops by hydration of obsidian on fracture surfaces that are exposed to moisture in the atmosphere or to meteoric water (groundwater). As the outer rind hydrates, it expands and separates along a crack from the nonhydrated substrate. Inward repetition of this process creates a sequence of concentric perlitic cracks that reflect light, creating the characteristic pearl-gray color. Perlitic hydration occurs by atomic diffusion so the rate of inward advance is described by $t = x^n/C,$ where t is time, x is distance, n is about 2, and C is a constant for a given area that is chiefly a function of mean ambient T (Friedman et al., 1966). Fast rates occur in hot climates and are about 1–2 micrometers/1000 years, according to measurements of the thickness of the hydrated rind on obsidian artifacts of known age. Once a calibration has been established in a particular climate, obsidian artifacts of unknown age can be dated by measurement of the thickness of the hydrated rind.

Devitrification, or delayed crystallization, of silicic glasses produces two crystalline textures, felsitic (discussed later) and **spherulitic** (Figures 7.6 and 6.15d).

Hydration and devitrification tend to be simultaneous in basaltic glass, producing the alteration product called **palagonite.** In thin section, this is isotropic to weakly birefringent, variably orange to brown, and commonly concentrically layered in colloform fashion. Relative to pristine basalt glass (Plate IIIa), palagonite is vastly but variably enriched in water, to as much as 30 wt.%; most of the Fe is oxidized to the ferric state; and most other elements have perturbed concentrations. Palagonite is a complex mixture of clay and zeolite minerals and hydrated ferric oxides.

7.1.2 Aphanitic Texture

Aphanitic texture (Figures 7.7–7.9), typical of surface extrusions and near-surface small intrusions of magma, consists of a mosaic of crystals too small to be identifiable by the naked eye without magnification by a hand lens or a microscope. Aphanitic texture implies high crystal nucleation rates relative to growth rates, such as occur during rapid reduction in T or water content of the magma system.

Relatively few aphanitic rocks are **aphyric,** or nonporphyritic. The presence of phenocrysts in most aphanitic rocks testifies to the fact that few magmas reaching near the surface of the Earth are superheated above liquidus temperatures.

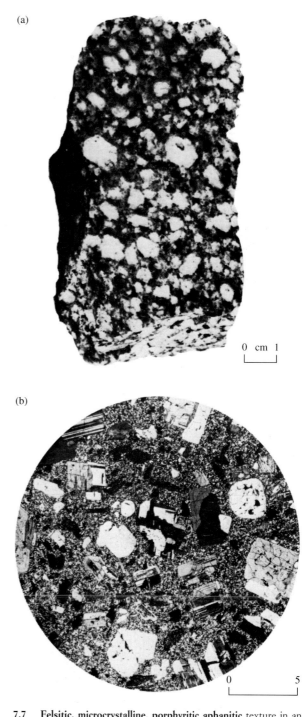

(a)

0 cm 1

(b)

0   5

7.7 **Felsitic, microcrystalline, porphyritic aphanitic** texture in an intrusive granodiorite porphyry. Photomicrograph under cross-polarized light. Weak **oscillatory zoning** in plagioclase phenocryst just below center of view.

Grains in **cryptocrystalline** texture are too small to be resolved optically but are visible with an electron microscope and can be identified by X-ray diffraction analysis. Larger grains in **microcrystalline** texture (Figures 7.7–7.9) can be discerned with a petrographic microscope. Microcrystalline rocks in which elongate rectangular grains of feldspar are dominant have **felty**

(a)

(b)

Plagioclase phenocryst
with corroded core

Pyroxene
phenocryst

Hornblende
phenocryst

Microlites
of feldspar
in felty matrix

Hornblende phenocrysts
with decomposed rims
rich in opaque Fe–Ti oxides

0   cm   1

0   mm 0.5

7.8   The most common texture in volcanic rocks is **porphyritic aphanitic,** as in this andesite from a lava flow. (a) Hand sample in which phenocrysts of white plagioclase and black hornblende lie in a gray aphanitic groundmass. (b) Photomicrograph of same rock in plane-polarized light showing phenocrysts surrounded by a **microcrystalline** matrix made largely of small birefringent **microlites** of euhedral plagioclase that are interwoven in a **felty** texture. For a magnified view of the partially decomposed hornblende phenocrysts see Figure 5.34a.

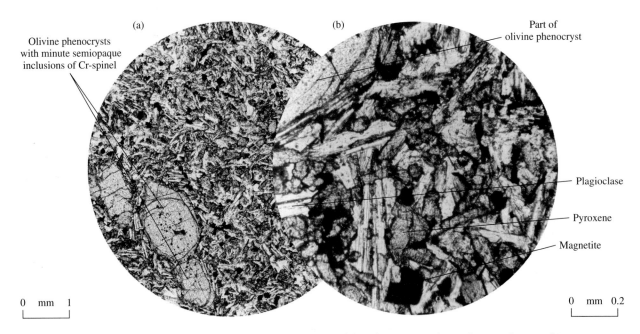

(a)

(b)

Part of
olivine phenocryst

Olivine phenocrysts
with minute semiopaque
inclusions of Cr-spinel

Plagioclase

Pyroxene

Magnetite

0   mm   1

0   mm   0.2

7.9   **Porphyritic aphanitic** basalt from a lava flow. **Intergranular** texture of the aphanitic groundmass, shown in the same thin section at two different magnifications in plane-polarized light, is an interlocking network of randomly oriented plagioclase **microlites** and abundant more equant pyroxene and Fe–Ti oxide grains. Euhedral to subhedral olivine phenocrysts may be present in some basalts. Intergranular texture resembles the felty texture of the andesite in Figure 7.8, except that mafic minerals are more abundant.

texture (also called pilotaxitic texture; Figure 7.8). The feldspar **microlites** in such rocks are large enough to display optical birefringence in polarized light. Felty texture is common in the aphanitic matrix of andesitic rocks and some basaltic rocks; in both, the feldspar is plagioclase. If feldspar microlites are oriented in a common direction the texture is trachytic (see Section 7.9.1). **Felsitic** texture is that of any aphanitic mosaic of mostly felsic minerals—feldspars and quartz—found commonly in rhyolite, dacite, and trachyte. Some devitrified glass has a felsitic texture. Basaltic rocks commonly have **intergranular** texture, wherein randomly oriented microlitic plagioclases and abundant, more equant pyroxenes, Fe-Ti oxides, and, in some rocks, olivines, form a tight interlocking mosaic (Figure 7.9).

Many aphanitic rocks appear to be holocrystalline in hand sample though examination of a thin section reveals the presence of minor glass. In many basaltic rocks, **intersertal** texture resembles intergranular texture except that brown glass is interspersed among the microcrystalline grains as a result of incomplete crystallization (Plate IIId).

7.1.3 Phaneritic Texture

Phaneritic texture occurs in rocks in which grains of major rock-forming minerals are all large enough to be identifiable with the unaided eye (Figures 7.10–7.14). Smaller Fe-Ti oxides and accessory minerals, such as zircon and apatite, are typically not visible without a microscope. Phaneritic rocks are typically found in magmatic intrusions and reflect crystallization at small degrees of undercooling, perhaps only a few degrees; nucleation rates are relatively low regardless of crystal growth rates. Magmatic intrusions worldwide of differ-

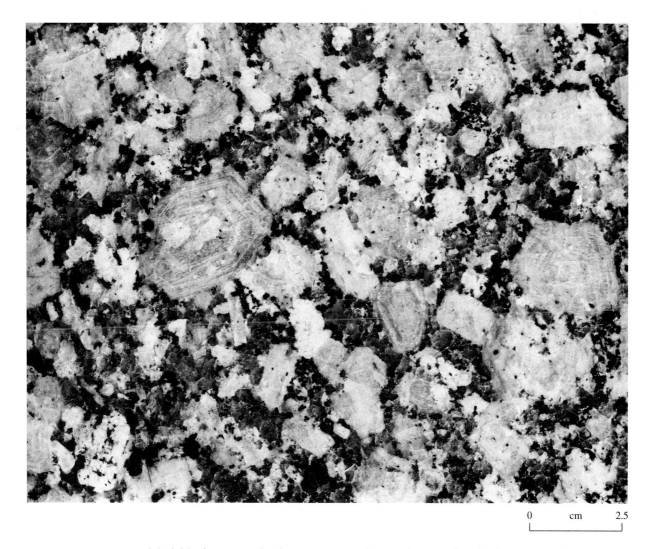

0   cm   2.5

7.10 **Seriate** texture in a polished slab of **inequigranular phaneritic** granite. White, mostly rectangular **subhedral** grains are plagioclase. Light gray, more equant grains are potassium-rich alkali feldspar; note the **oscillatory zoning** in the larger ones made evident by delicate alternating layers of whiter and grayer color concentric to their margins. Dark gray **anhedral** grains without apparent crystal faces are quartz. Black grains are biotite. Slight differences in nucleation and growth rates during crystallization of constituent minerals at small undercooling are responsible for this seriate fabric.

tion and growth rates. Other phaneritic rocks have an **inequigranular** texture in that they contain grains of conspicuously variable size. These include rocks of porphyritic texture in which there is a bimodal size population (large distinct from small; see below) and rocks having **seriate** texture in which grains have a more or less continuously ranging size (Figure 7.12). Many granites, such as shown in Figure 7.10, possess seriate texture made up of apatites and zircons visible only under the microscope, somewhat larger but generally <1-mm Fe-Ti oxides, larger mafic silicates, commonly still larger plagioclase and quartz grains, and alkali feldspars that are as much as 2–3 cm, or more. The

7.11 Pegmatitic fabric is exceptionally coarse, but variable seriate grain size. Most grains are >1 cm. Pegmatitic fabric is best discerned in outcrops, rather than hand samples, because grains are commonly as large as, or larger than, a typical hand sample. Giant crystals in some pegmatites attain extraordinary dimensions of 10 m or more. The photograph shows a quarry face in the Harding pegmatite, Taos County, New Mexico; the gallon jug beside the 0.6-m-long box in the lower left corner of photo attests to the giant size of the white crystals of spodumene (LiAlSi$_2$O$_6$). (Photograph courtesy of Richard H. Jahns. Reproduced with permission from Jahns RH and Ewing RC, 1977. The Harding Mine, Taos County, New Mexico, Min. Record 1977; March, April: 123.)

ent compositions, sizes, magma viscosities, and depths of emplacement (dictating cooling rates) all have a restricted range of grain sizes, generally 1–20 mm. This order-of-magnitude range suggests that nucleation and growth rates are not significantly different in different magmas at small degrees of undercooling. Otherwise, more variable grain sizes might be anticipated.

Phaneritic rocks have an **equigranular** (Figure 7.12), or simply a **granular,** texture if the grains are of similar size, within an order of magnitude, or so. Each crystalline phase must have experienced similar nuclea-

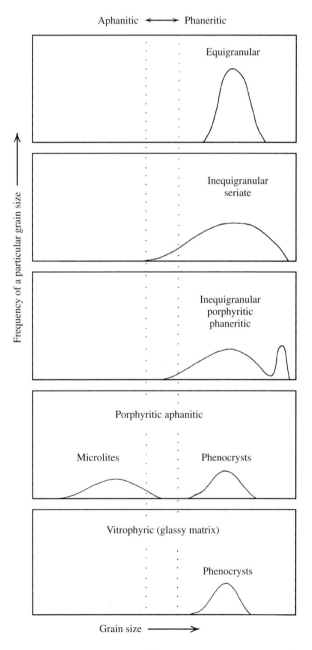

7.12 Schematic grain size populations in common magmatic rocks.

more or less continuous variation in grain size probably reflects differing ease of nucleation among the coprecipitating minerals in the slightly undercooled magma; Fe-Ti oxides nucleate readily, alkali feldspar nucleates at the slowest rate, and the other minerals in intermediate manner.

Small intrusive bodies and parts of larger ones made of exceptionally large, but heterogeneously sized, crystals whose dimensions are at least several centimeters and locally meters define **pegmatitic** fabric (Figure 7.11). Outcrops are generally required to identify this fabric, which must reflect limited nucleation and fast crystal growth rates.

7.1.4 Porphyritic Texture

Most aphanitic rocks, many glassy rocks, and some phaneritic rocks contain large, more or less euhedral **phenocrysts** embedded in a distinctly finer-grained or glassy **matrix,** or **groundmass.** This is **porphyritic** texture. Porphyritic aphanitic textures are shown in Figures 7.7–7.9; porphyritic glassy, or **vitrophyric,** texture in Figure 7.4 and Plate IIIa; and porphyritic phaneritic texture in Figure 7.13. Phenocrysts rarely constitute more than 50% of aphanitic and glassy rocks formed in extruded magma because abundant crystals immobilize magma and retard extrusive flow. In hand sample, phenocryst-rich (35–50%) aphanitic and glassy rocks may superficially appear to be phaneritic, careful examination with a hand lens or microscope will reveal otherwise.

Porphyritic textures originate in different ways: That is, they are polygenetic. Probably the most common

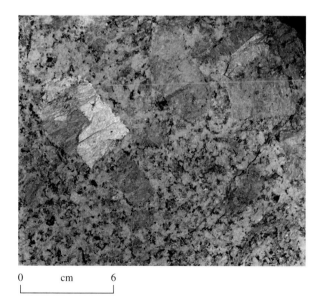

0   cm   6

7.13 Porphyritic phaneritic texture in granodiorite. The commonly pink, subhedral Carlsbad-twinned phenocrysts of K-rich alkali feldspar are as much as 6 cm in longest dimension and are surrounded by a finer but **phaneritic** matrix of gray quartz and feldspar and black biotite. One of the Carlsbad-twin individuals is oriented so as to reflect light off its cleavage.

origin for porphyritic aphanitic and vitrophyric textures involves a two-stage cooling history for the melt. An initial episode of slow cooling rate (small undercooling) yields few nuclei just below liquidus temperatures in a thermally insulated plutonic environment below the surface of the Earth. These grow to produce relatively large phenocrysts. After this partial crystallization, the magma experiences an episode of relatively rapid heat loss in a small intrusion in the shallow cool crust or in an extrusion onto the surface; both create the aphanitic or glassy matrix around the phenocrysts.

However, other kinetic paths can create porphyritic texture under more or less *uniform cooling* of plutonic magma bodies:

1. A two-stage cooling path cannot account for deep plutonic phaneritic rocks that must have cooled slowly at a rather uniform rate but contain large phenocrysts of one mineral in a finer matrix of others. In this case, different nucleation rates for different minerals may be involved. An example is euhedral alkali feldspar phenocrysts in porphyritic phaneritic granodiorite (Figure 7.13). These large alkali feldspars have provoked decades of debate as to their origin, but it is now believed that they nucleate more slowly just above solidus temperatures than other constituent minerals (Vernon, 1986). Many granitic magmas, such as granodiorite in Figure 5.29, reach alkali feldspar and quartz saturation within only a few degrees to tens of degrees above solidus temperatures. As much as half of the magma by volume may still be melt at this stage because of the considerable solubility of alkali feldspar and quartz. Therefore, growing alkali feldspars have ample space to produce a large crystal as they grow from sparse nuclei. Experiments suggest that alkali feldspars nucleate more slowly than quartz and plagioclase in granitic magmas (Swanson, 1977).

2. Porphyritic aphanitic texture can also develop in a shallow granitic intrusion that experiences an essentially isothermal reduction in water pressure. The texture of the resulting **porphyry** rock evolves through two levels of volatile concentration, not by two-stage cooling. Initial nucleation and growth of phenocrysts of plagioclase and other high-T minerals promote volatile oversaturation in the residual melt. If the exsolved fluid rapidly escapes from the system, such as by rupturing roof rocks and venting fluid through the cracks, the remaining melt becomes supersaturated with less soluble crystalline phases. In effect, the solidus of the system rotates clockwise in P-T space (Figure 5.2; see also Figures 5.11, 5.29, and 5.33), stranding the melt in the crystalline stability field, so a shower of nuclei develops, yielding an aphanitic matrix around the phenocrysts.

3. Bowen (1914) pointed out that continuous uniform cooling of some melts can produce porphyritic texture. Consider crystallization of a melt containing 90% of the $CaMgSi_2O_6$ component in the simple binary system $CaAl_2Si_2O_8$-$CaMgSi_2O_6$ (Figure 5.4). For some amount of undercooling below the liquidus, diopside nucleates. Growth of these pyroxene crystals at decreasing T occurs until, at the eutectic T, anorthite plagioclase begins to coprecipitate with diopside. Meanwhile, the near-liquidus diopside crystals have had a long growth period and could exist as large phenocrysts in a finer matrix of anorthite and possibly diopside crystals if more of these had nucleated near the eutectic.

Many aphanitic and glassy rocks contain clots, or polygranular aggregates, commonly of the same minerals of the same size that occur as isolated phenocrysts in the same rock. The clots in this cumulophyric texture (Figures 6.17 and 7.34) can originate in all of the ways that phenocrysts can. Clots originate as suspended crystals attach to each other, or they may be derived from breakup of the more crystallized wall of the magma chamber where precipitated crystals accumulated. Alternatively, some clots may be restite material dislodged from the site of magma generation.

7.1.5 Poikilitic and Ophitic Textures

In both ophitic and poikilitic textures, larger crystals enclose smaller, randomly oriented crystals. The larger crystals form from fewer nuclei than the smaller enclosed mineral grains. In **poikilitic** texture, large **oikocrysts** (literally, house crystals) completely surround many smaller grains. Poikilitic texture occurs in a wide range of rock compositions. For example, in phaneritic ultramafic rocks, oikocrysts of amphibole or pyroxene many centimeters in diameter enclose millimeter-size olivines, chromites, and other minerals (Figure 7.14). In some granitic rocks, near-solidus alkali feldspar oikocrysts surround minerals precipitated at higher T. **Ophitic** texture in rocks of basaltic composition is found in some aphanitic lava flows and more typically in slightly coarser, marginally phaneritic dikes of diabase. In this texture, large clinopyroxenes partially to completely enclose smaller euhedral plagioclases (Figure 7.15; see also Figure 7.22).

※7.2 FABRICS RELATED TO CRYSTAL-LIZATION PATH: GRAIN SHAPE

Most, but not all, aspects of the shape of magmatic crystals result from the way they crystallize from melts. Under near-equilibrium conditions at small amounts of undercooling, slow uninhibited crystallization of a melt yields **euhedral** crystals bounded by characteristic crystal-face forms. *Euhedral* is synonymous with **idio-**

0   cm   3

7.14 **Poikilitic** texture in peridotite, Stillwater Complex, Montana. A single large pyroxene **oikocryst,** with light reflecting off its cleavage, encloses numerous smaller, randomly oriented dark grains of olivine. A single pyroxene nucleus formed and grew in the residual melt, surrounding numerous earlier-formed olivine crystals.

morphic (Greek, "one's own form"). Euhedral crystals are commonly found as isolated phenocrysts in aphanitic and glassy rocks. Thus, euhedral olivines in Plate IIIa and Figures 7.3b and 7.9a are bounded by prism and pinacoid forms; prismatic pyroxenes, tabular feldspars, and platy micas can be seen in other figures in this chapter. Faster cooling rates (greater undercooling) yield progressively less compact grains of greater surface area relative to their volume. Figures 6.14 and 6.15 show a hierarchy of plagioclase grain shapes with increasing amounts of undercooling—from compact tabular crystals (see also Figures 7.15 and 7.16) to **skeletal,** to **dendritic** and **feathery,** to **spherulitic** grains.

As some crystals grow in a skeletal fashion they can completely encompass pockets of melt, forming **melt inclusions.** If the host crystal is rapidly quenched as the magma is extruded, inclusions solidify as glass rather

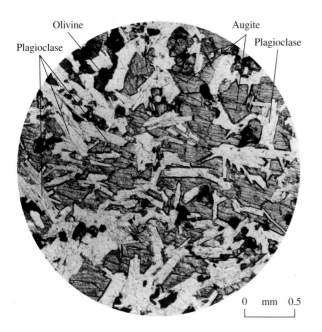

Olivine

Plagioclase

Augite

Plagioclase

0   mm   0.5

7.15 **Ophitic** texture in a **diabase** (coarse basalt) lava flow. Several randomly oriented, **euhedral** plagioclases are partially to completely surrounded by a single large crystal of clinopyroxene. Compare Figure 7.22.

than crystallizing. Some inclusions have an inverted crystallographic habit of their host crystal with which they are in equilibrium—a sort of "negative crystal" (Figure 6.18).

Less perfectly formed compact grains are **subhedral** if only partly bounded by crystal faces (plagioclases in Figure 7.17) and **anhedral,** or **xenomorphic** (foreign form), if not bounded by any characteristic crystal face (pyroxenes in Figure 7.16). These noneuhedral grains can form by restricted growth from melts in highly crystallized magmas where neighboring grains interfere with development of characteristic growth forms. Note that grains within cumulophyric clots (Figure 6.17) are commonly subhedral to anhedral, and if, for example, a clot breaks up during extrusion of the host magma, the resulting isolated phenocrysts are anhedral to subhedral. Other noneuhedral grains result from fragmentation processes discussed in Section 7.7 and from partial dissolution (resorption) of grains into the enclosing melt with which they are unstable (Plate II and Figures 6.19 and 6.20).

Phaneritic aggregates of grains showing varying euhedralism define two rock textures. In **hypidiomorphic-granular** texture there is a mix of euhedral, subhedral, and anhedral grains. It is seen in a wide range of rock types, from gabbro (Figure 7.16) to granite (Figure 5.18b and Plate IV). In granodiorites, Fe-Ti oxides, hornblendes, biotites, and plagioclases tend to be euhedral to subhedral, whereas quartz and alkali feldspar grains are typically anhedral. As discussed in relation to porphyritic phaneritic rocks, melts in grano-

dioritic magma systems become saturated with alkali feldspar and quartz within a few tens of degrees above solidus temperatures (Figure 5.29). Just above the solidus, only about half of the magma by volume has crystallized as free-formed, euhedral mafic minerals and plagioclase. Consequently, during near-solidus growth of quartz and alkali feldspar there is less freedom for growth among the existing higher-T euhedral grains and they become anhedral. If only sparse nuclei of alkali feldspar form, these grains become phenocrysts if earlier formed grains are moved aside or oikocrysts if they are enveloped.

In equigranular phaneritic rocks, exemplified by granite aplites consisting of generally fine leucocratic aggregates of alkali feldspar and quartz, virtually all grains are equant and anhedral to subhedral (Figure 7.17). This texture is appropriately known as **aplitic.** It appears likely that all of the grains crystallized essentially simultaneously from the melt and competed equally for space.

(a)

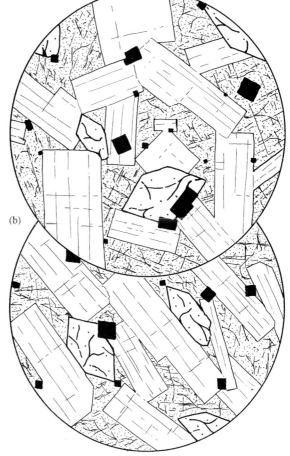

(b)

7.16 **Hypidiomorphic-granular** texture in gabbro. Euhedral magnetites (black), plagioclases (prominent cleavage), and olivines interspersed among **anhedral** pyroxenes. (a) **Isotropic** fabric. (b) Preferred orientation of tabular plagioclases defines **igneous lamination,** a type of **anisotropic** fabric.

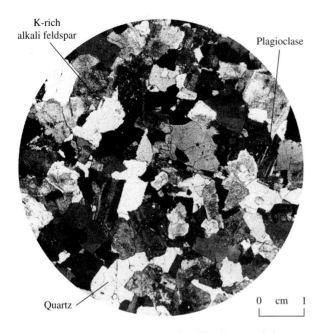

7.17 **Aplitic** texture in a granite aplite dike that intruded grano-diorite. Photomicrograph under cross-polarized light. The quartz and feldspar grains are of subhedral to anhedral shape and of similar size, so the rock is **equigranular**. This texture likely results from similar rates of nucleation and growth of the felsic minerals, all of which were growing more or less simultaneously. In hand sample, the texture appears sugary, like sandstone, but, unlike in sedimentary rock, the grains are somewhat interlocking and pore spaces or secondary cement is nonexistent.

❋7.3 FABRICS RELATED TO CRYSTALLIZATION PATH: INHOMOGENEOUS GRAINS

In a system at equilibrium, every phase must be homogeneous, including each mineral grain. But in most crystallizing magmas, sluggish reaction rates between melt and crystals lag behind rates of changing intensive parameters. Accordingly, many grains in magmatic rocks are inhomogeneous. Several types of zoned and composite grains manifest states of disequilibrium.

7.3.1 Zoned Crystals

A systematic pattern of chemical variation within a solid solution mineral is called **zoning**. It is a record of incomplete *continuous* reaction relations between a melt and the crystallizing solid solution as intensive parameters were changing in the magma system faster than kinetic rates could maintain equilibrium (Section 5.5.2). In some minerals, zoning forms a sector pattern that is hourglass shape in prismatic crystals, such as clinopyroxene. In feldspars, zoning is concentric to the exterior grain margin. Even in the most slowly cooled, hottest magmatic intrusions, plagioclases are **normally zoned** from calcic cores to more sodic rims, testifying to the very sluggish rates of diffusion of NaSi and CaAl ions during crystallization (Figure 5.14b). **Reverse zoning** is

rarely evident in plagioclases. **Oscillatory zoning,** especially widespread in plagioclases in intermediate composition magmatic rocks (Figure 6.17), most likely originates in the sluggish kinetics of crystal growth (e.g., Bottinga et al., 1966), but the exact process is not clearly understood. However, some oscillatory zoned feldspars might reflect cyclic changes in intensive parameters throughout the magma chamber, such as episodic emission of steam from an overlying volcanic vent, causing fluctuations in water concentration and consequent shifts in crystallization conditions (Figure 5.16).

7.3.2 Reaction Rims

Incomplete *discontinuous* reaction relations in fractionating magmas are recorded in a **reaction rim** that surrounds an anhedral, partially resorbed grain of another

7.18 **Rapakivi** feldspar in dacite, Clear Lake, California. Photomicrograph in cross-polarized light showing sanidine (high-*T,* K-rich alkali feldspar with a sector twin on left) surrounded by polysynthetically twinned oligoclase plagioclase. Stimac and Wark (1992) believe these mantled feldspars were produced by mixing of a sanidine-bearing rhyolite magma with basaltic andesite magma. (Photograph courtesy of James A. Stimac. Reproduced from Stimac JA and Wark DA. Plagioclase mantles on sanidine in silicic lavas, Clear Lake, California: Implications for the origin of rapkivi texture. Geol. Soc. Am. 1992; 104:728–744, with permission of the publisher, the Geological Society of America, Boulder, Colorado, USA. Copyright © 1992 Geological Society of America.)

mineral. Such reaction rims develop during fractionation in the model Mg_2SiO_4-SiO_2 system (Section 5.3.2) and are found in many magmatic rocks where rim and core phases are solid solutions. In Plate II, decreasing T and increasing water concentration in the crystallizing magma destabilized the clinopyroxene so that it reacted with the melt to form hornblende reaction rims. If magmatic conditions had changed more slowly so that the diffusion-controlled reaction relation could have been completed, achieving a state of equilibrium, all of the pyroxene would have been consumed in the reaction. In Figure 5.32, reduction in water pressure in the volcanic system promoted a reaction between the unstable resorbing hornblende and the enclosing melt, creating a reaction rim of stable pyroxene, plagioclase, and Fe-Ti oxides; the developing rim curtailed further resorption in the time available. In Figure 6.20, a xenocryst of quartz incorporated into basalt magma reacted with the silica-undersaturated melt, forming a reaction rim of stable clinopyroxene crystals. Their needlelike habit probably reflects rapid crystallization as the hot melt was quenched around the cool xeno-cryst. Rim crystals nucleated heterogeneously on the unstable core crystal.

Observations of volcanic rocks and complementary experiments suggest that **rapakivi** texture, in which an alkali feldspar grain is rimmed by plagioclase (Figure 7.18), can originate during mixing of magmas (Stimac and Wark, 1992). Alternatively, Nekvasil (1991) proposes that magma decompression during intrusion can create plagioclase-mantled alkali feldspars.

Another history of reactions with enclosing melt is revealed in some plagioclases (Figure 7.19), which contain an internal concentric zone crowded with irregular inclusions of minute mineral grains like those constituting the microcrystalline matrix.

7.3.3 Subsolidus Decomposition and Exsolution in Unstable Minerals

Partial to complete replacement of hydrous mafic phenocrysts is common in volcanic deposits (Figure 5.34). Because such replacements occur in the interior and especially upper part of the deposits but are lacking in

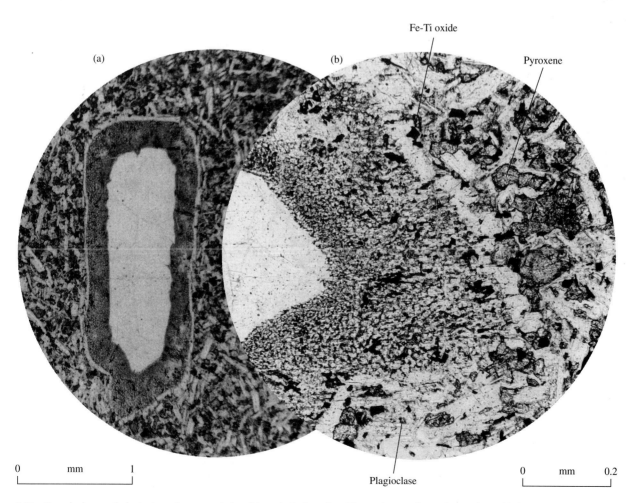

7.19 Complexly zoned plagioclase phenocryst in basaltic andesite lava flow. Photomicrographs in plane-polarized light at two different magnifications. Clear anhedral core is andesine and thin clear rim is labradorite. "Spongy" zone is crowded with irregular glass inclusions and minute mineral grains like those in groundmass of rock. (b) Enlarged view of bottom of (a) rotated 90° counterclockwise.

7.20 **Graphic** texture in hand sample of graphic granite. The scattered quartz grains in a single-crystal host of perthitic alkali feldspar resemble poikilitic texture. However, the quartz grains are more or less uniformly spaced and are not randomly oriented. They are all crystallographically continuous, as may be verified by their optical continuity viewed in thin section under cross-polarized light, and grew simultaneously with the alkali feldspar as an **intergrowth.**

the rapidly quenched, commonly glassy bases, it is inferred that these replacements developed after extrusion at subsolidus temperatures. As the interior and upper parts of the deposit cool and adjust toward the higher oxygen and lower water fugacity of the atmosphere, biotites and amphiboles are destabilized, and very fine-grained, anhydrous replacements that consist of pyroxenes, Fe-Ti oxides, and feldspars develop in their place. Partial replacement creates what may be called **decomposition rims,** but over time the entire grain can be replaced, or pseudomorphed. In the same volcanic rocks, pyroxenes are generally not decomposed. Olivines in mafic lava flows are commonly rimmed or completely pseudomorphed by red-brown "iddingsite," which is a cryptocrystalline mixture of ferric oxides and clay minerals.

Pyroxenes, magnetite-ulvöspinel solid solutions, and alkali feldspars that crystallized in slowly cooled plutons as more or less homogeneous minerals from the melt can exsolve at subsolidus temperatures into more stable coexisting crystalline phases. The most widespread **exsolution** texture is perthite (Figure

5.18a), an intergrowth of Na- and K-rich alkali feldspars formed by unmixing below the alkali feldspar solvus (Figure 5.17). Exsolution texture is also seen in some pyroxenes in gabbro (Figure 5.14b), where they unmixed below a solvus.

✳7.4 FABRIC RELATED TO TEXTURAL EQUILIBRATION: SECONDARY GRAIN-BOUNDARY MODIFICATION

In slowly cooled magma systems, textural equilibration may modify grain size and shape after the initial episode of crystallization. Textural equilibration is more likely if temperatures remain high so that movement of atoms can minimize the surface free energy of the crystalline aggregate (Section 6.4).

Puzzling intergrowths of quartz grains in an alkali feldspar host is called **graphic** texture (Figure 7.20). This appellation is suggested by the wedge- and hook-shaped quartz that calls to mind ancient (circa 3000 BC) cuneiform writing. It occurs in some leucocratic granite pegmatites, called **graphic granites.** A similar micrographic, or **granophyric,** texture (Figure 7.21) occurs in felsic rocks—usually intrusive—called **granophyres.** A long-favored origin, among many that have been proposed, involves simultaneous crystallization of quartz and feldspar from a melt; this process has been confirmed experimentally by Fenn (1986) and MacLellan and Trembath (1991). As a euhedral plagioclase begins to grow in a viscous granitic melt, a silica-

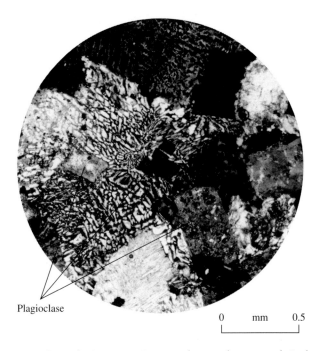

Plagioclase

7.21 **Granophyric** texture in granophyre under cross-polarized light. Note somewhat altered euhedral plagioclase at center of sprays of micrographic quartz and alkali feldspar. The suffix *-phyric* implies a porphyritic texture in which the fine-grained intergrowth surrounds a phenocryst.

and alkali-enriched boundary layer (Figure 6.16) forms around it. Continued crystallization may create a dendritic to spherulitic intergrowth of alkali feldspar and quartz, the former crystallographically continuous with and projecting radially outward from the euhedral plagioclase core. Some Ostwald ripening might coarsen the large-surface-area intergrowth, (Figure 6.21) producing granophyric texture. Whether phaneritic graphic granite might be created by further Ostwald ripening is a speculative possibility.

A superficially similar intergrowth in granitic rocks consists of an intergrowth of vermicular ("wormy") quartz in a sodic feldspar host typically in contact with K-rich alkali feldspar (Figure 5.18b). The origin of this **myrmekite** has been ascribed to direct crystallization or to replacement but Castle and Lindsley (1993) believe it may originate in subsolidus unmixing of K-bearing plagioclase in a ternary feldspar system open to excess Si.

*7.5 A WORD OF CAUTION ON THE INTERPRETATION OF CRYSTALLINE TEXTURES

Kinetic investigations of the dynamic behavior of melts in the laboratory, coupled with theoretical models and, as always, evaluated by observations of real rocks, provide insights into the origin of magmatic fabrics. However, the behavior of natural magmas is impossible to duplicate wholly by experiments and models. A deficiency in experiments is an appropriate handling of the time scale of rock-forming processes; the slow crystallization of plutonic magma bodies can never be duplicated in one researcher's professional career! In plutonic rocks, and to some extent in volcanic as well, the observed fabric is the cumulative product of a long period of changing geologic conditions and declining kinetic rates that took place above *and* below the solidus; the effects of overprinting processes are not easily unraveled. The spontaneous drive toward a state of textural equilibrium that potentially can modify grain boundaries is largely unexplored in slowly cooled magmatic rocks. The kinetic history of magmatic systems as indicated by the rock fabric is frequently ambiguous except in the simplest of cases.

To emphasize and illustrate this sense of caution we next briefly discuss a line of interpretive petrology that has been a part of the discipline virtually since its inception, namely, the determination of the order of crystallization of minerals from the magmatic rock texture.

7.5.1 Magmatic Rock Texture and Order of Crystallization

The sequence in which minerals crystallize with changing intensive parameters, most commonly declining T, between the liquidus and solidus of the magma, depends upon the magma composition, confining pressure, and fugacities of volatiles, especially water. It would be useful for the petrologist if inexpensive petrographic examination of rock fabric could determine the order of crystallization and thereby provide insight into the nature of changing intensive parameters without recourse to costly, tedious, less available laboratory equipment and techniques. Four textural criteria have been employed during the past decades to ascertain order of crystallization:

1. Comparison of phenocrystic and groundmass minerals in quenched magmas, usually volcanic rocks

2. Reaction rims

3. Mineral inclusions in larger grains

4. Relative euhedralism at mutual grain contacts

The latter two criteria are examined here (see also Flood and Vernon [1988, pp. 105–116] for a more thorough critique).

Inclusions of, for example, pyroxene poikilitically enclosed in a larger oikocryst of plagioclase are commonly interpreted to mean that pyroxene crystallized before plagioclase. But this interpretation is not necessarily correct in every case. At least three alternate possibilities are:

1. Late pyroxenes might have crystallized from melt inclusions entrapped within skeletal plagioclase after it ceased crystallizing.

2. Late pyroxenes might have crystallized from melt-filled cracks that cut thorough the plagioclase after it grew.

3. The inclusion-filled plagioclase might be a xenocryst or a restite grain derived from metamorphic rocks in the deep continental crust where the magma was generated.

The order of crystallization indicated by the relative euhedralism of grains in mutual contact can also be ambiguous (Figure 6.26). Ophitic texture (Figure 7.15) and hypidiomorphic-granular texture (Figure 7.16) have mutual grain contacts that tempt one to interpret order of growth. For ophitic texture, it is commonly assumed that the euhedral plagioclases grew freely in the melt, followed by pyroxene, which is anhedral because of the restricted available space within the network of earlier plagioclases. However, this is only one of at least three different possible orders of crystallization of plagioclase and pyroxene. In Figure 7.22a, a few nuclei formed in the melt allow early growth of pyroxene, followed by abundant nucleation of plagioclase and growth of both phases; the final resulting ophitic texture is shown on the right. In Figure 7.22b, pyroxene and plagioclase nucleate in the melt and grow simultaneously to create the final ophitic texture. In Figure 7.22c is the conventionally assumed order of crystallization in which plagioclase nucleates and grows before pyroxene does. In all three instances, pyroxene nucleates less abundantly than plagioclase; that can be taken as the only valid conclusion for ophitic texture. A

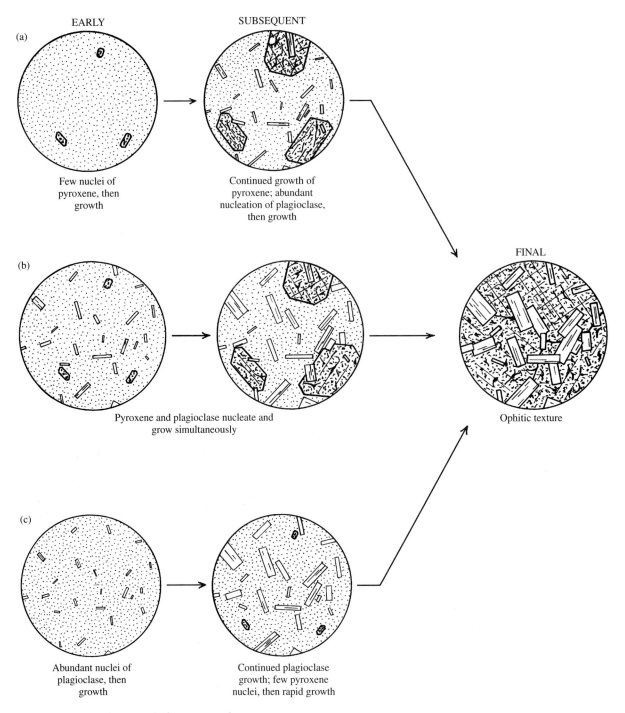

7.22 Three alternate kinetic paths for creating **ophitic** texture.

less important factor is the relative growth rates for the two crystalline phases that vary somewhat among the three scenarios.

One of the pitfalls in using relative euhedralism of two or more crystals to decide their order of crystallization is to assume that the later anhedral grain is molded onto the earlier euhedral one. However, the two grains could have grown simultaneously, or even in reverse order; the contrast in shape may simply reflect the lower surface free energy of rational faces of the eu-

hedral crystal that prevailed in minimizing the overall energy of the grain contact (Figure 6.26).

Interpretation of the order of crystallization for the hypidiomorphic-granular granodiorite previously discussed profited from a determination of the crystal melt equilibria in a similar model system. This is an example of how observations of real rocks *and* experimental data can be integrated to arrive at a more accurate conclusion than can be achieved by using only one source of information.

(a)

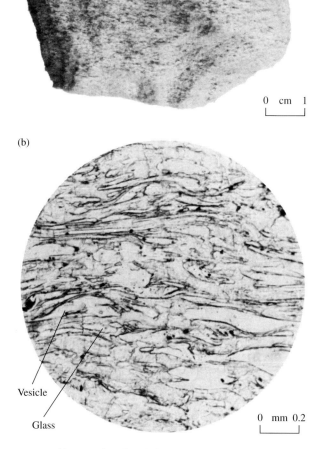

Volatile-fluid bubbles become **vesicles** as magma solidifies. The corresponding texture is **vesicular.** The surface tension of a static melt against a fluid phase makes bubbles spherical—the shape of least surface area relative to volume. Movement of either the bubble in the melt or the melt containing the bubble can distort this equilibrium shape. Elongate **pipe vesicles** (Figure 7.3a) are one result. Highly vesicular silicic glass, or **pumice** (Figure 7.23), has a **pumiceous** texture. **Scoria** and **scoriaceous** are parallel terms for andesite and basalt. Vesicles can be larger in basaltic lavas (to as much as 10 m or so) than in silicic, not necessarily because they contain more dissolved volatiles, but rather because basaltic melts have a less viscous nature, which allows the bubbles to expand and coalesce before the magma solidifies.

Some smooth-walled vesicles may be filled with secondary minerals precipitated from fluid solutions percolating through the rock, producing **amygdules** and **amygdaloidal** texture (Figure 7.24).

(b)

0   cm   1

Unfilled vesicles

Amygdules

0   cm   2

Vesicle

Glass

0   mm   0.2

7.23 Highly **vesicular** glass defining **pumiceous** texture in silicic pumice. Stretched **vesicles** are apparent in hand sample (a) and thin section photomicrograph in plane-polarized light (b).

✳7.6 FABRICS RELATED TO NONEXPLOSIVE EXSOLUTION OF VOLATILE FLUIDS

This section deals with fabrics produced by volatile exsolution from a melt (Sections 4.2.3 and 6.7) that did not result in fragmentation of the magma. Fragmentation is considered in Section 7.7.

7.24 **Amygdaloidal** texture in basalt. **Vesicles** filled with secondary minerals that precipitated from percolating aqueous solutions are **amygdules.** Filling minerals may be carbonate or zeolite minerals, or some form of silica, such as quartz, chalcedony, or opal. Amygdules are *not* phenocrysts.

In highly crystalline magmas, exsolved fluid can collect in angular **vugs** between crystals, creating **vuggy** fabric (Figure 7.25). Euhedral **vapor-phase** crystals, most commonly quartz in silicic magmas, nucleate on the crystalline walls of the vug and grow freely into the cavity. **Miarolitic cavities** are vugs in phaneritic rocks, especially in granitic rocks, that are widely spaced, many grain diameters apart. **Diktytaxitic** texture is found in some coarse-grained basalt (diabase) lava flows and consists of small angular vugs interspersed pervasively among the slightly larger plagioclase and pyroxene grains.

Hollow, bubblelike gas cavities in rhyolite lava flows and compacted tuffs are **lithophysae,** creating **lithophysal** fabric (Figure 7.26). In lava flows, lithophysae are subspherical masses generally a few centimeters in diameter composed of concentric onionlike shells of aphanitic material. Their origin is uncertain, but rhythmic exsolution and expansion of volatiles during crystallization may be involved. In compacted welded tuffs, lithophysae tend to be more irregular discoidal cavities commonly lacking the concentric structure seen in lava flows. They may form by collection of exsolving gas into pumice fragments.

7.26 **Lithophysal** fabric in **felsitic aphanitic** rhyolite. Lithophysae (singular, *lithophysa,* meaning "stone bubble") occur in partly glassy to aphanitic rhyolitic rocks and consist of onionlike masses a few centimeters in diameter with concentric shells of aphanitic quartz and alkali feldspar. (U.S. Geological Survey photograph courtesy of Robert L. Smith and Susan L. Russell-Robinson.)

Vapor-phase precipitates are common in lithophysae and include quartz, topaz (in F-rich metaluminous rocks), and specular hematite. Lithophysae are locally filled with secondary silica (quartz, chalcedony), forming geodes.

✳7.7 VOLCANICLASTIC FABRICS RELATED TO FRAGMENTATION OF MAGMA

The term **volcaniclastic** refers to a broad category of fragmental material, called **volcaniclasts,** composed at least in part of volcanic rock and formed by any particle-generating process and subsequently transported and deposited by any mechanism (Fisher and Smith, 1991). Volcaniclastic processes can operate over a wide range of scales. Therefore, their products are represented not only in rock fabrics but also in larger-scale stratigraphic field relations, described in Chapter 10. In this continuum of fabric and stratigraphic characteristics, distinquishing one from the other is somewhat arbitrary.

Volcaniclasts can be considered from four perspectives:

1. *Size* of fragments is the most fundamental attribute of volcaniclastic deposits, as it is in sedimentary deposits (Section 2.4.1): (a) **Ash** particles are <2 mm (corresponds to sand, silt, and clay sizes of sediment); (b) **lapilli** (singular, *lapillus*) are 2–64 mm (pebble and granule); (c) **blocks** and **bombs are** >64 mm (boulders and cobbles). Because work

7.25 **Vuggy** fabric in **felsitic aphanitic** rhyolite. Collection of exsolved volatiles into irregularly shaped pockets in a highly crystalline matrix produces vugs. They are commonly lined with euhedral **vapor-phase** crystals that grew freely in the entrapped volatile fluid. Vapor-phase precipitates include quartz (seen here), specular hematite, and, in some peraluminous rhyolites, topaz and Fe-Mn garnet.

(energy) is required to create surfaces by breaking a volume of rock or magma, clasts that have the largest surface-to-volume ratio, namely, ash and smaller lapilli, can only be created by the most energetic explosive processes. Blocks are angular clasts produced by lower energy fragmentation of solid material, such as pieces crumbled from steep margins of rigid extrusive lava flows and domes. Streamlined bombs are blobs of solidifying low-viscosity mafic magma shaped aerodynamically during ejection from an explosive vent.

2. *Composition:* Volcaniclasts can be (a) single crystals and crystal fragments that are typically of ash size; (b) rock fragments, or **lithic clasts,** of polygranular mineral aggregates that are typically lapilli and block size; (c) fragments of melt that quenched to glass, or **vitroclasts,** that are of any size; however, the term is usually applied to ash-size particles.

3. *Heritage:* **Juvenile** clasts, also referred to as **cognate** or essential clasts, are derived directly from the magma involved in the volcanic activity and consequently always consist in large part of glass formed by rapid quenching of the extruded melt. Crystals precipitated from the melt prior to extrusion or explosive eruption of the magma are usually present in a glassy matrix in larger vitrophyric clasts of lapilli and block/bomb size. During volcanic explosions, **accidental** clasts are derived from older rock torn from the vent walls or swept up from the ground surface by lava or pyroclastic flows. These foreign rock clasts can also be referred to as **xenoliths,** or, if the accidental material is in the form of individual crystals, **xenocrysts.**

4. *Process of fragmentation:* These include pyroclastic, autoclastic, and epiclastic processes. **Epiclasts** of a wide range of sizes are created by weathering and disintegration of volcanic rock—the same processes that produce sedimentary clasts. Transport and deposition of epiclasts, commonly in muddy **volcanic debris flows,** produce clastic deposits like the one shown in Figure 7.27.

7.7.1 Pyroclastic Processes

Pyroclastic processes explosively eject and aerially disperse **pyroclasts** of rock and magma from a volcanic vent. The terms **ejecta** and **tephra** are sometimes used synonymously for pyroclasts. In the most energetic explosions, fine ash ascends tens of kilometers above the vent, where it is entrained in world-circling air currents before eventually falling to the ground. Ground-hugging avalanches of ash and lapilli, locally with blocks, called **pyroclastic flows** can travel at hurricane-like speeds more than 100 km from the source. The wide range of pyroclast size, from finest ash to huge

7.27 Andesitic **epiclasts** in a **volcanic debris flow.** Note lack of sorting and stratification in the mixture of lithic blocks and lapilli and crystal-vitric ash. Camera lens cap for scale.

blocks many meters in diameter, reflects widely ranging explosive energies in three pyroclastic processes:

1. Exsolution of volatiles from melt and subsequent expansion and fragmentation of the bubbly magma (Figure 4.13)

2. **Hydromagmatic** interaction between near-surface magma and explosively vaporized external water in lakes, ocean, and pore spaces in rock and sediment

3. Combined exsolution and hydromagmatic interaction

The essential attribute of **pyroclastic fabric** is the presence of vitroclasts quenched from juvenile melt. The terms *vitroclastic fabric* and *pyroclastic fabric* are not always synonymous because subordinate crystals and lithic clasts are commonly present in pyroclastic material. The shape and size of vitroclasts (Heiken and Wohletz, 1985) depend on melt composition and viscosity and the manner of explosive process. Fragmentation of a highly vesiculated viscous silicic melt creates ash-size glass **shards,** many of which are angular Y shapes and arcuate slivers (Figure 7.28) that represent broken bubble walls. Unexploded silicic pumice, commonly with pipelike

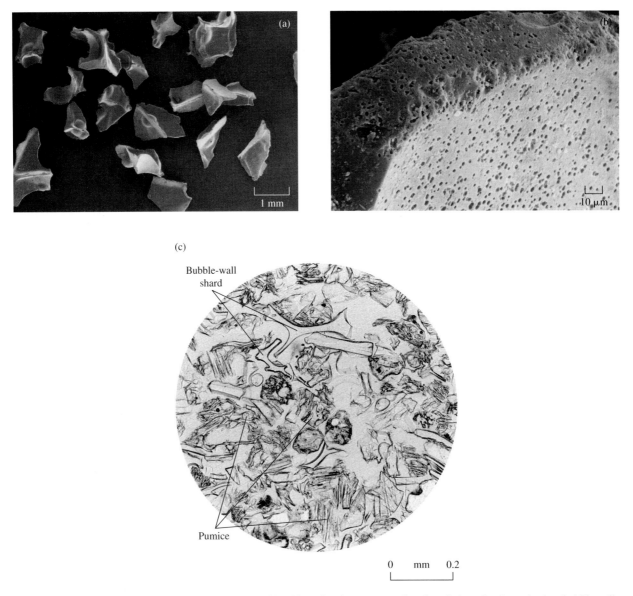

7.28 **Vitroclasts** produced by explosive disaggregation of bubbly melt. The arcuate and Y-shaped glass **shards** are broken bubble walls. (a) and (b) are scanning-electron photomicrographs at two different magnifications. (b) Highly magnified image of a shard in (a) showing minute **vesicles** in wall of larger vesicle. These two size populations are reminiscent of the two grain sizes in porphyritic aphanitic rocks. (c) Optical photomicrograph in plane-polarized light of thin section of Bishop ash (erupted from the Long Valley caldera in eastern California; see Figure 10.26) collected in southeastern Utah. Note "fibrous" pumice particles (Figure 6.27), which contain minute stretched vesicles, of same size as Y-shaped shards, testifying to the heterogeneity of bubble size in the erupting melt.

vesicles (Figure 6.27), which can range from ash to block size, can occur with the shards. Pyroclasts of low-viscosity basalt and some peralkaline rhyolite magmas are spindle- or ribbon-shaped bombs (Figure 7.29a). Smaller ash-size particles are smooth spheres, teardrops, and other aerodynamically streamlined shapes, such as basaltic Pele's tears and hair (Figure 7.29b). **Cinders** are lapilli of scoria. Cracked **breadcrust** surfaces on blocks and bombs resemble the crust on French bread, formed by expansion of gases trapped beneath the solidified exterior (Figure 7.30).

Hydroclasts, chilled juvenile vitroclasts produced hydromagmatically, are angular, poorly vesiculated ash-size granules bounded by conchoidal fractures.

In the turbulent ash-and-steam cloud associated with explosive eruptions, fine moist ash adheres to some sort of nuclei, such as crystals, building concentrically layered, spherical pea-sized **accretionary lapilli.**

Crystals are a common part of pyroclastic material. Some may be accidental xenocrysts, whereas most are euhedral to anhedral juvenile phenocrysts that

(a)

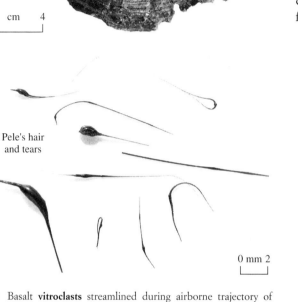

Bombs

0   cm   4

(b)

Pele's hair
and tears

0 mm 2

7.29 Basalt **vitroclasts** streamlined during airborne trajectory of clots of low-viscosity melt thrown from a lava fountain. (a) Volcanic **bombs** are spindle- or ribbon-shaped. (b) **Pele's hair** and **Pele's tears.**

7.7.2 Autoclastic Processes

Extrusions of all but the least viscous magma create block-size **autoclasts** on the margins of lava flows (see Figures 10.8 and 10.16). Autoclastic processes are self-inflicting, involve relatively small energy transfers so that fragments are generally short-traveled, and create essentially monolithologic, poorly sorted, relatively small-volume deposits of angular juvenile blocks. In a flowing lava, the cooler, crusted, more rigid margin breaks up as the hotter more mobile interior continues to move. Autoclasts may tumble off the toe of the advancing flow to be overridden in caterpiller-tread fashion so that the base of the lava flow becomes autoclastic as well. Talus piles around

0   cm   5

7.30 **Breadcrust** block of andesite pumice, Unzen volcano, Japan. Expansion of gas within block after the surface crusted caused it to crack in the fashion of French bread.

crystallized in the magma before eruption. The same phases occurring as isolated phenocrysts may also be found in small clotted cumulophyric aggregates. Pyroclastic deposits also commonly contain irregularly shaped fragments of quartz and feldspar crystals, called **phenoclasts** (Figure 7.31) whose shapes resemble pieces of a jigsaw puzzle. Some contain small inclusions of glass. In a decompressing magma exiting a volcanic vent, volatile-rich melt inclusions entrapped in crystals at higher pressure expand and rupture their host crystals, blowing them apart and creating phenoclasts (Best and Christiansen, 1997). In slowly extruded lava flows, entrapped melt inclusions have time to equilibrate with their host crystal and no phenoclasts form.

An example of a widespread type of pyroclastic fabric formed by explosive eruption of silicic magma is shown in Figure 7.32.

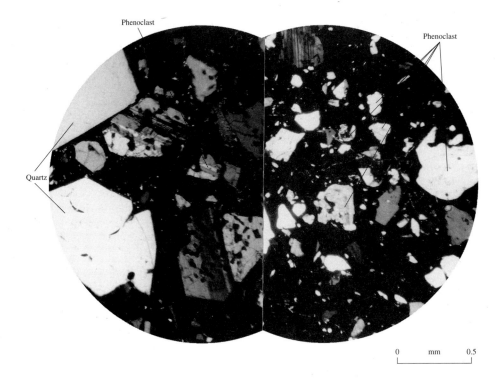

Phenoclast

Phenoclast

Quartz

0   mm   0.5

7.31 **Phenoclasts** of plagioclase in rhyolite tuff. These grains have shapes resembling jigsaw-puzzle pieces that were broken apart by expansion of volatile-bearing melt inclusions during extrusion of the decompressing magma. Some inclusions that did not burst are visible within phenoclasts on the left.

(a)

(b)

Broken crystal of feldspar

Glass shards

Pumice

0   cm   2

0   mm   1

7.32 **Pyroclastic** fabric in rhyolite lapilli tuff. Sample is from the Bishop pyroclastic flow deposit, eastern California (see Figure 10.26). (a) Hand sample with lapilli of fibrous pumice and smaller black lithic lapilli in a matrix of ash. (b) Photomicrograph under plane-polarized light showing ash matrix of glass **shards,** small pumice, and crystals.

7.33 **Autoclastic** talus blocks crumbled from margin of 600-year-old Glass Creek lava dome, eastern California. Blocks of **flow-layered** rhyolite are mingled black obsidian and gray felsitic rock visible in large block in foreground under black camera-lens cap.

highly viscous silicic domes emerging slowly from a vent are another type of autoclastic deposit (Figure 7.33).

☀7.8 FABRICS RELATED TO CONSOLIDATION OF VOLCANICLASTS INTO SOLID ROCK

Ash is consolidated into tuff, an ash and lapilli mixture into lapilli tuff, and blocks—usually with smaller intervening clasts—into breccia (Figure 2.7). In addition to the same consolidation processes that lithify sedimentary particles, such as cementation, a vitroclastic body can become consolidated by other means. For example, in appropriate geologic environments, glass is converted into more stable clay and zeolite minerals, which can bind initially loose vitroclasts together.

Another important lithification process that consolidates vitroclasts is **welding,** which takes place if they are still hot enough to stick together when deposited. Thus, molten spatter, mostly bombs, from a basaltic

lava fountain can weld together in an **agglutinate** deposit immediately surrounding the vent. Smaller ash-size vitroclasts dispersed by pyroclastic processes into the air lose heat and are too cool to weld upon deposition. However, silicic pyroclastic flow deposits can be emplaced as thermally retentive, ground-hugging avalanches of hot ash and lesser lapilli and locally blocks. Because emplacement temperatures are nearly that of the erupting magma, vitroclasts in the flow are still hot and sticky; therefore, as they come to rest, they can weld together into a **welded tuff.** Glass shards and pumice lapilli are flattened by compaction into disks oriented more or less parallel to the depositional surface, forming **eutaxitic** fabric (Figure 7.34). Flattening is more intense toward the base of the deposit as a result of the increasing weight of the overlying pyroclastic material. Locally, on hill slopes, the welding mass of hot vitroclasts may creep downhill by secondary, or **rheomorphic,** flow, stretching the vitroclasts into linear shapes or creating folds in the otherwise planar eutaxitic foliation.

Deposits of unconsolidated vitroclasts overridden by hot lava or pyroclastic flows can absorb sufficient heat that they become sticky and fuse near the contact, forming **fused tuff.** This secondary fusion produces a texture resembling that which develops in a hot pyroclastic flow as it cools.

Because silicic pyroclastic flow deposits are hot and contain entrapped gas at deposition, they can experience secondary crystallization through vapor-phase processes and more pervasive devitrification (Ross and Smith, 1961). This secondary crystallization can produce slight bonding of otherwise loose particles. Devitrification can produce spherulitic and felsitic textures (Figure 7.35) that can obliterate, preserve, or even enhance primary vitroclastic fabric.

Vapor-phase crystallization occurs as mineral components dissolved in the gas phase nucleate on walls of any cavity and grow into it.

☀7.9 ANISOTROPIC FABRICS

7.9.1 Descriptive Geometric Aspects

The fabric in some rock bodies looks different and has different properties in different *places* in the body. Obviously, this inhomogeneity, or heterogeneity, depends on the scale of observation. For example, an aphyric, holocrystalline basalt appears homogeneous on the scale of a hand sample, but, on a microscopic scale, a plagioclase in one place in the field of view is certainly not the same as a pyroxene in another; on this scale the basalt is inhomogeneous. On the scale of the outcrop, the basalt lava flow may have a glassy vesicular top, unlike the wholly crystalline, nonvesicular interior, so on this scale the flow is again inhomogeneous.

(a)

(b)

(c)

Biotite
phenocryst

Compressed
glass shard

Plagioclase
phenocryst

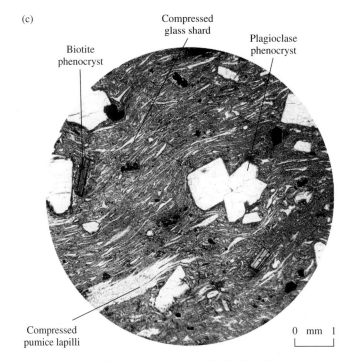

Compressed
pumice lapilli

0   mm   1

7.34 **Eutaxitic** fabric in compacted welded lapilli tuff. Compare with nonwelded lapilli tuff in Figure 7.32. (a) Hand sample showing fresh surface that appears to be featureless obsidian. (U.S. Geological Survey photograph courtesy of Robert L. Smith.) (b) Weathered surface of
the same hand sample as in (a) showing distinctive eutaxitic fabric that is defined by collapsed and flattened pumice lapilli, which have
distinctive frayed, flamelike terminations, hence the designation **fiamme** (Italian, "flame"). The lighter-colored matrix between the lapilli
is composed of compacted welded ash. (U.S. Geological Survey photograph courtesy of Robert L. Smith.) (c) Photomicrograph under
plane-polarized light of a more crystal-rich tuff than the hand sample in (a) and (b) which illustrates the smaller-scale details of the eutaxitic fabric of flattened shards and pumice lapilli. The flattened glass shards, which appear white against submicroscopic, fine gray dust
particles, are more tightly compressed around the **cumulophyric** clot of plagioclase phenocrysts (photo center), which behaved as a rigid
body during compaction of the surrounding soft glass fragments. Also note the collapsed pumice lapillus with its frayed or flamelike termination. Eutaxitic fabric is produced by gravity-induced compaction and elimination of pore spaces and not by flow; thus, it is *not* a
type of flow layering.

(a)

0   cm   2

(b)

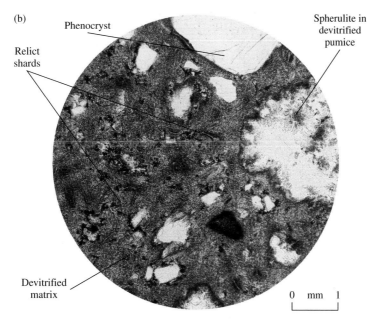

Phenocryst

Spherulite in
devitrified
pumice

Relict
shards

Devitrified
matrix

0   mm   1

7.35 **Devitrified** lapilli tuff from the nonwelded top of the Bishop pyroclastic deposit that was subjected to vapor-phase crystallization. Compare Figure 7.32. (a) Hand sample in which relict pumice lapilli are composed of strings of tiny delicate spherulites replacing the thin glass walls of the original fibrous pumice. (b) Photomicrograph in plane-polarized light of same rock. In the microcrystalline felsitic matrix most of the original vitroclastic fabric has been erased by the devitrification, but a few relict shards are still evident. Fibers of alkali feldspar and cristobalite nucleated on shard walls and grew inward normal to the wall, preserving its outline. Spherulites have replaced the glass walls of the pumice.

The fabric in some rock bodies looks different and has different properties in different *directions* in the body. This directional, or **anisotropic,** fabric can also depend on the scale of observation, as will become obvious. Bedded sedimentary rocks have anisotropic fabric, whereas most—but not all—bodies of granite are **isotropic** in that the fabric is random, appearing the same in any direction. Anisotropic fabric can be planar, like a stack of blank sheets of paper; or it can be linear, like a clutched handful of pencils; or it can be combined planar and linear, like the pages with printed lines in this textbook. Rocks with planar fabric are **foliated,** or possess **foliation;** the eutaxitic welded tuffs just described are foliated rocks. Rocks with linear fabric are **lineated,** or possess **lineation.** If both lineation and foliation are present in a rock, the lineation almost always lies within the foliation, like the lines of print running across the pages of this book. It is, therefore, necessary to examine the foliation surface directly for a lineation. Some of the ways in which anisotropic fabric is expressed in rocks are shown schematically in Figure 7.36.

An important attribute of foliation and lineation is that they are *penetrative* properties; that means the directional aspect is repeated throughout the rock and can be found in any part of it. If the "rock" blocks in Figure 7.36 were broken into many pieces, each piece would still show the anisotropic fabric, within some limits of size and number of pieces. Thus, widely spaced joints or dikes in a granitic body do not constitute a foliation.

A common point of confusion is to conclude that *any* linear aspect that is visible on a random rock surface manifests an intrinsic, penetrative lineation within the rock volume. However, if other surfaces of the rock oblique to the first also reveal the same linear aspect and all of these lie parallel to one another (with due allowance for warped and folded foliations), then the internal fabric within the rock is a foliation. The linear aspect seen on the rock surfaces are simply line traces of the foliation where it intersects the surfaces, like the lines on the sides of this textbook, which are the edges of the planar pages in it. (The fabric of the rock may also be lineated, but this pattern must be verified by close examination of the foliation surface itself.) Such *apparent* lineations, that is, traces of foliation on obliquely intersecting exposure surfaces, and true penetrative foliation and lineation can be distinquished by careful examination of Figure 7.36. Careful examination of a rock is required to discern the true nature of the anisotropic fabric, whether it is planar, linear, or a combination of the two. It generally helps to examine the rock on all scales—outcrop, hand sample, and thin section.

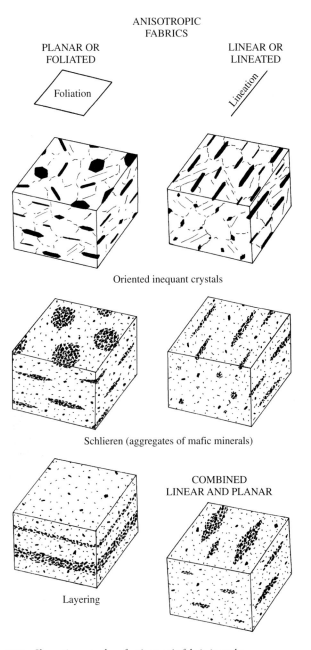

ANISOTROPIC
FABRICS

PLANAR OR
FOLIATED

Foliation

LINEAR OR
LINEATED

Lineation

Oriented inequant crystals

Schlieren (aggregates of mafic minerals)

Layering

COMBINED
LINEAR AND PLANAR

7.36 Shematic examples of **anisotropic** fabric in rocks.

It may be noted in the lower left of Figure 7.36 that the sides of the layered block have contrasting mineral bands, defining a banding that is the two-dimensional trace of the layers on the random oblique surfaces. However, the term **layers,** defining a **layering,** more accurately describes the actual penetrative three-dimensional fabric element within the rock than do the less appropriate terms *bands* and *banding.*

<u>Expressions of Foliation.</u> Figure 7.36 shows that foliation can be expressed or defined in different ways:

1. A common type of foliation is the layering just discussed. In sedimentary rocks and some volcaniclas-

7.37 Sequence of curved and lenticular **composition layers** in more felsic granodiorite, Alta, Utah. This is a type of **anisotropic** fabric.

tic deposits, layering is called *bedding,* or *stratification.* Layering is expressed by planar contrasts and heterogeneities in texture (grain and pore size and/or shape) or mineral modal proportions. Various names have been applied to layering that is defined by contrasts in modal proportions of minerals (e.g., Figure 7.37), but the nongenetic term **compositional layering** is appropriate.

2. Another expression of foliation are **inequant** mineral grains, whose longest dimensions are preferentially oriented within parallel planes. Oriented platy micas are a common expression of foliation. Tabular grains, such as feldspars, can have their two longest dimensions preferentially oriented in planar fashion, whereas columnar amphiboles can have their one longest dimension so oriented. Within that foliation plane, the longest dimensions of feldspars and amphiboles are randomly oriented if there is no accompanying lineation or are oriented in linear fashion to express a combined planar-linear fabric. Planar preferred orientation of tabular feldspars in phaneritic rocks, or **igneous lamination,** is shown in Figures 7.16b and 7.38. A similar fabric in an aphanitic aggregate is **trachytic.**

3. Yet another expression of foliation in granitic rocks are **schlieren** (singular, *schliere*). These are oriented wispy, diffuse concentrations of mafic minerals in a more leucocratic matrix (Figure 7.39) that may be planar disks, planar-linear blades, or linear pencil-like shapes. They can originate in different ways; some may be flattened or otherwise deformed mafic inclusions (Section 7.10).

7.9.2 Origin

It is one thing to discern and describe whether a fabric is isotropic or anisotropic, what its nature is, and how it is expressed, as in the preceding paragraphs. But deciding exactly how a particular anisotropic fabric in a rock body originated can be challenging and controversial. Many different forms of layering in magmatic rocks have many possible origins (Naslund and McBirney, 1996). A particular type of anisotropic fabric may be polygenetic. For example, igneous lamination could result from the following:

1. Tabular plagioclases crystallizing and settling to the floor of a magma chamber, much like large snowflakes settling to the ground

2. Compaction of a random array of tablets in a semicrystalline "mush" resulting from the weight of more crystals accumulating on top

3. Planar flow of the magma, orienting the tabular plagioclases; a process that would be expected also to create a linear preferred orientation of the tabular crystals within the foliation, unlike the first two processes, in which the crystals would be randomly oriented within the igneous lamination.

Flow Orientation and Flow Layering. Magmas convecting in staging chambers within the crust, moving through overlying conduits, and flowing onto the surface of the Earth create anisotropic flow fabrics.

Magma flow occurs by movement of viscous melt between rigid suspended crystals. Flow velocity is slowest along fixed boundaries of a moving magma body but can increase inward, creating a gradient in flow velocity (see Figure 8.13). The pattern of velocity gradients and how internal flow markers interact with it result in different forms of anisotropic fabric (Figure 7.40). Suspended rigid crystals, volatile-fluid bubbles, and distinctive mineral aggregates embedded in magma can serve as **flow markers.**

Consider, first, flow orientation of inequant grains and grain aggregates. If inequant grains of columnar shape are caught in flowing magma that has a velocity gradient, the columns *rotate* into a stable orientation parallel to the velocity vectors, as logs do in a river, forming a lineated fabric (Figure 7.40a). In the same velocity-gradient flow regime, tabular feldspars rotate into a stable orientation that imparts both foliation (igneous lamination) and lineation to the magmatic body (Figure 7.40b). Other anisotropic fabrics are possible for different combinations of flow pattern and markers, either in the form of independent isolated crystals or as aggregated concentrations of crystals. If a flowing, homogeneous melt contains no markers, then no anisotropic flow orientation can result. Likewise, if the flowing magma has no velocity gradients—that is, if flow is uniform—or if the magma is stationary, then again, no anisotropic flow fabric can develop, regardless of the nature of the contained flow markers. Turbulent, random flow of magma containing markers can also create isotropic fabric.

7.38 Weak **igneous lamination** in granodiorite, Kern Mountains, Nevada. This is a type of planar anisotropic fabric.

7.39 **Schlieren,** a type of planar **anisotropic** fabric, are concentrations of mafic minerals in a more felsic granitic host rock. Schlieren may be planar, linear, or planar-linear. Note pocket knife for scale.

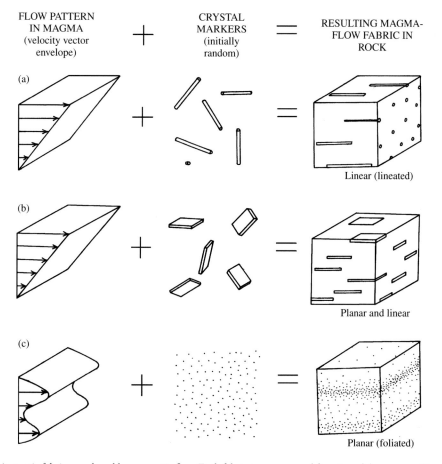

7.40 Schematic anisotropic fabrics produced by magmatic flow. Each fabric is a compound function of the pattern of laminar flow and associated velocity gradients in the solidifying magma and the nature of the crystal flow markers in it.

Consider, second, the origin of flow layering. Ridge-and-valley profiles of velocity in flowing magma can also *translate* grains of almost any shape toward the higher-velocity region (see Section 8.2.3). In three dimensions, if the velocity gradients are nonuniform in a planar sense (Figure 7.40c), then planar **flow layering,** expressed by layers of greater and lesser concentrations of crystals, can develop. Extruded, viscous magmas commonly have a flow layering (Figures 7.2, 7.4, and 7.41) that is expressed by varying planar concentrations or sizes of crystals or of vesicles that may have originated in this manner. Nelson (1981) suggests that thermal feedback in viscous flow can also play a role in development of flow layering in lavas. During flow, movement becomes localized in shear surfaces between more static layers; kinetic energy in the shear zones is transformed into thermal energy; the increased T lowers the viscosity, facilitating further flow in these zones and further increase in T (thermal feedback), reducing volatile solubility, increasing diffusion rates, and promoting crystal and bubble nucleation and growth. Mingling of contrasting magmas in the volcanic con-

duit is another way of producing flow layered lavas (Figure 7.42).

Field mapping of anisotropic flow markers in magmatic rock bodies can potentially provide valuable information concerning patterns of flow in the viscous magma during emplacement (Paterson et al., 1998). However, it is important to note that many anisotropic fabrics involving oriented mineral grains in plutons result not so much from magma flow as from the way in which the crystal-rich magma body was deformed, changing its shape, in the late stages of solidification. For example, almost wholly crystallized magma near a pluton margin can be squeezed by emplacement ("ballooning") of additional magma within the pluton interior, causing flattening of the margin and rotation of inequant mineral grains into an orientation more or less parallel to the pluton contact with its wall rocks; this foliation can resemble that created by magma flow along the wall rock contact during initial intrusion.

<u>Grain-Size-Graded Layering</u>. An example of rhythmic layers characterized by a gradation in grain size is seen

7.41 Folded **flow layering** in a rhyolite lava flow. Folded layers were formed where the flowing lava experienced local changes in the flow velocity regime. (Photograph courtesy of Glenn Embree.)

in the Duke Island complex in southeastern Alaska (Irvine, 1987). In this composite ultramafic intrusion, several structures (Figure 7.43) typical of sedimentary rocks, including cross- and graded-bedding and angular unconformities, testify to the action of currents of low-viscosity melt depositing suspended pyroxene and olivine crystals of varying size, in much the same way as streams of water deposit sand grains on top of larger pebbles.

Problematic Compositional Layers. It is tempting to infer a similar control on the origin of layers of contrasting mineral or modal composition: that is, compositional layering. If larger particles sink faster than smaller particles in a melt under the influence of gravity, then one would likewise expect that denser particles would sink faster than less dense particles. Chromite grains should sink more rapidly than less dense olivines of the same size. Many petrologists interpret compositional layers in terms of such gravitational sorting. However, in a study of compositional layers in the ultramafic Stillwater intrusion in Montana, Jackson (1961) found that small chromites near the base give way upwards into larger olivines. Because settling velocity (given by Stokes's law in Section 8.3.3.) is proportional to the first power of the density contrast

between crystals and melt and to the square of the grain size, the chromite-olivine layers are upside-down. For these and other reasons some petrologists have rejected any control by gravity and sought other explanations for compositional layers.

Rhythmic isomodal layers having a uniform modal composition that differs from that of adjacent layers (Figure 7.44) also have a problematic origin, as do rhythmic layers having gradational modal proportions from top to bottom (Figure 7.45b). Upward increasing plagioclase and decreasing mafic minerals are common in modally graded layers. Grain-size variations are generally absent but igneous lamination is common. Layers may be of uniform thickness, ranging from centimeters to meters, that extend for hundreds of meters and much more (Figure 7.45a) or they may be curved and lenticular, extending only several meters before disappearing into isotropic rock (Figure 7.45b). Numerous field photographs in Parsons (1987) illustrate the variety of layering in many intrusions.

0   cm   2

7.42 Mingled rhyolite-andesite pumice block from the pyroclastic flow that created the Valley of Ten Thousand Smokes, Alaska, in 1912. The crude **flow layering** was created as the two magma types mingled in the volcanic conduit beneath Novarupta volcano. (Sample courtesy of Richard V. Beesley.)

(a)

(b)

7.43 Sedimentarylike structures in the Duke Island ultramafic complex of southern Alaska. (a) **Size-graded layers** in which the base is dominantly large clinopyroxene crystals and minor smaller olivines; pyroxenes decrease in size upward. Pocket knife for scale. (b) Angular unconformity in layers of olivine clinopyroxenite.

7.44 "Inch-scale" **layering** in gabbro, Stillwater Complex, Montana. Incredibly uniform, rhythmic layers are mostly dark pyroxene alternating with mostly lighter plagioclase. Hammer for scale. (Photograph courtesy of A. R. McBirney.)

(a)

(b)

7.45 Rhythmic **modal layering** in the gabbroic Skaergaard intrusion, east Greenland. (a) Panoramic view of mountain face about 300 m high showing lighter plagioclase-rich layers. (b) Lenticular layers enriched in darker mafic minerals at base grading upward into lighter-colored, plagioclase enriched tops. (Photographs courtesy of G. M. Brown.)

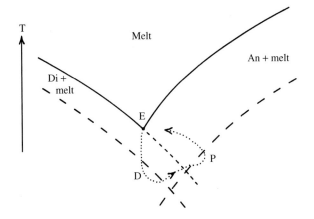

7.46 Possible nucleation phenomena in the binary system $CaMgSi_2O_6$-$CaAl_2Si_2O_8$ near its eutectic point. Solid lines are the liquidus surfaces in the binary system (Figure 5.4); dashed lines show schematically the undercooling needed to achieve optimal homogeneous nucleation of diopside and anorthite. The metastable liquidus of diopside is the fine-dashed line extending to the right of and below the eutectic point E. The track of a cooling melt with delayed nucleation of diopside and finally plagioclase is shown by the dotted line.

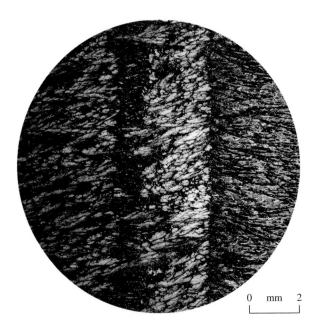

7.47 **Comb layering** in a diabase dike. Note curved, branching crystal morphologic characteristics in lighter-colored pyroxene-rich layers that alternate with olivine- and plagioclase-rich layers. (Photomicrograph of a sample provided by S. Agrell and H. Drever courtesy of G. E. Lofgren. Reproduced with permission from Lofgren GE, Donaldson CH. Curved branching crystals and differentiation in comb layered rocks. Contrib. Mineral. Petrol. 1975;49:309–319.)

The fact that the most widespread and best developed modal layering occurs in gabbroic intrusions and in some silica-undersaturated syenitic ones suggests that low melt viscosity is a controlling factor. Flow of the magma, gravity sorting of grains, or both, may be involved, possibly in concert with other processes. For layered gabbroic intrusions, gravitational sorting and settling of crystals of differing densities, possibly in conjunction with convective movement of the magma across the floor of the chamber, are controversial possibilities. Assertions made decades ago that each modal layer originates in intrusion of new magma can generally be rejected. This requires an unlikely repetition of special events—in some intrusions there are hundreds of layers in the rhythmic sequence—that does not explain the unique internal properties of each layer.

During his classic pioneering investigations of the Skaergaard intrusion, Greenland, Wager (1959) realized that different nucleation rates for different minerals precipitating from a melt might explain modally graded layers that appear to have originated through differential gravitative settling. Hawkes (1967; see also Maaløe, 1978) invoked this mechanism for rhythmic layers as thick as 100 m in the Freetown Complex of Sierra Leone in Africa, where the ideal sequence from the base upward in each layer is olivine + Fe-Ti oxides, olivine + plagioclase, olivine + pyroxene + plagioclase, and plagioclase alone. Multiply saturated gabbroic magmas can coprecipitate all of these minerals under equilibrium conditions. However, these phases, as listed, apparently have decreasing ease of nucleation, as discussed in Section 6.5.2. This can affect the crystallization history of the magma, possibly yielding a sequence of rhythmic modally graded layers.

Crystallization kinetics in a basaltic magma can be modeled by the simple binary system $CaMgSi_2O_6$-$CaAl_2Si_2O_8$, in which diopside and anorthite should coprecipitate from a eutectic melt under equilibrium conditions. The eutectic melt E in Figure 7.46 (see also Figure 5.4) may not begin to crystallize diopside until it has undercooled to D, where it experiences optimal nucleation. As diopside crystals grow, the melt becomes warmer in an adiabatic model system, because of release of latent heat, and richer in the anorthite component. Hence, the melt moves to P, where it nucleates anorthite rather than diopside. Growth of anorthite causes the melt now to track back toward the eutectic point E. Depending on possible upward expulsion of residual melt and gravity settling of crystals, the product of this crystallization excursion would be an upward increase in concentration of anorthite and decrease in diopside within each modally graded layer. Further cooling of the melt E could repeat the process, possibly creating alternating pyroxene- and plagioclase-rich layers like those in Figure 7.44.

7.48 Mafic inclusions, or **enclaves,** in granodiorite, Alta, Utah. Note thin subparallel leucocratic **aplite** dikes above senior author's wife.

<u>Comb Layering</u>. Long, branching, skeletal to feathery, subparallel crystals that are oriented perpendicular to a planar boundary of some sort define **comb layering** (Figure 7.47). Varieties of specific occurrences or mineral compositions include crescumulate texture, harrisitic layering, spinifex texture, Willow Lake layering, and orbicular fabric. Commonly, the comb layers are separated from one another by more granular isotropic textured rock. Apparently, these represent parts of the melt crystallized at low undercooling or supersaturation, whereas the comb layers are created by substantial supersaturation (Donaldson, 1977).

✳7.10 INCLUSIONS

Mafic inclusions, or **enclaves,** have attracted considerable attention and spawned long-lasting controversy. They are mafic aggregates embedded in a more felsic host rock and are especially common in granodiorite, quartz diorite, and tonalite (Figure 7.48). The minerals are essentially the same in the inclusion and its host; only their modal proportions differ. Inclusions are generally finer-grained than the host; grain shapes and mutual relations are variable. Inclusions are polygenetic and can originate in at least five ways:

1. As **xenoliths,** fragments of foreign country rock around the intrusion caught up into the magma. Existing minerals in the fragments recrystallize and equilibrate with the magma, forming new minerals like those stable in it

2. As **restite,** pieces of the crystalline residue of the magma-generating process in the lower crust carried in the ascending magma

3. As **cognate inclusions,** or **autoliths,** partially to wholly crystallized parts of the intrusion, perhaps lifted away from the margin of the intrusion

4. As cumulophyric accumulations of mafic minerals

5. As blobs of partially to completely recrystallized, mantle-derived basaltic magma mingled into the more felsic host magma; the basaltic magma may have provided the heat for felsic magma generation

In any of these possible origins, magmatic conditions were not conducive to complete assimilation, or digestion, of the mafic inclusion into the magma.

Summary

An infinite variety of kinetic paths during solidification of magmas create a virtually infinite variety of rock fabrics. Several time-dependent factors dictate the kinetic paths that are involved in crystallization (or the lack of it), textural equilibration, exsolution of volatile fluids, fragmentation of magma, and development of anisotropic fabrics. These factors include the rate of change in the intensive parameters of the magma system and the triad of transport processes in the magma—transfer of heat, transfer of mass by atomic diffusion, and transfer of momentum in viscous flow (Figure 6.1). A dominant role is played by widely ranging melt viscosity, which impacts rates of diffusion, crystal and bubble nucleation and growth, and magma ascent. Because of the complex interactions of these kinetic factors, accurate interpretations of the origin of fabrics must utilize careful and critical observations integrated with experimental and theoretical information.

Glass forms by rapid undercooling of melts, bypassing crystallization. All glasses are metastable at near-atmospheric conditions and, over millions of years, hydrate and devitrify into more stable crystalline phases.

Grain size in a magmatic rock unmodified by textural equilibration, fragmentation, or other secondary processes is proportional to the ratio of crystal growth rate to nucleation rate. Aphanitic grain size reflects a high rate of nucleation coupled with low growth rates appropriate to rapid cooling of extruded magmas and margins of shallow, small intrusions. Phaneritic grain size is characteristic of deeply emplaced, slowly cooled magmas that have low nucleation densities at small degrees of undercooling. The bimodal size distribution of porphyritic aphanitic and vitrophyric rocks typically reflects two different rates of cooling. However, a uniform cooling rate combined with contrasts in the nucleation rates of different minerals, loss of volatile fluid, and inheritance of xenocrysts and restite crystals can also be factors in the development of porphyritic texture.

Euhedral shapes of mineral grains—columnar, platy, tabular, and so on—grown slowly without restriction in slightly undercooled melts reflect their crystallographic lattice, whereas restricted growth in highly crystalline magmas creates subhedral to anhedral grains. Freely grown crystals in melts are increasingly less compact and have greater surface areas at increasing degrees of undercooling, or rate of cooling, creating skeletal, dendritic, feathery, and finally spherulitic morphologic characteristics.

After their initial growth, crystals can adjust their shapes and increase their sizes via Ostwald ripening, which minimizes the surface free energy of the aggregate. This textural equilibration is more likely to occur in more slowly cooled magmas.

Unstable crystals are dissolved in a melt. Partial dissolution creates anhedral, embayed grains that may be surrounded by a more stable reaction rim. Chemically zoned solid solutions, exemplified especially by plagioclase grains, are created in crystallizing magmas, where rates of diffusion are so slow that the the system cannot equilibrate as intensive parameters change.

Exsolution of volatiles from melts in decompressing and crystallizing magmas creates vesicular and vuggy fabric in rocks.

Volcaniclasts are produced by autoclastic breakup of solidifying margins of lava flows and by explosive pyroclastic fragmentation of magmas where they interact with external surface water (hydromagmatic activity) and where volatiles exsolve from the melt as the magma decompresses or crystallizes. Volcaniclasts are classified according to whether they are juvenile pieces of the magma or accidental rock and crystals; whether they are crystals, lithic fragments, or vitroclasts of quenched melt; and according to their size. Finer volcaniclasts (ash and lapilli) are produced in more energetic explosive pyroclastic eruptions, whereas less explosive activity yields larger volcaniclasts (blocks and bombs).

Deposits of vitroclasts that remain sufficiently hot undergo compaction and welding, in addition to possible devitrification and vapor-phase crystallization.

Anisotropic fabric is geometrically planar, linear, or both planar and linear and is expressed by preferred orientation of inequant mineral grains and mineral aggregates and by, for planar fabric, various types of rhythmic textural and compositional layering. Layering is polygenetic, and the origin of a particular sequence can challenge the petrologist's powers of observation and ability to conceive viable physical-chemical models. Possible mechanisms include magma flow, crystal sedimentation in low-viscosity melts, and crystallization controlled by differing ease of nucleation in different minerals.

CRITICAL THINKING QUESTIONS

7.1   Contrast texture, structure, fabric, and field relations.

7.2   Describe the different fabrics of glassy rocks and indicate their origin.

7.3   List all possible fabrics that can develop from a granitic (rhyolitic) magma. Comment on these.

7.4   Thoroughly discuss the origin of the fabric of the andesite in Figure 7.8.

7.5   Suppose there are no groundmass olivines in the basalt in Figure 7.9. Propose two possible origins for the porphyritic fabric. What bearing does your answer have on the order of crystallization of olivine from the magma with respect to the other constituent minerals?

7.6   In Figure 7.11, why might the spodumene crystals have a preferred subvertical orientation, rather than being random?

7.7   Propose an origin for the complex plagioclase in Figure 7.19.

7.8   What contrasts between the fabrics of rhyolites and basalts are dependent on the difference in the viscosity of their melts?

7.9   Critically discuss factors controlling grain shape in magmatic rocks. What evidence would allow the different factors to be distinguished?

7.10   What are possible mechanisms for the origin of igneous lamination? Do they differ from those for trachytic texture? Discuss evidence for contrasting mechanisms.

7.11   How does flow layering develop in extruded lavas? What evidence could be used to decide which process was involved in a particular rock?

7.12   How might oscillatory zoned plagioclases formed by kinetic processes in the growth of an individual crystal be distinguished from zoned crystals formed by cyclic fluctuations in intensive variables throughout the magma chamber?

7.13   How might reverse zoning develop in plagioclase?

7.14   Can a fused tuff and a welded tuff be distinguished in hand sample? In the field? Discuss.

7.15   Examine the variety of phaneritic fabrics evident in several samples of granitic rock. Polished rock slabs are an excellent means for study and can be examined on a "poor-man's field trip" in a cemetery, a commercial monument or tile establishment, and facades on buildings. Describe and sketch the variety of fabrics, and discuss how it might be possible for such a wide variety to originate between different granitic intrusions and even within a single intrusion (e.g. Clarke, 1992, Fig. 1.1).

CHAPTER

8

Physical and Thermal Dynamics of Bodies of Magma

FUNDAMENTAL QUESTIONS CONSIDERED IN THIS CHAPTER

1. How do magmas and rocks respond to geologic forces, formulated as states of stress?

2. What physical and compositional properties of bodies of magma are most important in controlling their dynamic behavior?

3. How do dynamic physical processes and heat transfer interact in magmatic systems?

4. How do these physical-thermal interactions influence the compositional evolution of magmas?

INTRODUCTION

Chapters 6 and 7 dealt mostly with atomic-scale chemical dynamics and the resulting kinetic paths in solidifying melts that control the evolution of magmatic fabric. This chapter and the next two deal with larger-scale physical and thermal dynamics of bodies of magma, including their response to states of stress, and the ways they flow, cool, and convect. The basic driving force of magma movement in the Earth is buoyancy. Acting against the buoyant force is viscous resistance to flow. As viscosity is partly dependent on magma temperature, the thermal evolution of the magma system becomes important.

Crystallizing and convectively cooling magma bodies evolve through time, unmixing into compositionally contrasting parts by fractionation of crystals and melt. Mixing of magmas of contrasting composition depends on their contrasting densities and viscosities. Magmas are contaminated by assimilation of wall rock if suffi-

cient thermal energy and time are available. Hence, chemical differentiation of magmas is intimately intertwined with their dynamic physical and thermal behavior.

✳8.1 STRESS AND DEFORMATION

8.1.1 Concepts of Stress

In buried rock and magma bodies, forces arise as a result of the weight of overlying rock material and of tectonism, mostly associated with lithospheric plates pushing together or a plate being stretched apart. **Stress** is the magnitude of a force, divided by the area over which it is applied. Hence, both stress and pressure have the same units of pascals or bars. When the same force acts over a smaller area, a greater stress or pressure results.

Suppose a body is acted upon by forces (Figure 8.1) so that it is in a **state of stress.** On any arbitrarily oriented planar surface through the body, there will be, in the general case, one **normal stress,** σ, perpendicular to the plane and two **shear stresses,** τ_1 and τ_2, perpendicular to one another and tangential (parallel) to the plane. These three orthogonal (mutually perpendicular) stresses constitute the components of the total stress on the plane. For a particular state of stress, the magnitude of these three surface stresses varies with respect to the orientation of the plane through the body. The general state of total stress in the body at a point, which may be considered to be an infinitesmally small cube, is specified by these three surface stresses on each of three orthogonal planes, parallel to the faces of the cube, or nine stresses altogether. However, it can be proved mathematically that it is always possible to orient a coordinate system so that no shear stresses exist

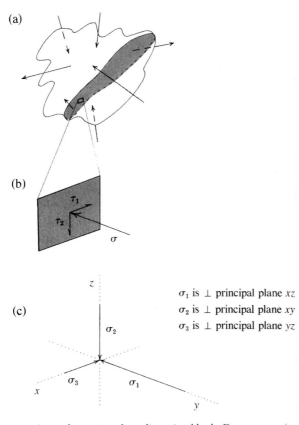

(a)

(b)

(c)

σ_1 is $\perp$ principal plane xz
σ_2 is $\perp$ principal plane xy
σ_3 is $\perp$ principal plane yz

8.1   State of stress in a three-dimensional body. Force vectors (arrows) acting on the body create the state of stress. On an arbitrarily oriented plane (shaded) through the body, forces acting on the area of the plane can be resolved into mutually perpendicular components of normal, σ, and shear stresses, τ_1 and τ_2. The normal stress, σ, is perpendicular to the plane, and the two shear stresses, τ_1 and τ_2, are parallel to the plane.

on the three orthogonal planes parallel to the faces of the cube, reducing the number of stresses needed to define the total state of stress to just three normal stresses. These *three normal* **principal stresses** *therefore represent the total state of stress at a point in the body* and are designated with subscripts according to their magnitude, $\sigma_1 > \sigma_2 > \sigma_3$. The specially oriented orthogonal planes on which only the principal stresses, and no shear stresses, prevail are called **principal planes.** Because shear stresses cannot be sustained on the interface between the solid Earth and its atmosphere, this more or less horizontal interface is a principal plane and the other two principal planes perpendicular to it must therefore be vertical. Hence, two principal stresses near the surface of the Earth are commonly considered to be horizontal and the third vertical.

Within the Earth, **compressive** states of stress press material together. **Tensile** stresses act in opposite directions to pull material apart.

A special state of total stress in which principal stresses are of equal magnitude in all directions, $\sigma_1 =$ $\sigma_2 = \sigma_3$, is referred to as a **hydrostatic state of stress,** because such states occur in bodies of water or other liquids. This hydrostatic stress, also called **pressure,** is approached at increasing depths in the Earth where rocks have diminishing strength and flow in the same way as viscous liquids, especially on long geologic time scales. This assumption is implicit in the formulation of **confining pressure,** P (Section 1.2), which is due to gravitational loading, which is referred to throughout this textbook as one of the three intensive thermodynamic variables that defines states of equilibrium in rock systems. This assumption was also made in calculating the geobaric gradient (equation 1.8).

Despite the obvious utility of the assumption of hydrostatic states of stress in rock systems, a nonhydrostatic state of stress is more typical and its effects on the dynamics of magma bodies are one focus of this chapter. In a **nonhydrostatic state of stress** the principal stresses are unequal, $\sigma_1 > \sigma_2 > \sigma_3$.

8.1.2 Deformation

Deformation is the way a body responds to applied stress. The character of the deformation depends on the state of stress and the properties of the body. Individual particles, such as mineral grains, within a stressed rock body are displaced or changed in shape or size until a new state of more stable equilibrium between the imposed forces and the body is attained, according to Le Chatelier's principle. The body, or parts of it, in its final deformed state has experienced one or more of the following components of deformation:

1. Translation (movement, or displacement) from one place to another with respect to Earth coordinates (e.g., in a thrust sheet)
2. Rotation, as in the limbs of a fold
3. Distortion, or **strain,** which can consist of a change in shape and/or volume

A nonhydrostatic state of stress can produce all of these components in folding, faulting, and other types of deformation of rocks in the Earth. On the other hand, a hydrostatic state of stress can generally only produce a change in volume; for example, a magnetite grain is compressed into a smaller grain of the same shape.

In rocks, strain can only be measured with respect to a reference marker whose shape and/or size is known in the initial, undeformed state. Reliable markers in rocks are difficult to find; a more or less ellipsoidal basalt pillow (Figure 7.3a) is a possible marker, but its exact shape and size are uncertain. Strain, ε, is measured in terms of the change in some dimensional or angular aspect of the final deformed body relative to the initial undeformed body. Examples of measures of strain include volumetric strain

8.1 $\varepsilon_v = \dfrac{(V - V_0)}{V_0} = \dfrac{\Delta V}{V_0}$

and linear strain, either an elongation or a flattening,

8.2 $\varepsilon_l = \dfrac{(l - l_0)}{l_0} = \dfrac{\Delta l}{l_0}$

where V_0 and l_0 represent the initial undeformed state and V and l the final deformed state. Note that strain is a pure number without units, such as 0.45, or, as a percentage change, 45%.

In Figure 8.2, a cubic volume of isotropic rock, which has uniform properties in all directions, is subjected to a nonhydrostatic state of stress, $\sigma_1 > \sigma_2 > \sigma_3$. Consequently, it is flattened perpendicular to σ_1 (the maximal compressive principal stress) and elongated parallel to σ_3 (the least compressive principal stress), which is the direction of least work of deformation. Note that a nonhydrostatically stressed body extends in the directions of least and perhaps intermediate principal stress but not in the direction of the maximal compressive principal stress, σ_1, as that would require the maximum of work to be done on the surroundings. Nature, being lazy and parsimonious, prefers the least-work alternative.

8.1.3 Ideal Response to Stress

Stress and strain are mathematical concepts that apply to any material. The response of *real* rocks and magmas to applied stress can be complex and depends upon many factors, as will be discussed in the next section. But before dealing with real behavior we first consider *ideal*, or "end-member" elastic, plastic, and viscous behavior, which have simple relations between applied stress and resulting strain.

Elastic behavior is the only ideal response to applied stress that is recoverable, or reversible; the deformed material returns instantaneously to its initial undeformed state when the stress is eliminated (Figure 8.3a). Rubber bands, springs, sheets of window glass, and rocks under certain conditions behave elastically. There is a linear, direct proportionality (Hooke's law) between stress and elastic strain, $\sigma = E\varepsilon$, where E, Young's modulus, is a property of the material. The value of E is the slope of the stress-strain line in Figure 8.3a. As the applied stress increases, the material reaches its **elastic strength,** σ_e, and the body ruptures or breaks, losing cohesion and causing permanent deformation. If a pane of window glass is subjected to a small applied stress, it bends slightly and reversibly, but if it is stressed more to exceed its elastic strength, it fractures into pieces.

Elastic strains of glass, minerals, and rocks are much smaller than stretching of a rubber band. Small elastic volumetric strains in geologic solids are a consequence of their very small elastic compressibility. This property, formally defined in Section 8.3.1, is a measure of how much a material body compacts under pressure; even large pressure changes yield only small changes in volume in most rocks, on the order of 1%.

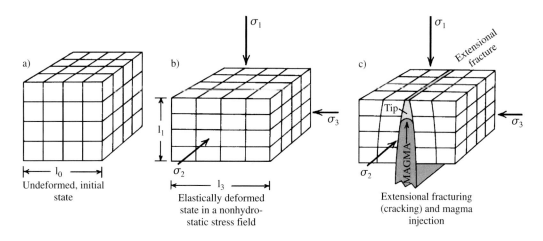

8.2   Deformation of a body under a nonhydrostatic state of stress, $\sigma_1 > \sigma_2 > \sigma_3$. (a) Undeformed body represented by a cubical stack of smaller cubes. (b) Principal stresses of unequal magnitude cause a change in shape—a **strain**. Actual recoverable elastic strain produces only a small percentage of strain, far less than is represented here for illustrative purposes. However, the relative amounts of strain are as expected, that is, flattening perpendicular to the maximal principal compressive stress, σ_1, and elongation, or extension, parallel to the least principal compressive stress, σ_3. Strain parallel to σ_2 can be either elongation or flattening or nil. (c) Nonhydrostatic stresses may exceed the brittle (essentially elastic) strength of a rock so that an **extensional fracture** forms perpendicular to the least principal compressive stress, σ_3. These open cracks may be subsequently filled with magma or fluid. The pressure exerted by the magma or fluid itself may be sufficient to create an extensional fracture by **hydraulic fracturing.** Just beyond the injecting fluid or magma that wedges apart the walls of rock is a tip cavity where transient low pressure can suck out volatiles dissolved in the magma or pore fluids lodged in the wall rock.

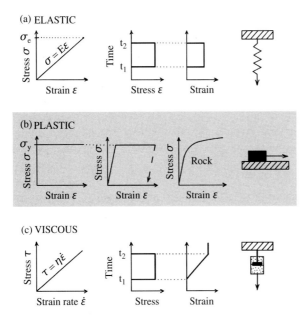

8.3 Definitions and comparisons of three ideal types of response to applied stress. For each type of ideal behavior, the relation between stress and strain is shown in the left-hand diagram and a mechanical analog is shown in the diagram on the right; the middle diagrams show special attributes. (a) **Elastic behavior** in which stress is linearly proportional to strain. If the stress exceeds the elastic limit, or **elastic strength**, σ_e, the material breaks permanently. Another attribute of elastic behavior is that strain is instantaneous and nonpermanent (reversible). This is shown in the middle diagrams, where a particular stress, σ, is applied between time t_1 and t_2 resulting in an immediate accompanying strain whose magnitude is σ/E. After t_2 when the stress is eliminated there is no longer any strain. Elastic behavior is modeled by a spring. (b) **Plastic behavior** in which no strain occurs until the applied stress reaches the **yield strength**, σ_y, after which permanent strain can accumulate indefinitely. The mechanical model of plastic deformation is a block resting on a table. An applied force cannot move (deform) the block until the frictional resistance (the yield strength) is overcome. Before the stress reaches the yield strength in a material, some amount of reversible elastic strain may result (second diagram); in this case, removal of the stress after an excursion through elastic and then plastic behavior is tracked by the dashed line. A typical response of a stressed rock (third diagram) shows initial elasticlike behavior grading continuously into more plasticlike behavior. The yield strength can only be approximated and the plastic-like region shows "strain hardening" in which increasing stress is required to induce more strain. (c) **Newtonian viscous behavior** in which the time rate of shear strain, $d\varepsilon/dt = \dot{\varepsilon}$, and shear stress, τ, are linearly proportional. The proportionality constant is the coefficient of viscosity, η. Any applied shear stress between time t_1 and t_2 results in an immediate, permanent strain that increases until the stress is removed. After t_2 when the stress is eliminated the strain that had accumulated at time t_2 remains permanently. A mechanical model of viscous response is the dashpot—a cylinder containing a viscous liquid, such as honey, in which moves a loose-fitting piston.

In contrast to elastic behavior, ideal **plastic** response involves a nonrecoverable, or irreversible, strain occurring at a stress equal to a critical value known as the **yield strength**, σ_y (Figure 8.3b). For applied stresses less than the yield stress, only reversible elastic strain

occurs. Under some conditions, real minerals and rocks behave more or less plastically, but the stress-strain response is continously curved and a yield stress can only be approximately determined.

The ideal **viscous** response to applied stress was discussed in Section 6.1 (see also Figure 6.1a) with reference to bodies of melt. Unlike elastic and plastic bodies, a viscous body has no strength and deforms permanently by flow under the smallest applied shear stress. This property is typical of liquids. As shear stress is applied, a proportional strain *rate* results (Figure 8.3c). Such linear flow behavior is referred to as **Newtonian viscosity** and the proportionality constant is the viscosity, η.

☀8.2 RHEOLOGY OF ROCKS AND MAGMAS

Real rocks and magmas generally deviate, commonly substantially, from the three ideal modes of deformational behavior just discussed. The real deformational behavior of rocks and magmas—their **rheology**—is a combination of the ideally elastic, plastic, and viscous end-member responses that are expressed either simultaneously or under particular conditions. For example, basalt magmas near their liquidus T where few if any crystals exist exhibit Newtonian viscous behavior, whereas with increasing crystallinity the magma rheology becomes non-Newtonian and the magma possesses yield strength; its behavior is a compound of viscous and plastic, or viscoplastic. Accordingly, the applied stress must exceed a critical yield strength in order for permanent viscous flow to occur. For an applied shear stress less than this critical yield strength no permanent deformation occurs. Highly crystalline magmas near their solidus temperatures can fracture more or less elastically as wholly crystalline rock does. Bodies of folded rock layers (Figure 8.4) exemplify compound viscous-elastic behavior. They are viscoelastic. At high T over long periods the compositional layers were folded as if they had low viscosity. Yet if it had been possible to strike the hot body with a hammer when the folds were developing, it would have fractured into pieces. Thus, rocks and magmas have a duality of behavior that differs with circumstances.

Viscoelastic behavior illustrates the importance of the **time rate of strain**, $\Delta\varepsilon/\Delta t$, which has units of reciprocal seconds (s^{-1}). A rock subjected to a large strain rate, in which strain accumulates quickly in a small time, such as during a blow with a hammer, results in an elastic response; the rock breaks into pieces. On the other hand, slower strain rates allow a greater amount of permanent but slowly developing viscous flow to occur. The viscoelastic mantle of the Earth transmits elastic seismic waves at speeds measured in kilometers per second (fast strain rate) yet also

8.4 Ductile-flow folds in metamorphosed bedded chert, Sutter Creek, California. This body of rock is mainly quartz and cannot have melted under the crustal conditions of its metamorphism and deformation. Hence, folding must reflect low-strength ductile behavior. (Photograph courtesy of L. E. Weiss. Reproduced with permission from Weiss LE. Minor structures of deformed rocks. New York: Springer Verlag, 1972. Copyright © 1972 by Springer Verlag.)

convects by viscous flow at rates of centimeters per year (very slow strain rate). Magmas experience strain rates as low as 10^{-8} s^{-1} in slow extrusions of lava to as fast as 30 s^{-1}, in explosive volcanic eruptions.

8.2.1 Rheology of Rocks

<u>Brittle Behavior</u>. Rocks near the surface of the Earth subjected to high strain rates where P and T are low respond essentially elastically by fracturing, breaking apart, and losing cohesion when the applied stress reaches the elastic limit; this is **brittle behavior.** Every geologist who has collected a hand sample from a rock outcrop by breaking off a piece with a hammer has witnessed brittle behavior. A more revealing exhibition is to squeeze a small cylinder of rock in a hydraulic testing machine (Figure 8.5). As the axially directed load, in this case the maximum principal compressive stress, σ_1, is increased under atmospheric pressure ($P = 1$ atm $= \sigma_2 = \sigma_3$), the rock cylinder shortens proportionately in an elastic manner to a maximum of about <1%. Once the load (σ_1) reaches the elastic strength of the rock, the cylinder splits along **extensional fractures** (Figure 8.6a), which are oriented subparallel to σ_1. These fractures open in the direction of least stress ($\sigma_2 = \sigma_3$) for minimal expenditure of work. Pushing parallel to the pages of this book causes the pages to separate in a manner resembling extensional fracturing.

The stress difference required to produce permanent deformation under specified conditions in a body is its **strength.** The brittle strength that was exceeded to produce the permanent deformation manifest in the extensional fractures in Figures 8.2 and 8.6a is equal to the **stress difference** $\sigma_1 - \sigma_3$; it can be as low as 5 MPa

(= 50 bars) for some weak rocks. It should be realized that open extensional fractures can develop in a volume of rock under a compressive state of stress where $\sigma_1 > \sigma_2 > \sigma_3 > 0$.

In a second experiment at somewhat greater P ($= \sigma_2 = \sigma_3 = 35$ bars), corresponding to a depth of about 0.13 km, the axially loaded rock cylinder continues to respond in a brittle way by forming a **fault,** or **shear fracture** (Figure 8.6b), along which parallel displacement occurs and that is inclined at an angle of approximately 30° to σ_1. Faulting occurs where the shear stress, τ, exceeds the frictional resistance on the potential fracture plane. Frictional resistance is the product of multiplying the normal stress, σ, acting on the fracture (Figure 8.1) by the coefficient of friction of the rock, μ. Friction of rocks does not vary appreciably, but the normal stress that can act on a fracture increases, as does the confining pressure, P, with increasing depth in the Earth. Hence, the frictional resistance, $\sigma\mu$, to faulting increases with depth, necessitating greater shear stresses to cause faulting. In other words, the brittle strength of rocks increases with depth.

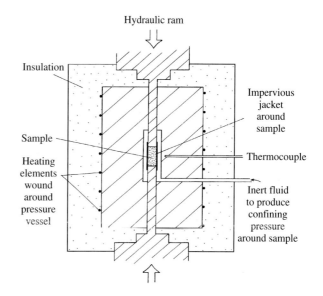

8.5 Laboratory deformation apparatus. The behavior of rocks under nonhydrostatic stress can be investigated in the laboratory by using a hydraulic squeezer, illustrated here schematically. Elevated temperatures, measured by the thermocouple, on the deforming sample are produced by the heating wires around the hardened steel pressure vessel (widely spaced diagonal lines). Alternatively, because steel is weak at high T, a small heater can be built immediately around the sample inside the pressure vessel, which may then be cooled by circulating water to maintain its strength. Such heaters may be deformed with the sample and are discarded after an experiment. A confining pressure, P, is created by pumping some inert fluid, such as compressed argon, around the sample. If this fluid pressure is less than the axially directed stress exerted on the ends of the sample cylinder by the hydraulically driven pistons, the axial stress is σ_1 and the confining pressure around the cylinder is $P = \sigma_2 = \sigma_3$, consequently, the sample will be shortened, or compressed, as in Figure 8.6.

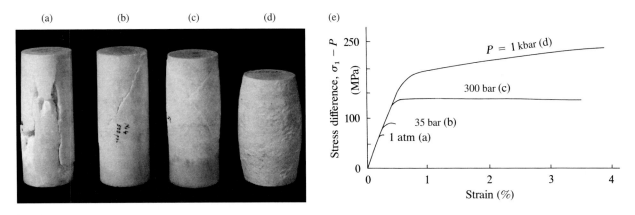

8.6   Cylinders of marble deformed in the laboratory show a transition between brittle and ductile behavior as a function of confining pressure, $P = \sigma_2 = \sigma_3$. Marble cylinders on the left were deformed at 25°C and P indicated in the stress-strain curves on the right. Note the increasing strength (stress difference required to produce permanent, "plasticlike" deformation) of marble with increasing P. See text for further discussion. (Photograph courtesy of M. S. Paterson. Reproduced with permission from Paterson MS. *Experimental rock deformation. The Brittle Field*. New York: Springer Verlag, 1978. Copyright © 1978 by Springer Verlag.)

Not only does increasing depth increase brittle strength, it also retards brittle behavior. This follows because fracturing creates openings in rocks where none was initially present, thus increasing the rock volume for a given mass. **Cataclasis,** creating **cataclastic fabric** (Figure 8.7), involves closely spaced (micro)fracturing, crushing, frictional sliding of broken fragments past one another, and rotation of grains. All of these dilatant processes increase the volume of the rock body because of the introduced open spaces, which are increasingly prohibited at increasing P.

Liquids play a profound role in deformation. The pressure of a fluid, such as water, or of a silicate melt, that pushes outward on the walls of open spaces in a rock counteracts the normal stress compressing the rock together. The **effective normal stress** is therefore $(\sigma - P_f)$, where P_f is the fluid or melt pressure. Fluid or melt pressure reduces the frictional resistance,

8.7   **Cataclastic** fabric. (a) In polished surface of granitic rock. Note irregularly oriented cracks and wider shear fractures that contain obvious broken mineral grains in a dark matrix, which is more finely crushed grains cemented by silica introduced after the rock lost cohesion during cataclasis. (b) Lunar anorthosite under cross-polarized light. Intense brittle fracturing and brecciation was produced by meteorite impact on the Moon. (National Aeronautics and Space Administration photograph 60215,13.)

$(\sigma - P_f)\mu$. If P_f is sufficiently large in some place in the crust and even in the upper mantle, extensional fracturing can occur parallel to σ_1 and σ_2 and perpendicular to σ_3. The magma or fluid can then penetrate into the cracks, forming dikes and veins. This phenomenon is called **hydraulic fracturing** (Figure 8.2c).

<u>Ductile Behavior</u>. In a third experiment, at still greater P, the marble cylinder in Figure 8.6c displays some permanent bulging or shortening along its axis in addition to faulting. In the fourth experiment, at the highest P ($= \sigma_3 = \sigma_2 = 1$ kbar), the bulging and shortening of the cylinder are greater and clearly involve a more or less uniformly distributed deformation mostly on the scale of individual mineral grains (Figure 8.6d); this is in contrast to the localized deformation along faults cutting many grains that occurred at lower confining pressures. Part of this distributed, permanent deformation in which virtually every grain has changed its shape is brittle microcracking (cataclasis). But another part that does not involve loss of cohesion and dilatancy involves two additional, atomic-scale mechanisms within individual mineral grains, the effects of which can be discerned microscopically. One mechanism involves recrystallization of grains, as atoms diffuse from more highly stressed parts to regions of lower stress. The other mechanism entails intracrystalline plastic slip and twinning, in which groups of atoms, or domains within the grain, move relative to neighboring domains. Both mechanisms produce changes in grain shape without any loss of cohesion, and both are promoted at elevated P and especially at the accompanying higher temperatures in the Earth. Hence, these two mechanisms that involve enhanced atomic mobility at elevated temperatures are said to be **thermally activated.**

The distributed, grain-scale permanent deformation just described is known as **ductile deformation;** because it macroscopically resembles viscous flow, it is also called **ductile flow.** It is manifested in folds (Figure 8.4) and flow of glaciers and salt domes.

Because of the thermally activated nature of ductile flow, the ductile strength of minerals and rocks decreases exponentially with depth as T increases (Figure 8.8). The ductile strength of felsic minerals and rocks decreases by several orders of magnitude in the continental crust. Higher temperatures combined with slower strain rates make long-term atomic mobility increasingly significant so that ductile rock strengths can be as small as a few megapascals. This vanishingly small strength, comparable to that in toothpaste and cake batter, in hotter rocks in the deep crust and mantle means that large stress differences cannot be sustained; consequently, states of stress are nearly hydrostatic. In Figure 8.8, note that the lesser geothermal gradients prevailing in stable cratonic continental regions cause minerals to be stronger at a particular depth because of lower temperatures, relative to hotter geologically active orogenic regions.

With increasing depth, increasing brittle strength is interrupted by exponentially decreasing ductile strength (Figure 8.9). This phenomenon, together with contrasts in the brittle and ductile strengths of different minerals and rocks, produces a stratification in strength within the lithosphere. Assuming the continental crust to be made of felsic rock and the upper mantle predominantly of olivine, the weak lower continental crust is embedded between the stronger brittle upper crust and uppermost mantle. In models in which the lower continental crust is made of mafic rock whose ductile strength lies between that of olivine and quartz-feldspar a weak middle crust results. Stratification in strength of the lithosphere has an important bearing on how magmas ascend through it and intrude a particular volume of crustal rock; these topics are considered in the following chapter.

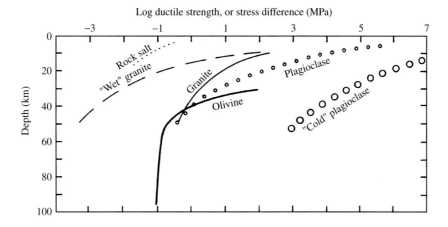

8.8   Ductile strength as a function of depth for some rocks and minerals. All curves are for a strain rate of $10^{-14}\,s^{-1}$ and a geothermal gradient appropriate to an active orogenic zone, except the stronger "cold" plagioclase curve, which is for a lesser geothermal gradient in a craton. (Redrawn from Twiss and Moores, 1992, Fig. 18.19.)

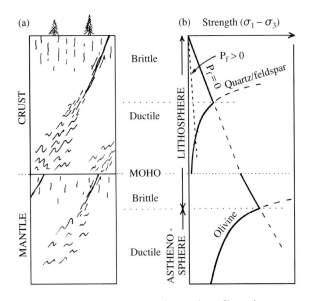

8.9 Generalized deformation and strength profiles with respect to depth for the continental lithosphere. (a) Cartoon showing that extensional fractures and faults near surface give way to ductile folds and flow in deeper crust. This same transition in behavior may occur in the upper mantle. (b) Schematic strength profiles for dominant rock-forming minerals in the continental lithosphere that define stratified brittle and ductile regimes. Brittle strength increases essentially linearly with depth, or P, independently of T. At some depth that is shallower for a greater geothermal gradient the brittle strength curve is truncated by the exponentially T- and mineral-dependent ductile strength curve, which is relatively insensitive to P. Ductile strength of the lower crust can be less than a few megapascals, whereas the brittle strength of the upper crust and mantle can be hundreds of megapascals. The solid lines for brittle and ductile strength are the maximum stress difference that can occur in the Earth. Note the dashed line labeled $P_f > 0$, which shows that a rock regime at a high fluid pore pressure has a smaller brittle strength and the realm of brittle fracturing extends to greater depth in the crust. Thus, otherwise ductile rocks can fracture if fluid pressure is high.

8.2.2 Non-Newtonian Rheology of Magma

For the Newtonian viscosity described thus far in this chapter and in Section 6.1, there is a linear proportionality between stress and strain rate (Figure 8.3c). This simple relationship appears to be valid for most crystal- and bubble-free melts, including metastable melts at subliquidus temperatures, which are subjected to moderate applied stress and relatively low strain rates. However, nonlinear behavior, or **non-Newtonian rheology,** prevails in melts subjected to high strain rates that exceed their characteristic relaxation time (see end of Section 6.1) and especially in magmas that contain suspended crystals or bubbles. For such behavior, the **apparent viscosity, η_a,** can be defined by the ratio of shear stress to strain rate at a particular set of conditions. The importance of non-Newtonian rheology in volcanic activity can be readily appreciated from the fact that erupting magmas are vesiculated, are com-

monly partially crystallized, and ascend rapidly at very high strain rates through a feeding conduit.

The following sections explore factors involved in non-Newtonian behavior.

<u>Yield Strength</u>. Many petrologists have noted that some homogeneous, bubble- and crystal-free melts possess a yield strength, σ_y, which makes viscous flow non-Newtonian. (For a contrary viewpoint see Lejeune et al., 1999.) The applied shear stress, τ must exceed σ_y before any viscous deformation can take place. This non-Newtonian viscoplastic rheology obeys the equation

$$8.3 \quad \tau = \sigma_y + \eta_a\left(\frac{dv}{dt}\right)$$

<u>Crystal-Rich Magma</u>. Another reason for non-Newtonian behavior is an increasing volume fraction of crystals (crystallinity) in magmas. Rigid crystals increase the apparent magma viscosity because a more tortuous shear path must be followed in the melt around neighboring crystals, even though they can rotate during flow. Observed yield strengths in crystal-bearing magmas appear to result from grain-to-grain interactions, especially as the crystalline network becomes continuous (Smith, 1997).

The Einstein-Roscoe equation can be adapted (e.g., Marsh, 1981) for the apparent viscosity, η_a, of a crystal-bearing non-Newtonian magma relative to the viscosity of the compositionally equivalent melt, η.

$$8.4 \quad \eta_a = \eta(1 - BX_v)^{-2.5}$$

where X_v is the volume fraction of crystals and $B = 1.35$ for uniformly sized rigid crystals, assumed to be spherical for simplicity. If the fraction of crystals is small, the viscosity is nearly that of crystal-free melt, but as the fraction of crystals increases, the apparent viscosity of the magma increases exponentially (Figure 8.10), until near $X_v = 1/1.35 = 0.74$ it approaches infinity. In real magmas that contain nonspherical crystals of various shapes and sizes, B is larger, about 1.7, so that solidlike behavior occurs where X_v is about 0.59. At greater degrees of crystallinity, magma acts as a solid and can fracture in a brittle manner. Dikes can be rapidly intruded into cracks in high-crystallinity magma as if it were a brittle solid, but subsequently the dike and its host magma can still be deformed, however slightly, by slow viscous flow; no increase in T or applied stress is required to allow flow, only more time.

As magma cools below the liquidus, an increasing non-Newtonian viscosity due to increasing concentration of crystals is compounded by increasing residual melt viscosity because of decreasing T and, in many systems, because of increasing silica concentration of the residual melt. However, in magmas where mostly anhydrous minerals are crystallizing, water concentration

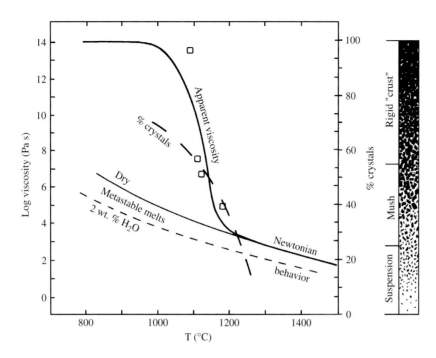

8.10   Dramatic exponential increase in apparent viscosity as the crystallinity of magma increases. Volume fraction (as percentage) of crystals shown by long dashes. Thick solid line is experimentally determined viscosity at 1 atm near the NNO buffer of the dacite lava that formed a dome within the crater of Mount Saint Helens after explosive activity waned in 1981. (From Murase et al., 1985.) Viscosity increases by a factor of 10^9 over less than 200°C as the crystal content increases from about 30% to 70%. Note the simple, nearly linear T-dependent Newtonian viscosity for crystal contents less than about 30%. Light lines show calculated Newtonian viscosities for metastable, crystal-free melts at low temperatures. Note also nearly constant viscosity near the solidus where the magma behaves effectively as a solid rock and has a plastic yield strength of 0.2 MPa. Open squares indicate apparent viscosity calculated by using equation 8.4 and $B = 1.7$. The classification of the magma as to crystal suspension, mush, and rigid "crust" is as occurs in a lava lake (Marsh, 1989.)

increases in the residual melt; thus the melt viscosity might actually decrease. If a residual melt becomes saturated in water so that it exsolves and forms bubbles, the viscosity of the melt increases because of its decreased water concentration, which in turn raises the liquidus T and can result in increased crystallinity, further elevating the viscosity. Exactly how bubbles themselves influence the rheology of the magma body is uncertain. Experiments have revealed no consistent results (Lejeune et al., 1999). These competing factors emphasize the danger of overgeneralization of magma rheology. Each evolving magma system must be considered individually.

Because the viscosity of the melt between crystals and any bubbles can be highly variable and because crystal sizes and shapes are also variable, no universal critical crystallinity value distinguishing between viscous and brittle behavior can be defined for all magmas. Rather, even in a particular magma, there is a crystallinity-dependent transition in behavior.

At any time a magma intrusion consists of different rheologic zones—mobile magma, rigid magma still above its solidus, and magmatic rock below the solidus (Figures 8.10 and 8.11). As the magma cools and crystallizes, a diminishing proportion remains mobile. It should also be noted, in Figure 8.11, that the mobile

part of the magma body is smaller than the volume that is still magma, and this magma volume is less than the magmatic intrusion, which includes solidified, wholly crystalline rock.

8.2.3 Deformation and Flow of Magma

Deformation of high-crystallinity magma bodies can produce dilatancy (volume increase) as crystals rotate, shuffle, and interact. This **dilatancy pumping** phenomenon (van der Molen and Paterson, 1979) may redistribute the melt into dikes oriented perpendicular to the least compressive principal stress, σ_3.

Shaw (1969) drew attention to the possible role of viscous heating and thermal feedback phenomena in magma systems. As a viscous body is sheared, kinetic energy is degraded into thermal energy in an analogous manner to the frictional heating of two solid objects sliding on one another. The T increase in the flowing system depends on the rate of strain, the apparent viscosity and dimensions of the system, and the rate of dissipation of heat out of the system. As T rises during viscous heating, the viscosity of the magma decreases, potentially localizing subsequent flow and heating, causing further reduction in viscosity, and so on, in a process called **thermal feedback.** Magma flow and feedback may become localized along discrete, thin

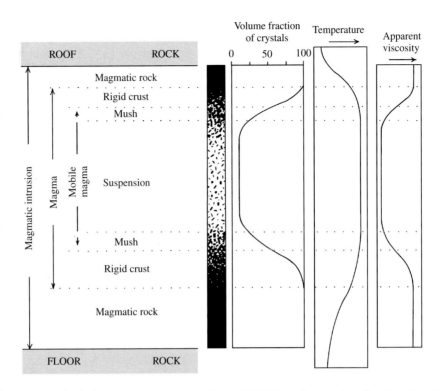

8.11 Correspondences among rheologic zones, magmatic properties, and definitions of magmatic realms in an intrusive system. Compare Figure 8.10.

surfaces intervening between thicker layers that remain unsheared. This phenomenon may have a bearing on the development of flow layering in viscous lava flows (Section 7.9.2).

In rapidly ascending bubbly magma in a subvolcanic conduit, localized shear may focus more rapid strain, exceeding the relaxation time of the melt (Section 6.1) so that, having made the transition into glass, it ruptures in a brittle manner. This sudden rupturing and release of excessive internal volatile pressure in the bubbles may well be the cause of explosive volcanic eruptions (Section 6.7.2).

Flow of Magma in Channels and Conduits. Rheologic properties of magma influence its movement in extruded lavas, in subterranean conduits such as tabular dikes and circular pipes, and in convecting intrusions.

The nature of the flow regime of a viscous magma, considered here for simplicity to be Newtonian, depends upon the magma properties and the boundary conditions surrounding it. Relevant properties are the viscosity, η, and the density, ρ. Boundary conditions of the magma body include the flow velocity, v, and the *hydraulic radius, D,* which is four times the cross-sectional area perpendicular to flow divided by the *wetted* perimeter of the flowing magma. For flow in a perfectly circular pipe, $D = 4\pi r^2/4\pi r = r$ where r is the pipe radius, and for a flow in a dike with a rectangular cross section, $D = 4lw/2(l + w) = 2lw/(l + w)$, where l and w are the length and width; if the dike width is very small compared to its length, $D \sim 2w$.

These properties and boundary conditions can be combined into a dimensionless **Reynolds number**

$$8.5 \quad \mathrm{Re} = \frac{vD\rho}{\eta}$$

Flow regimes dominated by a viscous resistance to flow, where η is relatively large and/or v small, have small Re, and the flow is said to be laminar. **Laminar flow** consists of movement in parallel sheets or "pencils" in pipes (see bottom of Figure 8.12). In contrast, flow regimes dominated by inertial forces at high velocities where viscous effects are relatively less significant have large Re and the flow is turbulent. **Turbulent flow** is a rapid and chaotic motion of particles in small eddies superimposed on the overall flow of the moving fluid.

The transition from laminar to turbulent flow depends somewhat on the configuration of the conduit or channel but begins at Re as low as about 500; flow may not be wholly turbulent until Re $\sim$ 4000 (Figure 8.12). In the transitional region, laminar flow is metastable and some disturbance may trigger turbulence. Despite the wide variation in magma viscosity, geologic boundary conditions generally seem to preclude turbulence, except possibly in some very low-viscosity komatiite and carbonatite lavas (Figure 6.2) and gas-charged kimberlite magmas. Fast-moving streams of water are turbulent.

Flowing magma invariably has a velocity gradient near the solid boundary against which it flows. This gradient arises from the viscous drag along the boundary

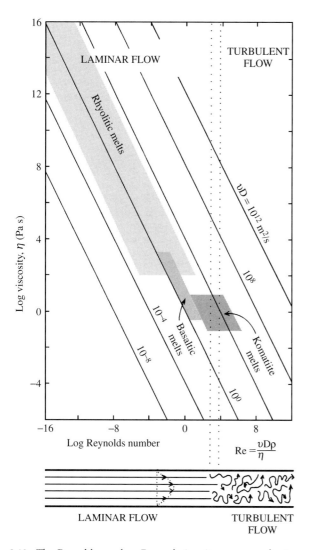

8.12 The Reynolds number, Re, and viscosity, η, govern laminar versus turbulent flow in Newtonian bodies. The transition between laminar and turbulent flows in a pipe, shown schematically at the bottom of the diagram, occurs at Re = 2000 − 4000 (=$10^{3.3}$ − $10^{3.6}$). The range of values of Re and η for komatiite magmas (dark shaded), basaltic magmas (intermediate shaded), and rhyolitic melts (light shaded) are plotted with respect to typical flow velocities and cross-sectional dimensions. Some komatiite lava flows may be turbulent.

where the velocity is nil. In flowing Newtonian magma, the velocity profile is parabolic, whereas flowing non-Newtonian (viscoplastic) magmas have a central stagnant "plug" zone where velocities are uniform and no gradient exists (Figure 8.13). Within the central plug, shear stresses are less than the yield strength of the magma so it is undeformed, whereas in flow margins shear stresses exceed the yield strength and the magma flows viscously. Plug flow is not restricted to lavas but is also observed in muddy volcanic debris flows in which the poorly sorted mixture of clasts as much as several meters in diameter impedes flow just as crystals do in magma. With due allowance for surface crusting of lava flows and other natural perturbations, the flow

profile is a good indicator of the rheology of the mobile material.

Grain dispersive pressure. Another consequence of a velocity-gradient boundary layer is evident in dikes that have a greater concentration of usually larger crystals nearer the center than at the margins (Figure 8.14). As magma containing suspended crystals flows in a laminar manner between the solid conduit walls, a **grain dispersive pressure** in the boundary layer forces grains into the interior of the dike (Komar, 1976). Mechanical interaction between suspended solids in a viscous flow has been studied by engineers in many laboratory experiments and can be observed, for example, in streams of water carrying pieces of wood or leaves. The dispersive pressure is created as more interior grains shear at greater velocity past more slowly moving, more exterior grains. Overtaking grains bump into slower grains in the boundary layer, and, since they can only

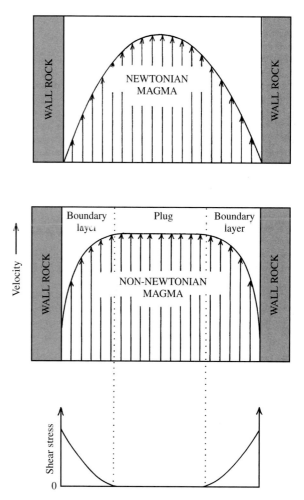

8.13 Velocity and shear stress gradients in confined magma flow. Profiles of a lava flow in a confined channel viewed from above or of a subterranean conduit viewed from the side. (a) Parabolic velocity profile (arrows) in flowing Newtonian magma. (b) Plug flow in non-Newtonian (viscoplastic) magma or volcanic debris flow. (c) Schematic profile of shear stress.

8.14 Effect of grain dispersive pressure in a subvertical dike intruded into Pliocene gravels 15 km south of Hoover Dam, Arizona. The highest concentration of largest black amphibole and white plagioclase crystals in the dike lies above and below the rock hammer, where the intrusive flow velocity gradient was negligible. Platy-weathering margin of dike is virtually free of crystals. Compare Figure 8.13. The inward concentration of larger crystals cannot be related to differential cooling rate across the dike because the crystals are anhedral and therefore did not grow in place, because the amphiboles were equilibrated under upper mantle P–T conditions, and because small mantle-derived peridotite xenoliths also occur in the crystal-rich interior of the dike. Thus, the flowing magma contained inward concentrated crystals and rock fragments derived from an upper mantle source as it stalled at this near-surface depth.

move *into* the channel or conduit, away from the fixed solid wall, there is a net component of grain motion into the dike. These mechanical interactions diminish into the interior, causing a gradient in the dispersive pressure. Higher concentration of crystals in the dike interior increases its apparent viscosity, enhancing plug flow. Concentrations of crystals in nonvertical dikes and pipes may be asymmetric as a result of gravitational sinking of denser crystals after flow has ceased but before solidification of the magma.

Other explanations of nonuniform concentration of crystals between the walls of pipes or dikes appeal to special circumstances such as strongly non-Newtonian flow and multiple intrusions.

Grain dispersive pressure, possibly coupled with thermal feedback, in flowing magmas may play a role in development of flow layering, as depicted in Figure 7.40c.

*8.3 DENSITY OF MAGMA AND BUOYANCY

Melt densities vary mainly between 2.8 and 2.2 g/cm^3. Crystal-bearing *magmas* have somewhat larger densities, but the most significant variation occurs in magmas that contain volatile bubbles, in which densities can be as small as 0.05 g/cm^3. From the smallest to largest density is a factor of about 50, compared to the enormous variation in viscosity of about 20 orders of magnitude. This would seem to relegate density to the category of insignificant magma properties. However, small variations in density *in large masses* of rock and magma can have significant effects. Unlike in viscosity, whose absolute value dictates flow phenomena, it is seemingly small *contrasts* in density between solids and melt, or between contrasting parcels of magma, that strongly affect dynamic magmatic behavior.

In addition to controlling the rise of magma from deeper sources in the mantle and crust, small density contrasts play major roles in the generation and diversification of magmas. If partial melts in mantle and deep crustal sources had the same densities as their source rock or greater densities, there would be no magma movement to shallower depths and consequently no intrusions and volcanism. Physical incorporation and chemical assimilation of chunks of wall rock into magma depend on their density contrasts. Mixing of contrasting magmas and convection in bodies of magma involve density differences.

8.3.1 Density Determinations

Densities are measured in grams per cubic centimeter or in the less familiar SI units as kilograms per cubic meter. The former is used here.

Melt densities over a range of *P* and *T* are generally measured by the falling sphere method using Stokes's law (discussed later).

Measured densities at one particular *P* and *T* can be corrected to another *P* and *T* by using compressibility and thermal expansion data for melts (Lange, 1994) and crystals (Clark, 1966). The **coefficient of isothermal compressibility, β,** expresses the change in volume or density as *P* changes with depth in the Earth at constant *T*

$$8.6 \quad \beta = -\left(\frac{1}{V}\right)\left(\frac{dV}{dP}\right)_T = -\left(\frac{1}{\rho}\right)\left(\frac{d\rho}{dP}\right)_T$$

where *V* is the molar volume and ρ is the density. Because *dV/dP* is negative β is positive; it has units of reciprocal pressure. β depends somewhat on *P* and *T* and differs slightly in different crystallographic directions in anisotropic crystals, so a hydrostatic state of stress can actually cause an anisotropic strain, or different changes in shape in different directions. For crystalline solids (Clark, 1966), β = 1–2 × 10^{-11} Pa^{-1}, and for melts, $\beta \sim 7 \times 10^{-11}$ Pa^{-1}. The volumetric **coefficient of thermal expansion,** α, expresses the change in volume or density as T changes at constant P

$$8.7 \quad \alpha = -\left(\frac{1}{V}\right)\left(\frac{dV}{dT}\right)_P = -\left(\frac{1}{\rho}\right)\left(\frac{d\rho}{dT}\right)_P$$

α has units of reciprocal degrees. Most minerals and rocks have a thermal expansion in the range of $1-5 \times 10^{-5}$/deg and for many silicate melts, $\alpha \sim 3 \times 10^{-5}$/deg.

8.3.2 Densities of Minerals and Melts

Densities of common rock-forming minerals (Figure 8.15) range from about 5 g/cm^3 for Fe-Ti oxides to about 2.6 g/cm^3 for felsic minerals.

Unlike the familiar, but atypical, liquid water–ice system in which the solid is less dense than liquid, silicate melts are 10–20% less dense than the compositionally equivalent crystalline solids (Figure 8.15). The densities of natural melts depend mostly on their chemical composition, especially the concentration of water, and to a lesser extent on P and T (Advanced Topic Box 8.1). Figure 8.15 shows that the change in density for a particular melt over a range of several hundreds of de-

grees Celsius is only a small fraction of the variation between melts of contrasting major-element composition, such as between basalt and rhyolite. The effect of dissolved water on melt density is also dramatic. For example, dissolving only 0.4 wt.% water in a basalt melt at 1200°C and 700 bars has the same effect as increasing T by 175°C or decreasing P by 2300 bars (Ochs and Lange, 1997). This significant control exerted by water follows because its coefficients of thermal expansion and isothermal compressibility (4.7×10^{-4}/deg and 1.7×10^{-5}/bar, respectively, at 1000°C and 1 kbar) are the most expansive and compressible of those of any component in a melt. It may also be recalled (Problem 4.2) that a modest weight percentage of water translates into a much greater amount on a molecular basis.

Water-oversaturated magmas that contain bubbles of exsolved water can produce even greater variations in density, to less than 0.05 g/cm^3 for some very highly vesiculated melts at low P.

Densities of compositionally changing residual melts during fractional crystallization of magma depend mostly on the concentrations of Fe and dissolved water. In fractionating tholeiitic mafic magma (Figure 8.16), the residual melt increases in density as a result of Fe

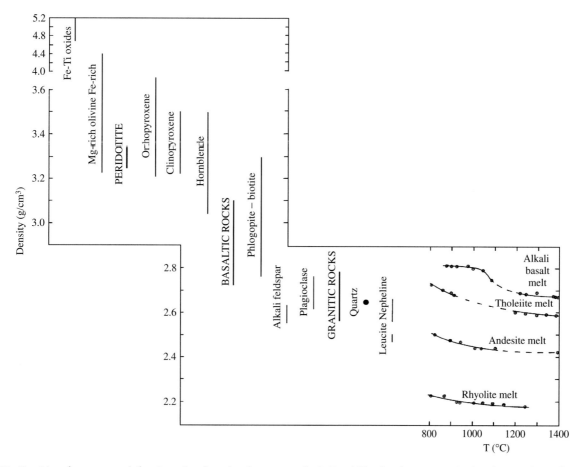

8.15 Densities of common rock-forming minerals and rocks at atmospheric P and T and melts at 1 atm. Higher densities for mafic solid-solution silicates are Fe-rich end members; lower densities are Mg-rich end members. Note change in density scale in upper left. Experimentally measured densities for crystal-free melts. (Redrawn from Murase and McBirney, 1973.)

Advanced Topic Box 8.1 Thermodynamic model for determining the density of silicate melts as a function of their bulk composition, P, and T

Bottinga and Weill (1970) found that the molar volume, V, of a silicate melt at a particular P and T can be approximated by simply adding the partial molar volumes of the constituent liquid oxide components for that P and T: $V = \Sigma_i X_i v_i$ where X_i is the mole fraction of the i oxide and v_i is the partial molar volume (Section 3.4.2) of the i oxide. It is assumed that the partial molar volumes of the oxides are independent of the melt composition. Refinements of this model using new experimental data have been made by I.S.E. Carmichael and coworkers (for example, Lange, 1994). The molar volume can now be calculated with good accuracy for almost any silicate melt and, most importantly, including ones that contain dissolved water, for a wide range of P and T in the lithosphere.

The equation for calculation is

$$V - \Sigma_i X_i \left[v_i + \frac{dv_i}{dT(T-1673K)} + \frac{dv_i}{dP(P-1\,bar)} \right]$$

where dv_i/dT is the thermal expansivity and dv_i/dP is the compressibility of the i oxide. Units for T are in degrees kelvin and for P in bars. Values for v_1 dv_i/dT, and dv_i/dP are given in Table 8.1.

Once the value for V is obtained, the density of the melt, ρ, is found from

$$\rho = \frac{\Sigma_i(X_i MW_i)}{V}$$

where MW_i is the molecular weight of the i oxide.

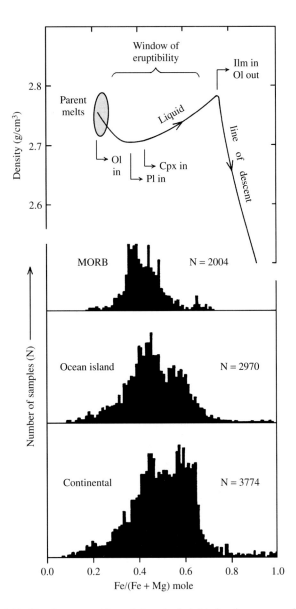

8.16 Density-composition relations in tholeiitic basalt magma and rocks. Top of diagram shows variation in density and Fe/(Fe + Mg) mole ratio of residual melts following a liquid line of descent as indicated minerals crystallize (e.g., Ol in, cpx in) and fractionate out of a mid–ocean ridge basalt (MORB). Note the minimum density where plagioclase (Pl) and clinopyroxene (Cpx) begin to precipitate. The three histograms below represent Fe/(Fe + Mg) mole ratios in 2004 analyzed samples of MORB, 2970 oceanic island basalts, and 3774 continental basalts. Stolper and Walker (1980) postulate that the density minimum in the residual melts constitutes a "window of eruptibility," which correlates with the most frequent basalt compositions observed in oceanic and continental crusts. (Reproduced with permission from Stolper EM and Walker D. Melt density and the average composition of basalt. Contr. Min. Petrol. 1980;74:7–12. Copyright © 1980 by Springer Verlag.

enrichment during crystallization of olivine, pyroxene, and plagioclase then decreases at lower T as Fe-Ti oxides precipitate and fractionate. In fractionating calc-alkaline magmas Fe enrichment is generally limited and residual melt densities diminish as silica, alkalies, and especially water become more concentrated.

8.3.3 Buoyancy

The reason that density contrasts are so important in magma dynamics is their interaction with gravity. In the presence of a gravitational field, density contrasts produce **buoyancy** forces. Bodies denser than their immediate surroundings experience a downward force— **negative buoyancy**—whereas less dense bodies experience an upward force—**positive buoyancy.** A body of the same density as its surroundings has a **neutral buoyancy** and no force prevails. Whether bodies of contrasting density actually sink or rise depends upon viscous resistive forces acting against buoyancy forces.

Many magmatic phenomena involve buoyant movement of a crystal, rock, or volatile bubble (hereafter simply referred to as a *particle*) within viscous melts and magmas of contrasting densities. Buoyancy is also relevant in movement of pyroclasts in volcanic plumes created from exploding volcanoes. Moving particles are driven by a buoyant driving force but retarded by a viscous drag force so that after a steady state is attained, the **terminal velocity** of upward or downward movement is constant. For a smooth, isolated sphere of radius r moving without interference through a Newtonian melt (or other liquid) of viscosity η the terminal velocity is given by **Stokes's law**

$$8.8 \quad v = \frac{2r^2 g \Delta\rho}{9\eta}$$

where $\Delta\rho$ is the contrast in density between particle and melt (or other surrounding media) driving buoyant movement and g is the acceleration of gravity. For slowly moving particles, departures from the smooth sphere model are generally insignificant. Thus, particle terminal velocity is chiefly proportional to the reciprocal of the viscosity, η, and the square of the particle radius, r, both of which can vary over several orders of magnitude. Density contrasts between melts and silicate crystals have less effect on velocity because they are generally small, $\Delta\rho < 1.0$ g/cm$_3$. However, density contrasts between melts and volatile bubbles are greater.

In crystal-rich magmas, movement of an individual particle can be hindered by neighboring particles. However, clumps of numerous particles may move rapidly en masse according to the dimension of the clump and its density contrast. This may be important in some intrusions where crystallization occurs along the walls and cascades of crystals slump into the chamber.

Stokes's law is not valid for non-Newtonain rheologies. Thus, the sinking velocity of dense blocks of roof rock into a chamber of partially crystallized magma cannot be determined. The Stokes model cannot apply to melts that possess a yield strength; crystals of insufficient mass cannot overcome this strength and would remain suspended indefinitely. Nonetheless, sinking of near-liquidus olivines in basaltic melts has been documented experimentally and in Hawaiian lava lakes, suggesting that such melts have a negligible yield strength.

Because of the somewhat greater compressibility of melts than of crystals, a density contrast at low P that makes a crystal negatively buoyant, causing it to sink, might be reversed at high P. One of the first to demonstrate this experimentally was Kushiro (1980), who showed that calcic plagioclase is more dense than tholeiite melt at low P but less than the melt at high P. If plagioclase can float deep in the continental crust in a mantle-derived basaltic melt of negligible yield strength, it may explain the origin of problematic Proterozoic anorthosite intrusions made largely of plagioclase. A similar crossover in densities (Figure 8.17) might occur for olivine in ultramafic melts in the mantle (Stolper et al., 1981). This has possible implica-

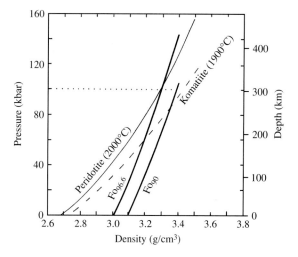

8.17 Density relations for ultramafic melts and olivine at high P and $T = 1900 - 2000°C$. Thicker solid lines represent two olivine compositions; two thinner lines represent ultramafic melts at two different temperatures. Neutral buoyancy between peridotite melt and equilibrium composition olivine, Fo$_{96.6}$, occurs at 100 ± 10 kbar. (Redrawn from Agee and Walker, 1993.)

tions for the earliest differentiation of Earth, when a primeval "magma ocean" is postulated to have existed.

☀8.4 CONDUCTIVE HEAT TRANSFER

As briefly discussed in Section 1.1.3, heat transfer occurs by radiation, conduction, advection, and convection. Lavas lose heat and cool rapidly because of radiation and convection of heat from the surface into the transparent and very low-viscosity atmosphere. In wet climates, vaporization of rain on the surface of the extrusion absorbs heat, enhancing cooling. In just a few hours, lava flows can form a crust rigid enough to walk on, but complete cooling of flows to ambient atmospheric temperatures takes several years to tens of years for thicker flows. Intrusions of magma have slower rates of heat loss for comparable volumes because rapid radiative and convective transfer does not occur. Instead, slow heat loss is dictated by a complex interplay of transfer processes, including internal convection within the magma, conduction into the enclosing rocks, and advective circulation of hydrothermal fluids that extract heat from near the intrusion and transfer it through the surrounding rock.

With regard to rock and magma properties and boundary conditions for the system, the rate of cooling and the T distribution within a magmatic body depend on its T at emplacement, volatile content, latent heat of crystallization, apparent viscosity, thermal conductivity, density, specific heat, dimensions, and shape, as well as the T, conductivity, specific heat, volatile content, and permeability of the surrounding rocks. It is impossible to evaluate all of these variables in any but the most ideal, simplified cases.

Section 6.3 introduced some concepts of conductive heat transfer. In this section, these concepts are applied to the conductive cooling of bodies of magma where heat is transferred into the adjacent rock. Advection and convection are discussed in Sections 8.5 and 8.6.

It is appropriate here to note that the rocks surrounding a magmatic intrusion are commonly referred to as **country rocks.** Depending on their position, these are wall rocks, floor rocks, and roof rocks.

8.4.1 Conductive Cooling Models

Because exact heat conduction, or thermal diffusion, equations are complex and many input variables cannot always be evaluated, approximation models are used.

One such approximation is $t = z^2/\kappa$ (equation 6.7), which indicates that a *thermal transfer time, t,* increases as the square of the dimension in the z direction of the body for a constant thermal diffusivity κ. Thus, doubling the dimension of a body increases its heating or cooling time in that direction by a factor of 4. The **thermal diffusivity,** $\kappa = k/\rho C$ (where k is the thermal conductivity, C is the specific heat, and ρ is the density), is the ratio of the ability of a material to conduct heat relative to its capacity to accumulate thermal energy. The diffusivity is a property of the conducting material and is about 10^{-6} to 10^{-7} m^2/s for melts and common dry rocks but is somewhat less for rocks that contain water or air in pore spaces (Delaney, 1987). Strongly anisotropic rocks such as schists have a greater diffusivity parallel to the foliation than across it.

A more accurate heat transfer approximation was formulated by Jaeger (1968). He proposed a *characteristic, or nondimensional, time,* $t = \kappa t/a^2$, where a is the radius of a sphere or cylinder or half the thickness of a sheet. For example, if $\kappa = 10^{-6}$ m^2/s, t is in years, and a is in meters, $t = 31.5t/a^2$ or

$$8.9 \quad t = \frac{ta^2}{31.5}$$

If $t < 0.01$, cooling is superficial; for $t \sim 0.1$, cooling will have penetrated to the center of the body; and for $t \sim 1$, there is substantial cooling at the center of the body and about as much heat has been lost to the country rocks as remains in the body: That is, its average T is roughly halved. For $t > 10$ heat transfer is practically complete.

Equation 8.9 can be used to approximate the time required for substantial cooling, $t \sim 1$, and loss of extrusive capability in a hypothetical static circular cylinder of magma 5 m in radius feeding a volcano. Substituting, $t = (1 \times 5^2)/31.5 = 0.794$ y $= 290$ days.

The spatial relations of T and t in a conductively cooling static magma body and its country rocks are conveniently represented by a family of **isotherms.** The configuration of these three-dimensional isothermal surfaces in real rock bodies can be represented as isothermal lines in two-dimensional sections through the body. Figure 8.18 shows evolving isotherms in and adjacent to a common intrusion—a basaltic dike—in which the wall rocks initially have a uniform $T = 50°C$. Consequences of this intrusive situation, which also applies to any sheetlike intrusion (Delaney, 1987), may be generalized as follows:

1. Within a day or two, all but the center of a 2-m-thick dike has cooled to <90% of its initial $T = 1150°C$. This cooler T (<1035°C) is near or below the solidus of the magma (Figures 5.10 and 5.11); therefore, any further magma flow is impossible. Hence, the time available for subsequent transport of magma is significantly less than overall dike cooling times. The observation that a basaltic fissure eruption commonly evolves into a focused central eruption after days to a few weeks of activity implies inward solidification of magma and sealing in thin feeder dikes, except in one subvertical cylindrical conduit, where magma transport is thermally most efficient. The efficiency follows from the smaller surface area/volume ratio in the cylinder relative to the dike (Section 6.3.1). The preceding calculation suggests that a 5-m-radius conduit can remain viable for several months.

2. Immediately after instantaneous intrusion, the maximum T of the wall rock at the contact is half the T of the magma, regardless of the thickness of the dike. However, if the latent heat of crystallization of the magma is taken into consideration, the peak wall rock T can be as much as 100°C greater and the rate of inward movement of solidification into the dike as much as three times less. Also, prolonged transport of new magma past a particular area of wall rock introduces more heat into it.

3. Because $t = z^2/\kappa$ (equation 6.7), the velocity of an isotherm moving away from a contact is $v = dz/dt = 0.5(\kappa/t)^{1/2}$. Therefore, shortly after emplacement, isotherms advance rapidly but then slow. Thus, margins of a magma body chill rapidly, but the interior cools at a slower rate, a conclusion in agreement with observed variations in grain size in relation to cooling rate in typical dikes. The outer few centimeters of the margin of thin basaltic dikes less than a few meters in thickness are typically quite glassy. Internal parts have partly glassy intersertal to wholly crystalline intergranular texture. Thicker dikes may be of ophitic-textured diabase.

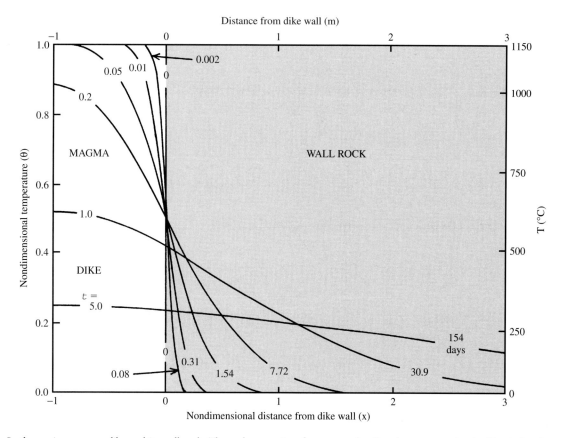

8.18 Isotherms in a magma dike and its wall rock. Thermal properties of magma and wall rock are assumed to be identical and constant. Latent heat released during crystallization of the magma is ignored. Upper and right axes are distance in meters and T in degrees, respectively, for a dike 2 m wide in which the magma initially was 1150°C and the wall rocks 50°C. The labeled curves show T with respect to distance from the dike contact at times of 1.54, 30.9, 7.72, and so on, days. Bottom and left axes are nondimensional distance and temperature, respectively, that permit the cooling history of a planar body of any width to be represented. For example, a dike whose width is 8 m has its center at $x = -1$ and a distance of 4 m into the wall rock is at $x = 1$. The nondimensional temperature, $\theta = (T - T_{wri})/(T_{mi} - T_{wri})$, where T_{wri} is the initial wall rock T and T_{mi} is the initial magma T, varies between 1 and 0. The curves are also labeled in nondimensional time, $t = t\kappa/a^2$ where a is the dike half width, κ is the thermal diffusivity, and t is time; therefore, for a dike 2 m in width at $t = 0.2$, $t = 7.72$ days, and for a dike four times that width (8 m) at $t = 0.2$, $t = 7.72 \times 16 = 123.5$ days. (Redrawn from Delaney, 1987.)

An important factor in the conductive cooling of a magma body is its shape (Section 6.3.1). Isotherms in Figure 8.19 illustrate the influence of the shape of a magmatic body on its conductive cooling. Heat is conducted away from outside corners faster than along planar sides because the corner volume has a considerable area (roughly twice that of the sides) through which heat can be conducted. Therefore, after some cooling, isotherms are located farther into the body near corners. In contrast, near reentrant corners where country rock projects into the magma body (Figure 8.19b), heat from the two adjacent sides conducts into the same mass of country rock, which thermally "saturates," reducing the country rock thermal gradient, thus impeding cooling of the magma in that vicinity; therefore, isotherms are crowded together. In an extrusive body (Figure 8.19c), more heat conducts from the surface than from the base; consequently the part of the body that remains hottest longest is not at the center but is displaced downward. Actually, the downward

displacement is more extreme in real lava flows, where radiation and convection also dissipate heat into the atmosphere. Beneath relatively thin lava flows, especially those having insulating rubbly bases, not much heat is conducted into rock material beneath the extrusion before it cools from above so that thermal effects on the substrate are minimal.

It must be emphasized that these conductive cooling models only apply to bodies of static, or motionless, magma and ignore other heat transfer processes. Nonetheless, these models serve as a valuable "baseline" against which to evaluate thermal histories of more dynamic advective and convective systems where movement of liquids facilitates heat transfer.

⁂8.5 ADVECTIVE HEAT TRANSFER

Rocks in the brittle upper crust are fractured. Any outcrop or roadcut is laced with cracks, usually of two or more orientations and spaced centimeters to no more

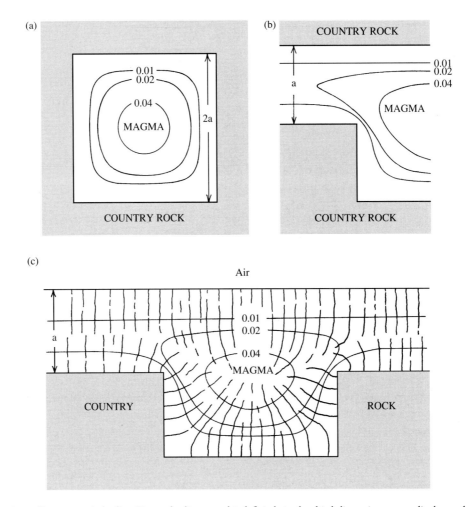

8.19   Isotherms in cooling magmatic bodies. Magma bodies extend indefinitely in the third dimension perpendicular to the page. Isotherms are $0.8T_0$ at nondimensional times (see caption Figure 8.18) $t = 0.01$, 0.02, and 0.04 after emplacement of magma initially at a uniform temperature T_0. The pattern of a family of isotherms at different temperatures at one instant of time would be similar. Country rocks (shaded) are initially at a uniform temperature. (a) and (b) are intrusions and (c) is an extrusion. In (a), a magma conduit that is square in cross section and 8 m on a side, the $0.8T_0$ isotherm has advanced more than two-thirds of the way into the conduit and is a circle (cylinder in three dimensions) after a time $t = ta^2/\kappa = [0.04 \times (4m)^2]/10^{-6} \ m^2/s = 6.4 \times 10^5 \ s \times 1.16 \times 10^{-5} \ days/s = 7.4 \ days$. In (c), contraction-induced joint columns oriented perpendicular to isotherms are shown schematically by irregular lines. (Redrawn from Jaeger, 1968.)

than a meter or so apart. These fractures are commonly filled with fluid, usually aqueous. Deep drilling and study of exhumed once-buried rocks reveal the presence of fluids in cracks to depths of at least 10 km. Even in the deeper crust and upper mantle, where rock flows as a viscous or ductile material, extensional fractures must still form by hydraulic fracturing, as evidenced by sheetlike magma intrusions and veins (Plate VI) found in such rocks now exposed at the surface.

The significance of these facts is this: Movement of magma and hydrothermal fluids in the Earth through interconnected openings in rock can transfer heat much more rapidly than by conduction alone. The dry country rock model assumed for simplicity in conductive calculations is inappropriate for many geologic environments. **Advection** of liquids through passageways in rocks "short-circuits" heat transfer from a cooling

magma intrusion via a lower resistance path into the cooler country rocks. Far traveling advecting liquids can move faster than heat can conduct through solid rock. Magma derived from a central intrusive mass can quickly invade fractures in the surrounding rock, forming dikes and sills. Swarms of such sheetlike intrusions are common over large plutons (see Figure 9.5) and testify that heat has been transferred far into the country rocks. Exsolved aqueous fluids expelled from a crystallizing magma intrusion might mix with larger volumes of meteoric water lodged in country rock openings to form huge advective systems (Section 4.3.3).

In the absence of preexisting fractures, or where preexisting fractures are not suitably oriented in the local stress field to be open, high-pressure magmas and hydrothermal fluids can move into self-generated hydraulic extensional fractures (Figure 8.2c). At shallow depths of the crust, wholesale brecciation can occur where roof rocks rupture over an intrusion if excessive fluid pressures develop (last part of Section 4.3.1). Magma and fluids advecting into widely distributed cracks and breccia openings rapidly lose heat by conduction to the large contacting surface areas of cooler rock.

Advective transfer of heat associated with magmas ascending from the mantle elevates geothermal gradients in the crust above subduction zones and in rifts. As a result, deeper crustal temperatures locally reach the solidus and cause magma generation.

Advection requires permeable rock, as either preexisting channels or self-induced hydraulic extensional fractures created by the advecting magma or fluid. **Permeability** is the ease with which fluid can move through interconnected openings. **Porosity** is the proportion of openings available in the rock to hold fluid: that is, pore volume/total rock volume. In a lava flow, vesicles and cracks formed during thermal contraction and perhaps by continued flow of the still-mobile interior create porosity. In a plutonic rock, such as granite, porosity lies in cracks and perhaps vugs produced by exsolved fluids. Porosity and permeability generally decrease with depth in the Earth, especially if related to cracks.

✳8.6 MAGMA CONVECTION

Convection is a more efficient mode of heat transfer, by one to two orders of magnitude, than conduction in cooling magma bodies. Convecting magma bodies transfer more heat to the roof rocks, where solidus temperatures can be exceeded, causing partial melting, especially above large mafic intrusions. Convection also influences the way a magma body crystallizes and unmixes into compositionally diverse parts, or remains homogeneous.

The essence of **convection** is transfer of heat within a body by buoyant movement of thermally contrasting parts of it. (In contrast, advection is movement of a liquid through openings in a solid.) In magma systems, convection may be initiated by injection of hotter magma into the base of a chamber filled with cooler, denser magma, causing a convective overturn to restore gravitational equilibrium. Convection also occurs during foundering of slabs of cooler, dense roof rock into a magma chamber, causing cool magma adjacent to the slab to be dragged into the hotter interior. The convection discussed in the remainder of this section is driven by internal density differences within the liquid body caused by contrasts in T and/or composition.

Purely thermal convection in a homogeneous melt is considered first to establish some concepts of con-

vection. Then follows a discussion of convection in more typical crystallizing magmas, where contrasts in both T and composition drive thermochemical convection.

8.6.1 Thermal Convection in a Completely Molten Body of Melt

Convective motion of water in a pot on a hot stove is a familiar example of thermal convection. Gravitational stabilization occurs when cooler, more dense fluid at the top sinks and hotter, less dense fluid at the bottom rises. The released gravitational potential energy is consumed in overcoming viscous resistance to flow.

As a consequence of conductive heat loss into the wall rock, the T of a melt decreases exponentially toward a vertical contact of an intrusion (Figure 8.20). This cooler and, therefore, denser melt in the thermal boundary layer is negatively buoyant and potentially able to sink. A gradient in the concentration of water in the melt near the contact may develop as a result of inward diffusion from water-bearing wall rock. This more hydrous melt is less dense, positively buoyant, and potentially able to rise. *Any* density contrast in a steeply inclined thermal or compositional boundary layer (Figure 8.20) is gravitationally unstable, but whether the buoyancy is sufficient to cause movement depends on melt rheology. If movement of the boundary layer does occur, return flow must necessarily occur farther into the closed body of melt. These complementary flows establish convection.

What happens at more or less horizontal contacts between melt and country rocks, such as at the margins of a flat sill (Figure 8.21)? For simplicity, assume that the body of melt, *initially,* has a uniform T above the liquidus. Self-compression in the insulated intrusion due to the downward increasing weight of the melt creates an adiabatic T gradient that is only a few tenths of a degree per kilometer. (Recall in Section 3.2.1 that compressive work done on a system transforms into thermal energy.) Whether gravitational instability develops in the melt near the roof as it cools and becomes denser depends on the contrast in T between roof melt, T_r, and floor melt, T_f, or $(T_f - T_r)$. More heat tends to be lost by conduction through the roof than the floor of a thick horizontal sheet because shallower rocks are cooler on an ordinary geotherm and conductive heat transfer is greater where the thermal gradient (equation 1.5) is greater. Additionally, for shallow intrusions, groundwater in fractured roof rocks absorbs heat and advects it away. After conductive, and possible advective, heat loss (Figure 8.21b), the cooler melt at the roof might be dense enough to overcome viscous resistance to flow, causing convective overturn (Figure 8.21c). Cooler, denser melt sinks and deeper, hotter, less dense

Gradients

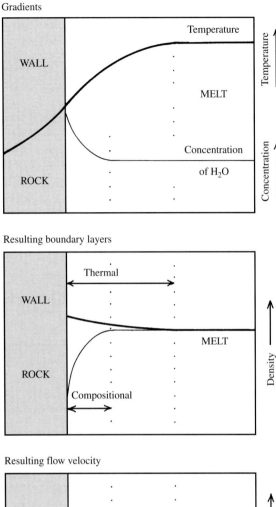

Resulting boundary layers

Resulting flow velocity

8.20 Thermal and compositional gradients create gravitationally unstable boundary layers along the vertical wall of a body of melt. Conductive transfer of heat to the wall rock creates a broad thermal gradient. Heavier lines in middle and bottom diagrams pertain to thermal boundary layer and thinner lines to compositional boundary layer. Actual boundary layers differ in breadth more than shown here. In thermochemical convection, the effect of decreasing T at the wall is counteracted by water enrichment in residual melts produced by sidewall crystallization, which creates a less dense, positively buoyant boundary layer that can float upward.

melt moves to the top of the body. The T gradient in the gravitationally restabilized body of melt is again adiabatic. Further heat loss through the roof rocks would initiate further convective overturn, either in episodes or in a steady state.

Gravitational stability in this sort of fluid body can be quantified in terms of $(T_f - T_r)$, with other relevant factors cast into a ratio of buoyancy forces driving convection (in numerator) to opposing resistive drag forces (in denominator)

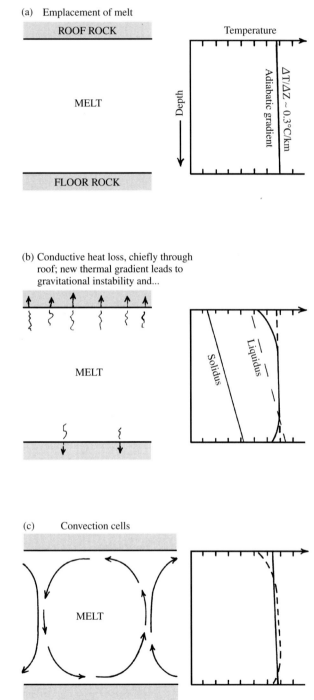

(a) Emplacement of melt

(b) Conductive heat loss, chiefly through roof; new thermal gradient leads to gravitational instability and...

(c)   Convection cells

8.21 Thermal convection in a horizontal slab of melt cooled mostly at the roof by conduction. The melt and roof and floor country rocks extend indefinitely to the right and left of the three vertical cross sections shown here. Thermal gradient shown by solid line, pre-existing gradient by dashed line.

$$8.10 \quad Ra = \frac{[\rho g \alpha (T_f - T_r) h^3]}{\eta \kappa}$$

where ρ is the density, g is the acceleration of gravity, α is the coefficient of thermal expansion, h is the vertical thickness of the convectable fluid, η is the viscosity, and κ is the thermal diffusivity. The **Rayleigh number,** Ra is a dimensionless number that prescribes whether convection occurs. For magma chambers, Ra must be > 500–2000, depending on the exact shape of the melt body. The larger the ratio of buoyant to resistive forces, the more vigorous is convection. Convection occurs in roughly equidimensional convection cells (Figure 8.21c). Occurrence and vigor of thermal convection are most sensitive to four factors:

1. Thickness of the melt body, h; doubling h increases Ra by a factor of 8.
2. $(T_f - T_r)$ the difference in T between the bottom and top of the magma body.
3. Viscosity, η, which ranges over many orders of magnitude.
4. Density, ρ, which is most sensitive to composition, especially the concentration of dissolved water or to the proportion of exsolved volatile bubbles in a volatile saturated magma.

Superliquidus bodies of homogeneous low-viscosity basaltic melt of virtually any vertical dimension convect. Thick bodies of more viscous water-rich granitic melts with large values of $(T_f - T_r)$ might convect, but dry granitic melt bodies less than a kilometer or so thick probably do not thermally convect.

Because the liquidus T of minerals increases about 3°C per kilometer depth whereas the adiabatic gradient in melts is about an order of magnitude less, crystallization occurs at the base of a uniform melt body of considerable vertical thickness (kilometers, rather than meters), even though most of the cooling occurs through the roof (Figure 8.21b). However, in magma chambers that have a vertical compositional gradient, the preferential bottom crystallization may not occur.

8.6.2 Thermochemical Convection in Crystallizing Magmas

Beginning with the pioneering studies of Shaw (1965), numerous theoretical and experimental studies have shown that compositional buoyancy is far more significant in driving convection than that resulting from thermal gradients alone (e.g., McBirney, 1980; Sparks et al., 1984). Whereas the variation in density from 800°C to 1200°C for a particular melt composition is only about 0.1 g/cm^3 or less, common volatile-free melt densities at 1 atm and, say, 1000°C range from 2.2 to 2.8g/cm^3 depending on composition (Figure 8.15). Density variations in volatile-bearing crystallizing magmas can be much greater.

In crystallizing magmas, the residual melt in equilibrium with precipitating crystals is always different in composition—and, therefore density—from the initial melt. (This principle is a central theme of Chapter 5.) For example, the residual melt just above the solidus in basalt magmas (see the Makaopuhi basalt in Plate IIId) is enriched in silica and alkalies and is approximately of rhyolite composition; the density contrast between rhyolite and basalt melts is about 0.4 g/cm^3. Residual melts in fractionating mid–ocean ridge basalt magma vary by more than 0.2 g/cm^3 (Figure 8.16) and in basaltic andesite magma by about 0.1 g/cm$_3$. If a melt becomes water-saturated, the bubbles of exsolved water can substantially lower the density of the vesicular melt.

Convection driven by compositional differences depends on a density difference between different parcels, A and B, of magma: that is, a gradient in density ($\Delta\rho = \rho_A - \rho_B$). However, since gradients in both T and composition occur in crystallizing magmas, the dynamic process is known as **thermochemical convection,** or, because the relative rates of thermal and chemical diffusion govern these dynamic systems, **double-diffusive convection.**

Magmas can be envisaged to crystallize in two end-member chamber shapes, namely, bottle-shaped vertical cylinders in which subvertical walls dominate the external contacts and flat slabs dominated by a sub-horizontal roof and floor. Two end-member magma compositions may also be considered:

1. Calc-alkaline magmas, such as andesite and dacite, in which residual melts are generally more enriched in silica, alkalies, and water, so that they have lower density and are positively buoyant relative to the initial parent magma. Dissolved water has the greatest effect in promoting buoyancy.
2. Basalt magmas, most commonly tholeiitic, in which residual melts tend to be more Fe-rich, more dense, and negatively buoyant.

Bottle-Shaped Magma Chambers. At near-vertical wall-rock contacts, cooler temperatures in the thermal boundary layer of the magma produce **sidewall crystallization.** In calc-alkaline magmas, less dense, positively buoyant residual melt can free itself from the crystal mush in the boundary layer and rise, collecting into a pool at the top of the magma chamber (Figure 8.22). The chamber, filled initially with what may be compositionally uniform magma, unmixes into contrasting parts, a cap that is enriched in silica, alkalies, and water and an interior that is less evolved in composition. Although this unmixing cannot be directly observed, the geologic record of countless compositionally zoned pyroclastic deposits (see, for example, Figure 10.38) is widely interpreted to have resulted by eruption of magma from a chamber subjected to thermochemical convection driven by sidewall crystallization. Moreover, model studies in tanks of room-T, multicomponent

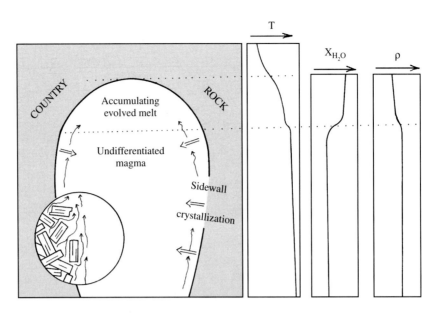

8.22   Thermochemical convection in a crystallizing bottle-shaped calc-alkaline magma chamber. Pronounced compositional stratification can be produced in an initially homogeneous magma chamber. **Sidewall crystallization** (double-line arrows) yields less dense, silica- and water-enriched residual melt that can separate from the associated mush of crystals adhering to the chamber wall (inset circular diagram at enlarged scale). This melt buoys upward and accumulates in a gravitationally stable, growing cap zone. Schematic properties of the stratified chamber are shown in three graphs on right. Continued heat loss from the magma body allows inward advance of the crystallizing wall so that the final solidified pluton can be concentrically and vertically zoned in composition. Explosively erupted silicic magmas are derived from the upper volatile-rich parts of such stratified chambers.

saline solutions show the phenomenon to be viable. Alternatively, one is left with the dilemma of how to create compositional differentiation in highly viscous magma chambers in which rates of atomic diffusion are exceedingly slow, purely thermal convection cells may be precluded because of high viscosities, and through-chamber crystal settling is very slow, or nonexistent, because of non-Newtonian viscosity.

In addition to, or in lieu of, sidewall crystallization, a positively buoyant compositional boundary layer at a vertical wall might originate in two other ways:

1. Hot magma may raise the wall rock T to above its solidus, generating a low-density partial melt that segregates and buoys upward.

2. Relatively dry melt can absorb water from wet wall rock, reducing the melt density. However, because the rate of chemical diffusion of water is orders of magnitude slower than thermal diffusion— about 10^{-10} m^2/s versus 10^{-7} m^2/s—the chemical boundary layer is thinner than the thermal (Figure 8.20).

A vertical compositional gradient in a magma chamber has implications not only for differentiation of magmas but also the way they erupt.

In bodies of mafic tholeiitic magmas with predominantly vertical walls, residual melts resulting from crystallization are enriched in Fe and, if not also too enriched in volatiles, are more dense (Figure 8.16). This residual melt, or possibly a dense crystal-laden magma, may sink en masse along the wall of the magma chamber and onto its floor as a density current, not unlike sediment-laden turbidity currents in standing bodies of water.

Flat Slabs. The dynamics of thermochemical convection differ in magma chambers lacking extensive sidewalls and dominated instead by a subhorizontal floor and roof. One possible situation is shown in Figure 8.23, which may be compared with Figure 8.21, where it was shown that most of the heat loss is through the roof but most of the crystallization may be at the floor. Cooling melt at the roof becomes denser and sinks, whereas floor crystallization of calc-alkaline magmas could release compositionally buoyant residual melt that might also be thermally buoyant because of the release of latent heat during crystal growth. Residual melts from fractionating tholeiitic basalt magmas could be more Fe-rich and more dense unless compensated by latent heating and water enrichment. Depending on the contrasts in viscosity of different magma parcels, varying degrees of magma mixing and homogenization may occur by ascending and descending plumes, retarding differentiation of the chamber magma. This contrasts with compositional differentiation, or magma "unmixing," which can occur in bottle-shaped magma chambers.

These contrasts in the convective dynamics of bottle-shaped and flat-slab magma chambers demonstrate how a seemingly irrelevant factor such as chamber shape can influence the compositional evolution of magmas (de Silva and Wolff, 1995).

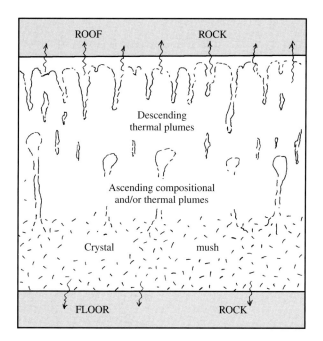

8.23 Thermochemical convection in a flat slab of crystallizing magma. Compare Figure 8.21. In contrast to magma unmixing in a bottle-shaped chamber (Figure 8.22), the magma in this slab tends to be homogenized and mixed by the descending and ascending plumes.

Special Interest Box 8.2 Controversial origin of layering in the Skaergaard Intrusion, Greenland

The Skaergaard Intrusion, magnificently exposed in fjords near the Arctic Circle in eastern Greenland, is briefly described in Section 12.4.2. The Skaergaard has served for decades as a supreme example of the effects of convection in a basaltic magmatic body, creating a wide variety of layering (Figure 7.45) together with an extreme compositional differentiation.

The classic investigation of Wager and Deer (1939) and subsequent studies by Wager and his associates concluded that the 2.5-km-thick sequence of subhorizontally layered rocks making up most of the intrusion was, for the most part, a result of magmatic sedimentation. They envisaged convection currents of crystal-laden magma descending from the roof and walls and sweeping across the floor, sorting and depositing the crystals according to their differing densities. Repeated currents were believed to have created the rhythmic modally graded layers (mafic minerals more concentrated downward and plagioclase upward in each layer) that dominate the layered sequence.

However, Bottinga and Weill (1970) pointed out that during the fractional crystallization of the Skaergaard magma the plagioclase crystals should have floated in the increasingly Fe-enriched residual

melts. McBirney and Noyes (1979; see also an updated discussion of the intrusion in McBirney, 1996) attempted to reconcile this paradox of how apparently floating plagioclases could form the major mineral constituent in the rhythmic layers in terms of a non-Newtonian magma. McBirney and coworkers proposed that other processes, mainly of a kinetic character (Section 7.9.2) in more or less static floor zones, were involved in creating the various types of layering.

It was not until the 1990s that sufficient data became available to determine the densities of hydrous melts of any composition with reasonable accuracy (Lange, 1994). Prior conclusions regarding plagioclase flotation were necessarily based on anhydrous melt models using less accurate data for partial molar volumes of condensed components (CaO, MgO, etc.). The influence of dissolved water on melt densities is significant, and it can be shown (Problem 8.13) that modest water concentrations in Fe-rich residual melts can allow sinking of plagioclase in the Skaergaard. But whether these modest concentrations actually prevailed cannot be answered because of the lack of independent information. The widespread absence of biotite and amphibole in the evolved Skaergaard differentiates only preclude water concentrations of less than about 3–4 wt.%.

The origin of compositional layering in the Skaergaard (Irvine et al., 1998) and, by implication, many other layered intrusions remains controversial.

8.6.3 Replenishment in Evolving Magma Chambers

Lifetimes of magma chambers are commonly extended by episodes of **replenishment.** New draughts of introduced magma are usually hotter, commonly denser, and nearly always less viscous than the resident evolving magma already in the chamber. Consequently, as the new magma is injected from below, its upward momentum and buoyancy carries it well into the chamber, leading to mixing and possible eventual homogenization. This scenario may occur at oceanic spreading ridges as primitive basalt magma from a mantle source replenishes somewhat evolved basalt magma already in the chamber (see Figure 12.12).

In another scenario, slowly moving new magma of more mafic composition rises into an evolving continental silicic chamber and spreads across the floor. Model experiments using a tank of aqueous solutions of contrasting density and viscosity (Snyder and Tait, 1995) reveal that the invading magma traps a layer of the less dense magma beneath it (Figure 8.24). The invading magma moves laterally as subparallel fingers,

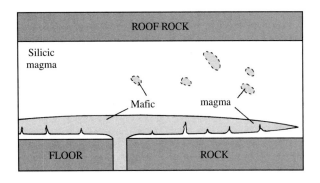

8.24 Replenishment of a silicic chamber by basaltic magma that has been introduced from a feeder below the chamber. The basalt magma (light shaded) forms a lens near the base. A thin underlying layer of less dense silicic melt is gravitationally unstable and has penetrated into the lens, initiating convective breakup. Overlying blobs of basalt from a previous episode of replenishment and convective disruption are dispersed through the silicic intrusion. After complete crystallization these blobs may be preserved as mafic inclusions in the granitic host.

resembling a hand inserted into some liquid. The gravitationally unstable layer of resident silicic magma entrapped at the base of the chamber subsequently forms buoyant plumes that pierce upward through the denser layer of new recharging magma, disrupting it into pillowlike blobs, a possible example of which is shown in Figure 8.25. Convective motion in the chamber due to the heating of the silicic magma by the hotter invaded basaltic magma disperses the blobs. Alternatively, or in addition, as the basaltic magma cools and crystallizes, its melt may become volatile-saturated, reducing magma density and causing convective mingling of basalt magma into the silicic magma. Whatever the dispersive mechanism, the blobs of basalt may become the

8.25 Pillowlike bodies of dark gabbro in syenite that appear to indicate coexisting gabbro and syenite magmas. If the syenite magma had intruded into fractured solid gabbro, pieces would be angular. The mingled magmas may have originated by replenishment of a syenite magma chamber by gabbro magma intruding as a basal lens and subsequent breakup of the lens, in the manner of Figure 8.24. (U.S. Geological Survey photograph courtesy of R. E. Wilcox and Louise Hedricks.)

ubiquitous mafic inclusions in granitic intrusions (Section 7.10). Hybridization and other effects of the replenishment significantly impact the compositional evolution of the silicic-mafic magma system (e.g., Wiebe, 1996).

SUMMARY

Forces acting on a body create a state of stress that can be conveniently represented by three orthogonal principal stresses, $\sigma_1, \sigma_2, \sigma_3$. Near the surface of the Earth, one of the principal stresses is vertical and the other two are horizontal. Hydrostatic states of stress, where $\sigma_1 = \sigma_2 = \sigma_3$, prevail in fluid bodies and in the deep ductile crust and asthenospheric mantle because rock strengths are small, especially at slow strain rates, and cannot sustain large stress differences of a nonhydrostatic state. The strain resulting from hydrostatic stress is a change in volume, whereas nonhydrostatic states of stress (where $\sigma_1 > \sigma_2 > \sigma_3$) produce changes in volume and shape of a body.

The rheologic response of most magmas and rocks to applied stress involves combinations of ideal elastic, plastic, and viscous behavior. Rocks and nearly crystallized magmas near their solidus deform by brittle (essentially elastic) fracturing at rapid strain rates. Extensional fractures form perpendicular to σ_3. Hydraulic fractures are self-generated by magmas and hydrothermal solutions, whose pressure counteracts P. At increasing depth in the crust the brittle strength of rocks increases, but the ductile strength diminishes exponentially as thermally activated mechanisms at the atomic scale take the place of fracturing. Viscouslike ductile flow is favored over brittle fracturing by low strain rates, which allow more time for atomic movement; by elevated T, which promotes atomic mobility; and by high P, which increases frictional resistance and impedes brittle dilatant deformation. Ductile strength under any of these conditions also depends upon the mineral composition of the rock and this varies with depth in the lithosphere; thus, rheology and strength are stratified in the lithosphere.

During crystallization of a body of magma its rheologic behavior progressively changes from Newtonian, through non-Newtonian with exponentially increasing apparent viscosity, to that of a brittle solid. The Newtonian viscosity of near-liquidus melts depends mostly on T and concentration of silica, water, and fluorine. More crystal-rich, non-Newtonian magmas possess a plastic yield strength, and applied stress below this cannot produce flow.

The rheology and flow velocity of most magmas result in laminar flow. Grain dispersive pressure concentrates suspended particles in a flowing magma into the interior of the channel or conduit. Plug flow occurs in viscoplastic magmas.

Compared to the many orders-of-magnitude range in viscosities of magmas, densities vary at most by only a factor of about 50. Most of this variation is in volatile-oversaturated magmas in which bubbles can reduce the density to as low as 0.05 g/cm^3, compared to 2.8 – 2.2 g/cm^3 in bubble-free melts. The small range in bubble-free melt densities is primarily a result of their composition, especially the concentration of dissolved water; thermal expansion and compressibility of melts are small. The density of magmatic rocks is 10–20% greater than their corresponding melts. Despite these relatively small density variations and correspondingly subtle contrasts in density between melt, magma, and rocks, they nonetheless provide a significant buoyant force in large volumes that can overcome viscous resistive forces to drive magma ascent, convection, mixing, unmixing, and other dynamic phenomena in magma systems.

The efficiency of conductive cooling is related to the surface area over which heat can be lost relative to the volume. For a particular shape of body, cooling times increase as the square of a critical dimension. Although conductive cooling (thermal diffusion) rates are so slow that even modest-size plutons a few kilometers in diameter may require tens of thousands of years to crystallize, chemical diffusion rates are orders of magnitude slower still; therefore, little chemical transport is possible in conductively cooling, static bodies.

Advection of liquids through openings in permeable country rock is an efficient heat transfer process that hastens cooling of intrusions. Advecting magma and hydrothermal fluids in country rock can produce economically viable ore deposits and geothermal reservoirs.

Thermal and especially compositional gradients within bodies of magma create internal density contrasts. Gravitational instability of contrasting density parcels can drive convection if viscous resistance can be overcome. Convecting bodies generally cool more rapidly than by conduction and transfer more heat into roof rocks, in some cases sufficient to melt the roof. Despite the generally greater heat loss at the roof, crystallization can occur primarily at the base of a vertically extensive body because the liquidus T increases with depth at a greater rate than the adiabatic gradient.

Thermochemical convection is likely to be more common than purely thermal convection because density contrasts in melts in cooling crystallizing magma are larger than can be produced by any differences in T. It is widely believed that thermochemical convection occurs in tall bottle-shaped calc-alkaline magma bodies by sidewall crystallization. This creates a buoyant boundary layer of silica- and water-enriched residual melt that can separate from the growing crystals and float upward, ponding at the top of the compositionally stratifying magma chamber. This more evolved, commonly water-oversaturated capping magma can erupt explosively. Other thermochemical convective systems are possible in magma chambers having different shapes, such as horizontal slabs, and different compositions, such as mafic tholeiitic magma that yields denser Fe-rich residual melts. Depending on magma composition and chamber shape, thermochemical convection can lead to magma differentiation and unmixing of an initially homogeneous magma or to stirring and mixing, precluding differentiation.

The physical and thermal dynamics of magmas as well as the geometry of their chambers play a significant role in the compositional diversity of igneous rocks.

CRITICAL THINKING QUESTIONS

8.1   Contrast shear, normal, and principal stresses and hydrostatic, nonhydrostatic, and total states of stress.

8.2   What is the difference between strain and strain rate? What are their units?

8.3   Distinguish among ideal elastic, plastic, and viscous responses to applied stress and composite behavior such as viscoelastic and viscoplastic behavior. Draw mechanical analogs, in the manner of Figure 8.3, for the two types of composite behavior.

8.4   Contrast between brittle and ductile behavior in rock with regard to mechanisms and geologic conditions.

8.5   How do extensional fractures develop? Discuss with respect to geologic conditions under which they form, especially the role of liquids, and to state of stress in the Earth.

8.6   Sketch how, and explain why, rock strength varies with respect to depth in the lithosphere.

8.7   Discuss factors governing the rheology of magma, and contrast Newtonian and non-Newtonian behavior.

8.8   Why do most magmas flow in a laminar manner?

8.9   What does plug flow tell about the rheology of the flowing material? Why?

8.10 How can flow of a partially crystallized magma impact its composition?

8.11 Discuss factors that govern melt density. Magma density.

8.12 How do viscosity and density interact in dynamic magma systems to control particle buoyancy?

8.13   Contrast the control on buoyant motion of large blocks of foreign rock immersed in a body of magma versus with buoyant motion of an isolated crystal.

8.14   Characterize factors governing heat transfer by conduction.

8.15   Discuss relevant factors and processes in advective heat transfer.

8.16   What drives convection and what retards it?

8.17   How do advective and convective transfer of heat differ?

8.18   How do thermochemical and thermal convection differ?

8.19   How can the geometry (size and shape) of a body of magma influence its chemical evolution?

PROBLEMS

8.1   In Figure 8.2a and 8.2b calculate the linear strain in the directions of σ_1 and σ_3 by carefully measuring lengths l_0, l_1, and l_3 with a millimeter scale. (*Partial answer:* In the σ_1 direction $\varepsilon_1 = -0.2 = -20\%$.)

8.2   What is the change in volume of an assumed isotropic crystal in moving it isothermally from the surface of the Earth to a depth of 35 km in the continental crust? (*Answer:* 1%.)

8.3   Show that the state of stress in the deep crust is virtually hydrostatic by comparing the ductile strength of "wet" granite at a depth of 30 km (Figure 8.8) with the magnitude of the confining pressure, P, at that depth. (*Answer:* The strength, which is the maximum stress difference possible at that depth, is 0.001% of P.)

8.4   Draw a stress-strain rate diagram for a viscoplastic body that obeys equation 8.3.

8.5   Using MELTS software (Advanced Topic Box 5.3) explore the counteracting effects on residual melt viscosity of increasing concentrations of silica and water in a fractionating andesite magma whose initial water content is 1.5 wt.%. Using equation 8.4 determine the apparent viscosity of the magma at T increments of 25°C below the liquidus. Discuss your results and any assumptions made.

8.6   Determine and discuss the flow regime (laminar versus turbulent) of (a) the upper mantle whose apparent viscosity is approximately 10^{21} Pa s; (b) a river of water whose velocity is as much as 20 m/s; (c) a river of basalt lava near its liquidus T flowing at a velocity of 8 m/s in a channel 2 m deep and 5 m wide. For each situation explain or justify your choice of viscosities and boundary conditions.

8.7   Calculate the approximate density of an andesite magma at 1000°C and 6 bars that contains 75 volume % water bubbles. Use values for water and melt density read from Figures 4.4 and 8.15.

8.8   Determine the laminar flow velocity of a pahoehoe "tongue" of basalt melt 1 m wide and 0.3 m thick. Justify assumptions made in this determination.

8.9   Compare the relative effect on the density of a silicate melt by increasing P versus increasing T in the crust of the Earth. Do melt densities increase or decrease with depth? Assume the geothermal gradient is $\Delta T/\Delta z = 25$°C/km and the geobaric gradient is $\Delta P/\Delta z = 27$ MPa/km. (*Answer:* The compressive effect of P per 1-km increase in depth is 2.5 times that of the expansive effect of T per kilometer.)

8.10   Compare the distance traversed by isolated crystals 1 mm in diameter of plagioclase ($\rho = 2.65$ g/cm^3) and pyroxene ($\rho = 3.3$ g/cm^3) in a granite melt ($\eta = 10^7$ Pa s; $\rho = 2.3$ g/cm^3) and a basalt melt ($\eta = 10^2$ Pa s; $\rho = 2.7$ g/cm^3). Indicate any assumptions made in your calculations. (*Partial answer:* The pyroxene would settle 400 m in 1 year in the basalt melt.)

8.11   Using the 1-atm density of tholeiite melt in Figure 8.15 and a 1-atm density for An$_{60}$ of 2.7 g/cm^3 determine the P at which the plagioclase crystals are neutrally buoyant at 1200°C. Indicate assumptions made in this determination.

8.12   A famous basalt flow on Hualalai volcano on the Island of Hawaii contains mantle-derived peridotite xenoliths that are as much as 70 cm in diameter. What minimum ascent velocity would have been required to lift these xenoliths to the surface of the Earth so they would not sink en route? For the acceleration of gravity use 980 cm/s^2 and note that 1 Pa = 10 g/cm s^2. Justify your choice of parameters and explain any assumptions made in the calculation.

8.13   Read Special Interest Box 8.2. Using the Lange density equations in Advanced Topic Box 8.1 and the partial molar volumes in Table 8.1 show that crystals of plagioclase An$_{58}$($\rho = 2.70$ g/cm^3) would float in a tholeiitic basalt melt equivalent to the LZb Skaergaard composition in Table 12.6 because its density at 1160°C and 2160 bars is calculated to be 2.76 g/cm^3. For this calculated melt density, partition the 12.84 wt.% total Fe as FeO listed in Table 12.6 into 2.14 wt.% Fe$_2$O$_3$ and 10.91 wt.% FeO and assume

Table 8.1. Partial molar volumes, thermal expansivities, and compressibilities of oxides at 1673 K and 1 bar. Data from Lange (1994) except for H_2O which is from Ochs and Lange (1997).

i	v_i cm³/mole	dv_i/dT 10^{-3} cm³/mole K	dv_i/dP 10^{-4} cm³/mole bar
SiO_2	26.90	0.00	−1.89
TiO_2	23.16	7.24	−2.31
Al_2O_3	37.11	2.62	−2.26
Fe_2O_3	42.13	9.09	−2.53
FeO	13.65	2.92	−0.45
MgO	11.45	2.62	0.27
CaO	16.57	2.92	0.34
Na_2O	28.78	7.41	−2.40
K_2O	45.84	11.91	−6.75
H_2O	27.75	10.86	−3.82

0.5 wt.% dissolved water. Also show that the same plagioclase would be neutrally buoyant in the same melt but contains about 1.2 wt.% dissolved water.

8.14 Using the "characteristic dimensionless time" of Jaeger (1968) in Section 8.4.1 compare the cooling history of a granite batholith 20 km in diameter, a granite stock 1 km in diameter, and a basaltic dike 0.2 m in width. Discuss assumptions made in this comparison. (*Partial answer:* The dike is almost entirely cooled after 28 hours.)

8.15 Calculate the Rayleigh number, Ra, in terms of $(T_f - T_r)$ for a 200-m-thick horizontal sill of uniform basalt melt. What vertical thermal gradient is necessary for thermal convection to occur? Is this reasonable? Discuss. Assume the values of $g = 9.8$ m/s², $\alpha = 3 \times 10^{-5}$/deg, and $\kappa = 10^{-6}$ m²/s, $\eta = 10$ Pa s, and $\rho = 2.6$ g/cm³.

9

Magma Ascent and Emplacement: Field Relations of Intrusions

FUNDAMENTAL QUESTIONS CONSIDERED IN THIS CHAPTER

1. How does magma rise from its site of generation in the deep crust and upper mantle and move to shallower depths in the solid lithosphere: That is, how does magma *ascend?*

2. How is room created in solid rock for magma to intrude: That is, how is magma *emplaced?*

INTRODUCTION

It has been recognized for at least a century that the fundamental driving force causing magma to rise is its buoyancy—the difference between the density of the magma and its surrounding country rock. Magma generated by partial melting of solid source rock in the deep crust or uppermost mantle is less dense than the surrounding rock and is, therefore, gravitationally unstable and capable of rising. But whether a mass of magma can actually ascend buoyantly depends on the relative magnitude of the resistive force that is dictated by the magma rheology, basically its viscosity. Large volumes of silicic magma are required to provide the necessary buoyant force to overcome the viscous resistive force at the margin of the magma body where it contacts solid country rock. Smaller volumes of less viscous basaltic magma can ascend with greater facility and more speed, even in large surface area dikes that have been created by tectonic forces or by the pressure of the magma itself.

Ascending magma constitutes a classic interaction of the two fundamental but opposing energy sources—thermal and gravitational—within the Earth. Melting at the magma source might proceed without bound until all of the source rock is melted and the causative mechanism (e.g., decompression, volatile influx) stops. But in a gravitational field, the partially melted rock or a segregated partial melt usually becomes sufficiently buoyant to rise out of the source before complete melting occurs. Gravity-driven ascent is commonly arrested as the magma loses heat to the country rocks, becomes more viscous, and stops flowing. Or magma may ascend all of the way to the surface of the Earth and exit as a volcanic extrusion.

During ascent and intrusion of magma, nature exercises its principle of parsimony, taking the path of least work that consumes the least energy.

Magma ascent and final emplacement cannot always be separated and distinguished; rather, they are a dynamic continuum. Thus, a vertical dike through which magma ascended from a deep source also constitutes an intrusion, unless somehow the magma drained out or the crack closed together to eliminate the filling magma.

The challenge for the petrologist is to try to understand the physical and thermal dynamics of ascent and emplacement of a magma body from the field relations, fabric, and composition of a partially exposed and long-dead cold intrusion.

Magma ascent constitutes a significant advective transfer of heat from deep levels of the lithosphere to shallower and, therefore, plays a major role in the thermal evolution of the cooling Earth.

✷9.1 MOVEMENT OF MAGMA IN THE EARTH

9.1.1 Neutral Buoyancy and the Crustal Density Filter

Mafic to ultramafic magmas in the *uppermost* mantle are less dense than the mantle peridotite in which they might have been generated (Figure 8.17). They are, therefore, positively buoyant and can potentially rise. However, above the Moho, crustal rocks are dominantly feldspathic rather than olivine-rich, and consequently their density is much less than the underlying mantle. Whether mantle-derived magmas are positively, negatively, or neutrally buoyant in the crust depends entirely on their T, their P, and especially their composition and the mineralogical and modal composition of the crustal rock (Figures 8.15 and 9.1). Subtle differences of a few tenths of a gram per cubic centimeter in the density of the magma or the crustal rock are sufficient to change the buoyancy from positive to negative.

The density of the oceanic basaltic crust (about 7 km thick) increases with depth because of progressively decreasing vesicularity, closure of pore spaces, and expulsion of water. Mantle-derived, olivine-rich basalt (picrite) magma and crust are of similar density at a depth of about 1–3 km which is, therefore, a **horizon of neutral buoyancy** at which magma may stagnate and accumulate (Ryan, 1994). Lateral enlargement of magma chambers at oceanic spreading junctures may occur at this horizon. Evolved, less dense magma can rise via dikes to the seafloor where it extrudes.

Continental crust is far more heterogeneous than oceanic crust, making generalizations of magma ascent less certain. Beneath a variably thick veneer of sedimentary rock whose density ranges from 2.2 to 2.7 g/cm³, crustal igneous and metamorphic rocks have densities ranging from about 2.6 to 2.9 g/cm³ and averaging approximately 2.7 g/cm³, corresponding to granodiorite to diorite bulk compositions (Figure 9.1). In many places, deeper continental crust has densities of approximately 2.9 g/cm³, corresponding to more

mafic compositions, such as amphibolite. Andesite and less dense silicic magmas are probably positively buoyant in crust of any composition and can rise all the way to the surface. Volatile undersaturated basaltic magmas have densities >2.7 g/cm³ and would not be expected to rise buoyantly through less dense continental crust of granite to granodiorite composition. There is strong geologic evidence (discussed in later chapters) for appreciable **underplating** of the lower continental crust, as well as intrusion into it, by basaltic magmas. Feldspathic crustal rocks can, therefore, serve as an effective **density filter** blocking ascent of denser mafic, mantle-derived magmas. These buoyantly blocked basaltic magmas are believed to be responsible for partially melting the already hot lower crust as they cool and transfer heat. During cooling, they also are likely to fractionate olivine, so their residual melts, which can be less dense (Figure 8.16), may be buoyant enough to rise. Fractionating magmas that contain dissolved, and especially exsolved, volatiles or magmas assimilating silicic crustal material may also have low enough density that they can rise higher into the overlying crust. Repeated stagnation and solidification of mafic magmas within the crust may densify it sufficiently that progressively denser, more primitive basalt magmas may subsequently ascend farther.

Yet, paradoxically, rather primitive, unevolved basalt magmas that have densities greater than that of continental rock are commonly extruded onto the surface as lavas in continental areas. Some magmas have even intruded through porous alluvium (Figure 8.14) whose density can be <2.0 g/cm³. How can this happen? One possibility is volumetric expansion upon melting in the source, which drives the magma upward

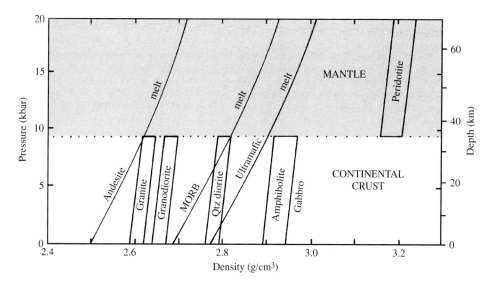

9.1   Density relations between some rock compositions and volatile-free melts in the continental crust and uppermost mantle. Range of densities for rock types is indicated in rectangular boxes. Note the smaller compressibilities of rocks compared to melts as P (depth) increases. MORB, mid-ocean ridge basalt. Amphibolite is a metamorphic rock composed of hornblende and plagioclase that is formed by recrystallization of mafic igneous rocks (basalt, gabbro) under hydrous conditions. (Redrawn from Herzberg et al., 1983.)

toward a lower P regime. However, it is unlikely that this effect is capable of driving the magma very far from its source. Other driving forces are exsolution and expansion of bubbles in volatile-saturated magmas, which can greatly reduce magma density, enhance buoyancy, and cause volcanic eruption. But magmas must have another driving mechanism, independent of volatile exsolution, that can commonly propel them upward through less dense rock. This driving force is magma overpressure.

9.1.2 Magma Overpressure

Consider a lens-shaped body of magma buoyantly blocked at a density contrast in the lithosphere after having ascended from deeper in the lithosphere (left side of Figure 9.2). The pressure in the magma, P_m, is equal to the weight of the rock column overlying the lens, $\rho_r g z_d$, which is the lithostatic (confining) pressure, P, at that depth (Section 1.2). In a deeper body of magma (center of Figure 9.2) at a depth $z_d + z_D$, the magma pressure, $P_m = \rho_r g z_d + \rho_r g z_D$. Now suppose a conduit is accessible from the deeper magma lens all the way to the surface. Ignoring viscous drag of the magma in the conduit and assuming the density of the magma, ρ_m is constant regardless of depth, how far can the magma rise? The P-z (pressure-depth) diagram on the right of Figure 9.2 indicates that the magma will rise to a height h above the surface, satisfying the equality between lithostatic and magma pressure, $P = P_m =$

$\rho_m g(h + z_d + z_D)$. In other words, the magma has an excess hydraulic head, h, due to the load of the *overall* denser rock column on the less dense body of magma at the base of the column. Provided conduit continuity between subterranean magma body and volcanic vent is maintained, less dense magma is pushed out of the ground by the weight of the overall more dense rock column. Theoretically, the deeper the magma column, the greater is the height to which a volcano can grow above the surface; this may have a bearing on the maximum summit heights of Hawaiian and other volcanoes.

In the P-z diagram on the right of Figure 9.2 note that the magma pressure, P_m, *at any depth* exceeds the confining (lithostatic) pressure, P. The magnitude of this **magma overpressure,** or hydraulic head, at any depth z is $(P - P_m)$. Magma overpressure can be sufficient to cause hydraulic fracturing in brittle rock (Figure 8.2c) or, if not, at least can drive magma into and through existing cracks, overcoming viscous resistance. Or overpressure (in effect, buoyancy) can slowly force magma through ductile overlying rock.

Thousands of Cenozoic basaltic extrusions worldwide in extensional tectonic regimes contain xenoliths of dense mantle peridotite. The basaltic magmas must have ascended rapidly (Problem 8.12) through essentially continuous passageways from a mantle source in order to have lifted the xenoliths. Additionally, copious volumes of magma extruded from fissures in continental and oceanic flood basalt plateaus testify to the

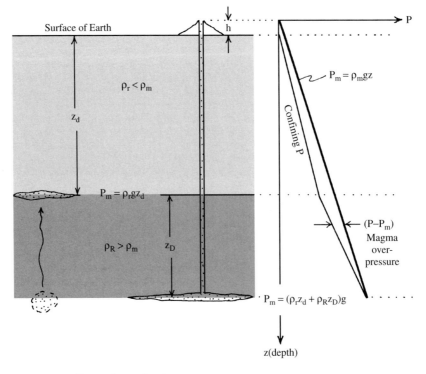

9.2   Origin of magma overpressure. See text for explanation.

tapping of huge subterranean reservoirs and efficient transport to the surface.

9.1.3 Mechanisms of Magma Ascent

Most magmas move upward through solid rock in basically two ways: as diapirs and as dikes. **Diapirs** are bodies of buoyant magma that push slowly through surrounding ductile, highly viscous country rock in the lower crust or mantle. Diapir originates from the Greek verb *diaperien,* "to pierce." The existence and nature of magmatic diapirs are inferred from examination of field relations of intrusive magma bodies, model studies of viscous fluids, and theory. Magma can also rise rapidly through subvertical cracks in brittlely fractured rock as **dikes.** Their existence is a matter of simple observation. Movement of magma through fractures has been tracked seismically. Table 8.1 compares these two end-member processes that operate at vastly different time scales and depend on contrasting magma and host rock rheologies. Ascent of a particular mass of magma can involve either mechanism at different depths. Shallow crustal processes including stoping and "drilling" of gas charged magma are considered further in the section on magma emplacement (Section 9.4).

☀9.2 SHEET INTRUSIONS (DIKES)

9.2.1 Description and Terminology

Sheet intrusions, as the name implies, are tabular bodies having very small **aspect ratios** of thickness/length, generally $10^{-2}-10^{-4}$. A **dike** is a sheet intrusion that cuts discordantly across planar structures, such as bed-ding, in its host rock. Dike also refers to a sheet intrusion hosted within massive, isotropic rock, such as granite. In contrast, a **sill** is a concordant sheet intrusion that parallels planar structures in its host rocks (Figure 9.3). Some geologists define a sheet intrusion, regardless of country rock concordancy, as a sill if horizontal, or nearly so, and as a dike if vertical, or nearly so.

Compositionally diverse dikes and sills are commonly associated together beneath volcanoes and near margins of larger intrusions (Figures 9.4 and 9.5) where they inflate the volume of rock into which they are intruded.

Dike swarms consist of several to hundreds of dikes emplaced more or less contemporaneously during a single intrusive episode. Dikes in swarms may be irregular in orientation, more or less parallel, or **radial,** arrayed in map-view-like spokes of a wheel from a central point (Figure 9.6a). Huge radial swarms (Ernst and Buchan, 1997) are believed to form above mantle plumes associated with continental extension and ocean opening (Figure 9.7). Individual dikes in such swarms can be >2000 km long and tens to rarely 100 meters in width. However, most common dikes are less than 10 m wide; 1- to 2-m-wide dikes are typical.

On a global perspective, most dikes are of basaltic composition and manifest ascent of a vast volume of mantle-derived magma through fractured lithosphere throughout Earth history.

Feeder dikes supply magma to connected sills or other intrusions and overlying volcanoes. The huge radial swarms just mentioned likely fed copious extrusions of lava, perhaps forming vast thick flood

Table 9.1 Comparison of Sheet Intrusion (Diking) and Diapirism in the Continental Crust

ASPECT	SHEET INTRUSION	DIAPIRISM
Most common magma composition	Basalt	Granitic
Rheologic behavior of country rock	Brittle (elastic)	Ductile or viscoplastic
Viscosity contrast between country rock and magma	Many orders of magnitude	A few orders of magnitude
Ascent velocity	0.1–1 m/s	0.1–50 m/y
Time for magma ascent	Hours to days	10^4–10^5 y
Factors controlling ascent velocity	Magma viscosity and density contrast with country rock; dike thickness	Country rock ductile strength and thickness of boundary layer around diapir
Effect of state of stress on path of magma transport	Sheet perpendicular to least principal stress, σ_3	Probably slight
Country rock deformation	Nil	Substantial penetrative ductile, chiefly in boundary layer
Nonmagmatic example	Hydrothermal quartz vein	Salt dome

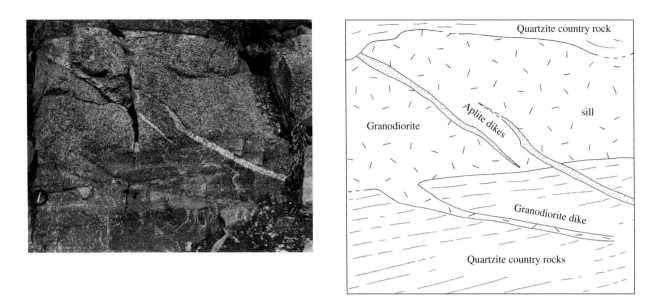

9.3   Sill and dikes. A **sill** of granodiorite intruded concordantly with layering in quartzite host rock and a smaller offshoot **dike** penetrating discordantly across layering. Thin subparallel dikes of leucocratic granite aplite cut across more mafic granodiorite and layered country rock. Note sharp contacts. Camera lens cap in lower left for scale. (a) Photograph. (b) Annotated sketch.

9.4   Sill-dike swarm in subhorizontally layered country rocks surrounding granodiorite pluton, Alta, Utah. Most of the leucocratic aplite-pegmatite granitic rock forms **sills** parallel to layering in darker-colored recrystallized shale, but smaller **dikes** cut discordantly across layers. The original stratigraphic section of shale is approximately doubled in thickness by the inflating sills. Pocket knife 8 cm long for scale in center of photograph.

(a)

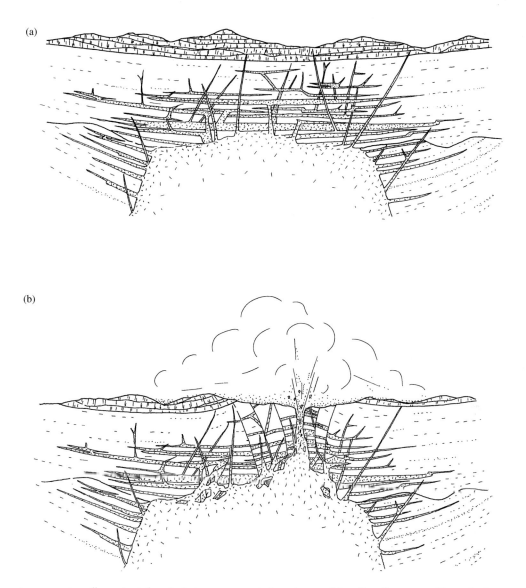

(b)

9.5   Schematic cross sections illustrating inferred relations among main intrusion, overlying dike-sill swarm, and layered volcanic rocks in the Miocene-Pliocene Tatoosh complex in Mount Rainier National Park, Washington. (a) In an early stage in the rise and emplacement of the intrusion, magma is lodged in an overlying dike-sill complex, advectively heating the roof of older volcaniclastic rocks. (b) In a later stage, the main mass of magma has continued to rise by **stoping** into its dike-sill complex and by **doming** the sills, their older host rock, and overlying volcanic rock layers. Note stoped blocks of roof rock in the main intrusion. Magma has broken through to the surface in an explosive eruption. A still later stage can be envisaged in which the still ascending magma intrudes its own volcanic cover. (Redrawn from Fiske et al., 1963.)

(a)   (b)

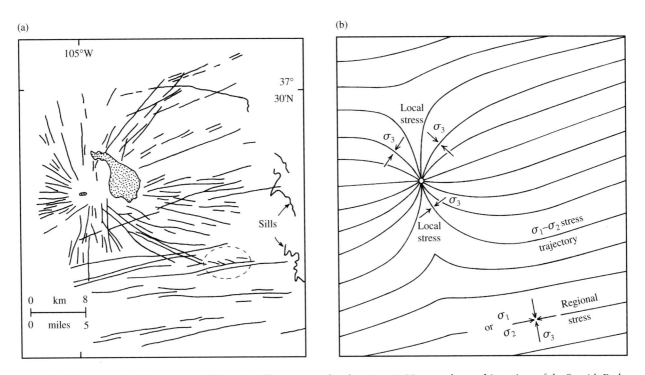

9.6   Radial and parallel dike swarms. (a) Subvertical dikes were emplaced at 28 to 20 Ma around **central intrusions** of the Spanish Peaks (stippled) in south central Colorado. Flow markers (aligned tabular phenocrysts, elongate vesicles) in the dikes indicate the central intrusions as the source of the radially diking magma. Most dikes consist of segments a few meters to several kilometers long; many segments are *en echelon* but cannot be shown on this small-scale map except some unusually well-expressed ones enclosed by the dashed-line ellipse. An origin for *en echelon* dikes is shown in Figure 9.9. (Redrawn from Smith, 1987.) (b) Theoretical stress analysis. Central intrusion (open circle) is responsible for a *local stress* field that allows for radial diking. The central intrusion perturbed a *regional stress* field that controlled emplacement of the mostly older swarm of subparallel east-northeast-striking dikes mainly of more mafic magma. Trajectory lines are traces (intersections) in the horizontal plane of vertical surfaces parallel to σ_1 and σ_2. Because these surfaces are perpendicular to σ_3, they are potential avenues for magma intrusion. Note that most radial dikes are oriented nearly parallel to the regional $\sigma_1 = \sigma_2$ trajectory. (Redrawn from Odé, 1957.)

basalt plateaus. Basalt dikes are also the means for growth of continental and oceanic island volcanoes (such as Hawaii, Figure 9.8) and for growth of oceanic-ridge submarine volcano systems related to seafloor spreading and, therefore, for growth of the entire seafloor itself. **Sheeted dike complexes** formed at ocean ridges in the extending oceanic crust consist of subparallel dikes intruded into older dikes. They testify to long-term prolific diking in the dilating ridge.

Some dikes never reach the surface of the Earth where the magma can "see" the light of day; they are "blind" intrusions.

Basalt **sill swarms** underlie flood-basalt fields and thick plateaus and can have volumes comparable to those of dike swarms. A huge swarm of Jurassic diabase sills, together with an accompanying feeder dike swarm, formed during the breakup of Gondwanaland. Segments of these swarms are found in Antarctica, South Africa (see Figure 13.20), and Tasmania, where individual sills are as much as 300 m thick and the swarm segment crops out over 25% of the 65,000-km² surface area of the island (Walker, 1993). The famous Carboniferous Whin sill in the British Isles has an areal

extent of >5,000 km² and an average thickness of 40 m. Sills accompany the dike swarm on continental margins in the mid-Atlantic (Figure 9.7b). Some sills preserve evidence of episodic replenishment during growth. Other sheet intrusions include cone sheets and ring dikes described later.

9.2.2 Some Thermomechanical Concepts Pertaining to Emplacement of Sheet Intrusions

Magma can invade existing fractures in shallow crustal rock if the normal stress perpendicular to the fracture is less than the pressure exerted by the magma and if that pressure is sufficient to overcome the resistance to viscous flow. However, most dikes, even in the mantle and ductile lower crust, are probably created as the magma itself rapidly stresses and fractures the rock (Shaw, 1980) and fills the propagating crack as it advances (Plate VI). This is the hydraulic fracture mechanism shown in Figure 8.2c. The work to fracture rock is less than that required to push magma through a crack, especially for viscous magma. The stress at the tip of the opening crack is sufficient to continue the fracturing process. Evidence for magma-generated

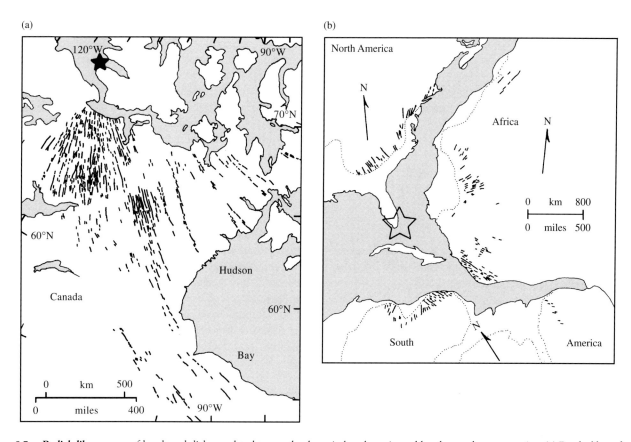

9.7   Radial dike swarms of basalt and diabase related to mantle-plume-induced continental breakup and ocean opening. (a) Basalt dikes of the gigantic 1.27-Ga radial Mackenzie swarm northwest of Hudson Bay, Canada, associated with opening of a middle Proterozoic ocean. Flow markers in the dikes show that the direction of magma transport was subvertical within 500 km of the postulated mantle-plume source (star) but subhorizontal at distances up to 2000 km from the source. Dikes near the source fed a flood basalt province and the large differentiated Muskox intrusion (see Figure 12.17). (Redrawn from Earth and Planetary Science Letters, v. 96, A. N. LeCherminant and L. M. Heaman, Mackenzie igneous events, Canada: Middle Proterozoic hotspot magmatism associated with ocean opening, pp. 38–48, 1989 with permission of Elsevier Science NL, Sara Burgerharstratt 25, 1055 KV Amsterdam, The Netherlands.) (b) Radial swarm of Triassic-Jurassic dikes along the margins of the North American, African, and South American continents. (Redrawn from May, 1971; see also Puffer and Ragland, 1992.) Continents have been restored to an early Mesozoic, predrift configuration. The swarm indicates a possible mantle-plume source at the southern tip of the Florida peninsula (star) (Ernst and Buchan, 1997). Dotted lines show the inland contact of post-Jurassic sedimentary deposits, which likely conceal many additional dikes.

fractures can be found in the time-space relations of dikes, such as a radial swarm centered around a contemporaneous central intrusion (Figure 9.6) and in the absence of similarly oriented joints in the country rocks far from the dikes.

The magma flow velocity through a dike is (Rubin, 1995)

$$9.1 \quad v = \left(\frac{w^2}{3\eta}\right)\left(\frac{dP_m}{dz}\right) = \frac{w^2 g \Delta\rho}{3\eta}$$

where w is the width of the dike, g is the acceleration of gravity, and η is the magma viscosity. The vertical magma pressure gradient, dP_m/dz, that drives ascent equals $\Delta\rho g$ where $\Delta\rho$ is the density contrast between magma and host rock. A pressure gradient of only 0.1 bar/km can maintain a flow velocity of almost 1 m/s (about as fast as a person can walk) in a 5-m-thick basalt dike (Spera, 1980). Doubling dike thickness al-

lows magma to flow four times as fast because there is proportionately less viscous resistance along dike walls for movement of a larger volume of magma.

The distance magma can be transported through a dike from its source depends critically on the competition between two rates:

1. The rate of magma cooling, and accompanying viscosity increase, by conductive (Figure 8.18) and advective heat transfer into the wall rocks
2. The rate of magma flow

Near source, the magma still contains most of its thermal energy and flows readily; but "downstream" farther from the source an increasing amount of heat has been lost to the wall rocks, increasing viscosity and arresting flow. One way to model the distance of magma transport, d, is by multiplying the characteristic conductive cooling time by the magma flow velocity (equation 6.7 by 9.1)

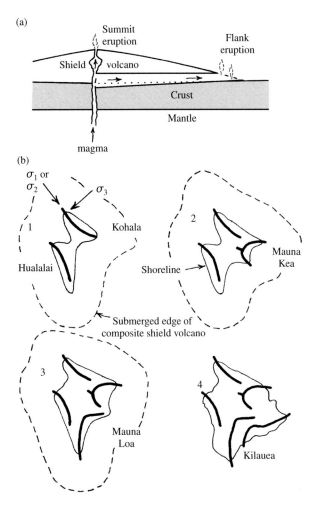

(a)

(b)

9.8 Evolving magma injection paths during growth of the five basalt shield volcanoes that the exposed part of the island of Hawaii comprises. (a) Schematic cross section of a Hawaiian basalt shield volcano built of porous lava flows and minor intercalated clastic deposits. Basalt magma rises in a subvertical central conduit through the uppermost mantle and lower crust. Some magma spreads laterally along subvertical extensional fissures in a **rift zone** and feeds a flank eruption. (b) Schematic growth of the five coalesced shield volcanoes from oldest to youngest (1 through 4) by rift-zone controlled eruptions (thick lines). After the composite Kohala and Hualalai shield had formed, the orientation of subsequent extensional rifts was governed by local rather than regional stresses. The flanks of the shield are gravitationally unstable and pull away from the center, forming arcuate extensional fractures, like gigantic landslides. (From Shaw, HR. The fracture mechanisms of magma transport from the mantle to the surface. In: Hargraves RB, ed. Physics of magmatic processes. Princeton, NJ: Princeton University Press, 1980:201–264. Copyright © 1980 by Princeton University Press. Reprinted by permission of Princeton University Press.)

$$9.2 \quad d = \left(\frac{w^2}{\kappa}\right)\left(\frac{w^2 g \Delta\rho}{3\eta}\right) = \frac{w^4 g \Delta\rho}{\kappa 3\eta}$$

Therefore, the distance of magma transport is sensitive to dike width, w (to the fourth power), and viscosity, η. Therefore, in terms of heat loss and solidification, doubling the thickness of a dike is equivalent to moving it

16 times closer to the source of the magma, all other factors being the same (Delaney and Pollard, 1982). Because magma viscosities range over many orders of magnitude, this factor also strongly influences the effectiveness of dike transport of magma, as higher viscosities reduce transport distance. But increasing viscosity can be compensated for by widening the dike. Wada (1995) found that the thickness of 44 dikes in Japan and Peru correlates with apparent viscosity (calculated from magma composition), but not exactly as predicted by equation 9.2. Mafic magmas whose viscosities were 10–100 Pa s formed dikes 1 m thick, whereas felsic magmas whose calculated apparent viscosities were 10^6–10^7 Pa s formed dikes 100 m thick. Crystal-rich, water-poor granitic magmas, which have viscosities several orders of magnitude greater than 10^6–10^7 Pa s, probably move distances measured in kilometers only if the conduit is several kilometers wide. Whether such magma intrusions can be called dikes at all is debatable, as this is the probable order of magnitude of diapir diameters (discussed later). Silicic dikes less than a meter or so in thickness are uncommon, except in the immediate vicinity of larger granite plutons. Granitic counterparts of the huge basalt dike swarms (Figure 9.7) do not exist.

Aplite dikes (Figures 7.48 and 9.3) are virtually ubiquitous within, or are closely associated with, more mafic granitic plutons and deserve special mention. Compositional and textural observations indicate that these fine-grained phaneritic and leucocratic dikes originate from minimum T (Figures 5.24–5.27) residual melts sucked into self-generated extensional fractures in the cooling and contracting, mostly crystalline host magma body. Although the dike walls are planar and the contact apparently sharply defined between the aplite and the coarser-grained, less evolved host, close examination reveals interlocking unfractured crystals across the textural-compositional contact (Hibbard and Watters, 1985). Hence, the dike host rock behaved in a brittle manner (Section 8.2.2) so it could fracture yet was actually a crystal-rich mush with a small percentage of interstitial residual melt. In terms of equation 9.2, the typical aplite dike thickness of a few centimeters is possible because the residual, virtually crystal-free melts are enriched in water, so their viscosities are not excessive, but also because the melts probably migrated through the dike no more than several meters.

9.2.3 Geometry and Orientation of Sheet Intrusions

The orientation of self-induced, sheetlike magmatic pathways through the crust is governed by the state of stress. According to nature's least work principle, hydraulic tensile fracturing by overpressured magma creates extensional fractures parallel to σ_1 and σ_2 (Figure 8.2c), which open in the direction of the least compressive principal stress, σ_3. These magma-filled dilatant cracks, therefore, serve as paleostress indicators. Magma transported upward from deep sources, such as mantle-derived basalt magma intruded into near-vertical dikes in the crust, implies that σ_3 is horizontal, the typical stress orientation above mantle plumes and in other regimes of tectonic extension. Swarms of subparallel, subvertical dikes exposed over large areas indicate a uniform regional state of extension in the crust at the time of intrusion.

Dikes commonly occur in segments that may be arrayed *en echelon,* one explanation for which is a shift in the orientation of σ_3 with depth (Figure 9.9).

Sheet intrusions are common in shallow crustal, subvolcanic environments, where they surround and overlie a more massive **central intrusion** (Figure 9.6a). These more or less upright, bottle-shaped or cylindrical central intrusions can perturb a uniform *regional*

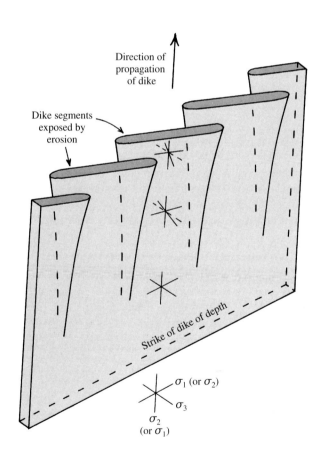

Direction of propagation of dike

Dike segments exposed by erosion

Strike of dike of depth

σ_1 (or σ_2)

σ_3

σ_2 (or σ_1)

9.9 Schematic three-dimensional form and origin of subvertical **en echelon dike** segments. At some depth the dike is an unsegmented sheet intrusion whose orientation is controlled by the nonhydrostatic state of stress indicated by the thin-line orthogonal principal stresses below the dike. At progressively shallower depths the orientation of the least principal horizontal stress, σ_3, rotates progressively counterclockwise, as shown in the upper part of the dikes. Consequently, the least-work dike configuration there is an *en echelon* system of dike segments. Note that the other horizontal principal stress must also rotate. (Redrawn from Delaney and Pollard, 1981.)

state of stress, which is dictated by tectonic environment, because of their buoyant magma pressure, compounded by thermal expansion of the wall rocks. The resulting superposed *local* state of stress is spatially variable in orientation around the intrusion. Sheet intrusions created in this perturbed stress regime by magma supplied from the central intrusion include radial dikes, cone sheets, and ring dikes.

Formation of Radial Dikes and Cone Sheets. The local state of stress laterally *around* a central magma intrusion differs from the state *above* it. Around an intrusion swelling by magma overpressure, wall rock is compressed so that the maximum horizontal compressive stress, σ_1, is oriented perpendicular to the wall rock-magma contact. The least compressive stress, σ_3, which can be tensile for high magma pressures, is oriented horizontally and tangentially to the contact in the stretched wall rock. Trajectories of planes parallel to σ_1 and σ_2 and perpendicular to σ_3 are accordingly arrayed vertically and radially around the subvertical central intrusion (Figure 9.6b) so that magma-filled extensional fractures form a **radial dike swarm** centered at the central source intrusion. Farther from the influence of the central intrusion, stress trajectories assume a regional orientation and, if this far-field state of stress is uniform, dikes become subparallel.

Above the central intrusion, trajectories of curved surfaces representing planes parallel to σ_1 and σ_2 define the orientation of potential concentric conical extensional fractures perpendicular to σ_3. Magma driven along these conical fractures from the apex of the central intrusion form one or more commonly concentrically nested **cone sheets** (Figure 9.10). Because of their inward dip, the magma intruded into a cone sheet elevates the segment of roof rock inside the cone. Considerable buoyancy-related magma pressure is required to accomplish this work against gravity.

The orientation of radial dikes and cone sheets is governed by two principles enunciated by Anderson (1951):

1. Both the surface of the Earth and the country rock–magma contact are solid-fluid interfaces that cannot support any shear stress; they are, therefore, principal planes, and local principal stresses must be normal to them (Figure 9.10).

2. Farther away from these two interfaces, principal stresses must bend to conform to the regional state of stress.

These geometrical constraints can be seen in Figure 9.10 and especially Figure 9.6b. The geometry of cone sheets differs from that of radial dikes because the magma pressure pushes upward against the free surface of the Earth above a central intrusion but sideways against confined rock around the intrusion.

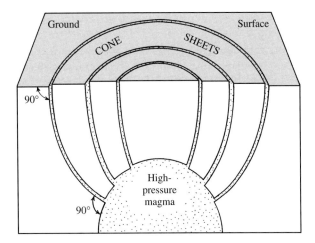

9.10 Idealized geometry of **cone sheets** above a shallow, forcefully intruded **central magma intrusion**. Inward-dipping cone sheets develop as high-pressure magma in the central intrusion invades conical extensional fractures that follow $\sigma_1 = \sigma_2$ trajectories above the apex of the intrusion. This geometry applies to situations in which the depth to the top of the intrusion is comparable to its width; in such cases, some of the magma commonly extrudes and the intrusive complex is referred to as *subvolcanic*. Many intrusions are emplaced farther beneath the surface and for this reason and other factors do not have cone sheets.

The fact that many central intrusions never create radial dikes or cone sheets must reflect something of the particular nature of the roof rock and magma overpressure in intrusions.

<u>Effects of Topography on Dike Configuration.</u> Regional states of stress can also be perturbed in topographically high land masses, such as large volcanoes. So-called gravitational stresses in the huge shield volcanoes that form the island of Hawaii create a reorientation of intravolcano stresses relative to regional intraplate stresses. Extensional fissures (called *rifts* by Hawaiian geologists) through which magma rises thus become reoriented during island growth (Figure 9.8).

<u>Ring Dikes.</u> During the history of a magma intrusion the pressure in the chamber is likely to vary. After an initial overpressured state sufficient to cause magma ascent and possibly produce radial dikes and cone sheets, pressure may decrease. In evolving magma chambers this may be the result of release of exsolved volatiles or simply contraction during cooling and crystallization. Partial intrusion into nearby country rock or extrusion of magma can create potential voids in the partially evacuated chamber. With loss of supporting magma pressure, the roof over the partially evacuated chamber can subside, creating a topographic depression known as a **caldera**. (This is sometimes called a **cauldron** and the foundering process **cauldron subsidence**.) The **ring fault** (in some places a ring-fault zone) bounding the subsiding roof slab more or less follows the outline of

the chamber. Magma can well up between the subsiding roof block and the undisturbed country rock, forming a **ring dike** (Figure 9.11). This is an arcuate, subvertical sheet intrusion that, in some cases, may form a complete 360° circular structure. In order for the denser roof rock to subside intact into the less dense, low-pressure magma, its margin must be vertical or dip outward. (Inward dips have been noted in some ring dikes, but such dips require special conditions for roof subsidence to compensate for diminished downward diameter of space into which the roof block can subside.)

More deeply eroded calderas expose increasing proportions of plutonic rock, including the ring dike, relative to extruded volcanic rock (see Figure 13.36).

9.2.4 Basalt Diking in Extensional Regimes

As a broad generalization, extruded basalt magmas are widespread in extensional tectonic regimes, whereas andesitic, or at least intermediate-composition calc-alkaline, extrusions predominate in compressional regimes. In the Basin and Range province of western North America, middle Tertiary andesitic volcanism was gradually supplanted by late Cenozoic basaltic activity as the tectonic regime changed from compressional to extensional. Magma extrusions are discussed in the next chapter, but the way basalt magmas ascend to the surface is relevant here.

Extensional and compressional tectonic regimes have fundamentally different states of stress. In extensional regimes, the lesser two principal stresses, σ_2 and σ_3, are horizontal and the greatest, σ_1, is vertical. In compressional regimes, the greater two principal

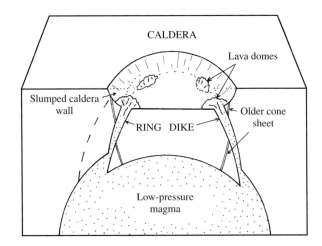

9.11 Hypothetical **ring dike** and **caldera** above a shallow magma chamber. Postcollapse caldera fill consists of landslide debris that is produced by slumping of the unstable caldera wall and epiclastic (sedimentary) deposits shed off the eroding caldera wall. Magma rising in the ring dike locally extrudes to form lava domes or flows; their vents mark the position of the usually concealed underlying ring dike and the ring fault it followed.

stresses, σ_1 and σ_2, are horizontal and the least, σ_3, is vertical. These contrasting states have different consequences for magma ascent and evolution in the lithosphere, as shown in Figure 9.12. In this figure, the vertical principal stress, σ_v, either σ_1 (in extensional regimes) or σ_3 (compressional), equals $\rho_r gz = P$ where the density of rock, ρ_r, is assumed to be constant with depth.

In a compressional regime, where σ_3 is vertical, the *minimum* horizontal principal stress, σ_{Hmin}, is σ_2. Its magnitude is fixed by the brittle and ductile rock strengths, as in Figure 8.9. In the lower part of the brittle crust in Figure 9.12a, the straight heavy line representing the magma pressure (compare Figure 9.2), Pm, is less than σ_2. Therefore, magma cannot invade this part of the crust by vertical diking and ascent is prevented. Magma can, however, spread horizontally in sills, lifting or inflating the overlying crust against σ_3 (Figure 9.12c). Sills of mantle-derived basalt magma can evolve into less dense differentiates and these evolved magmas can then ascend. Although vertical diking would be precluded in this depth interval for the relative values of σ_{Hmin} and P_m in the diagram, variations in rock strength, geothermal gradient, and magma density might locally permit vertical diking in compressional settings, as actually observed in some instances.

In contrast, in an extensional regime, with σ_1 vertical, the *minimum* horizontal principal stress, σ_{Hmin}, is σ_3. Its magnitude, fixed by the brittle and ductile rock strengths, is less than σ_v and much less than the magma pressure. Therefore, magma can vertically dike through this part of the crust (Figure 9.12b). The magma overpressure is sufficient to create extensional fractures, independently of any additional contributing motive force, such as might be provided by expanding exsolved volatiles, and regardless of how much low-density material, such as alluvium, might exist at the surface. These deductions agree with geologic observations.

In the ductile regime in Figure 9.12a, sustainable stress differences, that is, rock strengths, are small and the state of stress is essentially hydrostatic; thus, the two horizontal principal stresses equal the vertical, σ_v and $\sigma_1 = \sigma_2 = \sigma_3 = \sigma_v = P = \rho_r gz$. Fracturing can occur accompanied by vertical magma ascent.

Horizontal sills of basalt are commonly intermingled with essentially contemporaneous vertical dikes that must have served as magma feeders. Such coexisting dikes and sills would seem to defy the concept that state of stress dictates the orientation of sheet intrusions, as just discussed. However, prolonged vertical diking and magma inflation in vertical dikes can produce an interchange or switching of local principal stresses (Parsons et al., 1992), which allows emplacement of magma in horizontal sills (Figure 9.13). Alternatively, vertical dikes passing through layers of weak rock, such as shale sandwiched between stronger sandstone, can balloon into them, forming sills.

State of Stress and Petrotectonic Association.

In extensional stress regimes of continental rifts, basalt is one of the most common rock types, and in oceanic

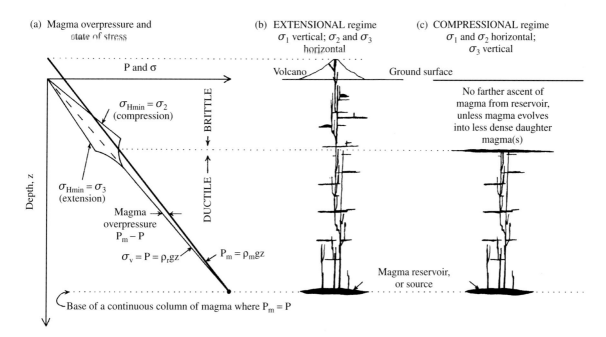

(a) Magma overpressure and state of stress

P and σ

$\sigma_{Hmin} = \sigma_2$ (compression)

$\sigma_{Hmin} = \sigma_3$ (extension)

Magma overpressure $P_m - P$

$\sigma_v = P = \rho_r gz$

$P_m = \rho_m gz$

Depth, z

BRITTLE

DUCTILE

Base of a continuous column of magma where $P_m = P$

(b) EXTENSIONAL regime
σ_1 vertical; σ_2 and σ_3 horizontal

Volcano

Ground surface

Magma reservoir, or source

(c) COMPRESSIONAL regime
σ_1 and σ_2 horizontal; σ_3 vertical

No farther ascent of magma from reservoir, unless magma evolves into less dense daughter magma(s)

9.12 Schematic relations between static magma pressure and state of stress in the lithosphere that govern magma ascent and stagnation. See also Marrett and Emerman (1992). It is assumed for simplicity that there is no reduction in magma pressure due to viscous loss that occurs during upward flow in a dynamic system and that the tensile strength of rock is nil.

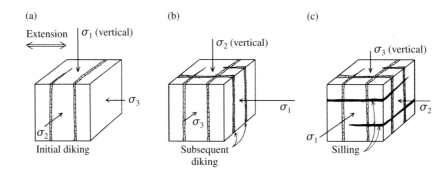

9.13 Intrusion of horizontal sills in an extensional tectonic setting after vertical diking. (a) Initial intrusions are vertical dikes perpendicular to horizontal least compressive principal stress, σ_3. Wedging of magma reinforced by thermal expansion of heated wall rocks increases the stress perpendicular to dikes so that σ_3 becomes σ_1 in (b). Relative magnitude of other two principal stresses remains the same: Vertical σ_1 becomes σ_2 and horizontal σ_2 becomes σ_3. In this new state of stress, additional magma is emplaced in vertical dikes perpendicular to initial ones. After this subsequent diking, magnitudes of principal stresses are again interchanged to yield a third state of stress in (c) where σ_3 is now vertical, allowing horizontal sills to develop as more magma is introduced. The sills lift the overlying crust against gravity. (Redrawn from McCarthy and Thompson, 1988.)

rifts basalt is virtually the only rock type. On the other hand, along continental margins subject to compressional states of stress at convergent plate junctures basalt is commonly subordinate to andesite and more silicic rock types, especially where the crust is thicker. Although reasons for this contrast will be considered further in later chapters, it can be noted here that in compressional regimes (Figure 9.12c) basalt magma that is stalled in the lower to middle crust has ample opportunity to diversify into more evolved, more silicic magmas, including andesite. Basalt magma can crystallize and yield more evolved residual melts, it can partially melt surrounding country rock as a result of the transferred heat and create silicic magmas from that rock, and it can assimilate this rock or mix with its partial melts. These diversification processes are less common, but by no means absent, in extensional settings, where the basalt magma tends to be erupted rather than stagnating in the crust.

Hence, states of stress related to plate tectonic motion have a profound impact on the associated types of magmatic rocks. This is one facet of the concept of **petrotectonic associations.**

✳9.3 DIAPIRS

Diapirs and plumes are terms used for bodies of buoyantly rising material. Mantle plumes are long-lived columns of ascending less-dense mantle *rock*. Diapir is used for columns of rock salt (also called *salt domes*) in sedimentary basins and for bodies of magma rising in the lithosphere. Diapirs of felsic magma ascending in the continental crust are emphasized here. Diapirs have been the subject of numerous laboratory model studies (e.g., Ramberg, 1981) and theoretical and numerical analyses. Field relations of some granitic intrusions are believed to be compatible with emplacement as diapirs (Figure 9.14).

Any layer of less dense material overlain by denser material is gravitationally unstable. The upper boundary of the less-dense unstable layer develops sinusoidal bulges, known as *Rayleigh-Taylor instabilities,* which grow until the density inversion is stabilized in some way (Figure 9.15). As the bulges extend upward, smaller ones may die whereas larger ones grow and separate from the "mother" layer, forming diapirs that continue buoying toward the surface. The rise of thunderclouds from heated near-surface air on hot summer days is an example of this buoyant instability. The wavelength of the bulges in the low-density layer depends upon the thickness of the layer and the viscosity and density contrasts with the overlying denser layer. Thus, diapir spacing provides insight into these parameters (Lister and Kerr, 1989).

The velocity of ascent and diapir longevity are complex functions of many parameters, not the least of which are diapir shape and size. A magma diapir rising through ductile crust must have a relatively large ratio of volume to surface area so that the buoyant body force—a function of its volume—is maximized, whereas the resistive drag force—a function of its surface area—is minimized. Therefore, the "ideal" diapir shape is a perfect sphere, which also happens to be the most thermally retentive for conductive heat loss. But a sphere is only approximated in nature, because, among other possible factors, the drag during ascent maintains a tail in the wake of the rising sphere. The relative magnitudes of the driving and resistive forces determine the ascent velocity of the diapir, as for solid particles in a viscous material (Stokes's law, equation 8.8). Thus, doubling the mean diapir diameter can increase ascent velocity by a factor of 4 if other factors remain constant. Resistive drag also depends on the rheology and thickness of a boundary layer of thermally perturbed country rock adjacent to the ascending diapir (Marsh, 1982). More heat transferred from the magma into the

9.14 Elliptical plutons (diapirs?) in the Archean Pilbara craton, Western Australia. Light colored 3.4 and 3.0 Ga granitic rock is surrounded by darker colored 3.5 to 3.0 Ga metamorphic greenstone belts. Overlying these rocks in marked angular discordance is a gently dipping cover sequence deposited 2.7 to 2.4 Ga visible on the east (right). Indian Ocean is to the north. Area is about 400 km wide. Image furnished through the courtesy of Clive A. Boulter, University of Southampton, UK, and provided by the Australian Centre for Remote Sensing (ACRES), AUSLIG, Canberra and SPOT Imagine Services, Sydney and digitally enhanced and produced by Satellite Remote Sensing Services, Department of Land Administration, Perth, Western Australia Copy Licence 629/2000.

boundary layer can reduce its ductile strength and possibly induce partially melting, allowing the diapir to slip through the country rock with greater ease. But the amount of available thermal energy in the diapir is finite. Small-volume diapirs and those with larger surface area would be expected to stall sooner than larger subspherical ones. Even large diapirs may stall at a crustal level where the ductile strength increases exponentially (Figures 8.8 and 8.9). In the overlying stronger brittle layer, other mechanisms of movement and emplacement of large bulbous masses of magma must come into play, as discussed later.

Because of the transferred heat into country rocks, an ascending diapir leaves in its wake higher-T, softened country rock. Subsequent diapirs can rise significantly faster and farther in this thermally perturbed, preconditioned column of rock. This may explain why major centers of magmatism commonly have lifetimes of several million years. As one ascending mass of magma is intruded and stalls, another follows in its wake.

Some geologists doubt that diapirs of highly viscous felsic magma are thermally and physically capable of ascending very far in the continental crust.

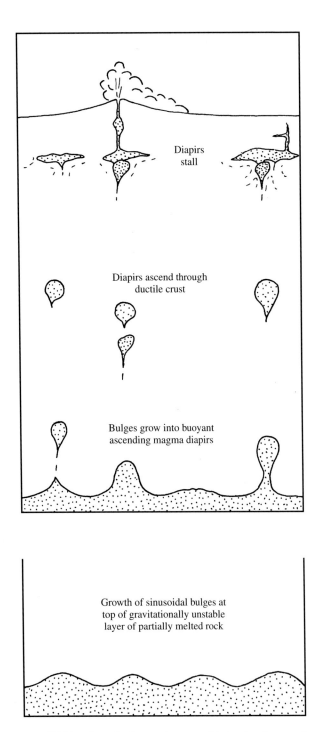

Diapirs
stall

Diapirs ascend through
ductile crust

Bulges grow into buoyant
ascending magma diapirs

Growth of sinusoidal bulges at
top of gravitationally unstable
layer of partially melted rock

9.15 Highly schematic diagram (not to scale) showing the growth, ascent, and stalling of buoyant magma **diapirs.** Beginning at bottom of diagram, a layer of partially melted rock (magma) in source region in upper mantle or lower crust of lesser density than the overlying rock develops sinusoidal Raleigh-Taylor instabilities. In next higher frame of diagram, these bulges grow and separate from the source layer, forming inverted "teardrop"-shaped diapirs of magma that ascend through denser ductile country rock, as do hot-air balloons rising into the atmosphere. Eventually (top of diagram), magma diapirs stall at a density barrier or where they encounter stronger brittle rock. Subsequent diapirs may follow in the wake of earlier ones. Some magma may erupt.

Doubts stem, at least in part, from the theoretical models, which depend strongly on parameters chosen and simplified boundary conditions assumed. Further doubt stems from the lack of unequivocal field evidence for vertical movement. Except in unusual instances, such as where a major sector of the crust has been tilted substantially after magma emplacement, the geologist can see only a subhorizontal section of limited vertical extent through an intrusion and its wall rock. Complete exposures from the top of an alleged diapir to its tail are not seen. In these vertically limited exposures (2 km of relief is exceptional), the only evidence for diapirs lies in highly strained country rock immediately surrounding the magma intrusion, separating the generally less deformed rock within it from more distant country rock. In overly simplified terms, the question facing the geologist is whether the ductile deformational fabric in the country rock (Cruden, 1990) is the result of passage of an ascending diapir or of ballooning of a body of magma that ascended in some other manner, such as by unexposed dikes. However, the paucity of felsic dikes at all levels of the crust but widespread occurrence of more equidimensional intrusions having small aspect ratios strengthen the case for felsic diapirs.

✳9.4 MAGMA EMPLACEMENT IN THE CRUST: PROVIDING THE SPACE

Once magma generated in deep sources has ascended to shallower depths, final emplacement occurs at a particular position within the lithosphere. The concern in this section is how space is provided for nonsheet intrusions, which are referred to as **plutons.**

On a lithospheric scale, mantle-derived magma can be emplaced into the crust by thickening it, displacing the Moho downward to replace the volume of melted mantle source rock and/or lifting the surface of the Earth. Room for magma generated in the lower crust and emplaced in the upper crust involves an exchange in position of material, rock for magma. Deep magma moves up and displaces shallow crustal rock. At an observed level of exposure, creation of the space that is now occupied by hundreds or tens of thousands of cubic kilometers of magmatic rock is a nontrivial **"room problem."** For example, the belts of batholiths that characterize orogenic subduction zones, such as the Cordillera of western North America (Figure 9.16), involve intrusion of a vast amount of granitic magma into the presently exposed crustal level, an estimated 10^6 km^3 in the case of the Mesozoic Sierra Nevada batholith of California (Paterson and Fowler, 1993). Intrusion into orogenic zones that are typically under a compressional state of stress only compounds the paradox.

However, the room problem diminishes if the entire intrusion is considered from a three-dimensional per-

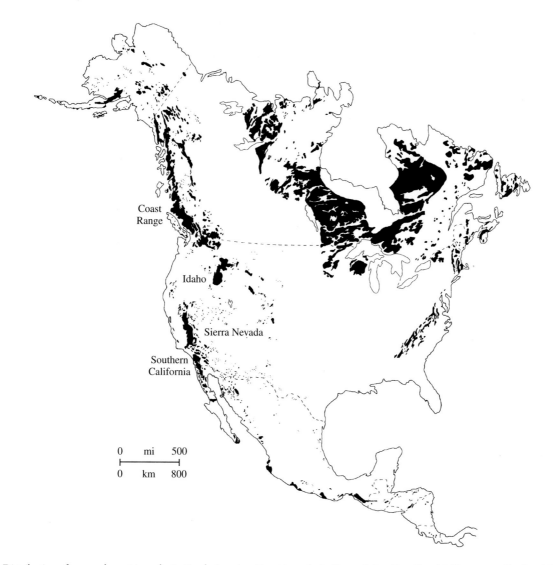

9.16 Distribution of exposed granitic rocks in North America. Granitic rocks in Precambrian Canadian shield are generalized and include some metamorphic rocks. Rocks in the Appalachian orogen along the U.S. East Coast are mostly Paleozoic. Labeled **batholiths** along the west coast in the Cordilleran orogen are mostly Mesozoic. (Redrawn from the Tectonic Map of North America, U.S. Geological Survey.)

spective. The view that plutons extend indefinitely downward with constant horizontal dimensions into the deeper crust, lacking a floor, is unquestionably flawed. Large areal extent at a particular level of erosion does not imply large vertical dimension (e.g., Figure 9.17). Perceived difficulties in accounting for the space occupied by a pluton at a particular level of exposure may be reduced if all three dimensions of it and the surrounding country rocks can be examined.

9.4.1 Some Aspects of Granitic Plutons

Careful studies reveal that very few, if any, plutons are truly homogeneous in composition. This internal inhomogeneity is expressed in composite and zoned plutons. **Composite intrusions** have compositionally and/or texturally distinguishable parts reflecting emplacement of two or more contrasting magmas. In

many instances, an appreciable time elapses between successive intrusions, as indicated, for example, by a chilled, finer-grained contact of the later intrusion against the earlier colder one. In other cases, subsequent magma may have been intruded before the first intrusion one cooled very far below its solidus T, so that their contact shows less evidence of a thermal contrast. **Zoned intrusions** have more or less concentrically arrayed parts of contrasting composition. In normally zoned plutons, more or less concentric parts are successively less mafic inward (Figures 9.18 and 9.19). Normal zoning might develop, for example, as a diorite magma diapir stalls at a particular level in the crust and, in its thermal wake, slower-moving, more silicic and viscous granodiorite and then granite magma diapirs are intruded, inflating the diorite envelope. These successive surges create gradational contacts due to in-

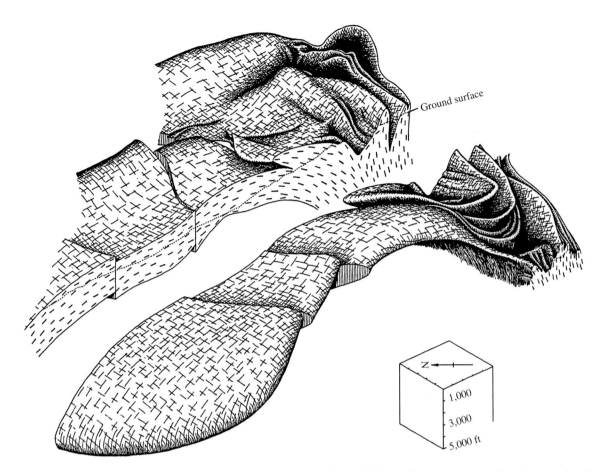

Z ← →
1,000
3,000
5,000 ft

9.17   Three-dimensional projection of the Rattlesnake Mountain granitic pluton, California. The projection is split and separated to show internal structure of the pluton and position of ground surface. The wrinkled parts of the intrusion are screens of hornblende-quartz diorite embedded in more felsic rock. (From MacColl RS. Geochemical and structural studies in batholithic rocks of southern California. Part I. Structural geology of the Rattlesnake Mountain pluton. Geol. Soc. Am. Bull. 1964;75:805–822. Reproduced with permission of the publisher, The Geological Society of America, Boulder, Colorado, USA. Copyright © 1964 Geological Society of America.)

teractions between the incompletely crystallized magmas. Normal zoning might alternatively develop by assimilation of mafic wall rock or by sidewall crystallization processes in a homogeneous magma in which higher-T mafic minerals crystallized preferentially near the cool wall rock margin, allowing the felsic constituents in the magma to concentrate upward and in some manner inward. In some cases, a reverse zoning is evident in more mafic pluton interiors.

A **batholith** is a pluton or commonly groups of separately intruded plutons exposed over generally tens of thousands of kilometers (Figure 9.16). The composite Sierra Nevada batholith in California consists of hundreds of intrusions emplaced over about 130 million years (late Triassic to late Cretaceous) that range in composition from gabbro to granite (mostly granodiorite). Individual plutons crop out over areas ranging from <1 km^2 to >10^3 km^2; their average volume is about 30 km^3.

Plutons smaller than batholiths, commonly consisting of only a single intrusion, are called **stocks;** their

outcrop area is generally <100 km^2. Features related to stocks and batholiths are shown in Figure 9.20.

9.4.2 Emplacement Processes and Factors

This section deals chiefly with emplacement of felsic plutons. Mafic magmas are mostly emplaced in sheet intrusions.

Several emplacement processes have been identified. Some space is created in country rocks through dewatering, removal of chemical constituents by migrating solutions, and growth of denser minerals during metamorphism of country rocks around the intrusion. However, these volume-reduction processes probably only contribute a small fraction of the total space occupied by an intrusion. Other more significant mechanisms (Figure 9.21) include the following:

1. Stoping: pieces of country rock that are physically incorporated into the magma, these xenoliths may be chemically assimilated ("digested") to varying degrees as well

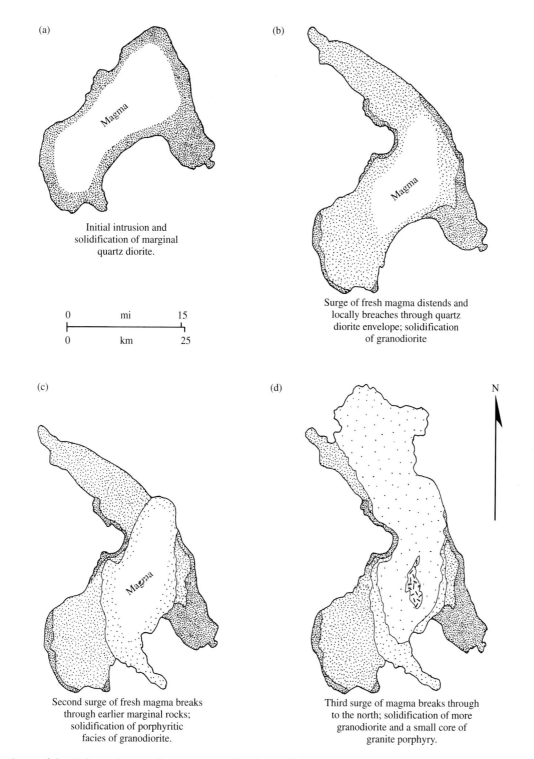

(a)

Magma

Initial intrusion and
solidification of marginal
quartz diorite.

0   mi   15
0   km   25

(b)

Magma

Surge of fresh magma distends and
locally breaches through quartz
diorite envelope; solidification
of granodiorite

(c)

Magma

Second surge of fresh magma breaks
through earlier marginal rocks;
solidification of porphyritic
facies of granodiorite.

(d)

N

Third surge of magma breaks through
to the north; solidification of more
granodiorite and a small core of
granite porphyry.

9.18 Evolution of the Tuolumne Intrusive Series, a compositionally **zoned pluton** within the Sierra Nevada batholith, California. Mantle-derived basalt magmas contaminated by increasing amounts of partial melts of the lower continental crust were intruded into the shallower crust over a time span of several million years (Kistler et al., 1986). See also Table 13.8. (a–d) Schematic sequence of events during growth of pluton. (e) Compositional variations along a west-east line across the pluton. (Redrawn from Bateman and Chappell, 1979.)

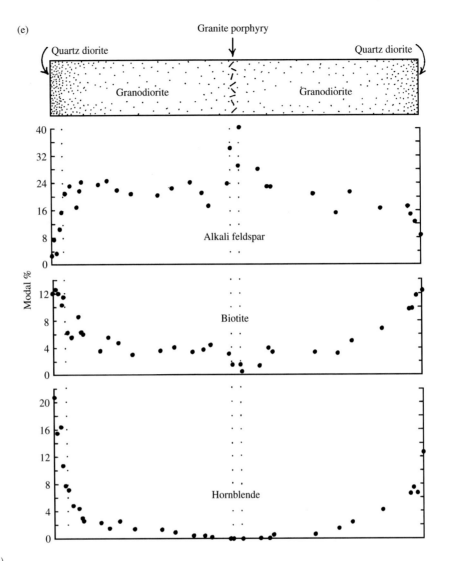

9.18 (*Continued*).

2. Brecciation and kindred phenomena involving expansion of a volatile fluid at generally low *P*

3. Doming: flexural and block-fault uplift of roof rocks

4. Ballooning: forcefully swelling magma that pushes aside ductile wall rocks and lifts roof rock

5. Magma invasion into tectonically favored "potential void" sites such as shear zones, fold hinges, and local extensional domains.

The emplacement history of a particular pluton has to be carefully evaluated on the basis of its field relations, fabric, and composition. A single pluton may be emplaced by more than one process, depending on the variable rheology of the country rocks and magma as its ascent slows and then stalls. Episodic emplacement and growth of a pluton might begin with brittle emplacement processes succeeded by more ductile ones as the country rocks are heated by the magma. Each new surge of magma can ascend higher, intruding former

roof rocks (Figure 9.5), and in some cases intruding its own volcanic cover. Emplacement processes in the brittle upper crust differ from those in the weaker ductile lower crust.

<u>Stoping</u>. The apt term *stoping* is derived from underground mining operations in which human-made caverns, called *stopes,* are created by removal of ore. As envisaged by R. A. Daly in the early 20th century (see Daly, 1968, Chap. 12) and by contemporary geologists, **stoping** is engulfment of pieces of brittle country rock by magma. In the usual case, fracture-bounded blocks of denser host rock, dislodged from the roof or wall, sink into the magma (Figures 9.5 and 9.21), allowing magma to move vertically upward or horizontally. Denser blocks of already fractured rocks can sink into the magma. In addition, new fractures in country rock can be created by the adjacent magma through thermal stressing of initially cool country rock (Marsh, 1982) and by hydraulic fracturing (Figure 8.2c). Magma can

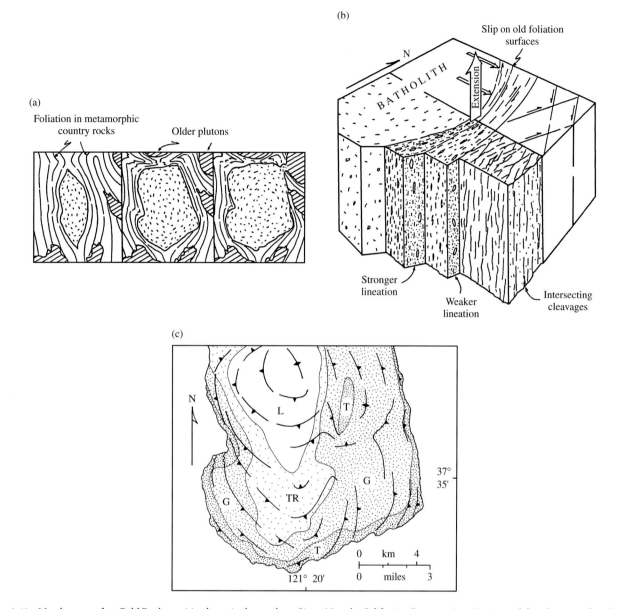

9.19 Mostly **concordant** Bald Rock granitic pluton in the northern Sierra Nevada, California. Compton (1955) estimated that about one-fourth of its area at the exposed level was gained by stoping and assimilation and the remainder by **forceful emplacement.** Ductile metamorphic wall rocks were pushed aside and dragged upward during intrusion. (Redrawn after Compton, 1955.) (a) Three small-scale maps show hypothetical stages in progressive growth of pluton. (b) Accommodation of pluton by subhorizontal flattening combined with vertical stretching in the metamorphic wall rock that created a strong vertical stretching lineation and flattening foliation. (c) Generalized map of **zoned pluton** showing concordancy between internal foliation in pluton and its contact with the wall rock. Note, however, that magmatic foliation cuts across most internal compositional contacts in zoned pluton. L, leucotrondhjemite; TR, trondhjemite; G, granodiorite; T, tonalite. (Redrawn from Paterson and Vernon, 1995.)

insinuate into country rock surrounding the main body of magma, producing a network of dikes and sills (Figure 9.5). Magma-bounded chunks of the denser rock can fall into the main magma body. Exsolved magmatic fluids and heated groundwater in cracks and other pore spaces in country rock can expand and wedge blocks into the magma. Confined water in the upper few kilometers of the crust held at constant volume experiences an increase in pressure of 1.5 MPa for each degree Celsius increase. Stoping mecha-

nisms are, thus, favored in shallower, cooler brittle crust.

Evidence for magma emplacement by stoping lies in the occurrence within the pluton of pieces of foreign country rock, or **xenoliths.** However, countless plutons neither have xenoliths nor offer evidence of alternate mechanisms of emplacement. For example, the strikingly discordant multiple intrusions at Mount Ascutney, Vermont (Figure 9.22), display no indication of emplacement by doming or ballooning. Thus it was

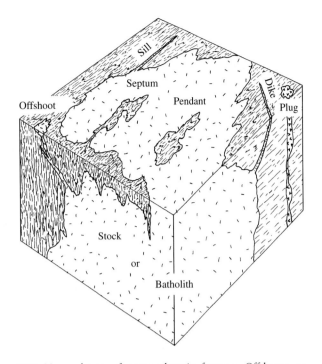

9.20 Nomenclature of some plutonic features. Offshoots, or **apophyses,** are any off-branching intrusions from the main pluton; they can be plugs or sheet intrusions such as dikes and sills. A **cupola** (not shown) is an upward-projecting part of the pluton into the roof rock. Erosional remnants of downward-projecting roof rocks completely surrounded by pluton are **roof pendants;** if they have peninsulalike connections to the main mass of roof rock, they are **septa,** or **screens.**

that Daly hypothesized the stoping mechanism: He believed the absence of xenoliths in these intrusions was due to their complete assimilation into the magma, leaving no trace of their former presence.

Stoping is an example of **passive emplacement** of magma. The intrusive rock tends to have an isotropic fabric that reflects dominance of crystallization over deformation, and the host rock is likewise little deformed.

<u>Ring-Fracture Stoping</u>. In contrast to the "piecemeal" stoping of outcrop-scale or smaller fragments of rock just described, much larger slabs of roof rock (measured in kilometers) may founder into a low-pressure magma chamber. An example is **ring-fracture stoping** (Daly, 1968), so called because the roof, initially supported by buoyant underlying magma, fails along a steeply dipping circular fault that follows the perimeter of the generally subcircular magma chamber (Figure 9.23). As the more or less intact slab sinks, less dense magma moves upward along the outward-dipping fault, invading the space between it and the wall rock, forming a **ring dike.** However, unlike in the situation shown in Figure 9.11, the ring fault and related dike do not reach all the way to the surface to create a caldera and magma extrusions. Rather, a fault-bounded roof slab detaches from higher roof rock, perhaps along a bedding plane or some other planar weakness. The resulting intrusion has the form of an inverted bowl in

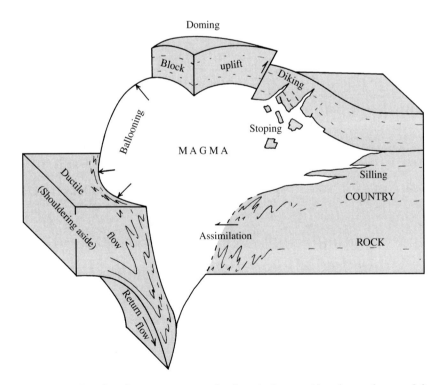

9.21 Highly schematic perspective view of emplacement processes of a pluton in the crust. Note deeper, downward-directed, return flow and forcefully emplacing ballooning in more ductile rocks in contrast to shallower, brittle processes of stoping and fault-block uplift. (Redrawn from Paterson et al., 1991.)

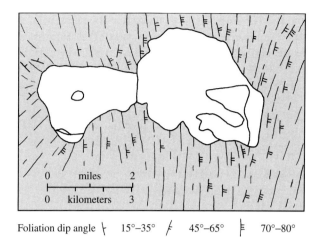

Foliation dip angle ⊦ 15°–35°   ⊦ 45°–65°   ⊨ 70°–80°

9.22 **Posttectonic plutons** with highly **discordant,** sharp contacts against their foliated metamorphic host rock at Mount Ascutney, Vermont. Note the lack of deflection of the foliation in the country rock (shaded) adjacent to the six Cretaceous intrusions, as if a gigantic cookie-cutter had removed metamorphic rock and allowed magma to move into its place. Compare with the concordant plutons in Figures 9.19 and 9.28 emplaced in large part by forceful pushing aside of ductile wall rock. (Redrawn from Daly, 1968.)

three dimensions. Erosion just into the top of the inverted-bowl-like pluton gives the appearance of a discordant, "cookie-cutter" pluton. Only by deeper erosion are the foundered roof slab and surrounding ring dike revealed.

It is not unusual for multiple episodes of ring-fracture stoping to occur at nearly the same location, developing multiple ring dikes and circular intrusions (Figure 9.24).

Brecciation. Slender, subvertical columns or funnel-shaped bodies of intrusive magmatic rock, commonly emplaced into the brittle upper crust, are called **pipes, or plugs.** They have an elliptical or circular cross section whose diameter is generally <1 km. Some merge with dikes and sills laterally and/or at depth, and some have fed volcanic extrusions, in which case the term **volcanic neck** is used.

Delaney and Pollard (1981) compared three geometric and energetic constraints on the movement of magma through a dike versus a cylindrical pipe in a *brittle* host rock:

1. The conduit most likely to grow by extensional fracturing of the host rock

2. The conduit that accepts the greatest volume of magma for a given increase in magma pressure at the source

3. The conduit around which the least work is done on the wall rocks to accommodate a given volume of magma

For all three of these aspects, a dike is favored over a plug. So how do plugs develop? Significant observa-

tions of plugs include an absence of wall rock deformation, textures and mineral compositions indicating fluid over-saturation in the magma, and abundant xenoliths of wall rock in the plug; **breccia pipe** is a more apt label for a xenolith-rich intrusion. In some breccia pipes wall rock fragments from a higher stratigraphic level are concentrated along the margin, whereas fragments from greater depth prevail in the

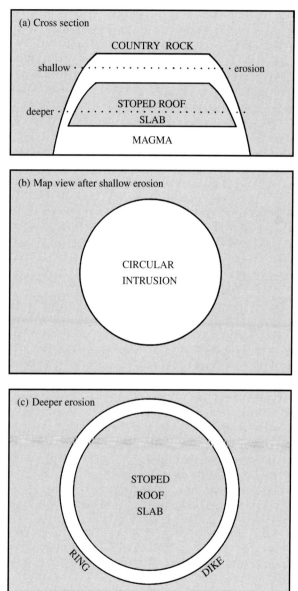

9.23 Idealized **ring-fracture stoping.** (a) Vertical cross section of magmatic body showing foundered slab of roof rock. (b) Hypothetical geologic map of subhorizontal surface exposure after shallow depth of erosion reveals a "cookie-cutter" circular intrusion. (c) Hypothetical geologic map of subhorizontal surface exposure after deeper erosion reveals a **ring dike** surrounding the foundered slab of roof rock. Compare actual circular intrusions and ring dikes in Figure 9.24. (Redrawn from Hall, 1996.)

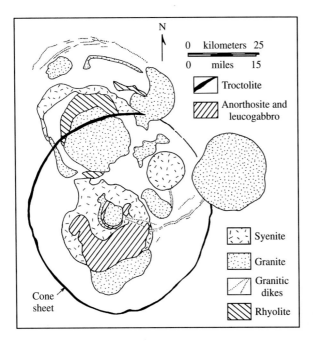

9.24 Ring complex and associated cone sheet in Niger, West Africa. (a) Simplified geologic map of Silurian **ring complex** of circular intrusions and unusually large, nearly perfectly circular, inward-dipping Meugueur-Meugueur **cone sheet** of troctolite (olivine-plagioclase rock). Unpatterned area is chiefly older basement rocks. (Redrawn from Moreau et al., 1995.)

center (Williams and McBirney, 1979, p. 52). These observations suggest an upward-convecting column of fluid-oversaturated magma, at the top of which rock fragments were stoped from the roof and carried downward along the cooler margin of the column. The upward "drilling" process is accomplished by efficient, concerted hydraulic and thermoelastic-stress fracturing of roof rocks at the top of the magma column. Thus, a plug may begin as a magma-filled brittle crack, but as volatile saturation occurs the drilling action tends to create a more circular, thermally efficient conduit. Some breccia pipes may be "blind," having never erupted through the surface of the Earth.

Diatremes are funnel-shaped breccia pipes apparently emplaced at low temperature that contain a high concentration of mantle- and/or crust-derived fragments, even to the exclusion of a recognizable magmatic matrix. They are commonly overlain by a shallow, dish-shaped explosion crater called a **maar** that has an encircling low pyroclastic ring (Figure 9.25). Many diatremes originate from alkaline, mafic to ultramafic kimberlite magmas that are exceptionally enriched in H_2O and CO_2 so that they probably become oversaturated with CO_2 at subcrustal depths. Above a "root" zone of dikes and sills that contain unfragmented kimberlite the highly overpressured, fluid-charged magma creates its own fractures and ascends

rapidly (10–30 m/s) and turbulently to the surface, carrying well-mixed dense mantle xenoliths and crustal fragments produced as the system drills upward through the lithosphere. The complexity of diatremes and their related crater and root zones makes generalizations difficult (for a description see Scott-Smith in Mitchell, 1996). Model experiments by Woolsey et al. (1975) have created many of the features of kimberlite diatremes (Figure 9.26).

Doming and Fault-Block Uplift. Clear evidence for overpressured magma that makes room for itself is seen

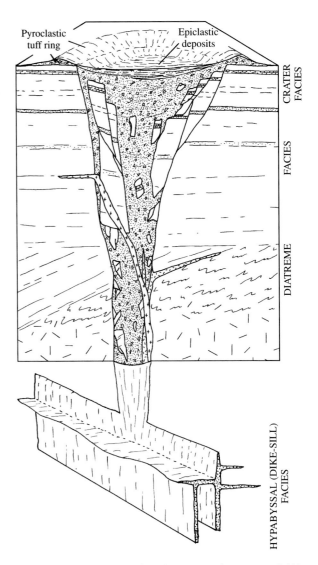

9.25 Idealized **diatreme** and overlying **maar.** The maar is a dishlike explosion crater surrounded by a ring of pyroclastic ejecta and is partly filled with lake sediments and alluvium (epiclastic deposits). Below this crater facies is the diatreme, here intruded by late dikes, which contains caved blocks of wall rock as much as 100 m long 1 km below their original stratigraphic level. Diagram not to scale; horizontal dimension (typically less than 1 km in diameter) exaggerated relative to vertical.

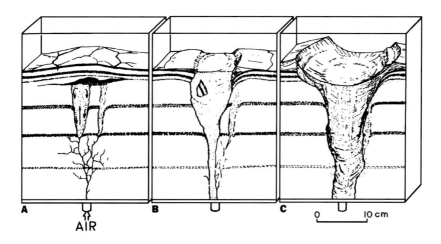

9.26 Diagrammatic summary of fluidization experiments bearing on the origin of a **diatreme.** (a) Introduced compressed air fractures cohesive clay layers and uplifts surface, forming ring and radial fractures in domed layers; fluidized convective cells develop in voids below uplifted layers, mimicking a blind diatreme. (b) Larger convective cell breaches surface; outcasted ejecta accumulate in a ring around growing maar. Saucerlike stratification and wall-rock fragments are evident in the diatreme. (c) Continued bubbling circulation further enlarges maar and underlying diatreme; bedded ejecta sags into center of diatreme and faint cross-cutting channels manifest where ascending gas has flushed out finer particles. (From Woolsey et al., 1975.) (Reprinted from Physics and Chemistry of the Earth, Volume 9. Woolsey TS, McCallum ME, Schumm SA. Modeling of diatreme emplacement by fluidization, pp. 29–42. Copyright © 1975, with permission from Elsevier Science.)

in domed and fault-block-uplifted roof rock overlying shallow crustal intrusions. Undisputed doming has occurred over the classic laccoliths first described in 1877 by G. K. Gilbert in the Henry Mountains, Utah (Figure 9.27; see also Corry, 1988; Jackson and Pollard, 1988). A **laccolith** is a flat-floored intrusion with a domical upper surface essentially concordant with the layered rocks into which the magma was forcefully inserted. Laccolith intrusions apparently begin as tabular sills, generally with only a few kilometers of covering roof rock, but then inflate upward—by as much as 2 km—as more magma is injected. During the growth of some laccoliths, the arching roof rock breaks along one or more steeply dipping faults so that the roof rises in a trap-door fashion.

In the view of Hamilton and Myers (1967), the relatively lesser volume of granitic batholiths in the Paleozoic Appalachian Mountain belt than in the Mesozoic Cordilleran of western North America (Figure 9.16) reflects the laccolithic shape of plutons (e.g., Figure 9.17). The present 5- to 8-km depth of erosion in the Cordilleran is believed to expose more of their alleged laccolithic extent than does the deeper erosion in the older Appalachian region.

Ballooning. Many felsic plutons have the following characteristics, seen in Figures 9.18, 9.19, and 9.28:

1. Roughly circular to elliptical shapes in map view

2. More or less concentric magmatic foliation within the pluton that is concordant with the contact and foliation in the wall rocks

3. Increasing strain toward the pluton-host rock contact, mainly evident in the wall rock but also locally seen in the intrusive rock; wall rocks flattened perpendicular to the contact and having steeply inclined stretching lineation parallel to it

4. Concentric compositional zonation within the pluton

A commonly invoked explanation of these characteristics is **ballooning,** a radially directed inflation of a magma chamber as additional magma is intruded. As an ascending magma diapir stalls, its "tail" continues to rise, inflating the surrounding cooler, more crystalline mass, producing a concentric foliation and a flattening strain within the pluton and the immediately adjacent, heated wall rocks. An important facet of the ballooning model is a return downward flow of the ductile wall rocks (Figure 9.21) as the ascending magma diapir stalls and inflates. This return flow can accommodate some of the lateral expansion and flattening of the wall rock that provide space for the pluton. However, Paterson and Vernon (1995) urge caution in several aspects of this model:

1. Concentric magmatic foliation patterns that cut across internal compositional contacts (e.g., Figures 9.19c and 9.28) suggest that magma flow may be less responsible than late-stage flattening strain of the largely crystallized magma.

2. Use of mafic inclusions as **strain markers** in the intrusive rock (Figure 9.28) can be misleading. Although commonly used to determine the amount

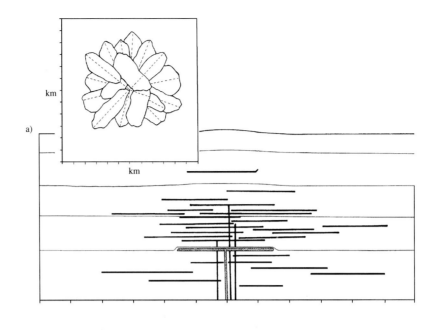

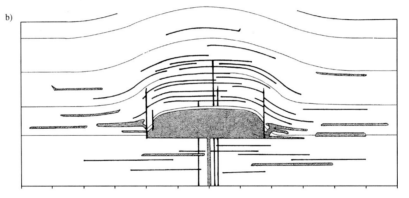

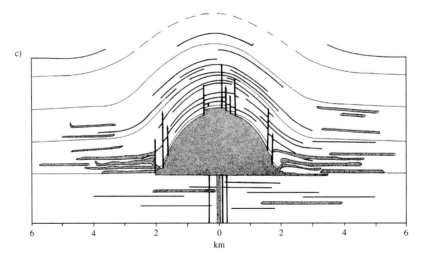

9.27 Growth of a **laccolith** by **doming** of overlying sedimentary strata in the Henry Mountains, Utah, according to Jackson and Pollard (1988). (a) Emplacement of sills and dikes (thick lines) and a thin protolaccolith (stippled). Inset is a plan view of the tongue-shaped, radial sills; tick marks at 1-km intervals. (b) Inflation of central laccolith induces bedding plane slipping and warping of overlying sills and 3.5-km thickness of Permian–early Tertiary strata. Additional diking and silling occur around the laccolith. (c) Increased inflation of the laccolith, now probably a composite intrusion formed of multiple pulses of magma, produces numerous faults (not shown) in overlying steepened, stretched strata. Sills below floor of main laccolith are conjectural. Horizontal and vertical scales are the same. (From Jackson MD, Pollard DD. The laccolith stock controversy: new results from the southern Henry Mountains, Utah. Geol. Soc. Am. Bull. 100:117–139. Reproduced with permission of the publisher, The Geological Society of America, Boulder, Colorado, USA. Copyright © 1988 Geological Society of America.)

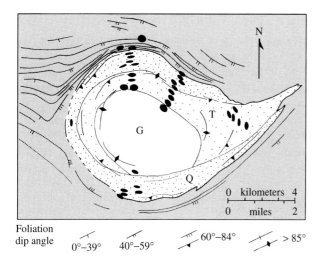

Foliation
dip angle   0°–39°   40°–59°   60°–84°   > 85°

9.28 **Zoned concordant** Ardara pluton, Donegal, Ireland, a result
of **forceful emplacement.** Black ellipses show increasing flattening of mafic inclusions in the intrusion and mineral grains
in the wall rock nearer the concordant contact. Dip of foliation in wall rocks (shaded) indicated by tick marks on strike
lines and in pluton by triangles. G, granodiorite; Q, quartz
monzodiorite; T, tonalite. (Redrawn from Pitcher, 1997.)

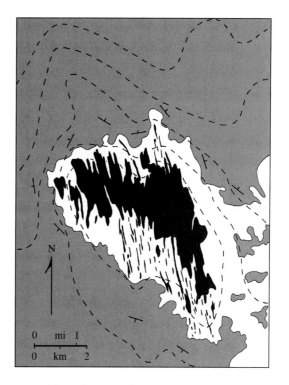

9.29 **Forcible emplacement** of granitic magma, Black Hills, South
Dakota. Magma (black) wedged along steeply dipping foliation
surfaces in country-rock schist (unshaded). Unconformably
overlying, initially subhorizontal, sedimentary strata (shaded)
were domed, as indicated by strike-dip symbols and by equal
elevation structure contour lines (dashed) spaced 500 feet
apart on the unconformity between the schist and sedimentary
rock. Many faults are omitted. (Redrawn from Noble, 1952.)

of strain during chamber inflation, the initial inclusion shape is rarely known with certainty
and the final shape can be dependent on several
factors.

3. Regional deformation may accompany pluton emplacement so that its fabric and that of the country
rock may bear an imprint of far-field tectonic
processes and these may well influence magma emplacement.

In a variant of the ballooning mechanism, magma
forcefully injected along steeply dipping foliation surfaces in schist can accumulate in such volume as to
dome unconformably overlying strata (Figure 9.29).

Ballooning is an example of **forceful emplacement**
of magma. Both host and intrusive rock, but especially
the former, display effects of deformation; their contact
tends to be concordant. Because of the inherently
higher *T* of the ductile host rock, relative to that in the
shallower crust, more chemical interaction occurs between country rock and magma, creating broad, compositionally complex border zones and contact aure-
oles (discussed later).

<u>Tectonically Created Room.</u> The notion that magma
emplacement is "structurally controlled" in orogenic
belts has been popular for decades. Dilatant fault zones
where rock has been broken up (brecciated) are favor-
able sites for magma emplacement. Bends along an active fault (Figure 9.30) provide a potential void for
magma emplacement. Hinge zones of actively forming
folds may also be a local site where magma might find
space.

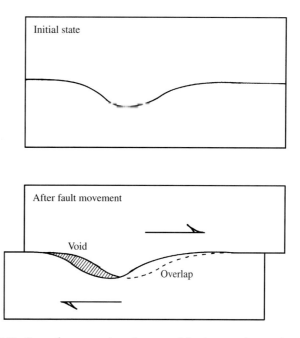

9.30 Space for magma intrusion created by jogs, or changes in
orientation, in a fault. As movement occurs, a potential void
into which magma may be emplaced is created. The fault may
be of any type—normal, reverse, or strike-slip.

Emplacement of magmas into the continental crust in subduction zones would seem to be difficult because of the compressive state of stress in this tectonic regime. Yet, paradoxically, this setting is where voluminous granitic batholiths have been emplaced during broadly concurrent plate convergence. Lithospheric plates usually converge obliquely, rather than everywhere perpendicular, a characteristic that actually creates large-scale possibilities for magma emplacement (Bouchez et al., 1997). A case in point is the relative convergence direction of the North American and oceanic plates to the west when the Mesozoic Sierra Nevada batholith (Figure 9.16) was forming. The right-lateral component of motion by the oceanic plates along the plate juncture created shear zones tens of kilometers long within the batholithic terrane. Domains of extension in this overall compressional/strike-slip tectonic regime can develop between adjacent subparallel shears and can provide potential room for magma emplacement (Figure 9.31).

9.4.3 The Intrusion–Host Rock Interface

Aspects of magma emplacement are recorded in the interface, or **contact,** between the intrusion and its host rock. Smaller intrusions, such as thin, rapidly cooled dikes, transfer heat but not matter across the intrusive contact and in this sense are closed systems. Larger, longer-lived intrusions tend to be open in that matter—chiefly volatile fluids—as well as heat move across the contact. Also, mechanical work of deformation can be done on the country rocks by the intrusive magma (Figures 9.5, 9.19, 9.21, 9.27, 9.28, 9.29). The **contact aureole** of perturbed country rock and the thermal, chemical, and deformational gradients recorded in it tell of emplacement processes and provide chronologic information (Paterson et al., 1991). Likewise, records of similar gradients may be evident within the margin of the intrusion, and these can provide additional insights into emplacement processes.

<u>Contacts and Border Zones.</u> The simplest interface is the **sharp contact.** On a particular scale of observation,

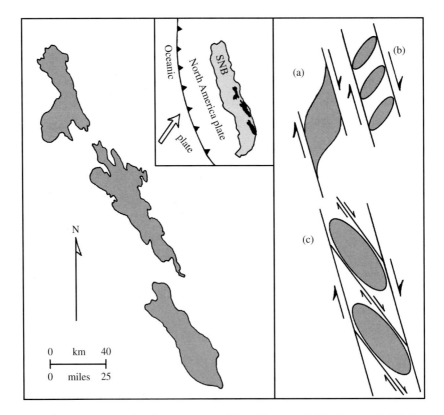

9.31 Possible tectonic control on making room for plutons within the Sierra Nevada batholith. Map on the left shows three composite plutons (shaded) comprising the Cathedral Range intrusive sequence emplaced at 92–81 Ma. Smaller-scale map in upper middle of figure shows position of the three intrusive masses (black) within the larger Sierra Nevada batholith (SNB; Figure 9.16), which was created during mostly oblique convergence (arrow) of oceanic plates beneath the North American continental plate during the Mesozoic. Panel on right shows three possible ways (a, b, c) in which local extensional domains (shaded ellipses) might be created in the right-lateral strike-slip regime in the continental plate during oblique plate convergence; these extensional domains could create room for magma emplacement. (c) Extensional regions bridging between secondary right-lateral shear zones seem to correspond most closely in shape and orientation with the Sierra intrusive masses. (Redrawn from Tikoff and Teyssier, 1992.)

the interface between magmatic and host rock can be pencil line–thin (Figure 9.32). Intrusion of magma against significantly cooler, usually shallow crustal, host rock leads to rapid conductive and perhaps advective cooling along a pronounced thermal gradient. Chemical and thermal interactions between magma and host rock are minimal in a narrow contact aureole. Grain size in the magmatic rock generally diminishes toward the contact. Sharp contacts without grain size reduction might reflect a more dynamic intrusive margin where flowing magma swept away the initial chilled contact material.

For many, generally larger intrusions, the interface between magmatic and country rock is gradational over several to as much as hundreds of meters. For such a **border zone,** significant thermal, chemical, and physical interaction occurred between magma and host rock. One type of physical interaction is pervasive injection of dikes and sills into country rock that is initially relatively cool and brittle (Figures 9.4, 9.5, and 9.33a), which allows dislodged fragments to be stoped into the magma. Or hotter country rock permeated by magma can be assimilated or heated sufficiently to melt partially, forming a contaminated border zone (Figure 9.33b, c).

The variety of magma-country rock interfaces is virtually limitless. Wide variations that can be seen in a single intrusion reflect contrasts in thermal, chemical, and rheologic processes and properties.

Chronology of Magma Emplacement Relative to Tectonism.

As challenging as elucidating the process of magma emplacement is determinating the time of intrusion relative to episodes of regional tectonism, such as characterize orogenic zones at convergent plate boundaries. Was the intrusion pretectonic: Was the magma emplaced prior to deformation of the rocks in a broad region around the pluton? Or was the intrusion syntectonic, as the plutons in the alleged Sierra Nevada shear zones of Figure 9.31? Or was the intrusion posttectonic, so that structures in the wall rocks are cut by the intrusion? Answers to these questions are often crucial in working out the chronologic evolution of orogenic belts because intrusive rock is commonly the material best suited to isotopic dating.

Determination of the relative age of forcefully emplaced intrusions is hindered by the fact that they deform aureole rocks and are themselves deformed, especially in their outer, more rigid margin. It must be determined whether deformation is localized in and around the intrusion or is of a regional character in which the pluton participated. A general difficulty in determining chronology of emplacement stems from the nature of the pluton itself. The lack of mechanically

(a)

(b)

9.32 Sharp intrusive contacts of magma against host rock. (a) Panoramic view of contact between Taboose Pass septum of dark-colored metamorphosed sedimentary rock (exposed on ridge and peaks) against lighter-colored Jurassic granite (in lower slopes) in the Sierra Nevada batholith, California. The contact with the septum (or screen) appears horizontal but actually dips steeply between the granite and Cretaceous granodiorite (on back side of ridge). Width of field of view is about 3 km. (Photograph provided by John S. Shelton and information by Clifford A. Hopson.) (b) Outcrop view of a **sharp, discordant contact** between granodiorite and overlying darker-colored, layered quartzite. Note hammer for scale.

contrasting layers, such as bedding in sedimentary rocks and foliation in metamorphic, precludes development of folds that manifest deformation. Other possible effects of deformation must be sought.

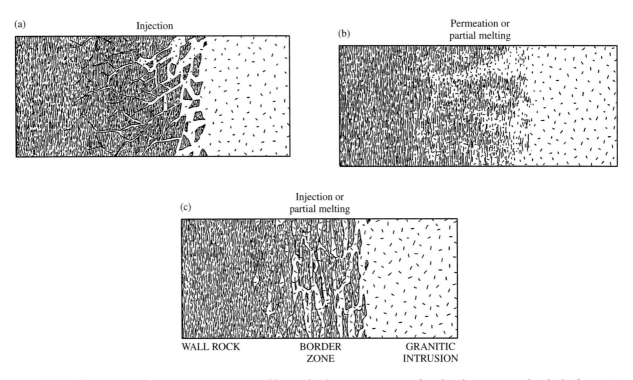

9.33 Idealized **border zones** between magma intrusion and host rock. These zones can range from less than a meter to hundreds of meters in thickness.

Determination of the relative chronology of pluton emplacement and regional tectonism is based on fabric and field relation criteria (Paterson et al., 1989). The strongest, but not the only, evidence for **pretectonic pluton emplacement** is an intrapluton foliation that parallels country rock foliation and is independent of the pluton margin, locally cutting across it at high angles. An example of a pluton and its country rock that were deformed together so as to develop a solid-state foliation continuous through both is shown in Figure 9.34. Evidence for **posttectonic pluton emplacement** can be found in the "cookie-cutter" Mount Ascutney

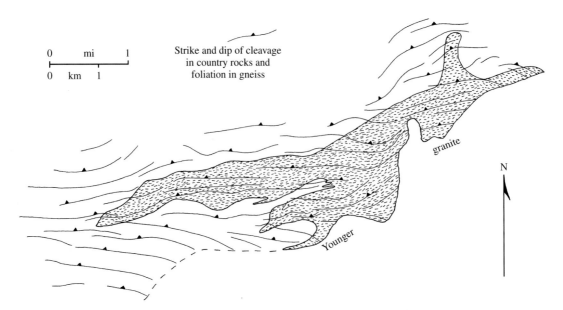

9.34 **Pretectonic pluton.** The original porphyritic granitic rock of the Saint Jean du Gard pluton north of Paris, France, is now a metamorphic augen gneiss in which the foliation is continuous with cleavage in the slate country rocks. Thus, both country rocks and pluton were deformed and metamorphosed as a single rock mass and possess similar anisotropic fabric. (From De Waard, 1950.)

intrusions (Figure 9.22), which are sharply discordant to country rock foliation and, therefore, were intruded after development of foliation in the metamorphic wall rocks. **Syntectonic pluton emplacement** can be the most difficult to ascertain. The most convincing occurs where the pluton contact is subparallel and continuous with the foliation in the country rocks beyond that of the immediate thermal contact aureole. Magmatic minerals, perhaps defining an internal foliation in the pluton, such as an igneous lamination (Section 7.9), have the same isotopic age as minerals defining the country rock foliation.

SUMMARY

Magma ascent through the lithosphere is a thermal-mechanical phenomenon governed by buoyant driving forces and viscous resistive forces related to the rheology of magma and lithospheric rock and to their contrast in density. Other governing factors include tectonic regime, state of stress, and stratification of the lithosphere with respect to strength and brittle versus ductile behavior. Energy for ascent is ultimately provided by gravity and internal heat in the Earth.

"Density-filtered" mantle-derived basaltic magmas can underplate and stagnate near the base of the lower-density continental crust. These buoyantly blocked magmas provide heat that can drive partial melting of the crust. Magma fractionation and assimilation of felsic country rock may reduce the density of evolved magmas, allowing farther ascent.

Subvertical basaltic dike swarms are crustal-scale features that testify to copious magma transport from a mantle source through the crust. In extensional tectonic settings, magmas follow self-generated extensional cracks oriented perpendicular to the least horizontal principal stress. If vertically continuous, overpressured basaltic magma can be driven upward tens of kilometers, even through less dense rock material. In compressional regimes, basaltic magmas are less able to rise all the way to the surface. Regional, far-field states of stress can be modified in the local environment of a central magma intrusion, allowing magma to intrude as cone sheets and radial dikes. Dike transport of magma is a trade-off between rate of flow and rate of solidification through conductive cooling; slowly flowing viscous magma dissipates heat to the wall rock and becomes immobile. Because distance of magma transport through a dike is proportional to the fourth power of its width and inversely proportional to magma viscosity, low viscosity basaltic magma can readily move a long distance through dikes a meter or so wide. In contrast, much more viscous granitic magma requires wider dikes through which to move.

Volatile-rich mafic to ultramafic magmas probably begin their ascent through the lithosphere via self-induced fractures. As overpressured magma follows the upward-propagating crack, it may be arrested if the supply is exhausted or continue as a fluid-oversaturated magma column that "drills" its way upward and finally may breach the surface, where it creates an explosive maar crater underlain by a diatreme.

Slow ascent of viscous granitic magma through ductile country rock as buoyant diapirs kilometers in diameter is an alternative mechanism to dike transport. The more or less spherical shape of the ascending diapir minimizes viscous drag because of minimal surface area relative to volume. This, in turn, maximizes buoyant lift and retention of thermal energy. Diapirs are most likely in the ductile, hotter lower crust, where wall rocks can readily be thermally "softened" to facilitate rise before stagnation at higher cooler crustal levels, where increased country rock strength arrests continued ascent. Subsequent diapirs are likely to ascend more easily in the thermally conditioned wake of preceding ones.

Magma is finally emplaced as an intrusion or pluton as a result of increased viscosity and loss of mobility upon cooling, insufficient buoyancy to carry it higher, or increased strength of the country rocks. Emplacement processes are most properly considered in three dimensions. However, the pluton–country rock system is usually observable only in surface exposures of generally less than a kilometer of surface relief. Careful and critical observation of field relations, fabric, and rock and mineral compositions is required to interpret the specific processes involved in magma emplacement for a particular pluton accurately. The relative contributions of different emplacement processes are variable, even for a particular pluton. Laccoliths and some other crustal plutons create room by upward forceful deflection of the overlying roof rocks. Stoping is favored in fractured shallow crustal rock but even with assimilation of country rock does not create new room for magma, therefore permitting only a transfer in position between these masses. The absence of country rock xenoliths in an intrusion does not necessarily rule out stoping and assimilation; piecemeal stoped blocks may have been completely assimilated into the magma, or foundered roof slabs may have sunk below the level of exposure. Some plutons appear to have forcefully pushed aside their ductile wall rocks as the magma ballooned laterally. Local extensional tectonic environments may create room for magma emplacement.

Criteria for determination of the time of magma emplacement relative to regional tectonism are based on fabric and field relations near the pluton margin. Such determinations can be crucial in unraveling complex orogenic chronology.

CRITICAL THINKING QUESTIONS

9.1   Discuss the role of buoyancy in the movement and stagnation of magma in the lithosphere.

9.2   Discuss the origin and consequences of magma overpressure. How is it a factor in ascent and emplacement of magma? How does high versus low overpressure control different mechanisms of intrusion? How does overpressure move magma to the surface, where it can erupt?

9.3   Contrast a local state of stress near a central intrusion and a regional state and the processes through which these control the ascent and emplacement of magma.

9.4   Contrast how different sheet intrusions are emplaced.

9.5   Compare and contrast ascent of magma from its source by diapirism and by diking with respect to thermomechanical factors and rheology of magma and country rock.

9.6   Characterize mechanisms of emplacement and making of room for intrusive magma, indicating specific evidence for each mechanism.

9.7   Discuss the so-called room problem for large intrusions: whether it is real or fictive and why.

9.8   Indicate how the relative age of regional deformation and processes of intrusion can be determined in features of an intrusion and its country rock.

9.9   In Figure 9.7a, propose a reason why most of the dikes in the Mackenzie swarm are oriented approximately north-south.

9.10 A long-standing observation is that the most common intrusive rock is granitic and the most common extrusive is basaltic. Propose reasons why this might be so.

PROBLEMS

9.1   If the change in melt density with depth is taken into account (Problem 8.9), how does this affect the magnitude of the excess magma pressure in Figure 9.2? Explain fully and indicate any assumptions made.

9.2   Derive the expression for magma overpressure, or hydraulic head, $\Delta \rho g z$. Indicate any assumptions made.

9.3   What is the velocity of basalt magma rising through a 1-m-wide dike if $\eta_{magma} = 100$ Pa s and $(\rho_{crust} - \rho_{magma}) = \Delta \rho = 0.1$ g/cm^3? (*Answer: 3.3 m/s.*)

9.4   Some 2-m-wide basalt dikes in Iceland are as much as 30 km long. All other factors being the same, including magma supply rates, a 6-m-wide dike in the Mackenzie swarm in Canada could be how long? (*Answer: 2430 km.*)

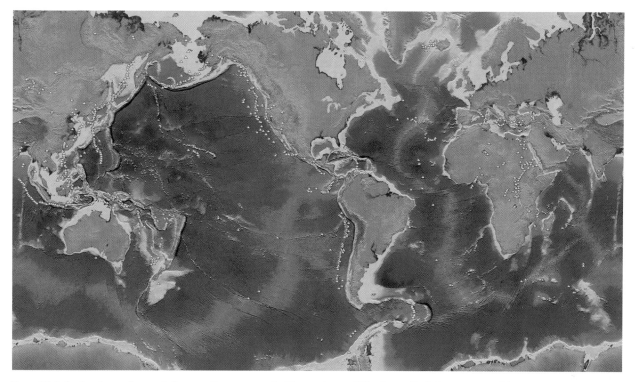

I   Global volcanism and tectonic features. Base map used with permission of Ken Perry, Chalk Butte Inc., Boulder, Wyoming. Representative volcanoes active in about the past 1000 years (yellow triangles) are from the Smithsonian Institution of Washington Global Volcanism Program Website.

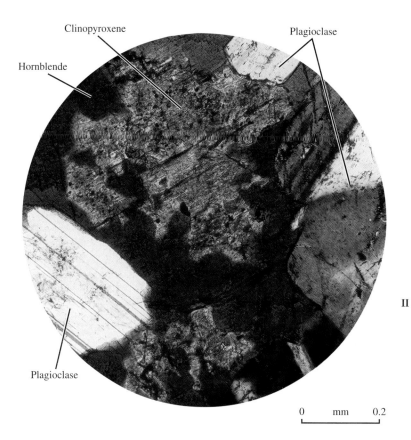

Clinopyroxene

Plagioclase

Hornblende

Plagioclase

0   mm   0.2

II   **Reaction rim.** Photomicrograph under cross-polarized light of an irregularly shaped (anhedral), partially resorbed clinopyroxene surrounded by a **reaction rim** of hornblende in a fine phaneritic diorite. The interference color of hornblende is essentially masked by its pleochroic dark brown color. Small black grains in clinopyroxene that has orange-red interference color are opaque Fe-Ti oxides. Note polysynthetically twinned plagioclase.

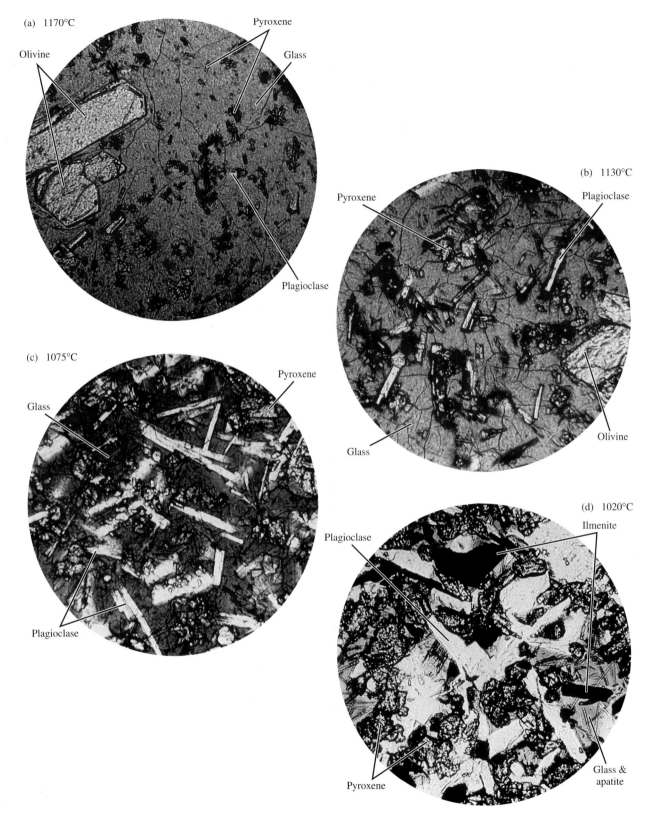

III　Sequence of photomicrographs in plane-polarized light shows progressive crystallization of Makaopuhi basalt from 1170°C to 1020°C. See also Table 5.1 and Figure 5.10. The diameter of each view is about 1 mm. (Photographs courtesy of Thomas L. Wright, U.S. Geological Survey.)

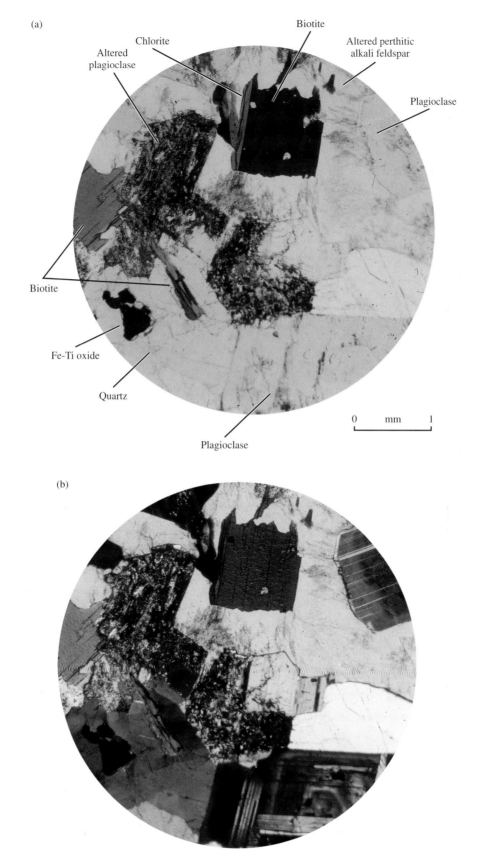

(a)

Chlorite

Biotite

Altered perthitic
alkali feldspar

Altered
plagioclase

Plagioclase

Biotite

Fe-Ti oxide

Quartz

0   mm   1

Plagioclase

(b)

IV Photomicrograph of an altered subsolvus granite in which primary high-T magmatic plagioclase and biotite have been partially decomposed by subsolidus breakdown reactions. (a) Plane-polarized light showing dark brown biotite partially replaced by green chlorite in thin lenses parallel to (001) and plagioclases on left replaced by a fine-grained phyllosilicates and possibly an epidote mineral. Also note clay-altered alkali feldspars that have a "turbid" appearance. (b) Cross-polarized light of same view showing highly birefringent replacement particles in upper plagioclase and euhedral form of fresh plagioclase against alkali feldspar and quartz in lower part of view. This contrast in euhedralism is interpreted to have originated as alkali feldspar and quartz crystallized near the solidus from a melt lying between the higher-T plagioclase crystals. Note undulatory extinction in strained quartz (lower left).

V Xenolith of mantle peridotite jacketed by vesicular basalt ("cored bomb"). Most of this lherzolite is made of pale green olivine about Fo_{90}. Lesser dark gray-brown orthopyroxene about En_{90} is more abundant than emerald-green Cr-rich diopside clinopyroxene. A small amount of minute grains of black Mg-Al-Cr-Fe spinel is not visible.

0 cm 3

Amphibole

Apatite

VI Xenolith of mantle peridotite with selvage of black amphibole plus accessory apatite (top) and basalt (bottom) from Dish Hill, California. This lherzolite is composed of pale green Mg-rich olivine (some stained by orange Fe-oxide) and lesser gray-brown Mg-rich orthopyroxene and emerald-green Cr-rich diopside. Note planar boundary of amphibole selvage that is interpreted to represent a vein emplaced in a brittle fracture in this mantle rock.

0 cm 2

CHAPTER

Magma Extrusion: Field Relations of Volcanic Rock Bodies

10

FUNDAMENTAL QUESTIONS CONSIDERED IN THIS CHAPTER

1. How is magma extruded onto the surface of the Earth?

2. What factors govern explosive eruptions versus quiet effusion of lava?

3. How can the fabric and field relations of volcanic rocks be used to decipher their manner of emplacement?

INTRODUCTION

Extrusive processes are more amenable to direct observation and study than processes of magma ascent and intrusion, which were discussed in Chapter 9. This is not to say, however, that geologists fully understand everything about extruding magmas and the rocks formed from them. Although some magma extrusions, such as low-viscosity, gas-poor basalt lavas, can be observed safely from within a few meters, one can only observe explosive eruptions from distances of many kilometers. Interpretation of the products of these exceedingly dangerous explosive eruptions can be challenging.

This chapter delves into processes of magma extrusion and the rock bodies and volcanic edifices so produced. The emphasis here is on the field relations of the entire volcanic rock body—the whole genetic entity. The rock fabrics that were considered in Chapter 7 and are observed in a single outcrop or on smaller scales than the whole body are not sharply distinct

from larger scale field relations. Despite their treatment in separate chapters, *both* fabric and field relations of volcanic rocks record the character of formative processes; they cannot and should not be divorced from one another if the petrologist is to understand fully the origin of the rock body. For example, identical porphyritic aphanitic fabrics could have originated in a thin dike or in the margin of a small stock, in a shallow crustal environment, or in a lava flow, or as a clast in a volcanic debris flow. Only by means of the field relations can the true origin be ascertained because the cooling rate (thermal kinetic path) was similar in each environment and as a result the same fabric developed.

In this textbook, only a very brief summary of the rapidly growing field of volcanology as it relates to rock-forming processes and the fabric and field relations of rock bodies can be provided. More extensive general treatments are those by Williams and McBirney (1979), Cas and Wright (1987), Francis (1993), and the monumental *Encyclopedia of Volcanoes* edited by Sigurdsson et al. (2000). Since Fisher and Schmincke (1984), an explosion in works on pyroclastic topics has been edited by Sparks et al. (1997), Freundt and Rosi (1998), and Sparks and Gilbert (1999). A useful brief summary is Walker (1993).

✳10.1 OVERVIEW OF EXTRUSION: CONTROLS AND FACTORS

Two magma properties are of supreme importance in processes and products of extrusion: dissolved volatile concentration in the melt fraction and rheology of the magma. Rheology is expressed in Newtonian and non-Newtonian viscosity and depends not only on the concentration of dissolved volatiles in the melt, but also on major element composition of the melt (especially concentration of silica), magma T, crystallinity, and strain rate. In this chapter, apparent viscosity (Section 8.2.2) is used to denote magma rheology.

Volatile phenomena and rheology are involved in moving magmas to the surface from buried chambers and feeding conduits; they are also involved in the processes of extrusion from the vent and emplacement of the magma onto the surface.

10.1.1 Moving Magma to the Surface: What Allows Extrusion

Basically two requirements must be satisfied if magma is to extrude, either directly from its source in the deep crust or upper mantle where it was generated or from a staging chamber in the shallower crust. First, there must be an opening to the surface from the buried magma body. Second, magma must be able to move and be propelled through the opening. These are not necessarily independent of one another and, in fact, are usually related. Several mechanisms, all of which depend on development of overpressure in the subterranean magma body, allow venting of magma:

1. Independently of any exsolving and expanding volatiles, a buried mass of magma may have the capacity to rise and even fracture the overlying rocks by virtue of its buoyancy. Thus, in extensional tectonic regimes basaltic magma in upper mantle or deep crustal reservoirs can invade subvertical fractures of its own making, ascend to the surface, and extrude. This mechanism probably accounts for most extrusions of basaltic magma.

2. After buoyant ascent from its source through dikes or as diapirs, magma stored for a time in a shallow crustal chamber can subsequently erupt once its volatile fluid pressure or buoyant force exceeds the tensile strength of the roof rock overlying the chamber, causing the roof to rupture and then allowing the gas-charged magma to erupt. This can happen in the following ways:

   (a) As a stationary magma cools and crystallizes feldspars, pyroxenes, and so on, the residual melt becomes saturated in volatiles. The resulting volatile fluid pressure or the buoyancy of the bubbly magma can drive eruption.

   (b) The magma may rise to still shallower crustal levels, causing more volatiles to exsolve and expand in the decompressing system. Exsolution and bubble growth may be retarded in rapidly ascending, viscous magmas so that eventual pressure release is greater and explosive eruption more violent. Instead of the magma's rising to shallower levels to cause decompression, a stationary magma system may be unroofed.

The catastrophic May 18, 1980, explosive eruption of Mount Saint Helens, Washington (Lipman and Mullineaux, 1981), furnishes an example. After 2 months of seismic activity, steam-blast explosions at the summit, and bulging of the northern summit and flank area at a rate of about 2 m/day, a magnitude 5+ earthquake triggered a massive landslide in the unstable bulge, unroofing the buried growing body of dacite magma (Figure 10.1). The sudden decompression of the overpressured magma system produced a violent explosion.

   (c) Mafic magma may be injected into the base of a chamber of cooler, less dense, more silicic magma (Figure 8.24; Sparks and Sigurdsson, 1977). Transfer of heat to the intruded resident silicic magma may create enough additional buoyancy to cause eruption. But probably more significantly, cooling and crystallization of the mafic magma cause volatile saturation and exsolution. Released volatiles float into the overlying silicic magma, oversaturating it and increasing the volatile pressure and magma buoyancy.

   (d) External water in the ground or in lakes or ocean may come into contact with buried magma, absorb heat, and expand explosively, blowing off the shallow cover over the magma body.

Changes in an extruding magma system can arrest further eruption. Deeper levels of evolved silicic crustal chambers tapped during continued extrusion are commonly poorer in volatiles and are more crystalline: both characteristics create greater apparent viscosity of the magma and less eruptibility. A decrease in ascent velocity, caused by whatever process, can allow more cooling, crystallization, and potential loss of exsolved volatiles through permeable wall rock.

Magma is extruded either from a central vent or from a fissure. In a **central eruption,** magma vents from a more or less subvertical cylindrical feeding conduit and builds a conical volcano (Figure 10.2). Other magmas, commonly of basaltic composition, extrude from a long crack in the crust and constitute a **fissure eruption** (Figure 10.3); the subterranean feeder is a subvertical dike. For thermal reasons (Section 8.4.1), eruptions that begin from a fissure commonly become localized into a central vent as extrusion continues.

10.1.2 Two Types of Extrusions: Explosive and Effusive

Depending on whether or not near-surface magmas blow apart into separate pieces, either of two types of extrusion, **explosive** or **effusive,** can result. These contrasts in the dynamics of extruding magma are linked to vesiculation phenomena that depend on volatile con-

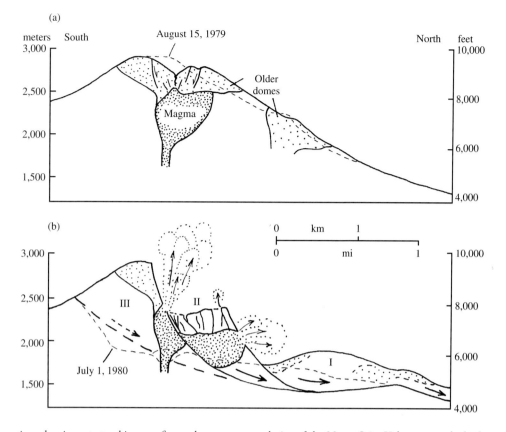

10.1 Cross sections showing catastrophic unroofing and consequent explosion of the Mount Saint Helens magma-hydrothermal system on May 18, 1980. (a) Situation just before the earthquake-induced 2.3-km³ rockslide. Intrusive dacitic magma had perceptibly bulged and destabilized the north side of the volcano (compare August 1979 topographic profile, dashed line). (b) Three successive unstable masses of rock slipped northward on May 18. Movement of I and II caused a lateral blast, pyroclastic surge, and plume (vertical eruption column) of ash and steam. Movement of III further beheaded the magma body. See Figures 10.20 and 10.21 for distribution of explosive deposits. The dashed line shows the July 1, 1980, topographic profile. (Redrawn from Moore and Albee, 1981.)

centrations in the magma and its rheology, and to whether magma comes into contact with external water.

In *exploding* magmas, juvenile particles of melt and crystals, together with possible accidental rock and single crystal fragments (xenoliths and xenocrysts, respectively), are blown from the volcanic vent, dispersed through a medium of air or water, and finally deposited on the surface of the Earth. All of these particles and fragments, collectively called **pyroclasts, tephra,** or **ejecta,** accumulate subaerially on dry ground and subaqueously on the floors of lakes and oceans.

Nonfragmented but commonly bubble-bearing magmas pour *effusively* from volcanic vents as coherent overflows of **lava.** The morphological characteristics and style of movement of lava reflect the ways magma composition and heat loss and gas loss impact magma rheology. Apparent viscosity dictates whether lava spreads as a thin sheet or stream with an **aspect ratio** (thickness/horizontal dimension) as small as 10^{-4} in the case of some basaltic lava flows (Figures 10.3 and 10.4) or as a bulbous dome with a ratio near 1 in the case of many silicic lavas (Figures 10.2 and 10.5). To some degree, the rate of discharge of lava from the vent

also influences aspect ratio; rapid discharge can reduce effects of cooling on lava mobility and lengthen the flow. Apparent viscosity as well as other factors such as the vent diameter and volume of the magma supply influence the rate of discharge of lava from a vent; less viscous lava generally extrudes faster. Figure 10.2 illustrates some of the variety of extrusive forms that can occur in one local volcanic field.

Some lava finds its way into topographic depressions, such as stream canyons, where it moves as a confined flow. On surfaces lacking pre-exsiting channels, lava moves as an unconfined flow and, because of a greater surface area exposed to the atmosphere, tends to cool faster by radiant and convective heat transfer. Locally, rootless lava flows may be produced by coalescence of molten blobs of magma falling around the base of lava fountains.

During the history of a particular long-lived volcano the mode of magma extrusion commonly fluctuates between explosive and effusive. Single episodes of volcanic activity after a period of repose that can last from months to hundreds of years commonly begin explosively and then, as the supply of more volatile-enriched

10.2 Mount Shasta, a **composite volcano** in the Cascade Mountains of northern California, and nearby lava extrusions. See also Figure 10.44. View is looking southeast up Whitney Creek. Extruded magmas with differing apparent viscosities created a range of lava-flow morphological characteristics. The most viscous lava produced Haystack dome in left foreground, which is 150 m high and 900 m in diameter, and snow-covered Shastina, a **parasitic cone** just to the right of the summit of Mount Shasta. Note the lobate tongues at the toe of the youthful flow to the right of Haystack and the well-developed **lava levées** along the flow margins that testify to viscoplastic flow behavior. Lava flows and volcanoclastic deposits built Mount Shasta, which rises almost 3 km above Haystack dome and about 16 km away. (Photograph from Shelton JS. Geology Illustrated, New York, W.H. Freeman, 1966. Used with permission of John S. Shelton, who holds the copyright.)

magma at the top of the preeruption chamber is exhausted, evolve into effusive activity. Large composite and shield volcanoes, such as Mount Saint Helens and Mauna Loa (Hawaii), respectively, are built on time scales of $<10^6$ y (Figure 10.6) by episodic eruption of magma at repose intervals of $1–10^2$ y. Such recurring activity reflects an interplay of many factors, including the rate of replenishment of the magma in the staging chamber from deeper sources; rate of cooling and crystallization of magma in the chamber, which depends largely on chamber size and shape; apparent viscosity of the magma and its composition, especially volatile content; and other factors.

In contrast to these long-lived, large, polygenetic volcanoes, some eruptions consist of only one episode, lasting perhaps months to years, after which activity ceases at that vent system. This **monogenetic** activity forms small simple volcanic edifices, such as a cinder cone and its associated lava flow (Figure 10.4) or a rhyolite dome nestled in its precursory pyroclastic crater (Figure 10.5).

Contrasts between explosive and effusive processes and products are strongly influenced by magma composition. Pyroclastic basaltic deposits are typically only of local, minor volume around source vents. Far more energetic and explosive activity, which reflects generally greater volatile contents and especially greater apparent viscosity in silicic magma, can create vast, thick pyroclastic deposits tens of kilometers distant from the vent and dispersal of finer tephra worldwide. Basaltic lavas a few tens of meters thick can spread tens, even hundreds, of kilometers from the vent, building enormous shield volcanoes (Figure 10.6) and larger flood-basalt plateaus. Some low-silica trachytic and phonolitic magma extrusions are comparable to basaltic extrusions with regard to mobility. Increasingly more silicic lava of greater apparent viscosity forms small thick flows and domes piled high over the vent (Figures 10.2 and 10.5). Intermediate-composition magma extrusions behave in some intermediate manner between rhyolitic and basaltic end members and yield more or less intermediate lava flow morphological characteristics.

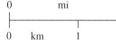

10.3 Youthful Kings Bowl basaltic **fissure eruption** in the eastern Snake River Plain, Idaho. Note subparallel fissures in older underlying basaltic flows (paler gray in photograph) on each side of major feeder fissure. Mantle of ash (lightest gray) covers the east-central part of the youthful flow. Irregular white line crossing photograph from east to west is a road. (Photograph by U.S. Department of Agriculture.)

☀10.2 EFFUSIONS OF BASALTIC LAVA

Basalt is the most widespread magmatic rock on Earth. More than half of the world's volcanoes are of basalt or include basalt. It forms most of the oceanic crust that covers about three-fourths of the Earth and huge continental flood-basalt plateaus as well as smaller local fields. Basalt is found in virtually all tectonic settings.

Effusions of basaltic magma from fissure and central vents vary in size, rate of discharge, surface morphological features, and internal structure. The aggregate volume of one subaerial lava flow, commonly composed of many gushes of lava extruded during a single eruptive event lasting hours to perhaps a year or more, is generally 0.01–1 km^3, although volumes on the order of 10^2–10^3 km have apparently occurred in a single plateau flood. A relatively high extrusion rate for the 12.3-km^3 Laki, Iceland, fissure eruption during

28 days in 1783 was about 5000 m^3/s (~0.5 km^3/day). Higher rates of extrusion allow lava to flow farther, as much as 40 km at Laki, because heat transfer and consequent cooling and immobilization are less important factors. Thickness of a single flow is generally 10–30 m but can be as little as a few centimeters for a low-viscosity lava.

10.2.1 Types of Basaltic Lava Flows

Basically, four end-member types of basaltic flows can be recognized among a wide spectrum of forms. Three are typically subaerial, for which two have names from the native Hawaiian tongue: *aa* and *pahoehoe* (pronounced "ah'-ah" and "pa-ho'-e-ho'-e"). The fourth flow type is subaqueous pillow lava. The main subaerial aa and pahoehoe flows differ in surface morphological characteristics and in the dynamics of emplacement.

Pahoehoe Flows. Low-viscosity lava, especially basalt but also carbonatite, can produce **pahoehoe flows** that consist of thin, glassy sheets, tongues, and lobes, commonly overlapping one another. A quickly congealed vesicular glassy skin insulates the interior and blocks the escape of exsolved gas bubbles from the typically slowly moving lava (10–100 m/h). Eruptive T and gas content can just about be maintained in the lava, even during flow over several kilometers. Restrictions in downslope flow cause the glassy skin of the flow tongue to wrinkle into ropelike festoons (Figure 10.7a). Downslope, lava pressure builds up within the rubbery skin, inflating the sheet or tongue and causing breakouts of a new tongue. The hotter interiors beneath the skin of pahoehoe flows form an intricate network of **lava tube** distributaries so the lava advances in multiple "fingers." Major feeding tubes in upstream parts of the

10.4 SP Mountain, a basaltic cinder cone and associated lava flow in the **monogenetic lava field** 45 km north of Flagstaff, Arizona. Sharply defined, symmetric **cinder cone** with crater at top lies at head of lava flow. Age is about 0.07 Ma. Older cinder cones and a **tuff cone** (upper right) are also visible in distance. (Photograph courtesy of John S. Shelton.)

10.5   Mono Craters silicic **lava domes** in east central California. Note circular **tuff cone** almost 1 km in diameter of pyroclastic material surrounding low lava dome in right foreground. Southward (to left) are additional low lava domes and lava flows or coulees. Sierra Nevada underlain mostly by a Mesozoic batholith in background. (Photograph from Shelton JS. Geology Illustrated, New York, W.H. Freeman, 1966. Used with permission of John S. Shelton, who holds the copyright.)

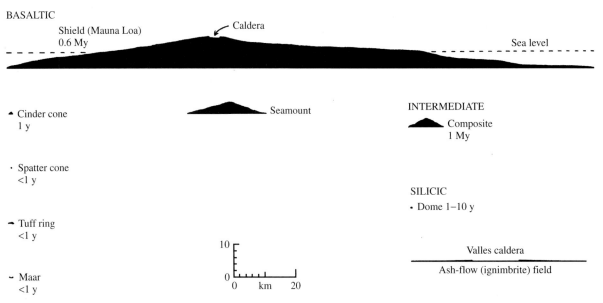

10.6   Comparative sizes of volcanoes and their life spans in years (y) and million years (My). Although composite volcanoes are impressive edifices that can tower some 3 km above their base, they are smaller than seamounts (submerged oceanic volcanoes) and about two orders of magnitude smaller than the large Hawaiian **shield volcanoes.** Monogenetic basaltic tephra volcanoes and silicic domes are two to three orders of magnitude smaller than composite volcanoes.

(a)

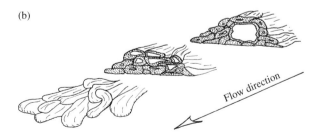

(b)

10.7   Pahoehoe lava flows. (a) Corded glassy pahoehoe tongues 1 m or so in width formed during the 1974 eruption of Mauna Ulu, Hawaii. An active **lava fountain** playing in the left background over the vent feeds actively moving incandescent lava flow tongues that are somewhat lighter colored in this photograph. (U.S. Geological Survey photograph courtesy of Robin T. Halcomb.) (b) Schematic serial sections through a tube-fed pahoehoe flow showing major upstream lava tube and multiple smaller downstream distributary tubes. (Reproduced by permission from Rowland SK, Walker GPL. Pahoehoe and aa in Hawaii: Volumetric flow rate controls the lava structure. Bull. Volcan. 52:615–628, 1990; copyright © 1990 by Springer-Verlag.)

flow can be several meters in diameter and many kilometers long. If, as commonly happens, the lava drains downslope from beneath the crusted skin, the tube becomes an open cavity; these characterize exposed subvertical sections through the interiors of solidified pahoehoe flows (Figure 10.7b). On gentle slopes, elongate **pressure ridges** and more equidimensional, domelike tumuli are elevated segments of pahoehoe flows inflated by more lava input or stranded highs between drained segments. Crestal clefts or tension gashes are commonplace.

Though commonly developed on land, pahoehoe flows can also form subaqueously.

Aa and Block Flows. **Aa flows** are thicker than pahoehoe and have exceedingly rough, treacherous surfaces of irregular, clinkerlike scoriaceous fragments (Figure 10.8). A complementary but thinner rubble layer lies at the base. Flow advance is faster than pahoehoe so that the tensile strength of the cool rigid crust is overcome by the applied stress, causing autoclastic breakup. Heat loss is also higher from the discontinuously exposed incandescent core, causing increased downslope crystallinity in the groundmass (Polacci et al., 1999), which is another striking contrast from glassy pahoehoe. Increasing crystallinity produces more irregularly shaped vesicles than the smooth-walled subspherical vesicles of more glassy pahoehoe. Non-Newtonian rheology in more crystalline lava may be manifested in plug flow (Figure 8.13) and lava levées along channel margins (Figure 10.2).

Block flows resemble aa but have a mantle of more regularly shaped polyhedral chunks rather than jagged, highly vesicular, scoriaceous clinker. Block flows range from basaltic to highly silicic obsidian.

The contrasting morphological characteristics, as well as observed transitions from pahoehoe to aa, *never the reverse,* suggest that greater apparent viscosity promotes development of aa (Wentworth and Macdonald, 1953). Any downslope loss of heat and gas (causing increasing crystallization) allows near-vent pahoehoe to become aa farther downstream. Degassing and increasing nucleation and crystallization are promoted by stirring; consequently lava flows plunging over steep escarpments and lava produced by more vigorous fountaining may change from pahoehoe to aa. More viscous aa that is flowing faster with greater strain rate (Peterson and Tilling, 1980) results in fragmentation of the surface. Calm eruption and slow advance result in pahoehoe.

10.8   Toe of **aa lava flow** in the Black Rock Desert monogenetic basalt field, west central Utah.

Pillow Lavas. Pillow lavas are usually of low-viscosity basaltic magma formed where it comes into contact with water or water-saturated sediment, even in shallow intrusive situations (Walker, 1992). Their most widespread occurrence is on the seafloor where they have developed by extrusion along spreading ridges and on seamounts. Although having the appearance in most exposures, such as roadcuts (Figure 10.9a), of a pile of discrete, independent ellipsoids of pillow shape and size, submarine pillow lavas in some outcrops and especially those viewed on the seafloor in three dimensions consist of a tangled mass of elongate, grooved, interconnected flow lobes that are circular or elliptical in cross section (Figure 10.9b). Flattening of the still hot, not quite rigid pillows produces convex upward tops and cusped bottoms that fill openings between underlying pillows. Such forms are useful indicators for the field geologist of stratigraphic "right-side-up" direction in deposits that have been tectonically tilted. Pillows may resemble pahoehoe toes in cross section; they can be distinguished by the lack of open ellipsoidal internal tubes and the presence of fewer vesicles and radial contraction cracks. Pillows typically have a concentrically zoned fabric (Figure 7.3) reflecting decreasing inward rate of cooling. They may also be progressively encrusted with Mn-Fe oxides as the glassy envelope "weathers" subaqueously to palagonite (Section 7.1.1). Shattering of the hot glassy rinds of pillows in water creates vitroclasts, discussed later.

Moore (1975) observed submarine pillows forming off the 20°-sloping coast of Hawaii in tens of meters of water. Toothpastelike protrusions of fresh lava squeeze out of trapdoorlike openings in an upslope pillow; the newly protruded pillow tongue may then detach and roll away or, if it remains connected, may bud a new protrusion. Pillow formation on flatter slopes has not been observed.

Smooth-surfaced sheet lava flows and lava lakes along oceanic rifts are apparently formed in deep water by more rapid extrusion rates than those that create pillow lavas.

10.2.2 Columnar Joints

All rocks are fractured, mostly because of tectonic forces. However, most tabular bodies of magmatic rock, especially aa lava flows and thin sheet intrusions of basalt, have uniformly spaced **columnar joints** formed by shrinkage during cooling (Figure 10.10). Thermal contraction creates tensile stresses that exceed the brittle strength of the rigid magmatic body. Resulting extensional cracks nucleate at more or less equidistant points on the upper and lower margins of a uniformly cooling tabular body. At each of these nucleation points, a randomly oriented planar crack, or perhaps a three-prong crack, forms and propagates away from its point of origin along the essentially isothermal surface of the flow until it intersects an extensional crack propagating from a neighboring nucleation point. The network of cracks so formed consists of rather regularly sized polygons with four, five, six, or seven sides that resemble dessication cracks in a thin layer of drying, shrinking mud. As cooling and contraction advance into the tabular body, the polygonal cracks likewise propagate inward, forming mostly hexagonal joint columns; these represent the "least-work" configuration of thermally induced tensile-stress fractures in the magma body.

Joint columns have more than aesthetic appeal. Because they develop perpendicular to isothermal cooling surfaces parallel to the margins of a cooling tabular body, the configuration of the body, though its defining margins are now possibly missing as a result of erosion,

(a)

(b)

10.9 Basaltic **pillow lavas.** (a) Roadcut at Nicasio Dam in the Mesozoic Franciscan Complex north of San Francisco. (Photograph courtesy of Mary Hill.) (b) Submarine pillows at a water depth of about 2 km along the Puna Ridge east of the island of Hawaii. (Photograph courtesy of Hank Chezar and D. J. Fornari.)

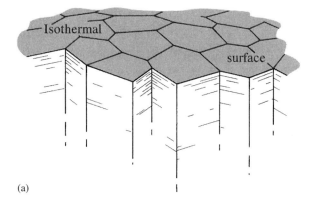

(a)

(b)

10.10 Columnar joints. (a) Schematic view of joint columns oriented perpendicular to an isothermal surface in a cooling magma body. See also Figure 8.19c. (b) Multitiered columnar joints in a remnant of a 100-m-thick, ponded basalt lava flow now exposed just above the Colorado River in the western Grand Canyon, Arizona. Abruptly overlying a basal "colonnade" of thick columns is the "entablature" of thinner columns. Though seemingly haphazard, two radially oriented arrays can be discerned on the left in the entablature. Many lava flows find their way into drainage channels and are subsequently overtopped by water, which penetrates down into the fractured cooling lava flow. Margins of the radial arrays may delineate where the water entered, locally depressing isothermal surfaces. (Photograph courtesy of W.K. Hamblin.)

can be inferred (Figure 8.19c). The pattern of jointing is useful in delineating individually emplaced cooling units in a succession of volcanic deposits.

10.2.3 Subaerial Lava Accumulations

On land, the three most important types of basaltic lava-built accumulations are, in terms of increasing volume, small basalt fields, shield volcanoes, and plateau-forming floods.

Basalt fields form on flanks of larger composite and shield volcanoes, within large calderas, and in other continental areas. Vents may be localized along exten-

sional faults. Generally small (<1 km³), monogenetic extrusions produce simple tonguelike lava flows and associated cinder cones (Figures 10.3 and 10.4). Where rising basalt magma encounters water-saturated sediment or surface bodies of water (lakes), hydromagmatic explosions produce tuff rings and maars (discussed later). In some places, a few lava flows may be superposed but no thick pile is produced as in basalt plateaus. The activity of a field may last a few millions of years and create hundreds of volcanic edifices.

Shield volcanoes are built by innumerable extrusions of low-viscosity basaltic lava flows from a central

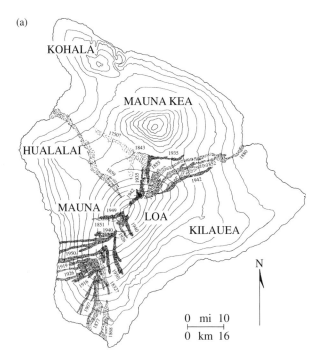

(a)

10.11 The island of Hawaii is built of seven coalescing **shield volcanoes.** See also Figure 9.8. Four of these—Hualalai, Mauna Loa, Kilauea, and Loihi—are active. Loihi seamount south of Kilauea has not yet emerged above sea level. Extinct Mahukona northwest of Hualalai has submerged isostatically (Moore and Clague, 1992). Each shield grew over about 0.6 My. (a) Aa and pahoehoe lava flows, such as the historic flows of Mauna Loa (patterned and labeled), build the subaerial parts of the shields. These flows are extruded from flank fissure (rift) systems and from a summit central vent complex. Deeply eroded older shield volcanoes on older islands reveal extensive feeder-dike swarms marking the fissure systems. Topographic contours are thousands of feet above sea level. (Redrawn from Stearns, 1966.) (b) Offshore bathymetry (depth contours in km) and major submarine slumps (dark shade), and debris avalanches (light shade). Double lines through subaerial shields are rift zones through which most of basaltic lava is extruded. AA′ line is line of cross section in (c). (c) Enlarged cross section along AA′ in (b) of slump south of east rift zone of Kilauea shield volcano showing subaerial lava flows (subhorizontal lines), hydroclastic debris (dashed lines), pillow lava (ellipses), sheeted feeder dikes (vertical lines), gabbro (dotted pattern), and magma (black). Dark shaded in lower right is mantle. (Redrawn from Moore et al., 1994.)

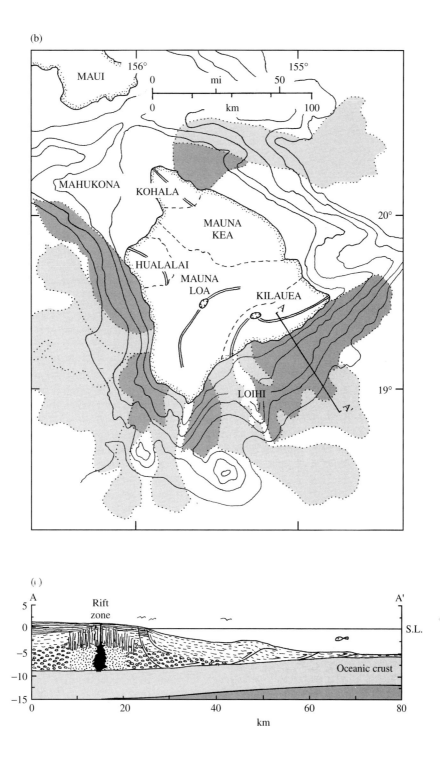

10.11 (*Continued*).

vent complex and locally one or more fissure systems radially disposed from it; the resulting edifice shape resembles a warrior's shield (Figures 9.8, 10.6, and 10.11). Small shields, as in Iceland and many continental areas, have diameters of a few kilometers. Shield volcanoes find their greatest development in the Hawaiian Islands. These gigantic edifices grew from the seafloor by submarine extrusions (discussed later) for tens to hundreds of thousands of years before even-

tually becoming subaerial. They are the largest volcanoes on Earth, have diameters of over 100 km, and rise as much as 10 km above their base on the seafloor. (Mount Everest is only 8.85 km above sea level!) So great is their weight that the oceanic lithosphere is flexed downward, resulting in greater ocean depths immediately around the isostatically subsiding island mass. The largest island, Hawaii (Figure 10.11), has grown over the past 0.6 My and is composed of seven coalesced shield volcanoes, the oldest of which is submerged to the northwest; the youngest, to the southeast, has not emerged above sea level. Gravitational instability of the flanks of the compound shield edifice causes recurrent landslides (Figure 10.11b, c). These are more or less coherent slumps of a flank sector whose slope is >3° as well as fast-moving chaotic debris avalanches on gentler slopes, some of which have gigantic volumes on the order of 5000 km³—the largest avalanches known on Earth.

Plateau-forming, fissure-fed **flood basalts** are the most voluminous subaerial lava extrusions of any composition known on Earth. Many **continental plateaus** occur around the globe, including the Jurassic Karroo-Ferrar in southern Africa and Antarctica, Paraná-Etendeka in South America and southwestern Africa, and the late Cretaceous-early Paleocene Deccan in India (Figure 10.12). The Columbia River Plateau of the northwestern United States (Figure 10.13) consists of more than 100 flows emplaced in a remarkably brief period of the Miocene—almost entirely 17–15 Ma (Reidel and Hooper, 1989). Their aggregate volume is about 180,000 km³, covering an area of 160,000 km² to a depth as much as 3 km. By comparison, the large Mauna Loa shield volcano on the island of Hawaii has one-sixth this volume.

Individual large lava flows, literally floods, whose stratigraphic correlation has been facilitated by "chemical fingerprinting" using particular element concentrations and ratios, range in volume from 90 km³ to perhaps as much as 3000 km³. Some flows traveled hundreds of kilometers down gentle slopes of 1 m in 10 km (1/10,000). On the basis of a model of lava transport in inflated pahoehoe flows, Self et al. (1997)

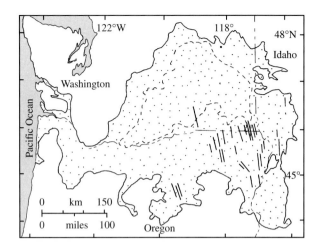

10.13 Columbia River **plateau flood basalts,** northwestern United States. Subparallel north-northwest-trending lines show the approximate location and orientation of known groups of feeder dikes that number in the thousands. Extent of Roza flows (Figure 10.14) indicated by dashed line. (Redrawn from Reidel and Hooper, 1989.)

propose that the gigantic compound Roza flow field (Figure 10.14) was emplaced over a period of 6–14 y and individual flows in 5–50 months. Lava was able to travel great distances because it did so under an insulating pahoehoe crust and the only place where the hottest mobile lava became exposed was at local breakouts feeding new pahoehoe tongues. Calculated effusion rates are about 4000 m³/s, which is comparable to that of the 1783 Laki, Iceland, fissure eruption cited previously. Though such discharge rates seem extraordinary, the rate for the peak 2-My plateau-forming episode is about 0.08 km³/Y, comparable to discharge rates of Hawaiian shield volcanoes and oceanic rifts.

In flood basalt plateaus, geologically rapid withdrawal of $10^2 - 10^3$ km³ of magma did not result in caldera collapse, as in silicic ash-flow eruptions of comparable volume (discussed later). Roofs over shallow crustal silicic chambers tend to be thinner than their spanning diameter and hence readily collapse. In contrast, roofs over magma reservoirs feeding basalt floods that lie in the upper mantle, possibly the lower crust, have sufficient thickness and strength to resist collapse and merely subside over time. Discharge of huge volumes of compositionally relatively uniform basalt magma from the mantle raises questions regarding magma generation and storage.

10.2.4 Submarine Basaltic Accumulations

Most submarine extrusions are of basaltic lava and account for most of global volcanism (Figure 1.1). Most submarine activity involves fissure eruptions at oceanic ridges, whereas more localized central eruptions have built in excess of one million basaltic volcanoes dotting the seafloor (Plate I). Most of these volcanoes are sub-

10.12 **Plateau flood basalts** exposed in the Western Ghat escarpment of the Deccan Plateau, India. Photograph provided courtesy of Peter R. Hooper. [Reproduced by permission from Hooper PR. Flood basalt provinces. In Encyclopedia of Volcanoes, Sigurdsson H, Houghton B, McNutt SR, Rymer H, Stix J, eds. 345–359:2000; copyright © 2000 by Academic Press (a division of Harcourt Brace and Company).]

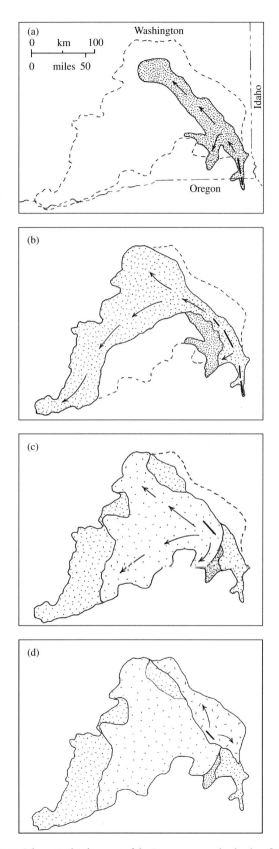

10.14 Schematic development of the Roza compound pahoehoe flow field in the Columbia River plateau (Figure 10.13). Arrows indicate direction of travel of basalt lava from fissure vent source (heavy line) beneath the insulating pahoehoe crust. (Redrawn from Self et al., 1997; Martin, 1989.)

merged **seamounts** >50−100 m high; others have grown from a submarine base (Figure 10.11c) into sub-aerially exposed volcanic islands such as Hawaii.

The confining pressure exerted by seawater and especially the relatively low volatile concentrations in basalt magmas preclude explosive eruptions in all but the shallowest water depths. Because hydrostatic pressure increases by about 1 bar ($= 10^5$ Pa) for every 10 m of water depth and because oceanic ridge basalt magmas typically have small water concentrations of <0.5 wt.%, exsolution is limited to water depths of <500 m (Figure 4.7). Ocean ridge basalt magmas may contain concentrations of CO_2 comparable to that of water but, because of its much lower solubility (Figure 4.10), may exsolve below the seafloor. Moreover, explosive fragmentation of magma due to bubble growth cannot take place until some critical bubble volume fraction develops, and this requires shallower depth for bubble expansion. After a volcano has grown from the deep seafloor to shallower depths of perhaps <200 m, explosive fragmental deposits can be formed.

Magmatism along the 65,000-km-long system of oceanic spreading ridges around the world (Frontispiece and Plate I) is the most prolific on Earth (Figure 1.1). In concert with seafloor spreading, the entire present-day oceanic crust has been produced in <200 My.

Extrusions along oceanic ridges at water depths of a few kilometers typically produce pillow lava and,

---

Special Interest Box 10.1 Speculative triggering of El Niño by submarine basalt extrusions

In the last two decades of the 1900s, a major anomaly in the weather pattern in some large continental areas surrounding the Pacific Ocean gained considerable attention. This so-called El Niño is caused by episodic perturbations at irregular recurrence intervals among sea surface temperatures, sea level anomalies, and atmospheric wind stress. Shaw and Moore (1988) have proposed a novel but controversial hypothesis that El Niños might be triggered by extrusion of large-volume submarine basalt flows along the East Pacific Rise off the northwest coast of South America. This is the fastest oceanic spreading ridge in the world and is the area where El Niños seem to be spawned. Shaw and Moore speculate that some of these submarine flows may have volumes and extrusion rates comparable to those of the flood basalt lavas that formed continental plateaus. Every 1 km^3 of submarine basalt that cools to ambient ocean temperatures within the time of an El Niño cycle, they calculate, accounts for about 1% of the anomaly.

where extrusion rates are greater, sheet lava flows, locally pahoehoe. Along the Mid-Atlantic Ridge, a crestal rift valley, 20 to 30 km wide and 2 km deep, has extensional fault-controlled terraces flanking a narrow inner valley only a few kilometers wide and 100 to 400 m deep. Topographically controlled lava flows on the inner valley floor are elongate (Figure 10.15). Along the faster-spreading East Pacific Rise, lava lakes have ponded on low-relief flanks. One youthful flow discovered by side-looking sonar covers 220 km² and has a volume estimated at 15 km³ (Shaw and Moore, 1988).

Counterparts of the huge continental flood-basalt plateaus were discovered during investigations of the ocean floors in the late decades of the 1900s (Mahoney and Coffin, 1997). **Oceanic plateaus** are broad topographic highs rising 1 km or so above the surrounding seafloor and underlain by a crust as much as 40 km thick, five to six times typical oceanic crust. Little is known of these plateaus.

✳10.3 EFFUSIONS OF SILICIC LAVA

10.3.1 Morphological Characteristics and Growth

The most crystalline and, especially, the most silica-rich lavas are the least mobile because of their high apparent viscosities. Accordingly, silicic effusions have much larger aspect ratios (thickness/length) than typically sheet-like basaltic lava flows. A spectrum of shapes can

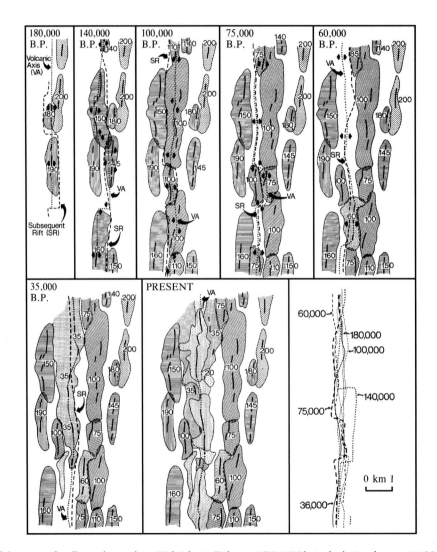

10.15 Evolution of the inner rift valley in the northern Mid-Atlantic Ridge at 36°N 50′N latitude during the past 0.2 My. Shaded areas show volcanic edifices built along the active volcanic axis (VA) that are later cut by a subsequent rift (SR). Lower right diagram combines all of these rifts. Small numbers give ages of volcanic edifices in thousands of years. (From Ballard RD and van Andel TH, Morphology and tectonics of the inner rift valley at lat 36° 50′N on the Mid-Atlantic Ridge. Geol. Soc. Am. Bull. 88:507–530, 1977; Reproduced with permission of the publisher, The Geological Society of America, Boulder, Colorado USA. Copyright © 1977 Geological Society of America.)

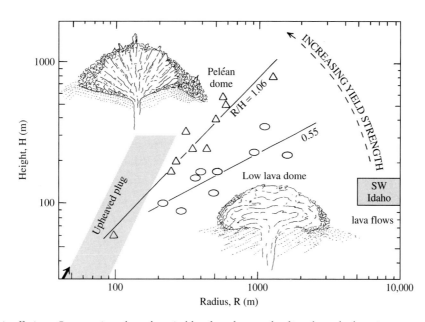

10.16 Shapes of silicic effusions. Cross sections through typical low **lava dome** and **peléan dome,** both resting on precursory pyroclastic deposits, shown as dotted lines. (Redrawn from Williams, 1932.) Note aprons of mostly block-size talus. Measured dome height, H, versus dome radius, R, of selected low lava domes (ellipses) and peléan domes (triangles) around the world are plotted with a best-fit line representing the indicated aspect ratio, H/R. (Redrawn from Blake, 1990.) The aspect ratio of peléan domes is similar to the tangent of angles of repose of the talus pile of unconsolidated fragments that typically surround them. Note that these talus accumulations are significantly less around low lava domes. Shaded rectangle on right represents large-volume (as much as 200 km³), high-T rhyolite lava flows in southwestern Idaho. (From Bonnichsen and Kauffman, 1987.) Shaded band on left represents possible aspect ratios for high-yield-strength upheaved plugs. As silicic lava emerges from a vent (arrow, lower left), with increasing height, H, it may grow into a lava flow, low lava dome, peléan dome, or upheaved plug depending on increasing yield strength.

be recognized (Figure 10.16). The most mobile with the smallest aspect ratio is the lava flow (Figure 10.17). The more viscous mushroom-shaped **lava dome** (Figures 10.2, 10.5, 10.18) has a larger aspect ratio. The still larger-aspect-ratio **peléan dome** resembles an artichoke and grows by expansion from within, pushing slabs of rigid lava out of and away from the vent in fan fashion or along sled-runner-shaped ramps. Although most silicic effusions have blocky fractured tops and surrounding aprons of blocky talus (Figure 7.33), these are more pronounced in peléan domes. Slender **spines** may be elevated above the remainder of the peléan dome before being shattered by steam explosions or thermal stresses or collapsing under their own weight. Least mobile, rigid lava is extruded as an **upheaved plug,** which is an elongate cylindrical mass roughly the diameter of the vent conduit so that its aspect ratio is >1; a plug resembles a cork in a narrow-necked wine bottle.

Many domes grow in craters (Figure 10.5) produced by preceding explosive eruption of more volatile-rich magma from the top of the supply chamber. Explosive activity may continue during the slow continuous or episodic effusion of lava, which advances from the vent at velocities of meters per hour to meters per day. The duration of lava effusion can be many decades in the case of large domes, such as the 1-km³ growth of Santiaguito, Guatemala, since 1922 (Rose, 1987). In some effusions it is clear that new lava is slowly added in an **endogenous** manner (Figure 10.19), inflating the interior beneath a more rigid carapace or cover of closely packed blocks of lava. Other effusions are **exogenous** and grow by addition of lava onto the surface.

Dome growth may be modeled in two ways, as a brittle shell enclosing pressurized magma or as viscoplastic lava. In the viscoplastic model (Blake, 1990; see also Section 8.2.2), the chilled carapace of a low lava dome has yield strengths of 10^5–10^6 Pa, whereas a steeper-sided peléan dome, whose form is dictated as much by its talus apron as by the rheology of the lava core, has a yield strength >10^6–10^7 Pa.

Autoclastic fragments, generally of block size, mantle flows and domes and form talus aprons around them (Figure 7.33). Fragments are created by movement of the rigid carapace, thermal stresses, or internal gas pressure, or by collapse under their own weight. As the effusion advances laterally, the clastic apron may be pushed aside or overridden and possibly engulfed within the flow. Larger-volume crumbling of sectors of the steep dome and plug margins on summits or on slopes of larger volcanoes creates avalanches of considerable runout, as in the Chaos Jumbles in Figure 10.18. More explosive shattering generates devastating pyro-

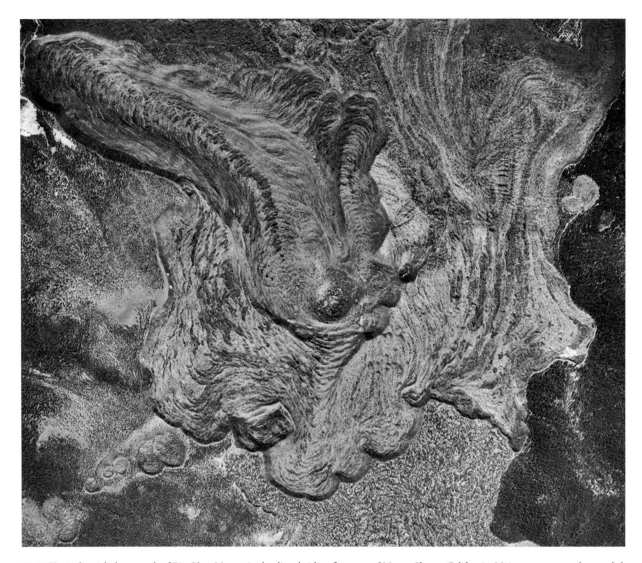

10.17 Vertical aerial photograph of Big Glass Mountain rhyolite obsidian flow east of Mount Shasta, California. Main tongue toward upper left is nearly 3 km long, has a thickness of about 75 m, and flowed down an approximate 9° slope from the vent near photo center. Arcuate **pressure ridges** in flow surface transverse to flow direction that are spaced about 60 m apart with amplitudes averaging 8 m possibly originated through compressional drag forces exerted in the more rigid flow surface by the underlying hotter, more mobile flowing interior. Note east-northeast trending line of extrusive vents beginning at lower left corner and ending at upper right that were probably controlled by a fissure. (Photograph by U.S. Forest Service.)

clastic avalanches that cascade with hurricane speed downslope.

10.3.2 Internal Fabric

Planar and linear fabric elements are widespread in silicic effusions and constitute markers of the pattern of internal flow in the body during emplacement. More or less planar flow layering (Figures 7.4 and 7.41) expressed by textural variations (crystal size, crystallinity, vesicularity) in a wide range of lava viscosities reflects laminar flow. Contorted and folded flow layers reflect local changes in flow velocity or drag, particularly near flow margins. Grooves and striations, which resemble slickenlines on fault surfaces, and associated transverse

open tension gashes also develop on shear surfaces in viscous lava.

Rhyolite flows and domes have a more or less consistent vertical internal zonation of fabric developed by variable vesiculation, fragmentation, and devitrification superposed on rheologic flow. This zonation serves as a useful field guide in the interpretation of poorly exposed old flows. Thicknesses of zones vary considerably, and a particular zone may not be everywhere present. Tops and bottoms of the extrusion are of autobreccia that is variably vesiculated, as are local internal seams representing brecciated flow margin material engulfed during flow or zones of rigid magma that were fragmented. Interiors are flow-layered. Young effusions

(a)

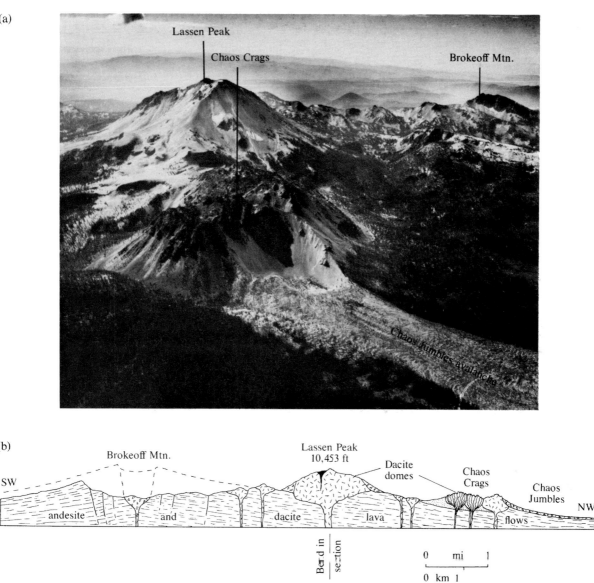

(b)

10.18 Aerial photo (a) and cross section (b) of a part of the Lassen volcanic area, northern California. Brokeoff Mountain is an erosional remnant of a large composite volcano (dashed lines). Collapse of a sector of Chaos Crags dacite peléan dome, possibly in the 19th century, produced the Chaos Jumbles avalanche. A prominant talus apron extends around the dome complex on each side of the sector that collapsed. A small eruption of dacite lava, black in (b), from the summit of Mount Lassen occurred in 1915 and spawned a major debris flow. (Cross section redrawn from Williams, 1932; photograph courtesy of John S. Shelton.)

less than a few million years old generally have obsidian interiors, but with increasing age, glass gradually hydrates and devitrifies, developing perlitic and crystalline textures, respectively. Early devitrification produces spherulites strung along flow layers, which, in cross sections, resemble beads on a string. In the more slowly cooled interiors of some thick flows, devitrification begins during effusion, whereas quenched flow margins are glass. Ultimately, all glass devitrifies and becomes a felsitic mass of aphanitic feldspar and

quartz. Interacting crystallization and release of volatiles creates lithophysal zones.

During flow of silicic to intermediate composition lavas, platy feldspar microlites in the groundmass become aligned to form trachytic fabric; volatile bubbles may be drawn out and similarly oriented. The foliation thus produced creates a pervasive parting that, particularly after accentuation by weathering, is manifested in platy fragments a few centimeters in thickness forming aprons of talus around flow margins.

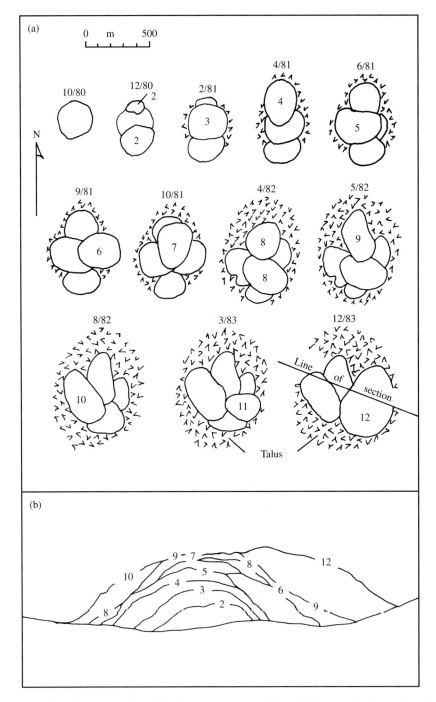

10.19 Mainly endogenous growth of the composite dacite **lava dome,** Mount Saint Helens, Washington. (Redrawn from Swanson et al., 1987.) The dome grew as new magma inflated the more rigid carapace. It is sited in a crater formed by the catastrophic explosions of May 18, 1980 (Figure 10.1) and was modified somewhat in succeeding months when numerous pyroclastic flows were erupted. Until August 1982 (8/82) the dome is portrayed in schematic map views (a) at end of each growth episode. During 1983 growth was continuous and only two arbitrary stages are represented in March 1983 (3/83) and December 1983 (12/83). (b) Schematic cross section at larger scale viewed southward shows topographic profiles of the composite dome at times when individual domes formed, from (a). Talus omitted.

✳ 10.4 EXPLOSIVE ERUPTIONS

The complex sequence of interrelated processes whereby magma near the surface of the Earth explodes and becomes a clastic deposit is not completely understood. For purposes of discussion, the continuum of interrelated processes can be considered in terms of the initiating explosive production of pyroclasts and their subsequent transport and deposition. Explosive discharge produces a volcanic plume from which pyroclasts are eventually deposited.

Despite accelerated research in the last two decades of the 20th century due to increasing application of fluid dynamic modeling and detailed observations of many eruptions, more new questions seem to have been created than old ones answered.

10.4.1 Explosive Mechanisms: Production of Pyroclasts

Explosive production of pyroclasts (tephra or ejecta) involves the expansion of volatiles—whether contained in magma or external water in the environment or in combinations of these (Section 7.7).

<u>Exsolution and Expansion of Dissolved Volatiles.</u> Fragmentation of magma caused by the volatiles dissolved within the melt follows a sequence of events beginning with exsolution of volatiles in an oversaturated melt. After nucleation, bubbles grow by continued exsolution and possible volumetric expansion, culminating in explosive rupture of the bubble walls, converting thermal energy into kinetic energy. The mechanism through which bubbles rupture their walls and explode is poorly understood (Figure 4.13; see also Section 6.7.2); it is obviously related to melt viscosity and/or water content because these are greatest in the most explosive silicic magmas. Excessive internal pressure in volatile bubbles may not be the only factor. The rate at which the bubbly melt expands upward in the volcanic conduit may play an important role; localized faster expansion strains the bubble walls faster so that they rupture as brittle glass rather than slowly stretching as a viscous melt.

There have been countless witnessed explosive eruptions. A well-documented one is that of Mount Saint Helens on May 18, 1980 (Figures 10.1, 10.20, 10.21). Though the pyroclastic material produced was small

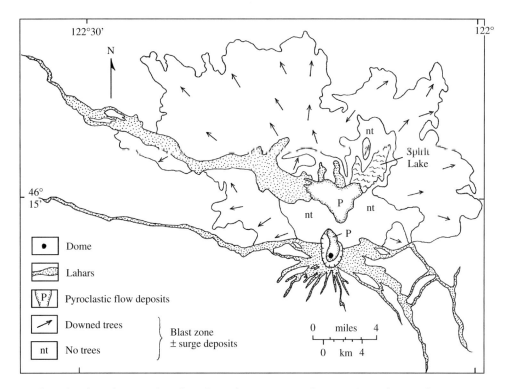

10.20 The variety of mainly volcaniclastic products from the explosive eruption of Mount Saint Helens, Washington, May 18, 1980 (Figure 10.1). Only deposits proximal to the volcano are shown. More distal ash-fall deposits are shown in Figure 10.21. In the 600-km² lateral blast zone where local minor surge deposits occur, trees as much as 2 m in diameter were blown down (arrows indicate tree orientation) and were completely blown away closer to the volcano in the "no tree" (nt) area. Some three dozen steam-blast explosion pits in the pyroclastic flow (P) are also not shown; they were produced as the hot deposit vaporized overridden bodies of water. Note that lahars were mostly confined to existing stream channels; on the upper steep slopes of the volcano on the east, south, and west, loose rock and soil were scoured away to feed lahars farther downslope. The lava dome emerged and grew months after the May 18 eruption (Figure 10.19) within the elliptical crater shown as line with tick marks. (Redrawn from Lipman and Mullineaux, 1981, Plate 1.)

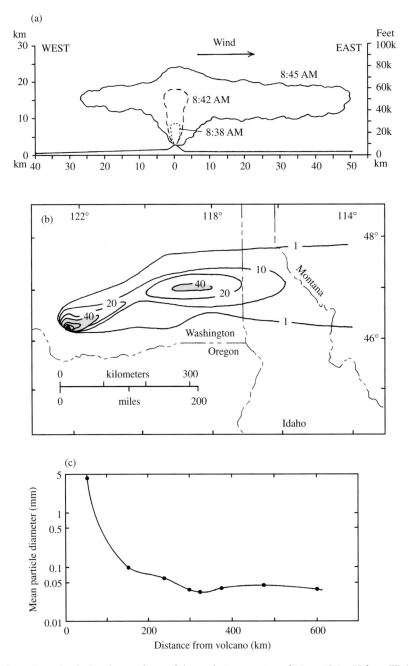

10.21 Ash-fall deposit from the 20-km-high volcanic plume of the explosive eruption of Mount Saint Helens, Washington, May 18, 1980. (a) Thirteen-minute growth of plinian plume beginning at 8:32 AM. (b) Distribution of ash. Isopach lines of constant uncompacted thickness in millimeters. Innermost two isopachs immediately surrounding the volcano are 100 and 200 mm. Anomalously thick area in eastern Washington probably reflects clumping of fine ash and premature fallout. (c) Mean particle diameter plotted on logarithmic scale against distance along axis of fallout in (b). Note change in horizontal distance scale from (a) to (b) and (c). (Redrawn from Sarna-Wojcicki et al., 1981.)

(<1 km³), the accompanying blast devastated a large area and debris flows (lahars) created considerable downstream damage. Airborne ash traveled around the Earth, and as much as 1 cm of ash was deposited 500 km distant.

Magma-Water Interactions. **Hydromagmatic explosions** can occur wherever magma contacts external water.

Because of the ubiquity of water on the surface of the Earth these explosions occur in a wide variety of geologic environments. Ascending magma can encounter shallow water along coasts of islands and continents as well as water-soaked ground, rocks containing water in fractures, and lakes (Figures 10.22 and 10.23). Because water is common in volcanic craters, recurrent rise of magma into volcanoes can trigger hydromagmatic ex-

10.22 Surtseyan eruption of Capelinhos volcano, Azores. The central black plume choked with basaltic ejecta is estimated to be about 400 m high. A ring-shaped **base surge** has formed at its base and a white steam cloud lies behind and to the right. (Photograph courtesy Richard V. Fisher from Othon R. Silveira of Horta, Azores; from Waters AC, Fisher RV. Base surges and their deposits: Capelinhos and Taal volcanoes. J. Geophys. Res. 76:5596–5614, 1971; Published 1971 by the American Geophysical Union.)

plosions. Lava extruded from subaerial vents can flow into the sea, lakes, and rivers and over water-soaked ground. On high volcanoes or at high latitudes, magma can contact snowfields and glaciers.

Because of their high T and large heat capacity, magmas contain a vast amount of thermal energy that can be transformed into PV energy (Section 3.2.2) as external water is vaporized to high-pressure steam. Explosive yields of rapidly vaporized water can be as much as one-third the yield of an equivalent mass of TNT (Francis, 1993). A kilogram of magma that contains 1.6×10^6 J of energy converting a fixed (confined) volume of water from 0°C to 1000°C produces a pressure of 500 MPa (Sparks et al., 1997, p. 14). This pressure exceeds the brittle breaking strength of rock by as much as two orders of magnitude. It is little wonder that some of the most explosive volcanic eruptions involve magma-water interactions.

However, not all magmas contacting water produce explosions. For example, lava entering the water at depths of less than a few tens of meters along the coast of Hawaii produces pillows (Moore, 1975) and only minor explosive activity. Other factors obviously control the intensity of hydromagmatic explosions. One appears to be "premixing" of large surface areas of magma with water. Explosive activity along the coast of Hawaii is more common where aa flows rather than pahoehoe enter the sea. The irregularly shaped, vesicular chunks of lava in aa provide for more heat transfer to water than smaller-surface-area pahoehoe and pillow lavas insulated by a smooth skin of glass.

It may also be that many hydromagmatic eruptions occur because magma contacting external, near-surface water is independently vesiculating and perhaps fragmenting because of exsolving volatiles in the magma; this is another way of premixing large-surface-area magma with water. Fine-scale fragmentation is apparently required to create high rates of heat transfer and energy release. This may happen (Wohletz, 1986) as an expanding layer of vaporized water develops at the interface between magma and water; in some, not well understood, manner, cyclic collapse and regeneration of this layer on short time scales (microseconds) explosively fragment the magma, typically into ash-size granules.

Processes and products in magma-water interactions range widely, depending chiefly on the water/magma ratio. Ratios of about 0.3–0.4, depending on magma composition, appear to be optimal for conversion of the thermal energy of the magma into the work of magma fragmentation, ejection/dispersal of pyroclasts, and possible crater excavation into underlying rock.

Hydroclastic refers to any fragmental material created by interactions between magma and water. Such deposits that are principally vitroclasts are referred to as **hyaloclastites**. These form by nonexplosive spalling and granulation of glassy rinds on pillow and pahoehoe flows in contact with water and by explosive hydromagmatic processes in a wide range of subaqueous and subaerial geologic environments. Where magma invades unconsolidated sediment near the surface of the

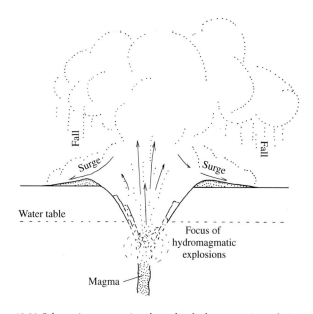

10.23 Schematic cross section through a **hydromagmatic explosion** system. Focus of explosion is the area where ascending column of magma contacts external water. Pyroclastic surge and fall deposits, including ballistic clasts, build a low **tuff ring** around the deepening **maar** crater.

Earth, water lodged in pore spaces expands and creates a situation where the magma can produce complex physical mixtures with sediment; the resulting rock is called **peperite.**

10.4.2 Pyroclasts in Volcanic Plumes

Most pyroclasts are ejected from a vent into a **volcanic plume**—a mixture of pyroclasts and hot expanding gas, chiefly steam, that is discharged explosively into the atmosphere (Sparks et al., 1997). On the basis of ballistic clast trajectories (discussed later), plumes exit volcanic vents at velocities as high as 600 m/s (a supersonic 2160 km/h). Higher velocities correspond with higher exsolved volatile concentrations in the erupting magma, whereas the height of the plume, to as much as 50 km, is controlled mainly by magma discharge rates (mass flux) from the vent. Discharge rates as high as about 0.1 km^3/h have been determined for historic eruptions, but they may have been at least an order of magnitude greater for colossal prehistoric eruptions. Discharge rates are in turn largely governed by the radius of the vent. (Most explosive vents can be considered to be more or less circular as the explosive process reams out fractured wall rock, eliminating inward projecting irregularities and creating a minimal area of circumferential surface.) Once formed, plumes can be sustained for hours to months if the magma supply is not exhausted and if the ascent rate of the material in the plume is less than the discharge rate. Other plumes accompany single instantaneous bursts.

Plumes are of many types. Those that are produced by hydromagmatic explosions are shown in Figures 10.22 and 10.23. Other types of plumes that depend on the water content of the erupting magma and vent radius are shown in Figures 10.24a and 10.24b. High-energy **plinian plumes** are created by blasting of gas-rich magma from smaller vents (Figure 10.25). (Pliny the Younger was an eyewitness to and described the 79 AD eruption of Vesuvius in southern Italy.) Above a gas-thrust region, turbulent plinian plumes engulf and heat atmospheric air, become buoyant, and rise convectively to tens of kilometers above the vent, forming the giant, visually impressive "cauliflower" ash-laden clouds accompanying explosive eruptions. Where the cloud becomes neutrally buoyant, it spreads horizontally, creating an umbrellalike form. Plinian plumes disperse pyroclasts over wide areas in ash-fall deposits. Lower-energy **collapsing columns,** resembling water fountains, are created by eruption of less-volatile-rich magma from larger vents. Discharge rate is so great that the plume contains more pyroclastic mass than can be lifted buoyantly; consequently the eruptive column collapses under its own weight. Collapsing columns produce ground-hugging pyroclastic flows and surges that move radially away from the base of the fountain at hurricane speeds. Such flows can themselves generate secondary **coignimbrite plumes** that are produced as fine ash is flushed out of the flow by buoyantly rising gas.

Some pyroclastic deposits (Figure 10.26b) indicate that low-energy fountains and high-energy plumes can alternate over a period of days to months from the same localized vent system. Other systems may begin, for example, with a plinian plume and end with a collapsing column.

10.4.3 Pyroclast Transport and Deposition

Pyroclasts blown from a volcanic vent, mostly in an explosive plume, are then transported by pyroclastic fall, flow, and surge and eventually deposited. The fabric and field relations of these three types of deposits are generally distinctive, but (as usually happens when humans impose a classification on nature) some pyroclastic deposits have hybrid aspects, emphasizing the need for caution and an open mind in interpreting them. For example, strong near-vent winds accompanying plinian fall can produce reworking of pyroclasts so that the deposit may resemble a surge deposit. In many localities, fall, surge, and flow deposits are interlayered in complex fashion (Figure 10.26b). Finally, nonvolcanic, or epiclastic, processes can rework tephra to produce features resembling those of primary pyroclastic deposits.

Pyroclastic flows are gravity-driven hot avalanches of mostly juvenile pyroclasts and gas that sweep downhill and across the landscape with hurricanelike speed; deposits are *unsorted* accumulations of ash and pumice lapilli and blocks that fill in topographic features as a flood of water does. Pyroclastic flows are of such great importance in the volcanic record that they are treated separately in Section 10.4.5.

<u>Pyroclastic Fall</u>. Gravity-induced fallout of ejecta from explosive volcanic vents, principally the overlying convecting volcanic plumes, creates **pyroclastic-fall deposits,** also called **ash-fall deposits.** Their extent, thickness, sorting, particle size parameters, especially maximal and mean sizes, and other characteristics depend on the nature of the preeruption magma chamber and conduit/vent geometric characteristics, discharge rate and duration, style of eruption, and nature of the associated eruption plume, especially its height, wind characteristics, and the aerodynamic properties of the pyroclasts. Careful measurements of fall deposit properties allow plume character and duration to be estimated (Sparks et al., 1997).

The largest fragments commonly ejected from volcanic vents are approximately >10 cm in the bomb- and block-size range and are called **ballistic clasts** because they are hurled on ballistic trajectories from the vent, resembling projectiles shot from a cannon. They can land as far as 25 km from the vent but most fall closer. Ballistic trajectories are essentially unaffected by

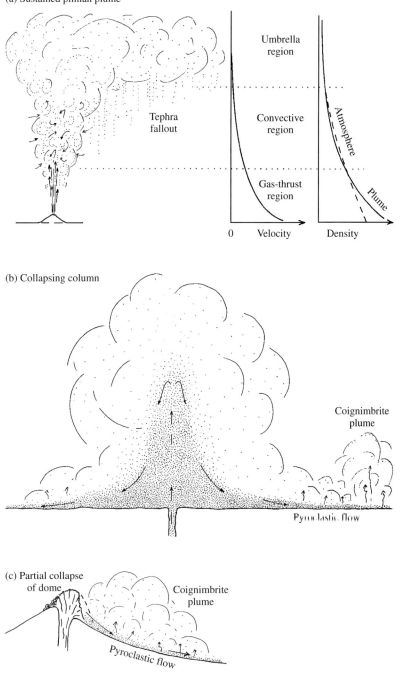

(a) Sustained plinian plume

Umbrella region

Tephra fallout

Convective region

Gas-thrust region

Atmosphere

Plume

0   Velocity   Density

(b) Collapsing column

Coignimbrite plume

Pyroclastic flow

(c) Partial collapse of dome

Coignimbrite plume

Pyroclastic flow

10.24 Volcanic plumes. Three diagrams are not at the same scale. (a) Sustained **convecting plinian plume** showing three dynamic regimes and their relation to column height, velocity, and density of plume and atmosphere. (Redrawn from Sparks et al., 1997.) In the gas thrust region, the expanding volatile fluid exsolved from the magma imparts upward-directed momentum to the plume, much as exploding gun powder propels shot from a gun barrel. In the convective region, thermal energy contained in the turbulent plume heats entrained atmospheric air, decreasing the density of the plume to less than that of the normal atmosphere and providing buoyant lift. In the umbrella region, the density of the expanding, convecting plume matches that of the density-stratified atmosphere so the neutrally buoyant plume spreads horizontally. Nonetheless, the upward momentum of the plume in the convective regime causes it to overshoot the level of neutral buoyancy so that the umbrella region can have a substantial thickness. Although the overall height of the plume is chiefly a function of magma discharge rate correlated with vent radius, other factors such as atmospheric T and humidity and wind velocity also influence height. **Ash-fall deposits** form by fallout from the plume hundreds to thousands of kilometers from the vent. (b) A **collapsing column** forms if the upward momentum of the plume exiting the vent is incapable of lifting it more than a few kilometers. The mass of ejecta is too great to be lifted buoyantly and, consequently, the pyroclasts fall back to the ground, where their kinetic energy gained during fall-back propels them away from the fountain as **pyroclastic flows.** However, an overlying convecting buoyant plume is formed by mixing of atmospheric air with fine pyroclasts in the outer part of the plume. A **coignimbrite plume** develops over the pyroclastic flows. Collapsing columns and plinian plumes may be less symmetric in nature than shown here. (c) Collapse of a sector of a peléan dome producing a small **block-and-ash pyroclastic flow** and overriding **coignimbrite plume.** Compare with Figure 10.36.

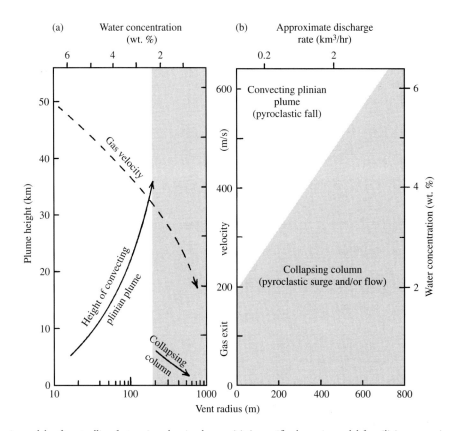

10.25 Fluid dynamic models of controlling factors in volcanic plumes. (a) A specific dynamic model for silicic magma in which *T* = 1200K, lithostatic pressure prevails in the conduit, and water is the only volatile. Decreasing water concentration in the erupting magma (from a zoned chamber in which the magma initially has about 6 wt.%) reduces the exit velocity of the plume gas at the vent. With increasing vent radius (due to erosion during explosive eruption), the height of the convecting plinian plume first increases as a result of increasing magma discharge rate and thermal energy available to heat entrained air to provide buoyancy; however, as this rate reaches a critical value at a vent radius of 200 m and water concentration near 2.4 wt.% the erupting mass is too great to be lofted higher by influx of heated air and the plume collapses (shaded part of diagram). (b) General model conditions showing that convecting plinian plumes are favored if the magma has a high water concentration and a high exit velocity at a small-diameter vent, whereas collapsing fountains occur for the opposite conditions. Note that a particular eruption can be stabilized or sustained in one or the other plume regime—either convecting or collapsing. A geologically likely transition is from a convecting plume to a collapsing column. (Redrawn from Wilson et al., 1980.)

wind or convection in the plume; thus, ejection velocities can be calculated. In contrast, smaller pyroclasts approximately 1–10 cm that include coarse lapilli and small blocks and bombs are mostly lofted by turbulent suspension in a convecting volcanic plume but will fall if their terminal velocity exceeds convective updraft in plume margins. Smallest pyroclasts are <1 cm and include fine lapilli and ash, which commonly account for most of the ejecta; they are also suspended by turbulence but can be dropped from the umbrella part of the plume as convective energy dissipates. Settling velocities of the finest ash particles may be smaller than wind currents, in which case they may circle the Earth several times before eventually settling. These can be responsible for multihued, pastel sunsets worldwide for years after a major pyroclastic eruption.

Particle size and density control terminal velocity (Stokes's law; Section 8.3.3) so that larger and denser clasts are preferentially dropped nearer the vent (Fig-

ure 10.21). Consequently, a pyroclastic fall deposit at any one location consists of particles of similar size: That is, deposits are well sorted. However, fine ash can clump into larger particles in the turbulent plume, if it is wetted by water condensed from cooling steam, forming accretionary lapilli. Other clumps form because of electrostatic attraction, forming porous "ash snowflakes." Both aggregates of fine ash fall closer to the vent. Fine ash can also fall prematurely if it is entrapped by falling raindrops or by larger falling particles. For these reasons, fallout deposits may not be as well sorted as expected, but the mean and maximal sizes of particles generally decrease with distance from the vent.

Fresh, unconsolidated pyroclastic-fall deposits of ash and highly vesiculated pumice fragments are best preserved in marshes, lakes, and deep oceans; where reworking by wind and water currents is limited; or where they are immediately covered by other deposits.

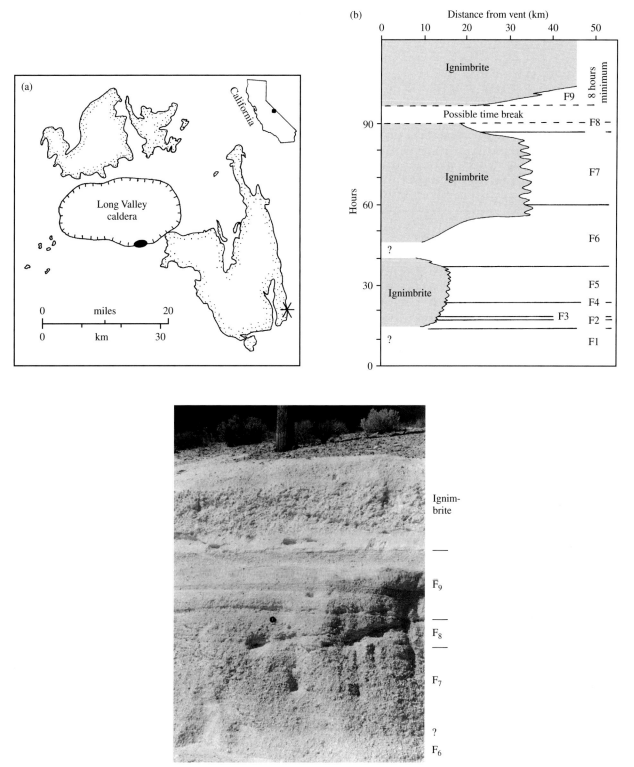

10.26 Pyroclastic deposits derived from the Long Valley magma system, eastern California, at about 0.76 Ma. (a) Distribution of the Bishop Tuff, consisting of ignimbrite and interlayered fall and surge deposits, chiefly in two lobes north and southeast of the Long Valley source caldera. (b) Stratigraphy of the interlayered proximal ignimbrite (shaded) and fall deposits (F1, F2, etc.) southeastward from the source vent (solid ellipse) on the caldera ring fracture in (a). Note that the total sequence of approximately 600 km^3 was deposited in about 100 h. (c) Proximal fall deposits of sorted pumice and ash capped by unsorted ignimbrite. See stratigraphic relations in (b). Photograph taken at locality marked by asterisk (*) in (a); Camera lens cap for scale. ([a] and [b] reproduced by permission from Wilson CJN, Hildreth W. The Bishop Tuff. New insights from eruptive stratigraphy. J. Geol. 105:407–439, 1997; Copyright © 1997 by the University of Chicago Press.)

Characterisitics of pyroclastic-fall deposits include the following:

1. Generally they are better sorted than surge and flow deposits.

2. Plane parallel beds form unless they are modified by erosion.

3. Unabraded vitric clasts form; if they are inequant in dimensions, they lie flat in the bedding plane.

4. **Mantle bedding** is created as the tephra showers uniformly over the ground surface, whether hill or valley, as in snowfall (Figure 10.27).

5. **Reverse-graded** beds form, in which increasingly coarser clasts occur upward, rather than normal-graded beds, in which the coarsest particles are at the base (Figure 10.28). Reverse grading might be caused by shifts in wind currents and speed, but another explanation stems from the dynamics of plinian plumes. A plume may rise to greater height, possibly because of increased magma discharge rate related to increased vent radius as the conduit is reamed out by the discharging magma. Consequently, larger pyroclasts are carried convectively to greater heights and distances before being released in fallout, in which their terminal fall velocities overcome convective lift.

Ash-fall beds are useful in **tephra chronology** because a widely dispersed ash-fall layer (Figure 10.21) can serve as a correlatable time stratigraphic horizon. Crystal and vitric particles have distinctive compositions inherited from the magma and can be dated isotopically. Tephra studies can be a valuable resource for

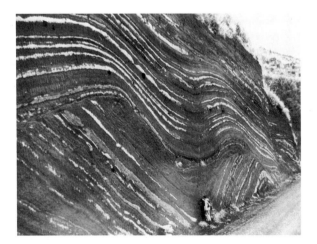

10.27 Mantle bedding of ash-fall layers, Oshima Volcano, Japan. The beds in the roadcut partly covered by snow are not folded, but dips are primary and mimic the configuration of the depositional surface on which they rest. Note unconformity where erosion cut into older underlying sequence. Geologist for scale at bottom of photograph. (Photograph courtesy of Jack Green.)

10.28 Reverse grading in **ash-fall deposit.** Note upward-increasing size of rhyolitic pumice fragments. Pocket knife 8 cm long.

archaeology. However, differential sorting due to contrasting size and density of the pyroclasts and derivation from different parts of compositionally zoned chambers precludes using the bulk composition of any fall deposit as an accurate indicator of preeruption magma composition.

Pyroclastic Surge. Like pyroclastic flows, **pyroclastic surges** are devastating mixtures of hot gas and solid particles that move laterally away from the base of a collapsing pyroclastic plume at hurricanelike speed (Figure 10.23). However, unlike flows, surges are dilute mixtures that have low concentrations of particles. Surges travel in turbulent manner to less than a few kilometers at most from a vent because they have less momentum, as they are mostly gas. Surges can develop bed forms similar to those in water- and wind-transported sediment. As in these modes of sediment transport, surges move particles by surface traction in a bed load and by turbulent suspension. Unlike water- and wind-transporting media, surges have density and viscosity that can vary during travel, thus creating variations in bedforms. **Surge bedforms** (e.g., Figures 10.29 and 10.30) include the following:

1. Poorly to moderately sorted, planar to pinch-and-swell strata that are 1 cm or so thick. Plane parallel beds may resemble pyroclastic fall deposits but can grade laterally into more typical surge bedforms, and flat clasts can be imbricated (dipping toward the source).

2. Low-angle, cross-bedded to wavy beds are common.

Flow

Surge

10.29 Silicic **pyroclastic deposit.** Poorly sorted **pyroclastic flow** deposit in Snake River Plain, Idaho, which contains light gray pumice blocks to as much as 12 cm, overlies **pyroclastic surge** deposit. **Surge bed forms** include climbing dune cross-beds and pinch-and-swell beds. Lowermost part sequence of thin-plane parallel beds may be also be surge material.

3. Climbing duneforms lie transverse to the surge direction.

4. Scoured bed contacts are due to local erosion.

5. **Bedding sags** are caused by impacting ballistic clasts depressing soft underlying layers; asymmetric sags can be interpreted to determine direction to vent.

6. Penecontemporaneous downslope slumps of water-saturated beds are also created.

Most surges are of the type known as **base surges** created by hydromagmatic explosions of mafic magma; gravitational collapse of a steam-saturated eruption column creates a ring-shaped surge traveling outward along the ground from the vent (Figures 10.22 and 10.23). Less common surges are related to steam blast eruptions, and some are jetted from toes of advancing pyroclastic flows. Whether surges are distinctly different from pyroclastic flows, or whether one grades into the other, is debatable. Some volcanologists consider a pyroclastic surge to be simply a dilute pyroclastic flow.

If most of the thermal energy in a hydromagmatic eruption is consumed in converting water to steam, the so-called wet surge may be near 100°C and consist of pyroclasts, steam, and water. Accretionary lapilli, soft-sediment deformation structures, and plastering of mud onto upright objects, such as trees that are not burned, indicate a wet surge. Other surges are hotter (able to carbonize trees) and are dry (steam only).

10.4.4 Explosive Style

Because of the myriad factors involved, there is a wide and continuous spectrum of **explosive style** that defies straightforward categorization. Styles range from small vents harmlessly "burping" low-viscosity basaltic spatter a few meters to the colossal, high-energy catastrophic explosions of silicic magma that create enormous convecting plumes or collapsing columns and deposit thick blankets of pyroclastic material over thousands of square kilometers and finer ash globally. During the course of a particular eruptive episode the style may change or alternate. Widely accepted names for particular styles are taken from geographic locales and famous exemplifying volcanoes. Associated with this spectrum of eruptive styles are a variety of volcanic plumes (Figure 10.31).

As with earthquakes, the frequency of volcanic explosions depends on their magnitude. The smallest explosions occur roughly every month somewhere on Earth, whereas the largest occur on a time scale of millions of years. Volcanologists employ various parameters to measure the **"explosiveness"** of a volcanic eruption; these include intensity (rate of magma discharged from the vent, or mass flux), magnitude (total volume or mass of material vented), explosivity index (ratio of pyroclastic deposits to all other volcanic material), and dispersive power or violence (area of dispersal of pyroclasts).

Eruptive styles are now discussed in order of increasing explosiveness. Characterizing properties of the resulting deposits are emphasized.

Hawaiian Eruptions. Typically, basaltic magma in **Hawaiian eruptions** takes the form of low viscosity lava flows and mildly effervesing **lava fountains,** less

10.30 Surge deposit 0.4 km northeast of Sugarloaf Mountain rhyolite dome, San Francisco Peaks volcanic field, Arizona. Climbing dune cross-beds indicate surge moved from right to left; note bedding sag to left of shovel below crest of dune. (From Sheridan and Updike, 1975; photograph courtesy of M. F. Sheridan.)

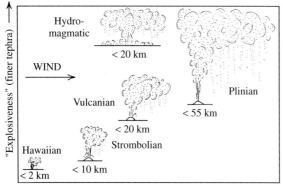

10.31 Highly generalized classification of **explosive eruption style** based on explosiveness and height of volcanic plume. Diagrams not to same scale. More explosive eruptions tend to have smaller mean size of pyroclasts. Higher plumes are capable of wider dispersal of fallout pyroclasts so that ash-fall deposits occur farther from the vent. (Redrawn from Cas and Wright, 1987.)

appropriately called "fire fountains." Expanding bubbles within a rising magma column propel fragments from the vent to form the incandescent lava fountain (Figures 10.7a and 10.32). The relatively large ejecta (centimeters to meters in diameter) retain their high eruptive T because the surface area for dissipation of heat is small compared to the enclosed mass. Hence, relatively little heat is transferred into the air, convective updrafts are minimal, and only the smallest pyroclasts are transported out of the fountain by the wind. Since most pyroclasts are large and unaffected by convecting air currents, they follow nearly ballistic trajectories in the collapsing fountain. Still hot, molten clots accumulate at the base of the fountain, where they form deposits of **welded spatter,** or **agglutinate.** Aerodynamic streamlining of the low-viscosity spatter during flight creates bombs and smaller Pele's tears and hair (Figure 7.29). Depending upon wind conditions, vent geometric features (central versus fissure), and possible obstructions in the vent that deflect the ejecta from the vertical, this welded spatter can form a more or less symmetric **spatter cone** around the vent or a less symmetric one-sided spatter rampart or mound. Spatter accumulations may be hundreds of meters in diameter, but most are smaller. Lapilli- and block-size fragments of scoria (vesicular basalt) that are cool and solid upon deposition accumulate as cinder deposits. Alternations of cinder and spatter create **cinder-and-spatter cones.** High-discharge-rate fountaining can produce sufficient accumulations of molten spatter at the fountain base that the mass recombines into a mobile lava that can move downslope as a rootless lava flow. Accumulation of lava in a depression can form a **lava lake.** Lava lakes also appear at the top of the magma column in the vent and can overtop a crater rim or undermine

and rupture a tephra cone, rafting away sectors of the cone on the flowing lava. Contemporaneous lava fountaining and lava effusion from the same or nearby vents can occur.

Strombolian Eruptions. More explosive **strombolian eruptions** that build monogenetic volcanoes of basaltic or andesitic magma involve bursting of large gas bubbles, as much as 10 m in diameter, near the top of a magma-filled conduit. Most of the ejecta are lapilli-size cinders and lesser larger blocks (Figure 10.33), which are solid upon deposition around the vent, forming a **cinder cone,** also called a **scoria cone** (Figure 10.4). Variable amounts of congealed spatter and streamlined bombs can be mingled with the cinders.

Vulcanian Eruptions. Vulcanian eruption activity occurred at Vulcano, another volcano on a Mediterranean island like Stromboli, but has taken place at many other subduction-related volcanoes erupting intermediate-composition magma. Typically, **vulcanian eruptions** begin with cannonlike, steam-blast explosions (discussed later) at intervals of minutes to hours that disintegrate rock plugging the vent over a magma-filled conduit. Blocks are ejected ballistically, whereas finer clasts fall out of convective plumes and accumulate in moderately sorted to well-sorted beds that are more widespread than those of strombolian eruptions. Once the rock cap on the magma column is removed, continued eruptions discharge juvenile pyroclasts ranging from vitric and crystal ash to bombs and breadcrust blocks. Pyroclastic surges and flows accompany

10.32 Incandescent 1959 Kilauea Iki, Hawaii, **lava fountain.** Diffuse plume of cooler black ash and cinder fanned from the fountain and carried downwind (to left) forms a crescent-shaped cinder rampart (not visible) and more distant beds of finer ash. Below the collapsing fountain, rapidly accumulating clots of magma merge into a rootless lava flow that feeds an incandescent lava river draining into a partly crusted **lava lake** that is barely visible in lower left corner of photograph. (U.S. Geological Survey photograph by G. A. Macdonald.)

10.33 Internal crude stratification and moderate degree of sorting in basalt **cinder cone.** Quarry face reveals blocks and mostly lapilli-size vesicular cinders that have been variably oxidized while hot in the oxygen-rich atmosphere. Essentially nonoxidized are black, whereas slight oxidation of the Fe-rich glass creates an iridescent coating. More thorough oxidation creates minute pervasive hematite grains that pigment cinders red-brown. Note hammer in lower left for scale.

some vulcanian eruptions. Effusion of highly viscous, less gas-rich lava commonly terminates the eruptive episode.

Steam-Blast Explosions. Water contacting hot rock is vaporized to expanding steam, which blows the rock to pieces in **steam-blast explosions.** Because the term *phreatic* refers to groundwater, explosions caused by contact between groundwater and hot rock may be called **phreatic explosions.** No juvenile magma is ejected, although in many cases it lies not far below and is responsible for the heating of the rock. Some explosions occur in areas of geothermal activity overlying active hydrothermal systems (Figure 4.12) where temperatures are increased as a result of magma recharge. Other explosions occur where hot lava or pyroclastic flows override or otherwise interact with bodies of water or water-soaked ground. Explosions produce steam-rich plumes laden with lithic ash as well as larger ballistic clasts. In certain cases, the nonjuvenile tephra has suffered mineral alteration as a result of prolonged prior hydrothermal and fumerolic processes. Many explosive volcanic episodes begin with steam blasts and later evolve into other styles of eruption.

Hydromagmatic Eruptions. One such style of activity that follows preliminary steam blasts is the **hydromagmatic eruption,** in which magma makes actual contact with external water (Figure 10.23). Basaltic eruptions in lakes and ocean are typical, such as that at Surtsey, Iceland, in 1963–1965; this eruptive style is appropriately called **surtseyan** (Figure 10.22).

Mostly juvenile fragments are ejected in hydromagmatic explosions. Shattering of quenched basaltic melt creates poorly vesicular, blocky juvenile vitric ash, which is caught up in a steam-rich plume. Finer ash in the convecting plume can aggregate into accretionary lapilli. Fallout from the high plume produces thinly bedded, sorted ash-fall deposits many kilometers from the vent in quiet water or on nearby land. Multiple surge beds numbering in hundreds to thousands, together with ballistic clasts, form wedge-shaped (in cross section) accumulations near the vent and tapering away from it. As the body of water is blocked from the vent by these encompassing accumulations, continued rise of magma erupts in strombolian or hawaiian style and may fill the crater with magma, forming a lava lake.

Monogenetic hydromagmatic eruptions produce low-rimmed edifices having bowl-shaped craters including tuff cones, tuff rings, and maars. Morphologically, these edifices form a continuum with cinder cones. In strombolian-generated **cinder cones,** the aspect ratio (edifice height/basal diameter) is approximately 1:3 and the constructional crater is small relative to the volume of the cone. In **tuff rings,** at the opposite end of the spectrum, the ratio is <1:5 and the volume of crater space is larger than the volume of ejecta (Figure 10.34). **Tuff cones,** the typical edifice formed by surtseyan eruptions, are intermediate in shape. Where lava erupted on land enters the sea, as on the island of Hawaii, hydromagmatic explosions may create a **littoral cone.** Such edifices are of unconsolidated tephra and can be called ash rings or ash cones, but most are fairly well cemented because of extensive and surprisingly rapid palagonitization (Section 7.1.1) of the warm wet vitric ash. A **maar** is a tuff ring in which the crater floor lies below the general elevation of the preeruption land surface. Maars form by excavation of older rock material by hydromagmatic explosions that occur just below the preeruption land surface (Figure 10.23). The surrounding rim of ejecta

10.34 MacDougal Crater in the Pinacate area of northwestern Sonora, Mexico. The **tuff ring** is about 1.5 km in diameter. (Photograph courtesy of John S. Shelton.)

therefore includes a significant proportion of accidental lithic material in addition to juvenile. In the Eifel, Germany, region, maars and cinder cones are closely associated, even along the same fissure system. But cinder cones tend to form on hills, whereas maars form in the valleys, where there was access to the shallow water table.

Diatremes are narrow, funnel-shaped masses of breccia underlying maars. They are discussed in Section 9.4.3 and shown in Figures 9.25 and 9.26. The origin of diatremes involves an ascending convecting fluid-rich magma system. However, some basaltic maars are underlain by what is interpreted to be a downward-growing diatreme that was produced by hydromagmatic processes (Lorenz, 1986).

<u>Plinian Eruptions.</u> Highly explosive plinian eruptions are characterized by plinian plumes (Figure 10.24a). **Plinian eruptions** involve volatile-rich silicic magmas (dacite-rhyolite) of high apparent viscosities, although andesitic and even basaltic eruptions have been documented. Eruptive velocities are hundreds of meters per second and eruptions last from tens of minutes to several days or intermittently for years where magma supply can be maintained. Plinian eruptions commonly, but not invariably, initiate silicic volcanic activity. Many of the most destructive eruptions of recorded history began as plinian, including that of Vesuvius, which inundated Pompeii and Herculaneum with several meters of a pumice fall in 79 AD; Krakatoa in 1883; and Mount Saint Helens on May 18, 1980.

In plinian eruptions, pyroclasts are as large as blocks, but lapilli-size pumice and vitric ash predominate. Most of the properties listed for pyroclastic fall deposits apply to characteristically sheetlike, moderately sorted to well-sorted plinian accumulations.

The 180 AD eruption of Taupo in New Zealand was the most powerful known plinian eruption, creating a layer as much as 12.5 cm thick of rhyolite ash 200 km from the vent; the plume is estimated to have had a height of >50 km. Some of the Taupo eruptions (130–186 AD) occurred where vesiculating silicic magma encountered lake water; such **phreatoplinian** eruptions are the silicic counterpart of basaltic surtseyan eruptions and produce widely dispersed, thin beds of very fine ash that has abundant fine vesicles and blocky shapes.

10.4.5 Pyroclastic Flows and Deposits: Overview

Large, widespread silicic pyroclastic-flow deposits in the western United States were an enigma to early geologists because of their superficial lavalike appearance but thin sheetlike aspect ratio (Figure 10.35) uncharacteristic of silicic lava flows. Beginning in 1902 with the tragic eruptions of Mount Pelée on Martinique (Figure 10.36) and La Soufrière on Saint Vincent, both islands in the Caribbean, and continuing with hundreds of similar eruptions in many parts of the world, volcanologists have gained much insight into the nature and origin of these puzzling deposits. Observations from safe distances have been integrated with laboratory experiments on model systems, computer simulations, and

10.35 Sheets of **ash-flow** tuff **(ignimbrite)** deposited outside the Valles caldera near Los Alamos, New Mexico (see Figure 10.41). Contrast aspect ratio of these sheets with a typical rhyolite lava dome (e.g., Figure 10.16). (U.S. Geological Survey photograph courtesy of Robert L. Smith.)

10.36 Pyroclastic eruption at Mount Pelée, Martinique. **Coignimbrite plume** rising above an inconspicuous pyroclastic flow dominates photograph. The remains of the town of Saint Pierre, devastated by an earlier eruption on May 8, 1902, which took 28,000 human lives, lie in the foreground. Compare Figure 10.24c. (Photograph courtesy of The Geological Museum, London.)

fluid dynamic studies to provide considerable insights, but no complete solutions. The classic exposition on pyroclastic eruptions and deposits in the western United States is by Ross and Smith (1961).

Nomenclature and Types. The French petrologist Alfred Lacroix, who was at the site of some of the 1902 Caribbean eruptions, called pyroclastic flows *nuées ardentes,* meaning "glowing" or "hot clouds." *Glowing avalanches* is a more accurate, sometimes used label for flows because the depositional agent is not a cloud of dispersed ejecta, as Lacroix believed, though these are invariably associated. **Ash flow** and **ash-flow tuff** are names commonly used by U.S. geologists for the eruptive agent and its deposit, respectively, even though pumice clasts of lapilli size are typically present with ash. Locally, blocks of pumice also occur, together with a wide size range of lithic clasts. **Welded tuff** is a nonporous rock made chiefly of vitric ash particles that are stuck together because of the high T at emplacement. For the Plio-Pleistocene rhyolite deposits on the North Island of New Zealand, Peter Marshall in 1935 coined the term **ignimbrite,** from Latin *ignis,* meaning "fire," and *bris,* meaning "cloud," hence, fiery cloud rock. The term *ignimbrite* has been widely adopted for pyroclastic flow deposits.

A **pyroclastic flow,** our preferred generic term, is a highly mobile, hot avalanche of pyroclasts and gas that is denser than ambient air and moves swiftly (as much as 300 m/s) along the ground surface away from its source. Resulting deposits are massive poorly sorted beds that can be hundreds of meters thick. Rheologic properties of a pyroclastic flow vary with respect to dis-

tance of transport from the vent. As it moves along the ground, denser particles may sink and lighter ones rise, buoyed up by the hot gas, forming an overlying dilute, turbulent ash cloud, commonly referred to as a *coignimbrite plume* (Figures 10.24b and 10.36). Flows tend to be confined to topographic lows but can cascade over tops of hills. Flows denser than water travel along the floor of lakes and oceans, whereas less dense ones travel over the water surface. Some pyroclastic flows have dilute ground surges propelled from their toe.

Many different types of pyroclastic flows have been recognized, but they fall essentially into two basic categories depending on the process of origin and character of the flow and flow deposit; these are block-and-ash flows and ash flows that are predominantly made of ash and lesser lapilli.

10.4.6 Block-and-Ash Flows

Relatively very small-volume avalanches produced by disintegrative collapse of growing andesitic to rhyolitic domes or thick flows on composite volcanoes produce **block-and-ash** flows (Figure 10.24c). Dome disintegration can be driven by exsolution of volatiles in the dome, causing explosive fragmentation; by magma-external water interactions in a water-filled crater, causing steam explosions; and by collapse of a gravitationally unstable dome. In any case, dislodged blocks cascade downslope, pulverizing one another in transit. Downslope flow is closely confined to topographic lows and canyons, diverging and turning according to slope configuration, resembling snow avalanches in mountain canyons. The accompanying upward-expanding, dilute ash-steam coignimbrite plume is less deflected by topographic features and lags behind the faster-moving flow. Block-and-ash flow runouts are less than a few kilometers and speeds are a few tens of m/s. Deposits are commonly tonguelike, levée-bounded, and a few meters thick or less and have volumes that are generally <0.1 km^3 to as small as 0.001 km^3. Deposits are unsorted, unwelded aggregates of ash and weakly vesicular blocks that are as much as a few meters in diameter; some blocks have radially arrayed or breadcrust cooling (Figure 7.30) cracks testifying to their emplacement at high T and cooling within the flow. All clasts have the same composition.

At Unzen volcano, Japan, in 1990–1995, tens to hundreds of very small block-and-ash flows occurred daily as an emerging dome episodically collapsed. Forty-three people, including three volcanologists, were killed in one flow.

10.4.7 Ignimbrite-Forming Ash Flows

Ash flows are made predominantly of ash and form mostly by collapsing pyroclastic columns (Figure 10.24b), although other mechanisms have been observed. If the proportion of lapilli (and possibly blocks) of pumice exceeds 50% in a matrix of ash, they are called **pumice flows.** Smaller flows, generally <1 km³, but some measuring in tens of cubic kilometers, are created by eruptions at composite volcanoes, commonly in subduction zones.

The largest ash-flow deposits, all prehistoric, are presumed to have been generated by collapse of eruptive columns formed by high rates of magma discharge. No preexisting conical volcanic edifice is associated with these eruptions in continental interiors which have been referred to as "erupted granitic batholiths." The volume of a single ignimbrite can be hundreds to thousands of cubic kilometers, equaling or exceeding the vast floods of plateau-forming basalt lava described in Section 10.2.3. Outflow sheets of ignimbrite can be tens to hundreds of meters thick, cover areas of tens of thousands of square kilometers, and reach more than 100 km from the source on negligible slopes. Topographical features are smoothed by the flows which flood depressions and thin over hills.

Flow Mobility: Fluidization. The mobility of ash flows is unquestioned; flows tens of meters thick and tens of kilometers from their source can surmount hills hundreds of meters high. The cause of their mobility, however, remains uncertain.

One mobilizing factor is the kinetic energy imparted at the vent. As a pyroclastic column collapses, gravitational potential energy of the fountain is transformed into kinetic energy that drives a horizontally moving flow. Higher, more massive collapsing columns would impart more kinetic energy and promote farther runout. The "energy line" concept of Sheridan (1979) indicates any topography can be surmounted if a straight line drawn from the top of the gas-thrust regime of the eruptive plume to the distal toe of the pyroclastic flow lies above the topographic feature.

The fact that pyroclasts in ash-flows are dispersed in a gas phase may enhance flow mobility by providing a "cushion" between the solids, reducing frictional and collisional particle interactions that would otherwise impede flow.

Gas-particle flows may be mobilized by the phenomenon called **fluidization,** a process used in industry for transport of solid particles without recourse to conveyors or vehicles. Upward-flowing gas passing through a mass of cohesionless (loose) particles lifts them apart at some critical velocity so the mass behaves as a frictionless fluid whose angle of repose is zero and whose overall density is less than that of individual particles. However, in the typically unsorted ash flows that contain a size range of ash and lapilli and perhaps blocks, overall fluidization is less effective because the gas permeability is less than in a mass of uniform-size particles; smaller particles clog spaces between larger. Additionally, larger or denser clasts cannot be lifted,

but smaller ones can be fluidized, and the smallest ones, whose terminal velocity is exceeded by the streaming gas, are entrained into it. This entrainment, or **elutriation,** of fine ash accounts for the universal dilute ash-steam clouds (coignimbrite plumes) observed over all pyroclastic flows, block-and-ash as well as ash flows (Figures 10.24b, c, and 10.36).

Nonetheless, this *partial* fluidization probably enhances ash-flow mobility and may produce local, subtle sorting of clasts (Figure 10.37). Fall deposits covering up to millions of square kilometers from coignimbrite plumes are enriched in fine vitric ash (glass shards) relative to the main ash flow. Thin ash flows possibly lose half their volume to winnowing (elutriation) of fine particles into the coignimbrite plume, whereas thicker flows probably lose a much smaller fraction.

Depositional and Cooling Units. Small volatile-rich silicic magma systems lodged in composite volcanoes explode for days to months during a particular eruptive episode, which can be separated by hundreds of years from other episodes. Large, caldera-forming eruptions

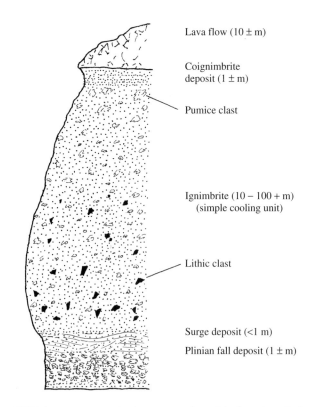

Lava flow (10 ± m)

Coignimbrite deposit (1 ± m)

Pumice clast

Ignimbrite (10 – 100 + m) (simple cooling unit)

Lithic clast

Surge deposit (<1 m)

Plinian fall deposit (1 ± m)

10.37 A common pyroclastic sequence that might form in a single eruptive episode. A basal plinian fall deposit of sorted ash and pumice is overlain by thinner surge deposits. The overlying simple ignimbrite **cooling unit** can be 1 m to more than 100 m thick. Note slight downward concentration of denser lithic clasts and upward concentration of less dense pumice fragments. The pyroclastic flow deposit is overlain by sorted fine ash deposited from the coignimbrite plume. Near the vent, a lava flow formed from largely degassed magma might be extruded over the pyroclastic sequence.

have repose times between eruptive episodes of as much as 10^5–10^6 y. Thus, ignimbrite sequences are built of multiple **depositional flow units;** each unit represents one explosive event. These ignimbrites may be separated by surge and fall deposits (Figures 10.29 and 10.37).

A **cooling unit** is a pyroclastic flow deposit nearly instantaneously laid down that cools as a thermal entity. A **simple cooling unit** can comprise one depositional unit, or it can be made of two or more that are emplaced nearly simultaneously so that there are no internal cooling breaks, such as intervening less welded tuff (discussed later). Discerning boundaries between depositional units within a simple cooling unit can be challenging; compositional discontinuities and intervening surge and fall deposits can be helpful. A **compound cooling unit** consists of a succession of flows emplaced closely in time so that only partial cooling breaks occur between depositional units.

Composition of Deposits. Most ignimbrites are rhyolite; fewer are dacite, trachyte, and phonolite; andesite is uncommon. Juvenile pyroclasts are predominantly ash-size vitroclasts and lesser crystals. Larger cognate pumice lapilli are typical, whereas lithic lapilli can be conspicuous in some deposits. Crystals include euhedral to subhedral intact phenocrysts formed in the preeruption magma chamber and phenoclasts of the same ancestry but broken during eruption. These primary crystals may be very sparse, <1% of the deposit, but can range to as much as about 50%. Some lithic clasts are cognate crystalline fragments related to the erupted magma, such as from the crystallized wall of the magma chamber. Accidental lithic fragments (xenoliths) can be chunks of rock torn from the enlarging conduit during explosive eruption, wallrock from the preeruption magma chamber, and loose rock fragments on the ground picked up and incorporated into a turbulent flow. The proportion of lithic fragments, pumice clasts, and phenocrysts in the deposit can vary widely, each from 0% to as much as 50% or so, whereas juvenile vitric ash particles are always in abundance.

Compositionally zoned ignimbrites were first well documented by Lipman et al. (1966; see also Hildreth, 1981). In some of these ash-flow deposits zonation is cryptic and can only be discerned by laboratory analyses, whereas in others it is quite conspicuous in the field (Figure 10.38a). Normally, the proportion of phenocrysts in the rock, as well as their sizes, increase stratigraphically upward in the sheet, as do FeO, MgO, and CaO, whereas the lower part of the ignimbrite sheet has a more evolved composition. A common zonation is a basal, high-silica rhyolite overlain by low-silica rhyolite, by dacite, or, in instances of strong zoning, by andesite. In some deposits, lateral zonation is

also evident from proximal parts near the source to distal parts tens of kilometers distant. Strong vertical and horizontal zonation in an ash-flow deposit may make it difficult to correlate isolated exposures of one depositional unit. Other useful correlation tools include precise isotopic dating and paleomagnetic analyses (Best et al., 1995).

Zoned ignimbrite deposits are derived by more or less systematic withdrawal from preeruption magma chambers that have compositional gradients (Figure 10.38b). The more evolved top of the chamber is erupted first; then successively deeper parts are erupted, producing, in the deposit, an *inverted* zonation of that in the chamber. It is obvious that a sample of bulk tuff in most instances does not accurately represent any of the preeruption magma; the tuff sample is an explosive mixture of pyroclasts derived from different compositional zones in the chamber. Moreover, the tuff has likely suffered some differential loss of fine vitric ash by elutriation from the ash flow. Unaltered pumice lumps hosted in the tuff, on the other hand, represent unmodified samples of the preeruption magma. Therefore, analyses of pumice best portray compositional variations in the preeruption magma body. Likewise, analyses of single glass shards, usually by electron microprobe, accurately reflect the melt composition.

Secondary Zonation after Deposition. Ash-flow deposits commonly have other zonal features that are superimposed on any primary compositional zonation just described. Secondary processes formed during cooling of the hot mass of pyroclasts and entrapped gas include (Figure 10.39):

1. Welding and compaction
2. Vapor-phase crystallization of minerals from the entrapped gas
3. Devitrification (delayed crystallization) of the glassy material

Welding is the bonding of hot glass particles. Because of the weight of overlying pyroclasts on these soft sticky particles, they are compacted together and trapped gas is squeezed out, collapsing pore spaces and vesicles within pumice fragments and reducing the bulk porosity of the tuff. In some deposits, interstitial gas that is largely steam may be partially resorbed into vitroclasts, reducing their viscosity and, hence, promoting welding more or less independently of compaction (Sparks et al., 1999). But generally the most intense welding and compaction occur in the lower portion of the flow, though not at the base, where faster cooling prevents much welding (Figure 10.39). Welding and compaction operate simultaneously to create the typical eutaxitic fabric of welded ash-flow tuffs, in which pumice lapilli and glass shards are flattened into

(a)

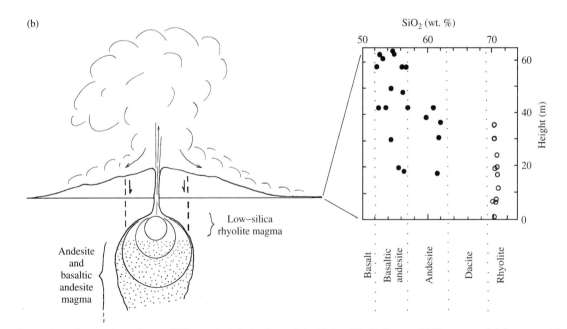

10.38 Compositionally zoned, poorly sorted Mazama ignimbrite, Crater Lake National Park, Oregon. (a) The upper, slightly more welded and erosionally resistant part of deposit is crystal-rich hornblende andesite; the underlying lighter-colored part into which it grades without a sharp cooling break is a crystal-poor, low-silica rhyolite. Altogether, the deposit is an inverted representation of the compositional zonation in the source magma chamber. (Photograph by Oregon State Highway Department along Wheeler Creek.) (b) Explanation of compositional zonation. On the right, silica concentrations in *pumice* clasts in the ignimbrite, plotted against height in the deposit, correspond to bulk magma compositions in the zoned preeruption magma chamber, but upside down. On the left, cartoon illustrates how successive, increasingly larger "spheres" of magma withdrawal from the zoned chamber beneath ancestral Mount Mazama first tapped only uniform low-silica rhyolite magma, followed by deeper withdrawal of more variable mafic magma. After caldera collapse along ring faults (dashed lines), the depression filled with water to form Crater Lake. (From Bacon and Druitt, 1988.)

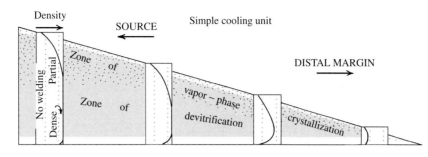

10.39 Idealized secondary zonation in a **simple ignimbrite cooling unit.** Vertical scale is exaggerated for clarity. Four upright rectangular panels show variation in density with respect to height in the unit and three degrees of related **welding and compaction**—dense, partial, and no welding and compaction. Dense welding occurs in lower two-thirds to half of unit near source. Zone of **vapor-phase crystallization** (stippled) is in upper part, and zone of **devitrification** (light shade) occupies most of cooling unit. All zonal boundaries and extents are highly variable in different deposits.

discoidal shapes more or less parallel to the depositional plane (Figure 7.34). The **compaction foliation** so expressed is enhanced by rigid inequant mineral grains, such as biotites and feldspars, that rotate in the soft glassy matrix during compaction and also become aligned in similar orientation to the flattened vitroclasts. Some densely welded tuffs with abundant lapilli and crystals, especially biotite, resemble foliated schists because of this well-developed planar fabric. Secondary **rheomorphic flowage** may occur where crystal-poor, high-T pyroclastic flows were deposited on slopes. Pumice lumps become extremely attenuated and lineated, even folded, particularly in alkaline, low-silica flows in which the glass is not highly viscous. Such **rheomorphic ignimbrites** may be difficult to distinguish from lava flows if the characterizing vitroclastic fabric is obliterated during flow.

Two types of secondary crystallization can occur simultaneously with welding and compaction in the hot ash-flow deposit. Trapped gas, which contains significant amounts of dissolved silica, alkalies, and other mobile chemical entities, migrates up through the permeable deposit and escapes into the atmosphere. Some gas collects into subvertical channels and vents at fumaroles (Figure 10.40). But **vapor-phase crystallization** takes place throughout the upper part of the deposit (Figure 10.39), where migrating gases cool and precipitate dissolved minerals in open pore spaces in the less compacted tuff. Minerals are mainly alkali feldspar, quartz, tridymite, and cristobalite. The other secondary crystallization that occurs in ash-flow deposits is **devitrification** of glass shards and pumice (Figures 7.35). Devitrification chiefly affects the middle to upper parts of the deposit; the lower boundary of devitrification against vitric tuff can be abrupt and sharp. In outcrop where this contact occurs within the welded zone, the devitrified tuff is red, pink, brown, or purple and the rock looks stony, in contrast to the black, glassy, underlying nondevitrified tuff (vitrophyre). In devitrified

tuffs, aphanitic quartz-feldspar intergrowths, locally spherulitic or lithophysal, replace glass, but delicate pumiceous and vitroclastic textures are faithfully preserved, even though the rock may be entirely crystalline. In other instances, devitrification completely erases fragment outlines; vitroclastic and eutaxitic fabrics are obliterated and the tuff assumes a massive, featureless aphanitic fabric similar to that of many lava flows. The presence of phenoclasts, however, can reveal its pyroclastic origin.

10.4.8 Calderas

The distinction between a **caldera** (Figure 10.41; sometimes called a *cauldron*) and a **crater** is succinctly expressed by Williams and McBirney (1979, p. 207): "A caldera is a large volcanic collapse depression, more or less circular or cirquelike in form, the diameter of

10.40 Fossil **fumeroles** in a 150-m-thick section of the densely welded Bishop tuff in Owens River Gorge, California, east of its Long Valley caldera source (Figure 10.26a). Normally subvertical columnar joints in ignimbrite sheet (left) curve and converge into fumarole conduit (partly in shadow, right). Remnant fumerole mound on top of sheet can be seen in left background.

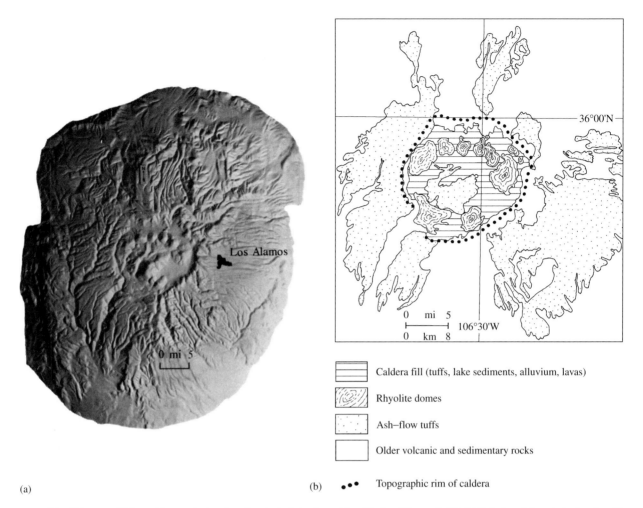

(a)

(b)

☰	Caldera fill (tuffs, lake sediments, alluvium, lavas)
▨	Rhyolite domes
⸬	Ash–flow tuffs
☐	Older volcanic and sedimentary rocks
●●●	Topographic rim of caldera

10.41 Plio-Pleistocene Valles **caldera** and related outflow ignimbrite sheet, New Mexico (Figure 10.35). (a) Relief model made by Stephen H. Leedom from U.S. Geological Survey relief maps. (b) Generalized geologic map. Faults have been omitted for clarity. The patchy distribution of the ignimbrite outside caldera is due to the uneven topographic features onto which it was deposited as well as to subsequent erosion. (Redrawn from Smith and Bailey, 1968.)

which is many times greater than that of any included vent. A crater may resemble a caldera in form but is almost invariably much smaller and differs genetically in being a constructional form rather than a product of destruction."

Contrary to widespread lay belief, calderas, such as the one that holds so-called Crater Lake, Oregon, did not form when the volcano "blew off its top," as Williams (1941) astutely observed. Had it done that, there would be a volume of accidental rock fragments in the pyroclastic deposit matching the volume of the caldera. This is definitely not the case, for the deposit is almost entirely juvenile vitric pyroclasts (Figure 10.38a).

Calderas form where a substantial volume of magma is withdrawn from a subterranean magma chamber in a geologically short time and the unsupported rock roof over the evacuating chamber collapses into the growing void (Figure 10.42). Calderas related to ash-flow eruptions are thought to begin where overpres-

sured magma fractures the roof, forming one or more extrusive conduits, possibly along arcuate ring fractures. Initial eruptions create an extracaldera outflow sheet. After some critical volume of magma has been vented, the unsupported roof subsides into the chamber. The sinking, denser roof rock possibly adds a driving force for continued expulsion of less dense magma that can accumulate to thicknesses of kilometers inside the caldera, depending on the amount of draw-down, forming an intracaldera tuff deposit.

The exact manner of caldera collapse (Lipman, 1997) is commonly difficult to establish because of incomplete exposures and erosion. Some roofs subside as more or less intact plates, or pistons, inside a circumscribing ring fault. Others fracture into blocks that subside in piecemeal fashion; still others, only partially circumscribed by a fault, subside in a hinged, or trapdoor, manner. Yet other roofs flex or downsag rather than subside along faults; these might form over deeper evacuating magma chambers.

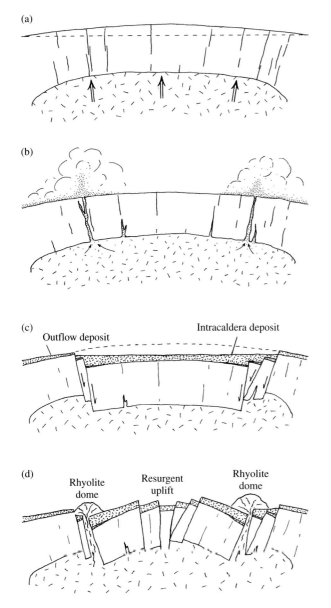

10.42 Schematic cross sections illustrating the generalized evolution of a **caldera** of the Valles type (Figure 10.41). (a) Doming of the roof over the intruding magma and formation of ring fracture system. Dashed line is original ground surface. (b) Initial eruption of ash flows from ring fracture(s) forms ignimbrite outflow sheet; partial evacuation of magma chamber. (c) Continued ash-flow eruption causes caldera collapse guided by existing ring fractures in roof and partially fills depression with intracaldera ignimbrite deposit. Collapse of steep caldera wall that forms intracaldera landslide breccias is omitted here for clarity. Additional postcaldera deposits in depression may consist of lake sediments and volcanic deposits from nearby sources. (d) **Resurgent uplift,** doming, and fracturing of central block due to renewed magmatic activity. Effusion of viscous rhyolite lava along ring fracture peripheral to central block forms an arcuate group of domes. (Redrawn from Smith and Bailey, 1968.)

During subsidence of fault-bounded calderas, high, steep, unstable walls along the caldera margin collapse in landslides to form lenses of breccia intercalated within the accumulating intracaldera tephra. Breccia

is commonly made of house-size blocks. Such wall-collapse breccias of older rock are definitive evidence that subsidence accompanied eruption of the tephra. A thick pyroclastic deposit with no interlayered breccia within a caldera could be subsequent filling from a later nearby eruption.

Sometime after collapse, a caldera floor may be uplifted, creating a **resurgent uplift** (Figure 10.42d). The amount of time involved and the mechanism of resurgence vary.

10.4.9 Subaqueous Pyroclastic Flows

Because pyroclastic eruptions most commonly occur in subduction zones along continental margins and island arcs it is inevitable that some ash flows either enter bodies of water from subaerial sources or have a subaqueous vent. Several questions follow from this inference. Given that ash flows have a density slightly more or less than that of water, what happens when they contact lakes or the sea from subaerial sources? To what extent do ash flows maintain their high eruptive T (as high as 600°C) in bodies of water? What water depth would suppress pyroclastic flow generation if eruption were subaqueous? Are there unique properties of subaqueous ash-flow deposits in the rock record that set them apart from strictly subaerial ones? Does deposition of a flow that consists of low-density pumice clasts, denser but smaller glass shards, and still denser crystals in water produce sorting not evident in subaerial deposits?

Subaerially generated ash flows from the August 1883 eruption of Krakatau in the Indonesian archipelago entered the sea around the island and created submarine deposits as much as several tens of meters thick that are virtually identical to corresponding subaerial ones on the island (Sigurdsson et al., 1991). Paleomagnetic studies indicate emplacement temperatures in cored samples of the submarine deposit to be 350°–550°C. Historical observations of other subaerially generated ash flows indicate a denser basal flow traveled beneath sea level and a more dilute and buoyant upper part swept across the ocean for tens of kilometers. Mesozoic ignimbrite hundreds of meters thick filling a partially preserved submarine caldera in a roof pendant in the Sierra Nevada batholith, California was deposited in water 150 m or less in depth (Kokelaar and Busby, 1992). Beneath a carapace of bedded and sorted ash-fall tuff, massive unsorted tuff is densely welded and eutaxitic, indicating emplacement temperatures of possibly >500°C.

Other submarine pyroclastic deposits are bedded and sorted and have been depleted in fine ash carried away in suspension. See Fisher and Schmincke (1984) and Cas and Wright (1987) for further discussion of subaqueous pyroclastic flows.

❋10.5 OTHER VOLCANICLASTIC DEPOSITS

Locally extensive subaerial deposits of fragmental volcanic rock owe their origin not to explosive processes but to the mobilizing effects of water and the downhill driving force of gravity. Some of these volcaniclastic phenomena and their related deposits are a direct consequence of volcanic activity that destabilizes rock material on a slope. Others are only indirectly related to volcanism, and still others are epiclastic in nature and involve weathering, transport, and deposition of rock and sediment, especially through the agency of running water.

10.5.1 Epiclastic Processes and Deposits

During repose intervals between extrusions of magma, volcanic material on slopes is subject to epiclastic processes. The building up of a volcanic edifice is counteracted by erosion wearing it down toward base level. Locally, some of this eroded material accumulates as an epiclastic, or sedimentary, deposit.

All of the explosive volcanic processes described so far in this chapter create deposits of loosely consolidated material susceptible to subsequent transport and deposition by wind and running water. Additionally, coarse autoclasts on margins of lava flows are amenable to transport by water on steep slopes. **Reworked volcanic deposits** display features typical of most fluvial epiclastic deposits, such as abrasion of clasts, cross-bedding, and lenticular beds. Because of these similarities, it is commonly difficult to distinguish between reworked deposits formed from pyroclastic material that was never consolidated and epiclastic deposits formed from fragments produced by weathering and disintegration of consolidated volcanic rocks. Deposits consisting chiefly of volcanic fragments, regardless of origin, can be simply classified on the basis of grain size and referred to, for example, as *volcanic sandstone*.

10.5.2 Volcanic Debris Flows: Lahars

Currently, the Indonesian word **lahar** has a dual usage, applied to

1. A mass of intimately mixed water and rock material moving under the influence of gravity down the slopes of a volcano, also referred to as a **volcanic debris flow**

2. The resulting deposit (Figures 7.27 and 10.43).

These lahars or volcanic debris flows generally consist of a wide range of unsorted blocks and lapilli suspended in a water-saturated mud (ash) matrix that imparts mobility to the body. Their viscoplastic rheology resembles that of lava flows. Thus, lahars move in plug manner (Figure 8.13b) with lateral levees and fairly steep margins; their yield strength enables large blocks to be transported. An important component of lahars

10.43 Lahars or **volcanic debris flows** on the flank of Mount Rainier, Washington (Fiske et al., 1963). The steep dips, to as much as 30°, are primary dips in this 50-m-high cliff face. Note crude vertical erosional columns and lenses of partly brecciated lava alternating with debris flows. Compare Figure 7.27. (Photograph courtesy of C. A. Hopson.)

is clay, most typically derived from hydrothermally altered rock exposed on the volcano. Clay-rich rock at the lahar source promotes generation by slope failure because clays can hold large amounts of water, adding to the weight of the mass. Wet clay also facilitates transformation of rock avalanches into debris flows as water is released from the clay into the flow.

Lahars can be generated in many different ways, and transformations in rheology and flow regime during downhill movement are typical (Smith and Lowe, 1991). Two end-member origins are dilution, whereby water is added to rock fragments, destabilizing and mobilizing the mass, and bulking, whereby fragments are added to water from the eroding bed. Hot lahars can be created as pyroclastic flows merge with external water. For example, about 4500 y ago a rhyolite pyroclastic flow was produced by catastrophic sector collapse on the side of Cotopaxi volcano, Ecuador (Mothes et al., 1998). The combination of the hot pyroclastic flow, high elevation (5890 m above sea level), covering thick ice cap, and 3 km of relief produced a 3.8-km³ lahar that descended river systems 326 km to the Pacific Ocean and >130 km to the Amazon basin. Lava flows and domes can also generate hot lahars as the lava contacts snowfields and glaciers typically found on slopes of lofty composite volcanoes (discussed later). Alternatively, rivers can be bulked with hot pyroclasts to create hot lahars. Autoclastic envelopes around the lava as well as fractured, rigid lava in the massive core can be swept up and move down-slope in a bulking stream. Sectors of cold domes may also collapse, forming rockfall avalanches of dry rock blocks (e.g., Chaos Jumbles in Figure 10.18). If avalanches ingest sufficient water they may transform, by dilution, into lahars. Pyroclastic fall deposits on volcano slopes can be diluted and mobilized by heavy rainfall during eruption as eruptive steam cools and condenses, or by chance concurrent torrential tropical storms (as at Pinatubo in June 1991), or at some time after eruption. Crater lakes at volcano summits can be breached and the flood waters bulked by picked-up loose rock debris. At the distal end of the lahar runout, commonly in confined downslope channels, water and fine particles may drain from the coarser flow mass to create hyperconcentrated flows, or mudflows, and these, in turn, can transform into more or less normal streams of sediment-laden water.

Unquestionably, the largest lahars originate from catastrophic collapse of unstable domes (Cotopaxi) or summit sectors of high composite volcanoes, such as occurred at Mount Saint Helens on May 18, 1980 (Figures 10.1 and 10.20). Unstable sector collapse also occurs on more gently sloping flanks of oceanic island shield volcanoes, as shown in Figure 10.11b, c.

Lahars are generally confined to existing topographic depressions (Figure 10.20). Lahars can be *monolithologic,* as clasts were derived from a single source, such as a lava flow that broke up as it entered a snowfield, or *heterolithologic,* where multiple sources fed the lahar. Near-source lahars are made of chaotic, extremely poorly sorted angular clasts. Farther traveled lahars tend to be better sorted and to be locally stratified deposits; clasts tend to show better rounding, either as a result of abrasion during transport or of accumulation of previously more rounded erosional rock debris. Farther transported clasts have a smaller mean fragment size. Nonetheless, huge blocks tens to hundreds of meters across can be rafted tens of kilometers from the source, forming the hummocky ground surface that is typical of lahars as well as rock avalanches.

Discriminating between a lahar and other volcaniclastic and epiclastic deposits (e.g., glacially deposited diamictite) can be challenging. Hot lahars can be identified by the presence of blocks in which radial cracks have formed by cooling and contraction during flow after incorporation from a hot source. Paleomagnetic analysis may disclose a common magnetization direction acquired in the geomagnetic field during cooling of clasts; had the clasts cooled and magnetized prior to incorporation into the flow their magnetization directions would be random.

10.5.3 Composite Volcanoes

Composite volcanoes are the lofty, more or less symmetric conical photogenic landmarks that most people consider to be volcanoes. Most active or recently active volcanoes in subduction zones around the margin of the Pacific Ocean, in the Caribbean, and in the Mediterranean are of this type, including famous ones such as Fujiyama in Japan, Vesuvius in Italy, Mayon in the Philippines, and Mount Saint Helens (Figure 10.1), Shasta (Figure 10.2), and Lassen Peak (Figure 10.18) in the Cascade Range of the Pacific Northwest of the United States. Many composite volcanoes reach great heights because they rise a few kilometers above an already elevated platform of older volcanic deposits and deformed basement rocks in the orogenic belt. Although it is imposing topographically, any one composite volcano (Figure 10.6) has a total volume that is less than might be anticipated. Fujiyama, one of the largest, has a volume of approximately 870 km^3.

A **composite volcano,** also called a **stratovolcano,** is built mostly of andesitic and dacitic magmas extruded from a central vent and consists of, as the names imply, innumerable alternating tongues of lava and volcaniclastic deposits, especially lahars (Figure 10.44). Until removed by erosion, a small crater lies at the summit. Locally, magma may be extruded from flanking central vents, forming **parasitic cones** (Figure 10.2). Magma solidified within feeder conduits and minor fissures as plugs, dikes, and sills forms a reinforcing skeleton for the edifice. The steep slope of composite volcanoes reflects the following compound factors:

1. Relatively small volume and low rates of extrusion of viscous magma that does not move far from the central vent summit

2. Near-vent ballistic ejecta resting at its angle of repose of about 30°–35°

3. Viscoplastic rheology of debris flows, which are a major component of any composite volcano

The effusive and volcaniclastic deposits that composite volcanoes comprise can be divided into central, proximal, and distal facies (Figure 10.44; see also Williams and McBirney, 1979, pp. 312–313). The central or near-vent facies (within about 2 km of the central vent) is a bewildering array of structures and both intrusive and extrusive rock that are commonly hydrothermally altered. Thin lava flows are subordinate to coarse, poorly sorted volcaniclastic deposits with steep initial dips. The proximal or flank facies (up to roughly 5 to 15 km from the central vent) comprises thick lava flows; lahars with subangular, coarse clasts; and some reworked clastic deposits. Zones of weathering and soil development may occur between layers of lava and volcaniclastic deposits. The distal facies comprises layers of rock with considerable lateral continuity formed of well-sorted and fairly well-bedded lahars and epiclastic deposits of rounded clasts; interstratified lake deposits may occur, and lava flows are restricted to less viscous types that flowed down valleys.

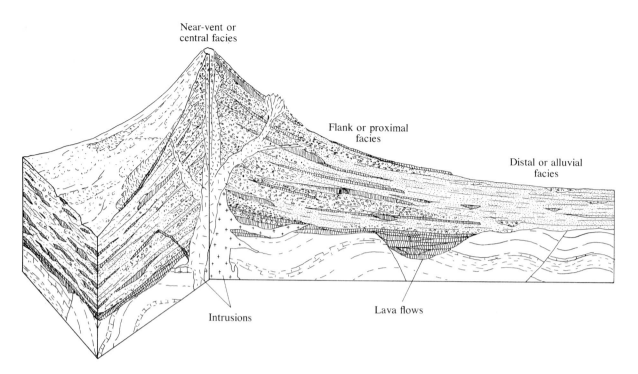

Near-vent or
central facies

Flank or proximal
facies

Distal or alluvial
facies

Intrusions

Lava flows

10.44 Idealized cross section through a **composite volcano** showing alternating layers of lava and volcaniclastic material cut by dikes, sills, and plugs, some of which feed surface lava extrusions. The platform on which the volcano rests is a hypothetical mass of folded and faulted sedimentary rocks capped by a sequence of basalt and andesite flows. Different facies of the composite volcano are discussed in the text. (Redrawn from Williams and McBirney, 1979.)

SUMMARY

The manner of magma extrusion is recorded in the field relations of layered volcanic deposits (stratigraphy) and their fabric. Whether extrusion is by explosive dispersal of pyroclasts or effusive flow of coherent lava depends on the dissolved volatile concentration in the melt, how the volatiles exsolve, possible interactions between external water and the magma, and the apparent viscosity of the magma. Volatile-poor magmas generally extrude as lavas. Low-viscosity lava flows, the most common of which are basaltic, are small-aspect-ratio (thickness/horizontal extent) sheets or streams of pahoehoe, aa, or subaqueous pillow lava that can travel tens or even hundreds of kilometers from their vent source on gentle slopes. Large apparent viscosities with a significant plastic yield strength, such as in high-silica and/or highly crystalline magmas, create large-aspect-ratio lava domes that pile high over the vent. High rates of discharge from the volcanic vent can decrease the aspect ratio.

Explosive eruptions, generally of volatile-rich silicic magmas, disperse pyroclasts mixed with hot gas in volcanic plumes. Plumes can be convecting plinian, or collapsing columns, or a combination of these two. Large concentrations of exsolving and expanding volatiles create high magma discharge velocities, especially where vent diameters are small, promoting high gas thrust that drives plinian plumes tens of kilometers above the vent. As upward momentum provided by expanding gas diminishes above the vent, convective heating of entrained air gives buoyant lift to the plume. Larger vent diameters and lower gas contents favor collapsing columns in which pyroclasts fall immediately around the vent for lack of upward momentum and/or lack of convective transfer to entrained air into massive columns. A common scenario is initial plinian activity that degrades into a collapsing column as the vent is reamed out, enlarging its diameter, and less volatile-rich magma is erupted. However, many possible combinations of plinian and collapsing fountain are possible during a particular eruptive episode, even concurrent play of both from a particular vent system.

Dispersed pyroclasts in plumes are transported to their site of deposition by three processes—fall, surge, and flow. Largest clasts follow ballistic trajectories and generally fall near the vent. Smaller clasts, generally of lapilli and ash size, fall vertically from plumes and are transported horizontally in surges and flows. Transport distance and depositional characteristics depend on many factors, including particle size, density, shape, trajectory (horizontal versus vertical), and concentration, which can fluctuate through the time of activity of the surge, flow, or fall process. Pyroclastic fall deposits are mantle-bedded, and are generally finer and increasingly better sorted with respect to particle size away from the vent. Beds may be normally-graded or reverse-graded.

In pyroclastic surge and flow horizontal transport occurs from the base of a collapsing column as a mixture of pyroclasts and hot gas. Surges have dilute particle concentrations and their moderately sorted deposits lie within 1 km or so of the vent. Plane-parallel surge beds are widespread, but other bedforms that can indicate direction of transport include wavy beds, low-angle cross beds, climbing dunes, and ballistic-clast sags. Near-vent accumulations of basaltic tephra built by pyroclastic surges and ballistic ejecta form tuff cones, tuff rings, and, if a below-ground-level crater is explosively excavated, maars and diatremes. Pyroclastic flows have high particle concentrations and produce massive unsorted deposits. Small block-and-ash flows, whose volume is generally <0.01 km³, are generated by collapse of a lava dome at the summit of a volcano and happen dozens to hundreds of times in an episode of activity at a subduction-sited composite volcano. Larger ash-flow, or ignimbrite, eruptions occur less frequently at a particular locale; deposits of ash and lapilli can be hundreds of meters thick (a few km in caldera depressions), extend over a hundred kilometers from source, and have volumes as much as thousands of cubic kilometers. The considerable mobility of hot pyroclastic flows results from their potential energy inherited from the collapsing massive column and their endowment of gas that produces partial fluidization. As some of the fluidizing gas escapes from the avalanche it carries with it the finer, mostly vitric ash and creates an overlying coignimbrite plume that produces downwind beds of fine ash. Escaping gas in deposits promotes vapor-phase crystallization in pore spaces, which accompanies devitrifaction and welding and compaction of the ignimbrite after deposition.

A wide spectrum of explosive style is dictated by the range in magma composition and volatile content. Least explosive are Hawaiian eruptions, in which basaltic lava fountains build near-vent spatter deposits and streams of lava, some formed as rootless flows of molten spatter, and flow away from the vent. Strombolian eruptions of a boiling top of the basaltic or andesitic magma column exposed in the vent throw out ballistic ejecta, forming a cinder (scoria) cone; lava flows away from the vent area. Vulcanian eruptions commonly begin with steam blast ejection of accidental clasts and then proceed to ejection of juvenile clasts as the ascending column of magma is exposed. At the most explosive end of the spectrum are plinian and ash-flow eruptions. Plinian eruptions create lofty plumes of ash and pumice lapilli, fallout from which creates bedded and sorted pyroclastic fall deposits.

Explosiveness and magma viscosity govern the character of volcanic edifices, or landforms. Steep, conical composite volcanoes are created by countless central eruptions over a million years or so of relatively viscous andesitic-dacitic lava and exploded ballistic ejecta accumulating near vent and pyroclastic flows and debris flows (lahars) sweeping farther downslope. Enormous floods of basaltic magma create oceanic plateaus on the seafloor. Fissure-fed floods create continental plateaus in just a few million years. Gently sloping, fissure- and central-vent-fed basaltic shield volcanoes grow in a half-million years or less. In contrast to these long-lived, focused extrusions, monogenetic extrusions of basaltic lava build strombolian cinder (scoria) cones and isolated lava flows in less than a few years, forming small basaltic lava fields.

CRITICAL THINKING QUESTIONS

10.1 Characterize conditions that allow explosive versus effusive extrusions of magma.

10.2 Describe and account for the contrasts in effusions of basaltic and silicic lavas.

10.3 Discuss the nature of explosive volcanic plumes and controlling factors in their development.

10.4 Summarize styles of explosive eruption in terms of explosiveness, associated volcanic plumes, and pyroclastic deposits that are produced.

10.5 Discuss the origin of volcanic edifices, explaining the contrasts in their shape and how they are built up above ground level or extend below.

PROBLEMS

10.1 Devils Tower in the northeastern corner of Wyoming is a mass of rock that rises 260 m above the surrounding plain and has spectacular subvertical columnar joints. It is often said to represent a volcanic neck, the feeder conduit of now-eroded overlying lava flows. Make a sketch of the tower and critique this explanation.

10.2 Make a sketch of Figure 10.10b showing how the radial arrays of joint columns in the entablature could have been produced by isotherms perturbed by water entering in widely spaced fractures.

10.3 Describe, or diagram, the transformations and transfers of conserved energy in a volatile-rich magma that take place from its generation by thermally induced partial melting in the deep crust, through explosive venting into a plinian plume, and final deposition in ash-fall beds.

10.4 In Figure 10.38b, why are there both rhyolite and mafic pumices in the Mazama ignimbrite deposit from heights of about 15 to 35 m, whereas there are only mafic above and rhyolite below?

10.5 Draw topographic profiles of hilly terrain on which you characterize and distinguish between the field relations and fabric of pyroclastic fall, surge, and flow deposits.

Generation of Magma

11

1. What is the role of the mantle in global magma generation?

2. How and where can solid rock be melted to generate magma?

3. How does the mineralogical and chemical composition of solid rocks in the upper mantle and lower continental crust dictate the composition of magmas generated by partial melting in these source regions?

4. How do partial melting conditions in source regions influence the composition of the generated magmas?

INTRODUCTION

The basic problems in understanding magmatic rocks are how, where, and why they are created and what accounts for their wide range of compositions, as exemplified in any variation diagram (e.g., Figure 2.4).

Clearly, active volcanoes manifest the occurrence of magma in the Earth. Basaltic lavas extruded from volcanoes have $T \sim 1200°C$, whereas rhyolitic magmas commonly erupt at 900°–700°C. These temperatures imply that magma has to be created at depths of *at least* the lower crust (45–35 km) or upper mantle (60 km), assuming a uniform geothermal gradient of 20°C and negligible cooling during ascent. But we know that the crust and mantle are generally solid rock all the way to the outer liquid metallic core because they transmit

seismic shear waves. So what are the anomalous conditions that explain how and why silicate magma is generated from solid silicate rock in the mantle and possibly the lower crust? Somehow, already hot but subsolidus rock must be perturbed to cause melting and magma generation.

Magma generation must be closely linked with global tectonics because active volcanoes occur where lithospheric plates are diverging and converging and where mantle plumes are actively rising, as under Hawaii (Figure 1.5 and Plate I). No active volcanoes occur in the central part of North America or Australia or in other huge areas around the globe.

✳11.1 MELTING OF SOLID ROCK: CHANGES IN P, T, AND X

Melting of a solid is conventionally associated with an increase in $T(+\Delta T)$. However, phase diagrams in Chapter 5 show that melting can also occur at virtually constant T (isothermally) by decompression $(-\Delta P)$ and by influx of volatiles $(+\Delta X_{volatiles})$ into already hot rock. In all three changes in state of the system, the rock already contains a significant store of thermal energy and becomes a melting **source rock** as $+\Delta T$, $-\Delta P$, and $+\Delta X_{volatiles}$ perturbations move the system above the solidus (Figure 11.1). Total melting of a source rock to its liquidus requires large, and geologically unrealistic, changes in these three intensive variables. Before total melting might occur, either the buoyant melt separates and moves out of the source or the partially melted rock becomes sufficiently buoyant to rise en masse, perhaps as a diapir, out of the source region and away from the perturbing changes.

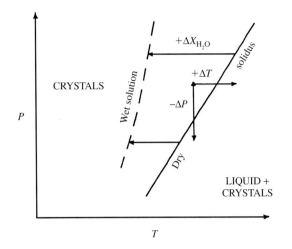

11.1 Perturbations in *P*, *T*, or X_{water} move a subsolidus rock to above the solidus, causing partial melting. Potential source rock just below solidus represented by solid triangle. Influx of water, $+\Delta X_{H_2O}$, depresses the solidus to a "wet" solidus position (not water-saturated), placing the rock in the stability field of liquid + crystals. Perturbations of $-\Delta P$ and $+\Delta T$ move potential source rock to above the solidus.

As the causes of these perturbations are examined next it should be kept in mind that they may not act independently. For example, increase in *T* may be accompanied by an increase in volatile concentration where mantle-derived basalt magma intruded into the lower continental crust heats it and adds volatiles exsolved during crystallization. Addition of volatiles lowers melting temperatures, enhancing the effect of heating in melt production.

11.1.1 Temperature Increase, $+\Delta T$

An increase in *T* sufficient to melt solid rock in the Earth can occur in several ways.

<u>Mass Movement of Rock or Magma</u>. Heat transfer associated with convective or advective movement of rock or magma is an important means of raising rock temperatures above the solidus in large rock volumes. Two global plate tectonic regimes where this takes place include (Figure 11.2):

1. The descending oceanic lithosphere in subduction zones absorbs heat from the surrounding hotter mantle. Consequently, partial melting of the less refractory basaltic crust might occur in young, still hot subducting lithosphere, as contrasted with subduction of old, cold lithosphere. Shear heating of rapidly subducted lithosphere can also promote higher temperatures (discussed later).

2. Already hot, deep continental crust can be heated in excess of its solidus *T* by juxtaposition of hotter mantle-derived magma above subducting plates and upwelling decompressing mantle. Density constraints indicate that ascending basaltic magma can

be arrested at the base of the less-dense feldspathic crust, **underplating** it, or perhaps stagnating not far into it above the Moho (Section 9.1.1). The thermal budget for melting crust can be roughly approximated by considering deep continental felsic rock at a depth of 30 km on an average geothermal gradient of 20°C and therefore at *T* = 600°C. To raise the rock *T* to its solidus of, say, 900°C and assuming the rock specific heat is 1.4 J/g deg requires (1.4 J/g deg) × (900 − 600) deg = 420 J/g (equation 1.4). Melting 1 g of rock requires an additional 300 J/g, its approximate latent heat of melting. Thermal energy to accomplish this heating and melting totals about 720 J/g and can be supplied by less than 1 g of basalt magma initially at 1300°C. This assumes that the latent heat of crystallization of the basalt magma supplies about 420 J/g and cooling to 900°C supplies about (1.4 J/g deg) × (1300 − 900) deg = 560 J/g. An approximate rule of thumb is that the mass of continental felsic rock melted is about the same as the mass of basaltic magma heating it. Melting consumes relatively large amounts of thermal energy, moderating increasing *T* within the source rock.

<u>Mechanical Work of Shearing</u>. Work can be transformed into thermal energy, *Q*, according to

$$11.1 \quad Q = \tau\left(\frac{d\varepsilon}{dt}\right)$$

where τ is the shear stress required for threshold deformation and *dε/dt* is the strain rate (Section 8.2). For a shear stress of 50 MPa and a representative geologic strain rate of 5 × 10^{-14}/s, such as occurs during shortening in a mountain belt or convective mantle flow, *Q* = 50 × 10^6 Pa [(kg/m s^2)/Pa] × [J/(kg m^2/s^2)] × 5 × 10^{-14}/s = 2.5 × 10^{-6} J/m^3 s. For 1 g of granite whose density is 2.7 × 10^6 g/m^3, *Q* = 6.8 × 10^{-13} J/g s ~ 2 × 10^{-5} J/g y. This heating is so small it would require 36 My to melt granite initially at 600°C (see previous paragraph), assuming the stress is maintained and no heat is dissipated from the system.

However, through a **thermomechanical feedback** phenomenon, shear heating might be sufficient to induce melting. Thus, shearing might become concentrated in certain zones, localizing more thermal energy production; the consequently higher *T* reduces the apparent viscosity, further localizing shear deformation, causing more heating and reduction of viscosity, and so on, until the melting *T* is reached. Such shear melting would tend to be localized in small volumes.

Local shear melting can occur at meteorite impact sites and along fault zones where strain rates are much higher.

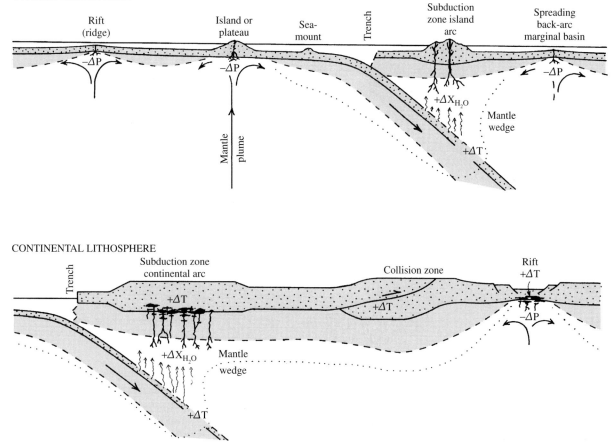

11.2 Geologically plausible plate tectonic settings at which partial melts may be generated from otherwise solid rock by perturbations in P, T, or $X_{volatiles}$. Compare Figure 1.5. Lithosphere is shaded; oceanic and continental crust is stippled. Scale is grossly distorted. Decompression ($-\Delta P$) of rising solid mantle beneath oceanic ridge, continental rift, and in mantle plume partially melts peridotite. In subduction zones, water released (wavy arrows) by dehydration of hydrated oceanic crust capping subducting oceanic lithosphere rises into overlying mantle wedge; resulting $+\Delta X_{water}$ in peridotite causes partial melting as solidus T is depressed. Locally, young, hot, hydrated oceanic crust partially melts as a result of $+\Delta T$ caused by absorbed heat from surrounding hot asthenosphere. Ascending mafic magmas from volatile-fluxed mantle wedge underplate continental crust or rise slightly above the Moho, where they stagnate (black blobs), transfer heat to the already hot lower crust, and cause $+\Delta T$ and partial melting. A similar $+\Delta T$ occurs in a continental rift where basaltic magmas are derived from upwelling mantle underplate crust. Where continental crust is greatly thickened in collision zones, such as the present-day Himalayan, $+\Delta T$ results from processes such as radioactive heating, shearing, and adjustment to the geothermal gradient. Dotted line is highly schematic isotherm perturbed to shallower depths by rising mantle (heat source) and depressed to greater depth in the colder subducting lithosphere (heat sink).

Decay of Radioactive Isotopes. Thermal energy is generated by slow decay of K, Th, U, and other less abundant radiogenic isotopes in rocks. However, the rate of heat production is very low (granite about 3.4×10^{-5} J/g y, basalt 5.0×10^{-6} J/g y, peridotite 3.8×10^{-8} J/g y), so that, again, as with distributed shear heating, tens of millions of years would be necessary to raise the T of deep rock to its solidus, even if the heat produced did not leak out of the body in that time. Additionally, the first melting episode would extract much of the incompatible K, Th, and U from the source rock, so that subsequent heat generation would be greatly retarded. Therefore, radioactive decay *alone*, in most instances, cannot produce sufficient perturbation in T to cause magma generation.

Tectonic Thickening. Thrusting and folding thicken continental crust to >50 km in active orogenic (mountain) belts in subduction zones (Figure 11.2). This can lead to supersolidus temperatures in the deep crust. Several factors are intertwined in complex ways in this heating. Radioactive heat production in the thickened continental crust is enhanced, as is the insulating effect on the flux of heat from the mantle. The 600°C temperature at the base of a 30-km-thick crust is subsolidus for most potential source rocks, but, after thickening to, say, 50 km and readjustment to the same average geotherm of 20°C/km, the T at the base of the crust is 1000°C, which is well above the solidus of many crustal rocks. Optimal "incubation" depends on the rate of thickening and may be curtailed by prompt

Special Interest Box 11.1 Highly localized melting of rock independent of tectonic setting

Heat generation by mechanical friction accompanying movement at very high strain rates ($>10^{-2}$/s) along fault zones is locally sufficient to cause melting of rock (equation 11.1). Sheets of melted rock are commonly associated with cataclastic fabric (Figure 8.7) in shear zones because cataclasis is an essential precursor to frictional melting. The sheets are generally <1 cm thick. Rapid heat loss quenches the melt sheet to a vein of typically black glass called **pseudotachylite.** (Tachylite is a term once used for basaltic glass; however, not all shear-generated glass is basaltic.)

Locally extensive melting occurs at meteorite impacts where strain rates are $>10^6$/s; high-pressure silica polymorphs—coesite and stishovite—are also created by the transient shock. Larger podlike or dikelike bodies of pseudotachylite as much as 1 km thick (Figure 11.3) are produced by the extreme comminution of fault walls during rebound and gravitational collapse of the transient impact cavity (Spray, 1998).

Many older pseudotachylites are devitrified to microcrystalline or cryptocrystalline aggregates. Devitrification textures can resemble those occurring in volcanogenic glass, for example, spherulitic texture. Fragments of unmelted rock are common within the glass; partially dissolved crystals are embayed. Flow layers, vesicles, and magmatic crystallization textures are locally evident.

An extremely rare and bizarre means of locally melting rock and unconsolidated sediment occurs through lightning strikes, producing **fulgurite.** These are irregularly shaped crusts or tubular glassy structures that can be as much as 40 cm long and 6 cm in diameter.

11.3 Dark-colored pseudotachylite surrounding blocks of unmelted granite at McCreedy West in the north range of the 250-km-diameter, 1.85 Ga Sudbury, Ontario, impact structure. After the impacting meteorite excavated the crater in Archean granite and granite gneiss, faulting accompanying rebound and gravitational collapse generated pseudotachylite melts by comminution and frictional heating at high strain rates along the fault walls. (Photograph and caption courtesy of John G. Spray.)

rapid uplift and erosion. Thermal modeling that takes into account these and other factors suggests optimal heating occurs on the order of a few tens of millions of years after orogenic thickening (e.g., Patiño Douce et al., 1990).

11.1.2 Decompression, $-\Delta P$

Because of the positive slope of the solidus curve of dry silicate systems in *P-T* space, adiabatic decompression of hot near-solidus rock can induce melting. Consider a solid parcel of mantle initially below the volatile-free solidus at *B* in Figure 11.4. As the parcel rises to shallower depth, it decompresses and cools adiabatically at a rate of about 0.3°C/km, or 1°C/kbar, along the path *BB′*, whose slope is steeper than the dry solidus. At *B′*, melting begins and the thermal energy required for the latent heat of fusion is drawn from the thermal energy

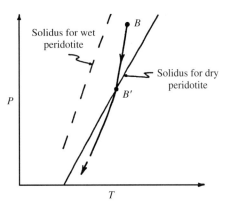

11.4 Schematic decompression melting. Note that decreasing depth is *downward* in diagram.

stored in the body. Hence, during continued decompression, the upwelling mantle cools at a more rapid rate at pressures less than B' than at pressures greater than B'. Thus, the P-T path of the ascending mantle closely follows the solidus. As the solidus has a slope of about 10°C/kbar, a decrease in P of 1 kbar (over about 3 km) results in an increment of T (corrected for adiabatic cooling) $\Delta T = 10 - 1 = 9°C$. Multiplied by a specific heat, C_P of 1.4 J/g deg, this gives (equation 1.4) 12.6 J/g, sufficient to melt about 3% of rock with a latent heat of fusion of 420 J/g (McBirney, 1993, p. 243).

Special Interest Box 11.2 Decompression melting: belated acceptance of a two-century-old idea

Almost three-fourths of the Earth is covered by rocks of basaltic composition; most of these rocks constitute the oceanic crust that is several kilometers thick. Most basaltic magmas are generated by decompression melting, which therefore is the dominant magma-generating mechanism in the planet. Decompression melting in upwelling mantle beneath the globe-encircling system of oceanic spreading ridges and in ascending mantle plumes is accepted today as one of the most fundamental concepts in petrology. Surprisingly, acceptance was slow (Sigardsson in Sigardsson et al., 2000).

John Playfair in 1802 followed by George P. Scrope in 1825 speculated that a change in pressure might generate magma within the Earth. In the early nineteenth century, the pioneering work of Carnot, Clausius, and Clapeyron in the infant science of thermodynamics resulted in an exact expression for the change of the melting T of a solid substance with respect to change in P—the Clapeyron equation (3.13), $dT/dP = \Delta V/\Delta S$. Nicholas Desmarest had recognized in 1763 that the prismatic joint columns in the Claremont area of France were of volcanic origin, and subsequent observers correctly inferred that they formed by contraction during cooling. Thus, because a volume (and entropy) increase accompanied melting, the Clapeyron equation predicted an increase in the melting T of rock with increasing depth in the Earth. Hence, decompression of hot solid rock could lead to melting.

As early as 1881, Osmond Fisher proposed that convection currents in a molten interior (as then believed) ascended beneath the oceans and descended beneath continents. After the discovery of radioactivity near the turn of the 20th century, Arthur Holmes was one of the first to apply it for dating of minerals. Also being aware of the heat produced by radioactive decay, Holmes embraced convection as

a means to cool the Earth and in so doing proposed in 1928 a global mechanism remarkably like the modern paradigm of plate tectonics. But convection in mantle known to be solid, like Wegener's proposal of continental drift in 1912, were insights too preposterous at the time to be widely accepted.

Without an alternative mechanism to relieve P in the interior of the hot Earth, petrologists in the first half of the twentieth century either equivocated or dismissed decompression melting as a viable process. As late as 1960, in their classic *Igneous and Metamorphic Petrology,* Francis J. Turner and John Verhoogen concluded that (p. 446) "Whether convection does occur in the mantle is not definitely known, nor is it known whether it could be effective in the upper part of the mantle where magmas are generated. Thus, although convection might lead to melting, it cannot be shown that it does, and the problem of generation of [basaltic] magma remains as baffling as ever."

During the Cold War (1950s to 1970s), the U.S. government generously funded Earth scientists in order to learn more about the oceans. The result was the serendipitous discovery of seafloor spreading and plate tectonics that in turn lead to the confirmation of mantle convection and the opportunity for decompression of mantle peridotite to generate basaltic magma.

Decompression partial melting of mantle rock is the principal mechanism by which huge global volumes of basalt magma are generated at ocean ridges and above mantle plumes (Figures 1.1, 1.5, 11.2).

Some geologists believe that decompression of uplifting continental rock can cause partial melting, but it is likely of minor importance.

11.1.3 Changes in Water Concentration, $+\Delta X_{water}$

Among the possible changes in chemical composition of a solid rock system that could induce melting on a global scale, increase in water concentration (or pressure), $+\Delta X_{water}$ ($+\Delta P_{water}$), is the most significant. Very locally, an increase in CO_2 or other volatiles may induce melting in already hot rock. Even small increases in water concentration can significantly depress the solidus of silicate systems (Figures 5.2, 5.11, 5.29). This perturbation is most significant in subduction zones (Figure 11.2), where the descending wet oceanic crust dehydrates, liberating water that advects into the overlying wedge of peridotite mantle. In the crust, water (and lesser amounts of CO_2) resides in pore spaces in sediment and in cracks in thicker basaltic rock. More significantly, basaltic rock has been variably metamorphosed into hydrous mineral assemblages by hy-

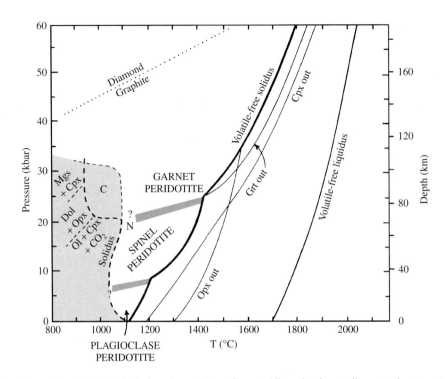

11.5 Phase relations in mantle peridotite. Volatile-free phase relations for spinel lherzolite from Kilborne Hole, New Mexico. (Sample KLB-1; redrawn from Takahashi et al., 1993.) Subsolidus assemblage of garnet peridotite is garnet (Grt) + olivine (Ol) + orthopyroxene (Opx) + clinopyroxene (Cpx), for spinel peridotite is Spl + Ol + Opx + Cpx, and for plagioclase peridotite is Pl + Ol + Opx + Cpx. Note subsolidus transitions between the three peridotite facies are "smeared" (shaded bands) over a range of T and P because of solid solution. Thin-line curves denoting complete dissolution of Opx and Cpx into the melt with increasing T correspond to approximately 40% melting of the lherzolite. On the left, short dashed lines show phase relations of volatile-bearing peridotite with 0.3 wt.% H_2O plus <5 wt.% $MgCO_3$ (Mgs, magnesite) or $CaMg(CO_3)_2$ (Dol, dolomite) beneath its solidus (heavy dashed line). Shaded area indicates where amphibole is stable. N, approximate P-T region for nephelinite partial melt in spinel peridotite. Between about 20 and 30 kbar a Mg-Ca-Na carbonatite melt is stable with an amphibole peridotite above the solidus in the field labeled C. (Redrawn from Wallace and Green, 1988.)

drothermal activity at spreading ridges. As these hydrous minerals are heated in the descending crust to depths of hundreds of kilometers, sequential dehydration reactions (Section 5.7.2, Figure 5.31) release water.

In the lower continental crust, dehydration mineral reactions accompanying metamorphism might provide water to induce partial melting. Underplating mantle-derived magmas might also give off water and CO_2 as they crystallize.

✳11.2 MANTLE SOURCE ROCK

Most magma generated in the Earth is basaltic. Its source is decompressing peridotitic mantle upwelling beneath global ocean ridges and to a lesser extent in ascending mantle plumes (Figures 1.5 and 11.2). Basaltic magmas in subduction zones are generated in the mantle wedge overlying the subducting lithosphere. In most continental tectonic regimes, significant proportions of heat as well as mass for magma generation are also directly or indirectly derived from the mantle. In the young Archean Earth, a large proportion of the sialic continents was created ultimately from partial

melts extracted out of the mantle. Hence: *The ultimate source of thermal energy and mass for production of magma in the Earth is the mantle.* Understanding the nature of the mantle source rock is therefore critically important.

Cosmological models relate the composition of the whole Earth to chondrite meteorites from which the Earth is presumed to have accreted at about 4.5 Ga. Their composition, on a volatile-free basis, is (Taylor and McLennan, 1985) as follows:

SiO_2	34.2 wt.%
TiO_2	0.11
Al_2O_3	2.44
FeO	35.8
MgO	23.7
CaO	1.89
Na_2O	0.98
K_2O	0.10
Total	99.2

Chondrites also contain water and carbon. Most of the Fe in the chondritic Earth segregated into its core shortly after or during accretion.

Measured seismic velocities in the upper mantle are compatible with a rock made of olivine, pyroxene, and garnet. These three phases are composed essentially of the five major chondritic chemical components: SiO_2, Al_2O_3, FeO, MgO, and CaO. Upper mantle densities constrained by isostatic calculations are about 3.35 g/cm^3 (increasing slightly with depth, Figure 1.3), which is more consistent with peridotite (about 3.3 g/cm^3) than with olivine-free, pyroxene-garnet rock (eclogite, 3.5 g/cm^3), although bodies of the latter might occur in minor amounts within the former.

Peridotite is an ultramafic rock made of Mg-rich olivine and lesser amounts of pyroxene, usually both Ca-Mg-rich clinopyroxene and Ca-poor, Mg-rich orthopyroxene; these three crystalline phases are a stable assemblage to a depth of about 410 km. Most peridotites contain more Al in the bulk rock than can be held in solid solution in pyroxenes and olivine, thus stabilizing a separate minor Al-rich phase, whose nature depends on P and less on T. At less than about 8 kbar (30-km depth), the stable Al-rich phase is plagioclase; from there to about 25 kbar (roughly 75 km, depending on crustal thickness), it is spinel; and at still higher P it is garnet (Figure 11.5). Transitions between these three peridotite assemblages are smeared over a range of P (depth) and T because the minerals are solid solutions. Garnetiferous peridotites equilibrated at depths greater than about 150 km and occurring as xenoliths in some kimberlitic rock also contain diamond. Diamonds commonly contain minute inclusions of minerals stable in the upper mantle (pyroxene, garnet). Extremely rare diamonds contain inclusions stable only at >670 km-depth in the deep mantle (Mg-Fe-Ca-Al perovskite, Figure 1.3), testifying to the depth from which at least some kimberlitic magmas carrying suspended diamonds are derived. At shallower depths where diamond is not stable, the stable C-bearing phase may be graphite if reducing conditions prevail or, if more oxidizing, a carbonate mineral or CO_2 found in fluid inclusions in mantle minerals.

Some petrologic information on the uppermost mantle is from dredged samples from upfaulted segments of the seafloor and from large tracts of *alpine* peridotite or peridotite *massifs* exposed in mountain belts. The bodies of peridotite are components of ophiolite—segments of oceanic lithosphere tectonically emplaced and exposed in mountain belts adjacent to oceanic trenches (see Section 13.6). Oceanic rocks commonly suffer overprinting metamorphic effects that have erased their primary character to varying degrees. However, ophiolites provide key stratigraphic and structural information that dredge samples and inclusions, described next, cannot.

11.2.1 Mantle-Derived Inclusions

Important compositional information on the mantle is provided by phaneritic **mantle-derived inclusions** of peridotitic rock, usually hosted in alkaline basaltic and kimberlitic rock in thousands of localities worldwide (e.g., Nixon, 1987). These typically fresh and unaltered dense inclusions, also called **nodules,** or **xenoliths,** are fragments of rock entrained from near the source of the host magma or plucked from shallower conduit walls and lifted to the surface during rapid ascent. Mineral geobarometers indicate equilibration in the upper mantle. Inclusions are of two types, classified according to their clinopyroxene: Cr-rich diopside or Al-Fe-Ti-rich clinopyroxene.

<u>Cr-Diopside Peridotite</u>. The most common mantle-derived inclusion consists mostly of olivine with subordinate pyroxenes and smaller amounts of either spinel (Mg, Fe^{2+})(Cr, Al, Fe^{3+})$_2$O$_4$ or pyropic garnet (Mg, Fe^{2+}, Ca)$_3$(Al, Cr)$_2$Si$_3$O$_{12}$. The clinopyroxene is essentially diopside ($CaMgSi_2O_6$) that contains only a small weight percentage of alumina and Fe-oxide, but because of a few tenths of weight percentage of Cr_2O_3 is characteristically an emerald-green color (Plate V). Although constituting <20% and generally <5% of these inclusions, this distinctive green Cr-diopside serves as a characterizing mineral. Olivines and orthopyroxenes contain about 90 mol% of the Mg end member. Some of these peridotites contain less than 1% or so of hydrous minerals, including phlogopite and pargasite or richterite amphibole (Appendix A). Modal proportions of olivine and pyroxenes classify most inclusions as lherzolite and harzburgite; fewer are dunite. A metamorphic fabric resulting from textural equilibration has created 120° triple-grain junctions (Section 6.4). Some inclusions show subtle, more pyroxene-rich layers. Some show a late imprint of shear deformation manifested in strained olivines that have *deformation bands* of contrasting optical extinction orientation in polarized light in thin sections.

Spinel-bearing peridotite inclusions are hosted almost exclusively in alkaline basalt, basanite, and nephelinite and very rarely in andesitic rocks in subduction zones. Host magmas are characteristically volatile-rich, are commonly explosive, and tend to ascend rapidly from their upper mantle source. The absence of inclusions in far more widespread tholeiitic basalts possibly stems from slower ascent to the surface and the more evolved nature of these magmas, which involves fractional crystallization of more primitive parents in crustal storage chambers; both factors inhibit lifting dense inclusions to the surface and promote their assimilation. Spinel peridotite inclusions are mostly of latest Cenozoic age. Whether from oceanic or continental regimes, they are not significantly different (McDonough, 1990), indicating that the uppermost mantle in these two global areas is similar.

Garnet peridotite inclusions derived from greater depths in the mantle are somewhat different chemically from spinel peridotites (Table 11.1). They are hosted in

Table 11.1   Average Worldwide Composition of Garnet Peridotite (Harzburgite) and Spinel Peridotite (Lherzolite) Inclusions

	GARNET	SPINEL						
SiO_2	45.00	44.22	Li	1.5	y	4.4	Mo	0.050
TiO_2	0.08	0.09	B	0.53	Zr	21	Ru	0.0124
Al_2O_3	1.31	2.28	C	110	Nb	4.8	Pd	0.0039
Cr_2O_3	0.38	0.39	F	88	Ba	33	Ag	0.0068
FeOt	6.97	8.47	S	157	La	2.60	Cd	0.041
MnO	0.13	0.14	Cl	53	Ce	6.29	In	0.012
MgO	44.86	41.60	Sc	12.2	Pr	0.56	Sn	0.054
NiO	0.29	0.27	V	56	Nd	2.67	Sb	0.0039
CaO	0.77	2.16	Cr	2690	Sm	0.47	Te	0.011
Na_2O	0.09	0.24	Co	112	Eu	0.16	Cs	0.010
K_2O	0.10	0.054	Ni	2160	Gd	0.60	W	0.0072
P_2O_5	0.01	0.056	Cu	11	Tb	0.070	Re	0.00013
Total	100.00	99.97	Zn	65	Dy	0.51	Os	0.004
			Ga	2.4	Ho	0.12	Ir	0.0037
Mg-value*	92.0	89.8	Ge	0.96	Er	0.30	Pt	0.007
Olivine	68	62	As	0.11	Tm	0.038	Au	0.00065
Opx	25	24	Se	0.041	Yb	0.26	Tl	0.0012
Cpx	2	12	Br	0.01	Lu	0.043	Pb	0.00016
Spinel	—	2	Rb	1.9	Hf	0.27	Bi	0.0017
Garnet	5	—	Sr	49	Ta	0.40	U	0.00012

Major element and calculated modal composition in the first three columns. Trace element composition (in ppm) of spinel periodiate in last six columns.
*Mg-value = 100[MgO/(MgO + FeO)] molecular ratio.
Data from McDonough (1990).

more alkaline, more silica-undersaturated magmas, including kimberlites and lamproites (Sections 11.5 and 13.12.2). These same host magmas also locally contain xenoliths of **eclogite,** a rock of basaltic chemical composition that consists of a high-*P* assemblage of mostly green clinopyroxene solid solution (jadeite-rich omphacite) and purple-red Mg-rich (pyropic) garnet (Appendix A). These xenoliths could be derived from basaltic melts generated and entrapped in the mantle and crystallized to eclogite or from remnants of subducted, recrystallized oceanic crust.

A Brief Digression on Volatiles in Mantle Minerals. In the mid-20th century, geologists began to accept the notion that Cr-diopside peridotite inclusions represent samples of the upper mantle. But as this mantle rock contains no volatile-bearing minerals, as then thought, Oxburgh (1964) realized that generation of typically volatile-rich alkaline basalt and kimberlite magmas from this source would be impossible. Further difficulties arise because of the very low concentrations of K, Na, and many other incompatible elements in peridotite (Table 11.1); only by very small degrees of partial melting could such magmas be generated. Obviously, a source of volatiles and incompatible elements was needed in the upper mantle. Alternatively, these Cr-diopside peridotite inclusions do not represent the source of such magmas.

Soon after Oxburgh's seminal paper, geologists began seeing previously overlooked or ignored amphibole, mica, apatite, and other hydrous and volatile-bearing minerals in mantle-derived inclusions. In addition to water, amphiboles and micas contain several tenths of 1 wt.% F and about an order of magnitude less Cl. And these minerals contain substantial amounts of the missing K, Rb, and other incompatible elements. Sulfur is sequestered in sulfides that occur in some inclusions. Diamonds and micas contain hundreds of parts per million of N. Virtually pure CO_2-fluid inclusions occur in mantle olivines. Infrared spectroscopy reveals that nominally anhydrous olivine, pyroxene, and garnet in upper mantle peridotite actually contain measurable amounts of structurally bound water as $(OH)^-$ or H^+ (Bell and Rossman, 1992). Garnets generally have <60 ppm water but may have as much as 200 ppm, pyroxenes mostly 100–600 ppm, and olivines 100 ppm, increasing with depth.

The oxygen fugacity of the upper mantle is commonly appropriate to stabilize carbonate minerals, such as magnesite, $MgCO_3$, and dolomite $CaMg(CO_3)_2$. During even the most rapid decompression, as in ascending explosive kimberlite magma, carbonate minerals quickly decompose with release of CO_2; consequently, inclusions generally lack this mineral and some that did have suffered disaggregation so that associated silicate grains have been dispersed as xenocrysts in the kimberlite.

Mantle volatiles could originate from two sources. Juvenile volatiles were derived from the primeval chondritic material from which the planet formed at 4.5 Ga. Volatiles liberated from subducting oceanic crust are introduced into the overlying mantle and perhaps, by some means, distributed more widely through the mantle. It is estimated that six times more water is introduced into the mantle over subducting oceanic crust than is delivered to the surface of the Earth in subduction-zone volcanism (Thompson, 1992).

<u>Al-Fe-Ti-Rich Clinopyroxene Inclusions</u>. Though not as abundant as Cr-rich diopside inclusions, Al-Fe-Ti-rich clinopyroxene inclusions are nonetheless widespread. They are especially significant in containing substantial amounts of volatile-bearing minerals enriched in incompatible elements (K, Rb, Ti, C, H, etc.). Some consist exclusively of volatile phases. These inclusions are petrographically more variable than Cr-diopside inclusions, into which they may locally grade, but are dominated by clinopyroxenes rich in Al, Fe, and Ti; because of their black, conchoidally fractured aspect in hand samples, these clinopyroxenes resemble obsidian. Modal proportions range widely among the following constituent minerals: clinopyroxene, olivine, orthopyroxene, high-Al/Cr spinel, magnetite, ilmenite, rutile, zircon, plagioclase (in low-P inclusions), carbonate minerals, Fe-sulfides, apatite, amphibole, and phlogopite. Note that the latter five contain volatiles and the latter three are hydrous. Wehrlites, olivine clinopyroxenites, and pyroxenites are common rock types (Figure 2.10b), together with mica- and amphibole-bearing varieties. Olivines are enriched in Fe and orthopyroxenes in Al, Fe, and Ti relative to these phases in Cr-diopside inclusions. Textures of inclusions are also variable, but magmatic ones are common, such as amphibole poikilitically enclosing other phases. Preservation of magmatic textures in inclusions suggests crystallization from melts not long before entrainment into the host magma; otherwise textural equilibration and development of metamorphic fabric would be expected to occur at the high T prevailing in the mantle.

Discrete **megacrysts,** up to several centimeters in diameter, of clinopyroxene, olivine, orthopyroxene, high-Al/Cr spinel, magnetite, ilmenite, rutile, zircon, amphibole, phlogopite, and plagioclase can occur with or without accompanying xenoliths in alkaline mafic host rocks.

Significantly, Al-Fe-Ti-rich clinopyroxene assemblages are locally found as veins, to as much as several centimeters thick (Plate VI and Figure 11.6a), in Cr-rich diopside peridotite inclusions (e.g., Wilshire et al., 1988). Mineral barometers indicate that some veined inclusions are derived from depths of at least 170 km in the mantle. Thicker veins may be the source of discrete megacrysts and xenoliths of the Al-Fe-Ti-rich clinopyroxene assemblage. Many veins are remarkably planar, indicating that the mantle host rock fractured in a brittle manner, probably hydraulically (Section 8.2.1), then the liquid hydraulic agent was emplaced into the fracture, forming the vein. Similar veins occur in large subaerial exposures of peridotite in ophiolite. Multiple generations of veins, one or more cutting earlier ones, are locally evident in both ophiolite and inclusions. Wall rock of Cr-diopside peridotite adjacent to a vein can be modified chemically and mineralogically for distances up to as much as several centimeters.

11.2.2 Metasomatized and Enriched Mantle Rock

Subsolidus modification of the chemical composition of a rock through the agency of invasive percolating liquids is called **metasomatism.** Rock volume can remain constant during this open-system replacive metasomatism. Two types may be defined: In **cryptic metasomatism** original solid-solution minerals remain but are changed in composition; for example, original clinopyroxenes may be made more Fe-rich, but subtle chemical changes may not be obvious, hence the designation *cryptic* (hidden) *metasomatism.* In **modal metasomatism** original minerals are replaced by entirely new minerals; an olivine-orthopyroxene assemblage (harzburgite) may be modally metasomatized to a new mineral assemblage of amphibole and clinopyroxene. Obviously, these two types of metasomatism can develop simultaneously in the same volume of rock; some original solid solution mineral compositions can be modified and other unstable minerals are replaced by new ones.

Metasomatism has been studied for at least a century in crustal metamorphic terranes and around ore deposits. However, Bailey (1970) first realized that xenoliths of Al-Fe-Ti-rich assemblages in alkaline volcanic rocks are records of mantle metasomatism and that they have a significant bearing on the generation of alkaline magmas. The rationale is that some highly alkaline mafic and ultramafic magmas cannot readily be derived from "normal" mantle of anhydrous, incompatible element-poor Cr-diopside peridotite. This sterile, or infertile, rock has to be metasomatically **enriched** in incompatible elements (Figure 11.6) before it can be a viable source of alkaline magmas that are enriched in such elements (see Section 11.5).

Enrichment details depend on the nature of the liquid, composition of the wall rock, and partition coefficients between the metasomatizing liquid and the minerals in the wall rock. Volatile-bearing minerals created by modal metasomatism—chiefly phlogopite, amphibole, apatite—harbor most of the incompatible elements—K, Ti, Al, Rb, Ba, Sr, H, F, Cl, and light rare earth elements. In their absence, incompatible elements are sequestered in clinopyroxene.

Possible metasomatizing liquids include volatile-rich silicate melts, carbonatite melts, and C-O-H fluids (see Section 4.2.1 for the distinction between fluid and melt). In most metasomatized mantle rocks it is uncertain whether a fluid or a melt was involved; in some cases it might have been both, or one type of liquid might have evolved into another. Some metasomatizing silicate melts are likely of Ti-K-rich alkaline basalt composition, judging from amphibole veins of roughly the same composition. Intraplate spinel peridotite xenoliths from the Kerguelen Islands in the southern Indian Ocean contain minute inclusions (a few tens of micrometers in size) in olivines and pyroxenes that consist of coexisting CO_2 fluid and silicate and carbonatite melt (now glass); still more minute quantities of minerals in these inclusions are Ti-rich kaersutite, diopside, rutile, ilmenite, and magnesite. Schiano et al. (1994) interpret these to be immiscible segregations of an initially homogeneous metasomatizing melt ($T \sim 1250°C$) that invaded the mantle beneath the islands. On the other hand, minute glass inclusions in strongly depleted, cataclastic harzburgite xenoliths from a Philippine arc volcano are dacitic and contain relatively high weight percentages of Na (3–5), K (2–4), and H_2O (4–5), as well as high Cl (1600–7300 ppm), and S (as much as 2500 ppm). Melts were incorporated into growing host minerals (primary olivine and enstatite, secondary metasomatic amphibole and phlogopite) at about 920°C (Schiano et al., 1995). Regardless of the nature of the percolating metasomatic liquid or its ultimate origin, it contains significant concentrations of incompatible elements that react with minerals in Cr-diopside peridotite, creating enriched concentrations. Mg-rich, Ti-Al-poor olivine and orthopyroxene are especially unstable in and susceptible to replacement by metasomatic liquids that contain relatively high concentrations of Fe, Al, alkalies, and volatiles.

Both cryptic metasomatism and modal metasomatism depend on diffusion of ions driven by concentra-

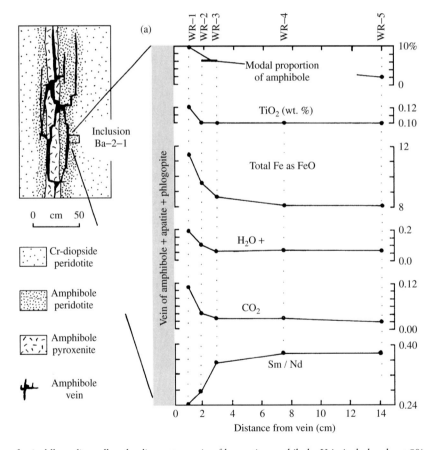

11.6 Metasomatism of spinel lherzolite wall rock adjacent to a vein of kaersutite amphibole. Vein includes about 5% apatite plus 5% phlogopite. Chemical and isotopic data from composite inclusion *Ba-2-1* from Dish Hill, California. (Redrawn from Nielson et al., 1993.) (a) Left, hypothetical spatial relations in veined mantle peridotite where 17 × 16 × 9.5 cm inclusion (small rectangle) might have been positioned (compare Plate VI). Right, select modal and elemental variations (in wt.%) in wall rock samples *WR-1*, *WR-2*, and so on, at indicated distances from vein. (b) Chondrite-normalized rare earth element (REE) diagram showing strong enrichment in vein amphibole relative to Cr-diopside peridotite, particularly of light REE, and progressively less enrichment in wall rock away from vein. (Redrawn from Menzies et al., 1987.) Shaded is range of Cr-diopside spinel lherzolites from Kilborne Hole, New Mexico (Redrawn from Irving, 1980.) (c) Isotopic ratios for Dish Hill sample shown by solid circles and connecting tie line. Wall rock and vein fields are for inclusion samples from California and Arizona. MORB, representative mid–ocean ridge basalt. OIB, ocean island basalt. (Redrawn from Wilshire et al., 1988).

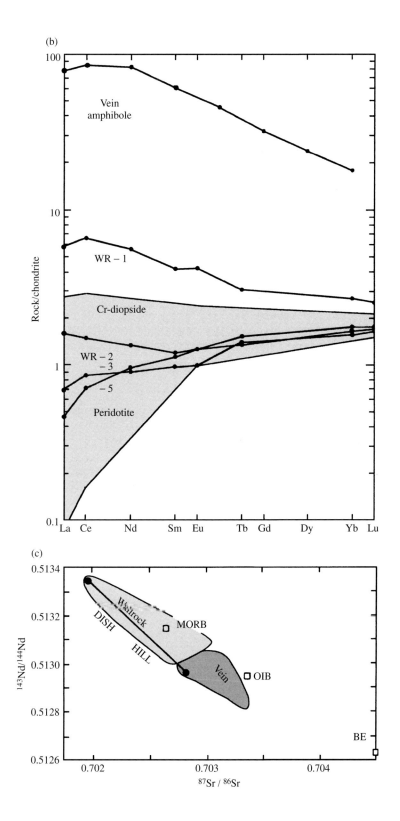

11.6 (*Continued*).

tion gradients. However, because it is slow, the effective distance over which metasomatism can occur solely by diffusion is limited. Metasomatism is greatly enhanced if there is advective flow of liquid along grain boundaries and especially through fractures, after which slower intragranular diffusion can take place.

The significance of liquid flow in granular media goes far beyond metasomatic processes. Obviously, migration of melts generated by partial melting of peridotite is a crucial step in their segregation into larger masses that ascend and intrude overlying lithosphere as magmas. Flow of fluids during metamorphism is another subject of considerable importance. For these reasons, an extended discussion of liquid movement between mineral grains follows.

Distribution and Migration of Liquids in Granular Aggregates.

Liquids can migrate through aggregates of mineral grains (rocks) by self-generated hydraulic fracturing (Section 8.2.1) or by **porous flow** (Section 8.5) that involves movement of the liquid along grain boundaries.

The principle of minimization of surface energy, or textural equilibration, is an important factor influencing the distribution and migration of very small volumes of liquid along grain boundaries in rocks (e.g., Watson et al., 1990). Governing principles are an extension of the discussion in Sections 6.4 and 7.4, which considered only crystal-crystal interfaces. Intersecting liquid/crystal interfaces (Figure 11.7) form an angle, θ, called the **dihedral (wetting) angle,** which depends on the relative magnitudes of the crystal-crystal interfacial energy, γ_{CC}, and the liquid-crystal interfacial energy, γ_{LC}

$$11.2 \quad \theta = 2 \arccos\left(\frac{\gamma_{CC}}{2\gamma_{LC}}\right)$$

Interfacial energies, which depend on liquid and crystal compositions and on *P-T* conditions, can be readily measured by experiment. Three cases are significant (Figure 11.7). If $\theta = 0°$, the liquid wets the entire crystal-crystal interface, forming a grain-boundary film. Apparently, no geologic phases exist for this case. If $\theta \neq 0$, no film exists, but interconnectivity of melt can still be maintained, not along grain surfaces, but rather along three-grain common edges, provided θ is less than or equal to 60°. The third case is for $\theta > 60°$, where the liquid exists as minute pools at isolated three-grain corners. These geometries assume that there is a monomineralic (one-phase) crystalline aggregate and that surface energies of grains are isotropic. However, in the more realistic case of a polyphase peridotite in which grains, especially of the dominant olivine, have a preferred crystallographic orientation as a result of mantle flow, melt distribution can be some-

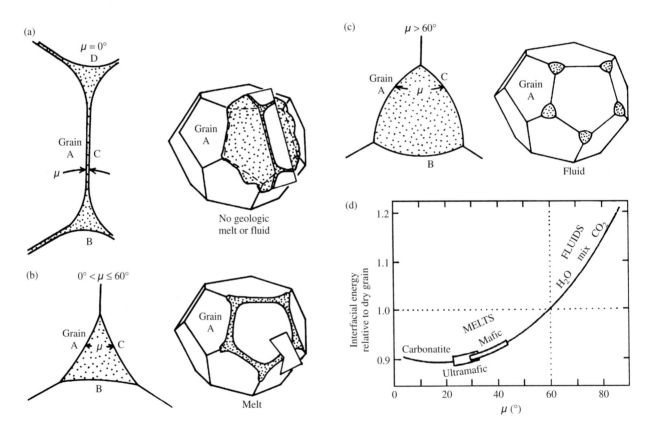

11.7 How very small volumes of liquid in a grain aggregate (rock) are distributed is a function of the relative interfacial energies of contacting phases (equation 11.2; see also Watson et al., 1990). The shape of reference grain A is dictated by the demands of minimal surface free energy in a one-phase grain aggregate (compare Figures 6.23 and 6.24). (a) For $\theta = 0°$ the liquid wets the entire surface of grain A and its neighboring grains B, C, D, and so on, in the form of a thin film. Though theoretically possible, this has not been found to occur in geologic systems. Left-hand diagram shows distribution of liquid along grain boundaries in a plane perpendicular to their common edges. (b) For $0° < \theta$ less than or equal to 60° melt is distributed in interconnecting strands with triangular cross sections that lie along three-grain edge junctions. The similarity of this melt distribution with thread-lace scoria (Figure 6.25) is remarkable, but not unexpected for textural equilibration driven by minimization of surface energy. (c) For $\theta > 60°$ small isolated packets of a fluid phase lie at grain corners. (d) Carbonatite melts and silicate melts of basaltic and ultramafic composition are stable along intergrain edges. H_2O-CO_2 fluids are stable along intergrain edges if in large volume whereas small volumes lie at grain corners.

what different from that depicted in Figure 11.7. Permeabilities (rates of liquid flow; Section 8.5) can be greater in anisotropic fabrics.

Figure 11.7d shows that CO_2 fluids are not interconnected until they make up about 8% of the aggregate as isolated pools merge. Water-rich fluids *may* have interconnectivity along three-grain edges. Mafic to ultramafic hydrous silicate melts have interconnectivity under virtually any upper mantle condition in peridotite even at melt fractions as low as about 1%. Carbonatite melts maintain interconnectivity at melt fractions as low as <0.1%.

Because of their low viscosity, melts are able to infiltrate along grain edges through mantle rock readily in the process of porous flow (McKenzie, 1985). The principal driving force, at least for larger volumes of melt, is the buoyancy of the less-dense melt. However, interconnecting strands of melt along three-grain edges can actually lower the interfacial energy in a grain aggregate, relative to the melt-free aggregate. Consequently, any available melt can be drawn spontaneously into an initially melt-free peridotite, reducing the energy of the system (Watson et al., 1990). Melts probably penetrate along grain boundaries by a coupled dissolution-precipitation process. Rates of porous flow are on the order of 1 mm/day for basaltic melts and perhaps a couple of orders of magnitude faster for aqueous fluids if their wetting angles are <60°. Carbonatite melts also move fast.

In sheared peridotite xenoliths, grain size has been reduced and grains stretched into lenticular shapes, creating a penetrative foliation. Any melt in ductilely deforming peridotite would tend to collect along shear surfaces where the rate of melt migration might be enhanced, that is, flow permeability increased, relative to a static, nondeforming rock system.

<u>Metasomatized Subcrustal Lithosphere.</u> It is quite possible that porous flow of liquids, perhaps assisted by the grain-scale deformation just described, occurs in the convecting asthenospheric mantle. Very slight partial melting (<1%?) of the asthenosphere might create metasomatizing melts. Where strands of ascending buoyant liquids collect to form overpressured volumes, hydraulic fracturing can occur, allowing faster ascent. But because moving liquid of either form carries little heat relative to the larger mass of wall rock, the melt or fluid readily freezes as veins, unless there is enough volume of melt to continue to ascend as magma. Mantle wedges overlying subducting oceanic plates (Figure 11.2) are likely extensively veined (Davies, 1999). Uppermost mantle beneath continental rifts and where mantle plumes are operative is also likely metasomatically veined. After liquid production has ceased beneath a particular plate, it may begin again during the course of changing plate motions. Thick, relatively cool (low-geothermal-gradient) mantle underpinnings of

continents, especially shields that may be as old as the early Proterozoic, might have been metasomatized to varying degrees by repeated veining episodes. The Archean mantle may have been too hot to allow freezing of veins. Younger lithosphere in oceanic areas has also been metasomatized. This metasomatized subcrustal or mantle lithosphere, laced as it is with metasomatic veins, is believed to play a very important role in the generation of alkaline magmas, especially the highly alkaline, highly potassic magmas (Section 11.5).

✳11.3 GENERATION OF MAGMA IN MANTLE PERIDOTITE

Chapter 5 introduced concepts of the melting of multiphase silicate rocks based on their crystal-melt equilibria. In this section, additional concepts, which relate to generation of a wide range of ultramafic and mafic melts from "normal" mantle sources, that is, Cr-diopside peridotite, are developed.

Although decompression melting is undoubtedly paramount in generation of basaltic melts from peridotite, it is convenient to examine partial melting that accompanies an increase in T using available T-X phase diagrams to elucidate pertinent concepts. The system Mg_2SiO_4-$CaMgSi_2O_6$-SiO_2 under water-saturated conditions at $P = 20$ kbar (Figures 11.8 and 11.9) will be used to model melting. The small amounts of Al, Ti, Fe, Cr, Mn, K, Na, and other minor and trace elements, mainly sequestered in solid solution in clinopyroxene, the small modal amounts of spinel or garnet in natural peridotites, and amphibole and mica in hydrous peridotite, are ignored at this point. The average worldwide spinel lherzolite (Table 11.1) with its somewhat depleted composition is used as a model of the source rock.

Generation of a melt that is less than the whole rock is referred to as **partial melting.** Different types of partial melting have been recognized, only two of which are considered here:

1. **Equilibrium (batch) partial melting** generates a melt that is in equilibrium with the **crystalline residue** as long as the two are in intimate communication and until the melt leaves. Equilibrium melting in a closed system is the reverse of equilibrium crystallization.

2. **Fractional partial melting** generates melt that is immediately isolated from the crystalline residue; no reaction relations occur between melt and crystals. As it is a disequilibrium process, it is not the reverse of fractional crystallization.

11.3.1 Equilibrium (Batch) Partial Melting of Lherzolite

In Figure 11.9a, the first melt to form with rising T in a lherzolite L has a composition at point M. This invariant point represents the only possible melt in the

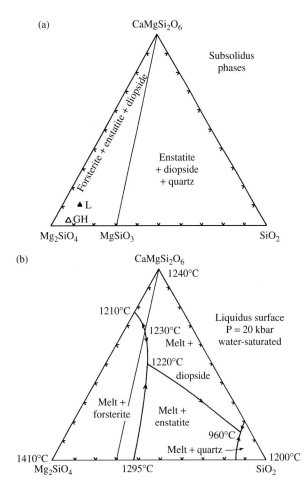

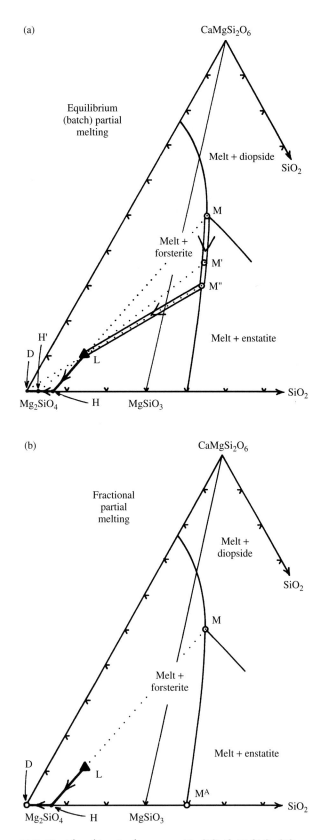

(b)

11.8 Ternary system Mg_2SiO_4-$CaMgSi_2O_6$-SiO_2. (a) Subsolidus assemblages. *Average worldwide spinel peridotite, which is a lherzolite, L, and garnet peridotite, which is a garnet harzburgite, GH, from Table 11.1. (b) Projected liquidus features at P = 20 kbar (2 GPa; depth about 70 km) water-saturated. (Redrawn from Kushiro, 1969.)*

ternary system that coexists in equilibrium with two pyroxenes plus olivine, the three major minerals constituting the lherzolite source rock. Point *M* at the confluence of three boundary lines in Figure 11.9a has a physical resemblance to the small pools of melt that form at mutual grain boundaries (Figure 11.7c) of two pyroxenes and olivine in a lherzolite; melts do not develop at olivine-olivine or pyroxene-pyroxene grain contacts. M is an incongruent melt that is substantially more enriched in silica and $CaMgSi_2O_6$ than the source. This reflects preferential melting of diopside and a lesser amount of incongruent melting of enstatite, which yields a more silica-rich melt plus residual olivine. If the melt M were to crystallize, it would form 7% quartz, 46% diopside, and 47% enstatite. Recalculation to 100% of the normative *Q, Di, Hy* (2.2, 17.9, and 21.5) in average Hawaiian tholeiitic shield basalt in Table 13.2 gives 5%, 43%, and 52%. This similarity in composition between the hypothetical model melt M and tholeiite magma ignores the neces-

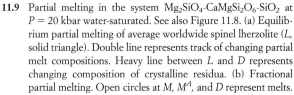

11.9 Partial melting in the system Mg_2SiO_4-$CaMgSi_2O_6$-SiO_2 at P = 20 kbar water-saturated. See also Figure 11.8. (a) Equilibrium partial melting of average worldwide spinel lherzolite (L, solid triangle). Double line represents track of changing partial melt compositions. Heavy line between L and D represents changing composition of crystalline residua. (b) Fractional partial melting. Open circles at M, M^A, and D represent melts.

sity that there be sufficient K, Na, Al, Fe, and Ti in the real source to stabilize plagioclase and Fe-Ti oxides in the real tholeiite magma at low pressures.

Continued melting of the source lherzolite simply yields more melt M, provided crystals of the two pyroxenes plus olivine are present in the crystalline residue. Invariant equilibrium persists and T and melt composition remain unchanged. In a T-increasing melting system, the latent heat of fusion consumes the increased thermal energy. The increased amount of melt is created by preferential dissolution of diopside, eventually shifting the composition of the crystalline residue to diopside-free harzburgite, H. The lever rule indicates that about 79% of the system is harzburgite residue and 21% is melt. That is, the **melt fraction** is 21%. With all of the diopside melted out of the source, so the residue is only olivine plus enstatite, further T increase can no longer generate a melt at the invariant point M. Instead the melt composition must track down the boundary curve on the liquidus surface between the stability fields of melt + forsterite and melt + enstatite, and away from the $CaMgSi_2O_6$ apex; this component in the melt is simply being diluted as enstatite dissolves. Melt M' on this boundary curve, for example, coexists in equilibrium with a harzburgite, H'. The tie line between the melt and crystalline residue *must* pass through point L, the composition of the initial source rock. The reason for this special geometric constraint is that in equilibrium melting the system remains closed and all melt-crystal tie lines are tethered at the bulk composition point. It may be noted that point M' is closer to this bulk composition than is M; the melt fraction in the system has increased to about 29%. Continued melting drives the melt composition to M'', where the melt fraction is about 33%; the crystalline residue is now pure forsterite, or dunite rock, D.

High-degree partial melts, 30%–40%, of an initial lherzolite source in the mantle contain high concentrations of dissolved olivine and are ultramafic picrite and komatiite. Low-degree partial melts, 5%–10%, are mostly of dissolved clinopyroxene (and spinel or garnet in real sources). The solid residue progressively loses spinel or garnet, clinopyroxene, and then orthopyroxene.

11.3.2 Fractional Partial Melting of Lherzolite

This manner of partial melting begins, as before, with generation of an initial melt M (Figure 11.9b). But each successive parcel of melt M generated is removed (fractionated) from the crystalline residue, which shifts directly away from L along a straight line projected from M to harzburgite, H, as before. However, once the crystalline residue becomes H, further partial melting yields different products from those in the equilibrium case. With a source rock at

H, the system has no "memory" of ever having been at L; it only "thinks for the moment" that the source is at H. With no diopside in H, only forsterite plus enstatite, the only possible melt that can coexist in the whole ternary system is melt M^A. But this melt requires a 75°C increase over melt M. Therefore, as more thermal energy is absorbed by the source rock after the last liquid M has been removed from the rock and as its last diopside is dissolved, no further melting can occur until sufficient heat has been absorbed to raise the source rock to 1295°C. At that T, melting resumes, generating melt M^A, chiefly by dissolving enstatite.

11.3.3 Factors Controlling Partial Melt Composition

Changes in P and volatile concentration significantly shift the position of cotectic boundary lines and invariant first-melt compositions in the model source systems shown in Figures 11.10–11.12.

Increasing P shifts anhydrous invariant-point first melts toward lower silica and higher Na (alkali) compositions. In Figure 11.10, from a silica-saturated melt (modeling a quartz-normative tholeiitic basalt magma) at 1 atmosphere, the equilibrium shifts to an olivine-saturated melt (alkaline basalt magma) at 15 kbar.

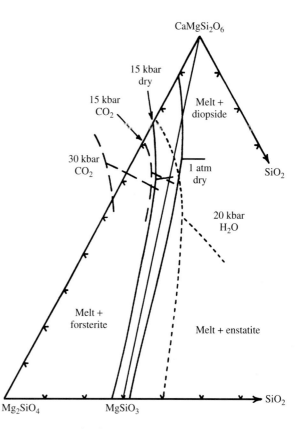

11.10 Beginning-of-melting invariant points and associated liquidus boundary lines in the system Mg_2SiO_4-$CaMgSi_2O_6$-SiO_2. Solid lines, volatile-free (dry); short dashed lines, water-saturated; long dashed lines, CO_2-saturated. (Redrawn from Eggler, 1974.)

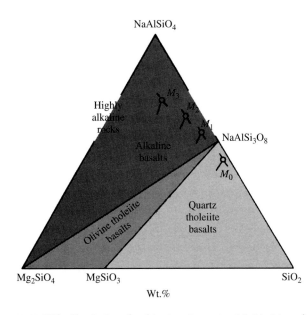

11.11 Shift of beginning-of-melting invariant points M_0, M_1, M_2, and M_3 at 1 atm, 1, 2, and 3 GPa, under volatile-free conditions. Basaltic melts modeled by this ternary system are indicated. (Redrawn from Kushiro, 1968.)

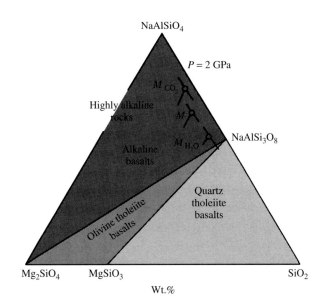

11.12 Shift of beginning-of-melting invariant points at 3 GPa under indicated volatile-saturated conditions. M is the volatile-free dry system. (Redrawn from Eggler and Holloway, 1977.)

In Figure 11.11, increasing P shifts first melts from silica-saturated (quartz tholeiite magma), to silica-undersaturated (alkaline basalt magma), and to highly silica-undersaturated (nephelinite and other highly alkaline magmas) at highest P. Partial melts contain more dissolved olivine at higher P. Hence, those developed from mantle peridotite at P in excess of about 3 GPa (about 100-km depth) have MgO content >15 wt.% and are picrites and komatiites, in contrast to the basaltic melts generated at lower P (Herzberg and O'Hara, 1998). Thus, when a high-P partial melt decompresses at shallower depths, olivine precipitates to maintain equilibrium. This phenomenon accounts, at least in part, for the common occurrence of olivine phenocrysts in extruded basalt magmas.

Increasing CO_2 pressure (fugacity) also decreases silica and increases alkali concentrations in first melts, reinforcing the effects of increasing P.

Increasing water pressure (fugacity) has the opposite effect, shifting invariant first-melt compositions to higher silica and lower alkali and $CaMgSi_2O_6$ concentrations.

Experimentally generated partial melts of a natural peridotite (e.g., Hirose, 1997) confirm the results of the simple ternary systems just described in shifting low-T first melts at increasing P to more silica-undersaturated compositions (Figure 11.13). But, in addition, the experiments on natural peridotite reveal the importance of yet another factor controlling melt compositions—the **degree of partial melting**, manifested by the melt fraction. Smaller-degree partial melts are nepheline-normative alkaline basalt compositions be-

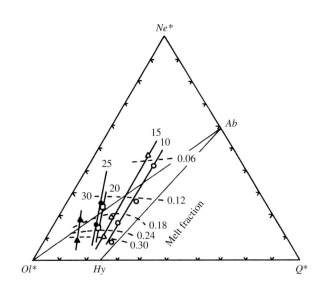

11.13 Factors controlling the composition of experimentally generated partial melts of anhydrous spinel lherzolite KLB-1 from Kilborne Hole, New Mexico. Partial melts (circles, squares, and triangles) are increasingly enriched in normative olivine (depleted in normative quartz) with increasing P (10, 15, 20, 25, and 30 kbar). Small extents of partial melting yield nepheline-normative melts but increasing melt fractions (0.06, 0.12, 0.18, 0.24, 0.30) generated at increasing T have more normative hypersthene. $Ne^* = ne + 0.6\,Ab$; $Q^* = Q + 0.4\,Ab$; $Ol^* = Ol + 0.75\,Hy$. Experiments by Falloon et al. (1999) on KLB-1 show that only the highest melting temperatures in these anhydrous melting experiments attained equilibrium; accordingly, they urged caution in interpreting melt compositions. However, the *pattern* of changing melt compositions with respect to degree of melting and P seems reasonable. (Redrawn from Hirose and Kushiro, 1993.)

cause they contain high concentrations of the incompatible elements Na and K as well as H, C, Ti, P, and others. These elements are not present in the simple model ternary systems just described but are sequestered in natural peridotite minerals, chiefly clinopyroxene. As the degree of melting increases at constant P, incompatible elements are diluted by increasing amounts of dissolved Si, Al, Fe, Mg, and Ca so that the melts become less alkaline, more silica-rich hypersthene- or even quartz-normative tholeiitic basalt compositions.

11.3.4 Modeling Partial Melting Using Trace Elements

Quantitative modeling of trace elements is another way of looking at the relation between melt composition and degree of partial melting. The composition of the source rock is also important in governing the trace element composition of partial melts, as may be appreciated from the bulk partition coefficient, D (equation 2.2 and Section 2.5.1).

Low degrees of partial melting corresponding to a small melt weight fraction, F, produce melts enriched in incompatible elements and depleted in compatible elements compared with the source rock. Larger degrees of melting reduce incompatible element concentrations but increase compatible element concentration until at $F = 1$ (100% melting) the melt has the same composition as the source. The type of melting, whether equilibrium (batch) or fractional, influences the trace element concentration in a partial melt, in addition to the source rock composition. Only equilibrium melting is considered here (but see Rollinson,

1993) because it may be the more likely case, but this point is controversial.

Equilibrium (batch) partial melting can be described by

$$11.3 \quad \frac{C_l}{C_0} = \frac{1}{(F + D - FD)}$$

and by

$$11.4 \quad \frac{C_s}{C_0} = \frac{D}{(F + D - FD)}$$

where C_0 is the initial concentration of an element in the solid source, C_l is the concentration of the same element in the partial melt, and C_s is the concentration in the unmelted solid residue. These equations are plotted in Figure 1.14 for several values of D as a function of F and melt, residue, and source compositions. For strongly incompatible elements, such as uranium in a basalt-peridotite system, D is very small (<0.001) and $C_l/C_0 \sim 1/F$; the concentration of the element in the partial melt depends essentially on F and approaches infinite concentration as F approaches 0. In the crystalline residue (equation 11.4), a highly incompatible element is strongly depleted with even very small degrees of partial melting. Relative to the source, small degree melts can have large changes in the ratio of two incompatible elements whose D differs significantly, say, 0.5 versus <0.01. Batch partial melting produces melts that are not extremely depleted in the compatible elements. The lowest concentrations are proportional to $1/D$ for very small F. For example, if $D_{Ni} = 5$, the lowest concentration of Ni (at very small F) that can oc-

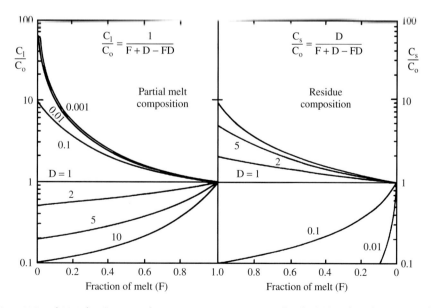

11.14 Plots of equations 11.3 and 11.4 showing trace element concentrations in partial melts (C_l) and residue (C_s) as a function of melt fraction, F, and bulk partition coefficient, D. Original concentration in source rock is C_0. On left, curves for $D = 0.01$ and 0.001 are almost coincident and appear to be a single, thick line.

cur in a single batch of partial melt is one-fifth of the initial concentration in the solid source.

Figure 11.15 shows primitive mantle-normalized rare earth element (REE) patterns (Section 2.5.3) for a garnet lherzolite source and the batch partial melts that could be derived from it over a range of melt fraction. Smallest-degree partial melts (low *F*) are most enriched in the most incompatible elements. Higher-degree partial melts are progressively more like the source. Remember that in these REE diagrams the most incompatible elements are on the left and less incompatible elements are on the right. Also note that the presence of garnet in the residue causes the patterns to have a markedly negative slope and high light REE to heavy REE ratio (Figure 2.22); in other words, La/Yb is large (Figure 2.23).

11.3.5 Characteristics of Primary Magma

Petrologists do not universally agree on the composition of **primary magmas** that have moved to the surface unmodified and undifferentiated from their mantle peridotite source. It is, nonetheless, important to establish a baseline reference against which differentiated magmas discussed in the following chapter can be compared.

Erupted magmas that contain dense mantle peridotite inclusions are commonly believed to be primary because their upward transport must be too rapid to allow differentiation of the magma and dissolution of the inclusion. However, entrainment of pieces of the mantle after some differentiation and subsequent to segregation from the source rock cannot always be ruled out. In any case, hosting of mantle inclusions is only permissive evidence; primary magmas need not contain inclusions.

Other criteria for recognition of primary magmas are chemical, though there is little consensus on the values to be used. Melts segregated from mantle peri-

dotite are in equilibrium with olivine, about Fo_{90}. The Fe^{2+}/Mg distribution coefficient between this olivine and an equilibrium melt, or $(Fe^{2+}/Mg)_{olivine}/(Fe^{2+}/Mg)_{melt}$, is *P*-dependent but is approximately 0.3 ± 0.03. This translates into an atomic ratio, called the **Mg number,** of $100\ Mg/(Mg + Fe^{2+}) = 68 - 75$, or a weight ratio $FeO/MgO = 0.4 - 0.7$ in the rock. These ratios for total Fe as FeO or Fe^{2+} are relatively insensitive to the degree of partial melting but are strongly influenced by fractional crystallization. In addition, 8 wt.% MgO, 400 ppm Ni, and 1000 ppm Cr are commonly used as lower limits for primary magmas.

A necessary (but again not sufficient) condition for a mafic volcanic rock to have solidified from a primary magma is that its composition lies at a beginning-of-melting invariant point (e.g., in Figures 11.9 and 11.10) at some mantle *P*. For a lherzolite source, this near liquidus high-*P* melt must be saturated or equilibrated with olivine, two pyroxenes, and either spinel or garnet. Most basaltic rocks have not solidified from magma so equilibrated, implying modifications after leaving the source, as recognized in early studies (Yoder and Tilley, 1962).

Metasomatized mantle sources do not necessarily contain olivine, or, if it is present, it is more Fe-rich than Fo_{90}. Primary magmas derived from such sources will not have the same chemical values as those derived from normal depleted peridotite.

Because of the uncertainties involved in establishing what is a primary magma, petrologists often attempt to decide which rock in a comagmatic suite under investigation solidified from a **primitive magma,** one whose composition was *least unmodified* by differentiation processes after leaving its source. Such a magma tends to have the largest Mg number and highest concentrations of Ni and Cr. However, caution must be exercised, for example, a basalt may have high Mg number, Ni, and Cr but petrographic examination might reveal unusually large amounts of olivine that could have accumulated by gravity settling, a type of differentiation process. Incompatible elements can be enriched in some differentiated magmas, but they can also be enriched in metasomatized mantle and in the primary magmas derived from them by partial melting.

Obviously, deciding whether a rock formed from a primary magma, or which rock solidified from a primitive magma, requires careful study.

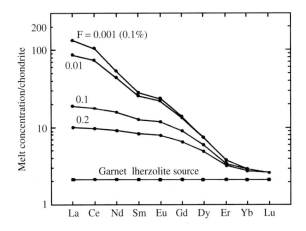

11.15 Chondrite-normalized diagram for rare earth elements in melts generated by varying degrees of partial melting of a garnet lherzolite source. (Data from Rollinson, 1993.)

✵11.4 MAGMA GENERATION IN SUBARC MANTLE WEDGE

After oceanic rifts, the most voluminous global magma production is in subduction zones (Figure 1.1). **Arc magmatism** occurs where oceanic lithosphere descends beneath overlying oceanic lithosphere in **island arcs** and beneath continental margins in **continental arcs.** Few, if any, mantle-derived magmas survive their journey through the crust, especially thick sialic continental crust, unscathed by contamination and other differentiation processes, making it difficult to understand their primitive character. This overprint, however, is highly variable, depending largely on the thickness of the crust. Island arc volcanic rocks, because they form on relatively thin and mafic crust, provide a logical starting point for inquiry into the nature of magma generation in the subarc mantle (see Davidson [1996] for a lucid summary).

Minor but widespread hornblende and biotite in volcanic arc rocks are stabilized by >3–5 wt.% water in magmas, in contrast with the essentially anhydrous mantle-derived magmas generated in other oceanic settings. As first pointed out by Coats (1962), dehydration ("drying out") of the oceanic slab as it subducts and heats up likely produces+ relatively "wet" arc magmas. The water liberated from the slab promotes generation of hydrous partial melts in the overlying wedge of hot mantle peridotite (Figure 11.2).

11.4.1 Dehydration of Subducting Oceanic Crust

Subducting oceanic lithospheric plates consist of mantle peridotite and overlying oceanic crust (see Figure 13.1). This crust has 6–7 km of basaltic rock and an overlying veneer of sediment, variable proportions of which are scraped off into the trench and accreted onto the edge of the overlying plate (Figure 11.16). The crust contains variable amounts of volatiles, chiefly H_2O and CO_2. Carbonate minerals occur in deep marine sediment and in pore spaces and veins in basaltic rocks; overall CO_2 concentrations in oceanic crust are probably about 0.1 wt.%. The overall concentration of water in the crust is 1–2 wt.%, part of which is physically entrapped in pore spaces and cracks. However, an important amount is chemically bound in hydrous silicate minerals formed by moderate-T seafloor metamorphism of the basalt and peridotite at oceanic spreading ridges and by subsequent long-term low-T submarine "weathering" as the plate moves away. Hydrous Mg-Fe-Ca-Al silicates that replace primary magmatic minerals in basalt include epidotes, micas, amphiboles, serpentines, chlorites,

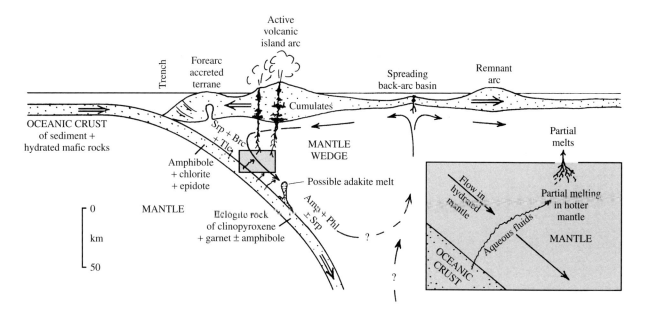

11.16 Highly schematic dynamics of an island arc-mantle wedge system. Double-line arrows indicate relative motion of crust (stippled). Subduction of hydrated oceanic crust liberates water in complex dehydration mineral reactions, forming ascending aqueous fluid solutions (wavy arrows). The forearc accreted terrane is built chiefly of scraped-off oceanic sediment and possible mafic-ultramafic rocks (ophiolite). Below this forearc, water liberated from the dehydrating crust hydrates the overlying peridotite in the mantle wedge to serpentine (± brucite ± talc). Some of this buoyant mass of low-density serpentinized peridotite rises as diapirs (see diapir in figure) into the accreted terrane and may be extruded onto the ocean floor. As the descending crust continues to heat up, more water is liberated, and it transforms into drier, high-P eclogite, but amphibole may still be present. In exceptionally young, hot subducting crust, dehydration partial melting of amphibole eclogite may generate adakite melts that rise through the mantle wedge, possibly mixing with andesitic partial melts, before intruding into the arc crust or erupting. Hydrated mantle peridotite overlying the subducting crust may be dragged down in a "corner flow" by viscous coupling, releasing water as the low-T hydrous minerals partially dehydrate into amphibole + phlogopite ± serpentine that are stable at higher P-T (Figure 11.18). Inset diagram in lower right (enlarged from the box in the main part of the diagram) shows that convective flow in mantle draws rising aqueous fluids laterally away from the crust into the hotter part of wedge where partial melting occurs. It should be noted in this inset diagram that temperatures increase in rocks from the lower left corner to the upper right: That is, there is an inverted thermal gradient.

Special Interest Box 11.3 Significance of the radiator on the front of the global magma engine

Sampling of oceanic lithosphere reveals widespread and locally intense alteration of the rocks at spreading ridges by seawater advecting through pervasive fractures. As it traverses through the hot crust, the seawater is heated and dissolves high concentrations of various mineral components from the hot rock. Where vented back into the ocean at T as much as 350°C these hydrothermal solutions form "black smokers" (Figure 11.17). Advective circulation of seawater at oceanic ridges has the following major impacts on global geology:

1. Circulation acts as a gigantic "radiator on the front of the global magma engine" (Sigurdsson et al., 2000, p. 5), providing for significant dissipation of heat from the interior of the Earth. It is estimated that seawater advects through the hot crust at a rate of about 1.5×10^{14} kg/y, or enough to recycle the entire mass of the oceans once every 5 My!

2. Seafloor hot springs support a unique biological community that is a factor in the global food chain.

3. The circulation system impacts the chemical composition of ocean water. Sinks and sources of mineral components are influenced by crust-seawater interactions. Components added by rivers and precipitated in marine sediments are not the only governing factors, as once thought.

4. Minerals, such as chalcopyrite ($CuFeS_2$) and sphalerite (ZnS), are deposited on the seafloor as the hydrothermal solutions cool at black smoker vents. Because of seafloor spreading, these potentially economic deposits are included in on-land ophiolite (Section 13.6) or are processed in the subduction "factory" where plates converge.

5. Primary anhydrous minerals in the mafic and ultramafic rocks of the oceanic lithosphere are transformed by seafloor metamorphism into hydrous, more oxidized minerals. As these hydrated rocks are heated in subducting slabs, the liberated water promotes magma generation in the overlying mantle wedge and is, therefore, chiefly responsible for global arc magmatism—the second most prolific on Earth (Figure 1.1).

11.17 Black smoker venting from a hot spring on the East Pacific Rise at a water depth of about 3 km. Dark-gray plume carries high concentrations of minute mineral precipitates, including Cu-Fe-Zn sulfides. Photograph taken by Robert D. Ballard from the deep submersible *Alvin* and furnished courtesy of Woods Hole Oceanographic Institution.

chloritoid, prehnite, pumpelleyite, talc, lawsonite, zeolites, and clay minerals. As the crust is compressed and slowly heated during subduction, physically entrapped water is first released at a depth probably not exceeding several kilometers. During further descent and heating, a series of endothermic, subsolidus dehydration reactions (Section 5.7.2 and Figure 5.31b) are "smeared" over a considerable range of T, P (depth), and volatile fugacities, progressively liberating water from the hydrous minerals (e.g., Liu et al., 1996). The smearing results from extensive solid solution in the hydrous phases, like the melting of plagioclase solid solutions that takes place over a range of T (Figure 5.13). P-T paths that a subducting slab take through time are highly variable, making generalizations difficult.

The major factor governing the depth of volatile mineral decomposition in subducting slabs is their thermal state (T distribution). For example, dehydration occurs at shallower depths in hotter slabs. This thermal state depends on many factors (Peacock et al., 1994):

1. The age of the incoming slab. Young lithosphere from a nearby spreading ridge is hotter.

2. The mass of previously subducted lithosphere. Previously cooled asthenosphere cannot heat the slab as fast.

3. Intensity of shear heating along the margins of the

more rigid lithospheric slab as it penetrates the more ductile asthenosphere. Faster-moving slabs create more heating (equation 11.1).

4. Slab dip: A slab that penetrates cooler mantle on a near-horizontal trajectory, as beneath the central Andes, does not heat as fast as a slab that penetrates hotter deeper mantle on a near-vertical path, as beneath the Mariana Islands in the western Pacific.

5. Vigor of convection in the mantle wedge: The basal part of the cooling wedge is partially coupled in a viscous manner with the descending slab and is dragged down with it, thus the slab does not contact hotter mantle (Figure 11.16).

6. Mineral reactions in the slab. Endothermic dehydration reactions cool the slab whereas exothermic hydration reactions heat it.

In short, older lithosphere created at more distant spreading plate junctures subducting at a slower speed without shear heating dehydrates at greater depths (Figure 11.18).

11.4.2 Magma Generation in the Mantle Wedge

Experiments on spinel lherzolite under water-undersaturated and water-saturated conditions at 10 kbar—conditions appropriate to the mantle wedge—yield basaltic andesite and andesite melts, respectively, for melt fractions of <0.23 at 1000°–1050°C (Hirose, 1997). Evidence for such degrees of partial melting is found in very rare, strongly depleted harzburgite xenoliths hosted in island arc rocks (Maury et al., 1992).

Trace Element Arc Signature. Primitive island arc rocks have a characteristic arc signature in normalized trace element diagrams (Section 2.5.3) that is "spiky" in contrast to the smooth pattern of MORB and most other oceanic rocks (Figure 11.19). Large-ion-lithophile elements (LILEs), such as Rb, Ba, Th, U, and K, are enriched relative to rare earth elements (REEs), La through Lu, and high-field-strength elements (HFSEs), such as Nb, Ta, and Ti. The relative

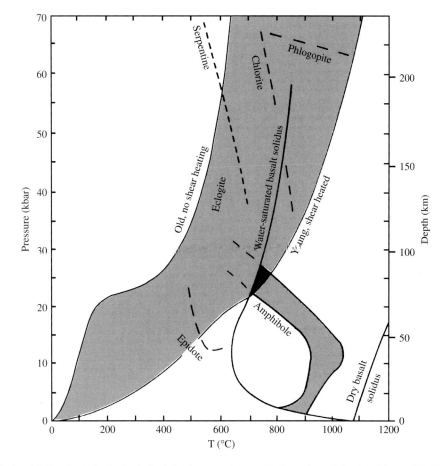

11.18 Range of calculated *P-T* paths (light shading) of subducting oceanic crust. End-member subducting slabs are (1) old, relatively cold upon entry into mantle, and slow-moving and (2) young, relatively hot, and fast-moving so shear heating occurs. Dark shading indicates *P-T* conditions for dehydration melting of hydrous basalt systems that contain amphibole (compare Figure 5.11). Note the small, wedge-shaped *P-T* "window" (black), where basaltic crust in young, shear-heated subducting slabs is predicted to experience dehydration melting. Basalt at *P* greater than about 25 kbar is composed of an eclogite assemblage of garnet + clinopyroxene ± amphibole. Upper *P-T* stability limit of minerals indicated by dashed lines. (Redrawn from Peacock et al., 1994.)

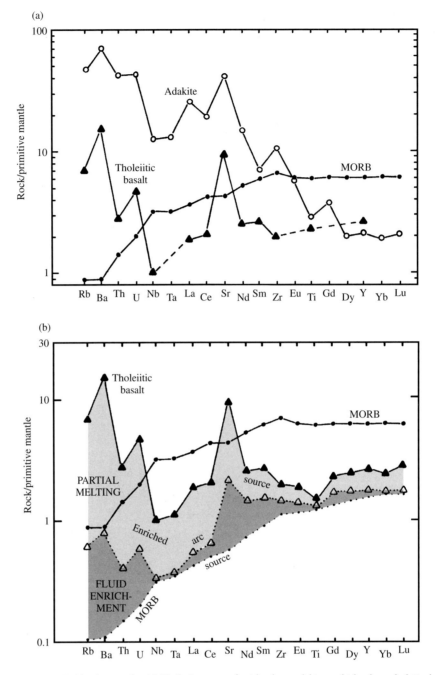

11.19 Trace-element arc signature in island arc rocks. (a) "Spiky" patterns for island-arc adakite and island-arc tholeiite basalt compared with smooth pattern for mid-ocean ridge basalt (MORB). (Data from Drummond et al., 1996; Wilson, 1987; Sun and McDonough, 1989.) (b) Differential enrichment of a hypothetical MORB source (lowermost dotted line) by fluid solutions liberated from the subducted oceanic crust (dark shading) produces an enriched mantle peridotite arc source. Partial melts of this arc source (light shading) yield the island-arc tholeiite basalt magma in (a). (Redrawn from Davidson, 1996.)

depletion in Nb, Ta, and/or Ti constitutes a **negative Nb-Ta-Ti anomaly** typical of arc rocks.

The origin of the arc signature has been controversial. The depletion in Nb-Ta-Ti has been explained by retention of a refractory phase in the source in which those elements are highly compatible, such as rutile. These HFSEs may also be sequestered in residual hornblende in the source (Drummond and Defant,

1990). Many petrologists now believe that aqueous fluids play a significant role in the arc signature. LILEs have low ionic potential (Figure 2.20) and are readily dissolved and transported in aqueous fluids at high *P* and *T* (Tatsumi and Eggins, 1995), whereas Nb, Ta, and Ti are less mobile. A basic working model for arc-magma generation, always subject to future revision as is any model, involves partial melting of subarc peridotite that has been metasomatically enriched in the more soluble ions (Figure 11.19b). Dissolved components might be from the slab with possible additions scavenged from the subarc wedge. Migrating silicate melts do not have these selective element preferences and therefore, if present, cannot dominate as metasomatizing agents.

Whether oceanic sediment contributes to arc magmas has also been controversial. However, significant amounts (to as much as 100 ppm) of boron (B) in arc rocks and the discovery of cosmogenic ^{10}Be (Section 2.6.3) in some of them have proved that at least some sediment is subducted (e.g., Leeman, 1996) and may be melted in some arcs. B is enriched in ocean-floor sediment and altered oceanic crust, whereas in mantle rocks it is <1 ppm. B is sequestered in clay and other phyllosilicate minerals and is quite mobile in aqueous fluids created as these minerals are heated in the subducting slab and liberate water. Be is like B in many ways and the near-uniformity of ^{10}Be/B ratios within a particular arc suggests that the two elements are governed by the same process that transfers matter from the slab to the mantle wedge. B abundances and ^{10}Be/B ratios decrease in arc rocks away from the trench, consistent with progressive scavenging of B and Be from the slab during its descent.

11.4.3 Partial Melting of Subducted Basaltic Oceanic Crust: Adakite

In the early days of the plate tectonic "revolution" it was believed by many geologists that melting of the subducted basaltic crust yielded the copious volumes of magmas manifested in this regime. However, it was soon realized that the overlying water-fluxed peridotite wedge provided a more viable magma source. Nonetheless, theoretical studies revealed a small "window" in P-T-time-composition space where partial melting of subducted basaltic crust might occur in young, hot slabs. Figure 11.18 indicates a range of P-T paths that subducting plates might take. Only the youngest and, therefore, hottest lithosphere descending rapidly, therefore inducing greatest shear heating, is expected to experience partial melting. At the other extreme, old lithosphere created at distant spreading ridges subducting at a slow speed without shear heating follows a path far below the T of even a water-saturated basalt solidus. The basaltic crust in which partial melting is likely to occur has been mostly dehydrated and converted to a high-P hydrous eclogitic assemblage of pyropic garnet, jadeitic clinopyroxene, and amphibole (Figure 5.11).

The search for candidate rocks in arcs where the subducting slab is <25 Ma and might represent partial melts of this amphibole-bearing eclogite assemblage has focused on **adakite,** named for Adak Island in the Aleutian arc. Adakite is basically a dacite, lo-

cally an andesite, having unusually high concentrations of Al_2O_3 (>17 wt.%), Na, Sr, and Eu, but low Mg, Ti, Nd, Y, Yb, and ^{87}Sr/^{86}Sr relative to the widespread andesite-dacite-rhyolite suite in subduction zones (Drummond et al., 1996). The high Sr and Eu suggest a lack of plagioclase in the source or no fractionation of this phase. The elevated light REE/heavy

---

Special Interest Box 11.4 Catalina schist: An exposed sample of the crust-mantle wedge interface

The discussion in Section 11.4 regarding hydration of the subarc mantle wedge, associated metasomatism, and partial melting in it and the basaltic crust is largely based on inference. However, exposures on Santa Catalina Island southwest of Los Angeles, California, are consistent with the nature of the inferred crust-mantle wedge interface and afford an opportunity for verification by real rocks (Sorensen, 1988).

The rocks show clear evidence for fluid migration, metasomatism, and partial melting, although mineral barometers indicate these processes occurred at shallower depths (approximately 30 km) than those generally inferred (Figure 11.18). The rocks are part of an accreted terrane in the Cretaceous forearc of coastal California formed where oceanic lithosphere was subducted beneath the continent (Figure 11.16) and subsequently uplifted to the surface. Structurally lowest rocks metamorphosed at lowest T are overlain by rocks metamorphosed at increasing T; this thermally inverted (relative to a normal geothermal gradient) sequence is presumed to be a sample of the crust-mantle wedge interface (see inset diagram in Figure 11.16). Metamorphosed gabbros and overlying seafloor clay rocks represent a segment of the oceanic crust. Overlying these is what appears to be a segment of the mantle wedge that consists of metasomatized peridotite and enclosed amphibole-eclogite blocks of tholeiitic basalt composition. The metasomatized peridotite consists of combinations of enstatite, anthophyllite, tremolite, talc, and quartz that indicate addition of water and Si and loss of Mg from the initial rock. (Widespread lower-T serpentine replaces these minerals.) The high-T hydration metasomatism culminated in partial melting of the basaltic rocks, producing thin dikes, stringers, and larger pods of plagioclase + quartz + muscovite. Had the partial melting occurred at T ~20 kbar or more (>70 km; see Figure 11.18), partial melts would have been adakitic (trondhjemitic) in composition.

REE ratios, specifically, high La/Yb or La/Y, unlike lower ratios and flatter REE patterns of most arc rocks (Figure 11.19a), are consistent with residual garnet in a source chemically like MORB (Figures 2.22 and 2.23). Whether partial melting is water-saturated or water-undersaturated is controversial, but the former allows for lower, more realistic temperatures and for rutile stability that would account for the observed HFSE depletion in adakites (Prouteau et al., 1999). On the other hand, the negative Nb-Ta-Ti anomalies and overall spiky aspect of the trace element pattern may indicate that the slab-derived adakite magmas were overprinted by an arc signature en route to the surface, likely involving some mixing with wedge-derived melts.

Some phaneritic rocks known as **trondhjemite** have a similar chemical signature to adakite and may have solidified from similar magmas intruded into the crust. Trondhjemites occur locally in Phanerozoic subduction zones but are most widespread in cratons of Archean rock.

❋11.5 GENERATION OF ALKALINE MAGMAS IN METASOMATICALLY ENRICHED MANTLE PERIDOTITE

The origin of silica-undersaturated alkaline magmas has been debated for many decades. According to early hypotheses primary alkaline magmas were not generated in the mantle or lower crust; they were created from primitive subalkaline basaltic magmas via differentiation overprints. One proposal invoked conversion of silica-saturated magma to silica-undersaturated magma by assimilation of limestone—a hypothesis based on the eruption of leucite-rich magmas from

Vesuvius, Italy. These silica-undersaturated lavas carry abundant carbonate xenoliths picked up during their ascent through about 3 km of Mesozoic rocks. Silica metasomatism of the xenoliths created Ca-Si minerals, such as diopside, whereas complementary contamination and desilication of the host magma were believed to create silica-undersaturated magma. However, this theory failed to explain many aspects of alkaline magma generation on a global basis, not the least of which is the absence of enough limestone in the oceanic crust to produce widespread alkaline rocks of volcanic islands. Alkaline magmas in many tectonic settings, including continental rifts, have primitive chemical compositions that are demonstrably not the result of crustal contamination.

Yet another untenable hypothesized process is low-P crystal fractionation from subalkaline basaltic magma; residual melts are more silica-rich, not silica-undersaturated (Yoder and Tilley, 1962).

Thus, the basic necessity for alkaline magmas is a means of generating them as primary magmas, probably in the mantle because so many of the less evolved magmas are quite mafic, even ultramafic. Subsequently, these primitive alkaline magmas may be differentiated to alkaline daughter magmas, as in oceanic islands such as Tristan da Cunha in the Atlantic (Figure 2.16).

Silica-undersaturated primary magmas can be generated at relatively high pressures and CO_2 concentrations and especially by small degrees of partial melting of "normal" spinel or garnet peridotite. Small-degree (e.g., F ~ 1%) partial melts that are believed capable of segregating from their source (Mckenzie, 1985) are enriched in the most incompatible elements by factors of about 100 (Figure 11.15).

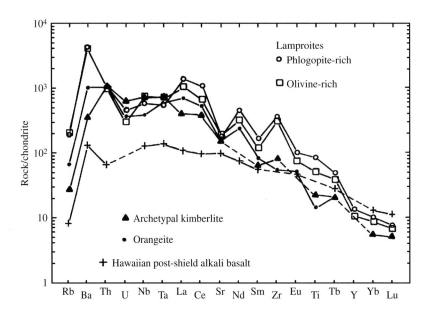

11.20 Chondrite-normalized trace element diagram for lamproites, archetypal kimberlites, and orangeites from Table 13.11 compared with oceanic island alkaline basalt from Table 13.2.

However, the enrichment of incompatible elements in some primitive magmas that carry mantle xenoliths, and hence probably suffered little modification from their source, is too extreme for an origin involving small degrees of partial melting to be valid. These magmas are represented by rare, highly alkaline, generally highly potassic, mafic, and ultramafic rocks found in continental rifts and stable cratons called kimberlites, lamproites, kamafugites, and phlogopite-rich minettes (Section 13.12). They commonly contain mantle-derived xenoliths and locally the first two are host to diamonds, indicating rather immediate transport from mantle depths. Normalized concentrations of the most incompatible elements in these rocks (Figure 11.20) are more enriched by about an order of magnitude than smallest-degree (F = 0.1%–1%) partial melts of a garnet lherzolite source (Figure 11.15).

11.5.1 The Metasomatized Mantle Connection

For most petrologists, a metasomatically enriched mantle is the most reasonable, or at the very least, a contributing, source for alkaline magmas and is likely the only source for highly alkaline kimberlites, lamproites, and kamafugites. The suite of ultramafic inclusions commonly occurring in these primitive rocks includes metasomatized mantle of the Al-Fe-Ti-rich clinopyroxene type that contains abundant phlogopite and K-rich richterite amphibole (1 and 2 in Table 11.2). Data in Table 11.3 indicate only modest enrichment factors (<10) between representative lamproite and metasomatic xenolith, which suggest that the latter could well be a source of partial melts of the former. A near-liquidus assemblage of olivine-diopside-phlogopite in a phlogopite-rich minette suggests a metasomatized mantle source (Esperanca and Holloway, 1987). Lloyd et al. (1985) used a common phlogopite clinopyroxenite inclusion from Uganda in the east African rift as a source rock and obtained, after 20%–30% partial melting under mantle conditions in the laboratory, a melt composition similar to that of the highly alkaline rocks hosting such inclusions.

The highly variable and exotic chemical compositions that are so typical of alkaline rocks are explained by the **vein-plus-wall-rock melting mechanism** (e.g., Foley, 1992). Alkaline magmas are viewed as partial melts of metasomatic veins that are enriched in volatiles and incompatible elements hybridized with variable amounts of partial melts from their less enriched peridotite wall rock. Initial melting of a mantle source laced with metasomatic veins generates very enriched, highly alkaline melts from the veins, as in the experiments of Lloyd et al. (1985). Because of extensive solid solution in the vein minerals, before the vein can be completely melted, components from the relatively depleted wall rock are dissolved and dilute the incompatible elements from the vein, producing a wide spectrum of melt compositions. The extreme composi-

Table 11.2 Chemical Analyses of Al-Fe-Ti-Rich Clinopyroxene Inclusions and Vein

	1	2	3	4
SiO$_2$	48.31	41.82	43.46	42.1
TiO$_2$	0.45	2.58	1.99	3.2
Al$_2$O$_3$	1.14	11.16	11.70	13.0
Fe$_2$O$_3$	6.88t	10.46t	2.62	—
FeO	—	—	7.19	8.1t
MnO	0.11	0.15	0.20	—
MgO	38.52	16.07	15.28	15.6
CaO	0.60	12.14	15.20	10.9
Na$_2$O	0.26	2.56	1.38	2.8
K$_2$O	1.17	2.01	0.32	1.3
P$_2$O$_5$	0.06	0.33	0.06	—
LOI	2.30	0.34	—	—
Total	99.80	99.62	100.58	97.29
Sc	2	25		
V	40	261		
Cr	1603	851		
Co	85	59		
Ni	1505	308		
Cu	10	41		
Zn	50	65		
Cs	0.29	0.4		
Rb	47	47		
Ba	154	1442		
Th	1.21	3.16		
U	—	1.4		
Nb	9	60		
Ta	0.82	4.18		
Ce	23.8	80		
Sr	116	747		
Nd	7.7	39		
Sm	0.87	7.93		
Zr	106	215		
Hf	3.05	5.15		
Eu	0.23	2.44		
Tb	0.08	0.89		
Y	4	19		
Yb	0.16	1.26		
Lu	—	0.18		

Column 1: Richterite (amphibole)-phlogopite peridotite, xenolith, Bultfontein, South Africa. (Column 1 data from Menzies et al., 1987.) Column 2: Phlogopite-apatite-amphibole xenolith, Pulvermaar, Eifel region, Germany. (Column 2 data from Menzies et al., 1987.) Column 3: Amphibole clinopyroxenite xenolith, Geronimo volcanic field, Arizona. (Column 3 data from Kempton in Menzies and Hawkesworth, 1987.) Column 4: Amphibole vein in composite xenolith, sample Ba-2-1, Dish Hill California. Analysis includes 0.29% Cr$_2$O$_3$. (Column 4 data from Wilshire et al., 1988.)

tional variability of alkaline magmas may, thus, reflect a widely ranging ratio of vein/wall rock components dissolved in the melts, compounded by the compositional variability of the metasomatic veins themselves.

Table 11.3 Comparison of Incompatible Element Concentrations (ppm) in Average Spinel Peridotite (Table 11.1). Metasomatic Phlogopite-Apatite-Amphibole Rock (Table 11.2), and Phlogopite-Rich Lamproite (Table 13.11)

	D	SPINEL PERIDOTITE	METASOMATIC XENOLITH	LAMPROITE	ENRICHMENT FACTOR LAMPROITE/ PERIDOTITE	LAMPROITE/ METASOMATIC XENOLITH
Rb	0.015	1.9	47	457	241	10
Ba	0.010	33	1442	10,607	321	7
Nb	0.020	4.8	60	147	31	2
K	0.010	8300	16,683	79,680	10	5
La	0.010	2.6		348	134	
Ce	0.010	6.3	80	629	100	8
Sr	0.020	49	747	1296	26	2
Ti	0.080	540	15,480	37,740	70	2

D is the approximate bulk partition coefficient for a peridotite-basalt system (Table 2.5), enrichment factors are ratios of indicated concentrations.

Generation of alkaline magmas is intimately related to a metasomatically enriched lithospheric mantle (e.g., Tingey et al., 1991). Subcontinental mantle lithosphere has been separated from the underlying convecting asthenosphere since probably as long ago as the Archean-Proterozoic transition (2500 Ma), whereas in suboceanic realms it has only been separated since about 200 Ma at most. An enriched lithosphere accounts for two additional characteristics of alkaline rocks: highly variable radiogenic isotopic ratios and occurrence of the most highly alkaline rocks in continental settings. Because of relative compatibilities (Section 2.6.2), metasomatically enriched peridotite has higher Rb/Sr and lower Sm/Nd ratios. The time-integrated effect of this metasomatism is, therefore, to increase $^{87}Sr/^{86}Sr$ and decrease $^{143}Nd/^{144}Nd$. The highest $^{87}Sr/^{86}Sr$ and lowest $^{143}Nd/^{144}Nd$ are in lamproites and orangeites that occur on Precambrian cratons (see Figure 13.42), whereas alkaline rocks of oceanic islands, such as Hawaii, are less radiogenic because the elapsed time since metasomatic enrichment of the lithospheric source is less. Small-volume alkaline magmas typically develop in intraplate settings and continental rifts where upwelling mantle, locally plumes, raise temperatures in the enriched lithosphere or cause it to decompress through uplift and tectonic thinning. Either or both perturbations (Figure 11.2) lead to partial melting.

✳ 11.6 MAGMA GENERATION IN THE CONTINENTAL CRUST

Vast volumes of felsic magmatic rocks composed mostly of two feldspars and quartz occur in continental arcs (e.g., western North and South America) but are only of trivial volume in oceanic island arcs (e.g., western Aleutians and Tonga). This simple fact emphasizes the need for preexisting continental crust in the generation of voluminous felsic magmas because the essential difference between these subduction zones bordering the Pacific Ocean is the nature of the crust in the overriding lithosphere. Because feldspars and quartz are near-liquidus phases in most felsic systems at crustal and upper mantle pressures (Figure 5.25), felsic magmas cannot be generated from peridotite in which olivine and pyroxenes are near-solidus phases. In addition, large volumes of felsic magma are not created by differentiation of vast amounts of basalt magma in oceanic areas.

Significant felsic magma generation can occur in three continental regimes, including continental arcs, where additional heat and possible fluxing aqueous solutions are available to perturb an otherwise subsolidus status in the lower crust. These three regimes are listed in order of decreasing volume of magma generation (Figure 11.2):

1. Continental arcs overlying subducting oceanic lithosphere where large volumes of ascending high-*T* mafic magma rising from the underlying mantle wedge are buoyantly blocked in and/or underplate the lower crust; additional thermal energy for partial melting is produced by this heat source (Section 11.1.1)

2. Continental rifts or areas above rising mantle plumes where underplating mantle-derived basalt magma provides additional heat for melting

3. Thickened crust in continent-continent collision zones, such as the Himalayan, heated by adjustments to the geothermal gradient

Magma generation might also occur as continental rock just below its subsolidus is decompressed as a result of rapid isostatic uplift and erosion or is fluxed by water liberated from nearby rocks experiencing metamorphic dehydration reactions, in (3), or from crystallizing mantle-derived magmas, in (1) and (2).

Many of the concepts of magma generation by partial melting for peridotite-basalt systems are applicable to continental rock-granite systems. However, the deep continental source is far more heterogeneous with respect to modal, mineral, and chemical composition than the mantle peridotite source, even allowing for its variable metasomatism. Moreover, silicic partial melts are cooler and orders of magnitude more viscous, making melt-residue segregation difficult, if not impossible, in reasonable geologic time scales. Consequently, there is greater opportunity for overprinting diversification processes to modify magmas slowly rising from their source. Source heterogeneity makes trace element modeling of the degree of melting and composition of the source rock more difficult and ambiguous (Harris and Inger, 1992).

11.6.1 Partial Melting of Continental Source Rocks

The continental crust is compositionally heterogeneous on most scales. Original rock types in the crust include a variety of sandstones, shales, and carbonate rocks plus rhyolite, dacite, andesite, basalt, and their phaneritic equivalents. All of these potential source rocks are metamorphosed because of deep burial and elevated temperatures where partial melting might take place. Constituent minerals are highly variable proportions of plagioclase, alkali feldspar, quartz, amphibole, micas, and generally lesser amounts of Fe-Ti oxides, carbonates, garnet, Al_2SiO_5 polymorphs, chlorite, epidote, and many other, less common metamorphic minerals. Among the major elements, most of the Mg, Fe, and considerable amounts of Al, Ca, and Na are sequestered in amphibole and to some extent in biotite, which together with muscovite harbors K. Significantly, these three minerals are hydrous. Accessory minerals, such as zircon and apatite, contain relatively large concentrations of trace elements, especially REEs.

Partial melting, sometimes called **anatexis,** of continental rocks generating felsic magmas can theoretically occur in strictly anhydrous systems that lack water as a separate phase as well as hydrous minerals. However, because solidus temperatures are relatively high, such melting is limited. But another reason why anhydrous melting rarely occurs is the widespread occurrence in potential source rocks of hydrous minerals. Partial melting of rocks that contain hydrous minerals in the absence of a separate aqueous fluid phase is referred to as **dehydration melting** (also called *fluid-absent, water-deficient, water-undersaturated, low-water fugacity,* or *low-water activity melting*). The only water in dehydra-

tion melting is chemically bound within micas and amphiboles, or other usually minor or accessory hydrous minerals, such as epidote and apatite. Partial melting of a source rock containing hydrous minerals results in their decomposition and liberation of water, together with other mineral components, forming a hydrous melt. Quartz, feldspar, and other anhydrous minerals are dissolved in this melt. A complementary less hydrous or anhydrous crystalline residue coexists with the melt. Partial melting may also occur, generally at relatively lower temperatures, under water-saturated conditions (also called *water-excess* or *high-water activity melting*). This extraneous water can be derived from, for example, nearby decomposing hydrous minerals below solidus temperatures or from crystallizing mantle-derived magmas that become water-saturated. There is no doubt that *water—in one form or another— is essential to generation of large volumes of felsic magma in the continental crust.*

<u>Source-Rock Fertility and Melt Fraction</u>. Because felsic melts are highly viscous, small melt fractions cannot readily segregate from the crystalline residue in geologically reasonable times (McKenzie, 1985). Also, small melt fractions cannot provide sufficient buoyancy to mobilize the entire melt-crystal mass so it can ascend, for example, as a magma diapir. Extensional tectonic environments or other circumstances may facilitate magma movement. A fundamental constraint in felsic magma generation in the deep continental crust is that large melt fractions, perhaps >20%, are necessary to provide for a buoyant volume sufficiently large that it is capable of ascending toward the surface. This contrasts strongly with the minimum melt fractions (1% or so) for segregation of basaltic melts. Therefore, source-rock fertility and intensive variables that enhance melt productivity are of paramount importance in creating a mobile mass of felsic magma capable of rising through the crust. **Source-rock fertility** is the potential amount of components available to yield a melt, in this case of felsic composition. For example, one way to compare relative fertilities is to compare the amount of minimum-T granite components (Figures 5.24–5.26; subequal proportions of normative quartz, orthoclase, and albite) in potential source rocks. In Hawaiian tholeiitic basalt (Table 13.2) it is ~7 wt.%, whereas in diorite (Table 2.2) it is ~31 wt.%.

Figure 11.21 shows melt fractions and modal compositions of crystalline residues resulting from dehydration melting of metamorphosed clay rocks (shales) and mafic rocks under presumed equilibrium conditions. These two rocks represent widespread sedimentary and magmxatic rock compositions, respectively, in the continental crust. Muscovite-rich, metamorphosed shale source rocks begin melting at 800°–825°C by

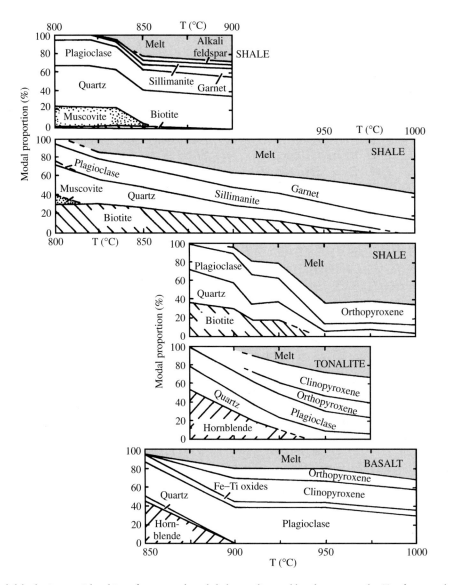

11.21 Experimental dehydration partial melting of metamorphosed shale, tonalite, and basalt source rocks. Top four panels at 10 kbar (about 37-km depth). (Redrawn from Patiño Douce and associates, e.g. Patiño Douce and Harris, 1998.) Bottom panel at 7 kbar (Redrawn from Beard and Lofgren, 1991).

breakdown of muscovite. By 850°C, all of the muscovite has been consumed, producing about 20% hydrous melt, by the reaction

muscovite + plagioclase + quartz

= hydrous felsic melt + alkali feldspar

+ sillimanite + garnet

In metamorphosed shales containing biotite, this phase decomposes over a wider and higher range of T than muscovite, yielding steadily increasing melt fractions. Two reactions are

biotite + quartz = hydrous felsic melt + garnet

and

biotite + plagioclase + quartz

= hydrous felsic melt + orthopyroxene

An example of the first reaction is shown in Figure 11.22.

Muscovite-free source rocks that contain abundant biotite or hornblende begin melting near 875°–900°C, consuming them by 950°C. Hornblende breaks down by the reaction

hornblende + quartz = hydrous felsic melt

+ clinopyroxene + orthopyroxene

Significantly increased melt productivity is produced by dehydration melting of an intimate mixture of a two-mica rock and a tonalite (Pl + Hbl + Qtz) compared to the melt fractions generated by melting each source rock alone. The enhanced fertility of a composite source, which may be quite realistic for the heterogeneous continental crust, apparently occurs because K migrates into the tonalite from the two-mica

11.22 Hand sample of pelitic **migmatite.** Dehydration melting of biotite and dissolution of quartz yielded garnet and hydrous felsic melt that is now crystallized to alkali feldspar and quartz.

rock and Na in the reverse direction, creating a composite source composition nearer the thermal minimum in the granite system (Figures 5.24–5.26). Granitic rocks and feldspathic sandstones can potentially yield large volumes of partial melt for the same reason, although dehydration melting of the generally minor amounts of biotite and amphibole in these rocks may not supply much water to flux dissolution of the felsic minerals.

<u>Melt Composition.</u> Incongruent felsic partial melts are generally relatively enriched in Si, Na, K, and water and depleted in Ti, Mg, Fe, and Ca relative to the source rock. These melts may be silica-oversaturated with normative Q, even though the source rock may be barely silica-saturated or even silica-undersaturated. Partial melts are peraluminous, metaluminous, or even peralkaline (Section 2.4.4), depending on the composition of the source, intensive parameters, and degree of melting. These same conditions dictate the composition of the crystalline residue, which in turn influences melt composition; for example, melts in equilibrium with residual garnet or plagioclase are less aluminous than if these Al-rich minerals are entirely dissolved and are depleted in either heavy REE (garnet) or Eu (plagioclase) because these trace elements are compatible.

Partial melting of source rocks that contain two feldspars and quartz just above their solidus predictably generates melts of granite composition at the thermal minimum in the granite system (Figures 5.24–5.26). Source rocks with only two, one, or none of these three crystalline phases yield partial melts further from the minimum composition.

Partial melting of basaltic rocks at $P = 5 - 32$ kbar under water-deficient conditions so that amphibole remains stable generates granite melts near the solidus, where melt fractions are less than a few percent but trondhjemite for melt fractions is near 10%.

At still higher degrees of melting, tonalite compositions are generated, essentially bypassing granodiorite (Figure 11.23). In contrast, progressive partial melting of basaltic rocks over about the same range in P but water-saturated conditions generates granite, granodiorite, and finally, at highest degrees of melting, tonalite, bypassing trondhjemite.

11.6.2 "Alphabet" Granitic Magmas: Contrasting Sources

Contrasting Devonian granitoid batholiths and minor associated felsic volcanic rocks in the Lachlan fold belt of eastern Australia are postulated to have had two distinctive sources (Chappell and White, 1992). **S-type granites** were derived from magmas generated by partial melting of sedimentary source rock composed in part of metamorphosed clay minerals. Weathering of feldspars and other rock-forming minerals at the surface of the Earth yields chemically differentiated products, chiefly Na in seawater, Ca and Sr in limestones, and Al-rich clay minerals in shales. Consequently, partial melting of metamorphosed shales that contain aluminous micas, Al_2SiO_5 polymorphs, cordierite, garnet, and so on (Figure 11.21, top three panels), generates strongly peraluminous magmas (Section 2.4.4). With >1 wt.% normative corundum, S-type rocks contain Al-rich minerals, including one or more of andalusite, sillimanite, garnet, tourmaline, cordierite, and muscovite in addition to biotite, two feldspars, and quartz.

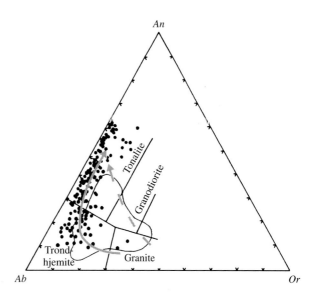

11.23 Experimentally generated partial melts of basaltic source rocks at 5–32 kbar plotted in terms of normative feldspar components (Figure 2.9). Full-line shaded arrow shows that smallest-degree dehydration partial melts (water-deficient conditions) are granite but more advanced melting generates trondhjemite and then tonalite melts. Water-saturated partial melts follow dashed shaded arrow with advancing degree of melting and are never trondhjemite. Outline encloses Archean tonalite-trondhjemite-granodiorite (TTG) suite. (Data from Rapp, 1997; Springer and Seck, 1997.)

Ilmenite rather than magnetite is the typical Fe-Ti oxide. Inclusions (enclaves) are of the same aluminous minerals but have higher concentrations of mafic phases. Whole-rock silica is generally >63 wt.%.

In contrast, **I-type granites** are believed to have been derived from magmas generated by partial melting of mafic and intermediate igneous rocks (Figure 11.21, bottom two panels). These mostly metaluminous granites contain more Na and Ca than S-type granites, which stabilize amphibole and commonly titanite, in addition to possible biotite, alkali feldspar, and quartz. Cogenetic rocks may be as mafic as gabbro, having as little as about 50 wt.% silica. Mafic inclusions are predominantly amphibole with plagioclase and lesser amounts of biotite, titanite, and clinopyroxene.

The postulated contrasting sources in the Lachlan fold-belt rocks spawned intense research on and debate about these and other granitoid rocks worldwide (e.g., Zen, 1988). Despite the important new focus on felsic magma sources, attempts to apply the eastern Australian S- and I-type labels elsewhere have caused considerable confusion (Miller et al., 1986). Their use implies that the source is known, but in practice this information is generally not verifiable. S-type is *not* synonymous with peraluminous because peraluminous magmas can also be generated from igneous source rocks. In addition, the composition of virtually every granitic rock reflects not only its source rock but a host of overprinting differentiation effects that acted on the primary magma. Collins (1996) finds that the Lachlan rocks have three distinct Sr-Nd-Pb-O isotopic sources: Mantle-derived mafic magma, late Proterozoic mafic arc rocks of the lower crust, and midcrustal metasedimentary rocks. I-type magmas were produced by mixing of the first two and S-type magmas by contamination of I-type magmas with metasedimentary partial melts.

A-type granites are a third, diverse group of felsics rock typically created in **anorogenic** areas and having diverse sources and origins (see Section 13.9).

11.6.3 Crystalline Residues

Partial melting of mica-amphibole-bearing metamorphic source rocks in the deep continental crust generates complementary hydrous felsic melts and refractory mafic to ultramafic crystalline residues.

In contrast with the simple olivine-pyroxene residue in mantle sources, continental residues are mineralogically heterogeneous and complex. Unlike mantle residues, new crystalline phases may be created during partial melting of felsic continental rocks (Figure 11.21). Many trace elements are not dispersed among major minerals as in peridotitic systems, for which equations such as 11.3 and 11.4 apply, but instead are sequestered in accessory minerals such as zircon, and monazite, whose stabilization depends upon sufficient activities of elements normally present in trace concen-

trations (e.g., Zr and Ce). These accessories may or may not be stable phases in the residue.

Three potential residues of partial melting have been discerned in deeply eroded roots of mountain belts and as xenoliths of this deep-seated rock in volcanic rocks. **Granulite** is a metamorphic rock that consists of an anhydrous assemblage of plagioclase + pyroxene ± quartz ± garnet ± sillimanite ± cordierite (compare Figure 11.21). **Migmatite** consists of layers, pods, and irregularly shaped masses of granite that are mingled with mafic metamorphic rock made largely of plagioclase, amphibole, biotite, pyroxene, and other refractory minerals (Figures 11.22 and 11.24). Migmatites are commonly interpreted to be partially melted rock in which the granitic partial melt has not segregated from the complementary residue. Some petrologists believe migmatites represent residues of production of granitoid plutons, but other possibilities exist. Garnet pyroxenites and eclogites, some containing amphibole, are the third residue.

Another possible relation between complementary partial melt and crystalline residue is represented in inclusions (granular aggregates) and single isolated crystals collectively called *restite* by Australian petrologists (e.g., Chappell et al., 1987). **Restite** crystals and inclusions are pieces of the residue caught up in a larger volume of melt. The whole mass forms a mobile body of magma that has sufficient buoyancy to ascend, perhaps diapirically, from the source into shallower crust. Magmas containing at least 60% restite might represent en masse mobilization of virtually the entire source volume, whereas magmas containing less restite might reflect some degree of melt-residue segregation.

Although intuitively reasonable, to what extent the restite concept actually applies to real felsic magmas is quite controversial and several questions may be posed. How much of a real felsic rock is restite? Zircon grains, commonly complexly zoned and having ages greater

11.24 Outcrop of deformed **migmatite**. Lower Proterozoic metamorphic belt, Tolstik Peninsula, Karelia, Russia. Felsic rock is concentrated in former shear zones. (Photograph courtesy of Michael Brown.)

than the age of the host magmatic rock, can be accepted as refractory restite, provided a xenocrystic derivation can be ruled out. But what of embayed calcic cores crowded with other mineral inclusions in plagioclase grains? Are these cores, and like ones occurring in other minerals, partially resorbed restite on which rim material was precipitated from the crystallizing melt? Or are they a result of mixing of magmas of contrasting composition? Are mafic inclusions of essentially the same minerals as the host rock, but in different proportions, pieces of restite equilibrated with the melt fraction? Or are they cognate inclusions ripped from the early crystallized margin of the magma chamber, equilibrated xenoliths of wall rock, recrystallized mantle-derived basaltic magma that powered the felsic magma system, or pieces of mixed and crystallized magma? Arguments in favor of the restite concept include the reluctance of restite material to segregate from melt of high apparent viscosity and the opportunity for restite grains to serve as sites for heterogeneous nucleation; widespread inclusions of relatively refractory apatite and zircon crystals in biotite and hornblende may be examples.

It should be noted that huge volumes of complementary mafic residues should underlie batholiths of felsic rock in continental mountain belts. However, geophysical evidence for these dense batholithic roots is often ambiguous. But Ducea and Saleeby (1998) present evidence from garnet pyroxenite xenoliths in Miocene volcanic rocks for a 70-km-thick keel of mafic-ultramafic rocks underlying the 30-km-thick Sierra Nevada, California, granitoid batholith (Figure 19.16; see also Section 13.16.2).

11.6.4 Melt Segregation

At least three factors that influence segregation of partial melts from residue can be identified:

1. Volume of melt produced: This is related to source fertility and melting temperatures. In pelitic sources, modest melt fractions, say, <10%–20%, from muscovite breakdown may produce granite segregations in migmatite, but larger melt fractions related to biotite and/or hornblende breakdown may be required for complete melt segregation from the source rock or en masse mobilization.

2. Melt process: Dehydration melting produces a larger volume increase than water-saturated melting and so is more effective in creating porosity in the source rock.

3. Deformation of the source rock during melting. Granite segregations are commonly found along shear zones (Figure 11.24) in deformed migmatites where mobile melt might have been drawn into local lower-pressure sites by dilatancy pumping (Section 8.2.3). It is difficult to be sure of cause and effect in such features because the shearing may be focused in the parts of the rock that contain granite melt.

11.6.5 Felsic Magma Generation and the Mantle Connection

The large volume of felsic magma in arc batholiths and ignimbrite fields (see Section 13.7) demands a significant advective input of heat from the mantle into continental source rocks to accomplish partial melting (Section 11.1.1). Production of granite magma by extreme fractional crystallization of basaltic magmas does locally occur, but the proportion (<10%) is too little to account for the observed volume, unless there are gigantic hidden volumes of complementary mafic-ultramafic rocks. In deep crust already at near-solidus temperatures of perhaps 500°–600°C, heat transferred from arrested intrusions and underplatings of mantle-derived basaltic magma at temperatures to as much as about 1400°C makes *partial melting thermally unpreventable*. Added to this heat input is the fluxing effect of volatiles liberated from these magmas as they become fluid-saturated during crystallization.

Evidence for a thermochemical mantle connection to felsic magma generation includes the following:

1. Isotopic compositions of many felsic rocks that indicate a significant mantle component derived from basaltic magmas that hybridize with partial melts of continental rock

2. Thick, tilted sections of lower crustal rock exposed in orogens that reveal kilometer-scale basaltic magma underplatings (e.g., Sinigoi et al., 1995)

3. Fragmented synplutonic mafic dikes (Figure 11.25)

4. Blobs of mafic rock in felsic hosts (Figure 8.25), which resemble shapes of lava pillows formed in water and suggest that hotter mafic magma invaded and quenched against cooler felsic magma

Widespread mafic inclusions (Figure 7.48) hosted in granodiorite, tonalite, and dioritic plutons (Didier and Barbarin, 1991) have different possible origins, i.e., are polygenic, but some are derived from (3) and (4).

SUMMARY

Global magma generation is a manifestation of a convecting mantle in a cooling still-hot Earth. The mantle and underlying core are the source of the thermal energy. Directly or indirectly, the mantle provides most of the mass for magma production. Convection in the cooling mantle via plumes and moving lithospheric plates allows magma generation by perturbing the *P, T,* and composition of potential solid source rock in the upper mantle as well as the lower crust. Incongruent

11.25 Fragmented synplutonic mafic dike. Pativilca pluton, Llanachupan, Pativilca Valley, Ancash, Peru. Height of cliff approximately 250–300 m. Dike intruded almost crystallized magma, which then became mobilized, disrupting the mafic sheet. (Drawing by K. Lancaster; Reproduced with permission of W. S. Pitcher from Pitcher, WS, 1997. The nature and origin of granite. New York, Chapman & Hall.)

partial melting generates melts enriched in lower-melting-T components and incompatible elements, leaving a more refractory crystalline residue. Extraction of basaltic partial melts from the peridotitic mantle over billions of years of geologic time is responsible for the differentiation of the crust from the mantle and continent growth.

The upper mantle is mostly Cr-diopside peridotite that has been locally metasomatized and veined by C-O-H fluids and silicate and carbonatite melts. These metasomatizing liquids have created assemblages of Al-Fe-Ti-rich pyroxenes, amphiboles, micas, apatite, and other materials, thereby enriching the relatively infertile, depleted peridotite in volatiles and incompatible elements. From this heterogeneous, variably enriched mantle a wide compositional spectrum of primary magmas has been generated by variable degrees of melting over a range of depth and volatile concentrations. Primary magmas range from ultramafic (komatiite, picrite, kimberlite) to mafic (alkaline and subalkaline basaltic) to intermediate (basaltic andesite, andesite). Tholeiitic basalt magmas generated by partial melting of upwelling decompressing mantle at the ocean ridge system constitute the largest volume of magma on Earth. Smaller volumes of more alkaline magmas are generated by lesser degrees of melting, likely of enriched source rocks. Highly alkaline magmas likely represent partial melts of metasomatic veins with little but variable wall rock contribution. Primary basaltic and andesitic magmas are generated by water-fluxed peridotite in subduction zones.

Source rocks in the continental crust are more heterogeneous than mantle rock. In addition to abundant feldspars and quartz, the presence of amphiboles, micas, and many alumino-silicates makes partial melting more complex. Melts and crystalline residues are more variable. Most melting probably occurs under water-deficient (dehydration) conditions in which the only water in the source-rock system is bound in hydrous minerals. Partial melt compositions range from granite to granodiorite to tonalite. These melts are highly viscous and are unlikely to segregate from their residuum as easily as more mafic mantle melts. Partially melted, buoyant crustal sources may rise en masse as diapirs.

CRITICAL THINKING QUESTIONS

11.1 Discuss the relative volumes of magma generated in the Earth by perturbations in P, T, and volatile concentrations (see Figure 1.1). Is increase in T responsible for the largest volume of magmas?

11.2 Critique the effectiveness of decompression melting of uplifted and eroded continental crust in comparison with decompression melting of upwelling mantle at ocean ridges and in plumes. What is the principal factor that allows melting to occur?

11.3 If T increases steadily into the Earth, why aren't magmas generated more abundantly at greater depths?

11.4 Color is generally an unsound basis for classification of rocks. Nonetheless, the two principal types of mantle-derived inclusions in basaltic rocks have strikingly different colors that can be used in identification. Discuss and justify the rationale for this distinction.

11.5 Explain how Al-Fe-Ti-rich veins can be the source of megacrysts and inclusions made of the same mineral assemblages.

11.6   Why are incompatible elements sequestered in some mantle minerals (amphiboles, clinopyroxenes) and not others (olivine)?

11.7   Why is there an "inverted" thermal gradient from the subducting crust into the overlying mantle in the area of the inset diagram in Figure 11.16?

11.8   Review the possible compositions of primary magmas that can be generated from mantle sources and indicate the conditions under which they form.

11.9   The oldest known alkaline rocks on Earth are about 2.7 Ga, whereas older rocks to as much as 4.0 Ga include tholeiitic basalts and a variety of silica-saturated felsic rocks. Suggest one reason for the apparent absence of alkaline magmas in the young Earth.

11.10 Discuss the differences between primary and primitive mantle-derived magmas. How can they be accurately identified?

11.11 Contrast partial melting in lower continental crust and upper mantle.

11.12 What distinctive major- and trace-element signatures, if any, might be used to distinguish dehydration partial melts of muscovite-bearing from hornblende source rock?

11.13 Explain why S-type granites typically have ilmenite rather than magnetite and have low Fe_2O_3/FeO ratios relative to I-type. (*Hint:* Is there something in shale source rocks that would cause their magmas to have low oxygen fugacity?)

11.14 How might juvenile primeval versus recycled surface water be distinguished in mantle minerals?

11.15 Section 11.2.1 indicated that eclogite xenoliths from the mantle might have two origins. How might these be distinguished in a particular xenolith?

11.16 Propose an alternate interpretation of the mafic blocks in granitoid pluton shown in Figure 11.25. What evidence might you seek to decide between your interpretation and that cited in the figure caption?

PROBLEMS

11.1   The minimal amount of water in the upper mantle to a depth of 670 km may be roughly estimated from just that contained as (OH^-) in nominally "water-free" crystalline phases. Calculate this amount. Compare it with a rough estimate of the amount of water in the oceans.

11.2   Estimate the amount of water cycled annually into the mantle via subducting oceanic crust that is entering the subduction zone at a rate of 5 cm/y. Compare this value with the amount of water exsolved from solidifying mantle-derived magmas intruded into and extruded onto the crust. Show all calculations and indicate and justify the numbers used. Discuss your results.

11.3   Draw the Mg_2SiO_4-SiO_2 binary phase diagram for 20-kbar anhydrous and for 20-kbar water-saturated from Figures 11.8b and 11.10. Compare with Figure 5.8. Discuss whether enstatite melts congruently or incongruently at 20 kbar.

11.4   Compare equilibrium and fractional partial melting of spinel lherzolite (Figure 11.9) by making diagrams in which T is plotted on the vertical axis and melt fraction on the horizontal axis. Indicate compositions of melts and crystalline residues at critical points in the diagrams. Discuss contrasts in these two styles of partial melting.

11.5   Partially melt, in both equilibrium and fractional manners, a rock composed of 40% quartz, 30% diopside, and 30% forsterite in the system in Figure 11.9. Describe in detail the compositions of crystalline residues and melts and plot their paths in the ternary.

11.6   Compare the time a basaltic melt would take to ascend buoyantly 100 km via porous flow through mantle peridotite versus through a dike 1 cm in width (see equation 9.1). Compare these times with the time for porous flow of an aqueous fluid rising 100 km above a subducting oceanic crust. Indicate all assumptions made and values of parameters chosen. Discuss the implications of your calculated times.

11.7   Write a mineralogical reaction whereby lherzolite is metasomatized by silica-bearing aqueous fluid to an assemblage of enstatite + anthophyllite + tremolite + talc + quartz. See Special Interest Box 11.4.

Differentiation of Magmas

12

FUNDAMENTAL QUESTIONS CONSIDERED IN THIS CHAPTER

1. Once magma has been generated in the lower crust or upper mantle, how can its composition be modified by subsequent differentiation processes?

2. What petrologic tools are available to unravel the effects of multiple processes that are involved in the origin and evolution of magmas?

INTRODUCTION

The origin of the broad compositional spectrum of magmatic rocks (e.g., Figure 2.4) has been one of the most fundamental problems in igneous petrology. In the previous chapter, it was demonstrated that variable source compositions and conditions allow generation of diverse primary magmas, but these fall short of accounting for the entire spectrum. Further modification of diverse primary magmas results from the overprinting effects of magmatic differentiation on primary magmas after leaving their source.

Near the surface of the Earth, rocks solidifying from undifferentiated, unmodified primary magmas are the exception rather than the rule. For example, most basaltic lavas have compositions that are not in equilibrium with their mantle peridotite source because olivine fractionation—one form of magmatic differentiation—takes place in the magma en route to the surface. Intrusions of basaltic magma that internally differentitate into layers of olivine (dunite rock), pyroxene (pyroxenite), and plagioclase (anorthosite) provide examples of the modifications that can happen.

The compositions of primary magmas ascending buoyantly out of mantle and deep crustal sources are modified to varying degrees by several processes of **differentiation.** Insights into this overprint can be gleaned from study of a suite of compositionally variable rocks that are closely related or associated in space and time: That is, they occur within a restricted geographic area and are of about the same age. Examples include rocks within a layered basaltic intrusion or a Hawaiian shield volcano. A necessary condition for a genetic relation between rocks and the magmas from which they solidified is that they are so related, but this is not sufficient to guarantee a **comagmatic,** or **cogenetic,** kinship via differentiation. Magmas from different sources or at least different evolving batches might, coincidentally, make up a composite intrusion or a long-lived composite volcano. Usually, a petrologist hypothesizes that a suite of rocks related in space and time are comagmatic, on the basis of some form of preliminary evidence, such as field relations and petrographic attributes. Additional chemical, isotopic, mineralogical, and modal compositions; trends in composition; precise isotopic dating; and other information can then be used to test for kinship. From these laboratory data, a **parent magma** is hypothesized; from it the more evolved rocks of the suite are assumed to have been derived by differentiation processes to be determined. A parent magma is usually not primary, or wholly unmodified from its source, but is at least primitive, having been modified less than other magmas that formed the comagmatic suite (Section 11.3.5). In many instances, an unseen parent must be hypothesized.

Differentiation processes can be outlined as follows:
I. Closed-system processes
  A. Crystal-melt fractionation
    1. Gravitational segregation

2. Flowage segregation

3. Filter pressing

4. Convective melt fractionation

  B. Physical separation of immiscible melts

  C. Melt-fluid separation

II. Open-system processes

  A. Assimilation of an initially solid contaminant

  B. Mixing of two or more contrasting magmas

Two or more of these processes commonly operate simultaneously or in tandem. For example, magma in a continental arc chamber may evolve from some primitive parent by crystal-melt fractionation while assimilating wall rock; this daughter magma may subsequently mix with another magma.

The plan of this chapter is first to present the basic principles of differentiation illustrated with simple examples, then to describe rock associations formed by more complex compound processes.

❊12.1 USING VARIATION DIAGRAMS TO CHARACTERIZE DIFFERENTIATION PROCESSES

Chemical variation diagrams that show coherent, regular compositional trends in a suite of comagmatic rocks can provide information on differentiation processes (Cox et al., 1979). Changing melt compositions (liquid lines of descent, e.g., Figures 5.23b and 8.16) during crystal-melt fractionation and proportions of two mixed magmas and amount of a solid contaminant can be determined. Graphical modeling techniques presented here, mostly for cartesian diagrams of two variables (x-y plots), can be readily adapted to a personal computer using widely available spreadsheet software.

Thorough sampling of fresh (unaltered) representative rocks and accurate laboratory analysis (Section 2.1) are crucially important in graphical modeling. So too is a careful petrographic study, as this is one means by which working hypotheses are formulated and tested by variation diagrams in which element concentrations and isotopic ratios are plotted. One diagram violating a hypothesis is sufficient to invalidate it, or possibly to indicate it needs some modification. For example, modification of the composition of the assumed, unseen parent magma may be required. Complete agreement with available data only indicates that the hypothesis hasn't been disproved.

Compositional patterns of suites of unaltered volcanic rocks are generally more amenable to unambiguous interpretation than plutonic rocks that might have suffered inconspicuous subsolidus modifications during slow cooling. Although the inverse procedure of inferring process from product always has some degree of ambiguity and uncertainty, volcanic suites may

closely represent extruded melts that evolved along liquid lines of descent. Volcanic rocks provide the opportunity to compare compositions of coexisting crystals and melt, represented in phenocrysts and surrounding glassy or aphanitic groundmass, respectively.

If some mass A is *added* to magma B in Figure 12.1, in which their compositions are represented by constituents x and y, the resultant hybrids plot along a straight line AB. Constituents x and y can be major elements or their oxides (in weight percentage) or trace elements (in parts per million); the latter generally have a greater range of variation within a comagmatic rock suite and serve as reliable indicators of differentiation processes. For a particular combination of A and B, say, C, the proportions of the two end members are given by the lever rule (equation 5.2). The mass of added A could be an assimilated contaminant into magma B, or A and B could be two mixed magmas, or B might be a parent magma from which precipitating crystals of uniform composition A accumulate a fixed amount by some means, forming a cumulate C. In the case in which A is *subtracted* from B a residue R is created. B might be a parent magma from which uniform crystals of A are fractionated out. The linear trends on this two-element variation diagram do not apply where x or y is a ratio of elements. Figure 12.2a–c shows two-element variation diagrams for a system in which more than one phase of fixed composition is fractionated from a parent magma.

Linear (straight line) trends on two-element diagrams are typically found in rock suites produced by mixing of two magmas and by assimilation of foreign rock into a magma because combinations of two compositionally fixed end members are involved. Additional information can indicate which of these two common possible differentiation processes operated.

Fractionating a solid solution of variable composition is a typical process in a crystallizing parent magma and yields a nonlinear (curved) trend in residual melts (Figure 12.2d).

Compositional trends for three constituents can be represented on ternary diagrams, but these provide no absolute concentration values.

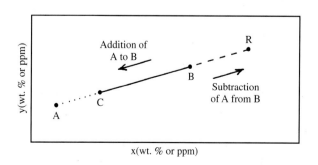

12.1 Hypothetical two-element variation diagram. The line $ACBR$ is a **control line.**

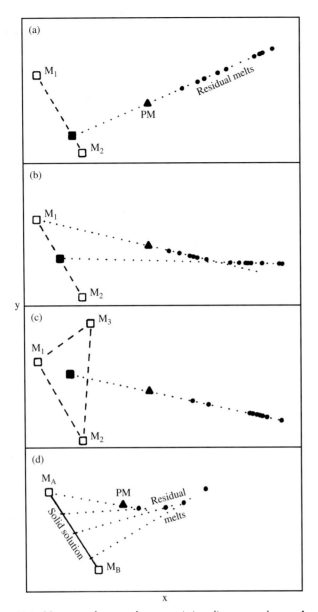

12.2 More complex two-element variation diagrams and **control lines** (dotted). (a) Two minerals, M_1 and M_2, in the proportions of 80:20 fractionate in varying amounts from a parent magma, PM, to produce variable residual evolved daughter melts (filled circles). (b) Mineral M_1 fractionates alone followed by cofractionation of equal proportions of M_1 and M_2. (c) Three minerals, M_1, M_2, and M_3, cofractionate in equal proportions from the parent magma. (d) Solid solutions between end members M_A and M_B fractionate successively, producing a curved residual melt trend; successive control lines are tangential to the curve at the intermediate "parent" melt that was produced as a residue of the previous fractionation.

❄12.2 CLOSED-SYSTEM MAGMATIC DIFFERENTIATION

With the recognition that magmas consist generally of melt plus crystals and locally an exsolved volatile fluid phase, three possible means of differentiation are logically possible in closed systems:

1. Separation of crystals and melt, called *crystal-melt fractionation, fractional crystallization,* or simply *fractionation*
2. Separation of two immiscible melts
3. Separation of melt and a fluid phase

12.2.1 Crystal-Melt Fractionation

Crystal-melt fractionation occupies a prominent status among all processes of magma differentiation and dominates in closed systems. Espoused as early as 1835 by Charles Darwin, fractionation was championed by Bowen (1928) through the first half of the 20th century (Young, 1998); Bowen referred to the process as *crystallization-differentiation* to emphasize that melt generally separates from growing crystals in dynamic magma systems.

Because of the contrast in chemical composition between any melt and its precipitating crystals, segregation of the two provides a powerful means of differentiating a parent magma into compositionally contrasting parts. For example, at 1020°C in the nearly crystallized Makaopuhi basalt magma (Plate III), a small proportion (about 10%) of silicic melt coexists with crystals of plagioclase, pyroxene, Fe-Ti oxides, and accessory apatite. Similar evolved melts that are enriched in silica, alkalies, and volatiles solidify to interstitial glass in many basalts (e.g., Table 12.1). Segregation of such a melt from the

Table 12.1 Chemical and Normative Compositions of a Whole-Rock Basalt and a Small Percentage of Residual Glass Lying between Crystals, Holocene Santa Clara Lava Flow Near St. George, Utah[a]

	BASALT	GLASS
SiO_2	51.1	66.5
TiO_2	1.7	1.3
Al_2O_3	14.3	13.9
FeO	10.4t	1.9t
MgO	8.1	0.1
CaO	9.2	0.9
Na_2O	3.1	3.3
K_2O	1.1	7.0
Total	99.0	94.9
Q	0.0	19.0
Or	6.5	41.2
Ab	26.2	27.7
An	21.5	2.4
Di	19.2	0.7
Hy	13.1	0.0
Ol	5.9	0.0
Mt	4.5	0.7
Il	3.2	2.4

[a] Note the low total for the glass that suggests about 5 wt.% volatiles, mainly water, in the microprobe analysis.

Special Interest Box 12.1 Picrites erupted in 1959 from Kilauea volcano, Hawaii: Mixing, not olivine fractionation

An example of a two-oxide variation diagram with a linear pattern, like the simple model in Figure 12.1, is shown in Figure 12.3. This diagram shows compositions of basaltic lavas erupted from the summit of Kilauea volcano, Hawaii, in 1959. Note that all plotted oxides in the lavas fit linear olivine control lines; this finding is consistent with the hypothesis, presented in the first edition of this textbook, that the lavas were differentiated by addition and subtraction of compositionally uniform olivine phenocrysts to and from a parent magma. However, it is puzzling why the fractionated olivines would be so uniform in composition during what was probably extensive crystallization of the magma; a curved trend of differentiation would be expected for fractionation of changing solid solutions from a changing magma (Figures 12.2 and 12.5). Subsequent investigations of this unusual eruption and its lavas, summarized by Helz (1987), who also presents significant new data, have shown that the olivine fractionation hypothesis is faulty.

Examination of thin sections of the picrites reveals seven textural types of olivines, rather than a single simple population of phenocrysts. In addition to euhedral and skeletal crystals (likely phenocrysts) there are blocky crystals as much as 12 mm long that contain planar extinction discontinuities due to strain, anhedral grains that appear to be strongly re-

sorbed, angular fragments(?), subhedral grains that contain sulfide inclusions, exceptionally large crystals (megacrysts as much as 20 mm in diameter), and metamorphic-textured aggregates. Most olivines regardless of textural type show evidence of partial resorption. Microprobe analyses reveal rather uniform compositions, Fo_{84-89}, but, in detail, indicate widespread disequilibrium in complex zoning.

Episodes of the November 14 to December 20, 1959, summit eruptions were unusual in recorded Hawaiian activity: Unusually extensive scoria deposits are of the most Mg-rich glass, erupted lavas were the hottest (1192°C), lava fountains were highest (580 m), and none of the summit lavas was also extruded on the flanks of the volcano, as commonly happens.

Altogether, the data suggest mixing of two magmas—a hot primitive, MgO-rich magma that ascended relatively rapidly from depths of 45–60 km in the mantle, passing through, and variably entraining a somewhat evolved magma stored in a subvolcanic chamber. In this model, texturally variable olivines represent crystallization products during decompression and cumulates in the crystallizing storage chamber. In addition, some mantle wall rock was assimilated into the ascending primitive magma.

This case history shows that although initial data can support a particular hypothesis (olivine fractionation), additional data can refute that hypothesis and a new one must be proposed (mixing of two magmas) to satisfy the whole body of information.

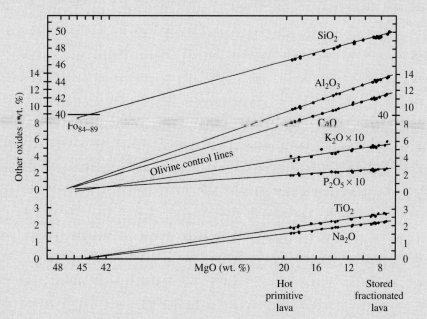

12.3 Composite two-oxide variation diagram for lavas erupted from the summit of Kilauea volcano, Hawaii, in 1959. Each analyzed sample is plotted as a solid circle with respect to MgO on the horizontal axis against the weight percentage of other oxides on the vertical axis. Olivine control line drawn through MgO-SiO_2 analyses passes through the range of composition of analyzed olivines Fo_{84-89} that have about 40 wt.% SiO_2 and 47–43 wt.% MgO (Helz, 1987). Other oxide control lines are near 0 wt.% at 47–43 wt.% MgO, reflecting their very small concentrations in the analyzed olivines. (Data from Murata and Richter, 1966.)

crystalline framework has been observed in some thick basalt flows and sills where it has collected into thin pods of granitic rock.

Thus, the effects of fractionation of a parent magma are manifested in complementary evolved residual melts and accumulated crystals. Rocks derived from residual melts have relatively evolved compositions and possibly phenocryst-free aphanitic or glassy texture.

Complementary accumulations of crystals that differ in composition not only from the separated melt but from the parent magma as well are called **cumulates.** Layers made virtually of only pyroxene, olivine, or plagioclase occur in some large, slowly cooled intrusions of basaltic magma. Such **monomineralic** layers of pyroxenite, dunite, or anorthosite, respectively, can only form by crystal accumulation because no magmas of equivalent composition are known to occur. Modest accumulation of a magnesian olivine near Fo_{90} (Appendix A), such as typically crystallizes from primitive basalt magmas, enriches the magma in Mg but depletes the residual magma in Mg and enriches it in Si, Ti, Al, Ca, Na, K, and P. Summit extrusions of basaltic lavas from Kilauea volcano, Hawaii, in 1959 ranged from olivine-rich picrites to olivine-poor basalts. Picrite lavas carrying as much as 30% olivine were erupted at unusually high rates of discharge from the vent, creating exceptionally high (580 m) lava fountains. Analyses of these two extreme lavas show the anticipated variations in major elements (Table 12.2; Figure 12.3).

Trace Element Modeling of Fractional Crystallization. In evolving magmas, the effect of fractional crystallization is more obvious in trace element than major element variations. For example, fractional crystallization drives a subalkaline magma toward the composition of a minimum melt in the granite system (Figure 5.27a). In such high-SiO_2 granite magma, the coexisting melt and solids (subequal proportions of plagioclase, alkali feldspar, and quartz) have nearly the same major element composition, and large amounts of fractionation yield only small changes in major elements of, perhaps, 2 wt.% in SiO_2. However, trace element concentrations can vary by several hundred percent. (Interestingly, in some such rocks Mg qualifies as a trace element.) Powerful constraints on the crystal-melt fractionation can be provided by trace elements.

During progressive crystallization of a magma, compatible elements are concentrated in the solids and incompatible elements are continuously enriched in the residual liquid. The extent of crystallization of a magma system is an important control on the trace element concentrations of the residual melt and the solids.

For a closed magma system undergoing perfect fractional crystallization, the relations among the bulk distribution coefficient, D (Section 2.5.1); melt fraction, F; and concentration ratio of a particular element in the residual melt to that in the parent magma, C_m/C_p, are given by the Rayleigh law

$$\frac{C_m}{C_p} = F^{(D-1)} \qquad\qquad 12.1$$

This equation is plotted in Figure 12.4 for several values of D. The extent of fractional crystallization is especially critical for those elements that have either very high or very low bulk distribution coefficients. For a compatible element, such as Ni ($D_{Ni} \sim 7$), in a basaltic magma precipitating olivine $\pm$ pyroxene, the concentration of Ni in the residual melt decreases to 0.5 that in the parent magma after only 10% crystallization ($F = 0.1$) and to 0.01 after 54% crystallization. In contrast, the concentration of an incompatible element, such as Rb ($D_{Rb} \sim 0.001$), in the residual melt of the same parent magma has only doubled after 50% crystallization, and advanced crystallization, about 90%, is needed to increase its concentration in the residual melt 10-fold.

Incompatible element enrichment in residual melts can set the stage for subsequent processes to enrich them further, possibly into economic concentrations in ore deposits. Beryllium (Be) is an example. Be ore deposits are typically associated with highly evolved granites and rhyolites. In granite magma systems, melt-fluid separation can concentrate the Be into pegmatitic minerals (e.g., beryl) that can be mined (discussed later).

If distribution coefficients and fractionating crystalline phases are known or can be assumed with some degree of certainty, crystal-melt fractionation models can be compared with trends from analytical data on rock suites. A model for a fractionally crystallizing magma system in which melt and solid solutions are continuously varying is fitted to analytical data on a rock suite in Figure 12.5. The variation trend for compatible versus incompatible element is strongly curved.

Table 12.2 Compositions of Most Olivine-Rich Basalt (Picrite) and Most Olivine-Poor Basalt Erupted from the Summit of Kilauea Volcano, Hawaii, in 1959

	PICRITE	OLIVINE-POOR BASALT
SiO_2	46.93	49.89
TiO_2	1.93	2.72
Al_2O_3	9.77	13.46
FeOt	11.74	11.35
MnO	0.18	0.18
MgO	19.00	8.00
CaO	8.29	11.33
Na_2O	1.58	2.25
K_2O	0.39	0.55
P_2O_5	0.19	0.27
Total	100.00	100.00

See also Figure 12.3.
(Data from Murata and Richter, 1966.)

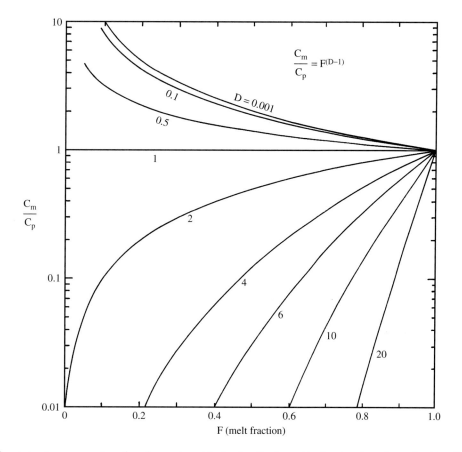

$$\frac{C_m}{C_p} = F^{(D-1)}$$

12.4 Theoretical variation in concentration of an element in residual melts, C_m, formed by fractionation of crystals with a distribution coefficient, D, from an initial parent magma, C_p. See equation 12.1.

The trace element pattern of residual melts can reveal important information about what phases are fractionating (e.g., Table 2.6). For example, in silicic melts, depletion of compatible Eu and Sr results from plagioclase fractionation and of Ba from mica or K-feldspar fractionation.

<u>Physical Mechanisms of Crystal-Melt Fractionation.</u> Given the chemical efficacy of crystal-melt fractionation, how is it physically accomplished? Four mechanisms have been identified:

1. **Gravitational segregation.** In a static body of melt, denser crystals might sink whereas less dense ones might float. This process has been widely accepted as a viable means of crystal-melt fractionation. However, except for the hottest mafic melts and largest crystals, the plastic yield strength of melts (ignored in the formulation of Stokes's law, which assumes Newtonian viscosity, Section 8.3.3) may preclude much movement of isolated crystals. Further experimental studies of the role of yield strength in crystal movement are needed.

2. **Flowage segregation:** In moving bodies of magma, grain-dispersive pressure pushes crystals and other solid particles into the interior of the flowing magma away from conduit walls where there are strong velocity gradients (Figure 8.13). This phenomenon has been documented in many dikes (e.g., Figure 8.14), sills, and extrusions.

3. **Filter pressing:** A cloth or paper filter passes liquid through it but not suspended solid particles, which are trapped by the smaller openings in the filter. Residual melt in partially crystallized magmas can be filter pressed from the interlocking network of crystals because of local gradients in pressure. On the floor of a crystallizing magma chamber, the weight of accumulating crystals may press some of the entrapped residual melt out of the compacting crystal mush into the overlying magma. Movement of small interstitial volumes of usually evolved, viscous silicic melt entangled in a network of crystals (Plate III, 1020°C) and separation from the network would, intuitively, appear to be unlikely. However, during crystallization, the melt is commonly also enriched in water, thereby reducing its viscosity; more importantly, it may also become water-saturated so that it exsolves a separate fluid phase. The increasing fluid pressure might drive the melt into regions of lower pressure. Sisson and Bacon (1999) recognize four situations in which gas-driven filter pressing might drive melt down a pressure gradient:

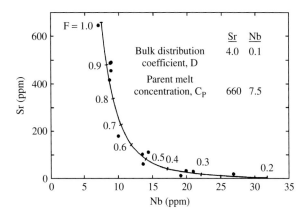

12.5 Trace element variation diagram for a fractionating magma system. Variation is for compatible Sr and incompatible Nb in pumice inclusions (filled circles) in a compositionally zoned rhyolite-dacite ignimbrite. The pumices are geologic samples of the zoned preeruption magma chamber (similar to that in Figure 10.38), which experienced fractional crystallization of two feldspars, quartz, biotite, and Fe-Ti oxides (present as phenocrysts in the ignimbrite) having the indicated bulk distribution coefficients. The curved line was constructed from equation 12.1, which was used to calculate the value of C_m for Sr and Nb as a function of F using the indicated bulk distribution coefficients and parent melt concentrations. Adjustment of these two model parameters and proportions of fractionating crystals allowed the curve to be fitted to the analytical data. It must be remembered that, no matter how good the fit, the model parameters constitute only a *possible* set of conditions in the actual magma system. That is, the agreement between the data and the model does not prove these parameters actually existed. (From Best et al., 1995.)

a. Migration of residual melt in crystallizing lava flows and shallow crustal intrusions.

b. Expulsion of melt from partly crystallized or melted rock inclusions into the host magma.

c. Expulsion of residual melt from mostly crystallized magma into nearby cracks, perhaps created by hydraulic fracturing (Section 8.2.1). Thin dikes of evolved leucocratic aplite in more mafic host granitoids (Figures 7.48 and 9.3) are a likely example. Merely opening the fracture creates a pressure gradient that might suck residual melt out of the host magma, but exsolved fluid in the melt would enhance the gradient and melt separation.

d. A crystallizing mafic underplate beneath or near the base of a crustal magma chamber contributes evolved melt that can mix with the chamber magma (Figures 8.24 and 8.25).

4. **Convective melt fractionation:** Residual melts may have sufficient buoyancy to move out of the enclosing crystalline network. This mechanism became well established in the 1980s, partly as a result of model laboratory experiments using tanks of salt solutions (e.g., McBirney et al., 1985) and partly as

a result of evidence for it found in rocks. Rather than gravitational compaction of a cumulate mat on the floor of a magma chamber, causing the interstitial residual melt to be pressed out, gravity may cause the melt to rise buoyantly out of the network of crystals because of its lower density. Bubbles of exsolved fluid would make them more buoyant. Ascending diapiric "fingers" of less dense residual melt, entraining some crystals, developed in Hawaiian lava lakes (Helz et al., 1989; see also Figure 8.23). Convective melt fractionation during sidewall crystallization in bottle-shaped intrusions is a well-accepted differentiation process in intermediate-composition magma chambers (Section 8.6.2; Figures 8.22 and 10.38).

12.2.2 Physical Separation of Immiscible Melts

Immiscible silicate melts were discovered in the early 1900s in the system $MgO\text{-}SiO_2$ (Figure 5.8). Further experiments as well as petrographic observations by the late 1900s had demonstrated the existence of immiscible silicate melts in other magmatic systems.

Many fresh, unoxidized tholeiitic basalts and some alkaline ones having high concentrations of FeO, P_2O_5, and TiO_2 and low MgO, CaO, and Al_2O_3 contain two compositionally distinct glasses lying interstitially between higher-T crystals. The two glasses, identifiable in thin section by contrasts in color and refractive index (Figure 12.6), are interpreted to have formed as two immiscible residual silicate melts (Philpotts, 1982). In slowly cooled intrusions, these two melts and any en-

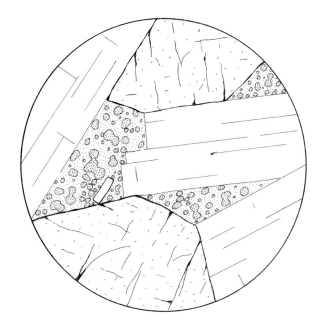

12.6 Immiscible silicate melts in basalt. Interstitial spaces between larger plagioclases and pyroxenes are of a **mesostasis** that was two immiscible melts but now consists of clear silicic glass hosting small globules of Fe-rich glass (darker stippling). Globule diameters are on the order of a few micrometers.

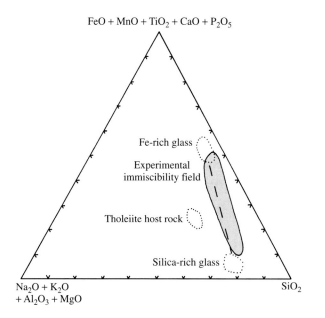

FeO + MnO + TiO$_2$ + CaO + P$_2$O$_5$

Fe-rich glass

Experimental
immiscibility field

Tholeiite host rock

Silica-rich glass

Na$_2$O + K$_2$O
+ Al$_2$O$_3$ + MgO   SiO$_2$

12.7 Compositions of immiscible melts in tholeiite basalt. Dotted
lines enclose analyzed immiscible melts (now glasses) connected by dashed tie line and host rock compositions. Shaded
area is field of immiscibility in synthetic model system
KAlSi$_2$O$_6$-Fe$_2$SiO$_4$-SiO$_2$. (Redrawn from Philpotts, 1982.)

Table 12.3 Chemical Compositions of Paired Glasses
and Rocks Formed by Separation of Immiscible Melts

	1	2		3	4
SiO$_2$	41.5	73.3		45.88	50.84
TiO$_2$	5.8	0.8		1.53	1.12
Al$_2$O$_3$	3.7	12.1		11.32	16.45
Fe$_2$O$_3$	—	—		6.60	4.40
FeO	31.0t	3.2t		7.56	5.04
MnO	0.5	0.0		0.15	0.12
MgO	0.9	0.0		6.50	3.78
CaO	9.4	1.8		11.42	7.28
Na$_2$O	0.8	3.1		2.06	2.97
K$_2$O	0.7	3.3		5.12	7.18
P$_2$O$_5$	3.5	0.07		1.84	1.07
Total	97.8	97.67		99.98	100.25
Q	13.7	37.8	Cs	0002.75	3.18
Or	4.1	20.0	Rb	0762	125
Ab	6.8	26.9	Ba	36,000	50,534
An	4.5	8.7	Th	00014.24	14.27
Di	17.2	0.0	U	3	2.81
Hy	21.8	2.5	La	824	600
Il	11.0	1.5	Ce	128.1	116.4
Ap	8.1	0.2	Sr	1591.7	1627.5
Mt	11.5	1.2	Sm	10.0	7.4
C		0.3	Hf	2.91	2.75
			Eu	2.00	1.68
			Dy	35.7	30.8
			Lu	0.32	0.25
			Co	26.25	13.74
			Cr	210.3	195.5
			Ni	622	39
			Sc	21.4	8.8
			V	1500.5	1437

1 and 2, Average oxide and normative composition of glasses in
mesostases in tholeiitic basalts. (Oxide data from Philpotts, 1982.)
3 and 4, Oxide and trace element compositions of shonkinite (3) and
"blob" syenite (4), Montana. Data from Kendrick and Edmond
(1981).

trained crystals might segregate, as a result of density
contrasts, into substantial volumes of magma. Crystallization could yield, on the one hand, an Fe-rich rock
composed essentially of Fe-Ti oxides, apatite, and Fe-
rich pyroxene and, on the other hand, a granitic rock
(Table 12.3 and Figure 12.7). Extremely rare nelsonite,
made mostly of Fe-Ti oxides and apatite, may also represent a segregated immiscible melt.

Field, fabric, and compositional relations indicate
some alkaline magmas in laccoliths in eastern Montana
unmixed into conjugate syenite and shonkinite (mafic
syenite) (Table 12.3). Each rock consists of clinopyroxene, olivine, biotite, and alkali feldspar of similar chemical composition, but in different modal proportions,
as would be expected if the two rocks formed from immiscible melts and suspended crystals that were all in
equilibrium. The syenite occurs as blobs within shonkinite, reflecting their original two-liquid status. These
two-liquid shapes together with trace element partitioning in the syenite and shonkinite rule out crystal-
melt fractionation as a mechanism to create the two
rock compositions.

Petrographic and experimental investigations have
unequivocally established the immiscibility of sulfide
and silicate melts. Concentrations of only a few hundred parts per million of S are sufficient to saturate
basaltic melts. Greater concentrations result in separation of a sulfide melt that is chiefly Fe and S with
minor Cu, Ni, and O that can ultimately crystallize to
pyrrhotite, chalcopyrite, and magnetite. Although trace
amounts of Ni are strongly partitioned into crystallizing olivine in basaltic melts, the uptake into an immis-

cible sulfide liquid is 10 times greater; for Cu it is 100
to 1000 times greater. Pb and Zn apparently tend to remain in the silicate melt. Sulfide-silicate melt immiscibility has significant implications for the genesis of
some magmatic and hydrothermal ore deposits.

Carbonatite. In several locales around the world,
small volumes of carbonatite occur with silica-undersaturated alkaline rocks. **Carbonatite** contains >50%
carbonate minerals, usually calcite. Since 1960 the
nephelinitic Oldoinyo Lengai volcano in Tanzania has
erupted alkali carbonate lavas and pyroclastics. Earlier
arguments whether carbonatite is truly a magmatic
rock were laid to rest by this eruption and discovery of
other carbonatite volcanic rocks. Table 12.4 reveals the
extreme composition of carbonatites; in alkali carbon-

Table 12.4 Chemical Composition of
Carbonatite Segregations in Nephelinitic Ash
Particles, the Two Representing Immiscible Melts,
Oldoinyo Lengai, Tanzania

	CARBONATITE	NEPHELINITE
SiO_2	3.17	43.97
TiO_2	0.10	2.34
Al_2O_3	1.05	7.96
FeOt	1.33	11.03
MnO	0.33	0.37
MgO	0.3	4.68
CaO	15.52	17.77
Na_2O	30.05	4.91
K_2O	5.35	1.57
P_2O_5	1.28	0.32

Data from Dawson et al. (1994).

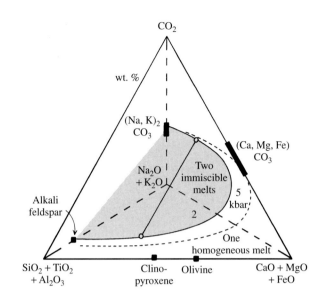

12.8 Immiscible melts in a carbonated silicate system. The shaded area represents the polythermal stability field of two immiscible melts below the arcuate 2-kbar solvus. Compositions of coexisting carbonate and silicate segregations in Oldoinyo Lengai volcanic ash from Table 12.4 shown by open circles connected by tie line. Segregation of these two immiscible melts from an initially homogeneous melt above (to the right of) the solvus may have occurred with decreasing T and $P < 2$ kbar. Solvus at 5 kbar is dashed line. Solid squares and rectangles represent compositions of minerals that might be stable in carbonated silicate systems. See also Lee and Wyllie (1997). (Redrawn from Kjarsgaard and Hamilton in Bell, 1989.)

atite, there is <3.2 wt.% SiO_2 + Al_2O_3 but major concentrations of SrO, BaO, SO_3, Cl, F, and, of course, CO_2. Carbonatites host no fewer than about 275 documented minerals in addition to carbonate minerals; other common minerals include apatite (repository of most REEs), magnetite, fluorite, pyrochlore (an Nb oxide that contains substituting Ta, Ti, Zr, Th, U, Pb, Ce, Ca, Sr, Ba), and alkaline amphibole. Most carbonatites are intimately associated with phonolite and more commonly with olivine-poor nephelinite.

Nd and Sr isotope ratios (Bell, 1989) unequivocally demonstrate that the ultimate source of carbonatite magmas is the mantle, probably long-lived, metasomatically enriched subcontinental lithosphere. Tangible samples of such a source are represented by rare garnet-phlogopite-calcite lherzolite inclusions in carbonatite tuffs. Experiments confirm that carbonatite partial melts can exist under special mantle conditions (Figure 11.5).

In June 1993 Oldoinyo Lengai erupted carbonatite lava and ash that contained nephelinitic spheroids that themselves contained alkali carbonatite segregations (Table 12.4). These are obviously immiscible silicate-carbonatite melts that separated from a carbonated silicate magma at low P (Figure 12.8).

12.2.3 Fluid-Melt Separation: Pegmatites

Experiments and observations of rocks indicate that aqueous and carbonatitic fluids in equilibrium with melts in magmatic systems contain significant concentrations of many chemical components, such as Si, Na, K, Fe, and many incompatible elements. Exsolution of fluid solutions from the coexisting melt is therefore a significant means of modifying its composition. In the terminal stages of the solidification of many intrusions

of granitic magma the creation of bodies of pegmatite is believed to be critically dependent on separation of an aqueous fluid phase from the residual water-saturated granite melt; this is the Jahns and Burnham (1969) model of pegmatite formation (see also, for example, Thomas et al., 1988).

Pegmatite is an unusually coarse-grained magmatic rock. Although giant crystals measured in meters occur in some (Figure 7.11), grain size is usually highly variable, and fine-grained phaneritic felsic rock (aplite) is commonly an intimate part (Figure 12.9). Large vugs (open cavities) are common. Most pegmatites are syenitic and granitic; more mafic ones also occur but ultramafic pegmatites are rare. Pegmatite bodies are relatively small—ranging from 1 m or less to a few hundred meters—and most occur as pods or lenses around the margins of nearby larger, deep-seated plutons, commonly extending from the pluton itself into the adjacent country rocks. In the famous Black Hills district of South Dakota (e.g., Norton and Redden, 1990), an estimated 24,000 pegmatite bodies are found over an area of 700 km². whereas the roof of the underlying comagmatic peraluminous granite is exposed over 100 km². **Simple pegmatites** consist essentially of a minimum-T granite composition (Figures 5.24–5.26) of

12.9 Pegmatite, San Diego mine, Mesa Grande district, San Diego County, California. Layered aplitic footwall underlies coarser pegmatitic upper part. Fringes of black tourmaline, especially evident at top of pegmatite, probably denote contemporaneous crystallization along footwall and hangingwall portions. Giant graphic microcline crystals radiate up from the footwall aplite, not down from the hangingwall as the Jahns-Burnhall model predicts. Scale card in center of photo is 9 cm long. (Photograph and caption courtesy of David London.)

albite, quartz, perthite, and possible minor muscovite, tourmaline, and Fe-Mn garnet. Rare (2% in the Black Hills) **zoned pegmatites** have an internal, layerlike zonation of fabric and mineralogical composition (Figures 12.9 and 12.10) that locally may be transected by late, metasomatic replacement bodies. A small proportion of these internally differentitated, zoned pegmatites, referred to as **complex pegmatites,** have relatively high concentrations of P, Cl, F, and B, as well as large-ion lithophile, rare earth, and other incompatible elements strongly partitioned into the residual melt during fractional crystallization of the parent granitic magma. For example, Li in the Black Hills granite averages about 30 ppm but in complex pegmatites can be as much as 7000 ppm, an enrichment factor of 233. These high concentrations stabilize minerals such as topaz (high concentrations of F), spodumene, lepidolite and amblygonite (Li), beryl (Be), columbite-tantalite and pyrochlore (Nb, Ta, Ce, Y), cassiterite (Sn), pollucite (Cs), monazite (Ce, La), zircon (Zr), and uraninite (U). Obviously, these complex pegmatites can be economically very valuable, but simple ones are also exploited for sheet muscovite (electrical and thermal insulators, such as in bread toasters) and large volumes of quartz and feldspar used in the glass and ceramic industries.

In the classic Jahns-Burnham pegmatite model, zoned pegmatites develop by inward solidification and differentiation of a lens-shaped body of water-saturated granite melt that produces the contrasting and commonly asymmetric mineralogical and textural zones. The outer margin, locally modally layered, is much like the host granite or syenite but abruptly coarser in grain

size, by as much as several orders of magnitude. Inward from the margin are graphic intergrowths of feldspar and quartz (Figure 7.20) and comb layers of crystals (Figure 7.47) that are commonly branching, inward-flaring, plumose, and locally of giant size and oriented perpendicular to pegmatite walls. Large internal portions are wholly quartz or feldspar or giant crystals of exotic minerals (Figure 7.11). Crystallization can continue to T as low as 300°C.

Many aspects of pegmatite development, particularly the large crystal size and other aspects of the internal fabric as just described, are controversial. Kinetic factors may be critically significant (London, 1992; Morgan and London, 1999).

✻12.3 OPEN-SYSTEM DIFFERENTIATION: HYBRID MAGMAS

12.3.1 Magma Mixing

If two or more dissimilar parent magmas blend together, a hybrid daughter magma compositionally intermediate between them is produced. Magmas can be derived from different sources, such as basaltic magma from the upper mantle and silicic magma from the deep continental crust, or they may have had a common parent magma but followed different evolutionary tracks, such as the contrasting magmas in a compositionally zoned chamber (Figure 10.38). Other scenarios are possible. Initially, dissimilar magmas are physically **mingled.** If solidification occurs soon afterward, the composite rock has layers, lenses, pillow-shaped blobs, or more irregularly shaped bodies in a dissimilar matrix (Figures 7.42, 8.25, 12.11). Rocks formed by mingling of magmas retaining their contrasting identity are evident on scales ranging from a thin section to large outcrops. After mingling, magmas may become **mixed** on an atomic scale by diffusion, if sufficient time and thermal energy are available, forming an essentially homogeneous melt. Homogenization and equilibration of crystals from the two batches of magma take a longer time. Hybridizing magmas can be as different as basalt and rhyolite or differ by as little as a few weight percentages in major elements.

Long after its first proposal by R. Bunsen in 1851, magma mixing became the subject of contentious debate in the 1920s and 1930s. C. N. Fenner advocated that the mixing of rhyolite and basalt magmas could produce the spectrum of compositions found in many magmatic rock suites, whereas N. L. Bowen persuasively advocated crystallization-differentiation (crystal-melt fractionation) as the dominant process of magmatic diversification. Writers of standard petrology textbooks of that time either made no mention whatsoever of mixing (Daly, 1968) or wrote only one brief sen-

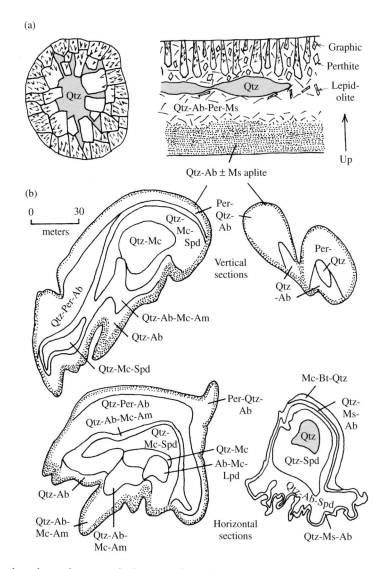

12.10 Schematic sections through zoned **pegmatite** bodies. (a) Body on left is several centimeters in diameter; on right is 1 to 2 m thick. Massive quartz is shaded; largest crystals are of quartz-perthite graphic intergrowth near pegmatite margin grading inward to perthite alone. (Redrawn from Jahns and Burnham, 1969.) (b) On the left are two sections of same body. Ab, albite; Am, amblygonite; Bt, biotite; Lpd, lepidolite; Mc, microcline; Ms, muscovite; Per, perthite; Qtz, quartz; Spd, spodumene. (Redrawn from Norton and Redden, 1990.)

tence (Grout, 1932): "Some mixing of magmas . . . may develop . . . but the process is probably rare." Since the late 1970s, however, petrologists have recognized widespread evidence for magma mixing so that it has now gained prominence as a significant differentiation process.

Compositionally dissimilar magmas are usually also dissimilar in *T* and physical properties, particularly apparent viscosity. Such contrasts are involved in many mechanisms of hybridization within intrusions and conduits feeding volcanic eruptions and even in flowing lava. The following are a few possibilities for mingling and possibly mixing:

1. During evacuation and flow through narrow conduits (e.g., Snyder et al., 1997): Erupted magmas form mingled lavas and pyroclasts.

2. Convective overturn and stirring in a magma chamber when underlying, initially denser mafic magma cools and exsolves volatile fluids, making it less dense than the overlying silicic magma: The underlying vesiculating magma develops m-scale Rayleigh-Taylor instabilities (compare Figure 9.15), causing detachment of buoyant blobs (Figure 8.24), some of which may become mingled mafic inclusions within the felsic host if solidification is rapid; longer residence in hot magma allows mixing.

3. Plumelike rising of hot, replenishing magma into an evolving chamber: This likely happens in basaltic oceanic-ridge systems as episodically replenishing primitive magma mingles and mixes with cooler magma that has evolved somewhat by crystal fractionation (Figure 12.12).

into the mafic magma in which clinopyroxene is a stable phase.

Compositional evidence for magma mixing in the history of a rock or rock suite includes the following:

1. Disequilibrium phases, such as coexisting quartz and Mg-olivine or calcic and sodic plagioclase.

2. Phenocryst-hosted glass (melt) inclusions that are compositionally unlike surrounding matrix glass.

3. Distinctive patterns on variation diagrams: Simple magma mixing produces straight-line trends on element-element variation diagrams (Figure 12.3 and Special Interest Box 12.1). However, because of limited thermal energy in most magma systems, only limited mixing occurs. Consequently, compositions of the rocks produced by mixing are generally only displaced slightly away from the parent magma composition toward the mixing magma composition.

If mixing proceeds to a homogeneous hybrid magma, its composition, C_h, can be expressed by a simple **mass balance equation** of the compositions of the

12.11 Mingled rhyolite and basalt lava, Gardiner River, Yellowstone National Park, Wyoming. In thin section, euhedral plagioclase and pyroxene phenocrysts from the basalt lava (darker colors) coexist with corroded quartz and sanidine phenocrysts from the rhyolite lava. (U.S. Geological Survey photograph courtesy of R. E. Wilcox and Louise Hendricks.)

Multiple lines of independent evidence can strengthen the mixing interpretation but are never sufficient to prove this process occurred. Disequilibrium texture in a rock that appears to be homogeneous but was produced by mixing of felsic and mafic magmas include the following:

1. Complex resorption and overgrowths in phenocrysts in quickly cooled volcanic systems, such as spongy zones within plagioclase (Figure 7.19) and rapakivi overgrowth of plagioclase on alkali feldspar (Figure 7.18): Normal fractional crystallization of ternary feldspars during cooling would not create rapakivi texture. However, these overgrowth textures can possibly be produced by changes in P or water fugacity in the magma system.

2. Anhedral, partially dissolved quartz rimmed by an aggregate of clinopyroxenes (Figure 6.20): The unstable quartz grains can be derived from felsic magma mingled with more mafic magma or from quartz-bearing rock assimilated (discussed later)

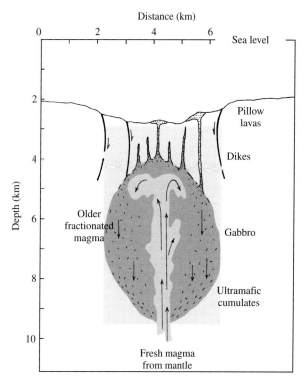

12.12 Schematic cross section through an episodically replenished oceanic rift magma chamber. Replenishing primitive magma from the mantle rises buoyantly into the chamber and mixes with cooler, denser, Fe-rich fractionated magma. Gabbro made of olivine, pyroxenes, and plagioclase crystallizes on the chamber walls; olivines and pyroxenes form ultramafic cumulates on the chamber floor. Evolving basaltic magmas are episodically injected through fissures in the overlying extending roof, forming feeder dikes for submarine extrusions, mostly pillow lavas. (Redrawn from Bryan and Moore, 1977.)

two end-member parent magmas, C_x and C_y; the mixing proportion, F_x, represents the weight fraction of one of the magmas

12.2 $C_b = C_x F_x + C_y(1 - F_x)$

Magma mixing, depending on how disparate the magmas are, may be accompanied by crystallization of new phases stabilized by the compositional and thermal properties of the hybrid system.

12.3.2 Assimilation

After leaving its source, a batch of ascending buoyant primary magma can encounter wall rock of different composition, especially basaltic magmas from mantle sources rising into sialic rocks of the continental crust and any silicate magma encountering Ca-rich limestone or Al-rich shale or their metamorphic equivalents. Magmas interact with their surroundings in an attempt to attain chemical and thermal equilibrium, especially where they slow or even stop in subterranean storage chambers. Hot country rocks are by no means inert to hotter, contrasting magma.

Incorporation of solid rock into a magma of different composition is the process of **assimilation;** it produces a **contaminated magma,** which is also hybrid, like mixed magmas. The contaminant can be country rock around the magma chamber or xenoliths within the magma. Assimilation may initially involve simple physical dispersal of xenoliths and xenocrysts into the magma, such as Precambrian zircons in Miocene rhyolite. Depending on magma and foreign material compositions and temperatures and available time, the foreign material chemically equilibrates with the melt to varying degrees. Minerals may selectively dissolve into the melt and contaminant ions incorporate into it by time- and T-dependent diffusion. Commonly, assimilation involves mixing with melts created by melting of the contaminant rock.

The thermal and chemical principles of assimilation were enunciated by Bowen (1928) many decades ago. Assimilation requires thermal energy, the source of which can only be the magma itself. Heat from the magma has two sources:

1.   That released during cooling to lower T
2.   The latent heat of crystallization

As few magmas appear to be superheated above their liquidus T, the available heat for assimilation is derived by concurrent crystallization and cooling of the magma below its liquidus. Section 11.1.1 indicated that the mass of lower continental crustal rock melted by a mass of intruded hotter basaltic magma is of the same order of magnitude, more of this magma would be required in the cooler shallower crust. Obviously, hotter, more mafic magmas have greater assimilative potential. But transfer of heat from a magma body into adjacent cooler rock leads to solidification at its contact, building an armor of solid magmatic rock that inhibits further assimilation. Pieces of stoped country rock (Section 9.4.3) within a body of magma afford a significantly greater surface area over which heat can be transferred and assimilative processes operate than the country rock. Volatile fluids liberated from heated country rock may contaminate volatile-poor magma with Si, K, Na, and other elements, as the fluid solution is absorbed into it.

The fate of assimilated crystals depends on their composition and that of the parent melt. Provided sufficient heat is available, a crystalline phase dissolves if the silicate melt is not already saturated with respect to that phase. Thus, quartz xenocrysts can dissolve on a time scale of days in basaltic melts in which the activity of silica is <1 (the usual case); melts so contaminated are enriched in silica. Alkali feldspar, biotite, and hornblende in granite assimilated by basalt follow a similar fate, but the details differ. Assimilation of granite in basalt magma promotes crystallization of some of the phases it would have normally precipitated and along similar lines of liquid descent; however, felsic derivatives are more abundant. Crystals react with a melt if they would have precipitated from the magma at higher T. Thus, physically ingested crystals of Mg-rich olivine into Makaopuhi basalt magma at $T = 1075°C$ and 1 atm (Plate III), where olivine is no longer stable, would induce precipitation of additional stable pyroxene by a reaction relation with the melt. Xenocrysts of clinopyroxene—perhaps derived from incorporated basalt xenoliths—in a granodiorite melt precipitating stable hornblende but not pyroxene would be expected to react with the melt, forming, by ionic diffusion, a reaction rim of hornblende surrounding and possibly eventually replacing the unstable clinopyroxene. Assimilation of quartz xenocrysts, perhaps from ingested blocks of sandstone, into a quartz-saturated granitic melt simply adds more modal quartz to the final granite. Assimilation of Al-rich minerals into basalt magma stabilizes calcic plagioclase at the expense of calcic clinopyroxene, so that leucocratic orthopyroxene gabbro (norite) magmas might form.

Evidence for magma contamination in the history of a rock is generally only permissive. The presence of xenocrysts (e.g., quartz in basalt, Figure 6.20) and xenoliths in a magmatic rock may suggest they are contaminants, but xenocrysts can also originate by mixing of dissimilar magmas, and foreign material can be incorporated late into the magma with minimal contamination of the melt. Strained xenocrysts that show undulatory optical extinction under cross-polarized light in thin section or other solid-state strain effects are especially useful in distinguishing assimilated solid material from phenocrysts in mixed magmas.

Sr and O isotopic signatures (as well as Nd and Pb) can also provide permissive evidence for assimilation of continental felsic crust into primitive mantle-derived basaltic magmas. Precambrian felsic rocks have relatively low Sr but high Rb; over time, ^{87}Rb decays and elevates the ^{87}Sr/^{86}Sr ratio (Section 2.6.2) to well above that of primitive mantle-derived partial melts, which are typically about 0.703–0.704 (Figures 2.26 and 2.27). If mantle-derived magmas assimilate old continental crust, their ratio is elevated. ^{18}O in mantle source rocks and their basaltic partial melts is ~6‰, in contrast to sedimentary rocks, in which δ^{18}O = 10‰−32‰ (Figure 2.24; Section 2.6.1). Consequently, mantle-derived magmas that assimilate sedimentary rocks are enriched in ^{18}O. Mantle-derived magmas that assimilate old felsic rocks *and* sedimentary rocks (or their metamorphic equivalents) are enriched in both ^{87}Sr and ^{18}O (Figure 12.13).

✳12.4 DIFFERENTIATION IN BASALTIC INTRUSIONS

Because of their relatively low viscosity and slow cooling, intrusions of basaltic magma potentially provide an opportunity to evaluate the role of crystal-melt fractionation in magmatic differentiation. In this section, three types of intrusions are examined to determine the extent to which fractionation does occur and the nature of the evolved fractionated magmas and whether intrusions behave as closed systems or other differentiation processes are also involved. If they can be accurately interpreted, these intrusions provide tests of petrologists' theoretical models and small-scale, time-restricted laboratory experiments on crystal-melt systems. Knowledge gained from mafic intrusions constrains models purporting to account for compositional variations in extruded lavas that erupted sequentially from a particular volcanic center and possibly derived from underlying differentiating storage chambers.

12.4.1 Palisades Sill

Differentiation in smaller subhorizontal sheets is considered first, using the Palisades sill as an example. The diabase-granophyre association in such sheets was the basis for the emphasis by Bowen (1928) that closed-system crystal-melt fractionation is the most important process of magmatic differentiation.

The Palisades diabase sill is a subhorizontal sheet intrusion exposed for at least 80 km along the Hudson River opposite New York City. It is one of ennumerable early Jurassic basaltic dikes and sills intruded along the east coast of North America during the breakup of the American and African continental plates (Figure 9.7b; see also Olsen, 1999). Whether or not it is everywhere a concordant sheet does not detract from the fact that it, like many others, displays more or less regular variations in bulk chemical, modal, and mineral composition and fabric throughout a vertical extent of hundreds of meters (Figure 12.14). Inward solidification of the sill progressed to a "sandwich horizon" near the top of the sill where the most evolved residual magma crystallized to patches of granophyric pegmatite. In this felsic differentiate, grain size is uneven, ranging

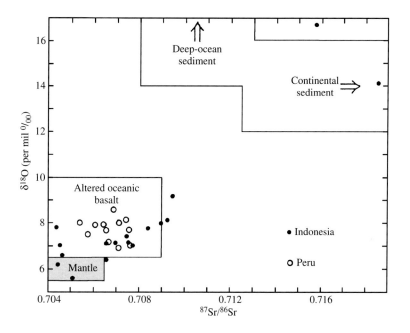

12.13 Oxygen and Sr isotopic composition of andesites from the Banda arc, Indonesia, and the Andean arc, southern Peru. Note two Indonesia samples in upper right. Range of composition of potential sources shown. (Redrawn from Magaritz et al., 1978.)

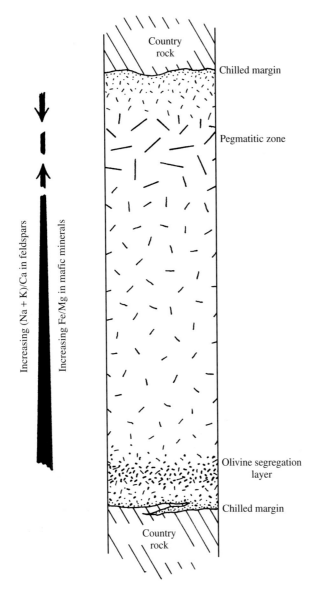

Increasing (Na + K)/Ca in feldspars

Increasing Fe/Mg in mafic minerals

Country rock

Chilled margin

Pegmatitic zone

Olivine segregation layer

Chilled margin

Country rock

12.14 Schematic cross section through an idealized diabase-granophyre sill. Lengthening dash lines indicate increasing grain size into pegmatitic granophyre zone.

from the micrographic quartz-alkali feldspar intergrowth typical of granophyre (Figure 7.21) to bladelike Fe-rich clinopyroxenes several centimeters long. An olivine-rich layer (10%–25% of rock) having abundant pyroxene is located 10–25 m above the floor of the sill.

The Palisades sill was interpreted for decades to be a simple product of closed-system differentiation driven by gravity settling of denser olivines and pyroxenes to the intrusion floor. However, several investigations in the late 1900s (see, for example, Gorring and Naslund, 1995) showed that this hypothesis is flawed. Studies of this and other similar differentiated sills indicate episodic recharge of fresh magma into the crystallizing fractionating bodies: That is, they were open magma systems.

The simple gravity settling hypothesis for the Palisades, wherein precipitating olivine crystals "rain down" through the crystallizing magma to form the olivine-rich layer, is rejected for several reasons:

1. The olivine-rich layer pinches and swells along the exposed extent of the sill and is locally absent.

2. The proportion of olivine increases abruptly into the layer from essentially zero above it, implying an unlikely very efficient gravitational settling "sweep" from above.

3. Olivines in the layer are far more Fe-rich (Fo_{55-70}) than expected from the olivine composition (Fo_{80}) in the finer grained, chilled margin, presumed to represent the quenched parent magma. Moreover, olivines in this chilled margin are embayed and were presumably unstable when the magma solidified.

4. Local internal "chilled" contacts within the sill and significant reversals or discontinuities in elemental trends in vertical sections (Figure 12.15) suggest multiple injection of additional magma into the fractionating magma system. One of these fresh draughts of magma is hypothesized to have carried a large proportion of suspended olivine that, during emplacement into the flat sheet, was smeared out as the olivine-rich basal layer.

The composite intrusion hypothesis seems intuitively probable because during the calculated several-hundred-year solidification time of the sill (Section 8.4.1) new ascending batches of magma feeding the overlying flood lavas may have intersected the horizontally widespread sill.

Despite the apparent magma replenishment and mixing, the Palisades magma has an overall vertical elemental variation that is readily explained by fractionation of plagioclase and clinopyroxene, the major minerals present throughout the sill. But the exact physical mechanism of the fractionation is uncertain. Because of possible restrictions imposed by the plastic yield strength of the melt (Section 8.2.2), gravitational settling of independent crystals of plagioclase and clinopyroxene through the magma may have been limited. But if this restriction was valid, how does one explain the highly asymmetric position of the late felsic differentiates near the top of the sill (Figure 12.14)? This seems to imply that gravity acted in some way. Perhaps large clumps of gravitationally unstable crystals growing near the roof fell to the floor of the crystallizing sill, augmenting the mat of crystals growing there. Buoyant residual melt could have escaped from the interstices of the mat of plagioclase and pyroxene grains, as a result of compaction under its own weight (filter pressing), or convective melt fractionation, or both. Displaced residual melt apparently accumulated near the roof of the sill, beneath downward crystallized magma, in the sandwich horizon, there forming the most-evolved Fe-rich granophyre.

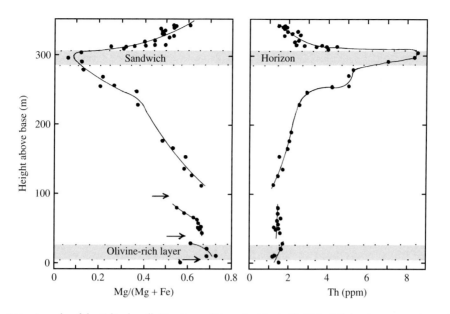

12.15 Chemical variation in rocks of the Palisades sill, New Jersey. Discontinuities in Mg/(Mg + Fe) ratio at arrows is consistent with influx of new, less differentiated magma into the sill. (Data from Shirley 1987.)

12.4.2 Layered Intrusions

Layered mafic intrusions have attracted the attention of petrologists for many decades (Cawthorn, 1996). Most of the large intrusions are of Precambrian age (Table 12.5), and the largest of these is the colossal Bushveld complex in South Africa which crops out over an area of about 65,000 km²; the estimated 0.5-million km³ of basalt magma that formed this 2-Ga intrusion was no doubt related to the head of a decompressing mantle plume. The volume of smaller intrusions, such as the early Tertiary Skaergaard intrusion in southeast Greenland (170 km³), is comparable to that

of some floods of plateau-forming basalt lava (Section 10.2.2).

<u>Layering and Cumulus Fabric.</u> Layering in these intrusions (Figures 7.43–7.45) is as pervasive and distinctive as stratification in sedimentary sequences. Single layers range from millimeters to hundreds of meters in thickness and from meters to tens, and even hundreds, of kilometers in lateral extent. In the Bushveld complex, the Merensky Reef—a pyroxenitic layer 1 to 5 m thick that is the chief global resource of platinum—extends along strike for nearly 150 km in the eastern part of the complex and 190 km along strike in the western part,

Table 12.5.   Large Layered Intrusions

NAME, LOCATION	AGE (MA)	REMARKS
Skaergaard, Greenland	55.7	55 km²; >3.5 km thick
Rum, Scotland	61–58	115 km²; >2 km thick
Duluth, Minnesota	1100	More than a dozen layered intrusions exposed over 5000 km²
Muskox, Northwest Territories, Canada	1270	Canoe-shaped, 11 × 150 km (Figure 12.17); associated with the Mackenzie dike swarm (Figure 9.7a)
Sudbury, Ontario	1850	1100 km²
Bushveld, South Africa	2050	Largest on Earth: 65,000 km² and 7–9 km thick; exceptional lateral continuity of individual layers
Jimberlana, Western Australia	2370	End-to-end canoe-shaped complexes averaging 1.5 km wide and 180 km long underlain by connecting dike
Great Dyke, Zimbabwe	2460	Four end-to-end canoe-shaped layered complexes 4–11 km wide and 550 km long underlain by a connecting feeder dike
Stillwater, Montana	2700	8 × 55 km and ~7 km thick; only basal ultramafic cumulates and overlying gabbroic and anorthositic rocks preserved
Windimurra, Western Australia	2800	35 × 85 km in area and 5–13 km thick; only slight differentiation

Data from Hatton and von Gruenewaldt (1990) and Cawthorn (1996).

about 300 km distant. Layering is most conspicuously defined by variations in relative proportions of minerals, defining modal layering. Gradational variations within a single layer, from top to bottom, may be obvious, forming graded modal layers. Some layers are graded in grain size. Rhythmic layering, defined as a sequence of recurring similar layers, is common.

Another aspect of the fabric of layered intrusions parallels that of clastic sedimentary deposits. In sandstones, a textural distinction between accumulated sand grains and secondary cement surrounding them is generally obvious. In cumulate rocks of layered intrusions, a similar distinction may be obvious between collected early formed **cumulus grains** and **intercumulus** mineral matter that formed later around them. This magmatic "cement" filling space between cumulus grains was precipitated from the intercumulus melt initially entrapped during accumulation of cumulus grains or from a later, modified melt percolating through the cumulate pile (Figure 12.16). On the floors of magmatic intrusions, cementation and overgrowth compete with compaction in "densifying" the cumulus aggregate, eliminating the intercumulus melt (Hunter, 1996). For a recent discussion of the validity of the cumulate paradigm see Morse (1998).

Cumulates on floors of layered intrusions do not necessarily reflect gravitational settling, as once widely believed. For example, the Archean Stillwater complex of Montana has layers of cumulus chromite grains (about 0.25-mm diameter, density 4.4 g/cm^3) near the base that grade upward into larger, less dense cumulus olivine grains (0.7–3.0 mm, 3.3 g/cm^3). As settling velocity, according to Stokes's law (Section 8.3.3), is proportional to the square of the particle size but only to the first power of the density contrast with the melt, the larger olivines should have settled faster than the smaller but more dense chromites. The modal grading, therefore, appears to be upside down. Jackson (1961) concluded from this and other observations that in situ bottom crystallization, not gravitational settling, created this layering. Kinetic factors may also play a role in development of modal layering (Section 7.9.2; Figure 7.46).

Evidence for upward migration of melt through the cumulus floor pile can be seen in mineral compositions in the Muskox intrusion (Figure 12.17). This layered intrusion, like the Stillwater, Duluth, Bushveld, and Great Dyke intrusions (Table 12.5), formed from episodic replenishment of basaltic magma during its crystallization. This open-system recharge is usually evident in upward reversals in compositional variation (Figure 12.18) and is logical in view of the multiple basalt dikes, sills, and lava flows spatially and temporally associated with the intrusions. Discontinuities in the chemical composition of cumulus chromite and olivine occur *up section* from the modal break marking

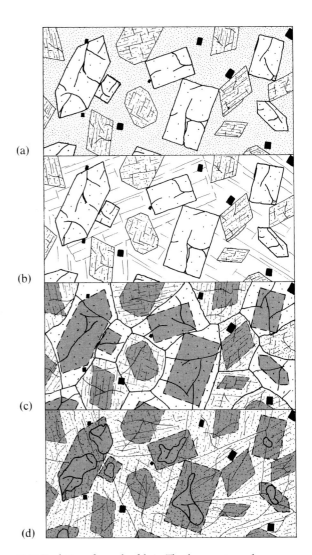

(a)

(b)

(c)

(d)

12.16 Evolution of **cumulus** fabric. The three **postcumulus** processes illustrated in (b–d) are ideal end members; real **cumulate rocks** generally form by some combination of these processes. See Hunter (1996) and Morse (1998) for additional discussion of cumulus textures. (a) Original cumulus grains of olivine (bold relief), pyroxene (with cleavage), and chromite (small black) in a melt (stippled). (b) Postcumulus **cementation** and development of poikilitic texture. The intercumulus melt surrounding the cumulus grains in (a) crystallized as large plagioclase grains, filling the intercumulate pore volume. Extraneous ions in the melt (Fe, Mg, Ti, etc.) not required by the growing plagioclase must be transferrred out of this space, perhaps by ionic diffusion. (c) Postcumulus **overgrowth** and development of equilibrated texture with approximately 120° triple-grain-boundary junctions. Intercumulus melt surrounding the cumulus grains (shaded areas enclosed by dashed outlines) in (a) crystallized as secondary enlargements on the cumulus olivines and pyroxenes until all pore space was eliminated. Plagioclase either was unstable or did not nucleate. Monomineralic dunites, pyroxenites, and anorthosites can originate by secondary enlargement of accumulations of olivines, pyroxenes, and plagioclases, respectively. Ions not needed during enlargement are transferred out of the local interstitial volume by some means. (d) Postcumulus **reaction replacement** combined with overgrowth. Intercumulus melt reacted with cumulus grains (shaded areas enclosed by dashed outlines) in (a), partly consuming olivines. Simultaneous secondary enlargement of the cumulus pyroxenes eliminated all pore spaces.

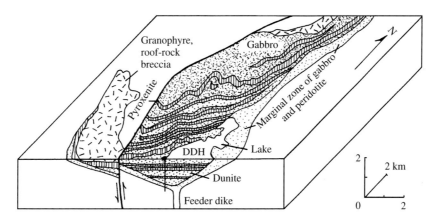

12.17 Schematic block diagram of the central segment of the Muskox layered intrusion, Northwest Territories, Canada. Regional tilting to the north of about 5° exposes most of this 1270-Ma trough-shaped intrusion. Feeder dike 150–500 m wide of picrite and gabbro extends tens of kilometers southward. The intrusion is as much as 11 km wide and at least 150 km long before disappearing under covering roof rock. Above the troughlike marginal zone of gabbro and peridotite is a 1800-m-thick series of 42 mappable layers individually ranging in thickness from 3 to 350 m that are predominantly dunite with a lesser proportion of pyroxenite. This layered series is overlain by gabbro and then granophyre formed from the evolved residual magma. DDH, diamond drill hole, whose core has been investigated in detail (Figure 12.18). (Redrawn from Irvine, 1980.)

the cyclic unit boundary. Irvine (1980) explains this discrepancy between modal and chemical breaks on the basis of compaction of the cumulate pile and upward transfer of the intercumetersulus melt out of it. Migration of melts having lower Mg/(Mg + Fe) and Ni/(Ni + Mg + Fe) ratios from the underlying cumulate unit into the overlying cumulus olivines and chromites causes them to reequilibrate via reaction relations. The wave of reequilibration for Mg and Fe extends farther into the overlying cyclic unit than does that of Ni because the melt contains about a third as much Mg + Fe as the cumulus crystals but only about a tenth as much Ni. Accordingly, adjustments in Mg and Fe are more pronounced.

In large, slowly cooled intrusions, a substantial fraction of the intercumulus melt can be transferred via filter pressing or convective melt fractionation from the compacting floor cumulates back into the uncrystallized central part of the intrusion. This is an effective crystal-melt fractionation that has in the past been attributed to crystal settling through the whole chamber.

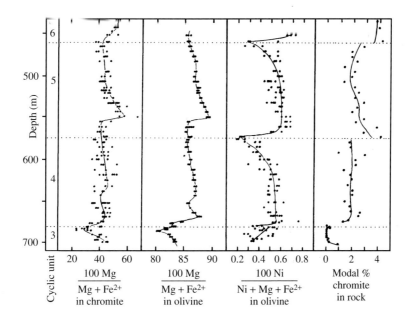

12.18 Analytical data on a part of the core from a drill hole into the Muskox intrusion (DDH in Figure 12.17). Boundaries of a cyclic unit are defined by discontinuities in modal proportion of cumulus chromite grains in right-hand panel. In the Muskox, each of the 25 cyclic units represents a magma recharge into the intrusion. Discontinuities in chemical composition of cumulus olivine and chromite occur above the modal boundaries. The top of unit 3 is chromite-free olivine clinopyroxenite; all other rock is chromite-bearing dunite. (Redrawn from Irvine, 1980.)

<u>Skaergaard Intrusion</u>. No discussion of layered intrusions would be complete without mention of the intensely studied Skaergaard intrusion, located in the remote southeast coast of Greenland just above the Arctic Circle. This intrusion and related dikes and overlying thick lava flows formed during the early Tertiary opening of the north Atlantic (see Figure 13.23). Ironically, the more that has been learned, from the earliest studies by Wager and Deer (1939) to later studies by A. R. McBirney and associates (e.g., 1996), the more controversy seems to develop (Irvine et al., 1998).

In contrast to many large differentiated sills and larger intrusions, the exposed part of the Skaergaard intrusion (Figures 12.19 and 12.20) reveals no unequivocal indication of multiple injections of magma that mixed with previously fractionated magma. For example, compatible element trends, such as for Ni, are continuous without major breaks (Figure 12.20). Therefore, the exposed part of the Skaergaard—above a hypothetical Hidden Zone—may have formed by fractionation in a closed magma system. Unfortunately, no unaltered "chilled" margin rock has been found that might represent the pristine, parent magma from which the intrusion fractionated.

Crystallization in the Skaergaard was dominantly upward from the floor, producing an uncertain amount of the Hidden Zone plus the exposed Layered Series of about 2500 m. This is much greater than the thickness of the downward crystallized Upper Border Series. These two series meet at the Sandwich Horizon, which represents the last-crystallized part of the intrusion,

and are enveloped within the still thinner Marginal Border Series, the part of the intrusion solidifying inward from the sidewalls. That this is indeed the pattern of inward crystallization of the intrusion is shown by variations in whole-rock chemical and mineral compositions and phase layering. **Phase layering** is the abrupt appearance or disappearance of a particular mineral, such as the vertical disappearance of olivine defining the lower boundary of the Middle Zone of the Layered Series and its reappearance in the Upper Border Series. **Cryptic layering** reflects systematic changes in the chemical composition of cumulus minerals, shown in the middle panel of Figure 12.20. Plagioclase becomes more sodic, to An_{25} in the Sandwich Horizon, from as much as An_{69} in the uppermost part of the intrusion and An_{66} in the lowest exposed part. Olivine ranges from Fo_{68} to pure fayalite Fo_0 and clinopyroxene to Mg-free hedenbergite in the Sandwich Horizon. These cryptic variations are more or less symmetric inward to the Sandwich Horizon, as are whole-rock chemical compositions (Figure 12.21).

The extreme enrichment in Fe and limited late enrichment in silica and alkalies have made the **Skaergaard trend** a reference standard for fractionation of tholeiitic basalt magma (Table 12.6; Figures 12.21 and 12.22). However, the physical mechanism of fractionation that operated in the Skaergaard is still controversial. Two competing hypotheses are briefly summarized in the following two paragraphs (see also Special Interest Box 8.1). The debate is yet another example that experienced petrologists can arrive at markedly contrasting opinions about the same rock.

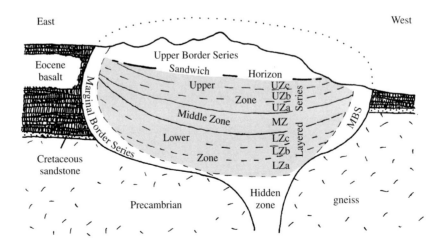

12.19 Schematic cross section through the Skaergaard intrusion, Greenland. Emplaced at 55.7 Ma, the intrusion measures about 6 × 11 km on the ground and, because of a regional northward tilt of 15°–25°, the 1200 m of glacially carved topographic relief exposes an almost continuous 3500-m stratigraphic section through the body. Configuration of the deeper unexposed part of the intrusion, the Hidden Zone, is based on geophysical measurements. The intrusion can be conveniently divided into four parts (see also Figure 12.20): (1) Marginal Border Series (MBS) consisting of an inner Layered Series and an outer Tranquil Zone; (2) Thick Layered Series (shaded), consisting, in ascending order, of the Lower Zone, which is divided into three parts (LZa, LZb, LZc); the Middle Zone (MZ); and the Upper Zone, which is divided into three parts (UZa, UZb, UZc); (3) Upper Border Series (UBS); (4) last crystallizing Sandwich Horizon. (Redrawn from McBirney, 1993.)

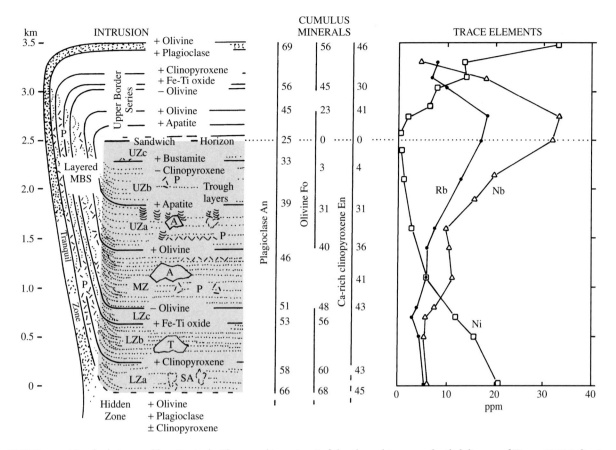

12.20 Compositional relations and layering in the Skaergaard intrusion. Left-hand panel is a more detailed diagram of Figure 12.19 indicating **phase layering,** which delineates the successive layer units (e.g., LZb, LZc, MZ). Also indicated are correlations of crystallization stages in the Layered, Marginal Border, and Upper Border Series. For example, after early crystallization of the texturally isotropic Tranquil Zone against the wall rock, an upper layer of the Upper Border Series crystallized concurrently with an outer layer of the Marginal Border Series (MBS) and the lowermost layer of the Layered Series, LZa. Modal graded layering is represented schematically by dotted lines. Note cross-bedding in modal layers at the junction of the subhorizontal Layered Series and the steeply inclined Marginal Border Series and trough layers in the UZa layer. P, schematically indicated patchy and strata-bound masses of coarse-grained mafic pegmatite; SA, schematically indicated masses of secondary anorthosite; T and A, respectively, schematically indicated blocks of troctolite (olivine + plagioclase rock) and anorthosite stoped off the roof of the intrusion. There are many more of SA, T, and A than shown. (Redrawn from McBirney, 1996; Irvine et al., 1998.)

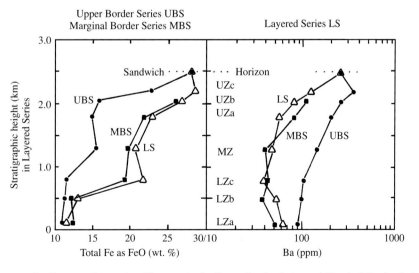

12.21 Chemical variations in the Skaergaard intrusion. Elements in the Upper Border Series and Marginal Border Series are plotted with respect to equivalent unit in Layered Series. (Redrawn from McBirney, 1996.)

Table 12.6. Average Compositions of the Units in the Layered Series, Sandwich Horizon (SH), and Granophyre, Skaergaard Intrusion

	LZ$_A$	LZ$_B$	LZ$_C$	MZ	UZ$_A$	UZB	UZ$_C$	SH	Grano
SiO$_2$	48.12	48.84	41.10	42.79	43.07	41.78	46.00	49.43	60.23
TiO$_2$	1.35	1.44	6.92	6.79	5.67	4.06	2.63	2.23	1.18
Al$_2$O$_3$	16.81	12.55	11.02	11.53	11.17	9.51	7.86	7.93	11.29
FeOt	11.13	12.84	21.10	20.00	22.52	26.64	28.67	27.87	14.08
MnO	0.16	0.21	0.26	0.26	0.31	0.41	0.65	0.25	0.24
MgO	9.42	10.13	7.61	6.24	5.62	3.41	0.38	0.09	0.51
CaO	10.11	11.57	9.77	9.87	8.62	9.36	10.14	8.23	5.11
Na$_2$O	2.52	2.13	1.97	2.23	2.55	2.59	2.42	2.72	3.92
K$_2$O	0.27	0.20	0.20	0.21	0.26	0.36	0.41	0.72	1.94
P$_2$O$_5$	0.11	0.09	0.05	0.08	0.22	1.88	0.84	0.53	0.27
Sr	285	219	199	218	233	244	263	450	
Zr	93	81	72	80	95	97	135	324	
La	5.62	4.68	3.84	3.09	3.49	14.57	22.68	57.8	
Sm	2.47	3.08	2.34	2.10	2.10	10.00	15.27	34.5	

Total oxides are 100.00.
Data from McBirney (1996).

In their classic study, Wager and Deer (1939) postulated that convection currents were responsible for the rhythmic modal layering, trough layering, igneous lamination, and local lenticular and cross bedding and slump structures. They envisaged the Layered Series as having formed by gravitational settling and sorting of crystals from convection currents that descended from the walls and swept inward across the cumulate floor. Depending on the vigor of convection, modally graded layers would be thin or thick or rhythmic. Irvine et al. (1998) reaffirmed and refined the convection-driven sedimentation hypothesis, documenting in considerable detail what they believe to be supporting evidence.

Beginning in the late 1970s, **magmatic sedimentation** via convection in the Skaergaard was rejected by McBirney and coworkers (1996), largely because plagioclases dominating the cumulate fabric should have floated, not sunk, in the increasingly denser Fe-rich melts. They appealed, instead, to fractional crystallization driven by compaction and convective melt fractionation within the pile of cumulate crystals on the intrusion floor. Because this fractionation is driven by gravity, there ought to be different fractionation effects in the Layered Series that accumulated on the floor versus in the Upper Border Series that was created near the roof. And both might differ from the Marginal Border Series. Figure 12.21 shows such differences. Upper Border Series rocks are depleted in Fe relative to Layered Series rocks because negatively buoyant Fe-rich residual melts, perhaps carrying some suspended crystals, drained out of the roof region and ponded in floor cumulates. Marginal Border Series rocks are slightly less Fe-rich than Layered Series rocks, again reflecting

draining out of Fe-rich melts, but not as efficiently as from the roof. The greater concentration of incompatible elements, such as Ba, in Upper Border Series rocks than in Layered Series is thought to reflect even later-stage, postpeak Fe-enrichment infiltration of buoyant incompatible-element-enriched residual melts derived from underlying magma. These late buoyant melts were relatively enriched in silica and alkalies and crystallized largely as scattered patches and dikes of granophyric rock throughout the intrusion but especially in the Upper Border Series. Some late mafic granophyres may have formed from immiscible melts. Other granophyres may be fused blocks of quartzofeldspathic gneiss country rock.

Magmatic Replacement in Large Layered Intrusions. Meter-scale, patchy masses of rock in the Skaergaard and other intrusions appear to have formed by replacement of original rock, facilitated by local concentrations of water. These masses include secondary anorthosite and mafic pegmatite that are commonly adjacent to one another (SA and P, respectively, in the left panel of Figure 12.20). The origin of the anorthosite (>90% plagioclase) by secondary, volume-for-volume metasomatic replacement is indicated by undisturbed continuity of layering from adjacent rock through the anorthosite. Irvine et al. (1998) suggest that locally higher concentrations of water in the intercumulus melt shift the olivine-pyroxene-plagioclase cotectic toward plagioclase, locally precipitating more of this phase and displacing mafic minerals into peripheral areas. Unusually coarse-textured gabbroic rock, or mafic pegmatite, some having additional biotite and amphi-

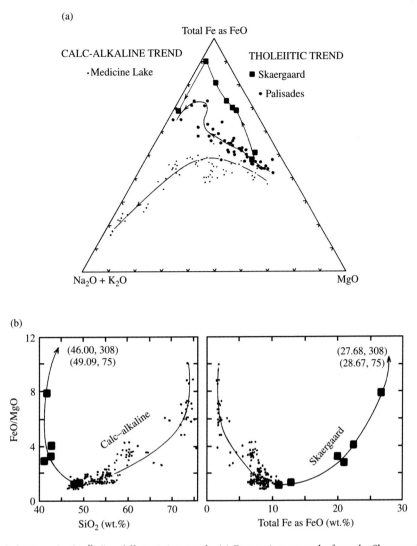

12.22 Comparison of tholeiitic and calc-alkaline differentiation trends, (a) Data points are rocks from the Skaergaard intrusion (data from McBirney, 1996; see also Table 12.6), the Palisades sill (data from Shirley, 1987), and the Medicine Lake volcanic field (data from Grove and Baker, 1984). (b) Palisades rocks are not plotted. The line with arrow through the Skaergaard rocks points in direction of more evolved rocks, the two most Fe-rich lying off the diagram at the indicated coordinates.

bole that are absent generally from the cumulates, is believed to have crystallized from locally more hydrous and buoyant intercumulus melts rising diapirically through the cumulus pile. The source of the water may be the intercumulus melt itself or perhaps water absorbed from country rock.

The origin of widespread thick layers of anorthosite in some layered intrusions, as much as 600 m in the Stillwater, is unresolved but cannot be due to secondary processes.

Meter-thick layers ("reefs") of pegmatitic pyroxenite in the Bushveld and Stillwater are major world repositories of platinum-group elements (PGEs), consisting of Pt, Pd, Rh, Ir, Ru, and Os. The traditional view has been that the PGE-bearing minerals were concentrated as dense immiscible sulfide melts on the intrusion floor. However, the intimate association of

the PGE minerals with pegmatite and hydrous minerals as well as exceptionally high concentrations of rare earth elements (REEs) in pyroxenes are interpreted in terms of redistribution of interstitial melt driven by compaction of the cumulate layers (Mathez et al., 1997).

12.4.3 Oceanic-Ridge Magma Chambers

After polybaric fractionation of olivine during their ascent from the mantle source, primary mid–ocean ridge basalt (**MORB**) magmas experience further differentiation at low P in crustal storage chambers beneath the rift.

Intuitively, it seems unlikely that the crustal chambers filled with magma would remain perfectly closed systems throughout their entire history of solidification. In the actively spreading rift environment, fresh draughts of more primitive magma would be expected to be episodically injected into the chamber of fractionating magma from the underlying mantle source (Figure 12.12). Three lines of evidence were found by early investigators (e.g., Rhodes et al., 1979) indicating chamber replenishment and mixing:

1. Disequilibrium phenocryst-melt compositions and textures: Many larger olivines and especially plagioclases have corroded, spongy cores (Fo_{85-90} and An_{83-86}) that are surrounded by more evolved crystalline material.

2. Glass inclusions within the high-*T* cores have compositions similar to the most primitive MORB and unlike the glass in the surrounding matrix.

3. Element variation trends follow those anticipated by fractionation of major phenocrystic phases combined with mixing of replenishing draughts of primitive magma (Figures 12.23–12.25).

Thus, the limited compositional variation in most MORBs reflects recurring replenishment and mixing of relatively primitive mantle-derived magmas with fractionating magma in rift chambers, moderating the effects of crystal-melt fractionation.

However, less common seafloor basalts found locally along the Galapagos rift and other sites on the East Pacific Rise follow a tholeiitic fractionation trend of extreme Fe-enrichment and limited enrichment in silica and alkalies resembling the Skaergaard trend (Figure 12.22). Apparently, in local more closed-system chambers, 60%–80% fractionation of olivine, plagio-

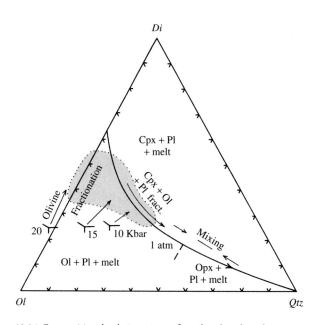

12.24 Compositional relations in seafloor basalts plotted in terms of normative *Di*, *Ol*, and *Q*. Plagioclase and/or spinel is stable in all assemblages. Experimentally determined beginning-of-melting invariant points in peridotite source rock at 20, 15, and 10 kbar, and 1 atm. Curved boundary line on 1 atm liquidus. Shaded area encompasses nearly 2000 analyses of basalts that are interpreted to have formed by polybaric fractionation of olivine from primary melts generated at 10–20 kbar. Subsequent low-*P* fractionation of plagioclase, clinopyroxene, and olivine from these primitive parent magmas within crustal magma chambers yields evolved magmas extending down the boundary line. Many of these derivative magmas have been modified by mixing with more silica-rich magmas. Significant enrichment in Fe in derivative magmas cannot be represented in this diagram. (Redrawn from Natland, 1991.)

clase, and lesser clinopyroxene produces ferrobasalts (13–16 wt.% total Fe as FeO) and less common ferrodacites (5–11 wt.% FeO, 64–71 wt.% SiO_2).

✳12.5 ORIGIN OF THE CALC-ALKALINE DIFFERENTIATION TREND

Several variation diagrams have been employed to distinguish between the tholeiitic differentitation trend of Fe enrichment with limited felsic derivatives and the calc-alkaline trend of limited Fe enrichment and abundant felsic derivatives (Figures 2.17 and 12.22; see also Miyashiro, 1974). The tholeiitic rock suite is typical of differentiated basaltic intrusions, as just described, as well as island arcs, whereas the calc-alkaline suite is typical of continental margin arcs.

These contrasting rock suites and their associated magmatic differentiation trends have been recognized, and their origin debated, for many decades. Prior to the mid-1900s, C. N. Fenner believed that crystallizing basaltic magmas evolved along a trend of Fe enrichment at nearly constant silica concentration, now des-

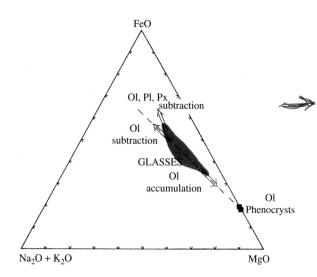

12.23 Crystal-melt fractionation in MORB magmas. Analyzed glasses from Bryan and Moore (1977) plot in shaded area. Filled squares are analyzed olivine (Ol) phenocrysts and dashed line is olivine control line. Accumulation or subtraction of olivine produces most of the compositional variation in the glasses, whereas limited fractionation of plagioclase (Pl) and pyroxene (Px) from low-MgO magmas shifts the trend slightly toward more FeO-rich compositions.

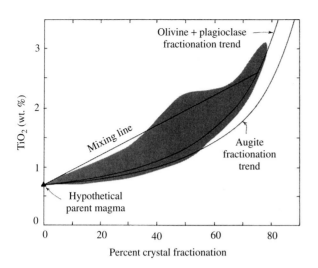

12.25 Effects of crystal-melt fractionation and magma recharge and mixing on the TiO$_2$ content of seafloor basalts. A hypothetical parent magma having 49.9 wt.% SiO$_2$ and 0.7 wt.% TiO$_2$ from which olivine, plagioclase, and augite of realistic compositions are fractionated produces residual melts along the curved fractionation-trend lines. Intermittent mixing of such residual melts with additional draughts of parental magma would produce magmas whose compositions lie within the shaded lens-shaped area bounded by the mixing line and the curved fractionation trend lines. The shaded area represents over 600 chemically analyzed seafloor glasses. (Redrawn from Rhodes et al., 1979.)

ignated the *tholeiitic trend* and exemplified by the extreme Fe enrichment of the Skaergaard intrusion. On the other hand, N. L. Bowen advocated a trend of limited Fe enrichment and greater production of felsic end products, exemplified by calc-alkaline rocks. It is now realized that there is a range of differentiation processes that can operate on primitive basaltic magmas, producing a continuous spectrum of rock suites and trends whose end members are the Skaergaard and the calc-alkaline trends. It should be recalled that this continuum and its two end members constitute the subalkaline suite (Section 2.4.5), which accounts for about 90% of global magmatism (Figure 1.1); hence, contrasts in the evolution of tholeiitic and calc-alkaline magmas merit special attention.

12.5.1 Tonga–Kermadec–New Zealand Arc

This continuous arc in the southwest Pacific Ocean developed above the subducting Pacific oceanic plate during the late Cenozoic; the volcanic rocks in the arc provide one of the clearest examples of contrasting tholeiitic and calc-alkaline suites or trends (Figures 2.17 and 2.18). Tholeiitic rocks in the youthful Tonga-Kermadec arc rest on oceanic crust, whereas calc-alkaline rocks in the along-strike Taupo volcanic zone of the North Island of New Zealand rest on continental crust (Figure 12.26). New Zealand obviously harbors a greater volume of silicic rocks than mafic-

intermediate composition rocks, whereas the reverse is true of the oceanic Tonga-Kermadec part of the arc. Relatively primitive basalts in the two arc segments have similar trace element concentrations and patterns, suggesting a similar ancestry in the subarc mantle wedge. However, the Taupo basalts have more radiogenic Nd and Sr isotopic ratios that are consistent with assimilation of an older sedimentary component resembling metamorphosed Mesozoic sedimentary rocks exposed to the east of the volcanic zone.

If the 16,000 km^3 of extruded rhyolite magma in the Taupo volcanic zone originated wholly by fractionation of parental basalt magma, about 10 times as much basalt magma would have been required. This questionable volume plus the relative minor amounts of contemporaneous intermediate andesite and dacite make such an origin unlikely. On the other hand, the high crustal heat flow and intense geothermal activity associated with the voluminous rhyolite volcanism suggest a major heat source in the crust; this is speculated to be underplated basalt magma from the mantle wedge. Compared to fractionation origin, rhyolite magma generation by partial melting of older continental rocks requires a significantly smaller mass of basalt magma, approximately equal to that of the lower crustal rocks being melted (Section 11.1.1). That the erupted basaltic magmas in the Taupo zone appear to show crustal contamination supports the hypothesized model.

12.5.2 Factors Controlling Development of the Calc-Alkaline Trend

Water is an essential component in arc magma generation, evolution, and typical explosive behavior. Water promotes partial melting in the hydrated mantle wedge overlying subducting wet oceanic crust. Subsequently, these hydrous primitive magmas evolve along distinct trends. High water fugacities stabilize amphiboles and biotite and depolymerize the melt, expanding the stability field of olivine while restricting plagioclase. Enhanced olivine crystallization reduces Mg, Fe, Ni, and Cr concentrations in residual melts. Restriction of plagioclase precipitation to calcic compositions, An$_{70-90+}$, enriches residual melts in Na and K. These factors are some of those responsible for development of the felsic calc-alkaline trend, rather than the Skaergaard trend of increasing Fe.

In continental arcs, at least three possible factors, not present in island arcs, might modify primitive magmas derived from the mantle wedge, producing greater proportions of felsic calc-alkaline rocks:

1. Sediment eroded off a high mountain terrain and deposited in the adjacent ocean trench, as well as tectonic slices carved off the edge of the overriding plate, may be subducted with the oceanic crust. This continental material may be partially melted,

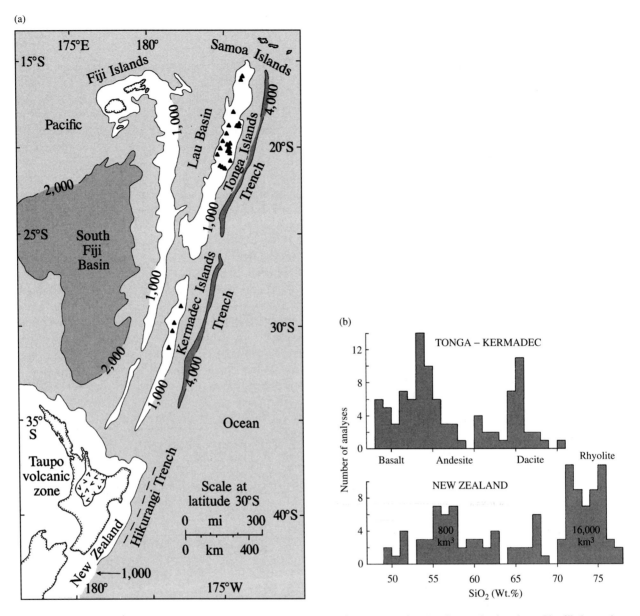

12.26 Tonga–Kermadec–New Zealand volcanic arc and arc rocks in the southwestern Pacific. (a) Volcanic islands indicated by filled triangles. Bathymetric contours are in fathoms (1 fathom = 1.83 m). For a more detailed, larger-scale map of the North Island of New Zealand see Figure 13.27. (b) Frequency distribution of silica in analyzed rocks from the Tonga-Kermadec island arc and from the North Island of New Zealand. Tonga is chiefly basalt and andesite; sparse dacite is overrepresented. In New Zealand, rhyolite is estimated to be about 20 times more voluminous than andesite. (Redrawn from Ewart et al., 1977.)

enhancing production of felsic magmas in the sub-arc environment.

2. Subcontinental lithospheric mantle may have been static (nonconvecting) and metasomatically enriched over long periods; ascending magmas scavenge these enrichments.

3. Contributions from the continental felsic crust itself are probably important. Where the continental crust is thicker there is a tendency for a greater proportion of K_2O at a given SiO_2 content, greater proportion of felsic rock types (dacite and rhyolite), and higher proportion of K_2O/Na_2O in andesite

(Figure 12.27); these compositional parameters indicate that the longer the path through which magmas rise in their journey to the surface of the continental crust the more they take on felsic continental characteristics. The first factor listed is independent of the path length and, thus, probably less significant. The "scatter" in the global data of Figure 12.27 likely reflects the compositional heterogeneity of the continental crust and other irregularities in differentiation processes.

The greater the path length that ascending magmas take through the continental crust the greater the op-

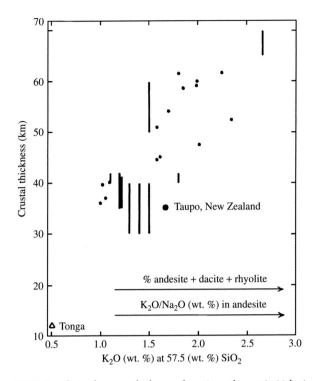

12.27 Correlation between thickness of continental crust (>30 km) and compositional parameters. The graphed parameter is the concentration of K_2O at 57.5 wt.% SiO_2 from a best-fit line through a K_2O versus SiO_2 wt.% variation diagram for a rock suite; for example, values of K_2O at 57.5 wt.% SiO_2 for the New Zealand suite and, for comparison, the Tonga suite are taken from Figure 2.18. Bars in diagram indicate variable crustal thickness for the K_2O value. Two other parameters not graphed also show a positive correlation with crustal thickness; these are the K_2O/Na_2O in andesite (57–63 wt.% SiO_2) and the proportion of andesite + dacite + rhyolite to basalt. (Data from Leeman, 1983; Hildreth and Moorbath, 1988.)

The first three parameters are commonly chosen to provide the best fit to the analytical data (Figure 12.28). This AFC model works best in cases in which the contaminant and parent magma compositions differ significantly.

The basalt-andesite-dacite-rhyolite suite in the Medicine Lake volcano in the northern California Cascade volcanic belt (Grove and Baker, 1984) is typical of many calc-alkaline suites. A general trend of increasing $^{87}Sr/^{86}Sr$ with increasing silica is consistent with progressive assimilation of felsic crust in mantle-derived magmas. Elevated incompatible element (e.g., Rb) concentrations in some evolved rocks indicate that fractional crystallization alone cannot be responsible, but combined fractionation and assimilation of felsic rocks can (Figure 12.29a). Mixing of basaltic and rhyolitic magmas also seems required for some rock types and is supported by mingled dacite and rhyolite lavas and by disequilibrium phenocryst assemblages in basaltic andesites and dacites; these assemblages include, in a single sample, Mg-rich olivine (Fo_{90}), calcic plagioclase (An_{85}), reversely zoned orthopyroxene with Fe-rich cores and Mg-rich rims, and reversely zoned plagioclase with andesine cores and labradorite

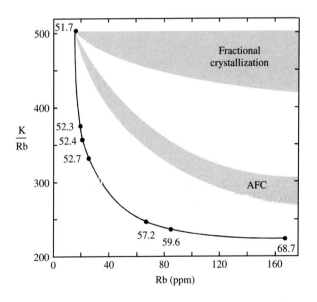

12.28 K/Rb versus Rb in the Volcanes Planchon-Peteroa-Azufre at 35°15'S latitude in the central Andes. Numbers alongside solid circles are weight percentage silica in rock samples that range from basalt to dacite. *For reference only* are approximate trends to be expected (1) if a parent magma, represented by the rock that has 51.7 SiO_2, were to evolve wholly by fractional crystallization of olivine, pyroxene, and plagioclase or (2) if fractional crystallization acted in concert with assimilation of granite (AFC). Note that fractional crystallization alone is inadequate to account for these diverse evolved rocks because of the similarity in partition coefficients of K and Rb. In contrast, assimilation of granite country rocks in concert with fractional crystallization is a more viable diversification process. (Redrawn from Hildreth and Moorbath, 1988.)

portunity for assimilation and fractional crystallization to produce felsic calc-alkaline magmas. Mantle-derived basaltic magmas intruded in the lower crust can assimilate felsic rock and mix with silicic partial melts generated by the heat from the crystallizing and usually fractionating basalt magma. The combined effects of simultaneous assimilation and fractional crystallization (**AFC**) on evolving magma systems were first formalized by Taylor (1980) and further quantified by De Paolo (1985) and Aitcheson and Forrest (1994). Their equations express isotopic ratios and element concentrations of hybrid daughter magmas to be expected from an assumed contaminant and parent magma based on the following:

1. Estimated bulk distribution coefficients
2. Ratio of the rates of assimilation to fractional crystallization
3. Remaining melt fraction
4. Effects of magma recharge and eruption

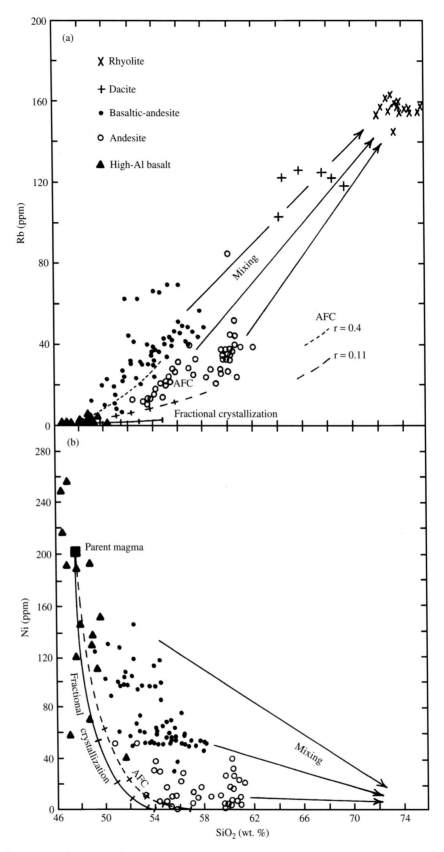

12.29 Evolution of Quaternary lavas at Medicine Lake, northern California, by fractionation, assimilation, and mixing. Fractionation of parental high-Al basalt magma modeled by extraction of olivine + plagioclase + clinopyroxene is shown by line with tick marks at intervals of 0.1 in melt fraction to $F = 0.5$. The fractionation model accounts for the observed variations in the high-Al basalts but not other magmas that must have originated by mixing of the evolved high-Al basalt magmas with rhyolite magma together with combined assimilation and fractional crystallization (AFC). The AFC model assumes the ratio (r) of the rate of assimilation of feldspathic sandstone and rhyolite to the rate of fractional crystallization was 0.11 (long dashes) or 0.4 (short dashes). Fractionation involving magnetite and hornblende could not reproduce the whole suite of lavas. (Redrawn from Grove et al., 1982.)

rims. Elevated compatible element (e.g., Ni) concentrations at a particular silica content in basaltic andesites and andesites were apparently produced by mixing of high-Al basalt and rhyolite magmas (Figure 12.29b).

The remarkably similar I-type granitic rocks of the Sierra Nevada batholith in California and the Coastal batholith of Peru are plotted in K_2O-Si_2O space in Figure 12.30a. These two plutonic calc-alkaline suites can be compared to the K_2O-Si_2O plot in Figure 12.30b, which shows trends of residual melts resulting from fractionation of the indicated minerals from parental basalt and andesite magmas. Parent basalt magma evolves toward andesite chiefly by fractionation of olivine. Fractionation of clinopyroxene, Fe-Ti oxides,

and hornblende from andesite magmas propels evolved melts along the calc-alkaline trend exhibited by the batholithic rocks in Figure 12.30a. Fractional crystallization of water-rich primitive magmas generated in the subarc mantle wedge causes residual melts to follow the calc-alkaline trend (Grove et al., 1982), in part by stabilizing hornblende, which is widely believed to be a major player in fractionating calc-alkaline magmas. Sidewall crystallization and accompanying convective melt fractionation (Figure 8.22) are probably important phenomena in production of calc-alkaline andesite-dacite-rhyolite magmas. Thick continental crust plays an important role in arresting or at least slowing mafic magmas during ascent, allowing them to fractionate.

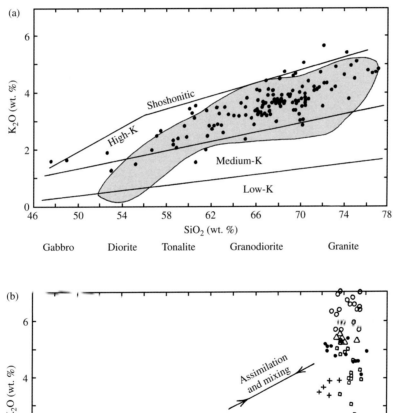

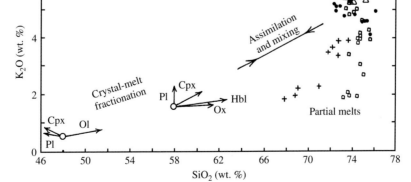

12.30 K_2O-SiO_2 relations in calc-alkaline plutonic arc rocks. (a) Composition of rocks in the Sierra Nevada batholith, California (filled circles; from data referenced in Bateman, 1992) and in the Coastal batholith of Peru. (Shaded area from Atherton and Sanderson, 1987.) (b) On the left, trends of residual melts resulting from fractionation of indicated minerals in parental basalt and andesite magmas (open circles). On the right, compositions of experimentally generated, dehydration partial melts of pelitic source rocks (open symbols), Pl + Hbl + Qtz rock (filled circles), and feldspathic sandstone (pluses). (Data from Conrad et al., 1988; Skjerlie and Johnson, 1993; Patiño Douce and Harris, 1998; Pickering and Johnson, 1998.) Calc-alkaline batholithic rocks can be created from fractionated basaltic and andesitic parental arc magmas that are mixed with partial melts of continental source rocks.

Thick continental crust also promotes assimilation of country rock into primitive magmas. If open magma systems are contaminated by surface-derived fluids and oxidized volcanic rocks, the relatively higher f_{O_2} hybrid magmas can precipitate magnetite early in their evolution, depleting residual melts in Fe and enhancing a calc-alkaline trend of silica and alkali enrichment. A high f_{O_2} is consistent with the lack of a negative Eu anomaly in many continental arc rocks. On the other hand, under f_{O_2}, reducing conditions, generally in magma systems open only to replenishment by primitive nonarc basalt magma, such as in MORB magma chambers, the Fe-enriched tholeiitic trend develops.

Also plotted in Figure 12.30b are experimental melts generated from pelitic (clay-rich progenitors) and mafic source rocks by dehydration partial melting (Section 11.6.1). These partial melts are silicic, mostly about 75 wt.% SiO_2, but range widely in K_2O concentration, reflecting corresponding variations in the source rock. Mixing of these partial melts with primitive as well as fractionated magmas is obviously an important process in production of calc-alkaline magmas.

Magmas developed in thick continental crust commonly have relatively radiogenic Sr and Nd isotope ratios created by contamination with ancient [87]Sr-enriched rock (Figure 12.31). However, the absence of an elevated [87]Sr/[86]Sr ratio cannot be taken as evidence for no crustal contamination because in some arcs, such as in the southern Andes, the crust is not very old and has similar isotopic composition to that

of the primitive magmas. In this instance, contamination can only be discerned with trace element data or $\delta^{18}O$ values if hydrothermally altered rocks were assimilated.

SUMMARY

A wide spectrum of magma-generating conditions in compositionally variable sources compounded with a wide range of differentiation overprints account for the wide spectrum of magmatic rock compositions found on Earth (e.g. Figure 2.4).

Despite their treatment in separate chapters, magma generation and differentiation in natural systems may be a continuum of processes, acting sequentially or possibly, to some degree, simultaneously. For example, it is reasonable to believe that, as basaltic melts segregate and begin to ascend buoyantly out of their mantle source, some precipitation of olivine might immediately occur during initial decompression and cooling. Or, as felsic magma is generated in the lower continental crust, it may simultaneously mix with basaltic magma that is providing heat for partial melting; assimilation of wall rock may also concurrently occur. To understand the origin of magmas, the petrologist must sort out signatures related to primary generation versus secondary differentiation in suites of comagmatic rocks, using their mineralogical, elemental, and isotopic compositions as well as their fabric and field relations.

Few magmas are likely to rise very far from their source without overprinting modifications in composition. Overprinting processes include closed-system differentiation—chiefly crystal-melt fractionation—and open-system mixing of magmas and assimilation of country rock. Crystal-melt fractionation by crystal settling, filter pressing, and convective melt fractionation is important in the differentiation of tholeiitic magmas emplaced in continental and oceanic crust, producing a trend of Fe enrichment with limited felsic differentiates. These basaltic intrusions may be open to replenishment by more draughts of primitive magma that mix with the crystallizing magma. In MORB chambers this replenishment and mixing apparently occur so frequently that modification of magma composition is limited; consequently, most MORB constituting the floor of the oceans is relatively uniform tholeiitic basalt.

Mantle-derived basaltic magmas can underplate or stall within the lower continental crust, cool, fractionally crystallize, and give off latent heat to allow assimilation of country rock (AFC). Mixing of mafic and more silicic magmas also occurs. Altogether, these processes produce calc-alkaline rock suites typical of continental margin magmatic arcs. The fractional crystallization ac-

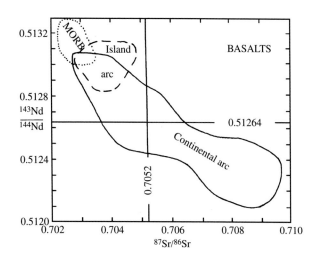

12.31 Nd and Sr isotope ratios in arc basalts compared to MORB. Basalts from western Pacific island arcs are slightly more radiogenic than MORB because of the slab-derived aqueous fluid component, whereas continental arc basalts (Japan, Philippines, New Zealand, Ecuador, Central America, Lesser Antilles) range to much more radiogenic ratios because of interaction with highly radiogenic old continental crust in which [143]Nd/[144]Nd can be 0.510 and [87]Sr/[86]Sr 0.900. (Redrawn from Tatsumi and Eggins, 1995.)

companying assimilation is particularly effective in producing the calc-alkaline trend if it occurs under elevated water and oxygen fugacities. Thicker continental crusts allow more opportunity for AFC and mixing, so that in thickest crusts, as in the central Andes of South America, little, if any, primitive mantle-derived basaltic magma is extruded and most shallow intrusions and extrusions are of rhyolite, dacite, and andesite.

CRITICAL THINKING QUESTIONS

12.1 Essentially monomineralic layers of pyroxenite, dunite, or anorthosite in large differentiated basaltic intrusions can only form by crystal accumulation because no magmas of equivalent composition are known to occur. What do you think is the evidence for no such magmas? (*Hint:* Consider the volcanic record.)

12.2 How do monomineralic layers (see preceding question) originate? Even the least "porous" cumulate still has a significant proportion of interstitial basaltic melt entrapped between the cumulus crystals.

12.3 Why are Sr and Nd isotopic ratios *not* used to evaluate details of crystal-melt fractionation in magma systems?

12.4 Explain the origin of the seven textural types of olivine grains in the 1959 summit eruption of Kilauea, Hawaii (Special Interest Box 12.1). Which are mantle-derived, cumulates, or precipitates of ascending magma?

12.5 Mixed magmas erupted from volcanoes can have an essentially homogeneous melt, but crystals are commonly markedly inhomogeneous and display disequilibrium textures and locally complex zoning. Why?

12.6 Why is assimilation probably insignificant in the evolution of MORB magmas crystallizing in oceanic crustal magma chambers?

12.7 Critically evaluate the origin of pegmatites, especially their distinctive fabric, in terms of the Jahns-Burnham model involving volatile-oversaturated magma systems versus the London model involving kinetic factors in an undercooled magma system.

12.8 Contrast primary, primitive, and parent magma—their differences and ways each can be discerned.

12.9 (a) Does the immiscible Fe-rich glass in mesostases of some tholeiitic basalts (column 1, Table 12.3) resemble any common silicate rock? (b) In what ways does it differ from other rocks of comparable silica content?

12.10 Examine Figures 12.30, 2.18, and 13.24. Is contamination of mantle-derived basaltic magmas with continental crust a viable means of creating highly potassic magmas that form the shoshonite series? Discuss.

12.11 What are the different ways that andesitic magmas can be produced in arcs? What trace element and isotopic attributes would distinguish among these?

12.12 Rocks originating by mingling of two magmas and by physical separation of immiscible melts may be superficially similar. What rock attributes would distinguish between these origins?

PROBLEMS

12.1 Evaluate the validity of hornblende fractionation from a parental andesite magma in producing rhyolite and dacite daughter magmas. Use compositions from Table 2.2 and Appendix A on two-element variation diagrams. Critically discuss your results.

12.2 In Figure 12.29a, what was the approximate proportion of rhyolite magma that mixed with andesite magma that contains 60 wt.% SiO_2 to create dacite that contains 69.5 wt.% SiO_2?

12.3 Write a stoichiometrically balanced reaction for assimilation of Al-rich minerals and rocks by basalt magma to yield anorthite and enstatite components in the hybrid contaminated magma at the expense of diopside in the basalt. Use Al_2SiO_5 to represent the aluminous material. What bearing might this assimilation have on the origin of some leuconorite?

12.4 Write balanced mineralogical reactions showing how: (a) An initially subalkaline, silica-saturated magma that contains normative enstatite and orthoclase could become silica-undersaturated with normative diopside and leucite by assimilating limestone. (b) Silica from the magma can metasomatize wall rock and xenolithic limestone and dolomite, producing calcium silicate minerals such as wollastonite and diopside. Discuss how effective these reactions are in modifying the degree of silica saturation in the original magma.

12.5 Examine the consequences of assimilation of crystalline material into magma using the binary system $NaAlSi_3O_8$-$CaAl_2Si_2O_8$ as a model. Consider a model magma that contains crystals of plagioclase An_{40} in equilibrium with melt into which crystals of cool An_{60} and of An_{20} are introduced. Compare and contrast the subsequent evolution of the two magmas as they crystallize.

12.6 Le Roex et al. (1990) show that the basanite-phonotephrite-tephriphonolite-phonolite sequence on Tristan da Cunha (Figure 2.16) was created by crystal-melt fractionation. This

Table 12.7. Concentrations of SiO$_2$, Rb, and Sc and Distribution Coefficients for Rb and Sc in Tristan da Cunha Lavas

	BASANITE								PHONOTEPHRITE							TEPHRIPHONOLITE					PHONOLITE		
	SP	A	A	A	SP	A	SP	A	P	A	A	A	P	P	P	A	P	P	A	A	SP	SP	P
SiO$_2$	44	45	46	46	46	46	44	46	50	48	51	48	50	48	48	55	57	56	55	55	61	58	59
Rb	73	64	75	73	78	71	74	76	112	90	95	80	119	88	87	129	151	153	139	128	176	151	165
Sc	18.1	17.1	14.5	14.7	13.9	12.7	17	10.3	4.4	10.1	6.2	9.3	5.4	6.5	7.8	3.1	2.7	2.6	2.5	3.4	1.3	1.6	1.3

	Cpx	TiMag	Pl	Amph	Alk feld
Rb	0.01	0.01	0.01	0.5	0.6
Sc	4	20	0.01	4	0.01

SiO$_2$ concentrations rounded to nearest whole number weight percentage; Rb and Sc, parts per million; A, aphyric; P, porphyritic; SP, sparsely porphyritic.
(From Le Roex et al., 1990.)

fractionation can be approximated by considering just two trace elements—Rb and Sc. (a) Are these elements compatible? Incompatible? (b) Using the data in Table 12.7, make a plot of Rb versus Sc for the sequence of rocks. (c) Which sample would be a reasonable parent magma for the sequence? (d) Determine bulk distribution coefficients for Rb and Sc that, when used in equation 12.1, yield a model Rb-Sc curve fitting the data points as closely as possible. Can the entire sequence be modeled by one set of coefficients? Discuss. (e) What weight proportions of clinopyroxene (Cpx), titaniferous magnetite (Timag), and plagioclase (Pl) account for differentiation of the basanite-phonotephrite part of the sequence? Discuss your results. (f) Can the tephriphonolite-phonolite suite be produced by fractionating the same minerals? Others? Discuss.

12.7 If the immiscible Fe-rich glass in column 1, Table 12.3, were to crystallize, of what minerals would it be composed and what would be their modal proportions? Assume an oxygen fugacity buffered by QFM (Figure 3.14).

12.8 For the shonkinite-syenite association in Table 12.3 plot the first 13 trace elements listed on a primitive-mantle normalized diagram (Table 2.7). How do the two patterns compare with those for other rocks in this textbook? Is it possible that these two rocks could be related by fractional cystallization, rather than by separation of immiscible melts? Critically evaluate whether fractionation of common rock-forming minerals might account for the trace element contents of the two rocks. For example, are the Cr and Ni contents consistent with olivine fractionation? Ti and V with Fe-Ti oxide fractionation? And so on.

Petrotectonic Associations

13

FUNDAMENTAL QUESTIONS CONSIDERED IN THIS CHAPTER

1. What rock types and rock suites form petrotectonic associations with particular tectonic settings?

2. How do trace element and isotopic compositions of petrotectonic associations, combined with other compositional, fabric, and field data, provide insights into magma evolution?

3. What role does global tectonic setting play in the way magmas are generated and differentiated?

INTRODUCTION

Global magma production is inextricably linked with movement of lithospheric plates and mantle plumes—the two manifestations of convection in the cooling mantle. Despite the stirring action of these two modes of convection, the mantle has a long-standing chemical and isotopic heterogeneity. Hence, mantle-derived magmas have varying elemental and isotopic signatures that depend on the part of the mantle where plate- and plume-related conditions allow partial melting. Contrasting mantle as well as continental sources, together with variable processes of magma generation and differentiation, depend upon the global tectonic setting in which they function. Consequently, different global settings yield suites of comagmatic rocks having distinctive compositional attributes, called **petrotectonic associations.** A striking contrast between two such associations is the copious production of tholeiitic basalt

magma along oceanic rifts versus the creation of calc-alkaline granitoid batholiths along continental margin subduction zones.

As in any categorization of geologic systems, some transitional associations are to be expected, such as the back-arc setting where rocks have attributes of both island arcs and oceanic spreading ridges. Additionally, some tectonic settings can evolve over geologic time from one to another. For example, a continental intraplate regime overlying a rising mantle plume might evolve into a continental rift and then into an oceanic rift; such is the well-documented continental breakup experienced by landmasses surrounding the Atlantic Ocean.

This chapter begins with associations of chiefly basaltic magmas generated by decompressing mantle where it wells up beneath oceanic spreading ridges and ascends in deep mantle plumes. Following sections deal with more varied arc associations where plates converge and finally with associations in continental rifts and stable cratons.

The intent of this chapter is to summarize the rock types and rock suites that occur in major global tectonic settings and the creation of the magmas in these petrotectonic associations. Trace element and isotopic data furnish insights into the origin of the magmas—their sources, partial melting conditions, and subsequent differentiation.

Plate tectonic settings where petrotectonic associations develop are displayed schematically in Figures 1.5 and 11.2. Additional details can be found in many books dedicated to specific associations, as referenced herein, and in overall summaries by Hess (1989), Wilson (1989), and McBirney (1993).

☀13.1 OCEANIC SPREADING RIDGES AND RELATED BASALTIC ROCKS

The 65,000-km-long system of oceanic spreading ridges encircling the Earth is by far the most significant expression of its magmatism (Figure 1.1). Acting in concert with seafloor spreading, magmatism at oceanic rifts has produced in <200 My, since the early Jurassic, the entire present-day oceanic crust, covering nearly 70% of the Earth. The continental rock record indicates that similar ridge magmatism and seafloor spreading occurred in pre-Mesozoic time, probably well into the Precambrian.

Spreading junctures in the oceanic lithosphere (Frontispiece and Plate I) are marked by broad rises that are thousands of km wide and stand about 2.5 km above the average level of adjacent abyssal plains. Beneath these oceanic rifts, upwelling, decompressing, and partially melting mantle generates basaltic magma, which rises and solidifies as **oceanic crust.** Seismic data, augmented by field study of subaerially exposed segments of the oceanic lithosphere (ophiolite; Section 13.6), indicate the typically 6–7 km thick crust is lay-

ered (Figure 13.1a). Beneath a veneer of sediment, which thickens away from the oceanic rift, are the following layers, from the top down:

1. Basaltic lava flows, commonly pillowed (Section 10.2.1).
2. Sheeted mass of their feeder dikes.
3. Massive gabbro that becomes more layered downward and merges into layered ultramafic cumulates; these form in crustal magma storage chambers.
4. Locally in sharp contact with the cumulates is mantle peridotite that has a high-T, solid-state strain fabric.

Advective percolation of seawater into the cooling, pervasively fractured basaltic rocks at oceanic rifts produces widespread and locally intense alteration, or seafloor metamorphism (Special Interest Box 11.3). Outcroppings of mantle peridotite on the seafloor, called **abyssal peridotite,** are also variably hydrated. In addition to advective cooling of the oceanic crust by circulating seawater, the lithosphere as a whole cools conductively so that farther from the rift it progressively cools and thickens (Figure 13.1d).

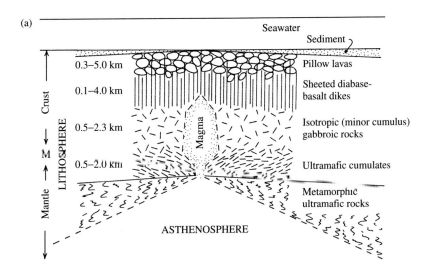

13.1 Oceanic lithosphere consisting of crust and underlying mantle. (a) Highly idealized cross section at a spreading ridge. Note that the seismic M-discontinuity between crust and higher-velocity ultramafic mantle is here drawn at the top of the ultramafic cumulates whereas the petrologic discontinuity between magmatic crust and metamorphic mantle lies 0.5–2.0 km deeper. (b) Heat flow from the seafloor as a function of its age. One heat flow unit (HFU) = 0.0418 W/m². Shaded area approximates distribution of observed values of heat flow perturbed by advective circulation of seawater through the crust. Solid line is the theoretical heat flow according to a model in which the lithosphere is cooling by conduction only. (c) Schematic geologic cross section through the upper lithosphere (ignoring surface topographic characteristics and structural complications) showing increasing thickness of sediment with age away from crest of ridge; deepening of seafloor away from ridge crest due to cooling and thermal contraction of lithosphere; approximately uniform thickness of layers 2 and 3 of basalt and gabbro, which together with sediment constitute the **oceanic crust;** and the underlying ultramafic rocks of the uppermost mantle. Advective circulation of seawater through the upper part of hot crust diminishes away from ridge crest. (d) The lithosphere beneath the basalt-gabbro crust is a thickening lid of cooling peridotitic mantle covering the hotter asthenosphere. Light lines are highly schematic isotherms suggesting that the base of the lithosphere, deepening with respect to age, is essentially an isothermal solidus surface. Dashed heavier lines indicate possible directions of flow of the asthenosphere away from axis of the rise: Stippled area under the crest is the approximate location of a zone of advanced partial melting (compare Figure 13.4). Note difference in depth scales in (c) and (d). (b–d Redrawn from Anderson et al., 1977.)

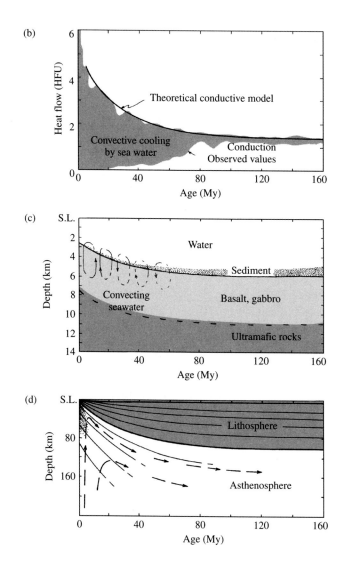

13.1 (*Continued*).

13.1.1 Mid-Ocean Ridge Basalt (MORB)

Basaltic rocks constituting the upper part of the oceanic crust have been called many names, including *abyssal tholeiite, seafloor* and *ocean-floor basalt, submarine basalt,* and *mid–oceanic ridge basalt,* or **MORB**. Although not all ridges are centrally positioned in oceans, the MORB label has become well entrenched. Although most of the oceanic crust is subalkaline tholeiite basalt, there are local, evolved rock types and rare alkaline basalts.

The fabric of MORB reflects rapid cooling of generally near-liquidus magma extruded in cool seawater as ponded sheet flows, pillow lavas, and associated hydroclastic breccias. Quenched glassy rinds on pillows (Figure 7.3) and vitroclasts have very sparse phenocrysts of bytownitic plagioclase with or without magnesian olivine, Fo_{80-90}, that contains minute Cr-Mg spinel inclusions. Glass is commonly altered to palagonite. Phenocrysts of augite are rare and usually confined to more extensively crystallized lavas that contain abun-

dant olivine and plagioclase. Glomeroporphyritic clots represent accumulations of fractionated crystals. Local disequilibrium attributes include embayed phenocrysts in chemically more evolved groundmasses, phenocrysts having corroded cores of higher-T solid-solution composition, and phenocrysts containing melt inclusions different from groundmass glass. These attributes likely reflect mixing of fractionated and more primitive magmas (Section 12.4.3).

The elemental composition of normal MORB is relatively uniform compared to that of other basaltic associations. Typical is subalkaline olivine or quartz tholeiite, which contains normative olivine and hypersthene or quartz and hypersthene, respectively (Table 13.1). The range of SiO_2 is only 47–51 wt.%. The most distinctive attribute is the low concentration of incompatible elements, including Ti and P, and large-ion-lithophile elements, such as K, Rb, Ba, Th, and U, compared to that of tholeiites in oceanic islands and continental plateaus (see Tables 13.2 and 13.5). On a

Table 13.1. Average Chemical and Normative Composition of N-MORB and Trace Elements (ppm) for N-MORB and E-MORB

	N-MORB		N-MORB
SiO$_2$	49.93	Or	1.00
TiO$_2$	1.51	Ab	22.06
Al$_2$O$_3$	15.90	An	31.14
FeO	10.43	Di	21.04
MnO	0.17	Hy	15.55
MgO	7.56	Ol	2.47
CaO	11.62	Mt	3.89
Na$_2$O	2.61	Il	2.87
K$_2$O	0.17	Ap	0.19
P$_2$O$_5$	0.08		

	N-MORB	E-MORB
Cs	0.007	0.063
Rb	0.56	5.04
Ba	6.3	57
Th	0.12	0.6
U	0.47	0.18
Nb	2.33	8.3
Ta	0.132	0.47
La	2.5	6.3
Ce	7.5	15
Sr	90	155
Nd	7.3	9
Sm	2.63	2.6
Zr	74	73
Eu	1.02	0.91
Gd	3.68	2.97
Dy	4.55	3.55
Y	28	22
Yb	3.05	2.37
Lu	0.455	0.354

Top, major oxides from McKenzie and O'Nions (1991); bottom, data from Sun and McDonough (1989).

primitive-mantle-normalized diagram, the trace element pattern is a relatively smooth, positively sloping line (Figure 13.2) that shows lesser concentrations of incompatible elements than, for example, oceanic island tholeiitic basalt.

Origin of MORB Magmas and the Mantle Source. Because ocean-ridge magmas ascend through the thinnest and compositionally least contrasting crust of any petrotectonic association, they potentially provide the clearest "window" into the mantle source. Normal MORB is commonly used as a reference composition for comparison with other mafic magmas. Nonetheless, most MORB cannot represent primary magma from a peridotite source (Section 11.3.5) for at least two reasons: First, MORB has only about 5–10 wt.% MgO

and <300 ppm Ni and most have Mg/(Mg + Fe) <0.7. Second, most MORB magmas are not multiply saturated near their liquidus with olivine + orthopyroxene + clinopyroxene at mantle pressures, as they would be if they were in equilibrium with a near-solidus peridotite source. Instead, compositions are displaced from 10- to 20-kbar beginning-of-melting invariant points in Figure 12.24 by apparent olivine fractionation. Compositional variations in fresh glasses corroborate olivine control (Figure 12.23). Primary magmas are believed to be olivine-rich picrite that on ascent and decompression loses olivine, reducing their MgO, Ni, and Mg/(Mg + Fe) to values observed in most MORB. Partial melting appears to occur in the stability field of spinel peridotite at depths of about 30–75 km (P = 10–25 kbr) because the trace element pattern of MORB (Figure 13.2) indicates that neither plagioclase nor garnet was a residual phase in the source; their absence is suggested by the lack of a negative Eu anomaly and low Sm/Yb ratio, respectively. However, if large melt fractions, >20%, were generated, these aluminous phases would be consumed and their associated geochemical indicators eliminated.

Large degrees of partial melting yield trace element signatures for the melts like that of the source (Figure 11.15). Indeed, spinel-bearing Cr-diopside peridotite xenoliths (Figure 11.6b) and normal MORB (N-MORB in Figure 13.2) have similar light rare earth element and incompatible element depleted patterns. This depletion in incompatible elements in the mantle source before the generation of MORB partial melts is conventionally interpreted to be a consequence of long-term (billions of years) extraction of enriched melts that formed the continents (Hofmann, 1988). One facet of this complementary relation between MORB and continental crust is their Sr and Nd isotope compositions (Figure 2.27).

The compositions of MORB and of mantle (abyssal) peridotite sampled at ocean ridges reflect varying degrees of partial melting and melt extraction, respectively. Global variations in MORB shown in Figure 13.3a reveal an inverse correlation between CaO/Al$_2$O$_3$ ratio and Na normalized at MgO = 8 wt.%. The ratio is essentially unaffected by fractionation of olivine, pyroxene, and plagioclase but increases with increasing degrees of partial melting as Ca-rich clinopyroxene is dissolved out of the peridotite source. Normalized Na decreases as the degree of partial melting increases: That is, Na behaves as an incompatible element in source peridotite. Modal variations in abyssal peridotites collected on the ocean floor are consistent with varying amounts of melted and extracted pyroxene components from this mantle source (Figure 13.3b).

After polybaric fractionation of olivine during their ascent from the mantle source, MORB magmas are

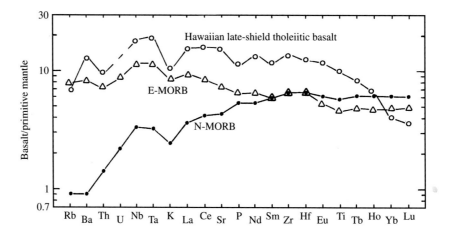

13.2   Primitive-mantle-normalized trace element patterns of oceanic tholeiitic basalts. Decreasing element incompatibility in mafic magmas from left to right. MORB data and normalizing values from Sun and McDonough (1989). Hawaiian tholeiite from Table 13.3.

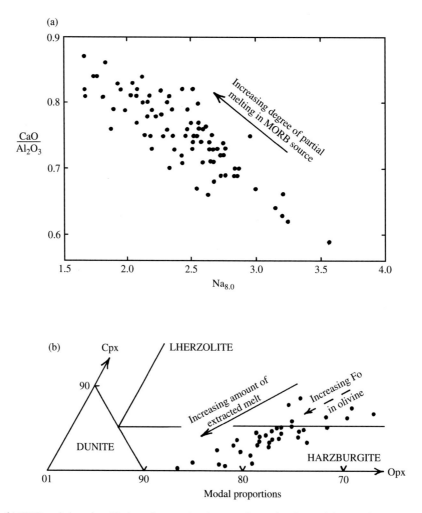

13.3   Compositions of MORB and abyssal peridotite reflect varying degrees of partial melting of the mantle source. (a) Global MORB composition. Variable on the *x* axis is the concentration of Na at 8 wt.% MgO on a Na-MgO variation diagram drawn for 84 sites at spreading ridges around the world (775 samples in all). The CaO/Al_2O_3 ratio is taken from a CaO/Al_2O_3-MgO variation diagram for samples at each site; intrasite variation in the ratio is negligible. (Redrawn from Klein and Langmuir, 1987.) (b) Mineral proportions in abyssal peridotite show that varying amounts of basaltic partial melt have been extracted. Data from Dick et al. (1984). Each point is the average modal proportion of Ol, Cpx, and Opx in 266 samples from 36 dredge-haul sites along rifts in the Atlantic Ocean, Indian Ocean, and Caribbean Sea. The classification diagram is for ultramafic rocks (Figure 2.10b). The most extensively melted peridotite, that is, the most sterile, has least Cpx and Opx and greatest Fo content of olivine (compare Figure 11.9).

differentiated at low P in crustal storage chambers beneath the rift. As discussed in Section 12.4.3, the limited compositional variation in most MORB reflects recurring replenishment and mixing of relatively primitive mantle-derived magmas with fractionating magma in rift chambers, moderating the effects of crystal-melt fractionation. However, uncommon seafloor basalts follow a fractionation trend of extreme Fe enrichment and limited enrichment in silica and alkalies.

<u>Configuration of Magma Bodies and Partial Melt Zones.</u> There has been considerable speculation, but few hard data, regarding the size, shape, longevity, and evolution of crustal magma chambers in the oceanic crust. Additional uncertainties relate to how chamber character and the shape of partial melt zones in the underlying mantle might explain compositional characteristics of MORB and subtle contrasts between MORB at slow and fast spreading rifts (Wilson, 1989). Slow spreading systems, such as the Mid-Atlantic Ridge, have rates less than about 5 cm/y, whereas the fast spreading East Pacific Rise has a spreading rate of more than about 8 cm/y. The zone of partial melting in the mantle beneath slow spreading rifts is presumed to be narrowly focused, perhaps only a few kilometers wide. This shape is believed to be governed by buoyancy-driven

flow in the slowly diverging asthenospheric mantle below the rift axis. On the other hand, below the faster spreading East Pacific Rise, a detailed geophysical investigation of a segment near 17°S latitude revealed a broad, partially melted zone in the upper mantle that is hundreds of kilometers wide and up to 150 km deep (Figure 13.4). This wide zone reflects passive upwelling driven by viscous drag of the separating lithospheric plates. In other words, convection in the asthenosphere is forced by plate movement, not the other way round. Significantly, there is no indication of asthenospheric upwelling flow from the deeper mantle, below the 410-km seismic discontinuity.

13.1.2 Iceland

Iceland is a large volcanic island positioned astride the actively spreading Mid-Atlantic Ridge and built of older, Tertiary olivine tholeiite lava flows flanking a central rift zone where more recent volcanism is focused (Figure 13.5). This relationship mimics that of the submarine oceanic ridge to the north and southwest of Iceland. Despite its topographic elevation (as much as 2 km above sea level) and unusual thickness (20–35 km), the Icelandic crust is not sialic as large continents are. The central rift zone has numerous northeast-trending, elongate extensional fault blocks

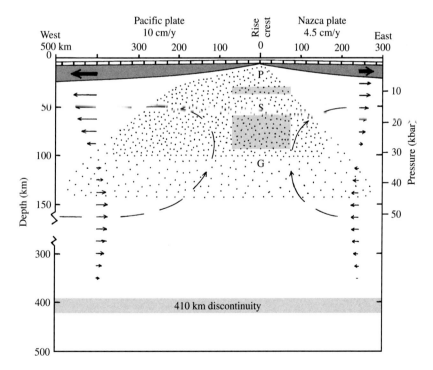

13.4 Geophysical cross section through the East Pacific Rise near 17°S. Electrical, magnetic, and seismic velocity data indicate a broad volume of slightly melted mantle peridotite (about 1–2% melt fraction) in the heavily stippled region and less in the lightly stippled region. Because there appears to be no deflection of the 410-km seismic discontinuity beneath the rise, it is believed that asthenospheric mantle flow (arrows) is confined to the shallow mantle above 410 km without ascending flow from the deeper mantle. Note the asymmetry of the partially melted zone, skewed to the west under the faster-moving (10 cm/y) Pacific plate relative to the slower (4.5 cm/y) Nazca plate. Lithosphere is shaded dark. P, depth to which plagioclase peridotite is stable; S, spinel peridotite; G, garnet peridotite; shaded bands indicate transition depths (Figure 11.5). (Redrawn from Forsyth et al., 1998.)

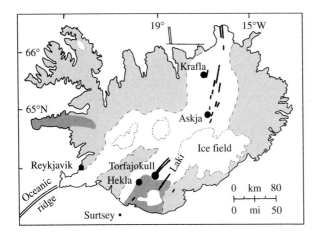

13.5 Generalized geological features of Iceland. Older (>3 Ma) volcanic rocks light shaded; younger (<3 Ma) rocks not shaded; alkaline basalts dark shaded. Eruptive fissures, including Laki, shown by heavy lines; major central volcanic complexes (Krafla, Askja, Hekla, Torfajökull) by large filled circles; glaciers enclosed by dotted lines. (Redrawn from McBirney, 1993.)

and fissures. The 12 km^3 of basaltic lava from the 25 km-long Laki fissure in 1783 was the largest recorded historic extrusion of lava, covering 565 km^2 in 8 months. Innumerable feeder dikes can be seen in the topographically lowest exposures, but their numbers diminish upward in subhorizontal lava flows. Picrites were erupted locally, and increasingly more alkaline basalts occur farther from the rift zone. More evolved tholeiitic-suite magmas—including quartz tholeiite, ferrobasalt, icelandite (low-Al, high-Fe andesite)—have erupted chiefly from numerous central volcano complexes. In these central volcanoes, as many as half of the rocks are silicic, commonly mildly peralkaline. The origin of the exceptionally abundant (for an oceanic island) silicic rocks is controversial. The significantly lower $\delta^{18}O$ values of the silicic rocks than of coeval basaltic rocks suggest partial melting of the hydrothermally altered, isostatically subsiding mafic Icelandic crust with heat supplied by intrusions of basalt magma along extensive rift-fissure systems (Gunnarsson et al., 1998).

An order-of-magnitude-higher rate of magma production in Iceland than in submarine oceanic rifts implies an unusually large volume of underlying hot decompressing mantle—a plume (Figure 1.5). Seismic imaging has indeed revealed a column of hotter, lower-velocity mantle rock beneath Iceland in which partial melting occurs to at least 410 km and probably deeper (Figure 13.6). Melting in an ascending plume of hotter mantle begins at greater depth than in upwelling mantle beneath an oceanic ridge and yields a greater melt volume to build thicker crust (Figure 13.7).

Many studies have found that MORB northward along the Mid-Atlantic Ridge, particularly primitive tholeiitic basalts in Iceland, is increasingly enriched in light REE and large-ion lithophile (LIL) elements. The implication of this trend is examined next.

13.1.3 Mantle Reservoirs

If the seismic images of the shallow convecting mantle beneath the East Pacific Rise (Figure 13.4) and the ascending plume from the deep mantle beneath Iceland (Figure 13.6) are credible, they support the existence of two fundamentally distinct **mantle reservoirs.** Widely espoused by geochemists, these mantle regions are the source of basaltic partial melts that have distinct trace element and isotopic signatures. One reservoir is relatively depleted in incompatible elements, has relatively nonradiogenic Sr and radiogenic Nd isotope ratios, and is the source of normal MORB, or **N-MORB;** this source corresponds to the upper mantle underlying the global oceanic spreading ridge system. The other reservoir is the deeper, relatively enriched mantle near bulk silicate Earth in its isotopic composition (Figure 2.27) that constitutes plumes. Many basalts from the Mid-Atlantic Ridge north of 30°N appear to be derived by mixing of the two mantle reservoirs or of partial melts from them. These mixed-source basalts are enriched in incompatible trace elements relative to N-MORB and are sometimes called **E-MORB** (Figure 13.2). LIL elements are enriched by about an order of magnitude, and the light-REE-to-heavy-REE ratio (e.g., La/Yb) is greater in E-MORB than in N-MORB.

A wide range of $^{87}Sr/^{86}Sr$ and $^{143}Nd/^{144}Nd$ ratios in oceanic basalts (Figure 13.8) can be produced by mixing of components from the two mantle reservoirs described. However, some ocean island rocks, such as those of Kerguelen in the Indian Ocean, have isotopic ratios extending to more enriched levels, even beyond bulk silicate Earth, implying derivation from sources with higher Rb/Sr and lower Sm/Nd ratios. Small degrees of partial melting of the mantle generate melts with higher Rb/Sr and lower Sm/Nd, because of the differing compatibilities of the element pairs (Section 2.6.2). Where the partial melts migrate and metasomatize the mantle, especially if the fixation is ancient (say, >1 Ga), $^{87}Sr/^{86}Sr$ is higher than in bulk Earth and $^{143}Nd/^{144}Nd$ lower.

Additional end-member sources are necessary to account for the entire range of isotopic ratios (including Pb and He) found in mantle-derived oceanic basalts (e.g., Hart et al., 1992).

✳13.2 MANTLE PLUMES AND OCEANIC ISLAND VOLCANIC ROCKS

Upwards of one million intraplate volcanoes are estimated to dot the ocean. Many are shields. Some emerge above sea level as islands, but the majority is

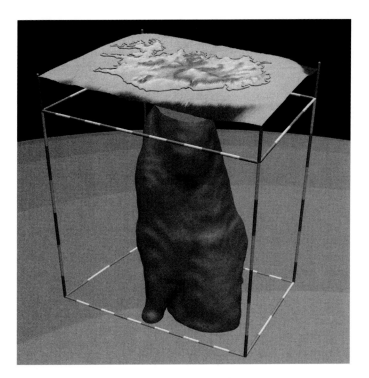

13.6 Seismic image of the Iceland mantle plume. (Courtesy of Cecily Wolfe.) The approximately 300-km-diameter plume to a depth of 410 km beneath Iceland. Other seismic data suggest a perturbed 660-km discontinuity that distinguishes the upper from the lower mantle (Figure 1.3), indicating that the plume ascends from the lower mantle, in contrast to the convective pattern beneath normal oceanic ridges (Figure 13.4).

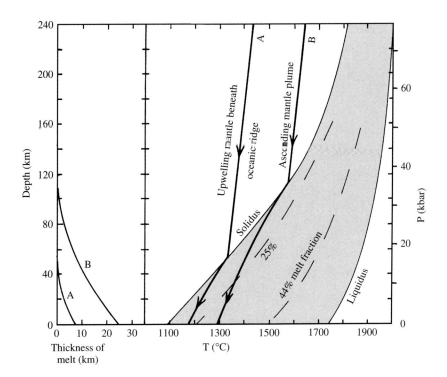

13.7 Partial melting of a model dry mantle peridotite. Compare Figure 11.4. Two hypothetical adiabatic *P* (depth)-*T* paths, A and B, of ascending, decompressing mantle are shown by heavy solid lines. The low-*T* path A, appropriate to upwelling mantle beneath oceanic ridges, reaches the solidus at a little over 50-km depth and about 1330°C and during continued ascent partially melts out the aluminous phase and clinopyroxene in the peridotite, yielding ultimately near the base of the crust about 20% melt whose *T* is about 1180°C. The subsidiary diagram on the left indicates that the generated melt forms a layer of melt about 7 km thick, matching the thickness of the oceanic crust. An ascending hotter mantle plume (path B) generates as much as 30% melt in the uppermost mantle at *T* ~1300°C, forming a layer about 25 km thick, matching that in the Hawaiian Islands. (Redrawn from McKenzie and O'Nions, 1991.)

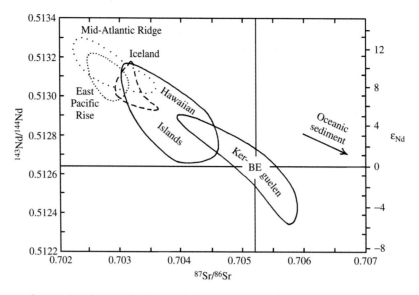

13.8 Isotope ratio diagram for oceanic volcanic rocks showing fields of oceanic island groups, East Pacific Rise MORB, Mid-Atlantic Ridge MORB, and oceanic sediment (off diagram). BE, isotopic ratios in bulk silicate earth. Compare Figure 2.27. (Redrawn from Wilson, 1989.)

submerged below sea level as **seamounts,** either because magma supply was limited and growth subdued or because, once emergent as islands, they were eroded off flat to sea level and then submerged as the underlying lithosphere cooled, contracted, and subsided (Figure 13.1c). Most volcanoes appear to form at or near spreading ridges but others, especially in the Pacific, lie in linear **volcanic chains** that have monotonically changing age along the chain (Plate I). The best known is the northwesterly trending, elbow-shaped Hawaiian Island–Emperor Seamount chain in the north Pacific. From its disappearance in the Kurile-Aleutian trench in the north, the chain includes >100 edifices culminating 6000 km to the southeast in the active volcanoes on the island of Hawaii (Figure 13.9). The rate of migration of volcanic activity along this chain has been about 8–9 cm/y. The proposal of Wilson (1963) that this chain originated by movement (at that rate) of the oceanic lithosphere over an essentially stationary mantle plume is now accepted by most geologists.

In contrast to the Hawaiian plume positioned beneath the interior of the Pacific plate, plumes under the Mid-Atlantic Ridge have produced paired chains of extinct volcanoes or aseismic ridges on each side of the divergent juncture (see Section 13.3.3). The V shape of the paired chains reflects a northerly component of plate motion during ocean opening. The displaced position of some supposed plume-related islands off the ridge axis, such as Tristan da Cunha, has been caused by westward drift of the ridge over the plume in the last few tens of millions of years.

13.2.1 Character of Volcanic Rocks

Oceanic island volcanoes are more diverse in composition than the predominantly tholeiitic basalts (MORB) of submarine ridges. Rocks range from tholeiitic to alkaline and from ultramafic to felsic. Whereas the Hawaiian Islands appear to have mostly tholeiitic and lesser alkaline underpinnings and small late alkaline cappings, subaerial exposures in other islands are wholly alkaline (e.g., Tahiti) or wholly tholeiitic (e.g., Galapagos).

Subalkaline tholeiites and alkaline basaltic rocks grade one into the other without any well-defined break. Mineralogical and modal contrasts (Wilson, 1989) are as follows: Olivine is exclusively a phenocrystic phase in tholeiites and is quite Mg-rich, Fo_{90-70}, whereas in alkaline basalts it is both a phenocrystic and a groundmass phase and is more variable in composition, Fo_{90-35}, even within a single hand sample. (Rims of olivines are commonly replaced by orange-brown "iddingsite"—a complex mixture of clay minerals and hydrous ferric oxides—in subaerially oxidized lavas, regardless of compositional affinity.) Plagioclase is more commonly phenocrystic in tholeiites than in alkaline basalts, in which it has higher K_2O content. The alkali concentration may be sufficient in alkaline magmas to stabilize alkali feldspar; this phase appears in the groundmass of alkaline basalts, together with feldspathoids or analcite in more silica-undersaturated rocks; vesicle linings may be of these same minerals. Tholeiitic lavas, on the other hand, may contain one or more silica polymorphs (quartz, cristobalite, and/or tridymite) in the groundmass and in vesicles. Tholeiites generally lack hydrous phases, but increasingly evolved alkaline rocks are more likely to contain Ti-rich amphibole (kaersutite) and phlogopite-biotite, either in the groundmass or as phenocrysts.

Much of the difference in bulk chemical composition between tholeiitic and alkaline basalts lies in compositional contrasts in the major pyroxene phase. Ca-rich clinopyroxene occurs in both basalt types, but its normative composition mirrors that of its host rock. Thus, clinopyroxene in alkaline basaltic rocks is silica-undersaturated because of relatively high concentrations of Ti and Al that replace Si. Titaniferous clinopyroxenes are slightly darker brown than those in tholeiites or are pale lavender-brown and have anomalous extinction colors under cross-polarized light in thin section. Some clinopyroxenes in alkaline rocks are enriched in Na, as the aegirine end member ($NaFe^{3+}Si_2O_6$), and these are greenish. Many Ca-rich clinopyroxenes are conspicuously zoned to brown (Ti-rich) or green (Na) rims. *Alkaline rocks never contain orthopyroxene.* However, it is common in tholeiites of oceanic islands, whereas rare in MORB. Many tholeiitic rocks also contain a third pyroxene phase—pigeonite (Ca-poor clinopyroxene; Figure 5.28).

Alkaline mafic rocks locally contain a variety of mantle-derived inclusions (Section 11.2.1) that are entirely lacking in tholeiites.

Olivine-rich picrite and olivine-pyroxene-rich **ankaramite,** some of which may have originated by accumulation of phenocrysts and may be ultramafic, are either alkaline or subalkaline.

Tholeiitic and alkaline parental magmas evolve into contrasting derivatives. Tholeiitic basalt magmas evolve through low-P crystal-melt fractionation into Fe-rich derivative ferrobasalt, ferroandesite, and so on. Mildly alkaline basaltic magmas in oceanic islands typically differentiate into the sodic lineage of hawaiite-mugearite-benmoreite-trachyte (Table 13.2; Figure 13.10) by multiphase fractional crystallization in crustal

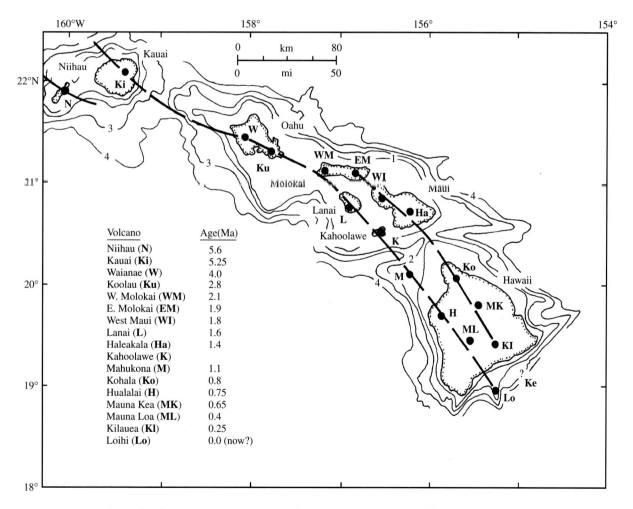

Volcano	Age(Ma)
Niihau (**N**)	5.6
Kauai (**Ki**)	5.25
Waianae (**W**)	4.0
Koolau (**Ku**)	2.8
W. Molokai (**WM**)	2.1
E. Molokai (**EM**)	1.9
West Maui (**WI**)	1.8
Lanai (**L**)	1.6
Haleakala (**Ha**)	1.4
Kahoolawe (**K**)	
Mahukona (**M**)	1.1
Kohala (**Ko**)	0.8
Hualalai (**H**)	0.75
Mauna Kea (**MK**)	0.65
Mauna Loa (**ML**)	0.4
Kilauea (**KI**)	0.25
Loihi (**Lo**)	0.0 (now?)

13.9 Hawaiian Islands and shield volcanoes on the Hawaiian Ridge. Age is approximate inception of tholeiitic shield building stage. A hypothetical volcano (Ke, Keikikea) is anticipated to emerge in the future off the southeast coast of Hawaii. Depth contours in kilometers. (Redrawn from Clague and Dalrymple, 1987; Moore and Clague, 1992.)

Table 13.2. Major Oxide Compositions of Hawaiian Volcanic Rocks

	PRESHIELD	SHIELD	POSTSHIELD						POSTEROSION		
STAGE	1	2	3	4	5	6	7	8	9	10	11
SiO$_2$	45.7	49.4	45.4	44.1	47.9	51.6	57.1	61.7	44.1	39.7	36.0
TiO$_2$	3.37	2.5	3.0	2.7	3.4	2.4	1.2	0.5	2.6	2.8	2.8
Al$_2$O$_3$	13.2	13.9	14.7	12.1	15.9	16.9	17.6	18.0	12.7	11.4	10.0
Fe$_2$O$_3$	3.50	3.0	4.1	3.2	4.9	4.2	4.8	3.3	3.6	5.3	5.7
FeO	8.55	8.5	9.2	9.6	7.6	6.1	3.0	1.5	9.1	8.2	8.9
MnO	0.16	0.2	0.2	0.2	0.2	0.2	0.2	0.2	0.2	0.2	0.1
MgO	7.08	8.4	7.8	13.0	4.8	3.3	1.6	0.4	11.2	12.1	12.0
CaO	12.00	10.3	10.5	11.5	8.0	6.1	3.5	1.2	10.6	12.8	13.0
Na$_2$O	3.13	2.2	3.0	1.9	4.2	5.4	5.9	7.4	3.6	3.8	4.1
K$_2$O	0.99	0.4	1.0	0.7	1.5	2.1	2.8	4.2	1.0	1.2	1.0
P$_2$O$_5$	0.45	0.3	0.4	0.3	0.7	1.1	0.7	0.2	0.5	0.9	1.1

1, Alkaline basalt; 2, tholeiitic basalt (200); 3, alkaline basalt (35); 4, ankaramite (9); 5, hawaiite (62); 6, mugearite (23); 7, benmoreite (5); 8, sodic trachyte (5); 9, basanite (11); 10, nephelinite (10); 11, melilite nephelinite (7).
Column 1, from Frey and Clague (1983), columns 2–11, averages (number of samples in parentheses) from Macdonald (1968).

chambers, rather than the potassic lineage of trachy-basalt-trachyandesite-latite (Figure 2.12). Some mildly alkaline magmas, or those transitional between tholeiitic and alkaline, evolve into silica-oversaturated peralkaline trachyte and rhyolite. More alkaline mafic magmas, such as basanite, can differentiate into silica-undersaturated, *Ne*-normative tephrite and phonolite, as on Tristan da Cunha (Figure 2.16). Another highly alkaline oceanic island association, not necessarily related via differentiation, is basanite-nephelinite-melilitite. (**Nephelinite** consists of nepheline + clinopyroxene ± olivine ± Fe-Ti oxides; **melilitite** is the same except that it includes melilite instead of nepheline.)

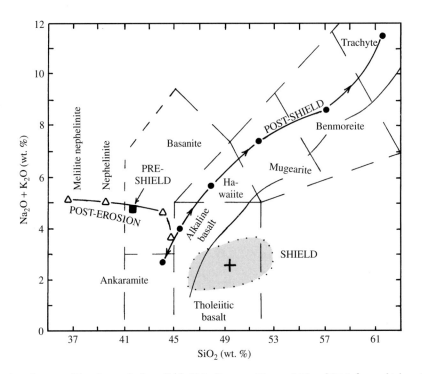

13.10 IUGS classification of average Hawaiian rocks from Table 13.2. Compare Figures 2.12 and 2.16, from which rock type names and thin curved dividing line between fields of alkaline and subalkaline rocks are taken. Filled square is preshield basanite. Shaded area encompasses 200 analyses of shield-forming tholeiitic basalt and large + is the average. Postshield alkaline suite evolved from parental alkaline basalt by multiphase fractionation denoted by filled circles. Posterosion, highly-alkaline mafic suite denoted by open triangles.

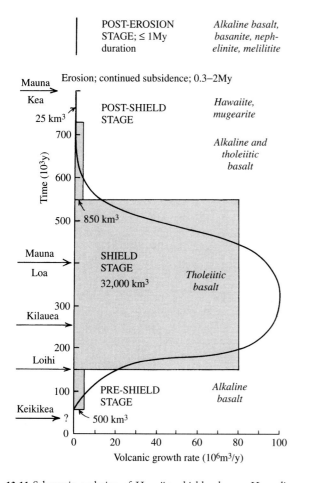

13.11 Schematic evolution of Hawaiian shield volcanos. Heavy-line, bell-shaped curve shows estimated variation in growth rate through time in the four stages of volcano evolution. A similar curve would represent the variation in degree of partial melting of the mantle source through time. Shaded rectangles show estimated volumes of rock. Rock types characterizing each stage listed on right in italics. Presumed stage status of some other Hawaiian volcanoes is indicated on the left, as follows: Mauna Kea has apparently just ended its postshield stage (last erupted about 4000 years ago). Kilauea activity is waxing in the shield-building stage (roughly, an eruption every year but some continue for >1 y); the activity of Mauna Loa, though waning (an eruption every 3.4 y on average), is still in the shield building stage; their shield stage is estimated to terminate about 300,000 and 150,000 years in the future, respectively. Loihi may be near the end of its preshield stage. Keikikea, a hypothesized volcano, has yet to be born. (Redrawn from Frey et al., 1990.)

13.2.2 Hawaiian Islands: Tholeiitic and Alkaline Associations

<u>Growth of Hawaiian Shield Volcanoes.</u> Hawaiian shield volcanoes grow from the ocean floor, or on the flank of an older shield, through three stages, lasting altogether about 0.6 My; a fourth evolutionary stage may follow after a period of inactivity and erosion. These stages, generalized in Figure 13.11, have been established principally by studies of the island of Hawaii (Figures 10.11 and 13.9; see also Moore and Clague, 1992). A summary follows:

1. Because the submarine parts of the island volcanoes are poorly known, the earliest, *preshield stage* of growth is highly speculative but is believed to be one of infrequent small-volume extrusions of alkaline basalt. The still submerged (about 1 km below sea level), active Loihi volcano is believed to be in transition from the preshield to shield stage of growth because its submarine rocks include both alkaline and tholeiitic lavas. In the preshield and early shield stages, magmas are extruded onto the ocean floor, forming submarine lava flows, commonly pillowed, and hydroclastic deposits formed by magma-water interaction (Figure 10.11c).

2. The main *shield building stage* is characterized by frequent, large-volume extrusions of tholeiitic basalt. As activity continues from the preshield stage, the volcano emerges above sea level and subaerial flows of thin pahoehoe and thicker aa lavas erupt from summit vents and from rift-controlled fissures (Figures 9.8 and 10.11). Pyroclastic deposits constitute <1% of the subaerial shield. A network of magma chambers and conduits within the volcano solidifies as coarser-textured, locally phaneritic gabbro in dikes and sills. Ultramafic cumulates and small volumes of evolved magmas are derived by crystal fractionation. Summit lavas are generally olivine-controlled (Figure 12.3). Less magnesian rift-zone lavas extruded on the shield flanks have experienced low-P multiphase fractionation, mainly of clinopyroxene and plagioclase, and mixing with more primitive magma recharging the fractionating chambers. As the shield grows, it sinks isostatically and massive landslides slough off gravitationally unstable flanks (Figure 10.11b, c). Mauna Loa is in shield-building decline, whereas Kilauea shield activity is waxing stronger.

3. The transition to the *postshield stage* is marked by waning tholeiitic activity and inception of infrequent, small-volume alkaline basalt extrusions; alkaline and tholeiitic lavas are commonly intercalated. Smaller shields have little or no postshield alkaline lava. Larger shields fed from robust magma generating plume systems have well developed alkaline activity (hawaiite-mugearite-benmoreite-trachyte) beginning before the culmination of shield building.

4. After a period of volcanic dormancy (as little as 0.3 My on Kauai to about 2 My on Oahu) and canyon cutting, the *posterosion stage* of activity in most shields involves commonly explosive extrusion of very small volumes of highly alkaline mafic magma. Rock types include alkaline basalt, basanite, nephelinite, and melilite nephelinite (Table 13.2; Figure 13.10).

Some postshield and posterosion magmas carry dense mantle-derived inclusions, testifying to their rapid ascent. However, other postshield magmas stall

Table 13.3. Trace Element Compositions of Select Hawaiian Volcanic Rocks

	1	2	3
Rb		18.6	16
Ba	50	304	791
Th	0.4	1.8	4.4
U	0.2		1.1
Nb		30.3	52
Ta	0.5	1.9	3
La	8.6	24.7	36.3
Ce	23.5	58.7	62.7
Sr	250	693	660
Nd	19	33.3	37.9
Sm	4.85	7.98	7.9
Zr	150	244	156
Hf	3.3	5.94	3.5
Eu	1.75	2.74	2.2
Gd			9.2
Tb	0.79	1.07	0.8
Y			25
Ho		1.2	0.8
Tm			0.3
Yb	2.03	2.21	1.5
Lu	0.28	0.29	0.05
Cr	260		576
Ni	125		402
Sc	30		13
V	282		309

1, Shield tholeiitic basalt, Mauna Loa volcano, Hawaii; 2, postshield alkaline basalt, Haleakala, volcano Maui; 3, posterosion nephelinite, Kauai.
Data from references in Watson (1993).

within the crust, where they differentiate. More voluminous, rapidly erupting shield-stage tholeiitic magmas experience limited differentiation before eruption.

Trace Element and Isotopic Characteristics of Hawaiian Magmas.

Besides the differences through time in the four stages of Hawaiian volcanism, variations in the trace element concentrations and isotopic ratios of the magmas reflect the degree of melting and character of their sources, respectively (Table 13.3).

Compared with MORB at similar MgO, Hawaiian shield-building tholeiites are enriched in TiO_2, K_2O, P_2O_5, and other incompatible elements (Figure 13.2). Alkaline basaltic rocks are even more enriched than tholeiites (Figure 13.12). Nearly all Hawaiian rocks have higher $^{87}Sr/^{86}Sr$ and lower $^{143}Nd/^{144}Nd$ than N-MORB (Figure 13.13). On Haleakala (east Maui) volcano (Figure 13.9), lavas of the three latest stages show differences in isotopic ratios. Shield tholeiites have the highest $^{87}Sr/^{86}Sr$ and lowest $^{143}Nd/^{144}Nd$ whereas posterosion alkaline lavas have the opposite ratios and

partly overlap N-MORB. Postshield alkaline lavas lie between these two. Preshield lavas are impossible to sample on Haleakala, but on the submarine Loihi volcano, alkaline lavas from this early stage have Nd-Sr isotopic compositions between those of the tholeiitic and alkaline lavas of Haleakala (Figure 13.13). This finding leads to the unexpected conclusion that Rb/Sr and $^{87}Sr/^{86}Sr$ are higher in the sources of the tholeiites than in the alkaline lava sources. And Sm/Nd and $^{143}Nd/^{144}Nd$ are lower in tholeiite than alkaline sources. In other words, tholeiitic basalts are derived from more enriched sources than those of incompatible-element-enriched alkaline lavas.

Geochemists try to explain this paradox in terms of different degrees of partial melting in different mantle sources. Shield tholeiites are largely generated from relatively enriched mantle near bulk Earth composition but mixed with depleted upper mantle similar to that of the MORB source. The relatively low Rb/Sr ratios and incompatible element concentrations in tholeiites indicate relatively high degrees of partial melting. Extensive melting dilutes the more incompatible Rb (and Nd) relative to Sr (Sm) while retaining the higher $^{87}Sr/^{86}Sr$ (lower $^{143}Nd/^{144}Nd$) of the enriched source in the melt. The generation of alkaline magmas is just the opposite. Small degrees of partial melting creates higher Rb/Sr ratios but from a source that is depleted in incompatible elements and is more like that of MORB. Posterosion lavas must also have a MORB-like source, since their isotopic ratios are like those of MORB, but are generated by even lower degrees of melting, so their incompatible element concentrations are highest; this source might lie in a metasomatically enriched lithosphere (Section 11.5.1).

Dynamics of Hawaiian Magma Generation: A Moving Plate above an Ascending Plume.

A long-standing problem concerning petrogenesis of Hawaiian magmas has been explaining the transitions between alkaline and tholeiitic magmas and back again that occur during the four stages of volcanic evolution. These sequential transitions are not readily explained by differentiation processes. To account for them by magma generation in a static mantle source there would have to be corresponding changes in degrees of melting, perhaps at different depths and volatile conditions, for which no reasonable independent explanation exists. However, the mantle beneath the Hawaiian Islands is not a static source, as Wilson (1963) first postulated, but is rather a rising plume beneath the moving Hawaiian lithosphere. This dynamic mantle source that is coupled to the moving overriding lithospheric plate explains time-dependent variations in eruption rate and magma composition (Figure 13.14). In such a thermally zoned plume, potential sources include the plume itself of enriched mantle, a sheath of entrained shallow and depleted asthenosphere through which it

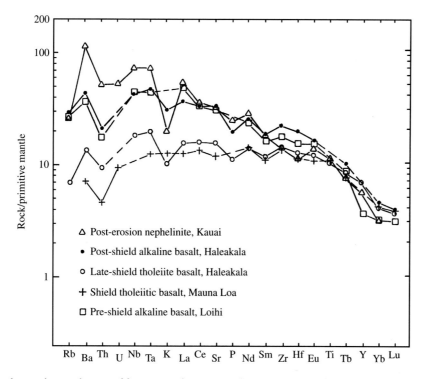

13.12 Normalized trace element diagram of four stages of Hawaiian volcanic rocks. (Data from references in Watson, 1993.)

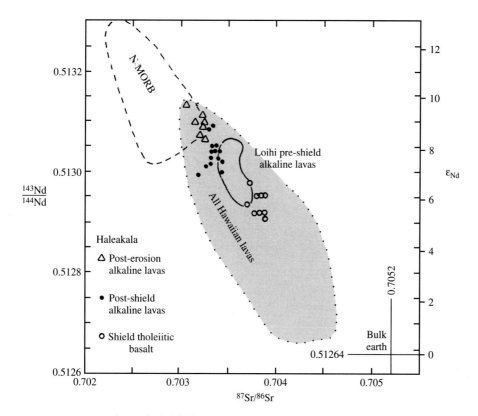

13.13 Isotopic ratios in Hawaiian lavas (shaded field) compared with Pacific N-MORB. (Redrawn from Chen and Frey, 1985.)

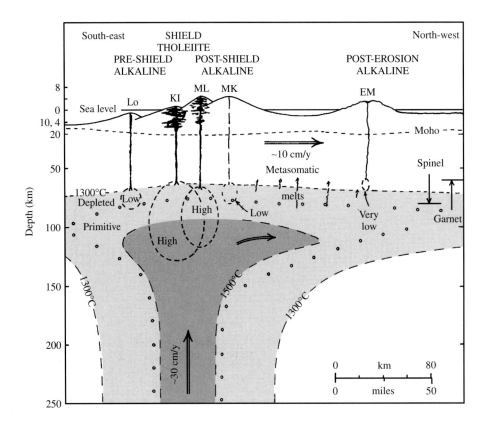

13.14 Schematic cross section through the southeast end of the Hawaiian Islands and underlying plume (see also Figure 1.5 and Watson and McKenzie, 1991). Note expanded vertical scale above ocean floor to show island topographical features. Horizontal scale for entire cross section is same as deeper vertical scale. Lithosphere lies above upper 1300°C isotherm. Asymmetry of plume is due to viscous drag of overlying lithosphere. Deep mantle part of plume inside line of small open circles has bulk Earth isotopic ratios; outside is depleted shallow mantle. Transition from spinel- to garnet-peridotite occurs at about 60- to 80-km depth but perhaps at 85–95 km in the hotter core of the plume. Preshield alkaline magmas (represented by a youthful Loihi volcano) are generated by relatively *low* degrees of partial melting (dashed-line circle) of depleted, decompressing shallow asthenospheric mantle entrained along margin of the plume. Relatively *high* degrees of partial melting (dashed-line ellipses), mainly in the hotter core of the enriched plume, generate tholeiitic magma in the shield-building stage; frequent eruptions occur mostly after storage, recharge, and fractionation in crustal magma chambers above the Moho (black treelike ornament beneath KI and ML). Peak magma production rates from the core of the plume whose *T* is >1500°C to 1300°C yield a layer of magma about 25 km thick—the Hawaiian crust (see path B in Figure 13.7). Mauna Kea (MK) apparently finished its postshield activity about 4000 years ago. Postshield alkaline magmas are again generated in depleted mantle by *low* degrees of partial melting. During this stage, the rate of magma production has dropped significantly so that some small crustal magma chambers cool and fractionate without frequent recharge, forming hawaiite, mugearite, and locally more fractionated daughter magmas. After a hiatus of as much as 2 My and about 160 km "downstream" of shield-building activity, a *very low* degree of partial melting of depleted MORB-like mantle that has perhaps been metasomatically enriched may locally segregate and ascend from near the base of the lithosphere. These posterosion nephelinitic magmas are most like MORB isotopically of any stage and reflect a decline in heat input from the passed-by plume.

moved, and heated overlying depleted lithosphere. Isotopically, the latter two correspond to the normal MORB source.

Blichert-Toft et al. (1999) find signatures of old crust and marine sediments in Pb, O, Hf, and Os isotopes in Hawaiian lavas, suggesting the plume source has somehow entrained old subducted crustal components, perhaps from the slab "graveyard" atop the core (Figure 1.4). They furthermore contend that Hf-Nd isotope trends preclude melting of depleted upper mantle sources. Obviously, much has yet to be learned about the origin of Hawaiian magmas, global recycling of the differentiated crust, and the process through which the mantle convects.

13.2.3 Highly Alkaline Rocks on Other Oceanic Islands

Unlike the Hawaii Islands, most intraplate oceanic islands have little, if any, exposed tholeiitic basalt. Whether the dominance of alkaline rocks is simply the result of their concealing the underlying largely tholeiitic shield, as in the older Hawaiian Islands, or whether volcanic islands other than the Hawaiian are dominantly alkaline throughout is difficult to say because of lack of subsurface information. However, in the plume model, degrees of magma generation and overall magma production rate are lower where the lithosphere is thicker because the ascending decompressing mantle encounters the rigid, nonconvecting

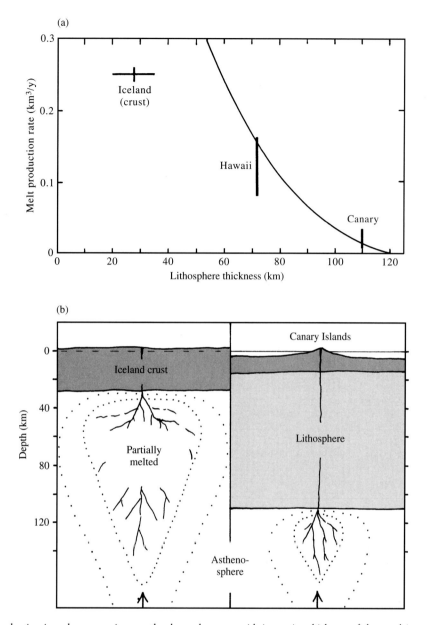

13.15 Partial melt production in a decompressing mantle plume decreases with increasing thickness of the overlying cooler mechanical lid (lithosphere). In the Canary Islands located on lithosphere about 110 km thick, tholeiitic basalt is scarce and the predominantly alkaline rocks represent magmas generated by small degrees of partial melting in a plume that could not decompress much above solidus temperatures. In contrast, the plume beneath Iceland experiences far more decompression before encountering thin lithosphere (essentially all crust), with the result that continuously high degrees of partial melting occur, generating copious amounts of tholeiitic magma. See also Watson and McKenzie (1991) and White (1993).

lithosphere before much partial melting can occur (Figure 13.15). This model suggests that oceanic islands founded on thicker lithosphere, such as the Canaries, have no major tholeiitic underpinnings and are predominantly alkaline basaltic rocks. Iceland, at the thin end of the lithospheric thickness spectrum, is predominantly tholeiitic.

On Gran Canaria, the largest of the seven volcanic islands in the Canary chain off the northwestern coast

of Africa in the Atlantic Ocean, late Cenozoic volcanism created only sparse tholeiite basalt but widespread nephelinite, peralkaline trachyte and rhyolite, phonolite, and less common picrite, basanite, tephrite, alkaline basalt, hawaiite, and melilite nephelinite.

The Azores are another group of presumed plume-generated islands just east of the Mid-Atlantic Ridge and the triple junction of the North American, African, and Eurasian plates. The islands delineate the African-

Eurasian plate juncture, which is a slow, obliquely spreading rift. Terceira, near the middle of the group, exposes mildly silica-undersaturated to -oversaturated basalt, hawaiite, mugearite, benmorite, and peralkaline trachyte lava flows and pyroclastic deposits making up >50% of the island.

Last erupted in 1961, Tristan da Cunha is the largest of a group of three islands in the South Atlantic about 430 km east of the Mid-Atlantic Ridge at the southwest end of the aseismic Walvis Ridge—a presumed plume track feature. The island is built of highly silica undersaturated basanite and lesser amounts of phonotephrite, tephriphonolite, and phonolite (Figure 2.16). Most of the evolved alkaline magmas—phonolites, trachytes, benmoreites, and others—in these islands originated through fractionation of plagioclase and mafic phases in parental mafic magmas.

❋13.3 PLUME HEADS AND BASALT FLOOD PLATEAU LAVAS

Prodigious floods of mainly basaltic lava have been extruded during relatively brief episodes throughout most of the history of the Earth. These **large igneous provinces** (Mahoney and Coffin, 1997) occur as submarine oceanic plate1aus as well as continental flood basalt plateaus (Figure 13.16).

13.3.1 Oceanic Plateaus

Oceanic plateaus are broad topographic prominences rising several hundred meters above the surrounding seafloor. Little is known about these submerged features, which are estimated to cover about 3% of the seafloor, because they are so inaccessible. Shipboard drilling, dredging off the seafloor, and sampling of rare, and sometimes questionable exposures on continental margins reveal that they are made mostly of thick sequences of basaltic lava flows. Some plateaus are underlain by crust as much as 40 km-thick—at least five times the thickness of normal oceanic crust.

The apparent magma production rate in the largest oceanic plateau—the Ontong-Java of the western Pacific—may be compared to that of oceanic ridge systems and oceanic island chains. The Alaska-size Ontong-Java is 30–43 km thick and has a volume of about 6×10^7 km^3. If it was produced in about 6 My, as available chronologic data suggest, the magma production rate was about 10 km^3/y, compared to the 21-km^3/y rate of the entire global ridge system (Figure 1.1) and 100 times the peak rate of a Hawaiian shield volcano (Figure 13.11).

One possible phenomenon that could produce such localized, brief bursts of tholeiitic magma production is very large-scale decompression melting of the head of a hot mantle plume. Plumes probably begin their ascent as Rayleigh-Taylor instabilities in the buoyant D″ layer

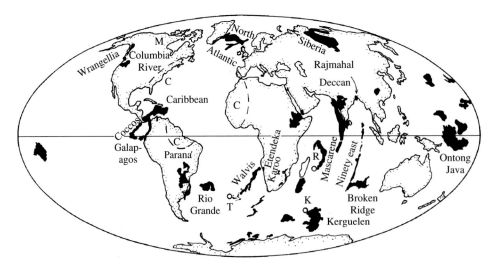

13.16 Large igneous provinces, chiefly basaltic. (See also Table 13.4.) Position of some currently active plumes indicated by open circles. Plume-caused flood basalt plateaus and tail-related island chains or ridge tracks can accompany opening of oceans. Examples include the following: (1) North Atlantic province (<60 Ma) related to currently active Iceland plume. (2) Northward opening of the South Atlantic about the Tristan da Cunha plume (T) since about 135 Ma; the Walvis Ridge tracks to the Etendeka province in Africa and the Rio Grande Plateau lies on the way to the Paraná Plateau in South America. (3) The Réunion (R) hotspot in the western Indian Ocean tracks northward along the Mascarene Plateau, thence along the Chagos-Maldive-Laccadive oceanic ridges to the Deccan Plateau (66 Ma) in western India. (4) The Kerguelen (K) hotspot formed the Kerguelen Plateau and Broken Ridge and Ninety east Ridge and the Rajmahal Plateau in eastern India (117 Ma). The Indian Ocean tracks have been modified by recent seafloor spreading. The huge Mackenzie basalt dike swarm in northern Canada (M; see also Figure 9.7a) formed above a 1.27-Ga plume head. The Central Atlantic igneous province, now mostly dikes and sills (C; see also Figure 9.7b), formed above a plume head at about 200 Ma near the Triassic-Jurassic boundary just preceding drifting apart of the American continents from Africa and Europe. (Redrawn from Coffin and Eldholm, 1994.)

of the lowermost mantle just above the core (Figure 1.5). Laboratory model experiments in tanks of viscous fluids reveal that a buoyant plume rises with a large-diameter head and a smaller-diameter tail. Large amounts of mantle wall rock are viscously drawn into the ascending plume and heated by it, particularly in the head so that temperatures there are perhaps 1350°–1400°C compared to 1550°C in the tail (Hill et al., 1992). As the plume head nears the rigid lithosphere, it flattens to as much as 2500 km in diameter, compared to the order-of-magnitude-smaller-diameter tail. Plume tails are thought to produce magma at a lower rate than plume heads and for tens of millions of years, creating linear volcanic island chains like the Hawaiian on moving lithospheric plates.

Because of their thickness and buoyancy relative to the mantle, oceanic plateaus riding on plates converging into island arcs or continental margins cannot wholly subduct. Consequently, parts are sliced off and tectonically emplaced, or **accreted,** onto the overriding plate. The Wrangellia terrane of coastal British Columbia and southern Alaska (Figure 13.16) is one such segment of an oceanic plateau formed in the Pacific and accreted onto the North American plate margin during the Mesozoic. More complete sections through a late Jurassic-Cretaceous ocean plateau are represented by widespread thick exposures of tholeiitic basalt and ultramafic rocks in the Caribbean area and western Colombia (Figure 13.17). Exposures on the small island of Gorgona off the west coast of Colombia are especially noteworthy as they are the only known Phanerozoic (post-Precambrian) occurrence of ultramafic lava flows called **komatiite.**

The high MgO (>18 wt.%) of komatiites and the general absence of phenocrysts, which precludes a cumulate origin, imply relatively high degrees of partial melting of a peridotite source. The necessarily high T for komatiite generation is consistent with derivation from a plume. So, too, is the somewhat enriched trace element pattern for the tholeiitic basalt lavas that apparently constitute most of oceanic plateaus (Figure 13.18; Table 13.5). In terms of their isotopic ratios, these lavas lie closer to bulk Earth than MORB and Hawaiian lavas (Figures 13.19), again consistently with a plume origin.

13.3.2 Continental Flood Basalt Plateaus

The best known flood basalts on continents are Mesozoic and Cenozoic, but some are Proterozoic (Figure 13.16; Table 13.4). Swarms of basalt sills and feeder dikes are associated with lava flows, particularly in the older, more deeply eroded plateaus (Figure 13.20). In a few locales, unusually large volumes of basalt magma were trapped within the crust and differentiated to form major layered intrusions (Table 12.5).

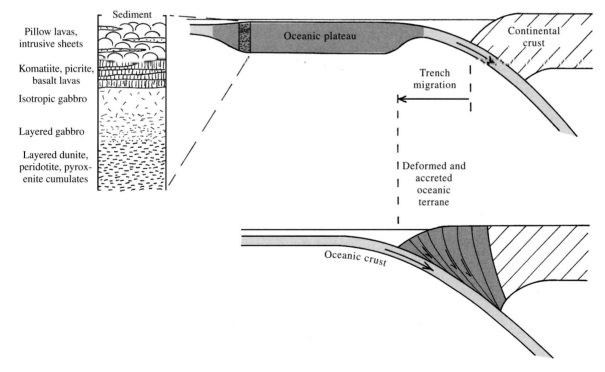

13.17 Highly schematic **accretion** of an oceanic plateau onto a continental margin in a subduction zone. This is one form of continental growth. Expanded stratigraphic section through the Caribbean Plateau on the left from Kerr et al. (1997). The lack of intercalated sediment in the sequence of extrusive rocks suggests rapid accumulation. The lack of sheeted feeder dikes, typical of the MORB association seen in ophiolite (Figure 13.1a), indicates no concurrent crustal extension occurred during extrusions of magma.

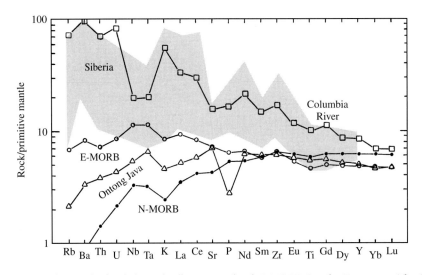

13.18 Normalized trace element diagram for flood plateau basalts compared with E-MORB. See also Figure 13.2. The Columbia River sample (Table 2.1) characterizes the more enriched continental flood basalt plateaus; note strong negative Nb-Ta anomaly. Nine Siberia samples (shaded; from Sharma, 1997) range to a less enriched pattern similar to E-MORB. Ontong Java (Table 13.5) is less enriched in more incompatible elements than some other oceanic plateau rocks.

Flood basalt plateaus consist of large numbers of superposed, fissure-fed, sheetlike floods of lava (Figures 10.12–10.14), at least some of which were extruded in a compound manner as successive lobes. The enormous volume of flood plateaus (generally $\sim10^6$ km³) and short duration of peak activity ($<$ a few million years and apparently ~1 My for well-dated ones) are compatible with copious magma generation in the head of a hot decompressing plume.

Lavas are chiefly tholeiitic basalt, but more evolved, Fe-rich lavas occur in some places and, in the Karoo, Ethiopia, and Paraná plateaus, substantial amounts of silicic lavas were also extruded. Picrites are uncommon except in the Karoo and Deccan. Basalts are commonly aphyric; porphyritic ones have plagioclase as the usual phenocryst. The groundmass typically contains plagioclase, Ca-rich and Ca-poor clinopyroxenes, relatively abundant Fe-Ti oxides, local glass, and, rarely, olivine.

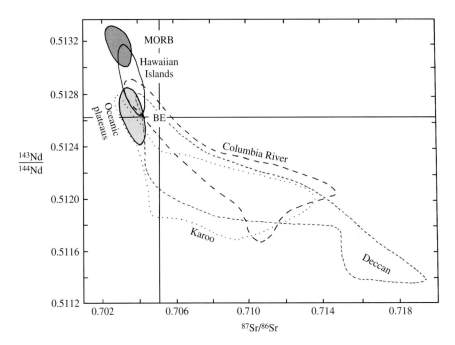

13.19 Sr and Nd isotopic ratios in flood basalts compared to MORB (dark shade) and Hawaiian Islands (Figure 13.8). Oceanic plateaus (light shade) are from Kerr et al. (1997). Continental Columbia River plateau from Hooper (1997), Karoo from Cox (1988), and Deccan from Mahoney (1988). Compare Figure 2.27. BE, bulk silicate Earth.

Table 13.4. Continental Flood Basalt Provinces

	AGE (MA)*	VOLUME (10^6 KM^3)	MAX. THICKNESS (KM)	AREA (10^6 KM^2)	COMMENTS
Columbia River	17–6 (16)	0.17	4+	0.16	300 Lava flows ranging from 700 to 3000(?) km³
Ethiopia	25±	0.35	3	0.75	Multiple episodes of activity in Cenozoic; upper part of plateau includes large volume of peralkaline silicic rocks
Deccan	(66)	2.6	2	>1.5	100–500 lava flows ranging to 10⁴ km³; possible cause of faunal extinctions near K-T boundary
NorthAtlantic	59±	6.6	6	1.3	
Paraná-Etendeka	138–120 (134–129)	1.5	1.7	1.2	Contemporaneous alkaline magmatism; late silicic lava flows
Karoo	184–179 (183 ± 1)	2.5	9	2	Covered much of southern Africa but now extensively eroded, exposing abundant sills and randomly oriented feeder dikes (Figure 13.20), correlative basalts in southern Australia and Antarctica (Ferrar province); in southeastern Africa includes voluminous silicic flows
Central Atlantic	200 ± 4	2	0.3 +?	7	Mostly sills and dikes in northwestern Africa, eastern North America, and northeastern South America; may have caused Triassic-Jurassic extinctions
Siberia	(248)	>2.5	3	>1.5	Large amounts of intercalated basaltic tuff; not clearly connected with an active plume; may have caused Permian-Triassic extinctions
Keweenawan	1095 ± 5	1.3	5	>2	>350 flows in Great Lakes region of North America
Coppermine River and Mackenzie	1267 ± 2	0.2	5	0.01	Mackenzie radial feeder dike swarm probably the largest in the world (Figure 9.7a); scattered flows in Coppermine River area

See Figure 13.16 for locations.
*Peak activity in parentheses.
Data from Sigurdsson et al. (2000), White and McKenzie (1995), and other sources.

Chemically, the dominant rock type is slightly silica-saturated quartz tholeiite more evolved than MORB in that Fe, Ti, P, and K concentrations are greater and Mg concentration less. MgO is commonly 5–8 wt.%, Ni < 100 ppm, Mg/(Mg + Fe^{2+}) < 0.65 (Table 13.5; Figure 13.18).

Origin of Magmas. These attributes indicate substantial crystal fractionation at low P from more primitive mantle partial melts. Indeed, Cox (1993) argues that primitive (picritic?) magmas experienced polybaric olivine removal during ascent, reducing compatible element concentrations and MgO to ~7 wt.%. Subsequently, these fractionated magmas experienced additional 30%–40% removal of olivine, clinopyroxene, and plagioclase in near-Moho storage chambers, where they were buoyantly blocked from further ascent. This fractionated gabbroic assemblage has about the same composition as the parent magma with regard to many major elements, and, hence, little indication of its removal may be apparent. However, Ni and Cr contents are low. Independent evidence for this extensive fractionation, at least in some provinces, lies in the voluminous contemporaneous silicic lavas, which require that large subterranean magma chambers either fractionate or supply heat to partially melt country rocks or do both. Underplating of perhaps 5 km of gabbroic fractionates is implied by the 1 km of permanent post-Karoo uplift in southern Africa: That is, 5 km of new crust was added by the magmatic activity.

Normalized trace element patterns for some continental flood basalts are similar to those of ocean island tholeiites and E-MORB (compare Figures 13.12 and 13.18), implying a similar plume source. However,

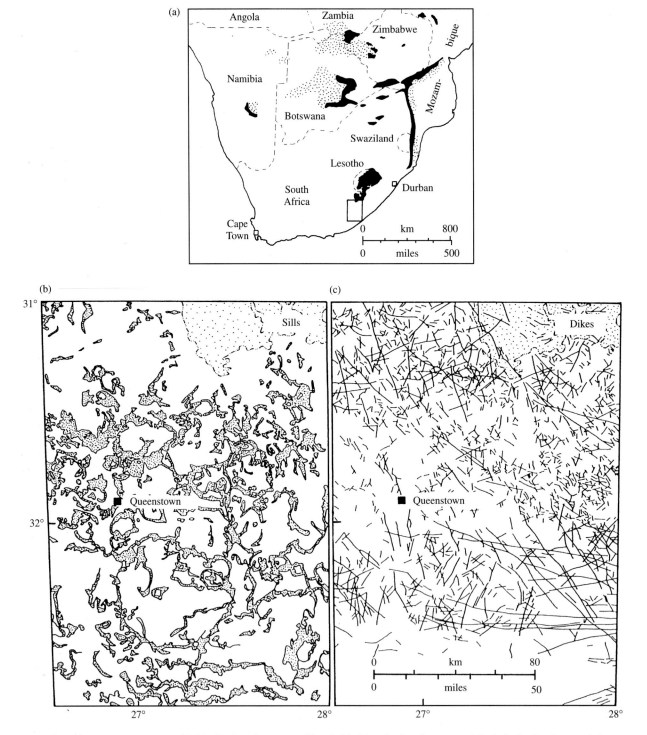

13.20 Karoo large igneous province. (a) Distribution of outcrops of basalt (black) and subsurface extent (stippled). Outlined rectangle is area in (b), showing sills, and (c), showing dikes. (From Marsh et al., 1997.)

other basalts are more enriched in incompatible elements. A continuing debate is whether this enrichment stems from contamination by the continental crust or by partial melting of metasomatically enriched continental lithospheric mantle (e.g., Wilson, 1989; Lassiter and De Paolo, 1997). Contamination is an expected companion to extensive crystal fractionation in crustal magma chambers and to generation of voluminous contemporaneous rhyolitic lavas, as in the Karoo and Paraná plateaus. Some continental flood basalts have distinct negative Nb-Ta anomalies (Figure 13.18). Such an anomaly is typical of arc magmas (Figure 11.19; see also Section 11.4.2) and continental crust built of them, suggesting contamination by continental crust. Many

Table 13.5. Compositions of Select Plateau Flood Basalts

	OCEANIC	CONTINENTAL		
	1	2	3	4
SiO$_2$	49.25	52.76	49.97	51.10
TiO$_2$	1.20	2.325	3.07	1.32
Al$_2$O$_3$	14.00	13.24	8.22	14.45
Fe$_2$O$_3$t	12.69	14.58	12.02	13.26
MnO	0.21	0.219	0.16	0.18
MgO	7.75	4.15	15.52	6.30
CaO	12.35	8.18	7.07	10.61
Na$_2$O	2.05	2.65	1.43	1.96
K$_2$O	0.14	1.28	2.10	0.53
P$_2$O$_5$	0.06	0.44	0.45	0.19
Rb	1.32	33	55	7.5
Ba	23.6	503	917	235
Th	0.33			
U	0.09			
Nb	3.76	13	19	3.7
Ta	0.27			
La	3.66	21.80		
Ce	10.55	47.34		
Sr	153	314	1000	185
Nd	8.26	30.51		
Sm	2.67	7.51		
Zr	69.3	210	402	117
Eu	0.97	2.34		
Gd	3.40	8.56		
Dy	3.90	8.60		
Y	22.6	53	28	31
Yb	2.28	4.66		
Lu	0.34	0.76		
Cr	133.0	22	804	156
Ni	98.1	19	827	57
Sc	46.5	39	21	
V	314	402	204	285

See also Table 2.1.
1, Tholeiitic basalt, average of 21 samples, Malaita, Ontong Java plateau. Data from Neal et al. (1997). 2, Grande Ronde basalt, Columbia River Plateau. Aphyric lava flows like this constitute about 85% of plateau. Data from Hooper (1988). 3, Average of 19 samples of olivine-rich basalt, Letaba Formation, north Lebombo, Karoo province; represents possible near-primary mantle-derived magma. Data from Cox (1988). 4, Average of five samples of basalt. Transvaal, Karoo province. Data from Cox (1988).

surrounding mantle domes the lithosphere upward, creating kilometer-scale uplifts. Extensional stresses are induced in the domed rigid plate, allowing intrusion of radial dike swarms that are thousands of kilometers in diameter (e.g., Figure 9.7). These swarms are exposed in older large igneous provinces where covering flood lavas have been removed by erosion. In some cases, plume activity has caused breakup of a large continental plate, leaving plateau basalt segments and/or dike-sill swarms stranded on passive continental margins adjacent to the opened ocean. Other apparently plume-related flood volcanism, as in Siberia and in the Columbia River Plateau (Figure 13.16), did not produce continental breakup.

Remnants of continental flood lavas on passive continental margins flanking the Atlantic provide an important record of plume initiation, continental breakup, and rifting.

Breakup of Gondwana began in the south Atlantic during the Cretaceous (Figure 13.21). Most of the ~1.5 × 10^6 km^3 volume of lavas making up the Paraná and Etendeka flood plateaus (Figure 13.22) bordering the Atlantic Ocean in eastern South America and southwestern Africa, respectively, were extruded about 134–129 Ma ago. These flood lavas are distinctly bimodal in composition. Most mafic rocks, usually aphyric tholeiite, have 48–59 wt.% SiO$_2$ whereas most silicic rocks have 63–72 wt.% SiO$_2$. Very few unaltered rocks have 59–63 wt.% SiO$_2$. Two distinct groups of silicic rocks have different sources. High-Ti porphyritic rhyolites were apparently derived from partial melts of earlier basalts underplated or lodged in the lower crust as succeeding basalt magmas supplied heat for melting. Low-Ti aphyric rhyolites have trace element and isotopic ratios indicating derivation by AFC—assimilation of crust into fractionally crystallizing basalt magmas. The aphyric rhyolite flows occur as remarkably homogeneous individual units lacking compositional zonation despite their large volume (commonly >1000 km^3). Some flowed distances of >300 km and, because of unique compositional attributes, have been correlated on opposite sides of the Atlantic Ocean! Their low aspect ratios, as low as 1/2000, are unusual for lava flows, and it has been suggested they were emplaced as ignimbrites, but the distinctive pyroclastic fabric is lacking, possibly because it was obliterated by rheomorphic flow. Alkaline magmatism that was broadly contemporaneous with Paraná flood volcanism includes relatively small local occurrences of alkaline gabbros, syenites, phonolites, and carbonatites.

The North Atlantic igneous province (Figure 13.23) developed during continental breakup in early Tertiary time as Greenland and eastern North America drifted away from northwestern Europe after initial rise of the Iceland plume.

flood basalts have greatly elevated ^{87}Sr/^{86}Sr and depressed ^{143}Nd/^{144}Nd ratios (Figure 13.19), which are consistent with extensive contamination with old continental crust.

13.3.3 Continental Breakup

The upward momentum of the ascending plume together with the thermally expanding rock in it and the

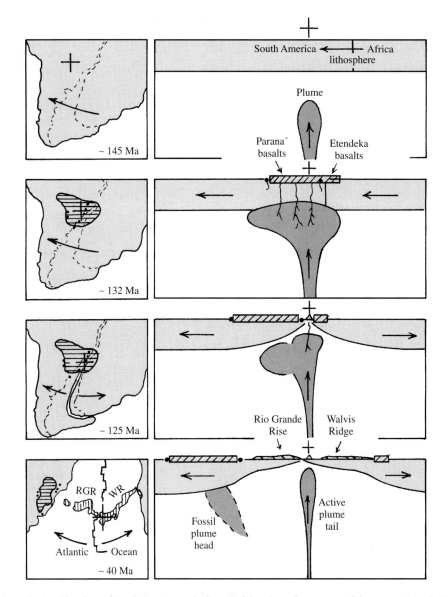

13.21 Plume activity and related breakup of South America and Africa. Left-hand panels are maps of the two conjoined (Gondwana) and then rifted-apart continents. Right-hand panels are enlarged highly schematic cross sections. Plus sign in both panel sets overlies plume axis. While conjoined, the two continents both were drifting westward. At about 132 Ma, the decompressing, broadening plume head partially melted, yielding basalt magmas that formed the Paraná-Etendeka flood basalt plateau and local alkaline magma centers (small filled circles). At about 125 Ma, the Gondwana protocontinent began to rift apart; the western part of the plume head that was attached to the South American lithosphere began to separate from the main part of the plume. Subsequently, the decompressing plume tail has generated basalt magmas that have formed the Walvis Ridge (WR) and the Rio Grande Rise (RGR) as the two continental plates drifted east-northeast and west-northwest. Since about 40 Ma the plume tail has underlain Tristan da Cunha just east of the Mid-Atlantic Ridge. Seismic studies indicate a 300-km-diameter low-velocity cylinder in the mantle to a depth of about 600 km beneath Brazil that is interpreted to be a still abnormally hot fossil plume. Most of the new basaltic crust at the Mid-Atlantic Ridge is being added to the South American plate, causing the ridge and the Nazca-South American plate juncture (Frontispiece) to migrate westward, opening the Atlantic and closing the Pacific. (Redrawn from Van Decar et al., 1995.)

❋13.4 ARC MAGMATISM: OVERVIEW

Compared to the chiefly basaltic petrotectonic associations of preceding sections, associations in subduction zones are far more variable. Diverse magmas are generated in multiple sources and are subsequently subjected to virtually the whole gamut of differentiation processes.

Arc magmatism at continental and oceanic plate margins overriding subducting oceanic lithosphere is second only to oceanic rift magmatism in terms of volume of magma produced (Figure 1.1), frequency of eruption, and longevity. Marked by the inclined Wadati-Benioff zone of earthquake foci, subduction regimes are the most complex of all global geologic systems. The complex dynamic interplay between

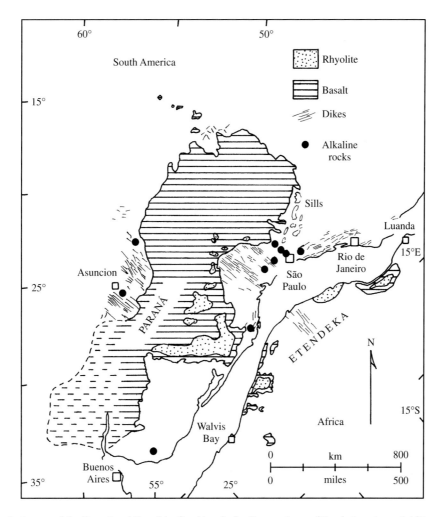

13.22 Simplified geologic map of the Paraná and Etendeka flood basalt-rhyolite provinces of South America and Africa restored to their pre-rift position before about 138 Ma. Paraná basalts in southwest are concealed beneath younger sedimentary rocks. (Redrawn from Peate, 1997.)

transfer of matter and energy—chiefly gravitational and thermal—is manifested not only in magmatism and seismicity, but also in mountain building or orogeny (forming mountain belts or orogens), crustal growth, sedimentation, metamorphism, and ore deposition.

Some of the transfer of matter in subduction zones is unidirectional and irreversible. Thus, arc magmatism adds partial melt extracts from the mantle to the crust, as does ocean ridge and plume activity; such additions may have been a significant factor in crustal growth throughout Earth history. Certain unidirectional changes in the composition of the mantle with time are complementary to crustal growth. Other transfers of matter in subduction zones involve recycling, the most obvious of which is cycling of volatiles, especially water. Physically entrapped and chemically bound water in hydrothermally altered subducting oceanic crust is liberated into the overlying mantle wedge. Partial melting generates hydrous magmas, some of which are extruded from volcanoes. Exsolved

water from erupting magmas is released into the atmosphere and condenses as rain and snow that fall to Earth and collect into rivers that flow back into the ocean, its original source. Still controversial as to its extent, is the cycling of oceanic rock. Subducted seafloor sediment may be partially melted and added to arc magmas building the crust. Some subducted slabs that include seafloor mafic rock converted to dense eclogite appear to sink to the bottom of the mantle (Figure 1.4), where matter may be caught up in ascending plumes that partially melt near the surface, creating magmas that are again added to the crust. Erosion of subaerial crustal rock dumps matter back into the ocean, completing the mass transfer circuit.

☀13.5 OCEANIC ISLAND ARCS

It is logical to begin a discussion of arc magmatism with oceanic island arcs. The effects of the "crustal filter" through which mantle-derived magmas ascend toward the surface are limited relative to continental arcs,

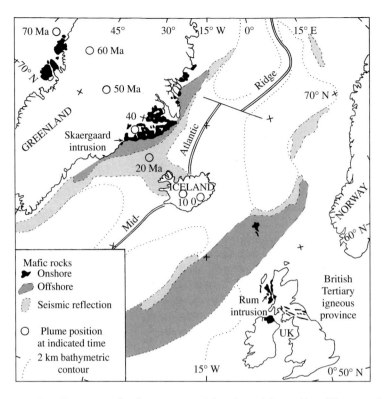

13.23 North Atlantic igneous province. Large areas of mafic magmatic rock have been delineated by offshore geophysical surveys and deep sea drilling. (Redrawn from Saunders et al., 1997.)

which have thicker crust of sialic rather than mafic composition.

Island arcs built on oceanic crust (Plate I) can be viewed as embryonic continents that grow and mature as the crustal thickness and area increase over millions of years with continued magmatic input. Although rarely exposed, intrusive rocks make up part of these volcanic edifices. Island arcs are linear groups of volcanoes several thousands of kilometers long and no more than 200–300 km wide. The width decreases with increasing dip of the subducting slab. They are bordered on one side by an **ocean trench,** as much as 11 km deep, that is the topographic expression of subducting lithosphere (Figures 11.16 and 13.17). Scraped-off ocean crust forms an **accretionary prism** (or forearc accreted terrane) on the edge of the overriding plate. The trenchward limit of volcanoes—the **volcanic front**—is 50–300 km from the trench, depending on slab dip, but is consistently about 110 km above the top of the subducting slab (Tatsumi and Eggins, 1995), as defined by earthquake foci. The volume of extruded magma decreases away from the volcanic front. In some arcs, a spreading back arc or marginal basin (Figure 11.16) lies on the opposite side from the trench.

13.5.1 Rock Associations

Whereas basalt is the typical rock of the ocean floor, island chains, and plateaus, the most common rock of convergent plate settings, whether island or continental

arc, is andesite (Gill, 1981). Although rocks with the same range of SiO_2 as andesite (57–63 wt.%) occur in virtually all petrotectonic associations, other element concentrations differ in these other associations, such as lower Al and higher Fe in icelandite and higher alkalies in benmoreite. Continental crust of intrusive and metamorphic rock is approximately andesite in average chemical composition, implying that andesitic magmatism has played a role in continental growth.

Arc rocks range widely and continuously in chemical composition on each side of andesite (Table 13.6 and Figure 13.24). Most are subalkaline, silica-saturated to silica-oversaturated. Rock types typically include basalt, basaltic andesite, andesite, dacite, and rhyolite. A large spectrum of K_2O concentration is subdivided into four rock series in Figures 2.18 and 13.24, including low-K (tholeiitic), medium- and high-K (calc-alkaline), and shoshonitic. High-K and shoshonitic rocks may be alkaline and silica-undersaturated. Complex variations along and across strike and through time within a single arc as well as variations between arcs preclude accurate generalizations on a global scale. Nonetheless, a few tendencies can be cited, as follows.

Low-K (tholeiitic) arc rocks are mostly basalt, basaltic andesite, and lesser amounts of andesite, rarely dacite and rhyolite, that form low mounds and shields around near-trench vents, chiefly in young arcs founded on thin crust. Lava flows are commonly

Table 13.6. Chemical Compositions of Island Arc Volcanic Rocks

	1	2	3	4	5	6	7	8	9
SiO_2	49.49	53.29	65.78	60.48	53.07	58.74	63.89	49.68	48.07
TiO_2	0.68	0.91	0.65	0.81	0.73	0.23	0.61	0.85	0.73
Al_2O_3	18.71	17.13	15.53	16.30	17.50	11.22	17.40	19.39	15.64
Fe_2O_3	2.75	3.47	1.63	7.90t	9.30t				2.34
FeO	7.14	7.83	3.39			8.53t	4.21t	9.33t	6.98
MnO	0.18	0.20	0.13	0.19	0.17	0.17	0.08	0.14	0.15
MgO	6.24	4.72	1.76	2.57	3.72	12.09	2.47	5.60	12.68
CaO	12.76	9.87	4.71	6.35	7.76	6.77	5.23	10.97	10.76
Na_2O	1.73	2.21	3.82	3.94	3.46	1.69	4.40	2.66	2.14
K_2O	0.28	0.30	0.44	1.25	3.93	0.47	1.52	0.82	0.48
P_2O_5	0.05	0.06	0.13	0.21	0.56	0.09	0.19		0.12
Total	100.00	100.00	97.97	100.00	100.20	100.00	100.00	99.44	100.099
Mg/(Mg+Fe)	0.57	0.47	0.44	0.56	0.61	0.72	0.51	0.52	0.74
Cs						0.71	0.49	1.19	0.52	0.06
Rb	5.6	4.9	6.2	18	75	11	30		
Ba	94	111	253	419	684	43	485	261	241
Th	0.39	0.36	1.27		1.2	0.40	3.52	1.33	1.42
U			0.49		0.5	0.19	0.99	0.54	0.70
Nb	0.53	0.76		2.4	1.8	2.2	8.3		
Ta					0.15	0.07	0.53	0.19	0.09
La	2.3	2.3		39.7	8.8	2.07	17.55	5.99	7.30
Ce	5.7	6.2		31.2	21	4.58	34.65	13.86	18.15
Sr	237	174	184	352	1545	100	869		
Nd				27.2	14	2.67	20.14	10.66	11.38
Sm				24	3.8	0.75	3.15	2.77	2.69
Zr	29	39	72	111	71	32	117		
Eu				21.1	1.2	0.28	0.97	0.911	0.834
Gd				22.2		1.06	2.25		
Dy				18.1		1.35	1.43		
Y	13	21		35	26	6.1	9.5		
Yb				15	2.5	0.89	0.91	1.91	1.52
Lu				15		0.15	0.15	0.239	0.223
Cr	66	27	22		9	969	54	32	51
Ni	33	16		8	7	223	39	22	272

1, Low-K (tholeiitic) basalt, Kermadec arc. Data from Ewart and Hawkesworth (1987).
2, Low-K (tholeiitic) basaltic andesite, Kermadec arc. Data from Ewart and Hawkesworth (1987).
3, Low-K (tholeiitic) dacite, Kermadec arc. Data from Cole (1982).
4, Medium-K (calc-alkaline) andesite, Mariana arc. Data from Woodhead (1989).
5, Shoshonite, Tavua, Fiji, subaerial Lau Ridge back arc behind Tonga arc. Data from Gill and Whelan (1989).
6, Boninite, average of 134 analyses. Data from Drummond et al. (1996).
7, Adakite, average of 140 analyses. Data from Drummond et al. (1996).
8, High-Al, low-Mg basalt, Aleutian arc. Data from Kay and Kay (1994).
9, High-Mg, low-Al basalt, Aleutian arc. Data from Kay and Kay (1994).

aphyric, especially where more mafic, but phenocrysts include olivine, calcic clinopyroxene and plagioclase, locally Fe-Ti oxides, and subcalcic clinopyroxene (pigeonite). In addition, these rocks usually have high Fe/Mg ratios that justify their designation as tholeiitic (Figure 2.17).

Medium- and high-K arc rocks of the **calc-alkaline series** are chiefly porphyritic andesites that contain phenocrysts of calcic plagioclase, ortho- and clinopyroxene, Fe-Ti oxides, and commonly minor amounts of hornblende and biotite. Basalt, basaltic andesite, dacite, and rhyolite are generally minor rock types; the latter two contain the same phenocrysts as andesites, plus quartz and sanidine in the high-K series. More silicic viscous lavas form steep-sided edifices, including composite volcanoes, and the greater water content of

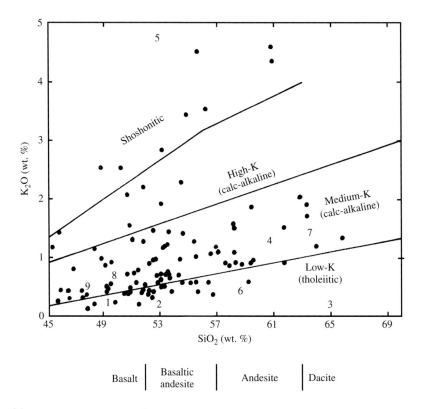

13.24 Volcanic rocks of the Mariana oceanic arc. Numbers 1–8 are analyses from Table 13.6. Compare with volcanic rocks in the Tonga oceanic arc that are low-K tholeiitic (Figure 2.18). (Data from Woodhead, 1989 and Bloomer et al., 1989.)

the magmas promotes explosive eruptions, forming pyroclastic deposits. Calc-alkaline rocks are typical in more mature arcs and in extrusions farther from the trench. This tendency implies that thicker crust favors development of calc-alkaline magmas that are mostly more evolved with higher alkalies and silica (Section 12.5.2). Apparently where the oceanic crust is still thin, not much more than average in the open ocean (6–7 km), basaltic magmas can rise and readily extrude. As the crust thickens from the top by the basaltic extrusions and from within by intrusive underplating, mantle-derived magmas are more likely to stagnate within the crust, where they differentiate, mainly by crystal fractionation, becoming more silicic, potassic, and hydrous.

Most arcs, especially the Aleutians, have low- to medium-K basalts that are highly porphyritic, containing 25%–40% phenocrysts of calcic plagioclase, olivine, and clinopyroxene. Chemically, these **high-Al basalts** have >17 wt.% to as much as 22 wt.% Al_2O_3 and low MgO (<6 wt.%) and compatible elements (Table 13.6). Less abundant high-MgO (>9 wt.%) basalts generally contain only olivine phenocrysts and have high Ni and Cr but lower Al_2O_3.

Shoshonitic series rocks (Figure 2.18), including **absarokite** (essentially a very-high-K basalt), **shoshonite** (very-high-K basaltic andesite), and **banakite** (very-high-K andesite), are less common than the other rock series and tend to form in mature arcs that have thick crust, but some have been found to be related to back-arc rifting. Plagioclase and two pyroxenes are typical phenocrysts in an alkali-rich groundmass.

Two less common rock types (Table 13.6) are defined independently of K_2O-Si_2O relations. **Boninite** is a high-MgO (>8 wt.%), low-TiO_2 (<0.5 wt.%), and very incompatible-element-depleted basaltic andesite or andesite belonging to the low- or medium-K series. It is commonly glassy, and typical phenocrysts include both orthorhombic and monoclinic enstatites (~En_{90}), Mg-rich olivine (~Fo_{90}), and calcic clinopyroxene (Crawford, 1989). **Adakite** is a dacite-andesite rock that has high La/Yb ratios relative to those of the more widespread calc-alkaline rocks (Figure 11.19).

13.5.2 Magma Evolution

The low MgO, Ni, Cr, and Mg/(Mg + Fe) content of most arc rocks (Table 13.6) indicates that magmas are not primary and have experienced extensive differentiation after leaving their source. Their evolved compositions can be appreciated in a CaO-MgO diagram (Figure 13.25), which shows that only the most Mg-rich arc rocks might represent relatively primitive magmas generated from hydrous peridotite in the subarc mantle wedge. The bend in the CaO-MgO variation trend of arc rocks at about 12 wt.% MgO likely reflects a transition from a fractionating assemblage dominated

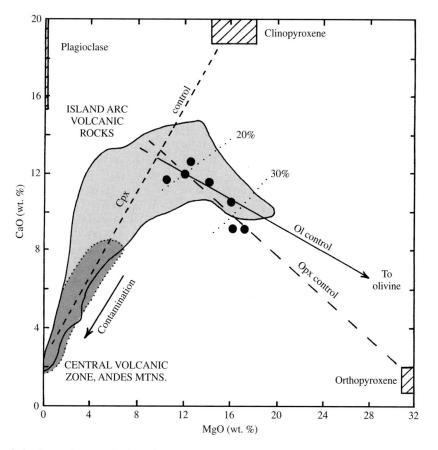

13.25 Compositions of island arc volcanic rocks from the west Pacific and Lesser Antilles in the Caribbean (light shading). For comparison, continental arc rocks from the central Andes are shown in dark shading. Compositions of fractionating crystals indicated by diagonally ruled boxes with representative control lines. More Mg-rich magmas probably experienced olivine fractionation; less Mg-rich magmas fractionated by clinopyroxene ± plagioclase removal. Filled circles are experimentally produced partial melts (18%–38% melt fraction) at 10 kbar and 1200°–1350°C of hydrous lherzolite (Hirose and Kawamoto, 1995). Most arc rocks solidify from evolved magmas that have experienced considerable differentiation, including fractionation, contamination with felsic crustal rocks, and magma mixing. (Redrawn from Davidson, 1996.)

by olivine ± orthopyroxene that operates on more primitive magmas to fractionation of mostly clinopyroxene + plagioclase at lower temperatures on previously olivine fractionated magmas.

Widespread, though generally minor, hornblende and biotite in arc rocks indicate that magmas contain at least 3 wt.% or so H_2O, implying that this volatile is an essential component in the evolution of arc magmas. High water fugacity expands the stability field of olivine and reduces plagioclase stability so that Mg, Fe, Ni, and Cr are reduced and Na and K increased in residual melts. These factors promote development of the felsic-rich calc-alkaline trend, rather than the Skaergaard trend of increasing Fe (Section 12.5.2). Primitive magmas ascending into more mature arc crust can also be contaminated by felsic rock, further accentuating the calc-alkaline tendency.

Differential solubility of incompatible elements in slab-derived hydrous fluids is believed to produce the typical spiky arc signature in normalized trace element diagrams (Figure 11.19).

Boninite and adakite magmas are generated in young arcs overriding young hot subducting lithosphere. Boninite is believed to be generated by partial melting of hydrated harzburgite in the mantle wedge and adakite by partial melting of hydrated basaltic crust in the descending slab (Section 11.4.2).

The origin of the high-Al basalts, whether by partial melting of the subducted oceanic crust or by fractionation from parental high-MgO basalt magmas, remains highly controversial, and both may be factors. High-MgO basaltic magmas, which may be parental to many arc rocks, cannot be derived from subducted basaltic crust and their origin remains problematic.

13.5.3 Back-Arc Basins

Back-arc or marginal basins are areas of spreading oceanic crust adjacent to many island arcs on the opposite side from the deep trench (Figure 11.16). They are especially common in the western Pacific, such as the Lau Basin behind (west of) the Tonga island arc (Hawkins, 1995). Crustal extension, confirmed by symmetric linear magnetic anomalies, is paradoxical and controversial; it may be caused by some sort of counterflow in the subarc mantle wedge or "roll back" of a gravitationally sinking slab within the overall convergent plate regime. Basalts forming new ocean floor along back-arc spreading ridges resemble MORB but have distinct arc affinities, such as negative Nb-Ta anomalies and elevated water content. Somehow, aqueous fluids liberated from the subducting slab are transferred into the MORB-like magma source beneath the back-arc spreading axis.

✻13.6 OPHIOLITE

Accreted terranes of oceanic rocks emplaced along margins of continental and oceanic plates overriding subducting slabs (Figures 13.17 and 11.16) commonly include a distinctive *sequence* of rocks called ophiolite. Most of the hundreds of ophiolite sequences recognized worldwide are dismembered to varying degrees so that only parts are exposed. Unusually well-exposed, complete ophiolites include the Semail along the northeast coast of the Arabian Peninsula in the Sultanate of Oman (Searle and Cox, 1999), the Bay of Islands in western Newfoundland, and the nearly complete Troodos ophiolite on the island of Cyprus in the Mediterranean (Moores, 1982).

Contact thermal metamorphic effects on country rocks are limited or absent, even though a high-*T* magma is implied by the high content of Mg-rich olivine in the ultramafic members of the ophiolite sequence. Instead, contacts between the ophiolite package and surrounding rocks, as well as internal lithologic contacts, are tectonic, marked by breccia and slickensides; hence, ophiolites appear to have been emplaced as subsolidus masses, commonly along fault zones.

13.6.1 Characteristics

Ophiolite is a distinctive sequence of magmatic, sedimentary, and metamorphic rocks formed in an oceanic environment and made up of the oceanic crust and uppermost mantle rocks (Figure 13.1a; see also Moores, 1982). (Ophiolite is from the Greek, *ophite*, referring to a serpent, in allusion to the widespread scaly green serpentine in the sequence.) Always variably deformed, recrystallized, and hydrated, a complete ophiolite sequence consists of the following, from the top down:

1. Marine sedimentary rock: Thinly layered (centimeters-scale) Fe-Mn-rich chert and shale are common, but deep ocean (pelagic) red limestone occurs in some. In many ophiolites, volcaniclastic deposits intercalated with turbidite sequences indicate nearby explosive volcanism and development of deep-sea fans essentially contemporaneous with underlying magmatic rocks; such deposits are typical of island arcs rather than the open-ocean, intraplate environment where chert, shale, and limestone form. Moores (1982) argues that the unconventional inclusion of the sedimentary component in the definition of ophiolite provides a geologic criterion for deciding the oceanic environment in which ophiolites are created.

2. Extrusive magmatic rock, chiefly basaltic: Pillow lavas predominate, but sheet flows and breccias are common. Apparently arc-related volcanic debris flows and silicic lavas are found in some ophiolites. Sills are locally common, as are dikes, which increase downward.

3. **Sheeted dike complex:** Dikes mostly of basalt and slightly coarser diabase (dolerite) are generally 1–3 m thick. In the Oman ophiolite, dikes have a uniform strike over an exposed distance of 400 km! Dikes intruded into dikes, without any other wall rock, are convincing proof of formation in actively extending crust. Anastamosing zones of brecciation on scales from microscopic to hand sample are widespread. Their irregular disposition is incompatible with a tectonic origin and suggests instead shattering due to advective entry of cold seawater into the hot dikes.

4. Massive (isotropic) gabbro: Below the depth of advective water penetration, magma intrusions cool more slowly by conduction and convection and solidify in the extending crust by plating crystals onto the walls, forming bodies of gabbro with isotropic fabric. Amphibole in local dioritic rocks testifies to high concentrations of water in the fractionated tops of crustal magma chambers. More felsic differentiates, which occur as irregularly shaped masses in diorite and gabbro and as thin dikes intruded into basalt, constitute 5%–10% of the plutonic part of ophiolite. These **plagiogranite** (also called *albite granite, trondhjemite,* or *granophyre*) differentiates consist of granophyric aggregates of quartz and strongly zoned oligoclase-andesine (Table 13.7); K-rich feldspar is notably absent and minor primary mafic minerals are altered to chlorite and actinolite.

5. Layered ultramafic-mafic cumulates: These are accumulations of fractionated crystals on the floors of gabbroic magma chambers. Olivine and pyroxene cumulates (dunite, peridotite) occur at the base, succeeded upward by Ol + Cpx + Pl cumulates (gabbro). Cyclic mineral and phase layering is common. A general lack of intrusive contacts within the gabbroic and ultramafic cumulate parts of ophiolites may be created by intermittent recharge of primitive magma into crystallizing magma before complete solidification occurred in the actively extending oceanic crust.

6. Deformed ("tectonized") peridotite: This metamorphic-textured mantle rock below the magmatic

Table 13.7. Composition of Plagiogranite from the Bay of Islands Ophiolite, Newfoundland

SiO_2	76.99	Co	4.4
TiO_2	0.13	Cr	43
Al_2O_3	13.07	Ni	16
Fe_2O_3	1.12t	Sc	2.06
MnO	0.01	Ba	92
MgO	0.39	Th	4.67
CaO	1.99	Ta	0.90
Na_2O	5.53	La	21
K_2O	0.12	Ce	47.3
P_2O_5	0.00	Sr	82
Total	99.35	Sm	6.79
		Zr	182
		Hf	7.43
		Eu	1.12
		Tb	1.24
		Y	62
		Yb	6.89
		Lu	1.04

Data from Elthon (1991).

cumulates, locally in sharp contact with them, is variably depleted in basaltic components. The deformed peridotite ranges from lherzolite to harzburgite to dunite with increasing degrees of depletion. This component of ophiolite is generally the most prominent, in some locales cropping out over thousands of square kilometers. Because of their occurrence in orogenic belts, such as the Alps (Figure 13.26), they have been referred to as **alpine peridotite.** They are variably hydrated, or serpentinized, form serpentinite rock, and show effects of pervasive, grain-scale deformation.

13.6.2 Origin and Emplacement

The initial formation of ophiolite in an oceanic extensional environment is unquestioned. But, in what kind—ocean ridge, spreading back-arc basin, or possibly an early extensional episode in island arc evolution? The overwhelming extent of ocean ridges worldwide no doubt led to the early opinion in the 1960s that they were the environment from which ophiolites formed. However, in the mid-1970s A. Miyashiro noted that some rocks in the classic Troodos ophiolite have arclike chemical affinities rather than normal MORB character. Criticism was immediately raised that the widespread and locally intense metamorphism likely invalidated any conclusions based on chemical composition of the rocks. Nonetheless, fresh volcanic glass subsequently found in the Troodos has arc attributes, and rocks from other ophiolites have also been found to have typical arc signatures in relatively immobile elements Th, Nb, Ta, and REEs.

A dwindling proportion of ophiolites, if any, appears to be of ocean ridge origin. This has significant implications for reconstructions of continental evolution by accretion of oceanic terranes.

The second basic question regarding ophiolites is how variably dismembered slices of dense oceanic lithosphere are accreted (tectonically emplaced) onto margins of less dense continents and onto island arcs in subduction zones, rather than being subducted. In one possible model, oceanic mantle lithosphere overlying subducting slabs is more buoyant because of widespread serpentinization from released water, compared to dry ocean-ridge mantle. (Density of olivine Fo_{90} is 3.3 g/cm^3 versus serpentine 2.6 g/cm^3.) Testifying to the buoyancy of hydrated ultramafic rock are serpentinite diapirs that have breached the ocean floor to create seamounts in the Mariana forearc (O'Hanley, 1996, p. 229). If ophiolites are emplaced within several million years of their creation, as some chronologic data seem to indicate, then the still relatively hot young oceanic lithosphere would have additional buoyancy. Lithosphere made buoyant by its youth and serpentinization may allow nearby converging continental margins of approximately similar, or slightly greater, density (about 2.7 g/cm^3) to be underthrust. Cool, nonhydrated, ocean-ridge lithosphere may only be very locally scraped off, if at all.

❋13.7 CALC-ALKALINE CONTINENTAL MARGIN MAGMATIC ARCS

During much of the Mesozoic and Cenozoic, the western coasts of the American continents were sites of arc magmatism as oceanic lithosphere subducted beneath them. Arc magmatism is also found on smaller continents, such as New Zealand and Japan, and islands in the Caribbean and Mediterranean.

The same rock types (basalt to rhyolite) and rock suites (low-, medium-, and high-K and shoshonitic) can be found in both island and continental arcs. However, in continental arcs there is a greater *proportion* of silicic calc-alkaline rocks—dacite, rhyolite (Figure 12.26b), and their plutonic granodiorite and granite counterparts. Quartz and alkali feldspar are widespread in silicic continental rocks but rare in island arc rocks. Vast sheets of silicic ignimbrite and huge composite granitic batholiths are unique to continents. Concentrations of large-ion lithophile elements, including K, Rb, Ba, Th, and U, are greater in continental rocks at a given silica content. In addition, they generally have higher $^{87}Sr/^{86}Sr$ and lower $^{143}Nd/^{144}Nd$ ratios (Figure 12.31). These compositional contrasts are basically a result of the thicker and more sialic crust in continental arcs into which primitive mantle-derived magmas are intruded and experience differentiation.

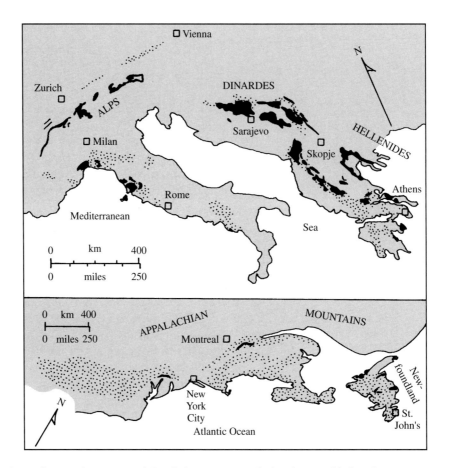

13.26 Ophiolite in the northern Mediterranean and Appalachian regions. Ophiolite shown in black and areas containing ophiolite or deep oceanic sedimentary rock are stippled. (Redrawn from Moores, 1982.)

13.7.1 Volcanic Arcs on Continental Margins

Four areas in the circum-Pacific provide a glimpse of the nature and the variety of continental volcanic arcs.

<u>North Island of New Zealand</u>. This continental segment of the Tonga–Kermadec–New Zealand arc in the southwest Pacific Ocean was previously cited in Section 12.5.1 to demonstrate the contrast between oceanic island and continental arc associations. Most of the volcanic rocks in the Taupo volcanic zone of the North Island of New Zealand (Figure 13.27) are calc-alkaline andesite and rhyolite; basalt and dacite are rare. The volcanic zone is a complex volcano-tectonic depression created by crustal extension and filled with as much as 3 km of volcanic rock. Eruption of poly-genetic andesite magmas since about 2 Ma continues today at composite volcanoes north and south of the central rhyolite ignimbrite field. This is the most frequently active and productive Quaternary rhyolite field on Earth, where an estimated 16,000 km^3 of rhyolite magma has been explosively erupted in the past 1.6 My in 34 caldera-forming ignimbrite eruptions whose volumes range from 30 to 300+ km^3 (Graham et al., 1995).

<u>Andes of Western South America</u>. This is the classic example of continental arc magmatism, consisting of thousands of Mesozoic-Cenozoic volcanoes as well as local deeply eroded plutonic roots. The overall relatively shallow dip (<30°) of the subducting oceanic lithosphere along the arc is attributed to its relatively rapid rate of convergence, ~10 cm/y, and youthful buoyancy. Numerous contrasts along the arc are significant (Figure 13.28). Active Quaternary volcanoes are concentrated in three segments—northern, central, and southern—of differing crustal thickness, age, and composition that lie above lithosphere subducting at angles of 20°–30°. Between these three segments, along-arc flexures in the subducting slab allow dips of only 5°–10°. These nearly horizontally subducting slabs appear to be buoyed up by thick aseismic ocean ridges, causing uplift of the overlying crust and erosional unroofing of older plutonic rocks. Because there is little or no subarc mantle wedge that could otherwise generate magma, no active volcanism occurs.

Virtually the whole spectrum of volcanic edifices (Chapter 10) are developed along the Andean arc, but andesitic composite volcanoes dominate. Single volcanic debris avalanches cover as much as 100 km^2.

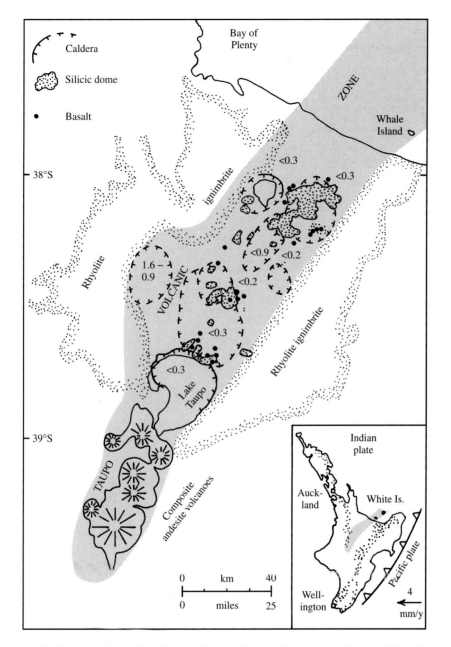

13.27 Taupo volcanic zone (shaded) on the North Island of New Zealand. See also Figure 12.26a. White and Whale Islands are active andesite composite volcanoes in the Bay of Plenty northeast of the caldera-ignimbrite field. Exposed metamorphosed sedimentary rocks are stippled on small-scale map lower right. Time of caldera-forming eruptions indicated in millions of years (e.g., <0.3, 1.6–0.9). (Redrawn from Graham et al., 1995.)

However, in the central volcanic zone, where the crust is oldest (includes Precambrian metamorphic rocks) and thickest (as much as 70 km), late Cenozoic rocks are predominantly silicic ignimbrites exposed over an area of about 5×10^5 km². The topographically high central Andes, which include the Altiplano-Puna (somewhat smaller than the Tibetan plateau in central Asia), is exceptionally arid, barren, and unpopulated. Consequently, very little is known of this remote central region, other than from a few reconnaisance studies (see de Silva and Francis, 1991). Huge resurgent

calderas, as much as 35×60 km, are related to voluminous (>1000 km³) dacite ash-flow eruptions apparently spawned by crustal thickening and mafic magma underplating since the late Miocene.

Southwestern North America Ignimbrite Provinces. A remarkable facet of Cenozoic volcanism in southwestern North America was an ignimbrite flareup during the Oligocene (Best et al., 1989). In <10 My, at least 10^5 km³ of ignimbrite blanketed parts of Nevada, Utah, Colorado, Arizona, New Mexico, and Mexico

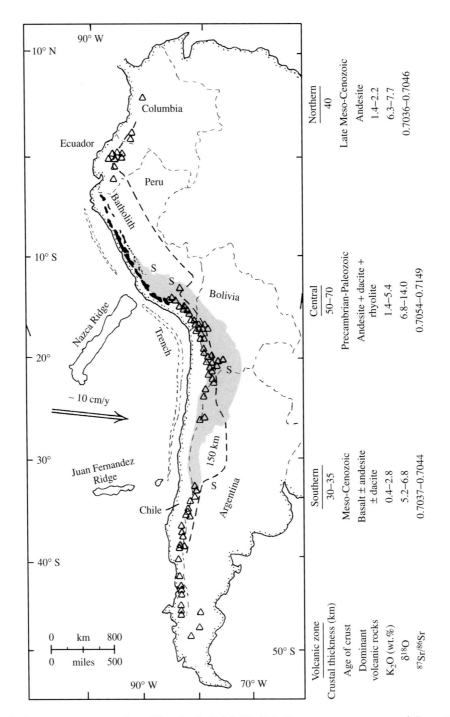

13.28 Western South America magmatic arc. Coastal Peruvian batholith (black) is Cretaceous next to coast and Cenozoic inland. Triangles, Holocene (active) volcanoes. Heavy dashed line, position of 150 km-depth contour on top of Wadati-Benioff seismic zone showing two segments of arc where slab subducts only 5–10°, precluding active volcanism. Shaded, area of average elevation >3 km and to as much as about 7 km, including Altiplano in Bolivia and southern Peru and Puna in northern Chile and Argentina. The notes along the right side contrast the three active volcanic arc segments. S, shoshonitic volcanic centers. (Redrawn from Wilson, 1989; Jordan et al., 1983.)

(Figures 13.29 and 13.30). Although most of the hundreds of caldera-forming ash-flow eruptions that created this ignimbrite blanket were of rhyolite magma derived from compositionally zoned crustal chambers, the very largest were of crystal-rich dacite magma. After disgorging thousands of cubic kilometers of dacite ejecta to depths of as much as 500 m, huge, 50–70 km-diameter calderas formed above probably slablike

chambers as similar volumes of ejecta continued to be erupted into the subsiding depression.

It may be surprising that subduction-related ignimbrites having typical arc trace element signatures occur so far inland from the North American plate margin, as far as southwestern Colorado. However, east-west crustal extension since the ignimbrite flareup has increased this distance by perhaps 200–300 km in

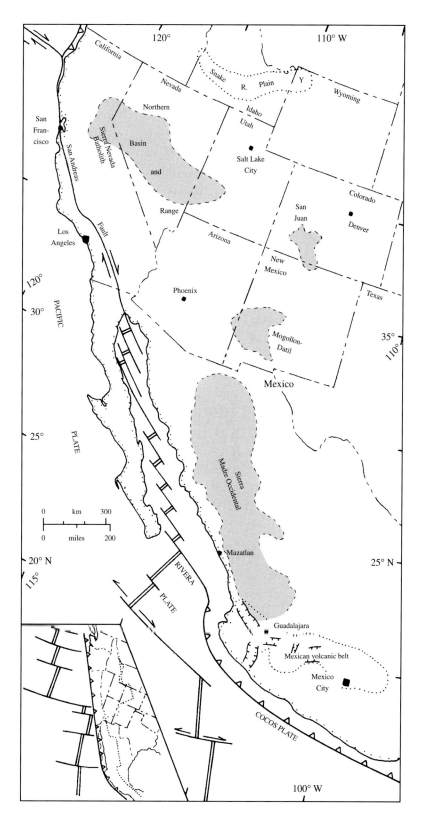

13.29 Some major Cenozoic volcanic provinces in the southwestern part of the North America plate and current plate boundaries. Cenozoic volcanic fields to the north of this map shown in Figure 13.31. Plate configuration at 30 Ma in small-scale inset lower left shows that a subduction zone extended continuously along the western margin of North America, and the eastern edge of the "viable" subducting slab (approximately located by dotted line) extended far inland. (Redrawn from Severinghaus and Atwater, 1990.) In most of California this convergent plate boundary has now been replaced by the San Andreas transform fault. Shaded areas are major ignimbrite fields emplaced during the middle Tertiary. Hundreds of source calderas in these ash-flow fields are not shown. Extent of the Sierra Madre Occidental ignimbrite plateau in Mexico (Figure 13.30) is highly generalized because so little is known of this region. A thin veneer of basalt flows erupted <17 Ma in the Snake River Plain covers thick rhyolite lava and ash flows that emerge from beneath the veneer at currently active Yellowstone (Y) (Hildreth et al., 1991). Rhyolite calderas monotonically decrease in age to the northeast, possibly because of the southwestward passage of the continental plate over a mantle plume, the head of which may have created the Columbia River Plateau off the northwest corner of this map (Figure 13.31). Late Cenozoic normal faults (lines ornamented with ticks) have formed rift grabens in the western part of the Mexican Volcanic Belt.

13.30 Panoramic view of the Sierra Madre Occidental ignimbrite plateau east of Mazatlan, Mexico. The entire rock sequence is rhyolitic ignimbrite; individual layers are as much as hundreds of meters thick. (Photograph courtesy of Gary J. Axen.)

places. Also, plate reconstructions for the Tertiary of western North America disclose a rapidly converging, northeastward-moving, low-dip oceanic lithosphere that advanced far inland (Figure 13.29, inset). This low-dip slab is reminiscent of the post-late Miocene character of the central Andes. Indeed, the compositions of Oligocene volcanic rocks in the northern Basin and Range province are most like those of the central Andes of any continental arcs and suggest that this Oligocene ignimbrite province was a high plateau underlain by crust thickened during Mesozoic compressional orogeny and probably by middle Tertiary mafic underplating.

<u>Cascade Range</u>. The linear arc of composite Quaternary volcanoes from southern British Columbia through Washington, and Oregon and into northern California (Figure 13.31) is one of the better known active volcanic arcs in the world. Included in the arc are, from north to south, the recently active Mount Saint Helens (Figures 10.1 and 10.20), Crater Lake and ancestral Mount Mazama (Figure 10.38), Mount Shasta (Figure 10.2), and Lassen Peak (Figure 10.18). Several attributes of the Cascades should dispel any notions that this arc is like all others and that all continental margin volcanic arcs are the same.

The subducting Juan de Fuca oceanic plate beneath the arc is relatively thin and warm because the ocean spreading ridge is near the trench. Volumetrically minor composite Quaternary volcanoes have been built within the past 1 My on a platform of older basaltic and andesitic rocks produced by more diffuse and intermittent volcanism during the Cenozoic. Deep erosion has exposed the plutonic roots of some of these magma systems (Figure 9.5).

While focused magma systems were creating andesite-dacite composite volcanoes, numerous sur-

rounding vents were erupting mainly basaltic lavas, forming monogenic cinder cones, lava flows, and shields. Many of these basaltic lavas in southern Washington (Leeman et al., 1990) are primitive, ranging from calc-alkaline through tholeiitic to alkaline. Significantly, only a few have typical arc signatures. Instead, most have trace element and isotopic compositions of oceanic island basalts, a feature noted in some rocks of other arcs. In the Cascades, the subducting young and still warm Juan de Fuca slab may have been virtually dehydrated before reaching conditions at which typical arc magmas could be generated in the subarc mantle wedge. Mount Saint Helens, for example, has erupted adakite magmas generated in the basaltic subducted slab. This postulated "dry" wedge, together with contamination by accreted Mesozoic and early Cenozoic oceanic and island-arc lithosphere, may explain the diverse and primitive basalt magma compositions. Quaternary composite volcanoes at the northern and particularly the southern end of the Cascades have a tendency to erupt magma of more silicic composition. This may reflect their construction on older continental crust, as opposed to the more oceanic character of the young mafic crust beneath the central Cascades.

13.7.2 Plutonic Arcs on Continental Margins: Granitic Batholiths

Felsic plutonic and volcanic rock-forming systems were once considered to be genetically separate. However, many observations and arguments clearly indicate that volcanoes have plutonic roots; however, the opposite is not necessarily true because not all intrusive magma vents to the surface. The volcanic-plutonic connection follows from the reasonable inference that erupting magma chambers do not completely empty themselves and is confirmed by the continuities in exposure between volcanic and plutonic bodies at some locales (e.g., Figure 9.5; also Figure 13.36). Along strike of the Cascade volcanic arc (Figure 13.31) and the Andean volcanic arc (Figure 13.28) locally deep erosion has stripped off the volcanic cover, exposing plutons of much the same age and compositions as the volcanic rocks.

Batholiths (Section 9.4.2) made predominantly of tonalite and granodiorite, but ranging from granite to gabbro, are a typical feature of every orogenic belt associated with subduction of oceanic lithosphere beneath a continental margin. In the western North American Cordilleran orogen (Figure 9.16), these mostly Mesozoic batholiths include the Coast Range, Idaho, Sierra Nevada, and Southern California–Peninsular Ranges. Additional Mesozoic-Cenozoic batholiths occur along the west coasts of Central and South America (Kay and Rapela, 1990).

<u>Sierra Nevada Batholith, California</u>. As no two batholiths are exactly the same, any description of one omits

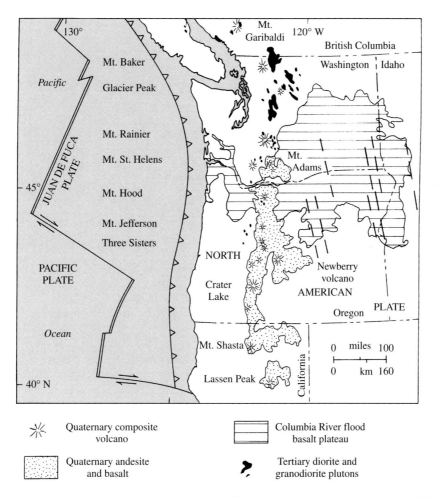

Quaternary composite volcano

Columbia River flood basalt plateau

Quaternary andesite and basalt

Tertiary diorite and granodiorite plutons

13.31 Selected Cenozoic magmatic rocks of southern British Columbia, Washington, Oregon, Idaho, and northern California. Juan de Fuca oceanic plate subducts beneath the continental margin. For clarity, names of most Quaternary composite volcanoes are placed off coast. Hundreds of known exposed feeder dikes in Columbia River Plateau are schematic.

features of others. However, because the central one-fourth of the Sierra Nevada Batholith (Bateman, 1992) is well-documented, it will be used as an example. Emplacement of the approximately 120 separate mapped plutons began in the late Triassic at about 210 Ma and continued episodically into the late Cretaceous until about 85 Ma (Figure 13.32). For uncertain reason, much of the topographic uplift of this mountain range did not occur until the late Cenozoic. Pluton host rocks are strongly deformed and mildly metamorphosed Paleozoic-Mesozoic volcanic and sedimentary rocks; in the west they include accreted oceanic terranes.

Most plutons in the Sierra are tonalite, granodiorite, and granite (Figures 2.6 and 12.30). Among mafic minerals, biotite and Fe-Ti oxides are ubiquitous; hornblende and titanite (sphene) are common. Many plutons are composite or zoned in fabric and/or composition and constitute comagmatic suites. Zones can be gradational from one to another or in sharp contact with one another, implying some cooling before the next intrusion of magma. The Tuolumne Intrusive Series (Figure 9.18; Table 13.8) is a classic example of a concentrically zoned pluton that has zones of younger and progressively more felsic rock that crystallized at lower T inward from the pluton margin. This comagmatic sequence, which appears to have been emplaced within a few million years near 87 Ma, spans the entire spectrum of rock types constituting the whole batholith, with the exception of the most mafic rock types; hence, understanding its origin can provide insight into the evolution of the entire Mesozoic batholith.

Competing models for zoned plutons include the following:

1. Contamination of the outermost part of the magma body with mafic wall rocks

2. Multiple intrusion

3. Convective melt fractionation driven by sidewall crystallization (e.g., Bateman and Nokleberg, 1978; Sawka et al., 1990; Bateman, 1992, pp. 67–82; see also Section 12.2.1)

This model has received widespread acceptance as the probable mechanism of differentiation of

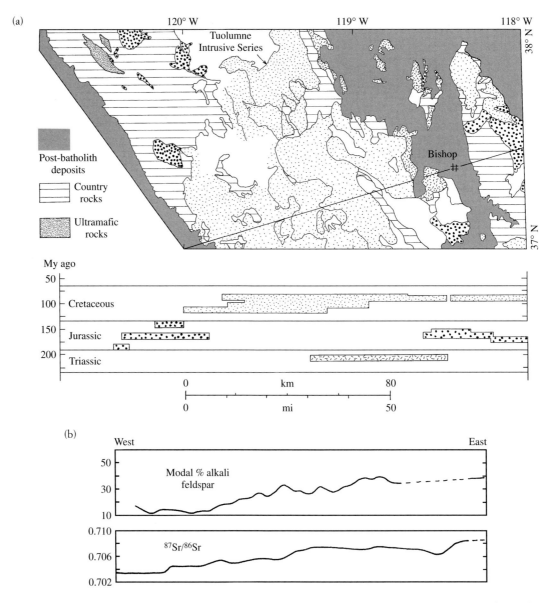

13.32 (a) Generalized geologic map across the central Sierra Nevada batholith. Shown are 18 outlined intrusive suites of multiple plutons of similar character that are believed to have originated in a distinct magmatic episode. Note space-time relations, including the predominantly Cretaceous eastward-migrating magmatism that split the older Jurassic arc. More detailed map of the Tuolumne Intrusive Series is shown in Figure 9.18. Late Cenozoic extensional faulting dropped valley-forming grabens, such as the one in which the town of Bishop is located, separating the main part of the batholith on the west from an eastern segment. Blank areas within batholith were unmapped (as of 1981). (Redrawn from Stern et al., 1981.) (b) Variation in modal percentages of alkali feldspar and $^{87}Sr/^{86}Sr$ across the batholith along the east-northeast trending line through Bishop shown in (a). Redrawn from Bateman (1992).

andesite-dacite-rhyolite magma chambers that erupt well before total crystallization to produce compositionally zoned pyroclastic deposits (Figure 9.38). Crystallization of higher-*T* mafic minerals and plagioclase along the cooled margin of the magma chamber allows buoyant residual melts enriched in water and more felsic components to escape, collecting into the interior and eventually rising to the top of the magma chamber. Because of limited vertical exposure, few plutons can be examined to verify this model (but see Cornejo and Mahood, 1997).

For the Tuolumne series, Kistler et al. (1986) find elemental and isotopic evidence that mantle-derived basalt magmas stagnated in the lower crust, fractionated, and mixed with variable amounts of more felsic and radiogenic anatectic melts generated by their heat. These magmas created by AFC were emplaced at the present level of exposure in at least four surges (Figure 9.18), each more felsic than the preceding one. Once emplaced, the magmas experienced limited convective melt fractionation as hornblende and plagioclase crystallized.

Table 13.8. Compositions of Granitic Rocks from the Sierra Nevada Batholith

	1	2	3	4	5	6
SiO_2	(61.52)	58.59	65.61	67.83	69.76	71.65
TiO_2	(0.73)	0.90	0.54	0.49	0.37	0.24
Al_2O_3	(16.48)	16.93	15.44	15.44	15.49	14.87
Fe_2O_3		2.26	1.76	1.55	1.24	0.84
FeO	6.18t	4.31	2.38	1.44	0.95	0.81
MnO	0.11	0.11	0.80	0.06	0.05	0.04
MgO	(2.80)	3.26	1.80	1.03	0.66	0.38
CaO	(5.42)	6.25	4.10	3.22	2.52	1.87
Na_2O	3.30	3.53	3.62	4.02	4.33	3.98
K_2O	2.41	2.38	3.11	3.65	3.72	4.19
P_2O_5	(0.25)	0.22	0.16	0.17	0.13	0.08
H_2O^+	(1.04)	1.01	0.78	0.61	0.45	0.58
H_2O^-	(0.2)	0.14	0.14	0.1	0.09	0.17
Total	100.44	99.89	99.52	99.61	99.76	99.70
Rb	79.8	108	138	129	137	158
Ba	715	720	515	905	740	1170
Th	6.37	16.6	21.8	18	18.4	16.8
U	2.89	4.2	7.4	5.2	8.8	3.8
Nb		8	7	7	8	9
Ta	1.27					
La	30.2	21	23	25	29	26
Ce	53.9	47	38	44	49	49
Sr	452	574	451	658	621	484
Nd	21.7	18	11	13	21	15
Sm	6.12	4	2.5	3	3	2
Zr	241	126	116	119	126	138
Hf	5.23					
Eu	1.52	1	0.7	0.9	0.8	0.6
Tb	0.98	0.7				
Dy	3.41					
Y		18	9	8	7	8
Ho		8	5.5	5	6.6	5.2
Yb	1.77	1.3	1.3	0.6	0.6	0.6
Lu	0.31	0.2	0.17	0.1	0.1	0.1
Cr	22.8	17	9	6	1	<1
Ni	11	11	5	2	2	2
Sc	9.66	16	8	4	3	2

Columns 2–6 from the Tuolumne Intrusive Series represent the five intrusive stages of the zoned intrusion shown in Figure 9.18a–d. (From Bateman and Chappell, 1979; Frey et al., 1978.)

1, Bass Lake Tonalite. Data from Dodge et al. (1982). Values in parentheses are average tonalite. From Le Maitre (1976). 2, Quartz diorite north of May Lake. 3, Equigranular Half Dome Granodiorite. 4, Porphyritic Half Dome Granodiorite. 5, Cathedral Peak Granodiorite. 6, Johnson Granite Porphyry.

A similar sort of interaction with the preexisting crust, but on a larger scale, is manifested in a regular compositional variation across the Sierra Nevada batholith (Figure 13.32). This pattern is mostly independent of pluton age. Tonalite is the most common rock type in the west, granodiorite along the batholith axis, and granite in the east. Alkali feldspar increases in modal proportion from about 4% in the west to about 30% in the east. Whereas SiO_2 is fairly constant, large-ion-lithophile elements (K, Rb, Ba, U, Th) and light REEs increase eastward, as does $^{87}Sr/^{86}Sr$, whereas $^{143}Nd/^{144}Nd$, CaO, MgO, and total Fe decrease. Tonalites on the west likely formed from fractionated mantle-derived magmas, possibly contaminated only by young mafic crust. Progressively eastward, these primitive magmas were increasingly contaminated and mixed with partial melts of older felsic continental crust.

A belt of widely scattered Mesozoic plutons extends several hundred kilometers inland of the North America batholith arc (Miller and Barton, 1990). Magmas producing these mostly Cretaceous plutons had an increasing crustal component through time. In some areas, latest Cretaceous plutons are strongly peraluminous two-mica granites that were probably derived entirely by anatexis of ancient crust. Unlike the S-type granites of eastern Australia, the North American peraluminous granites contain no cordierite. Nonetheless, high $^{87}Sr/^{86}Sr$ (to as much as 0.737) and $\delta^{18}O$ (to 13‰), and low $^{143}Nd/^{144}Nd$ (to 0.5118) indicate significant amounts of magma from a metasedimentary source. Many petrologists have hypothesized that crustal anatexis was energized by internal heating effects in the crust thickened by orogenic contraction (Section 11.1.1). Even if the thickened crust did not itself provide suffcient internal heat for anatexis, it would promote more contamination of ascending magmas because of the greater path length and residence time of ascending more primitive magmas.

❋13.8 GRANITES IN CONTINENT-CONTINENT COLLISION ZONES

Ancient examples of continent-continent collisions are the Paleozoic Hercynian orogen in southern Europe and the Proterozoic Grenville orogen in eastern Canada. In more recent geologic time, since about 55 Ma, the Indian and Asian continental plates have converged, forming the highest mountains on Earth—the Himalaya. An estimated 1000–1500 km of crustal shortening and doubling of crustal thickness to about 80 km have taken place in the collision zone that includes the Himalaya Mountains and the Tibetan Plateau farther north in the Asian plate. A volcanic arc formed before continent-continent collision but was extinguished as subduction of oceanic crust stopped. Consequently, no volcanic rocks are directly associated with this collision. However, highly potassic mafic lavas and local peraluminous rhyolites < 13 Ma old occur in the central Tibetan Plateau that may be related to local extensional tectonism in the gravitationally unstable high landmass.

Early Miocene (~20 Ma) magmatism in the Himalaya orogen is represented by generally sheet-like granitic plutons and swarms of innumerable smaller dikes, sills, and irregular intrusive bodies. They form a chain of intrusions for about 2000 km along the crest of the High Himalaya (Figure 13.33). Unlike in other granite associations where a range of rock types coexist, the Himalayan rocks are relatively uniform S-type leucogranites composed of a peraluminous assemblage of quartz (about 31 modal %)

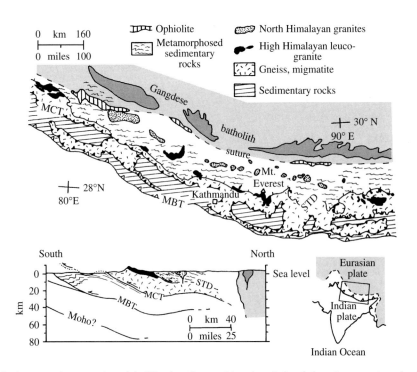

13.33 Generalized geologic map and cross section of the Himalaya Orogen. Note three belts of plutonic magmatic rocks: (1) The northernmost is the Gangdese batholith (dark shade) emplaced from the late Cretaceous to Eocene and generated during subduction of the oceanic Indian plate before continental collision. The North and High Himalayan granites were emplaced in thickened continental crust in the early Miocene after continental collision. The suture between the Eurasian continent (light shade) and the northward-colliding Indian continent is marked by lenticular slices of seafloor ophiolite scraped off the Indian oceanic lithosphere. The crustal thickening and north-south shortening are produced by an imbricate stack of thick thrust sheets that moved along the Main Central Thrust (MCT) and Main Boundary Thrust (MBT). SDT, Southern Tibetan Detachment, a low-angle normal fault that formed late in the topographically elevated orogen. (Redrawn from France-Lanord and Le Fort, 1988.)

+ alkali feldspar (22%) + sodic plagioclase (35%; An$_{10-20}$) and either muscovite (9%) + biotite (2%) ± tourmaline or tourmaline + muscovite (e.g., France-Lanord and Le Fort, 1988). Mn-Fe garnet, zircon, monazite, and apatite are common accessory minerals. The two-mica leucogranites commonly experienced subsolidus metamorphic deformation, whereas the apparently later, cross-cutting tourmaline-muscovite granites are not foliated. Leucogranites are underlain by a thick sequence of highly metamorphosed sedimentary and magmatic rocks (quartz-feldspar-mica-sillimanite gneisses and migmatites) of Precambrian to lower Paleozoic age and are overlain by another thick sequence of metamorphosed sedimentary rocks.

There is general agreement that the peraluminous leucogranite magmas are near-minimum-T (Figures 5.24 and 5.25) partial melts of metamorphosed micaceous sedimentary rocks. No mantle component is present. The following justify this consensus:

1. Isotope ratios of granites (^{87}Sr/^{86}Sr = 0.73–0.83; ^{143}Nd/^{144}Nd = 0.5121–0.5118; δ^{18}O = 9‰–14‰) that resemble those in the underlying metasedimentary quartzofeldspathic mica gneisses.

2. Lack of spatially and temporally associated magmatic rocks that have a mantle component.

3. Similarity to experimentally generated partial melts of quartz-feldspar-mica source rocks at water-undersaturated conditions (dehydration melting; Table 13.9): Under water-saturated conditions, partial melts dissolve larger concentrations of sodic plagioclase from source rocks, and these high-Na/K, Sr/Rb, and Sr/Ba trondhjemitic melts are unlike Himalayan leucogranites. Nonetheless, some variations in water fugacity and in source rock composition are indicated by the somewhat variable tourmaline/biotite ratio. Tourmaline is stabilized under low water concentrations and biotite under high. As for potential Himalayan source rock, metamorphosed graywackes have lower B (<16 ppm), whereas metamorphosed shales have higher B (~100 ppm), compared to 225 ppm B in average leucogranite. Either could be the source, depending on bulk distribution coefficients and degree of melting.

Anatexis could have been triggered by release of aqueous fluids from heated footwall rocks of the Lesser Himalayas as the hotter gneisses were emplaced over them along the Main Central Thrust (Figure 13.33); this would likely have been water-saturated melting. On the other hand, if water-undersaturated (dehydration) partial melting prevailed, it may have been caused by shear heating along the thrust (e.g., Nabelek and Liu, 1999) or by decompression of already hot, thick crust (Section 11.1.1) as the High Himalaya was tectonically unroofed.

Table 13.9. Average major and trace element composition of Himalayan S-type leucogranites compared to range of partial melt compositions generated experimentally by dehydration melting of quartz + plagioclase + muscovite + garnet ± biotite source rock at 6–10 kbar and 750°–900°C

	1	2
SiO$_2$	73.64	73.68–75.60
TiO$_2$	0.10	00.06–00.29
Al$_2$O$_3$	14.87	14.95–16.17
Fe$_2$O$_3$	0.83	
FeO	0.47	00.73–01.08t
MnO	0.03	00.02–00.07
MgO	0.11	00.17–00.39
CaO	0.47	00.42–01.15
Na$_2$O	4.05	03.07–04.92
K$_2$O	4.56	03.40–05.19
P$_2$O$_5$	0.13	
Total	99.27	
B	225	
Cl	35	
Co	96	
S	61	
Rb	286	
Ba	213	
W	4.4	
Th	6	
U	8	
La	11.5	
Ce	19	
Sr	76	
Nd	8	
F	1020	
Sm	2	
Eu	0.5	
Sn	19	
Gd	2	
Dy	2.4	
Li	170	
Y	14.5	
Er	1	
Yb	1	
Lu	0.1	

1, Average Himalayan leucogranite. Data from France-Lanord and Le Fort (1988). 2, Range of experimental partial melts. Data from Patiño Douce and Harris (1998).

✳ 13.9 ANOROGENIC A-TYPE FELSIC ROCKS

The anorogenic tectonic regime in which A-type rocks occur is noncompressional. Included are oceanic islands, generally near spreading ridges (Réunion, Ascension, Azores, Canary); apparently stable cratons;

and intraplate (within-plate) extensional continental rifts (discussed further in Section 13.11). Some A-type granite occurrences are said to be postcollisional or postorogenic, but these categories refer to a time frame rather than a specific tectonic regime. Examples of A-type rocks include many mid-Proterozoic (1.4–1.1 Ga) granites all over the world, Permian peralkaline trachytes ("rhomb porphyries") in the Oslo graben of Norway, Jurassic White Mountains syenite-granite intrusions (including some deeply eroded caldera ring complexes) of New Hampshire, and many late Cenozoic silicic volcanic rocks in the western United States.

A-type felsic rocks were only recognized in the last two decades of the 1900s as a distinct class that is different from other felsic rocks. In addition to appearing in *anorogenic* tectonic settings, magmas are more *alkali-rich* and so crystallize more abundant *alkali feldspar;* some are relatively *anhydrous,* compared with the generally wetter, more calcic magmas in continental arcs (Pitcher, 1997).

13.9.1 Characteristics

Chemically, A-type rocks have many distinctive attributes. With respect to arc felsic rocks, they have moderately high total alkalies; high K/Na, (K + Na)/Al, Fe/Mg, Ga/Al; halogens (F, Cl); high-field-strength (HFS) cations (Zr, Nb, Ta, Zn); and REEs (except Eu) but low Ca, Ba, Cr, Co, Ni, Sc, Eu, and Sr. High concentrations of incompatible elements form abundant accessory minerals, including zircon (in wholly crystalline rocks) and, in biotite granites, allanite. Cassiterite is found in some rocks. F combines with incompatible elements in fluorapatite, pyrochlore, cryolite, fluorite, topaz, and astrophyllite. In some occurrences these minerals are in sufficient quantity to warrant mining. Disruption of Si-O and Al-O polymers in the atomic structure of alkali-rich silicate melts allows for greater solubility of HFS cations through ionic complexing. Though alkalies are highly concentrated, so too is silica; consequently, there are no silica-undersaturated rocks containing feldspathoids. Molecular proportions of K, Na, Ca, and Al are close to the discriminating ratios for metaluminous, peraluminous, and peralkaline rocks (Section 2.4.4); consequently, both peralkaline rocks that contain alkaline amphiboles and peraluminous rhyolites that contain topaz ($Al_2SiO_4(OH,F)_2$) are A-type felsic rocks (Table 13.10).

Magma temperatures are elevated, commonly >900°C, compared to ±800°C for arc magmas. Low water fugacity is typical, so that hypersolvus perthitic feldspars (Section 5.5.3) are common in granitoids and anhydrous mafic minerals are more widespread. One discontinuous reaction series in which crystallization of hydrous mafic minerals is limited to near-solidus residual

Table 13.10. Chemical Composition of Average Peralkaline and Topaz Rhyolites

	1	2	3
SiO_2	74.0	71.2	75.6
TiO_2	0.21	0.37	0.14
Al_2O_3	11.6	9.11	12.8
Fe_2O_3	1.25	2.38	1.12t
FeO	1.88	4.52	
MnO	0.08	0.21	0.06
MgO	0.04	0.09	0.15
CaO	0.36	0.45	0.83
Na_2O	5.35	6.44	3.73
K_2O	4.46	4.40	5.04
P_2O_5	0.02	0.05	0.00
F	0.37	0.30	0.33
Cl	0.24	0.28	0.06
ASI*	0.81	0.563	0.98
Cs	2.8		11.3
Rb	140		423
Ba			41
Th	17.5		54.8
U	4.8		21.6
Nb	72		53
Ta	3.33		5.6
La	82		39
Ce	169		88
Sr			28
Nd	68		39
Sm	11.4		6.6
Zr	660		129
Eu	0.13		0.3
Sn			30
Tb	1.46		1.3
Li			50
Yb	6.0		8
Lu	0.77		1.2

Compare with average rhyolite in Table 2.2.
1, Peralkaline rhyolite. (Average comendite, major elements from Macdonald, 1974; typical trace elements from Mahood, 1981.)
2, Peralkaline rhyolite (average pantellerite). (Data from Macdonald, 1974.) 3, Average topaz rhyolite, Thomas Range, Utah. Data from Christiansen et al. (1986). *Alumina saturation index (ASI) = molecular $Al_2O_3/(K_2O + Na_2O + CaO)$.

melts is fayalitic olivine-hedenbergite (clinopyroxene)-ferrohastingsite (amphibole)-annite (Fe-biotite). Other alkaline magmas might crystallize riebeckite-arfvedsonite (Na-Fe amphiboles) and aegirine (Na-Fe clinopyroxene) (Appendix A). Unlike ternary-minimum (Figures 5.24–5.26) collision and arc granitic rocks, A-type granitoids include alkali-feldspar-rich syenite, quartz syenite, and alkali-feldspar granite (Figure 2.8). These phaneritic rocks are commonly linked in space, time, and genesis to extrusive rocks of similar composition, especially trachytes and rhyolites. In fact, the plutonic-volcanic link is as clear as in any magmatic association (Figure 13.36).

A recurring association of A-type rocks with tholeiitic and mildly alkaline more mafic rocks is petrogenetically significant. These mafic rock types include basalt, trachyandesite, hawaiite, mugearite, and benmoreite or their intrusive phaneritic equivalents. Proportions of the mafic and felsic rock types vary. In some locales, a distinct compositional gap exists between them, resulting in a **bimodal association** of basalt-rhyolite or basalt-trachyte. Some intrusive associations include anorthosite and granite, especially in Proterozoic terranes.

13.9.2 Petrogenesis

A-type magmas appear to be polygenetic: No single process creates all of them.

Fractionation of calcic, Al-rich plagioclase from parental, slightly alkaline basaltic magma has been advocated for production of peralkaline residual magmas. However, Rb and Cs are not as high as expected in every case, and crystallization of calcic plagioclase would be more likely in a hydrous magma than in the typical low-water-fugacity A-type magmas. The compositional gap (bimodality) between felsic and mafic rocks in many locales is also difficult to reconcile with fractionation, which would tend to yield a continuum of intermediate-composition daughter magmas.

A compilation of Y/Nb and Yb/Ta ratios (Figure 13.34) indicates the presence of both mantle and continental crust components in A-type felsic rocks. Many fall in the field of oceanic island basalts, and others lie on a trend toward average continental crust. Still others fall between island arc basalt and continental crust.

The mantle component could be derived from basalt magmas having an oceanic island, plumelike signature that were lodged in the lower continental crust or underplated beneath it. During prolonged magmatism, still hot basalts intruded early in the activity could be partially melted by later intruded basalt magmas. In the Yellowstone Plateau volcanic field, Wyoming (Figure 13.29), about 6000 km³ of high-silica rhyolite was extruded since 2.2 Ma along with only 100 km³ of tholeiitic basalt lavas. Exposures of Archean rocks peripheral to the volcanic field and presumed to constitute the deep crust beneath it are far more radiogenic ($^{87}Sr/^{86}Sr \sim 0.858$) than the rhyolites ($\sim 0.710$), thus precluding an origin by crustal anatexis (Hildreth et al., 1991). However, rhyolite magmas may have been generated as partial melts of voluminous earlier intruded basalts that were variably fractionated to Fe-rich daughters (e.g., ferrodiorite) and only slightly contaminated with continental crust. Late Cenozoic volcanism and accompanying crustal attenuation in the Snake River Plain–Yellowstone volcanic field are believed to reflect migration of the North American plate over a

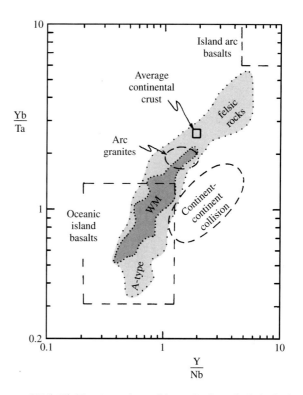

13.34 Y/Nb-Yb/Ta ratios in A-type felsic rocks shown by light shading. Dark shaded area is Jurassic A-type granitic, syenitic, and minor more mafic rocks in the White Mountains (WM) of New Hampshire. Compositions of rocks from various tectonic settings indicated by dashed line fields. Element ratios, which are not strongly influenced by differentiation processes, serve as indicators of magma source. Oceanic island basalt or its mantle source appears to be a major component in A-type rocks along with continental crust. (Redrawn from Eby, 1990.)

mantle plume, possibly the one that created the Columbia River flood basalt plateau to the west.

Frost and Frost (1997) point out that the Snake River Plain–Yellowstone rhyolites closely resemble rapakivi granites that are widespread in mid-Proterozoic terranes. Both can be considered A-type end members in having high $K_2O \sim 5$ wt.% at 70 wt.% SiO_2, K/Na > 1, and Fe/(Mg + Fe) > 0.9. Low water fugacity is indicated by the occurrence of fayalitic olivine ± Fe-rich pyroxenes and near-solidus precipitation of amphibole and biotite, if any. Oxygen fugacity is significantly lower than that of the typically more oxidized, magnetite-bearing arc granites.

Anatectic melts of continental crust heated by intrusive mafic magma from the mantle are believed by some petrologists to be an important component in many A-type magmas. For example, Collins et al. (1982) postulate magma generation from source rocks that were subjected to previous dehydration melting, leaving a residuum containing F- and Cl-enriched biotite and amphibole. However, other petrologists believe that the crystalline residue after dehydration melting is incapable of yielding additional felsic partial melts. They instead propose, as did T.F.W. Barth in the mid-1900s for the Oslo, Norway, rocks, dehydration melting of tonalite or granodiorite, which could have been produced in an immediately preceding, or in an ancient, episode of arc magmatism. Patiño Douce (1997) finds that high-*T* anatexis in the shallow crust (*P* less than or equal to 4 kbar) of arc source rocks containing only limited amounts of biotite and amphibole yields melts that have major-element compositions similar to that of A-type granitoids. Because the residue is mostly calcic plagioclase and orthopyroxene, partial melts contain virtually all of the F and Cl from the source and have high Ga/Al (Ga is incompatible in the residual minerals), Fe/Mg, K/Na, and (Na + K)/Ca but low compatible Ca, Sr, and Eu. Low Ni, Cr, and Co in the melts reflect low concentrations in the source. However, the partial melts of arc rocks would have a negative Nb-Ta anomaly and Rb/Nb would be high—neither of which is characteristic of A-type felsic rocks.

13.9.3 Anorogenic Ring Complexes in Nigeria and Niger

The classic A-type rocks lie within a diffuse north-south belt of ring complexes in Nigeria and Niger in central Africa (Figure 13.35). An imperfect southward migration from Silurian to Jurassic together with local doming might suggest movement of the African plate over a mantle plume. However, some geologists working in Africa (e.g., Bowden et al., 1987) believe that the belt originated as a result of reactivation of Precambrian zones of weakness—faults—in the craton. The reason for the overall southward migration of activity in the belt for 250 My and the migration of activity in a particular string of complexes, as well as the magma heat source, remains uncertain.

Although basaltic rocks occupy only a small proportion of exposures in the ring complexes, the role of basalt magmas in the evolution of the bimodal mafic-felsic association cannot be overlooked. There are early basalt lavas and basalt inclusions in granites—some of which are pillowlike, indicating mingling of basaltic and granitic magmas—and minor late basalt dikes. The amazing 65-km-diameter Meugueur-Meugueur cone sheet consists of troctolite (olivine + plagioclase gabbro) and encompasses a 900-km^2 anorthosite-leucogabbro body intruded by A-type granites and syenites (Figure 9.24). The petrogenetic linking of basalt, anorthosite, and granite/syenite seems strong. In other ring complexes, most precursory basalt magmas may have lodged deep in the crust, where they promoted generation of felsic magmas, some of which ascended high enough to erupt along ring fractures and vent as ash flows (Figure 13.36). The collapsed calderas outlined by the ring dikes partially filled with ignimbrite. Late biotite granite magmas rose into the volcanic roots.

✳13.10 GRANITES AND GRANITES

It should be obvious by now that a variety of granitic rocks develop in several different tectonic settings, including the following:

1. Granophyric plagiogranite in ophiolite sequences (Section 13.6.1); most (all?) sequences appear to have formed in arc settings.

2. Granophyre in differentiated basalt sills (Section 12.4.1) and layered intrusions (Section 12.4.2), both likely associated with mantle plumes.

3. Dioritic differentiates in island arcs.

4. Granodiorite and lesser associated gabbro, tonalite, granite, and dioritic rock in huge batholiths of continental arcs (Section 13.6.2). Both I- and S-types are represented.

5. S-type leucogranite in continent-continent collision zones (Section 13.8).

6. A-type granite and syenite in anorogenic regimes (13.9); these are especially significant in the mid-Proterozoic.

7. A chiefly Archean association of tonalite-trondhjemite-granodiorite and generally younger Archean and Proterozoic granitic rocks.

The first three occurrences have only minor volumes of granitic rock, but locally large volumes of leucogranite occur in continent-continent collision zones. The largest volumes of granitic rock are found in continental arcs and in the Precambrian occurrences.

Since Read (1957) and other early workers wrote of "granites and granites," petrologists have considered and debated granites from several different perspectives—all intending to cast some light on understanding the origin of these multifaceted, polygenetic rock masses (e.g., Clarke, 1992; Pitcher, 1997). In addition to tectonic setting, granites have been categorized with respect to

1. Time of emplacement relative to regional deformation or tectonism (pre-, syn-, and posttectonic)

2. Several chemical and isotopic attributes (e.g., peralkaline, peraluminous, and metaluminous; radiogenic and nonradiogenic; high and low $\delta^{18}O$)

3. Modal composition (leucogranite, granite, granodiorite, and so on)

4. Fabric related to degree of water saturation (hypersolvus and subsolvus) or other special conditions of magma evolution (porphyry, rapakivi)

5. Mechanism of ascent through the crust (diapir versus dike)

6. Depth of final crystallization (for example, stabilized epidote at high *P*)

7. Source (S- and I-type) and degree of crustal contamination

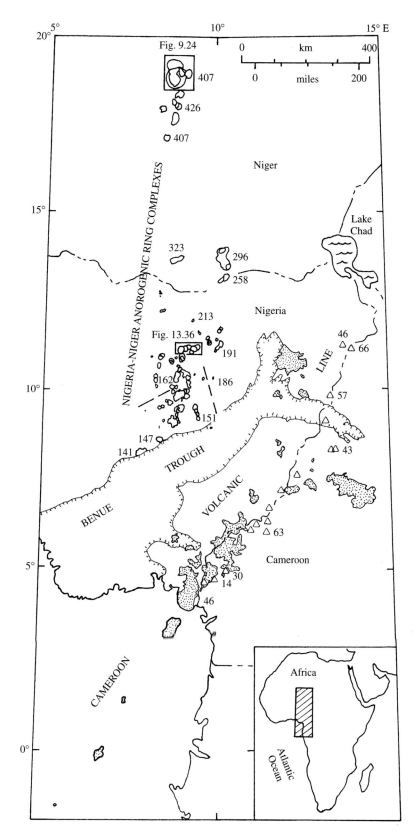

13.35 Generalized geologic map of the Nigeria-Niger anorogenic intrusive ring complexes and the Cameroon volcanic line in Africa. The ring complexes are generally younger toward the south, as shown by ages in millions of years. (Redrawn from Rahaman et al., 1984). Dashed lines in center of map indicate the "grain," essentially old fault zones, in Precambrian host rocks. There is no monotonic age progression of magmatic activity along the Cenozoic Cameroon line, where ages in millions of years are shown for intrusive complexes (open triangles) related to extrusive rocks (stippled). The Cameroon line is the only known alkaline province straddling continental and oceanic lithosphere; alkaline basaltic rocks are no different in chemical or isotopic composition along the line, suggesting derivation from asthenosphere underlying the contrasting lithospheres, whereas evolved rocks are peralkaline rhyolite on the continent and phonolite in the ocean islands. The Benue trough (filled with Cretaceous sedimentary rocks) and Cameroon line are considered to be a failed rift arm related to opening of the Atlantic Ocean along the actively rifting Mid-Atlantic Ridge. (Redrawn from Fitton, 1987.)

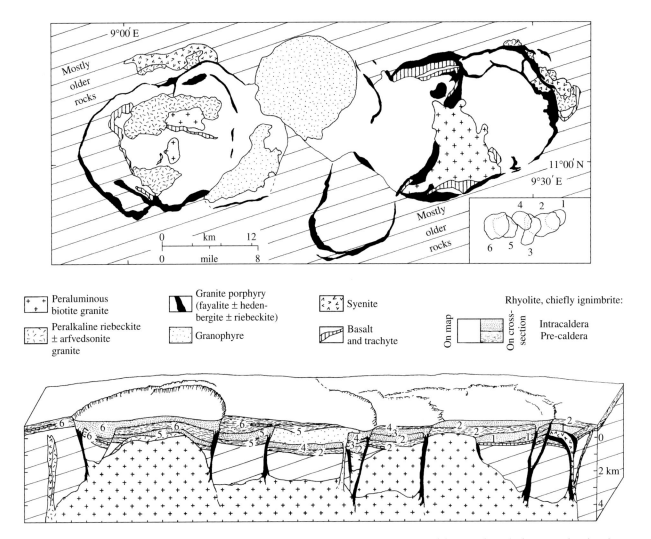

13.36 The Ningi-Burra ring complex of eroded calderas, Nigeria. See Figure 13.35 for location of this complex, which was emplaced at about 183 Ma. The generalized geologic map (top) and cross section (bottom) show six overlapping intrusive-extrusive centers that migrated from east to west. They are delineated by ring dikes of peralkaline fayalite ± hedenbergite ± riebeckite granite porphyry. Numbers on cross section indicate the center where the pre- and intracaldera deposits of chiefly peralkaline rhyolite ignimbrite were derived. (Redrawn from Turner and Bowden, 1979.)

Many petrologists have attempted to characterize magmatic rocks, particularly granites, in known tectonic setting with respect to their trace element concentrations (Figure 13.37). The rationale behind this characterization is that granitic magmas that have different petrotectonic associations should have contrasting evolutionary paths involving different sources and melting conditions and subsequent ascent and differentiation histories that together yield distinctive compositional signatures. However, certain components of these complex evolutionary paths may not be unique to a particular tectonic setting; for example, source rocks of much the same composition can be partially melted in continental arcs, rifts, and collision zones. Hence, the products of evolutionary paths in different tectonic settings tend to be compositionally "smeared" in most variation diagrams. This lack of natural discontinuity in

most variation diagrams of magmatic rocks was emphasized in Section 2.3.2; discriminating boundary lines can only delineate most, and seldom all, rocks in artificial subdivisions. Nonetheless, such discrimination diagrams are useful in trying to understand the origin of granitic magmas and trying to infer tectonic settings for rocks produced before the "modern" period of plate activity.

✳13.11 CONTINENTAL RIFT ASSOCIATIONS: BIMODAL AND ALKALINE ROCKS

Ancient and active modern continental rifts, of which there are an estimated 100 worldwide, are generally elongate sectors of the crust that have experienced extensional normal faulting. Grabens typically lie in a re-

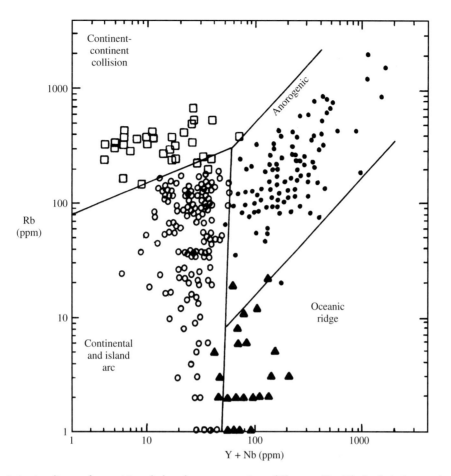

13.37 Tectonic discrimination diagram for granitic rocks based on concentrations of Rb versus Y + Nb. Symbols show modern rock suites used to subdivide the diagram. Arc granites broadly correspond to I-type and collision granites to S-type. (Redrawn from Pearce et al., 1984.)

gional uplift where the crust and mantle lithosphere are unusually thin and surface heat flow is high. These attributes are probably the consequence of upwelling hot mantle, a plume in some instances. Whether upwelling, plume-related asthenosphere is the cause and diverging plate motion the consequence (so-called active rifting), or vice versa (passive rifting), is difficult to ascertain in every instance. Spreading rates of oceanic rifts (generally 1–10 cm/y) are one to two orders of magnitude faster than that of continental rifts: Rhinegraben, western Germany, ~0.01 cm/y since the late Cretaceous; East African rift, Ethiopia, 0.06 cm/y; Malawi, 0.03 cm/y; Basin and Range, western North America, ~0.5 cm/y since the Miocene.

Some continental rifts evolve into oceanic rifts accompanying separation of the landmass (Section 13.3.3), generally along two arms of a three-arm rift geometry. (This three-arm rift geometry is clearly evident where the Arabian Peninsula has rifted from northeastern Africa; see Section 13.11.2 and Figure 13.38.) Rifting along the third arm, including some as old as Precambrian (Wilson, 1989, Figure 11.1), slowly dies, forming failed rifts. An example of a currently active rift destined to "fail" is the East African system de-

scribed later. Others are the northeast-trending Benue trough in Africa and the paralleling Cameroon Volcanic Line—a chain of mostly alkaline Cenozoic volcanoes extending 1600 km from the Atlantic Ocean into the African interior (Figure 13.35; see also Fitton, 1987). This crustal extension and chain of alkaline magmatism is not related to passage of the African plate over a mantle plume because there is no systematic age progression along the chain, as in the Hawaiian island chain.

Magmatic rocks in continental rifts range as widely as in any tectonic regime. Mafic rocks include subalkaline basalt and more often silica-undersaturated alkaline basalt, basanite, and nephelinite. Large volumes of associated felsic rocks, including rhyolite, trachyte, and phonolite, that occur with the mafic rock types define bimodal associations; intermediate compositions are subordinate or absent. Some rifts, especially the slower-spreading ones, have highly alkaline rock types that contain abundant feldspathoids, locally to the exclusion of both plagioclase and alkali feldspar. Silica-poor, feldspar-free carbonatite is a rare, extreme rock type associated with feldspathoidal rocks. In some deeply eroded rifts, A-type granitic rocks are

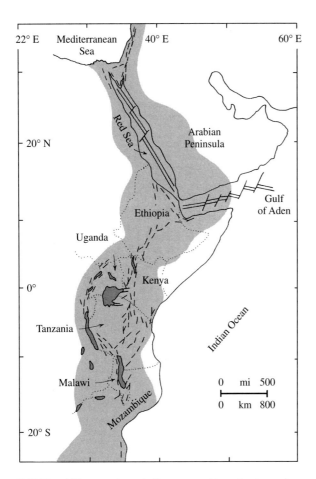

13.38 East African continental rift system and its union in northern Ethiopia with the Red Sea and Gulf of Aden oceanic rifts. These rifts lie within a chain of coalesced broad domical upwarps (light shaded). High-angle normal faults are shown by dashed lines. Double lines denote oceanic rifts cut by transform faults. Dotted lines outline countries mentioned in text. Note several lakes (dark shaded) in the grabens. (Redrawn from Gass, 1972.)

prominent, as described earlier, in others, such as the Precambrian Gardar province in southernmost Greenland, plutonic complexes of nepheline syenite-syenite-alkali granite are exposed. Many highly alkaline rocks are unusually enriched in incompatible elements; thus, Zr, Nb, Ta, Rb, Th, and others, can be major elements stabilizing major rock-forming minerals such as zircon, tantalite, and thorite. Associated complex pegmatites may be economically feasible resources of these exotic metals.

13.11.1 Transitions from Continental Arc to Rift Associations in Western North America

Long-lived, contractional continental arcs commonly evolve in their more mature stages into a regime of crustal extension. Although reasons for this tectonic transition vary and may never be certain, there is no question that the associated magmatic rocks experience a change in composition accompanying the shift in tectonism. Two examples suffice to demonstrate the variation in this transition.

<u>Basin and Range Province</u>. Since about 30 Ma in the middle Tertiary, subduction along the North American plate margin in the western United States has been progressively supplanted by transform plate motion. Inland of this growing transform system, which includes the San Andreas fault, east-west crustal extension has been forming the Basin and Range Province (Figure 13.29). Although the exact cause of the extension remains controversial, its inception has been associated with a transition from typical continental arc rocks to a variety of rift-related volcanic rocks.

In the northern Basin and Range, early and middle Tertiary subduction-related volcanism was characterized by extrusion of high-K arc lavas that are mostly andesite and, in the middle Tertiary, of rhyolite-dacite ignimbrite (Section 13.6.1). After about 24 Ma, increasing proportions of true basalt (IUGS classification) and rhyolite magmas were extruded, forming a bimodal association. The absence of intermediate-composition magmas in this association is conventionally interpreted to reflect the "ease" of extrusion of mantle-derived basalt magmas in an extensional stress regime (Figure 9.12) and the generation of anatectic rhyolite magmas from continental crust and/or basalt lodged earlier in the crust. Many Basin and Range rocks became more sodic, as manifested especially in local peralkaline rhyolites, whereas other rhyolites contain topaz. Both have A-type affinities.

In a study of 750 mafic lava samples from the southwestern United States, Fitton et al. (1991) found that all basaltic lavas with MgO >4 wt.% in the central Basin and Range Province <5 Ma have a distinct trace element signature like that of oceanic island basalt (OIB) generated by an asthenospheric source. Other basalt fields, as well as all fields >5 Ma, appear to have been generated in lithospheric mantle enriched by subduction-derived fluids, both of recent (early Cenozoic) and ancient (Proterozoic) vintage. This space-time contrast in source suggests that the lithosphere beneath the Basin and Range Province had been thermally eroded by upwelling, hotter asthenosphere, and in some parts removed completely, by 5 Ma. Sparse basaltic lavas lacking an arc signature (i.e., having no negative Nb anomaly) but having OIB affinities had appeared by about 20 Ma and rather primitive OIB-like basalt by about 16 Ma in the Basin and Range Province. Apparently, some unmodified asthenospheric magmas had begun to make their way to the surface long before the lithosphere had been eliminated, or at least made ineffective, in magma generation.

<u>Western Mexican Volcanic Belt</u>. The 1100-km-long Mexican Volcanic Belt trends east-west across Mexico opposite the actively subducting Rivera and Cocos

oceanic plates (Figure 13.29). Since the early Pliocene (about 5 Ma), and especially during the Quaternary, minor volumes of magma not generally found in arcs have been extruded from small rifts within the volcanic belt. These atypical extrusions are intimately associated with those of calc-alkaline arc affinity apparently differentiated from primitive, *Hy*-normative basalt magmas. The atypical rocks, similar to those found in many continental rifts, include OIB-type (Section 13.2.1) *Ne*-normative mugearite, benmoreite, trachybasalt, trachyte, and peralkaline rhyolite as well as leucitite and fresh lava flows of lamprophyre. (**Leucitite** is a glassy to aphanitic rock composed of leucite, clinopyroxene, and variable amounts of olivine; lamprophyre is described later.) The lamprophyres have an arclike trace element signature (depleted Ta and Nb) and high concentrations of incompatible elements (K, P, Ba, Sr, light REEs) typical of alkaline rocks. Luhr (1997) believes these attributes resulted from generation of magma in incompatible-element-enriched mantle veined by metasomatic phlogopite, amphibole, and apatite. Extensional fractures in the rifted crust allowed these low-degree partial melts of enriched composition to ascend to the surface, whereas larger-degree partial melts diluted in the vein component yield the more voluminous calc-alkaline magmas.

13.11.2 Magmatism in the East African Rift System

Broadly concurrent magmatism, crustal uplift, and extensional faulting are well expressed in the East African rift system, which extends some 3700 km from Mozambique in central eastern Africa northward through Ethiopia, where it splits into the Gulf of Aden and Red Sea oceanic rifts (Figure 13.38). The rate, amount, and time since inception of extension decrease, scissorlike, southward. In its central part, the system bifurcates into eastern and western branches. Grabens within these continental rifts have dropped as much as 3 km. Major crustal upwarp occurred in the early Tertiary, probably in relation to an underlying plume head, forming the broad Ethiopia dome and flooding nearly 10^6 km^2 in Ethiopia and southwest Arabia with alkaline basaltic lavas to depths of as much as 3 km (Table 13.4). Breakup of the continental crust created the Gulf of Aden in the Miocene and the Red Sea in Pliocene time. Typical MORB has been since extruded from these oceanic rifts.

In Ethiopia, Kenya, and northern Tanzania, Miocene and younger continental extension has been accompanied by extrusion of vast floods of transitional alkaline-tholeiitic basaltic lavas as well as peralkaline rhyolite, trachyte, and phonolite lava and pyroclastic flows. Individual phonolite flood lavas have volumes of as much as 300 km^3, and their aggregate volume is about 50,000 km^3. The origin of such a vast volume of compositionally uniform flood lavas poses a petrogenetic dilemma because of the brevity of activity during the Miocene and the absence of intermediate rock types extruded after the earlier-erupted alkaline basalt lavas. However, experiments by Hay and Wendlandt (1995) show that, under lower crustal pressures, crystalline phases at the liquidus of flood phonolite match the near-solidus assemblage of alkaline basalt; this suggests that the phonolite magmas may represent partial melts of earlier basalts underplating the crust below the rift.

The region between western Uganda and Zaire in the western branch of the rift has been famous for many decades because of its highly alkaline, ultrapotassic (K$_2$O > 3 wt.%), ultramafic, silica-undersaturated volcanic rocks, known as **kamafugites.** The label for these globally very rare lavas and pyroclastic deposits of post-Pliocene age has been coined from the three dominant rock-type names—*katungite, mafurite* (Table 13.11), and *ugandite.* These consist of a Si-poor alkalic mineral that is predominantly melilite, kalsilite, or leucite, respectively, in addition to olivine, clinopyroxene, Fe-Mg mica, Ti-rich magnetite, and perovskite. Kalsilite, virtually unknown outside Italy and Uganda, is essentially KAlSiO$_4$ and can only crystallize in the most extremely Si-Na-poor, K-rich magmas. Partial melting of mantle rock veined with phlogopite is probably the source of parental kamafugite magmas (Edgar, 1996).

Carbonatite-Nephelinite Association. The southern part of the African rift system in Tanzania and Malawi and including parts of the western rift (Figure 13.38) harbors volcanic and small shallow intrusive bodies of carbonatite. About one-half of the known 330 carbonatite occurrences worldwide (Bell, 1989) are on the African plate, including the only two on ocean islands (Canary and Cape Verde just west of the African coast). Most carbonatites occur in continental rifts and upwarps; the Paraña-Etendeka (Figure 13.22) is another region where they occur. Though very small in total worldwide area (a few 100 km^2), carbonatites are economically valuable and a great petrologic curiosity. No fewer than five books dealing wholly with carbonatite have been published.

Carbonatite contains >50% carbonate minerals, usually calcite. However, since 1960 the nephelinitic Oldoinyo Lengai volcano in Tanzania has erupted alkali carbonate lavas and pyroclastics. Earlier arguments whether carbonatite is truly a magmatic rock were put to rest by this eruption and discovery of other carbonatite volcanic rocks. Table 13.11 (columns 1 and 2) reveals the extreme composition of carbonatites; in alkali carbonatite, there is <0.2 wt.% SiO$_2$ + Al$_2$O$_3$ but major concentrations of SrO, BaO, SO$_3$, Cl, F, and, of course, CO$_2$. Relative to the mantle and continental crust, carbonatites are also strongly enriched in REEs, especially light ones, Y, Pb, Th, U, and Nb, and depleted in Sc, V, Cr, Co, Ni, Rb, Zr, and T (Barker,

Table 13.11. Chemical Composition of Carbonatites and Highly Alkaline Rocks

	1	2	3	4	5	6	7	8	9	10
SiO_2	0.16	13.53	42.20	39.06	48.36	47.4	53.64	42.31	35.09	30.0
TiO_2	0.02	1.94	2.48	4.36	0.70	1.82	6.29	3.75	1.06	1.8
Al_2O_3	0.0	2.40	12.04	8.18	16.80	9.3	8.13	3.92	2.55	2.6
Fe_2O_3	0.28t	12.96t	13.97t	4.61	2.55	9.82t			7.78t	9.4t
FeO				4.98	4.40		6.78t	8.27t		
MnO	0.38	0.45	0.27	0.26	0.13	0.14			0.15	0.2
MgO	0.38	8.45	6.22	17.66	6.57	16.4	7.82	24.42	29.02	29.4
CaO	14.02	35.33	15.17	10.40	9.85	9.32	3.23	5.00	3.49	10.9
Na_2O	37.22	0.87	3.93	0.18	1.30	2.14	0.49	0.50	0.18	0.3
K_2O	8.38	0.07	2.54	6.98	8.33	2.41	9.60	4.01	2.91	1.2
P_2O_5	0.85	3.27	0.88	0.61	0.61	1.27	1.23	1.59	0.68	1.6
H_2O^+	0.56	5.83					2.64	6.07		7.3
CO_2	31.55	11.64								5.4
SO_3	3.72	0.28								
LOI					4.76				11.76*	
F	25,000						3100			
Cl	34,000						127	110		
Cr				801	316	1022	373	1014	1852	1398
Ni		26	29	515	74	699	343	968	1253	1018
Rb	178		45	154	636	132	457	471	159	66
Ba	14,900	3400	1211	2407	1202	621	10,607	10,584	2442	915
Th	3.77		12		46	8.7	30	57	30	30
U	10.6				10.2	3.1	3.8	2.4	3	5
Nb	28	483	111	166	14	26	147	186	97	168
Ta	0.0				0.58	1.8	7.9	9.9	9	11
La	545		92	170	88	58	348	242	168	100
Ce	645		158		202	127	629	414	324	239
Sr	12,000	8000	910	1559	1812	678	1296	1325	1127	1145
Nd	102		61		90	70	212	146	115	
Sm	7.80				17.5	16.8	25	18	12.6	10
Zr	<10	845	193	202	266	682	1401	1167	214	308
Eu	1.62				3.2	5.33	5.8	4.3	3.00	3
Tb	0.32				1.5	2	1.8	1.4	0.82	0.8
Y	7		25	17	26	47	21	16		
Yb	0.46				2.2	2.26	1.6	1.5	0.99	1
Lu	0.058					0.28	0.19	0.17	0.13	0.13

1, Alkali carbonate lava, Oldoinyo Lengai volcano, Tanzania. Composed of alkali carbonate and minor MnS. Data from Barker (1996). 2, Calcic carbonatite lava, Fort Portal field, Uganda. Phenocrysts (and/or xenocrysts) of phlogopite, magnetite, clinopyroxene; groundmass of calcite, spurrite, apatite, periclase, perovskite, barite, pyrrhotite. Data from Barker (1996). 3, Olivine-poor nephelinite, Mt. Elgon, border of Uganda and Kenya. Data from Le Bas (1987). 4, Mafurite, Uganda. Data from Edgar (1996). 5, Leucite phonotephrite, Italy. Data from Conticelli and Peccerillo (1992). 6, Mica-rich minette lamprophyre, Wasatch Plateau, Utah. Data from Tingey et al. (1991). 7, Phlogopite-rich lamproite, average of 228 samples, West Kimberly, Western Australia. Contains <0.5 wt.% CO_2 and modal leucite but 8.8% normative quartz! Data from Mitchell and Bergman (1991). 8, Olivine-rich lamproite, average of 105 samples, West Kimberly, Western Australia. Data from Mitchell and Bergman (1991). 9, Orangeite (micaceous, or group II kimberlite), Sover, South Africa. Data from Mitchell (1995). 10, Archetypal (Group) kimberlite, average of 30 dike, sill, and pipe intrusions in Kimberly area, South Africa. Data from Scott-Smith (1996). *Mostly $H_2O + CO_2$.

1996a). Carbonatites host no fewer than about 275 documented minerals in addition to carbonate minerals; other common minerals include apatite, magnetite, fluorite, pyrochlore (an Nb oxide that contains substituting Ta, Ti, Zr, Th, U, Pb, Ce, Ca, Sr, Ba), and alkalic amphibole.

Most carbonatites are intimately associated with phonolite and more commonly with olivine-poor nephelinite (Table 13.11, column 3; nephelinites of oceanic islands, such as Hawaii, Table 13.2, are olivine-rich). Although young carbonatite-nephelinite associations make up composite volcanoes of lavas and volcaniclastic rocks, most older associations are sub-volcanic, shallow plutonic complexes that contain **ijolite,** the phaneritic nepheline-clinopyroxene plutonic counterpart of nephelinite. These complexes (Figure 13.39) are typically circular or elliptical in map view, and the constituent rock units form plugs, arcuate ring dikes, and cone sheets. Brecciation is widespread. Rock types include, in addition to carbonatite and ijolite, alkaline pyroxenite, phonolite, nepheline syenite, and, most characteristic of all, **fenite.** This is a metasomatic rock (Figure 13.40) produced by solid-state transformation of older wall rocks by infiltration of alkaline hydrothermal solutions derived from the nephelinitic or carbonatitic magmas. The combination of brecciation, multiple intrusion, assimilation, and fenitization produces complicated overprinted fabrics.

Nd and Sr isotope ratios (Bell, 1989) unequivocally demonstrate that the ultimate source of carbonatite magmas is the mantle, probably long-lived, metasomatically enriched continental lithosphere. Tangible samples are represented by rare garnet-phlogopite-calcite lherzolite inclusions in carbonatite tuffs. Experiments confirm that carbonatite partial melts can exist under special mantle conditions (Figure 11.5).

13.40 Hand sample showing fenite reaction zone adjacent to thin dike of carbonatite in granite host rock. Note acicular alkali amphiboles in carbonatite and their concentration as aphanitic grains in the **fenite.**

In June 1993 Oldoinyo Lengai erupted carbonatite lava and ash that contain nephelinitic spheroids that themselves contain alkali carbonatite segregations (Table 12.4). These are obviously immiscible silicate-carbonatite melts that separated from a carbonated silicate magma at low P (Figure 12.8).

❊13.12 ALKALINE ORPHANS, MOSTLY IN STABLE CRATONS

After alkaline basalts, nephelinites and phonolites (Barker, 1996b) are the most widespread and voluminous alkaline rocks. However, all these have received only a fraction of the attention given to lamprophyres, lamproites, orangeites, and kimberlites. These four are, collectively, generally primitive silica-undersaturated and potassic, mafic to ultramafic volcanic and shallow intrusive volatile-rich rocks. Subsolidus alteration to clay and carbonate minerals is common because of the

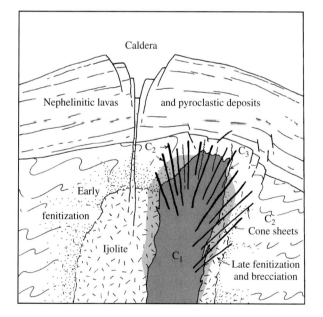

13.39 Idealized spatial relations of a subvolcanic carbonatite-nephelinite association in east Africa. Intrusion of nephelinite (ijolite) magma is accompanied by early fenitization of wall rocks. Eruption of magma to the surface builds a composite volcano that subsequently experiences caldera collapse. Carbonatite magma (C_1, dark shaded) intruded into the ijolite and older country rocks and doming the roof rocks is accompanied or preceded by a wave of fenitization and brecciation (light shaded). Resurgence of the magma locally breaches its fenite envelope. A swarm of later carbonatite cone sheets (C_2) is then emplaced, followed by still later carbonatite dikes (C_3). (Redrawn from Le Bas, 1987.)

high H_2O and CO_2 fugacities of the magmas. Consequently, rocks weather readily, are notoriously poorly exposed, and tend to underlie soil- or water-filled topographic depressions. The four are petrologic "orphans" in that they do not have well-constrained chemical, mineralogical, nor in some instances textural definitions, as do common magmatic rock types such as basalt or granodiorite. Each of the four is, in fact, not a rock type at all, but a diverse **clan** of rocks loosely related in some way. Their tectonic settings are also rather diverse. Most are found in continental regions of crustal extension or crustal stability (cratons), of which the African plate is the supreme example; but there are also occurrences of lamprophyre lava flows in continental arcs, such as the Western Mexico volcanic belt just described.

Were it not for their intriguing attributes and the potential insights they provide on mantle sources, magma generating processes, and tectonics, these rocks could be ignored. An added factor in their interest is economic—three are the sole known primary source of diamonds.

13.12.1 Lamprophyres

As implied by their name, lamprophyre rocks are porphyritic (phyric), and the most common variety—**minette**—contains abundant Fe-Mg mica (Greek *lampros,* means "glistening"). Traditionally, lamprophyres have been thought of as exclusively dike rocks, and that characteristic is commonly cited as one defining attribute, but local occurrences of lava flows (Luhr, 1997) and possibly moderate-size plutons of phaneritic rock have also been recognized. Petrologists only began to study lamprophyres seriously in the last decade or so of the 1900s. Thus, much has yet to be learned of their true nature, the best way to classify them, the origin of magmas that form lamprophyres, and even whether they should continue to be recognized as a distinct clan of kindred rocks. A *working definition* of **lamprophyres** can be proposed as follows (see also Le Maitre, 1989, p. 11; Mitchell, 1994, p. 142): a diverse group of polygenetic rocks crystallized under volatile-rich conditions and characterized by abundant mafic phenocrysts of biotite-phlogopite and/or amphibole together with lesser amouts of clinopyroxene and/or melilite; feldspars are confined to the matrix along with, either singly or in various combinations, feldspathoids, carbonate, monticellite, melilite, mica, amphibole, pyroxene, olivine, perovskite, Fe-Ti oxides, and glass. Strongly zoned and corroded minerals indicate a lack of equilibrium during final crystallization.

Lamprophyres in continental arcs commonly occur as dikes in tonalite-granodiorite plutons. This association is so widespread that some petrologists recognize

it as a trilogy, the third component of which are mafic inclusions and larger bodies of texturally and modally diverse but hornblende-rich rock called **appinite** (Pitcher, 1997, Chap. 10).

Mica-rich minettes (Table 13.11) and some kindred lamprophyres have arc affinities in their relatively depleted Nb and Ta contents, high Ba/Ti, spiky normalized trace element diagrams, and bulk chemical similarities to the absarokite-shoshinite-banakite series rocks of subduction zones. Shoshonitic rocks, composed mostly of olivine, two pyroxenes, and plagioclase and formed under low water fugacity, can be considered as heteromorphs of minettes that form under high water fugacity, destabilizing plagioclase and anhydrous mafic minerals. This heteromorphism accords with the proposal of Mitchell (1994) that lamprophyres should be considered as a **facies**—a diverse group of rocks crystallized under similar conditions, in this case, high water fugacity.

The high incompatible *and* compatible element concentrations of primitive lamprophyres, together with their somewhat radiogenic Sr and Nd isotopic ratios, suggest derivation from low degree partial melts of previously metasomatized mantle. In this vein-plus-wall-rock melting mechanism (Section 11.5.1), veins of phlogopite-rich assemblages could have been introduced by aqueous solutions rising into the mantle wedge above the subducting slab, in some cases as long ago as the Proterozoic (Tingey et al., 1991).

13.12.2 Lamproite, Orangeite, and Kimberlite Clans

From the late 1800s it was believed that the only primary source of diamonds was kimberlite; a secondary source was alluvium derived from kimberlite. However, in the late 1970s in the West Kimberly area of Western Australia, rich diamond concentrations were discovered in poorly exposed volcaniclastic deposits of lamproite. Better exposures of lamproites in Western Australia were made famous in studies in the mid-1900s by R. T. Wade and R. T. Prider.

Unlike lamprophyres, the lamproite-orangeite-kimberlite clans never contain plagioclase, nepheline, or melilite. All are markedly low in Al_2O_3 relative to alkalies so most are also peralkaline. In contrast to other rather primitive mafic and ultramafic rocks, such as picrite and komatiite, which have high concentrations of compatible Cr and Ni, rocks of these clans are also unusually enriched in incompatible elements (Table 13.11; Figure 11.20). Xenocrystic olivine may in part account for the high compatible element concentrations. Diamond is not an essential defining mineral. It is only present in some occurrences and in extremely minute concentrations; for example, in the famous Kimberly mine in South Africa where kimberlite was christened, 24 million tons of rock yielded only 3 tons of diamond, or 0.125 ppm!

Each clan ranges widely in composition and fabric, far more than common rock types. Additionally, there are overlaps in most any rock property. Instead of tight definitions, lengthy characterizations of each clan and contrasts between them must suffice.

Lamproite Clan. Of the 24 known worldwide locales of clustered lamproite bodies, altogether <100 km^3, the main locales are in southeastern Africa (Jurassic), Western Australia (Proterozoic and Miocene), and the Leucite Hills of Wyoming (Quaternary). All lie in tectonically stable plate interiors, although prior subduction or rifting appears to have metasomatized the thick continental lithosphere through which magmas ascended. In Western Australia, lamproite occurs in narrow pipes that flare at the surface into subcircular craters. These are filled with a variety of volcaniclastic deposits and are intruded by dikes and sills feeding surface flows. The <2-Ma Leucite Hills consist of lava flows and small cinder cones.

For comparison only (not as defining attributes), chemical analyses in Table 13.11 indicate that, relative to the orangeite and kimberlite clans, **lamproites** are relatively rich in SiO_2, TiO_2, Ba (commonly >5000 ppm), La (>200), and Zr (>500) and poor in CO_2. When compared to orangeites-kimberlites that contain abundant carbonate minerals, fresh lamproites generally have <0.5 wt. % CO_2. They are ultrapotassic (molecular $K_2O/Na_2O > 3$) and their peralkalinity is manifested in Fe^{3+}-Ba-rich, Al-poor leucite and sanidine; Ti-rich, Al-poor phlogopite; and Ti-K-rich richterite amphibole. Also present are Mg-rich olivine (both phenocrysts and groundmass in some), diopside, glass, and accessory amounts of apatite, ilmenite, and exotic minerals, which include wadeite ($K_2ZrSi_3O_9$), priderite (K, Ba, Fe^{3+}, Ti oxide), perovskite (Ca, Na, Fe^{2+}, REEs, Ti, Nb oxide), and several other minerals (Scott-Smith, 1996). Not all of these phases necessarily occur in a single sample and some samples can be mineralogically unique, unlike others in the clan. This attribute, together with dissimilarites between lamproites in different locales and the fact that some lamproites contain mantle-derived xenocrysts, imposes difficulties in accurate characterization of the whole clan.

Orangeite and Kimberlite Clans. The orangeite and kimberlite clans are hybrid rocks emplaced as explosive breccia in carrot- and funnel-shaped diatremes fed from intrusive, nonfragmental dike-sill complexes (Figures 9.25 and 9.26). Rocks are exceptionally rich (>12 wt.%) in $CO_2 + H_2O^+$, which are chiefly sequestered in primary calcite and in secondary serpentine that replaces olivine and orthopyroxene. The true composition of the primitive magmas remains uncertain because of differentiation and alteration overprints and especially because of contamination by abundant xenoliths and xenocrysts derived from both crust and mantle. Xenoliths of crustal rock (shale, granite, and others) and of mantle rock (eclogite and a wide variety of peridotitic rocks, commonly metasomatically altered; see Section 11.2) are readily identified. On the other hand, mantle-derived xenocrysts are difficult, and commonly impossible, to distinguish from phenocrysts that precipitated from the kimberlite melt at different places in its excursion from depths of perhaps 200 km or more, where diamond is stable (Figure 11.5). The degree of euhedralism as a guide to discriminating foreign versus cognate crystals may be misleading because deep euhedral precipitates might be resorbed at shallower depths or physically abraded in the explosive diatreme. Rocks are characteristically inequigranular because the polygenetic megacrysts (>5 mm in diameter) are hosted in an aphanitic, never glassy groundmass of calcite, secondary serpentine, phlogopite, Mg-spinel, and apatite. Most megacrysts, commonly about 25 modal %, are Mg-olivine, but in archetypal kimberlite, lesser megacrysts include Mg-rich ilmenite, phlogopite, enstatite, purple-red pyrope (Mg-rich) garnet, and green Cr-rich clinopyroxene; the latter two serve as colorful "indicator" minerals in the regolith overlying weathered kimberlites and in downstream stream and glacial deposits.

Early work just after the turn of the 1900s revealed two types of diamondiferous rocks in South Africa. In subsequent decades, this distinction became confused, but it was again clarified by Mitchell (1995). The two clans are:

(1) **Archetypal kimberlites** (also called basaltic or group I kimberlites)
(2) **Orangeites** (also called micaceous or group II kimberlites), which are mineralogically, isotopically, and, with regard to origin, more similar to the lamproite clan than to archetypal kimberlites, although orangeites contain more CO_2 and compatible Ni and Cr but less Rb, Ba, Zr, and Ti than lamproites

The only systematic chemical contrast between the kimberlites and orangeites lies in higher K_2O/TiO_2 in the latter. The most striking contrast is significantly higher $^{87}Sr/^{86}Sr$ and lower $^{143}Nd/^{144}Nd$ in orangeites than in archetypal kimberlites (Figure 13.41). Mineralogical contrasts among the lamproite, kimberlite, and orangeite clans are set out in Table 13.12.

All known (200+) orangeite bodies lie in the Archean Kaapvaal craton in South Africa (Figure 13.42), where they occur as early Cretaceous (125–110 Ma) and early Jurassic (165–145 Ma) swarms of dikes and diatremes. Worldwide, there are >5000 known bodies of archetypal kimberlites having an estimated volume >5000 km^3; most of these are in Precambrian cratons. They typically occur in clusters <40 km across of a few dozen intrusions (Nixon, 1995). In southern Africa, kimberlite emplacement has occurred episodically several times since 1600 Ma.

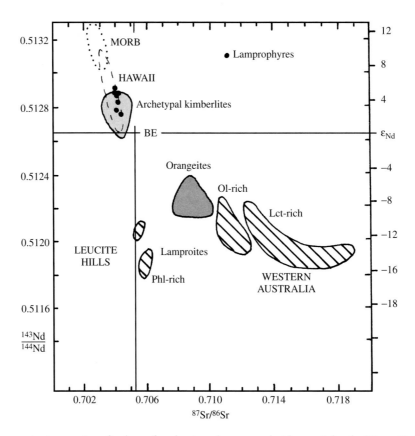

13.41 Radiogenic isotope ratios in potassic mafic-ultramafic volcanic rocks compared with oceanic basalts (Figure 13.8). Note that orangeites (dark shade) are far more enriched than worldwide archetypal kimberlites (light shade) and are isotopically more like lamproites (diagonally ruled). Note also provinciality of lamproites; those in Western Australia are more enriched in [87]Sr than those in Leucite Hills of Wyoming. BE, bulk Earth. (Redrawn from Mitchell and Bergman, 1991.)

Table 13.12. Mineralogical Comparison of Lamproites, Archetypal Kimberlites, and Orangeites

	LAMPROITES	ARCHETYPAL KIMBERLITES	ORANGEITES
Olivine	mm, ppp	mmmm, ppp	mm, pp
Phlogopite	mmm, ppp, ggg	mm, pp, ggg	mmm, ppp, ggg
Leucite	ppp, ggg		g
Ti-K richterite	ppp, ggg		gg
Sanidine	ppp, ggg		ggg
Diopside	ggg		ggg
Monticellite		ggg	
Spinel	gg	gggg	gg
Perovskite	gg	ggg	gg
Apatite	ggg	ggg	gggg
K-Ba-titanites	ggg	g	ggg
Zr-silicates	ggg	g	ggg
Mn-ilmenite	g	gg	ggg
Calcite		gggg	ggg
Pyrope		mmm	

Blank, absent; x, very rare; xx, rare; xxx, common; xxxx, abundant; g, groundmass; p, phenocryst; m, megacryst.
(Data from Mitchell, 1996.)

Implications of Diamonds and Inclusions in Them. Diamond-bearing kimberlites and orangeites occur only on Archean cratons (>2.5 Ga), whereas barren kimberlites lie in peripheral terranes that were subjected to post-Archean tectonism. On the other hand, diamond-bearing lamproites occur on Archean and Proterozoic cratons. Diamonds commonly contain inclusions of many different minerals (visible with a 10X hand lens) that provide significant information not only on their origin but also on the origin of the host magma and perhaps even the tectonic history of the planet (Haggerty, 1999).

Sm-Nd and Rb-Sr ages of garnet and pyroxene inclusions, and the assumed syngenetic host diamonds, are 900 to as much as 3300 Ma. Hence, most diamonds are much older than their host rocks (1600–90 Ma in southern Africa). The chemical composition of inclusions indicates that diamonds crystallized and were stored and preserved in stable craton roots at depths of mostly 120–200 km and temperatures <1200°–1500°C (Helmstaedt and Gurney, 1995; Gurney and Zweistra, 1995). This implies an unusually low geothermal gradient (8–10°C/km), especially for what is generally believed to have been a hotter planet during the Precam-

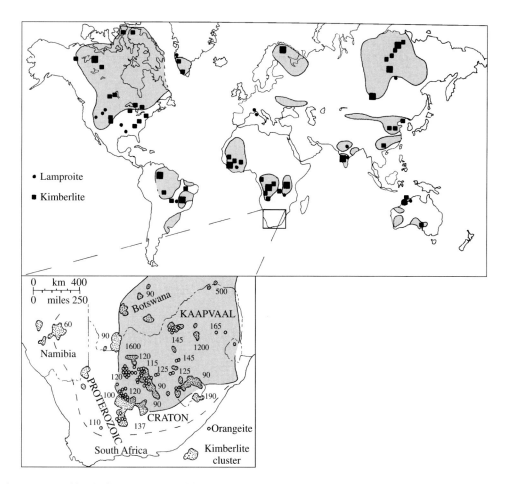

13.42 Global occurrence of kimberlite, orangeite, and lamproite. Generalized extents of Archean cratons shaded. Major diamond deposits shown by larger circles (lamproite) and squares (kimberlite). Lower left is enlarged southern Africa Kaapvaal craton, showing clusters of kimberlite and orangeite diatremes with ages of emplacement ranging from 1600 to 60 Ma. (Redrawn from Mitchell, 1995.) Diamond-bearing orangeite and kimberlite occur only on Archean cratons, but diamondiferous lamproites are found in both Proterozoic and Archean cratons. (Redrawn from Nixon, 1995; Haggerty, 1999.)

brian. Very rare inclusions even appear to be from the mantle transition zone at a depth of 410–670 km, and some may have crystallized in the deeper, lower mantle (e.g., Harris et al., 1997). After storage for long periods in relatively cool, presumably nonconvecting mantle, diamonds were caught up in ascending, younger, genetically unrelated magmas. Pieces of the rock in which the diamonds were originally lodged now occur as diamond-bearing xenoliths of eclogite and peridotite (chiefly depleted harzburgite) in the host lamproites, orangeites, and kimberlites. Disaggregation of these xenoliths into the much larger volumes of entraining magmas produced abundant megacrysts of indicator minerals—olivine, pyroxene, garnet, and others—as well as the orders-of-magnitude-more-diluted *xenocrystic* diamond population.

Note the hierarchy of petrologically significant inclusions: Minute mineral grains that serve as isotopic clocks and as geothermobarometers are hosted in diamonds that occur as grains in mantle eclogite and

harzburgite which are included as xenoliths in lamproites, and kimberlites, which are hosted in Precambrian craton country rock.

<u>Origin of Magmas and Relation to the Diamond Source</u>. Genesis of lamproite, orangeite, and kimberlite magmas must take into account their chemical and isotopic compositions, episodic production since the late Archean, and presence of entrained diamond-bearing eclogite and peridotite derived from deep, cool, nonconvecting lithospheric mantle, mostly of Archean age. The normalized trace element diagram (Figure 11.20) for lamproite and kimberlite is similar to the humped pattern for oceanic island basalts (OIBs) but the more incompatible elements are more enriched by about an order of magnitude than in OIB. This suggests derivation of magmas by smaller degrees of melting of an OIB-like, depleted mantle source or by melting of a metasomatically enriched source. The high ratio of light REEs/heavy REEs suggests garnet in the source

Table 13.13. Principal Factors in the Origin of Two Petrotectonic Associations

Tectonic Setting	Cause of Magma Generation	Magma Source(s) and/or Components	Differentiation Process(es)	Rock Type(s)	Characteristic Chemical and Isotopic Composition	Fabrics and Field Relations
Oceanic rift	Decompression of upwelling mantle	Mantle peridotite at depths of 40–80 km previously depleted in incompatible elements	Polybaric olivine fractionation of magmas ascending from source; low-P fractionation of Pl + Cpx + Ol in crustal magma chambers but moderated by mixing with draughts of more primitive mantle-derived magma	Relatively uniform tholeiitic basalt (MORB) and gabbro containing Q and Hy; rare ferrobasalt, ferrodacite, and plagiogranite; peridotite	Relatively low concentrations of incompatible elements; low $^{87}Sr/^{86}Sr$ and high $^{143}Nd/^{144}Nd$	Submarine basalt lava flows, commonly pillowed, and associated clastic deposits; sheeted dikes; rare dikes of plagiogranite; isotropic gabbros; layered cumulate gabbros and peridotites; altered metamorphic textured (deformed) peridotites
Continental arc	Reduction of solidus T in subarc mantle wedge by increase in water concentration; increase in T and perhaps in volatile concentration of lower crust due to intruded mantle-derived mafic magmas	Aqueous solutions and partial melts from subducting oceanic basalt crust; depleted upper mantle peridotite hydrated and metasomatized by aqueous solutions and melts; continental crust	Concurrent fractional crystallization of primitive mantle-derived magmas and assimilation of crustal rocks (AFC); magma mixing	Andesite, dacite, rhyolite, dioritic rocks, tonalite, granodiorite, granite; uncommon adakite, basalt, and gabbro	Evolved (nonprimitive) chemical compositions; high incompatible, low compatible and HFS elements (Ta, Nb), spiky trace element arc signature; variable and commonly relatively high $^{87}Sr/^{86}Sr$ and $\delta^{18}O$ and low $^{143}Nd/^{144}Nd$	Composite volcanoes; lava and debris flows; widespread pyroclastic deposits; calderas; typically huge batholiths made of hundreds of individual plutons emplaced over tens of millions of years

residue, agreeing with the deep source implied by the xenocrystic diamonds. Figure 13.41 shows that magmas creating the archetypal kimberlite clan have a source that is near bulk Earth and like that of asthenospheric, plume-related OIB, exemplified by Hawaiian magmas. In striking contrast, orangeite and lamproite clans have higher $^{87}Sr/^{86}Sr$ and lower $^{143}Nd/^{144}Nd$, so their magmas must have a major component from mantle lithosphere metasomatically enriched during the Precambrian.

It is possible that nonconvecting mantle underlying old cratons is a remnant of ancient mantle plumes (Herzberg and O'Hara, 1998; Haggerty, 1999) and has experienced an unusual long-term history of metasomatism and a younger episode of magma generation, as suggested by markedly different Sr isotope ratios of lamproites in Figure 13.41. Whether the diamondiferous eclogites in the subcratonic mantle represent frozen high-P basaltic melts or remnants of ancient subducted oceanic crust (Helmstaedt and Gurney, 1995) is controversial. Likewise, the source of the carbon for diamonds and the CO_2 dissolved in the kimberlitic magmas—whether from ancient subducted crust or primeval mantle—is also controversial. These questions bear on the problematic origin of Archean continents and lithosphere.

SUMMARY

Relating the wide diversity of magmatic rocks to their tectonic setting provides rich insights into the way the Earth works—a central theme of this textbook. Interpretive models of magma origin can constrain global models proposed by geophysicists and tectonic geologists, and vice versa. Of fundamental concern is the thermal budget of the Earth, the way its vast store of thermal energy is dissipated through motion of lithospheric plates and rise of mantle plumes and the magmatism associated with these convective processes.

From a narrower perspective, study of petrotectonic associations focuses attention on specific magma sources, conditions of generation, and subsequent processes of differentiation of primitive magmas that have produced distinct rock types and rock suites in specific tectonic settings. Rather than a lengthy summary of all of the associations described in this chapter, only two strongly contrasting associations are set out in Table 13.13.

CRITICAL THINKING QUESTIONS

13.1 Prepare a detailed summary of each of the petrotectonic associations described in this chapter using the format in Table 13.13. Two are provided as examples.

13.2 How does the composition of oceanic rocks constrain models proposed by geophysicists concerning the nature of the mantle and convection in it?

13.3 In the complementary Sr and Nd isotope compositions of MORB and continental crust (Figure 2.27) why is MORB closer to the mantle array than the continental crust?

13.4 How does the composition of rocks in ophiolites constrain their tectonic interpretation?

13.5 What was the state of stress during emplacement of the Karoo dikes and sills shown in Figure 13.20b?

13.6 What special geologic conditions in Iceland compared to other oceanic islands might be responsible for the exceptionally large proportion of silicic rocks?

13.7 Explain the paradox of the mica-rich lamproite in Table 13.11, column 7, that contains normative quartz but modal leucite (see also problem 13.2). How would you classify this rock? Is it silica-oversaturated or silica-undersaturated?

13.8 What do kimberlitic and lamproitic rocks tell us about the history of cratons (long-term stable sectors of continents unaffected by tectonism) and their thermal state?

13.9 Diamonds contain mineral inclusions that crystallized deep in the mantle at high T as long ago as 3 Ga. How is it possible for these to be preserved for our study?

13.10 Why does orthopyroxene never crystallize under equilibrium conditions in silica-undersaturated alkaline magmas?

13.11 From phase diagrams in Chapters 5 and 11 how might you account for the contrasting compositions of olivines in Hawaiian tholeiitic and alkaline basalts (Section 13.2.1)?

PROBLEMS

13.1 Make primitive-mantle-normalized trace element diagrams of granitic and rhyolitic rocks from Tables 13.7, 13.8, 13.9, and 13.10. Compare these patterns and comment on what they tell about the source of magmas and overprinting differentiation processes.

13.2 Calculate the normative composition of the plagiogranite in Table 13.7 and the phlogopite-rich lamproite in Table 13.11. How do these norms compare with the corresponding modes? See also critical thinking question 13.7.

APPENDIX A

Table A.1. Chemical Analyses of Representative Major Rock-Forming Minerals in Magmatic Rocks

	OLIVINE			GARNET	PYROXENE								
	1	2	3	4	5	6	7	8	9	10	11	12	13
SiO_2	41.23	38.95	30.25	41.55	58.13	52.52	51.52	55.08	51.5	41.5	48.48	51.92	54.76
TiO_2	0.02	0.04		0.11	0.04	0.71	0.18	0.07	1.2	4.8	0.08	0.77	0.44
Al_2O_3	0.03	0.02	0.00	20.28	0.79	1.10	0.32	2.24	2.2	10.4	0.3	1.85	8.33
Cr_2O_3	0.02	0.02		4.8	0.33	0.06		2.03	0.0	0.0			
Fe_2O_3									4.3	7.4		31.44	1.29
FeO	7.50t	18.16t	61.88t	6.93t	4.53t	18.0t	26.96t	2.21t	5.2	1.7	26.97t	0.75	2.70
NiO	0.45	0.01		0.01	0.11			0					
MnO	0.08	0.20	6.14	0.37	0.12	0.33	2.06	0.09	0.0	0.1	2.57		0.04
MgO	50.76	42.37	1.40	20.88	35.39	25.1	17.69	16.40	15.4	10.6	2.46		11.59
CaO	0.02	0.32	0.19	5.04	0.46	2.15	1.30	19.65	19.6	22.8	18.16		16.35
Na_2O	0.03	0.02		0.03	0.13	0.03		0.05	0.4	0.9	0.59	12.86	4.63
K_2O	0.00			0.00	0.00	0.00		0.01	0.1	0.1		0.19	0.05
Total	100.15	100.11	99.86	100.10	100.01	100.00	100.03	99.88	99.9	100.3	99.61	99.78	100.18

1, Mg-rich olivine ($Fo_{92.3}$) in coarse garnet lherzolite xenolith in kimberlite, Bulfontein, Kimberly, South Africa. From same rock as analyses 4, 5, and 8. (Data and chemical and modal compositions of the whole rock from Cox et al., 1987.) **2, Mg-rich olivine** ($Fo_{80.6}$), rim of phenocryst in basalt scoria, 1959 eruption of Kilauea volcano, Hawaii. (From Helz, 1987.) **3, Fayalite** ($Fo_{3.5}$) phenocryst in rhyolitic Bandelier Tuff, Valles caldera complex, New Mexico. From same rock as analyses 7 and 11. (Data from Warshaw and Smith, 1988.) **4, Pyropic garnet** in coarse garnet lherzolite xenolith in kimberlite, Bulfontein, Kimberly, South Africa. From same rock as analyses 1, 5, and 8. (Data and chemical and modal compositions of the whole rock from Cox et al., 1987.) **5, Enstatite** ($Fn_{93.2}$) in coarse garnet lherzolite xenolith in kimberlite, Bulfontein, Kimberly, South Africa. From same rock as analyses 1, 4, and 8. (Data and chemical and modal compositions of the whole rock from Cox et al., 1987.) **6, Hypersthene** ($En_{68.3}$) in groundmass in the prehistoric Makaopuhi lava lake, Kilauea volcano, Hawaii. (Data from Evans and Moore, 1968.) **7, Hypersthene** ($En_{52.1}$) phenocryst in rhyolitic Bandelier Tuff, Valles caldera complex, New Mexico. From same rock as analyses 3 and 11. (Data from Warshaw and Smith, 1988.) **8, Cr-rich diopside** in coarse garnet lherzolite xenolith in kimberlite, Bulfontein, Kimberly, South Africa. From same rock as analyses 1, 4, and 5. (Data and chemical and modal compositions of the whole rock from Cox et al., 1987.) **9, Augite** groundmass grain in tholeiitic basalt, Santa Clara lava flow, Saint George, Utah. (Data from Best and Brimhall, 1974.) **10, Ti-rich augite** groundmass grain in basanite lava, Uinkaret plateau, northwestern Grand Canyon, Arizona. (Data from Best and Brimhall, 1974.) **11, Fe-rich augite** phenocryst in rhyolitic Bandelier Tuff, Valles caldera complex, New Mexico. From same rock as analyses 3 and 7. (Data from Warshaw and Smith, 1988.) **12, Aegirine** in peralkaline A-type granite, south Kigom Hills, Nigeria. From same rock as analysis 17. (Data from Greenwood, 1951.) **13, Omphacite** in eclogite xenolith, Kao kimberlite pipe, Basutoland, South Africa. (Data from Nixon et al., 1963.)

Table A.1. *(Continued)*

	AMPHIBOLE						MICA	
	14	15	16	17	18	19	20	21
SiO_2	44.48	44.6	40.96	51.01	55.4	48.51	35.93	40.62
TiO_2	1.69	1.2	3.92	0.96	0.46	1.10	4.27	2.27
Al_2O_3	7.47	13.4	15.35	0.80	0.99	1.08	13.00	11.26
Cr_2O_3		2.7	0.27		0.48			
Fe_2O_3	3.91		4.82	16.41		17.53	2.61	
FeO	17.70	3.4t	3.30	17.62	2.25t	14.31	20.81	5.42t
MnO	0.26		0.05	0.48		0.66	0.11	0.06
MgO	9.34	18.1	15.24	0.22	22.9	1.31	9.26	24.67
CaO	10.72	10.1	10.42	0.19	6.74	5.43	0.41	0.13
Na_2O	1.46	3.4	2.56	7.98	3.82	7.87	0.10	
K_2O	0.97	1.3	1.74	1.80	4.63	1.48	10.14	10.13
H_2O^+	1.47		1.00	0.91		1.48	2.69	3.09
F	0.31		0.02	1.70			0.65	
Cl			0.12					
Total	99.78	98.2	99.77	100.09	97.67	100.76	99.98	98.94

14, Hornblende in quartz monzodiorite, Gaudalupe igneous complex, California. From same rock as analysis 20. (Data from Best and Mercy, 1967.) **15, Cr-rich pargasite** as interstitial grains in spinel lherzolite xenolith in basanite, Uinkaret plateau, Arizona. (Data from Best, 1974.) **16, Kaersutite** isolated anhedral megacryst in basanite, Uinkaret plateau, Arizona. (Data from Best, 1974.) **17, Riebeckite** in peralkaline A-type granite, south Kigom Hills, Nigeria. From same rock as analysis 12. (Data from Greenwood, 1951.) **18, Richterite;** mean of 27 in veined metasomatized garnet peridotite xenoliths, Bultfontein, South Africa. (Data from Dawson, 1987.) **19, Arfvedsonite** in A-type syenite, Morotu, Sakhalin Island, Russia (north of Japan). (Data from Yagi, 1953.) **20, Biotite** in quartz monzodiorite, Gaudalupe igneous complex, California. From same rock as analysis 14. (Data from Best and Mercy, 1967.) **21, Phlogopite** in mica-rich minette lamprophyre dike, Wasatch plateau. Utah. Analysis of whole rock in Table 13.11, column 6. (Data from Tingey DG, unpublished MS thesis, Brigham Young University, Provo, Utah, 1989.)

	FELDSPAR					FELDSPATHOID AND FELDSPATHOID-LIKE				FE-TI OXIDE	
	22	23	24	25	26	27	28	29	30	31	32
SiO_2	67.84	62.7	52.42	44.17	65.5	44.35	54.62	47.23	42.81	0.07	0.03
TiO_2	0.00		0.09				0.00		0.12	3.9	48.6
Al_2O_3	19.65	23.27	29.7	34.95	19.03	32.22	22.93	26.72	6.59	2.7	0.07
V_2O_3										1.0	0.05
Cr_2O_3										0.08	0.05
Fe_2O_3	0.03	0.19t	0.36	0.56	0.09t	0.47t	0.26	1.29t	1.90	58.1	8.8
FeO	0.02		0.13	0.08			0.26		3.06	34.0	39.8
MnO									0.08	0.28	0.44
MgO	0.04		0.08	0.00		0.23	0.00		7.48	0.7	1.9
BaO		0.02			0.16						
CaO	0.00	4.35	12.65	18.63	0.18	0.54	0.08	1.40	33.27	0.03	0.04
Na_2O	11.07	8.58	4.01	0.79	3.84	19.69	0.66	14.70	3.75		
K_2O	0.29	1.23	0.21	0.05	11.19	1.93	21.02	0.73	0.33		
H_2O^+	0.56			0.84		0.53	0.12	8.23	0.43		
Total	99.50	100.34	99.65	100.07	99.99	99.96	99.95	100.30	99.82	99.87	99.78

22, Albite in pegmatite, near Court House, Amelia County, Virginia. $An_{0.3}$, $Ab_{98.0}$, $Or_{1.7}$. (Data from Kracek and Neuvonen, 1952.) **23, Oligoclase** in rhyolitic Bishop Tuff, California. $An_{20.4}$, $Ab_{72.7}$, $Or_{6.9}$. From same rock as analysis 26. (Data from Hildreth W, unpublished Ph.D thesis, University of California, Berkeley, 1977.) **24, Labradorite** isolated megacrysts in basaltic scoria, Crater Elegante, Sonora, Mexico. $An_{63.1}$, $Ab_{35.7}$, $Or_{1.2}$. (Data from Gutman and Martin, 1976.) **25, Anorthite** in olivine norite, Grass Valley, California. $An_{92.6}$, $Ab_{7.1}$, $Or_{0.3}$. (Data from Kracek and Neuvonen, 1952.) **26, Sanidine** in rhyolitic Bishop Tuff, California. $An_{0.9}$, $Ab_{33.9}$, $Or_{65.0}$. From same rock as analysis 23. (Data from Hildreth W, unpublished Ph.D thesis, University of California, Berkeley, 1977.) **27, Nepheline** in theralite (olivine-nepheline-analcite gabbro), Square Top intrusion, New South Wales, Australia. From same rock as analysis 29. (Data from Wilkinson, 1965.) **28, Leucite** in leucitite, Lake Kivu National Park, Zaire. (Data from Sahama, 1952.) **29, Analcite** in theralite (olivine-nepheline-analcite gabbro), Square Top intrusion, New South Wales, Australia. From same rock as analysis 27. (Analcite is actually a zeolite but occurs in some rocks as if it were a feldspathoid.) (Data from Wilkinson, 1966.) **30, Melilite** in melilite nephelinite, Honolulu, Oahu, Hawaii. Whole rock analysis in Table 13.2, column 7. (Melilite is not a feldspathoid, but its occurrence in some rocks is like that type of mineral. (Data from Neuvonen, 1952.) **31, Magnetite** in groundmass in the prehistoric Makaopuhi lava lake, Kilauea volcano, Hawaii. From same rock as analysis 32. (Data from Evans and Moore, 1968.) **32, Ilmenite** in groundmass in the prehistoric Makaopuhi lava lake, Kilauea volcano, Hawaii. From same rock as analysis 31. (Data from Evans and Moore, 1968.)

Calculation of the CIPW Norm

In this normative calculation, devised by W. Cross, J. P. Iddings, L. V. Pirsson, and H. S. Washington, constituents from the chemical analysis of a rock are allocated in a prescribed sequence to hypothetical normative minerals. These have simple end-member compositions and all are volatile-free.

The normative minerals and their abbreviations, chemical formulas, and formula weights are shown in Table B.1. Examples of the calculation are shown in Tables B.2 and B.3. During the calculation, numbers should be rounded to four significant digits.

The oxidation state of Fe in the chemically analyzed rock can profoundly affect the degree of silica saturation in the norm. Because this degree of silica saturation is one of the primary reasons for making the calculation, the oxidation state must be standardized in some way. This is particularly important for hot extrusive rocks that oxidized in the oxygen-rich atmosphere or hydrosphere. Cox et al. (1979) suggest using a standard molecular ratio of $Fe_2O_3/FeO = 0.15$, which is equivalent to an atomic ratio of $Fe^{3+}/Fe^{2+} = 0.30$. The Fe^{3+}/Fe^{2+} ratio of the low-K basaltic andesite in Table 13.6, whose norm is calculated in Table B.2, is standardized as follows:

Oxide	Wt.% Oxide	Formula Wt. Oxide*	Formula Proportion**	Atomic Proportion	Total Fe Atoms
Fe_2O_3	3.47	159.69	0.0217	0.0434	
					0.1524
FeO	7.83	71.846	0.1090	0.1090	

* Sum of atomic weights.
** Wt.% oxide divided by formula wt. of oxide.

Simultaneous solution of the two equations in atomic amounts

$$Fe^{3+}/Fe^{2+} = 0.30$$
$$Fe^{3+} + Fe^{2+} = 0.1524$$

gives $Fe^{2+} = 0.1172$ and $Fe^{3+} = 0.0352$. As a check, $Fe^{3+}/Fe^{2+} = 0.3003$.

The standardized formula proportion of $FeO = 0.1172$ and of $Fe_2O_3 = 0.0352/2 = 0.0176$. These two proportions are used in the calculation in Table B.2.
The following calculation procedure ignores several constituents seldom found in analyses of common rocks. If the analysis includes NiO, add its formula proportion to FeO, add SrO and BaO to CaO, allocate F to Ap ($3P_2O_5 + 0.33F$), and if in excess, to make

Table B.1. Data for normative minerals

Name	Abbreviation	Chemical Formula	Formula Weight
Quartz	Q	SiO_2	60.08
Corundum	C	Al_2O_3	102.0
Orthoclase	Or	$K_2O \cdot Al_2O_3 \cdot 6SiO_2$	556.6
Albite	Ab	$Na_2O \cdot Al_2O_3 \cdot 6SiO_2$	524.4
Anorthite	An	$Ca_2O \cdot Al_2O_3 \cdot 2SiO_2$	278.2
Leucite	Lc	$K_2O \cdot Al_2O_3 \cdot 4SiO_2$	436.5
Nepheline	Ne	$Na_2O \cdot Al_2O_3 \cdot 2SiO_2$	284.1
Kaliophilite	Kp	$K_2O \cdot Al_2O_3 \cdot 2SiO_2$	316.3
Acmite	Ac	$Na_2O \cdot Fe_2O_3 \cdot 4SiO_2$	462.0
Sodium metasilicate	Ns	$Na_2O \cdot SiO_2$	122.1
Potassium metasilicate	Ks	$K_2O \cdot SiO_2$	154.3
Wollastonite	Wo	$CaO \cdot SiO_2$	116.2
Diopside	Di	$CaO \cdot FeO \cdot 2SiO_2$	248.1
		$CaO \cdot MgO \cdot 2SiO_2$	216.5
Hypersthene	Hy	$FeO \cdot SiO_2$	131.9
		$MgO \cdot SiO_2$	100.4
Olivine	Ol	$2FeO \cdot SiO_2$	203.8
		$2MgO \cdot SiO_2$	140.7
Dicalcium silicate	Cs	$2CaO \cdot SiO_2$	172.2
Magnetite	Mt	$FeO \cdot Fe_2O_3$	231.5
Ilmenite	Il	$FeO \cdot TiO_2$	151.7
Hematite	Hm	Fe_2O_3	159.7
Titanite	Tn	$CaO \cdot TiO_2 \cdot SiO_2$	196.0
Perovskite	Pf	$CaO \cdot TiO_2$	135.9
Rutile	Ru	TiO_2	79.87
Apatite	Ap	$3.3 CaO \cdot P_2O_5$	327.0

normative fluorite after step 5, Cl is allocated to normative halite after step 2, ZrO to normative zircon, and Cr (and FeO) to normative chromite before making normative ilmenite and magnetite.

1. Calculate the formula (molecular) proportions of the oxides in the chemical analysis by dividing their weight percentage (wt.%) by their corresponding formula weight, as was done for Fe_2O_3 and FeO.
2. Add the proportion of MnO to that of FeO.
3. Allocate an amount from the formula proportion of CaO equal to 3.3 times that of the formula proportion of P_2O_5 to make normative apatite (Ap).
4. Allocate an amount of FeO equal to that of the proportion of TiO_2 to make normative ilmenite (Il).
5. If there is an excess of TiO_2 over available FeO, allocate an amount of CaO equal to the excess to

Table B.2. Calculation of the Normative Composition of the Low-K Basaltic Andesite in Table 13.6

	Wt.% Oxide	Formula Weight	Formula Proportion	Ap	Il	Or'	Ab'	An	Mt	Rem	Di'	Hy'	Q
SiO₂	53.29	60.08	0.8870			0.0192	0.2136	0.2582			0.0912	0.1625	0.1423
TiO₂	0.91	79.87	0.0114		0.0114								
Al₂O₃	17.13	102.0	0.1679			0.0032	0.0356	0.1291					
Fe₂O₃	3.47	159.7	0.0176*						0.0176				
FeO	7.83	71.85	0.1172*		0.0114				0.0176	0.0910	0.0199	0.0711	
MnO	0.20	70.94	0.0028										
MgO	4.72	40.30	0.1171							0.1171	0.0257	0.0914	
CaO	9.87	56.08	0.1760	0.0013				0.1291		0.0456	0.0456		
Na₂O	2.21	62.00	0.0356				0.0356						
K₂O	0.30	94.20	0.0032			0.0032							
P₂O₅	0.06	141.9	0.0004	0.0004									
Wt.% normative mineral				0.13	1.73	1.78	18.67	35.92	4.07		10.50	18.56	8.55

* Proportions adjusted so that the molecular Fe_2O_3/FeO ratio = 0.15. FeO/MgO = 0.7771 in *Rem*.

Table B.3. Calculation of the Normative Composition of the Orangeite in Table 13.11

	Wt.% Oxide	Formula Weight	Formula Proportion	Ap	Il	Or'	Ks	Ac	Mt	Rem	Di'	Hy'	Ol'	Lc	Or
SiO_2	35.09	60.08	0.5841			0.1500	0.0059	0.0116			0.0928	0.7269	0.3631	0.0324	0.0324
TiO_2	1.06	79.87	0.0133		0.0133										
Al_2O_3	2.55	102.0	0.0250			0.0250								0.0196	0.0054
Fe_2O_3	7.78t	159.7	0.0113*					0.0029	0.0113						
FeO		71.85	0.0749*		0.0133				0.0113	0.0524	0.0032	0.0492	0.0493		
MnO	0.15	70.94	0.0021												
MgO	29.02	40.3	0.7201							0.7201	0.0432	0.6769	0.6768		
CaO	3.49	56.08	0.0622	0.0158						0.0464	0.0464				
Na_2O	0.18	62.00	0.0029					0.0029							
K_2O	2.91	94.20	0.0309			0.0250	0.0059							0.0196	0.0054
P_2O_5	0.68	141.9	0.0048	0.0048											
Wt.% normative mineral (Step. 30.)				1.57	2.02	0	0.91	1.34	2.62		10.14	0	52.63	8.56	3.01

Step 22. $D = 0.4023$

23. $D > Hy'/2 = 0.7261/2 = 0.3631$; $Ol' = 0.3631$; $Hy = 0$; $D_1 = 0.4023 - 0.3631 = 0.0392$

24. $D_2 = D_1 = 0.0392$

25. $D_2 > 4Ab' = 0$; $Ne = Ab' = 0$; $D_3 = D_2 - 4Ab' = 0.0392$

26. $D_3 < 2Or' = 0.0500$; $Lc = D_3/2 = 0.0392/2 = 0.0196$; $Or' - D_3/2 = 0.0250 - 0.0196 = 0.0054$

* Proportions adjusted so that the molecular Fe_2O_3/FeO ratio = 0.15. $FeO/MgO = 0.0728$ in *Rem*.

make provisional titanite (Tn'), but only after the allocation of CaO to make normative anorthite (step 12).

6. If there is still an excess of TiO_2 over available CaO, this latter excess is used to make normative rutile (Ru).

7. Allocate an amount of Al_2O_3 equal to the proportion of K_2O to make provisional normative orthoclase (Or').

8. If there is an excess of the proportion of K_2O over that of Al_2O_3, this excess is made into normative potassium metasilicate (Ks). The rock is peralkaline.

9. From any excess of Al_2O_3 in step 5, an equal amount is allocated to the proportion of Na_2O to make provisional normative albite (Ab').

10. If there is an excess of Na_2O, it is allocated, together with an equal amount of Fe_2O_3, to make normative acmite (Ac). The rock is peralkaline.

11. If there is still an excess of Na_2O from step 10, it is made into normative sodium metasilicate (Ns). The rock is peralkaline.

12. If there is an excess of Al_2O_3 over that used to make provisional normative Or' and Ab' in steps 7 and 9, this excess is allocated to an equal amount of CaO to make normative anorthite (An).

13. If there is an excess of Al_2O_3 over that required for An in step 12, it is made into normative corundum (C). The rock is peraluminous.

14. To the proportion of Fe_2O_3 (or the excess of Fe_2O_3 over that used in making Ac in step 10) an equal amount of FeO is allocated for normative magnetite (Mt).

15. If there is an excess of Fe_2O_3 over that required for Mt in step 14 it is made into normative hematite (Hm).

16. All of the FeO and CaO remaining from the previous allocations, together with the proportion of MgO, are entered into a remainder column (Rem). Calculate the FeO/MgO ratio.

17. Allocate to the remaining CaO an equal amount of (FeO + MgO) to make provisional normative diopside (Di'), *the relative amounts of FeO and MgO allocated must be in the ratio in which they occur in the Rem column.*

18. If there is an excess of CaO required to make Di' in step 17, this excess is made into provisional normative wollastonite (Wo').

19. If there is an excess of (FeO + MgO) required to make Di' in step 17, this excess is made into provisional normative hypersthene (Hy'). The relative amounts of FeO and MgO so allocated should be in the same ratio as in the Rem column.

All of the oxides except SiO_2 have now been allocated to actual or provisional normative minerals. In the next steps, SiO_2 is distributed to silicate normative minerals. If there is an excess in this distribution, the rock is silica-oversaturated. If there is a deficiency of SiO_2, the rock is silica-undersaturated, and adjustments must be made in the amounts of one or more provisional normative minerals so that there is sufficient SiO_2 to go around.

20. Allocate the necessary amounts of SiO_2 to the silicate minerals. For example, allocate six times the amount of K_2O for Or', an equal amount to that of CaO + FeO + MgO for Di', and so forth. Let this necessary amount of SiO_2 be Y.

21. If Y is less than the available formula proportion of SiO_2 calculated in step 1, the excess is made into normative quartz (Q). Go to step 30.

22. If Y is greater than the available SiO_2, the deficiency must be compensated for by reducing the amounts of silica-saturated provisional normative minerals (feldspars, pyroxenes) and creating new silica-unsaturated normative minerals (feldspathoids, olivine). Set the deficiency, D, equal to Y minus available SiO_2.

23. If $D < Hy'/2$ make normative olivine $Ol = D$ and make normative hypersthene $Hy = Hy' = 2D$. The relative amounts of FeO and MgO in Ol and Hy should be in the same ratio in which they occur in the Rem column. The deficiency in SiO_2 should now be zero. If $D > Hy'/2$, make provisional normative olivine $Ol' = Hy'/2$ and $Hy = 0$. Let $D_1 = D - Hy'/2$.

24. If no Tn' was made in step 5, let $D_2 = D_1$ and go to step 25. If Tn' was made in step 5 and $D_1 < Tn'/2$, make normative titanite, $Tn = Tn' - D_1$, and make normative perovskite, $Pf = D_1$. The deficiency in SiO_2 should now be zero. If $D_1 > Tn'/2$, make normative perovskite, $Pf = Tn'$ and $Tn = 0$. Let $D_2 = D_1$.

25. If $D_2 < 4Ab'$, let normative nepheline $Ne = D_2/4$ and normative albite $Ab = Ab' - D_2/4$. The deficiency in SiO_2 should now be zero. If $D_2 > 4Ab'$, let $Ne = Ab'$ and $Ab = 0$. Let $D_3 = D_2 - 4Ab'$.

26. If $D_3 < 2Or'$, let normative leucite, $Lc = D_3/2$, and normative orthoclase, $Or = Or' - D_3/2$. The deficiency in SiO_2 should now be zero. If $D_3 > 2Or'$, let $Lc = Or'$ and $Or = 0$. Let $D_4 = D_3 - 2Or'$.

27. If no Wo' was made in step 18, let $D_5 = D_4$ and go to step 28. If Wo' was made in step 18 and $D_4 < Wo'/2$, let normative dicalcium silicate $Cs = D_4$ and normative wollastonite $Wo = Wo' - 2D_4$. The deficiency in SiO_2 should now be zero. If $D_4 > Wo'/2$, let $Cs' = Wo'/2$ and $Wo = 0$. Let $D_5 = D_4 - Wo'/2$.

28. If $D_5 < Di'$, add an amount equal to $D_5/2$ to the amounts of Cs' and Ol' already determined in steps 27 and 23, respectively. Let $Di = Di' - D_5$. The deficiency in SiO_2 should now be zero. If $D_5 > Di'$, add an amount equal to Di'/2 to the amounts of Cs and Ol already determined in steps 27 and 23, respectively. Let $Di = 0$ and $D_6 = D_5 - Di'$.

29. Let normative kaliophilite $Kp = D_6/2$ and $Lc = Lc' - D_6/2$. The deficiency in SiO_2 should now be zero.

30. All of the normative amounts are converted into the weight percentage (wt.%) by multiplying them by the formula weights of the normative minerals in Table B.1. For example, the weight percentage of *Or* in the low-K basaltic andesite in Table B.2 is $0.0032 \times 556.6 = 1.78$; the weight percentage of *Di* is $(0.0218 \times 216.5) + (0.0238 + 248.1) = 4.72 + 5.90 = 10.62$. The final weight percentages of normative minerals are rounded to three significant digits. Rounding in the calculations may cause the normative total to differ somewhat from the oxide total.

REFERENCES CITED

Abbott DH, Hoffman Jr SE. 1984. Archean plate tectonics revised. I. Heat flow, spreading rate, and the age of subducting oceanic lithosphere and their effects on the origin and evolution of continents. Tectonics 3:429–448.

Agee CB, Walker D. 1993. Olivine flotation in mantle melt. Earth Planet. Sci. Lett. 114:315–324.

Aitcheson SJ, Forrest AH. 1994. Quantification of crustal contamination in open magmatic systems. J. Petrol. 35:461–488.

Anderson EM. 1951. The dynamics of faulting and dyke formation. Edinburgh, Oliver and Boyd.

Anderson GM. 1996. Thermodynamics of natural systems. New York, John Wiley & Sons.

Anderson JL, Smith DR. 1995. The effects of temperature and f_{O_2} on the Al-in-hornblende barometer. Am. Mineral. 80:549–559.

Anderson RN, Langseth MF, Sclater JG. 1977. The mechanisms of heat transfer through the floor of the Indian Ocean. J. Geophys. Res. 82:3391–3409.

Atherton MP, Sanderson LM. 1987. The Cordillera Blanca batholith: A study of granite intrusion and the relation of crustal thickening to peraluminosity. Geol. Rund. 76:213–232.

Atkins PW. 1978. Physical Chemistry. San Francisco, W.H. Freeman.

Bacon CR. 1989. Crystallization of accessory phases in magmas by local saturation adjacent to phenocrysts. Geochim. Cosmochim. Acta. 53:1055–1066.

Bacon CR, Druitt, TH. 1988. Compositional evolution of the zoned calcalkaline magma chamber of Mount Mazama, Crater Lake, Oregon. Contrib. Mineral. Petrol. 98:224–256.

Bailey DK. 1970. Volatile flux, heat focussing and the generation of magma. Geol. J. Special Issue 2:177–186.

Bailey EH, Stevens RE. 1960. Selective staining of K-feldspar and plagioclase on rock slabs and thin sections. Am. Mineral. 45:1020–1025.

Baker DR. 1996. Granitic melt viscosities: Empirical and configurational entropy models for their calculation. Am. Mineral. 81:126–134.

Baker DR, Vaillancourt J. 1995. The low viscosities of F + H_2O-bearing granitic melts and implications for melt extraction and transport. Earth Planet. Sci. Lett. 132:199–211.

Ballard RD, van Andel TH 1977. Morphology and tectonics of the inner rift valley at lat 36° 50′N on the Mid-Atlantic Ridge. Geol. Soc. Am. Bull. 88:507–530.

Barker DS. 1996. Carbonatite volcanism. In: Mitchell RH, ed. Undersaturated alkaline rocks: Mineralogy, petrogenesis, and economic potential. Min. Assoc. Can. Short Course Ser. 24:45–62.

Barker DS. 1996. Nephelinite-phonolite volcanism. In: Mitchell RH, ed. Undersaturated alkaline rocks: Mineralogy, petrogenesis, and economic potential. Min. Assoc. Can. Short Course Ser. 24:23–44.

Barker F, ed. 1979. Trondhjemites, dacites, and related rocks. New York, Elsevier.

Barnes HL. 1979. Geochemistry of hydrothermal ore deposits, 2nd ed. New York, John Wiley & Sons.

Bateman PC. 1992. Plutonism in the central part of the Sierra Nevada batholith, California. US Geol. Surv. Prof. Paper 1483.

Bateman PC, Chappell BW. 1979. Crystallization, fractionation, and solidification of the Tuolumne intrusive series, Yosemite National Park, California. Geol. Soc. Am. Bull. 90:465–482.

Bateman PC, Nokleberg WJ. 1978. Solidification of the Mount Givens granodiorite, Sierra Nevada, California. J. Geol. 86:563–579.

Beard JS, Lofgren GE. 1991. Dehydration melting and water-saturated melting of basaltic and andesitic greenstones and amphibolites at 1, 3, and 6.9 kb. J. Petrol. 32:365–401.

Bell DR, Rossman GR. 1992. Water in Earth's mantle: The role of nominally anhydrous minerals. Science 255:1391–1397.

Bell K, ed. 1989. Carbonatites. Boston, Unwin Hyman.

Best MG. 1974. Mantle-derived amphibole within inclusions in alkalic-basaltic lavas. J. Geophys. Res. 79:2107–2113.

Best MG, Brimhall WH. 1974. Late Cenozoic alkalic basaltic magmas in the western Colorado plateaus and the Basin and Range transition zone, USA, and their bearing on mantle dynamics. Geol. Soc. Am. Bull. 85:1677–1690.

Best MG, Christiansen EH. 1997. Origin of broken phenocrysts in ash-flow tuffs. Geol. Soc. Am. Bull. 109 63–73.

Best MG, Christiansen EH, Deino AL, Grommé CS, McKee EH, Noble DC. 1989. Eocene through Miocene volcanism in the Great Basin of the western United States. New Mex. Bur. Mines Mineral Resources Mem. 47:91–133.

Best MG, Christiansen EH, Deino AL, Grommé CS, Tingey DG. 1995. Correlation and emplacement of a large, zoned, discontinuously exposed ash-flow sheet: The $^{40}Ar/^{39}Ar$ chronology, paleomagnetism, and petrology of the Pahranagat Formation, Nevada. J. Geophys. Res. 100:24,593–24,609.

Best MG, Mercy ELP. 1967. Composition and crystallization of mafic minerals in the Guadalupe igneous complex, California. Am. Mineral. 52:436–474.

Bickle MJ. 1990. Mantle evolution. In: Hall RP, Hughes DJ, eds. Early Precambrian basic magmatism. London, Blackie: 111–135.

Blake S. 1990. Viscoplastic models of lava domes. In: Fink JH, ed. Lava flows and domes. New York, Springer-Verlag: 88–126.

Blichert-Toft J, Frey FA, Albarede F. 1999. Hf isotope evidence for pelagic sediments in the source of Hawaiian basalts. Science 285:879–882.

Bloomer SH, Stern RJ, Fisk E, Geschwind CH. 1989. Shoshonitic volcanism in the northern Mariana arc. 1. Mineralogic and major and trace element characteristics. J. Geophys. Res. 94:4469–4496.

Boettcher AL, Burnham CW, Windom KE, Bohlen SR. 1982. Liquids, glasses, and the melting of silicates to high pressures. J. Geol. 90:127–138.

Bohlen SR, Peacor DR, Essene EJ. 1980. Crystal chemistry of a metamorphic biotite and its significance in water barometry. Am. Mineral. 65:55–62.

Bonnichsen B, Kauffman DF. 1987. Physical features of rhyolite lava flows in the Snake River Plain volcanic province, southwestern Idaho. Geol. Soc. Am. Special Paper 212:119–145.

Bottinga Y, Kudo A, Weill DF. 1966. Some observations on oscillatory zoning and crystallization of magmatic plagioclase. Am. Mineral. 51:792–806.

Bottinga Y, Weill DF. 1970. Densities of liquid silicate systems calculated from partial molar volumes of oxide components. Am. J. Sci. 269:169–182.

Bottinga Y, Weill DF. 1972. The viscosity of magmatic silicate liquids: A model for calculation. Am. J. Sci. 272:438–475.

Bouhez JL, Hutton DHW, Stephens WE. 1997. Granite: From segregation of melt to emplacement fabrics. Boston, Kluwer.

Bouška V. 1993. Natural glasses. New York, Ellis Horwood.

Bowden P, Black R, Martin RF, Ike EC, Kinnaird JA, Batchelor RA. 1987. Niger-Nigerian alkaline ring complexes: A classic example of African Phanerozoic anorogenic mid-plate magmatism. In: Fitton JG, Upton BGJ, eds. Alkaline igneous rocks. Geol. Soc. Special Publication 30:357–380.

Bowen NL. 1914. The ternary system: Diopside-forsterite-silica. Am. J. Sci. 188:207–264.

Bowen NL. 1917. The problem of the anorthosites. J. Geol. 25:209–243.

Bowen NL. 1921. Diffusion in silicate melts. J. Geol. 29:295–317.

Bowen NL. 1928. The evolution of the igneous rocks. Princeton N.J., Princeton University Press.

Bowen NL, Anderson O. 1914. The binary system MgO-SiO_2. Am. J. Sci. 37:487–500.

Boyd FR, England JL. 1961. Melting of silicates at high pressures. Annu. Rept. Dir. Geophys. Lab. Yrbk. Carn. Inst. Wash. 60:113–125.

Brandeis G, Jaupart C. 1987. The kinetics of nucleation and crystal growth and scaling laws for magmatic crystallization. Contr. Mineral. Petrol. 96:24–34.

Brimhall GH, Crerar DA. 1987. Ore fluids: Magmatic to supergene. In: Carmichael ISE, Eugster HP, eds. Thermodynamic modeling of geologic materials: Minerals, fluids, and melts. Rev. Mineral. 17:235–321.

Brown WL. 1993. Fractional crystallization and zoning in igneous feldspars: Ideal water-buffered liquid fractionation lines and feldspar zoning paths. Contrib. Mineral. Petrol. 113:115–125.

Bryan WB, Moore JG. 1977. Compositional variations of young basalts in the Mid-Atlantic Ridge rift valley near 36° 49′ N. Geol. Soc. Am. Bull. 88:556–570.

Buddington AF, Lindsley DH. 1964. Iron-titanium oxide minerals and synthetic equivalents. J. Petrol. 5:310–357.

Buening DK, Buseck PR. 1973. Fe-Mg lattice diffusion in olivine. J. Geophys. Res. 78:6852–6862.

Burnham CW. 1979. The importance of volatile constituents. In: Yoder HS, ed. The evolution of the igneous rocks. Princeton N.J., Princeton University Press: 1077–1084.

Burnham CW. 1979. Magmas and hydrothermal fluids. In: Barnes HL, ed. Geochemistry of hydrothermal ore deposits, 2nd ed. New York, John Wiley & Sons: 71–136.

Burnham CW. 1985. Energy release in subvolcanic environments: Implications for breccia formation. Econ. Geol. 80:1515–1522.

Burnham CW, Davis NF. 1971. The role of H_2O in silicate melts. I. P-V-T relations in the system $NaAlSi_3O_8$-H_2O. Am. J. Sci. 270:54–79.

Burnham CW, Davis NF. 1974. The role of H_2O in silicate melts. II. Thermodynamic and phase relations in the system $NaAlSi_3O_8$-H_2O to 10 kilobars, 700°C–1100°C. Am. J. Sci. 274:902–940.

Burnham CW, Holloway JR, Davis NF. 1969. Thermodynamic properties of water to 1000°C and 10,000 bars. Geol. Soc. Am. Special Paper 132.

Carmichael ISE, Turner FJ, Verhoogen J. 1974. Igneous petrology. New York, McGraw-Hill.

Carroll MR, Webster JD. 1994. Solubilities of sulfur, noble gases, nitrogen, chlorine, and fluorine in magmas. In: Carroll MR, Holloway JR, eds. Volatiles in magmas. Rev. Mineral. 30:231–279.

Cas RAF, Wright JV. 1987. Volcanic successions: Modern and ancient. London, Allen and Unwin.

Cashman KV. 1988. Crystallization of Mount St. Helens 1980–1986 dacite: A quantitative textural approach. Bull. Volc. 50:194–209.

Cashman KV. 1990. Textural constraints on the kinetics of crystallization of igneous rocks. In: Nicholls J, Russell JK, eds. Modern methods of igneous petrology: Understanding magmatic processes. Rev. Mineral. 24:259–314.

Cashman KV, Mangan MT. 1994. Physical aspects of magma degassing. II. Constraints on vesiculation from textural studies of eruptive products. In: Carroll MR, Holloway JR, eds. Volatiles in magmas. Rev. Mineral. 30:447–478.

Castle RO, Lindsley DH. 1993. An exsolution silica-pump model for the origin of myrmekite. Contr. Mineral. Petrol. 115:58–65.

Cawthorn RG. 1996. Layered intrusions. New York, Elsevier.

Chappell BW, White AJR. 1992. I- and S-type granites in the Lachlan fold belt. Trans. R. Soc. Edin. Earth Sci. 83:1–26.

Chappell BW, White AJR, Wyborn D. 1987. The importance of residual source material (restite) in granite petrogenesis. J. Petrol. 28:1111–1138.

Chen C-Y, Frey FA. 1985. Trace element and isotopic geochemistry of lavas from Haleakala volcano, East Maui, Hawaii: Implications for the origin of Hawaiian basalts. J. Geophys. Res. 90:8743–8768.

Christiansen EH, Sheridan MF, Burt DM. 1986. The geology and geochemistry of Cenozoic topaz rhyolites from the western United States. Geol. Soc. Am. Special Paper 205.

Clague DA, Dalrymple GB. 1987. The Hawaiian-Emperor volcanic chain. US Geol. Surv. Prof. Paper 1350:5–84.

Clague DA, Straley PF. 1977. Petrologic nature of the oceanic Moho. Geology 5:133–136.

Clark SP, Jr. 1966. Handbook of physical constants. Geol. Soc. Am. Mem. 97.

Clarke DB. 1992. Granitoid rocks. New York, Chapman & Hall.

Cliff RA. 1985. Isotopic dating in metamorphic belts. J. Geol. Soc. London 142:97–110.

Coats RR. 1962. Magma type and crustal structure in the Aleutian arc. Am. Geophys. Union Monogr. 6:92–109.

Coffin MF, Eldholm O. 1994. Large igneous provinces: Crustal structure, dimensions, and external consequences. Rev. Geophys. 32:1–36.

Cole JW. 1982. Tonga-Kermadec-New Zealand. In: Thorpe RS, ed. Andesites. New York, John Wiley & Sons: 245–258.

Collins WJ. 1996. Lachlan fold belt granitoids: Products of three-component mixing. Trans. R. Soc. Edinb. Earth Sci. 87:171–181.

Collins WJ, Beams SD, White AJR, Chappell BW. 1982. Nature and origin of A-type granites with particular reference to southeastern Australia. Contr. Mineral. Petrol. 80:189–200.

Compston W, Pidgeon RT. 1986. Jack Hills, evidence of more very old detrital zircons in Western Australia. Nature. 321:766–769.

Compton RR. 1955. Trondhjemite batholith near Bidwell Bar, California. Geol. Soc. Am. Bull. 66:9–44.

Conrad WK, Nicholls IA, Wall VJ. 1988. Water-saturated and -undersaturated melting of metaluminous and peraluminous crustal compositions at 10 kb: Evidence for the origin of silicic magmas in the Taupo volcanic zone, New Zealand, and other occurrences. J. Petrol. 29:765–803.

Conticelli S, Peccerillo A. 1992. Petrology and geochemistry of potassic and ultrapotassic volcanism in central Italy: Petrogenesis and inferences on the evolution of the mantle sources. Lithos 28:221–240.

Cornejo PC, Mahood GA. 1997. Seeing past the effects of re-equilibration to reconstruct magmatic gradients in plutons: La Gloria Pluton, central Chilean Andes. Contr. Mineral. Petrol. 127:159–175.

Corry CE. 1988. Laccoliths: Mechanics of emplacement and growth. Geol. Soc. Am. Special Paper 220.

Cox KG. 1988. The Karoo province. In: Macdougall JD, ed. Continental flood basalts. Boston, Kluwer: 239–272.

Cox KG. 1993. Continental magmatic underplating. Phil. Trans. R. Soc. London A 342:155–166.

Cox KG, Bell JD, Pankhurst RJ. 1979. The interpretation of igneous rocks. Boston, George Allen & Unwin.

Cox KG, Smith MR, Beswetherick S. 1987. Textural studies of garnet lherzolites: Evidence of exsolution origin from high-temperature harzburgites. In: Nixon PH, ed. Mantle xenoliths. New York, John Wiley & Sons: 537–550.

Crank J. 1975. The mathematics of diffusion, 2nd ed. London, Oxford University.

Crawford AJ, ed. 1989. Boninites. Boston, Unwin Hyman.

Crisp SA. 1984. Rates of magma emplacement and volcanic output. J. Volc. Geotherm. Res. 20:177–211.

Cruden AR. 1990. Flow and fabric development during the diapiric rise of magma. J. Geol. 98:681–698.

Daly RA. 1968. Igneous rocks and the depths of the Earth. New York, Hafner. (Originally published in 1933.)

Davidson JP. 1996. Deciphering mantle and crustal signatures in subduction zone magmatism. In: Bebout GE, Scholl DW, Kirby SH, Platt JP, eds. Subduction top to bottom. Am. Geophys. Union Monogr. 96:251–262.

Davies JH. 1999. The role of hydraulic fractures and intermediate-depth earthquakes in generating subduction-zone magmatism. Nature 398:142–145.

Davis BTC, England JL. 1963. Melting of forsterite, Mg_2SiO_4, at pressures up to 47 kbar. Annu. Rept. Dir. Geophys. Lab. Yrbk. Carn. Inst. Wash. 62:119–121.

Davis MJ, Ihinger PD. 1998. Heterogeneous crystal nucleation on bubbles in silicate melt. Am. Mineral. 83:1008–1015.

Dawson JB, Pinkerton H, Norton GE, Pyle DM. 1990. Physicochemical properties of alkali carbonatite lavas: Data from the 1988 eruption of Oldoinyo Lengai, Tanzania. Geology 18:260–263.

Dawson JB, Pinkerton H, Pyle DM, Nyamweru C. 1994. June 1993 eruption of Oldoinyo Lengai, Tanzania: Exceptionally viscous and large carbonatite lava flows and evidence for coexisting silicate and carbonatite magmas. Geology 22:799–802.

Decker R, Decker B. 1998. Volcanoes. New York, W.H. Freeman.

Deer WA, Howie RA, Zussman J. 1962. Rock-forming minerals, 5 Volumes. London, Longman.

Deer WA, Howie RA, Zussman J. 1997. An introduction to the rock-forming minerals, 2nd ed. London, Longman.

De Graff JM, Aydin A. 1993. Effect of thermal regime on growth increment and spacing of contraction joints in basaltic lava. J. Geophys. Res. 98:6411–6430.

Delaney PT. 1987. Heat transfer during emplacement and cooling of mafic dykes. In: Halls HC, Fahrig WF, eds. Mafic dyke swarms. St. John's, Newfoundland. Geol. Assoc. Can. Special Paper 34:31–46.

Delaney PT, Pollard DD. 1981. Deformation of host rocks and flow of magma during growth of minette dikes and breccia-bearing intrusions near Ship Rock, New Mexico. US Geol. Surv. Prof. Paper 1202.

Delaney PT, Pollard DD. 1982. Solidification of basaltic magma during flow in a dike. Am. J. Sci. 282:856–885.

De Paolo DJ. 1985. Isotopic studies of processes in mafic magma chambers. I. The Kiglapait intrusion, Labrador. J. Petrol. 26:925–951.

de Silva SL, Francis PW. 1991. Volcanoes of the central Andes. New York, Springer-Verlag.

de Silva SL, Wolff JA. 1995. Zoned magma chambers: The influence of magma chamber geometry on sidewall convective fractionation. J. Volc. Geotherm Res. 65:111–118.

De Waard D. 1950. Koninklijke Nederlandse Akademic van Wetenschappen Proc. 53.

Dick HJB, Fisher RL, Bryan B. 1984. Mineralogic variability of the uppermost mantle along mid-ocean ridges. Earth Planet. Sci. Lett. 69:88–106.

Didier J, Barbarin B. 1991. Enclaves and granite petrology. New York, Elsevier.

Dodge FCW, Millard HT, Jr., Elsheimer HN. 1982. Compositional variations and abundances of selected elements in granitoid rocks and constituent minerals, central Sierra Nevada batholith, California. US Geol. Surv. Prof. Paper 1248.

Donaldson CH. 1977. Laboratory duplication of comb layering in the Rhum pluton. Mineral. Mag. 41:323–336.

Dowty E. 1980. Crystal growth and nucleation theory and the numerical simulation of igneous crystallization. In: Hargraves RB, ed. Physics of magmatic processes. Princeton, N.J., Princeton University Press: 419–551.

Drummond MS, Defant MJ. 1990. A model for trondhjemite-tonalite-dacite genesis and crustal growth via slab melting: Archean to modern comparisons. J. Geophys. Res. 95:21,503–21,521.

Drummond MS, Defant MJ, Kepezhinskas PK. 1996. Petrogenesis of slab-derived trondhjemite-tonalite-dacite/adakite magmas. Geol. Soc. Am. Special Paper 315:205–215.

Ducea MN, Saleeby JB. 1998. The age and origin of a thick mafic-ultramafic keel from beneath the Sierra Nevada batholith. Contr. Mineral. Petrol. 133:169–185.

Eby GN. 1990. The A-type granitoids: A review of their occurrence and chemical characteristics and speculations on their petrogenesis. Lithos 26:115–134.

Edgar AD. 1996. Kalsilite-bearing volcanics (kamafugites). In: Mitchell RH, ed. Undersaturated alkaline rocks: Mineralogy, petrogenesis, and economic potential. Min. Assoc. Can. Short Course Ser. 24:153–174.

Eggler DH. 1974. Volatiles in ultrabasic and derivative rock systems. Annu. Rept. Dir. Geophys. Lab. Yrbk. Carn. Inst. Wash. 73:215–224.

Eggler DH, Holloway JR. 1977. Partial melting of peridotite in the presence of H_2O and CO_2: Principles and review. Oregon Dept. Geol. Mineral Indust. 96.

Elthon D. 1991. Geochemical evidence for formation of the Bay of Islands ophiolite above a subduction zone. Nature: 354:140–143.

Ernst RE, Buchan KL. 1997. Giant radiating dyke swarms: Their use in identifying pre-Mesozoic large igneous provinces and mantle plumes. In: Mahoney JJ, Coffin MF, eds. Large igneous provinces: Continental, oceanic, and planetary flood volcanism. Am. Geophys. Union Geophys. Monogr. 100:297–334.

Ernst WG. 1976. Petrologic phase equilibria. San Francisco, W.H. Freeman.

Esperanca S, Holloway JR. 1987. On the origin of some mica-lamprophyres: Experimental evidence from a mafic minette. Contr. Mineral. Petrol. 95:207–216.

Eugster HP, Wones DR. 1962. Stability relations of ferruginous biotite, annite. J. Petrol. 3:82–125.

Evans BW, Moore JG. 1968. Mineralogy as a function of depth in the prehistoric Makaopuhi tholeiitic lava lake, Hawaii. Contr. Mineral. Petrol. 17:85–115.

Ewart E. 1982. The mineralogy and petrology of Tertiary-Recent orogenic volcanic rocks with special reference to the andesite-basaltic composition range. In: Thorpe RS, ed. Andesites. New York, John Wiley & Sons: 25–87.

Ewart A, Brothers RN, Mateen A. 1977. An outline of the geology and geochemistry, and the possible petrogenetic evolution of the volcanic rocks of the Tonga-Kermadec-New Zealand island arc. J. Volc. Geotherm. Res. 2:205–250.

Ewart A, Hawkesworth CJ. 1987. The Pleistocene-Recent Tonga-Kermadec arc lavas: Interpretation of new isotopic and rare earth data in terms of a depleted mantle source model. J. Petrol. 28:495–530.

Falloon TJ, Green DH, Danyushevsky LV, Faul UH. 1999. Peridotite melting at 1.0 and 1.5 GPa: An experimental evaluation of techniques using diamond aggregates and mineral mixes for determination of near-solidus melts. J. Petrol. 40:1343–1375.

Faure G. 1986. Principles of isotope geology, 2nd ed. New York, John Wiley & Sons.

Fenn PM. 1977. The nucleation and growth of alkali feldspars from hydrous melts. Can. Mineral. 15:135–161.

Fenn PM. 1986. On the origin of graphic granite. Am. Mineral. 71:325–330.

Fisher RV, Schmincke HU. 1984. Pyroclastic rocks. New York, Springer-Verlag.

Fisher RV, Smith GA. 1991. Volcanism, tectonics, and sedimentation. In: Fisher RV, Smith GA, eds. Sedimentation in volcanic settings. SEPM (Soc. for Sed. Geol.) Special Publication 45:1–5.

Fiske RS, Hopson CA, Waters AC. 1963. Geology of Mount Rainier National Park, Washington. US Geol. Surv. Prof. Paper 444.

Fitton JG. 1987. The Cameroon line, West Africa: A comparison between oceanic and continental alkaline volcanism. In: Fitton JG, Upton BGJ, eds. Alkaline igneous rocks. Geol. Soc. Special Publication 30:273–291.

Fitton JG, James D, Leeman WP. 1991. Basic magmatism associated with late Cenozoic extension in the western United States: Compositional variations in space and time. J. Geophys. Res. 96:13,693–13,711.

Fitton JG, Upton BGJ, eds. 1987. Alkaline igneous rocks: Geol. Soc. Lond. Special Publication 30. Boston, Blackwell Scientific.

Flood RH, Vernon RH. 1988. Microstructural evidence of orders of crystallization in granitoid rocks. Lithos 21:237–245.

Foley S. 1992. Vein-plus-wall-rock melting mechanisms in the lithosphere and the origin of potassic alkaline magmas. Lithos 28:435–453.

Forsyth DW and 15 others. 1998. Imaging the deep seismic structure beneath a mid-ocean ridge: The MELT experiment. Science 280:1215–1218.

France-Lanord C, Le Fort P. 1988. Crustal melting and granite genesis during the Himalayan collision orogenesis. Trans. R. Soc. Edin. Earth Sci. 79:183–195.

Francis P. 1993. Volcanoes: A planetary perspective. New York, Oxford.

Freer R. 1981. Diffusion in silicate minerals and glasses: A data digest and guide to the literature. Contr. Mineral. Petrol. 76:440–454.

Freundt A, Rosi M., eds. 1998. From magma to tephra: Modeling physical processes of explosive volcanic eruptions. New York, Elsevier.

Frey FA, Chappell BW, Roy SD. 1978. Fractionation of rare-earth elements in the Tuolumne intrusive series, Sierra Nevada batholith, California. Geology 6:239–242.

Frey FA, Clague DA. 1983. Geochemistry of diverse basalt types from Loihi Seamount, Hawaii: Petrogenetic implications. Earth Planet. Sci. Lett. 66:337–355.

Frey FA, Wise WS, Garcia MO, West H, Kwon S-T, Kennedy A. 1990. Evolution of Mauna Kea volcano, Hawaii: Petrologic and geochemical constraints on post-shield volcanism. J. Geophys. Res. 95:1271–1300.

Friedman I, Smith RL, Long WD. 1966. Hydration of natural glass and formation of perlite. Geol. Soc. Am. Bull. 77:323–328.

Frost CD, Frost BR. 1997. Reduced rapakivi-type granites: The tholeiitic connection. Geology 25:647–650.

Frost BR, Lindsley DH. 1992. Equilibria among Fe-Ti oxides, pyroxenes, olivine, and quartz. II. Application. Am. Mineral. 77:1004–1020.

Fuhrman ML, Lindsley DH. 1988. Ternary-feldspar modeling and thermometry. Am. Mineral. 73:201–215.

Gardner JE, Thomas RME, Jaupart C, Tait S. 1996. Fragmentation of magma during plinian volcanic eruptions. Bull. Volc. 58:144–162.

Gass IG. 1972. Proposals concerning the variation of volcanic products and processes within the oceanic environment. Phil. Trans. R. Soc. Lond. A 271:131–140.

Ghiorso MS. 1997. Thermodynamic models of igneous processes. Annu. Rev. Earth Planet. Sci. 25:221–241.

Ghiorso MS, Carmichael ISE. 1987. Modeling magmatic systems: Petrologic applications. In: Carmichael ISE, Eugster HP, eds. Thermodynamic modeling in geologic systems: Minerals, fluids, and melts. Rev. Mineral. 17:467–499.

Ghiorso MS, Sack RO. 1991. Fe-Ti oxide geothermometry: Thermodynamic formulation and the estimation of intensive variables in silicic magmas. Contr. Mineral. Petrol. 108:485–510.

Gill JB. 1981. Orogenic andesites and plate tectonics. New York, Springer-Verlag.

Gill JB, Whelan P. 1989. Early rifting of an oceanic island arc (Fiji) produced shoshonitic to tholeiitic basalts. J. Geophys. Res. 94:4561–4578.

Gorring ML, Naslund HR. 1995. Geochemical reversals within the lower 100 m of the Palisades sill, New Jersey. Contr. Mineral. Petrol. 119:263–276.

Govindaraju K. 1989. Geostandards newsletter. Special Issue, July, XIII.

Graham IJ, Cole JW, Briggs RM, Gamble JA, Smith IEM. 1995. Petrology and petrogenesis of volcanic rocks from the Taupo volcanic zone: A review. J. Volc. Geotherm. Res. 68:59–87.

Grand SP, van der Hilst RD, Widiyantoro S. 1997. Global seismic tomography: A snapshot of convection in the Earth. Geol. Soc. Am. Today no. 4;7:1–7.

Green TH. 1980. Island arc and continent-building magmatism: A review of petrogenetic models based on experimental petrology and geochemistry. Tectonophys. 63:367–385.

Green TH. 1982. Anatexis of mafic crust and high pressure crystallization of andesite. In: Thorpe RS, ed. Andesites. New York, John Wiley: 465–487.

Greenwood R. 1951. Younger intrusive rocks of Plateau province, Nigeria, compared with the alkalic rocks of New England. Geol. Soc. Am. Bull. 62:1151–1178.

Griggs DT, Blacic JD. 1965. Quartz: Anomalous weakness of synthetic crystals. Science 147:292–295.

Grout FF. 1932. Petrography and petrology. New York, McGraw-Hill.

Grove TL, Baker MB. 1984. Phase equilibrium controls on the tholeiitic versus calc-alkaline differentitation trends. J. Geophys. Res. 89:3253–3274.

Grove TL, Baker MB, Kinzler RJ. 1984. Coupled CaAl-NaSi diffusion in plagioclase feldspar: Experiments and applications to cooling rate speedometry. Geochim. Cosmochim. Acta 48:2113–2121.

Grove TL, Gerlach DC, Sando TW. 1982. Origin of calc-alkaline series lavas at Medicine Lake volcano by fractionation, assimilation and mixing. Contr. Min. Petrol. 80:160–182.

Guilbert JM, Park CF, Jr. 1986. The geology of ore deposits. New York, Freeman.

Gunnarsson B, Marsh BD, Taylor HP, Jr. 1998. Generation of Icelandic rhyolites: Silicic lavas from the Torfajökull central volcano. J. Volc. Geothermal Res. 83:1–45.

Gurney JJ, Zweistra P. 1995. The interpretation of the major element compositions of mantle minerals in diamond exploration. J. Geochem. Explor. 53:293–310.

Gutman JT, Martin RF. 1976. Schweiz. Min. Petrol. Mitt. 56:55–64.

Haggerty SE. 1999. A diamond trilogy: Superplumes, supercontinents, and supernovae. Science 285:851–860.

Hall A. 1996. Igneous petrology, 2nd. ed. Essex, England, Longmans.

Hall RP, Hughes DJ, eds. 1990. Early Precambrian basic magmatism. London, Blackie.

Halls HC, Fahrig WF, eds. 1987. Mafic dyke swarms. St. John's, Newfoundland: Geol. Assoc. Can. Special Paper 34.

Hamilton W, Myers B. 1967. The nature of batholiths. US Geol. Surv. Prof. Paper 554-C.

Hanson GN. 1978. The application of trace elements to the petrogenesis of igneous rocks of granitic composition. Earth Planet. Sci. Lett. 38:26–43.

Harris J, Hutchison MR, Hursthouse M, Light M, Harte B. 1997. A new tetragonal silicate mineral occurring as inclusions in lower mantle diamonds. Nature 387:486–488.

Harris NBW, Inger S. 1992. Trace element modeling of pelite-derived granites. Contr. Mineral. Petrol. 110:45–56.

Hart SR, Hauri EH, Oschmann LA, Whitehead JA. 1992. Mantle plumes and entrainment: Isotopic evidence. Science 256:517–520.

Haskins LA. 1983. Petrogenetic modeling—use of rare earth elements. In: Henderson P, ed. Rare earth element geochemistry: Developments in geochemistry. Amsterdam, Elsevier 2:115–152.

Hatton CJ, von Gruenewaldt G. 1990. Early Precambrian layered intrusions. In: Hall RP, Hughes DJ, eds. Early Precambrian basic magmatism. London, Blackie: 57–82.

Hawkes DD. 1967. Order of abundant crystal nucleation in a natural magma. Geol. Mag. 104:473–486.

Hawkins JW. 1995. Evolution of the Lau Basin—insights from ODP Leg 135. In: Taylor B, Natland J, eds. Active margins and marginal basins of the western Pacific. Am. Geophys. Union Monogr. 88:125–173.

Hay DE, Wendlandt RF. 1995. The origin of Kenya rift plateau-type flood phonolites: Results of high-pressure/high-temperature experiments in the systems phonolite-H_2O and phonolite-H_2O-CO_2. J. Geophys. Res. 100:401–410.

Heiken G, Wohletz K. 1985. Volcanic ash. Berkeley, University of California Press.

Helmstaedt HH, Gurney JJ. 1995. Geotectonic controls of primary diamond deposits: Implications for area selection. J. Geochem. Explor. 53:125–144.

Helz RT. 1987. Diverse olivine types in lava of the 1959 eruption of Kilauea volcano and their bearing on eruption dynamics. US Geol. Surv. Prof. Paper 1350:691–722.

Helz RT, Kirschenbaum H, Marinenko JW. 1989. Diapir transfer of melt in Kilauea Iki lava lake, Hawaii: A quick, efficient process of igneous differentiation. Geol. Soc. Am. Bull. 101:578–594.

Henley RW, Truesdell AH, Barton PB, Jr, Whitney JA. 1984. Fluid-mineral equilibria in hydrothermal systems. Rev. Econ. Geol. 1.

Herzberg CT, Fyfe WS, Carr MJ. 1983. Density constraints on the formation of the continental Moho and crust. Contr. Mineral. Petrol. 84:1–5.

Herzberg C, O'Hara MJ. 1998. Phase equilibrium constraints on the origin of basalts, picrites, and komatiites. Earth Sci. Rev. 44:39–79.

Hess K-U, Dingwell DB. 1996. Viscosities of hydrous leucogranitic melts: A non-Arrhenian model. Am. Mineral. 81:1297–1300.

Hess PC. 1989. Origins of igneous rocks. Cambridge, Mass., Harvard University Press.

Hess PC. 1995. Thermodynamic mixing properties and the structure of silicate melts. In: Stebbins JF, McMillan PF, Dingwell DB, eds. Structure, dynamics, and properties of silicate melts. Rev. Mineral. 10:145–189.

Hibbard MJ, Watters RJ. 1985. Fracturing and diking in incompletely crystallized granitic plutons. Lithos 18:1–12.

Hildreth W. 1981. Gradients in silicic magma chambers: Implications for lithospheric magmatism. J. Geophys. Res. 86:10153–10192.

Hildreth W. 1983. The compositionally zoned eruption of 1912 in the Valley of Ten Thousand Smokes, Katmai National Park, Alaska. J. Volc. Geotherm. Res. 18:1–56.

Hildreth W, Halliday AN, Christiansen RL. 1991. Isotopic and chemical evidence concerning the genesis and contamination of basaltic and rhyolitic magma beneath the Yellowstone Plateau volcanic field. J. Petrol. 32:63–138.

Hildreth W, Moorbath S. 1988. Crustal contributions to arc magmatism in the Andes of central Chile. Contr. Mineral. Petrol. 98:455–489.

Hirose K. 1997. Melting experiments on lherzolite KLB-1 under hydrous conditions and generation of high-magnesian andesitic melts. Geology 25:42–44.

Hirose K, Kawamoto T. 1995. Hydrous partial melting of lherzolite at 1 GPa: The effect of H_2O on the genesis of basaltic magmas. Earth Planet. Sci. Lett. 133:463–473.

Hirose K, Kushiro I. 1993. Partial melting of dry peridotites at high pressures: Determination of compositions of melts segregated from peridotite using aggregates of diamond. Earth Planet. Sci. Lett. 114:477–489.

Hofmann AW. 1980. Diffusion in natural silicate melts: A critical review. In: Hargraves RB, ed. Physics of magmatic processes. Princeton, N.J., Princeton University Press: 385–417.

Hofmann AW. 1988. Chemical differentiation of the Earth: The relationaship between mantle, continental crust, and oceanic crust. Earth Planet. Sci. Lett. 90:297–314.

Holland HD, Malinin SD. 1979. The solubility and occurrence of non-ore minerals. In: Barnes HL, ed. Geochemistry of hydrothermal ore deposits, 2nd ed. New York, John Wiley: 461–508.

Holloway JR, Blank JG. 1994. Application of experimental results to C-O-H species in natural melts. In: Carroll MR, Holloway JR, eds. Volatiles in magmas. Rev. Mineral. 10:187–230.

Hooper PR. 1997. The Columbia River flood basalt province: Current status. In: Mahoney JJ, Coffin MF, eds. Large igneous provinces: Continental, oceanic, and planetary flood volcanism. Am. Geophys. Union Monogr. 100:1–28.

Huebner JS, Turnock AC. 1980. The melting relations at 1 bar of pyroxenes composed largely of Ca-, Mg-, and Fe-bearing components. Am. Mineral. 65:225–271.

Hunter DR, Hamilton PJ. 1978. The Bushveld Complex. In: Tarling DH, ed. Evolution of the earth's crust. New York, Academic Press.

Hunter RH. 1996. Textural development in cumulate rocks. In: Cawthorn RG, ed. Layered intrusions. New York, Elsevier: 77–102.

Hunter RH. 1987. Textural equilibrium in layered igneous rocks. In: Parsons I, ed. Origins of igneous layering. Boston, Reidel: 473–503.

Huppert HE, Sparks RSJ. 1985. Komatiites I: Eruption and flow. J. Petrol. 26:694–725.

Hurwitz S, Navon O. 1994. Bubble nucleation in rhyolitic melts: Experiments at high pressure, temperature, and water content. Earth Planet. Sci. Lett. 122:267–280.

Ingrin J, Hercule S, Charton T. 1995. Diffusion of hydrogen in diopside: Results of dehydration experiments. J. Geophys. Res. 100:15,489–15,499.

Irvine TN. 1980. Magmatic infiltration metasomatism, double-diffusive fractional crystallization, and adcumulus growth in the Muskox intrusion and other layered intrusions. In: Hargraves RB, ed. Physics of magmatic processes. Princeton, N.J., Princeton University Press: 325–383.

Irvine TN. 1987. Layering and related structures in the Duke Island and Skaergaard intrusions: Similarities, differences, and origins. In Parsons I, ed. Origins of igneous layering. Boston, Reidel: 185–245.

Irvine TN, Andersen JCØ, Brooks CK. 1998. Included blocks (and blocks within blocks) in the Skaergaard intrusion: Geologic relations and the origins of rhythmic modally graded layers. Geol. Soc. Am. Bull. 110:1398–1447.

Irving AJ. 1980. Petrology and geochemistry of composite ultramafic xenoliths in alkalic basalts and implications for magmatic processes within the mantle. Am. J. Sci. 280-A:389–426.

Jackson ED. 1961. Primary textures and mineral associations in the ultramafic zone of the Stillwater complex, Montana. US Geol. Surv. Prof. Paper 358.

Jackson MD, Pollard DD. 1988. The laccolith-stock controversy: New results from the southern Henry Mountains, Utah. Geol. Soc. Am. Bull. 100:117–139.

Jaeger JC. 1968. Cooling and solidification of igneous rocks. In: Hess HH and Poldervaart A, eds. Basalts: The Poldervaart treatise on rocks of basaltic composition. New York, Wiley Interscience: 503–536.

Jahns RH, Burnham CW. 1969. Experimental studies of pegmatite genesis. I. A model for the derivation and crystallization of granitic pegmatites. Econ. Geol. 64:843–864.

Jayaraman A. 1984. The diamond-anvil high-pressure cell. Sci. Am. 4;250:54–62.

Jeanloz R, Romanowicz B. 1997. Geophysical dynamics at the center of the Earth. Physics Today, August:22–27.

Jenner GA, Dunning GR, Malpas J, Brown M, Brace T. 1991. Bay of Islands and Little Port complexes, revisited: Age, geochemical and isotopic evidence confirm suprasubduction-zone origin. Can. J. Earth Sci. 28:1635–1652.

Johannes W. 1978. Melting of plagioclase in the system Ab-An-H_2O and Qz-Ab-An-H_2O at P_{H_2O} = 5 kbars, an equilibrium problem. Contr. Mineral. Petrol. 66:295–303.

Johannes W, Holtz F. 1996. Petrogenesis and experimental petrology of granitic rocks. New York, Springer-Verlag.

Johannsen A. 1931, 1932, 1937, 1938. A descriptive petrography of the igneous rocks, 4 volumes. Chicago, University of Chicago Press.

Johnson MC, Anderson AT, Jr., Rutherford MJ. 1994. Pre-eruptive volatile contents of magmas. In: Carroll MR, Holloway JR, eds. Volatiles in magmas. Rev. Mineral. 10:281–323.

Jordan TE, Isacks BL, Allmendinger RW, Brewer JA, Ramos VA, Ando CJ. 1983. Andean tectonics related to geometry of subducted Nazca plate. Geol. Soc. Am. Bull. 94:341–361.

Jost W. 1952. Diffusion in solids, liquids, and gases. New York, Academic.

Jurewicz SR, Watson EB. 1985. The distribution of a partial melt in a granitic system: The application of liquid phase sintering theory. Geochim. Cosmochim. Acta 49:1109–1121.

Kay SM, Kay RW. 1994. Aleutian magmas in space and time. In: Plafker G, Berg HC, eds. The geology of Alaska. Geol. Soc. Am. G-1:687–722.

Kay SM, Rapela CW, eds. 1990. Plutonism from Antarctica to Alaska. Geol. Soc. Am. Special Paper 241.

Kendrick GC, Edmond CL. 1981. Magma immiscibility in the Shonkin Sag and Square Butte laccoliths. Geology 9:615–619.

Keppler H. 1999. Experimental evidence for the source of excess sulfur in explosive volcanic eruptions. Science 284:1652–1654.

Kerr AC, Tarney J, Marriner GF, Nivia A, Saunders AD. 1997. The Caribbean-Colombian Cretaceous igneous province: The internal anatomy of an oceanic plateau. In: Mahoney JJ, Coffin MF, eds. Large igneous provinces: Continental, oceanic, and planetary flood volcanism. Am. Geophys. Union Monogr. 100:123–144.

Kirby SH, Stein S, Okal EA, Rubie DC. 1996. Metastable mantle phase transformations and deep earthquakes in subducting oceanic lithosphere. Rev. Geophys. 34:261–306.

Kirkpatrick RJ. 1981. Kinetics of crystallization of igneous rocks. In: Lasaga AC, Kirkpatrick RJ, eds. Kinetics of geochemical processes. Rev. Mineral. 8:321–398.

Kirkpatrick RJ. 1983. Theory of nucleation in silicate melts. Am. Mineral. 68:66–77.

Kistler RW, Chappell BW, Peck DL, Bateman PC. 1986. Isotopic variation in the Tuolumne Intrusive Suite, central Sierra Nevada, California. Contr. Mineral. Petrol. 94:205–220.

Kjarsgaard BA, Hamilton DL. 1989. The genesis of carbonatites by immiscibility. In: Bell K, ed. Carbonatites. Boston, Unwin Hyman: 388–404.

Klein EM, Langmuir CH. 1987. Global correlation of ocean ridge basalt chemistry with axial depth and crustal thickness. J. Geophys. Res. 92:8089–8115.

Klug C, Cashman KV. 1994. Vesiculation of May 18, 1980, Mount St. Helens magma. Geology 22:468–472.

Klug C, Cashman KV. 1996. Permeability development in vesiculating magmas: Implications for fragmentation. Bull. Volc. 58:87–100.

Kokelaar P, Busby C. 1992. Subaqueous explosive eruption and welding of pyroclastic deposits. Science 257:196–201.

Komar PD. 1976. Phenocryst interactions and the velocity profile of magma flowing through dikes or sills. Geol. Soc. Am. Bull. 87:1336–1342.

Kracek FC, Neuvonen KJ. 1952. Thermochemistry of the plagioclase and alkali feldspars. Am. J. Sci. Bowen Volume:293–309.

Kress VC, Carmichael ISE. 1991. The compressibility of silicate liquids containing Fe_2O_3 and the effect of composition, temperature, oxygen fugacity, and pressure on their redox states. Contr. Mineral. Petrol. 108:82–92.

Kretz R. 1966. Interpretation of the shape of mineral grains in metamorphic rocks. J. Petrol. 7:68–94.

Kushiro I. 1968. Compositions of magmas formed by partial zone melting of the Earth's upper mantle. J. Geophys. Res. 73:619–634.

Kushiro I. 1969. The system forsterite-diopside-silica with and without water at high pressures. Am. J. Sci. 267-A:269–294.

Kushiro I. 1980. Viscosity, density, and structure of silicate melts at high pressure, and their petrologic applications. In: Hargraves RB, ed. Physics of magmatic processes. Princeton, N.J., Princeton University Press: 93–120.

Lange RA. 1994. The effect of H_2O, CO_2, and F on the density and viscosity of silicate melts. In: Carroll MR, Holloway JR, eds. Volatiles in magmas. Rev. Mineral. 30:331–369.

Lange RA, Carmichael ISE. 1990. Thermodynamic properties of silicate liquids with emphasis on density, thermal expansion, and compressibility. In: Nicholls J, Russell JK, eds. Modern methods of igneous petrology: Understanding magmatic processes: Rev. Mineral. 24:25–64.

Lasaga AC. 1998. Kinetic theory in the earth sciences. Princeton, N.J., Princeton University Press.

Lassiter JC, De Paolo DJ. 1997. Plume-lithosphere interaction in the generation of continental and oceanic flood basalts: Chemical and isotopic constraints. In: Mahony JJ, Coffin MF, eds. Large igneous provinces: Continental, oceanic, and planetary flood volcanism. Am. Geophys. Union Monogr. 100:335–356.

Le Bas MJ. 1987. Nephelinites and carbonatites. In: Fitton JG, Upton BGJ, eds. Alkaline igneous rocks. Geol. Soc. Special Publication 30:53–83.

Le Bas MJ, Le Maitre RW, Woolley AR. 1992. The construction of the total alkali-silica chemical classification of volcanic rocks. Mineral. Petrol. 46:1–22.

LeCheminant AN, Heaman LM 1989. Mackenzie igneous events, Canada: Middle Proterozoic hotspot magmatism associated with ocean opening. Earth Planet. Sci. Lett. 96:38–48.

Lee W-J, Wyllie PJ. 1997. Liquid immiscibility in the join $NaAlSiO_4$ -$NaAlSi_3O_8$-$CaCO_3$ at 1 GPa: Implications for crustal carbonates. J. Petrol. 38:1113–1135.

Leeman WP. 1983. The influence of crustal signature on compositions of subduction-related magmas. J. Volcan. Geotherm. Res. 18:561–588.

Leeman WP. 1996. Boron and other fluid-mobile elements in volcanic arc lavas: Implications for subduction processes. In: Bebout GE, Scholl DW, Kirby SH, Platt JP, eds. Subduction top to bottom. Am. Geophys. Union. Monogr. 96:269–276.

Leeman WP, Smith DR, Hildreth W, Palacz Z, Rogers N. 1990. Compositional diversity of late Cenozoic basalts in a transect across the southern Washington Cascades: Implications for subduction zone magmatism. J. Geophys. Res. 95:19,561–19,582.

Lejeune AM, Bottinga Y, Trull TW, Richet P. 1999. Rheology of bubble-bearing magmas. Earth Planet. Sci. Lett. 166:71–84.

Le Maitre RW. 1976. The chemical variability of some common igneous rocks. J. Petrol. 17:589–637.

Le Maitre RW. 1982. Numerical petrology. New York, Elsevier.

Le Maitre RW, ed. 1989. A classification of igneous rocks and glossary of terms. Oxford, Blackwell Scientific.

Le Roex AP, Cliff RA, Adair BJI. 1990. Tristan da Cunha, South Atlantic: Geochemistry and petrogenesis of a basanite-phonolite lava series. J. Petrol. 31:779–812.

Lesher CE, Walker D. 1991. Thermal diffusion in petrology. In: Ganguly J, ed. Diffusion, atomic ordering, and mass transport: Selected topics in geochemistry. New York, Springer-Verlag: 396–451.

Levi BG. 1997. Earth's upper mantle: How low can it flow? Physics Today, August:17–20.

Lindsley DH. 1983. Pyroxene thermometry. Am. Mineral. 68:477–493.

Lipman PW. 1997. Subsidence of ash-flow calderas: Relation to caldera size and magma chamber geometry. Bull. Volc. 59:198–218.

Lipman PW, Christiansen RL, O'Connor JT. 1966. A compositionally zoned ash-flow sheet in southern Nevada. US Geol. Surv. Prof. Paper 524-F.

Lipman PW, Mullineaux DR, eds. 1981. The 1980 eruptions of Mount St. Helens, Washington. US Geol. Surv. Prof. Paper 1250.

Lister JR, Kerr RC. 1989. The effect of geometry on the gravitational instability of a buoyant region of viscous fluid. J. Fluid Mech. 202:577–594.

Liu J, Bohlen SR, Ernst WG. 1996. Stability of hydrous phases in subducting oceanic crust. Earth Planet. Sci. Lett. 143:161–171.

Lloyd FE, Arima M, Edgar AD. 1985. Partial melting of a phlogopite-clinopyroxenite nodule from south-west Uganda: An experimental study bearing on the origin of highly potassic continental rift volcanics. Contr. Mineral. Petrol. 91:321–329.

Lofgren GE. 1974. An experimental study of plagioclase crystal morphology. Am. J. Sci. 274:243–273.

Lofgren G. 1980. Experimental studies on the dynamic crystallization of silicate melts. In: Hargraves RB, ed. Physics of magmatic processes. Princeton, N.J., Princeton University Press: 487–551.

Lofgren G. 1983. Effect of heterogeneous nucleation on basaltic textures: A dynamic crystallization study. J. Petrol. 24:229–255.

Lofgren GE, Donaldson CH. 1975. Curved branching crystals and differentiation in comb-layered rocks. Contrib. Mineral. Petrol. 49:309–319.

London D. 1992. The application of experimental petrology to the genesis and crystallization of granitic pegmatites. Can. Mineral. 30:499–540.

London D, Morgan GB VI, Hervig RL. 1989. Vapor-undersaturated experiments with Macusani glass + H_2O at 200 MPa and the internal differentiation of granite pegmatites. Contr. Mineral. Petrol. 102:1–17.

Lorenz V. 1986. On the growth of maars and diatremes and its relevance to the formation of tuff rings. Bull. Volc. 48:265–274.

Luhr JF. 1997. Extensional tectonics and the diverse primitive volcanic rocks in the western Mexican Volcanic Belt. Can. Mineral. 35:473–500.

Maaløe S. 1978. The origin of rhythmic layering. Mineral. Mag. 42:337–345.

Maaløe S. 1985. Principles of igneous petrology. New York, Springer-Verlag.

MacColl RS. 1964. Geochemical and structural studies in batholithic rocks of southern California. Part 1. Structural geology of Rattlesnake Mountain pluton. Geol. Soc. Am. Bull. 75:805–822.

Macdonald GA. 1968. Composition and origin of Hawaiian lavas. Geol. Soc. Am. Mem. 116:477–522.

Macdonald R. 1974. Nomenclature and petrochemistry of the peralkaline oversaturated extrusive rocks. Bull. Volc. 38:498–516.

MacLellan HE, Trembath LT. 1991. The role of quartz crystallization in the development and preservation of igneous texture in granitic rocks: Experimental evidence at 1 kbar. Am. Mineral. 76:1291–1305.

Mader HM, Zhang Y, Phillips JC, Sparks RSJ, Sturtevant R, Stolper E. 1994. Experimental simulations of explosive degassing of magma. Nature 372:85–88.

Magaritz M, Whitford DJ, James DE. 1978. Oxygen isotopes and the origin of high-$^{87}Sr/^{86}Sr$ andesites. Earth Planet. Sci. Lett. 40:220–230.

Mahoney JJ. 1988. Deccan traps. In: Macdougall JD, ed. Continental flood basalts. Boston, Kluwer: 151–194.

Mahoney JJ, Coffin MF, eds. 1997. Large igneous provinces: Continental, oceanic, and planetary flood volcanism. Am. Geophys. Union Monogr. 100.

Mahood G. 1981. Chemical evolution of a Pleistocene rhyolitic center: Sierra La Primavera, Jalisco, Mexico. Contr. Mineral. Petrol. 77:129–149.

Mangan MT, Cashman KV. 1996. The structure of basaltic scoria and reticulite and inferences for vesiculation, foam formation, and fragmentation in lava fountains. J. Volc. Geotherm. Res. 73:1–18.

Marrett R, Emerman SH. 1992. The relations between faulting and mafic magmatism in the Altiplano-Puna plateau (central Andes). Earth Planet. Sci. Lett. 112:53–59.

Marsh BD. 1981. On the crystallinity, probability of occurrence, and rheology of lava and magma. Contr. Mineral. Petrol. 78:85–98.

Marsh BD. 1982. On the mechanics of igneous diapirism, stoping, and zone melting. Am. J. Sci. 282:808–855.

Marsh BD. 1988. Crystal size distribution (CSD) in rocks and the kinetics and dynamics of crystallization: I. Theory. Contr. Mineral. Petrol. 99:277–291.

Marsh BD. 1989. Magma chambers. Annu. Rev. Earth Planet. Sci. 17:439–474.

Marsh JS, Hooper PR, Rehacek J, Duncan RA, Duncan AR. 1997. Stratigraphy and age of Karoo basalts of Lesotho and implications for correlations within the Karoo igneous province. In: Mahoney JJ, Coffin MF, eds. Large igneous provinces: Continental, oceanic, and planetary flood volcanism. Am. Geophys. Union Monogr. 100:247–272.

Martin BS. 1989. The Roza Member, Columbia River Basalt Group: Chemical stratigraphy and flow distribution. In: Reidel SP, Hooper PR, eds. Volcanism and tectonism in the Columbia River flood-basalt province. Geol. Soc. Am. Special Paper 239:85–104.

Mathez EA, Hunter RH, Kinzler R. 1997. Petrologic evolution of partially molten cumulate: The Atok section of the Bushveld Complex. Contr. Mineral. Petrol. 129:20–34.

Maury RC, Defant MJ, Joron J-L. 1992. Metasomatism of the sub-arc mantle inferred from trace elements in Philippine xenoliths. Nature 360:661–663.

May PR. 1971. Patterns of Triassic-Jurassic diabase dikes around the North Atlantic in the context of predrift position of the continents. Geol. Soc. Amer. Bull. 82:1285–1292.

McBirney AR. 1980. Mixing and unmixing of magmas. J. Volc. Geotherm. Res. 7:357–371.

McBirney AR. 1993. Igneous petrology, 2nd ed. Boston, Jones & Bartlett.

McBirney AR. 1996. The Skaergaard intrusion. In: Cawthorn RG, ed. Layered intrusions. New York, Elsevier: 147–180.

McBirney AR, Baker BH, Nilson RH. 1985. Liquid fractionation. I: Basic principles and experimental simulations. J. Volc. Geotherm. Res. 24:1–24.

McBirney AR, Murase T. 1984. Rheological properties of magmas. Annu. Rev. Earth Planet. Sci. 12:337–357.

McBirney AR, Noyes RM. 1979. Crystallization and layering of Skaergaard intrusion. J. Petrol. 20:487–554.

McCarthy J, Thompson GA. 1988. Seismic imaging of extended crust with emphasis on the western United States. Geol. Soc. Am. Bull. 100:1361–1374.

McDonough WF. 1990. Constraints on the composition of the continental lithospheric mantle. Earth Planet. Sci. Lett. 101:1–18.

McKenzie D. 1985. The extraction of magma from the crust and mantle. Earth Planet. Sci. Lett. 74:81–91.

McKenzie D, O'Nions RK. 1991. Partial melt distributions from inversion of rare earth element concentrations. J. Petrol. 32:1021–1091.

McMillan PF. 1994. Water solubility and speciation models. In: Carroll MR, Holloway JR, eds. Volatiles in magmas. Rev. Mineral. 30:131–156.

Means WD, Park Y. 1994. New experimental approach to understanding igneous texture. Geology 22:323–326.

Menzies M, Rogers N, Tindle A, Hawkesworth C. 1987. Metasomatic and enrichment processes in lithospheric peridotites, an effect of asthenosphere-lithosphere interaction. In: Menzies MA, Hawkesworth CJ, eds. Mantle metasomatism. New York, Academic Press: 313–361.

Merzbacher C, Eggler DH. 1984. A magmatic geohygrometer: Application to Mount St. Helens and other dacitic magmas. Geology 12:587–590.

Miller CF. 1986. Comment and reply on "S-type granites and their probable absence in southwestern North America." Geology 14:804–806.

Miller CF, Barton MD. 1990. Phanerozoic plutonism in the Cordilleran Interior, U.S.A. Geol. Soc. Am. Special Paper 241:213–231.

Mitchell RH. 1994. The lamprophyre facies. Mineral. Petrol. 51:137–146.

Mitchell RH. 1995. Kimberlites, orangeites, and related rocks. New York, Plenum.

Mitchell RH, ed. 1996. Undersaturated alkaline rocks: Mineralogy, petrogenesis, and economic potential. Mineral. Assoc. Can. Short Course 24.

Mitchell RH, Bergman SC. 1991. Petrology of lamproites. New York, Plenum.

Mitchell RS. 1985. Dictionary of rocks. New York, Van Nostrand Reinhold.

Miyashiro A. 1974. Volcanic rock series in island arcs and active continental margins. Am. J. Sci. 274:321–355.

Moore G, Righter K, Carmichael ISE. 1995. The effect of dissolved water on the oxidation state of iron in natural silicate liquids. Contr. Mineral. Petrol. 120:170–179.

Moore G, Vennemann T, Carmichael ISE. 1998. An empirical model for the solubility of H_2O in magmas to 3 kilobars. Am. Mineral. 83:36–42.

Moore JG. 1975. Mechanism of formation of pillow lava. Am. Sci. 63:269–277.

Moore JG, Albee WC. 1981. Topographic and structural changes, March–July, 1980—photogrammetric data. In: Lipman PW, Mullineaux DR, eds. The 1980 eruptions of Mount St. Helens, Washington. US Geol. Surv. Prof. Paper 1250:123–134.

Moore JG, Clague DA. 1992. Volcano growth and evolution of the island of Hawaii. Geol. Soc. Am. Bull. 104:1471–1484.

Moore JG, Normark WR, Holcomb RT. 1994. Giant Hawaiian landslides. Annu. Rev. Earth Planet. Sci. 22:119–144.

Moores EM. 1982. Origin and emplacement of ophiolites. Rev. Geophys. Space Phys. 20:735–760.

Moreau C, Ohnenstetter D, Diot H, Demaiffe D, Brown WL. 1995. Emplacement of the Meugueur-Meugueur cone sheet (Niger, West Africa), one of the world's largest igneous ring-structures. In: Baer G, Heimann A, eds. Physics and chemistry of dikes. Rotterdam, Balkema: 41–49.

Morgan GB, London D. 1999. Crystallization of the Little Three pegmatite-aplite dike, Ramona district, California. Contr. Mineral. Petrol. 136:310–330.

Morgan WJ. 1971. Convection plumes in the lower mantle. Nature 230:42–43.

Morse SA. 1970. Alkali feldspars with water at 5 kb pressure. J. Petrol. 11:221–253.

Morse SA. 1980. Basalts and phase diagrams. New York, Springer-Verlag.

Morse SA. 1998. Is the cumulate paradigm at risk? An extended discussion of the cumulate paradigm reconsidered. J. Geol. 106:367–370.

Mothes PA, Hall ML, Janda RJ. 1998. The enormous Chillos Valley lahar: An ash-flow generated debris flow from Cotopaxi volcano, Ecuador. Bull. Volc. 59:233–244.

Murase T, McBirney AR. 1973. Properties of some common igneous rocks and their melts at high temperatures. Geol. Soc. Am. Bull. 84:3563–3592.

Murase T, McBirney AR, Melson WG. 1985. Viscosity of the dome of Mount St. Helens. J. Volc. Geotherm. Res. 24:193–204.

Murata KJ, Richter DH. 1966. The settling of olivine in Kilauean magma as shown by lavas of the 1959 eruption. Am. J. Sci. 264:194–203.

Mysen BO. 1988. Structure and properties of silicate melts. Amsterdam, Elsevier.

Nabelek PI, Liu M. 1999. Leucogranites in the Black Hills of South Dakota: The consequence of shear heating during continental collision. Geology 27:523–526.

Naslund HR, McBirney AR. 1996. Mechanisms of formation of igneous layering. In: Cawthorne RG, ed. Layered intrusions. New York, Elsevier: 1–44.

Natland J. 1991. Mineralogy and crystallization of oceanic basalts. In: Floyd PA, ed. Oceanic basalts. New York, Van Nostrand Reinhold: 63–93.

Neal CR, Mahoney JJ, Kroenke LW, Duncan RA, Petterson MG. 1997. The Ontong Java plateau. In: Large igneous provinces: Continental, oceanic, and planetary flood volcanism. Mahoney JJ, Coffin MF, eds. Am. Geophys. Union Monog. 100:183–216.

Nekvasil H. 1991. Ascent of felsic magmas and formation of rapakivi. Am. Mineral. 76:1279–1290.

Nekvasil H, Lindsley DH. 1990. Termination of the 2 feldspar + liquid curve in the system Ab-Or-An-H_2O at low H_2O contents. Am. Mineral. 75:1071–1079.

Nelson SA. 1981. The possible role of thermal feedback in the eruption of siliceous magmas. J. Volc. Geotherm. Res. 11:127–137.

Neuvonen KJ. 1952. Thermochemical investigation of the åkermanite-gehlenite series. Bull. Comm. Géol. Finlande 26:1–13.

Nicholas A. 1992. Kinematics in magmatic rocks with special reference to gabbros. J. Petrol. 33:891–915.

Nielson JE, Budahn JR, Unruh DM, Wilshire HG. 1993. Actualistic models of mantle metasomatism documented in a composite xenolith from Dish Hill, California. Geochim. Cosmochim. Acta 57:105–121.

Nixon PH, ed. 1987. Mantle xenoliths. New York, John Wiley & Sons.

Nixon PH. 1995. The morphology and nature of primary diamondiferous occurrences. J. Geochem. Explor. 53:41–72.

Nixon PH, von Knorring O, Rooke JM. 1963. Kimberlites and associated inclusions of Basutoland: A mineralogical and geochemical study. Am. Mineral. 48:1090–1120.

Noble JA. 1952. Evaluation of criteria for the forcible intrusion of magma. J. Geol. 60:34–57.

Nockolds SR, Knox RWO'B, Chinner GA. 1978. Petrology for students. Cambridge, Cambridge University Press.

Norton JJ, Redden JA. 1990. Relations of zoned pegmatites to other pegmatites, granite, and metamorphic rocks in the southern Black Hills, South Dakota. Am. Mineral. 75:631–655.

Ochs FA, III, Lange RA. 1997. The partial molar volume, thermal expansivity, and compressibility of H_2O in $NaAlSi_3O_8$ liquid: New measurements and an internally consistent model. Contr. Mineral. Petrol. 129:155–165.

Odé H. 1957. Mechanical analysis of the dike pattern of the Spanish Peaks area, Colorado. Geol. Soc. Amer. Bull. 68:567–578.

O'Hanley DS. 1996. Serpentinites. New York, Oxford.

Oluon PE. 1999. Giant lava flows, mass extinctions, and mantle plumes. Science 284:604–605.

Oxburgh ER. 1964. Petrologic evidence for the presence of amphibole in the upper mantle and its petrogenetic and geophysical implications. Mineral Mag. 101:1–19.

Park Y, Hanson B. 1999. Experimental investigation of Ostwald-ripening rates of forsterite in the haplobasaltic system. J. Volc. Geotherm. Res. 90:103–113.

Parsons I, ed. 1987. Origins of igneous layering. Boston, Reidel.

Parsons T, Sleep NH, Thompson GA. 1992. Host rock rheology controls on the emplacement of tabular intrusions: Implications for underplating of extending crust. Tectonics 11:1348–1356.

Paterson MS. 1978. Experimental rock deformation—the brittle field. New York, Springer-Verlag.

Paterson SR, Fowler TK, Jr. 1993. Re-examining pluton emplacement processes. J. Struct. Geol. 15:191–206.

Paterson SR, Fowler TK, Jr., Schmidt KL, Yoshinobu AS, Yuan ES, Miller RB. 1998. Interpreting magmatic fabric patterns in plutons. Lithos 44:53–82.

Paterson SR, Vernon RH. 1995. Bursting the bubble of ballooning plutons: A return to nested diapirs emplaced by multiple processes. Geol. Soc. Am. Bull. 107:1356–1380.

Paterson SR, Vernon RH, Fowler TK, Jr. 1991. Aureole tectonics. In: Kerrick DM, ed. Contact metamorphism. Rev. Mineral. 26:673–722.

Paterson SR, Vernon RH, Tobisch OT. 1989. A review of criteria for the identification of magmatic and tectonic foliations in granitoids. J. Struct. Geol. 11:349–363.

Patiño Douce AE. 1997. Generation of metaluminous A-type granites by low-pressure melting of calc-alkaline granitoids. Geology 25:743–746.

Patiño Douce AE, Harris N. 1998. Experimental constraints on Himalayan anatexis. J. Petrol. 39:689–710.

Patiño Douce AE, Humphreys ED, Johnson AD. 1990. Anatexis and metamorphism in tectonically thickened continental crust exemplified by the Sevier hinterland, western North America. Earth Planet. Sci. Lett. 97:290–315.

Peacock SM, Rushmer T, Thompson AB. 1994. Partial melting of subducting oceanic crust. Earth Planet. Sci. Lett. 121:227–244.

Pearce JA, Harris NBW, Tindale AG. 1984. Trace element discrimination diagrams for the tectonic interpretation of granitic rocks. J. Petrol. 25:956–983.

Pearce TH, Kolisnik AM. 1990. Observations of plagioclase zoning using interference imaging. Earth Sci. Rev. 29:9–26.

Peate DW. 1997. The Paraná-Etendeka province. In: Mahoney JJ, Coffin MF, eds. Large igneous provinces: Continental, oceanic, and planetary flood volcanism. Am. Geophys. Union Monogr. 100:217–245.

Peterson DW, Tilling RI. 1980. Transition of basaltic lava from pahoehoe to aa, Kilauea volcano, Hawaii: Field observations and key factors. J. Volc. Geotherm. Res. 7:271–293.

Petford N. 1996. Dykes or diapirs? Trans. R. Soc. Edinb. 87:105–114.

Phaup WE. 1973. Geological Society of South Africa Special Publication 3.1.

Philpotts AR. 1982. Compositions of immiscible liquids in volcanic rocks. Contrib. Mineral. Petrol. 80:201–218.

Philpotts AR, Lewis CL. 1987. Pipe vesicles—an alternate model of their origin. Geology 15:971–974.

Pickering JM, Johnson AD. 1998. Fluid-absent melting behavior of a two-mica metapelite: Experimental constraints on the origin of Black Hills granite. J. Petrol. 39:1787–1804.

Pitcher WS. 1997. The nature and origin of granite. New York, Chapman & Hall.

Piwinskii AJ. 1968. Experimental studies of igneous rock series: Central Sierra Nevada batholith, California. J. Geol. 76:548–570.

Polacci M, Cashman KV, Kauahikaua JP. 1999. Textural characterization of the pahoehoe-aa transition in Hawaiian basalt. Bull. Volc. 60:595–609.

Prouteau G, Scaillet B, Pichavant M, Maury RC. 1999. Fluid-present melting of ocean crust in subduction zones. Geology 27:1111–1114.

Puffer JH, Ragland PC, eds. 1992. Eastern North America Mesozoic magmatism. Geol. Soc. Am. Special Paper 268.

Putnis A, McConnell JDC. 1980. Principles of mineral behavior. Boston, Blackwell Scientific.

Rahaman MA, van Breemen O, Bowden P, Bennett JN. 1984. Age migrations of anorogenic ring complexes in northern Nigeria. J. Geol. 92:173–184.

Ramberg H. 1981. Gravity, deformation, and the earth's crust. New York, Academic Press.

Rampino MR, Self S, Strothers RB. 1988. Volcanic winters. Annu. Rev. Earth Planet. Sci. 16:73–99.

Rapp RP, Watson EB, Miller CF. 1991. Partial melting of amphibolite/eclogite and the origin of Archean trond-hjemites and tonalites. Precamb. Res. 51:1-25.

Rautenschlein M, Jenner GA, Hertogen J, Hofmann AW, Kerrich R, Schmincke H-U, White WM. 1985. Isotopic and trace element composition of volcanic glasses from the Akaki Canyon, Cyprus: Implications for the origin of the Troodos ophiolite. Earth Planet. Sci. Lett. 75:369–383.

Read HH. 1957. The granite controversy. London, Murby.

Reidel SP, Hooper PR, eds. 1989. Volcanism and tectonism in the Columbia River flood-basalt province. Geol. Soc. Am. Special Paper 239.

Rhodes JM, Dungan MA, Blanchard DP, Long PE. 1979. Magma mixing at mid-ocean ridges: Evidence from basalts drilled near 22°N on the Mid-Atlantic Ridge. Tectonophys. 55:35–61.

Richet P, Bottinga Y. 1995. Rheology and configurational entropy of silicate melts. Rev. Mineral. 32:67–93.

Robie RA, Waldbaum DR. 1968. Thermodynamic properties of minerals and related substances at 298.15 K (25.0°C) and one atmosphere (1.013 bars) pressure and at higher temperatures. US Geol. Surv. Bull. 1259.

Robinson P. 1991. The eye of the petrographer, the mind of the petrologist. Am. Mineral. 76:1781–1810.

Roedder E. 1984. Fluid inclusions. Rev. Mineral. 12.

Rollinson HR. 1993. Using geochemical data: Evaluation, presentation, interpretation. New York, John Wiley & Sons.

Rose AW, Burt DM. 1979. Hydrothermal alteration. In: Barnes HL, ed. Geochemistry of hydrothermal ore deposits, 2nd ed. New York, John Wiley & Sons: 173–235.

Rose EI. 1987. Volcanic activity at Santiaguito volcano, 1976–1984. In: Fink JH, ed. The emplacement of silicic domes and lava flows. Geol. Soc. Am. Special Paper 212:17–27.

Ross CS, Smith RL. 1961. Ash-flow tuffs: Their origin, geologic relations, and identification. US Geol. Surv. Prof. Paper 366.

Rowland SK, Walker GPL. 1990. Pahoehoe and aa in Hawaii: Volumetric flow rate controls the lava structure. Bull. Volcan. 52:615–628.

Rubin AM. 1995. Propagation of magma-filled cracks. Annu. Rev. Earth Planet. Sci. 23:287–336.

Rutherford MJ, Hill PM. 1993. Magma ascent rates from amphibole breakdown: An experimental study applied to the 1980–86 Mount St. Helens eruptions. J. Geophys. Res. 98:19,667–19,685.

Ryan MP. 1994. Neutral-buoyancy controlled magma transport and storage in mid-ocean ridge magma reservoirs and their sheeted-dike complex: A summary of basic relationships. In: Ryan MP, ed. Magmatic systems. New York, Academic Press: 97–138.

Ryan MP, Blevins JYK. 1987. The viscosity of synthetic and natural silicate melts and glasses at high temperatures and 1 bar (10^5 pascals) pressure and at higher pressures. US Geol. Surv. Bull. 1987.

Sahama TG. 1952. Leucite, potash nepheline, and clinopy-roxene from volcanic lavas from southwestern Uganda and adjoining Belgian Congo. Am. J. Sci. Bowen Volume:457–464.

Sarna-Wojcicki AM, Shipley S, Waitt RB, Jr., Dzurisin D, Wood SH. 1981. Areal distribution, thickness, mass, volume, and grain size of air-fall ash from the six major eruptions of 1980. In: Lipman PW, Mullineaux DR, eds. The 1980 eruptions of Mount St. Helens, Washington. US Geol. Surv. Prof. Paper 1250:577–600.

Saunders AD, Fitton JG, Kerr AC, Norry MJ, Kent RW. 1997. The North Atlantic igneous province. In: Mahoney JJ, Coffin MF, eds. 1997. Large igneous provinces: Continental, oceanic, and planetary flood volcanism. Am. Geophys. Union Monogr. 100:45–94.

Sawka WN, Chappell BW, Kistler RW. 1990. Granitoid compositional zoning by side-wall boundary layer differentiation: Evidence from the Palisade Crest intrusive suite, central Sierra Nevada, California. J. Petrol. 31:519–553.

Schiano P, Clocchiatti R, Shimizu N, Maury RC, Jochum KP, Hofmann AW. 1995. Hydrous, silica-rich melts in the sub-arc mantle and their relationaship with erupted lavas. Nature 377:595–600.

Schiano P, Clocchiatti R, Shimizu N, Weis D, Mattielli N. 1994. Cogenetic silica-rich and carbonate-rich melts trapped in mantle minerals in Kerguelen ultramafic xenoliths: Implications for metasomatism in the oceanic upper mantle. Earth Planet. Sci. Lett. 123:167–178.

Scott-Smith B. 1966. Kimberlites. In: Mitchell RH, ed. Undersaturated alkaline rocks: Mineralogy, petrogenesis, and economic potential. Min. Assoc. Can. Short Course Ser. 24:217–244.

Searle M, Cox J. 1999. Tectonic setting, origin, and obduction of the Oman ophiolite. Geol. Soc. Am. Bull. 111:104–122.

Self S, Thordarson T, Keszthelyi L. 1997. Emplacement of continental flood basalt lava flows. In: Mahoney JJ, Coffin MF, eds. Large igneous provinces: Continental, oceanic, and planetary flood volcanism. Am. Geophys. Union Monogr. 100:381–410.

Severinghaus J, Atwater T. 1990. Cenozoic geometry and thermal state of the subducting slabs beneath western North America. Geol. Soc. Am. Mem. 176:1–22.

Shannon RD. 1976. Revised effective ionic radii and systematic studies of interatomic distances in halides and chalcogenides. Acta Cryst. A. 32:751–767.

Sharma M. 1997. Siberian traps. In: Mahoney JJ, Coffin MF, eds. 1997. Large igneous provinces: Continental, oceanic, and planetary flood volcanism. Am. Geophys. Union Monogr. 100:273–296.

Shaw HR. 1965. Comments on viscosity, crystal settling, and convection in granitic magmas. Am. J. Sci. 263:120–152.

Shaw HR. 1969. Rheology of basalt in the melting range. J. Petrol. 10:510–535.

Shaw HR. 1972. Viscosities of magmatic silicate liquids: An empirical method of prediction. Am. J. Sci. 272:870–893.

Shaw HR. 1974. Diffusion of H_2O in granitic liquids. Part I. Experimental data: Part II. Mass transfer in magma chambers. In: Hofmann AW, Giletti BJ, Yoder HS, Jr., Yund RA, eds. Geochemical transport and kinetics. Carn. Inst. Wash. Publ. 634:139–170.

Shaw HR. 1980. The fracture mechanisms of magma transport from the mantle to the surface. In: Hargraves RB, ed. Physics of magmatic processes. Princeton, N.J., Princeton University Press: 201–264.

Shaw HR, Moore JG. Nov. 8, 1988. Magmatic heat and the El Niño cycle. EOS, Am. Geophys. Union Trans., 1553.

Shelton JS. 1966. Geology illustrated. New York, W.H. Freeman.

Sheridan MF. 1979. Emplacement of pyroclastic flows: A review. Geol. Soc. Am. Special Paper 180:125–136.

Sheridan MF, Updike RG. 1975. Sugarloaf mountain tephra—a Pleistocene rhyolitic deposit of base-surge origin in northern Arizona. Geol. Soc. Am. Bull. 86:571–581.

Shirley DN. 1987. Differentiation and compaction in the Palisades sill, New Jersey. J. Petrol. 28:835–865.

Sigurdsson H, Carey S, Mandeville C, Bronto S. 1991. Pyroclastic flows of the 1883 Krakatau eruption. EOS, Trans. Am. Geophys. Union 72:377–381.

Sigurdsson H, Houghton BF, McNutt SR, Rymer H, Stix J, eds. 2000. Encyclopedia of volcanoes. New York, Academic.

Silver LA, Ihinger PD, Stolper E. 1990. The influence of bulk composition on the speciation of water in silicate glasses. Contr. Mineral. Petrol. 104:142–162.

Sinigoi S, Quick JE, Mayer A, Demarchi G. 1995. Density-controlled assimilation of underplated crust, Ivrea-Verbano zone, Italy. Earth Planet. Sci. Lett. 129:183–191.

Sisson TW, Bacon CR. 1999. Gas-driven filter pressing in magmas. Geology 27:613–616.

Skjerlie KP, Johnson AD. 1993. Fluid-absent melting behavior of an F-rich tonalitic gneiss at mid-crustal pressures: Implications for the generation of anorogenic granites. J. Petrol. 34:785–815.

Smith AL, Carmichael ISE. 1969. Quaternary trachy basalts from southeastern California. Am. Mineral. 54:909–923.

Smith GA, Lowe DR. 1991. Lahars: Volcano-hydrologic events and deposition in the debris flow-hyperconcentrated flow continuum. In: Fisher RV, Smith GA, eds. Sedimentation in volcanic settings. SEPM (Society for Sedimentary Geology) Special Publication 45:59–70.

Smith JV. 1997. Shear thickening dilatancy in crystal-rich flows. J. Volc. Geotherm. Res. 79:1–8.

Smith RL. 1960. Zones and zonal variations in welded ash flows. US Geol. Surv. Prof. Paper 354-F.

Smith RL. 1979. Ash-flow magmatism. Geol. Soc. Am. Special Paper 180:5–27.

Smith RL, Bailey RA. 1968. Resurgent cauldrons. Geol. Soc. Am. Mem. 116:623–662.

Smith RP. 1987. Dyke emplacement at Spanish Peaks, Colorado. In: Halls HC, Fahrig WF, eds. Mafic dyke swarms. St. John's, Newfoundland: Geol. Assoc. Canada Special Paper 34:47–54.

Snyder D, Crambes C, Tait S, Wiebe RA. 1997. Magma mingling in dikes and sills. J. Geol. 105:75–86.

Snyder D, Tait S. 1995. Replenishment of magma chambers: Comparison of fluid-mechanic experiments with field relations. Contr. Mineral. Petrol. 122:230–240.

Sobolev AV, Chaussidon M. 1996. H_2O concentrations in primary melts from supra-subduction zones and mid-ocean ridges: Implications for H_2O storage and recycling in the mantle. Earth Planet. Sci. Lett. 137:45–55.

Sorensen SS. 1988. Petrology of amphibolite-facies mafic and ultramafic rocks from the Catalina Schist, southern California: Metasomatism and migmatization in a subduction zone metamorphic setting. J. Metam. Geol. 6:405–435.

Sparks RSJ, Barclay J, Jaupart C, Mader HM, Phillips JC. 1994. Physical aspects of magma degassing. I. Experimental and theoretical constraints on vesiculation. In: Carroll MR, Holloway JR, eds. Volatiles in magmas. Rev. Mineral. 30:413–445.

Sparks RSJ, Bursik MI, Carey SN, Gilbert JS, Glaze LS, Sigurdsson H, Woods AW. 1997. Volcanic plumes. New York, John Wiley & Sons.

Sparks RSJ, Gilbert JS, eds. 1999. The physics of explosive volcanic eruptions. Geol. Soc. Lond. Spec. Publ. 145.

Sparks RSJ, Huppert HE, Turner JS. 1984. The fluid dynamics of evolving magma chambers. Phil. Trans. R. Soc. London, A. 310:511–534.

Sparks RSJ, Sigurdsson H. 1977. Magma mixing: A mechanism for triggering acid explosive eruptions. Nature 267:315–318.

Sparks RSJ, Tait SR, Yanev Y. 1999. Dense welding caused by volatile resorption. J. Geol. Soc. Lond. 156:217–225.

Spear FS. 1993. Metamorphic phase equilibria and pressure-temperature-time paths. Mineral. Soc. Am.

Spera FJ. 1980. Aspects of magma transport. In: Hargraves RB, ed. Physics of magmatic processes. Princeton, N.J., Princeton University Press: 264–323.

Spray JG. 1998. Localized shock- and friction-induced melting in response to hypervelocity impact. In: Grady MM, Hutchison R, McCall GJH, Rothery DA, eds. Meteorites: Flux with time and impact effects. Geol Soc. Special Publication 140:195–204.

Springer W, Seck HA. 1997. Partial fusion of basic granulites at 5 to 15 kbar: Implications for the origin of TTG magmas. Contrib. Mineral. Petrol. 127:30–45.

Stacey FD. 1992. Physics of the earth, 3rd ed. Brisbane, Brookfield.

Stearns HT. 1966. Geology of the state of Hawaii. Palo Alto, Calif., Pacific Books.

Stern TW, Bateman PC, Morgan BA, Newell MF, Peck DL. 1981. Isotopic U-Pb ages of zircon from the granitoids of the central Sierra Nevada: US Geol. Surv. Prof. Paper 1185.

Stimac JA, Wark DA. 1992. Plagioclase mantles on sanidine in silicic lavas, Clear Lake, California: Implications for the origin of rapakivi texture. Geol. Soc. Am. Bull. 104: 728–744.

Stolper EM, Walker D. 1980. Melt density and the average composition of basalt. Contr. Mineral. Petrol. 74:7–12.

Stolper EM, Walker D, Hager BH, Hays JF. 1981. Melt segregation from partially molten source region: The importance of melt density and source region size. J. Geophys. Res. 86:6261–6271.

Sugioka I, Bursik M. 1995. Explosive fragmentation of erupting magma. Nature 373:689–692.

Sun S, McDonough WF. 1989. Chemical and isotopic systematics of oceanic basalts: Implications for mantle composition and processes. In: Saunders AD, Norry MJ, eds. Magmatism in the ocean basins. Boston, Blackwell Scientific: 313–345.

Swamy V, Saxena SK, Sundman B, Zhang J. 1994. A thermodynamic assessment of silica phase diagram. J. Geophys. Res. 99:11,787–11,794.

Swanson DA, Dzurisin D, Holcomb RT, Iwatsubo EY, Chadwick WW, Jr., Casadevall TJ, Ewert JW, Heliker CC. 1987. Growth of the lava dome at Mount St. Helens, Washington, (USA), 1981–1983. In: Fink JH, ed. The emplacement of silicic domes and lava flows. Geol. Soc. Am. Special Paper 212:1–16.

Swanson SE. 1977. Relation of nucleation and crystal-growth rate to the development of granitic textures. Am. Mineral. 62:966–978.

Swanson SE, Fenn PM. 1986. Quartz crystallization in igneous rocks. Am. Mineral. 71:331–342.

Symonds RB, Rose WI, Bluth GJS, Gerlach TM. 1994. Volcanic-gas studies: Methods, results, and applications. In: Carroll MR, Holloway JR, eds. Volatiles in magmas, Rev. Mineral. 10:1–66.

Takahashi E, Shimazaki T, Tsuzaki Y, Yoshida H. 1993. Melting study of a peridotite KLB-1 to 6.5 GPa, and the origin of basaltic magmas. Phil. Trans. R. Soc. Lond. A 342:105–120.

Tatsumi Y, Eggins S. 1995. Subduction zone magmatism. Boston, Blackwell Science.

Taylor HP, Jr. 1974. The application of oxygen and hydrogen isotope studies to problems of hydrothermal alteration and ore deposition. Econ. Geol. 69:843–883.

Taylor HP, Jr. 1980. The effects of assimilation of country rocks by magmas on $^{18}O/^{16}O$ and $^{87}Sr/^{86}Sr$ systematics in igneous rocks. Earth Planet. Sci. Lett. 47:243–254.

Taylor HP, Jr., Sheppard SMF. 1986. Igneous rocks. I. Processes of isotopic fractionation and isotope systematics. In: Valley JW, Taylor HP, Jr., O'Neil JR, eds. Stable isotopes in high temperature geological processes. Rev. Mineral. 16:227–271.

Terry RD, Chilinger GV. 1972. Charts to aid the visual estimation of modal proportions of minerals in rocks. Am. Geol. Inst. Data Sheet 6.

Thomas AV, Bray CJ, Spooner ETC. 1988. A discussion of the Jahns-Burnham proposal for the formation of zoned granitic pegmatites using solid-liquid-vapour inclusions from the Tanco pegmatite, S.E. Manitoba, Canada. Trans. R. Soc. Edin. Earth Sci. 79:299–315.

Thompson AB. 1992. Water in the Earth's upper mantle. Nature 358:295–301.

Thompson JB, Jr. 1970. Geochemical reaction and open systems. Geochim. Cosmochim. Acta 34:529–551.

Tikoff B, Teyssier C. 1992. Crustal-scale, *en echelon,* "P-shear" tensional bridges: A possible solution to the batholith room problem. Geology 20:927–930.

Tilling RI. 1989. Volcanic hazards and their mitigation: Progress and problems. Rev. Geophys. 27:237–269.

Tingey DG, Christiansen EH, Best MG, Ruiz J, Lux DR. 1991. Tertiary minette and melanephelinite dikes, Wasatch Plateau, Utah: Records of mantle heterogeneities and changing tectonics. J. Geophys. Res. 96:13,529–13,544.

Turner DC, Bowden P. 1979. The Ningi-Burra complex, Nigeria: Dissected calderas and migrating magmatic centres. J. Geol. Soc. London 136:105–119.

Tuttle OF, Bowen NL. 1958. Origin of granite in the light of experimental studies in the system $NaAlSi_3O_8$-$KAlSi_3O_8$-SiO_2-H_2O. Geol. Soc. Am. Mem. 74.

Twiss RJ, Moores EM. 1992. Structural geology. New York, W.H. Freeman.

van der Molen I, Paterson MS. 1979. Experimental deformation of partially melted granite. Contr. Mineral. Petrol. 70:299–318.

Van Decar JC, James DE, Assumpção M. 1995. Seismic evidence for a fossil mantle plume beneath South America and implications for plate driving forces. Nature 378:25–31.

Van der Plas L, Tobi AC. 1965. A chart for judging the reliability of point counting results. Am. J. Sci. 263: 87–90.

Verhoogen J. 1980. Energetics of the earth. National Acad. Sci. 139.

Vernon RH. 1986. K-feldspar megacrysts in granites—phenocrysts, not porphyroblasts. Earth Sci. Rev. 23:1–63.

Wada Y. 1995. Reply (on the relationship between dike width and magma viscosity). J. Geophys. Res. 100:15,543–15,544.

Wager LR. 1959. Differing powers of crystal nucleation as a factor producing diversity in layered igneous intrusions. Mineral. Mag. 96:75–80.

Wager LR, Deer WA. 1939. Geological investigations in East Greenland. III. The petrology of the Skaergaard intrusion, Kangerlugssuaq, East Greenland. Meddelelser om Grønland 105:1–346.

Walker GPL. 1992. Morphometric study of pillow-size spectrum among pillow lavas. Bull. Volc. 54:459–474.

Walker GPL. 1993. Basaltic-volcano systems. In: Prichard HM, Alabaster T, Harris NBW, Neary CR, eds. Magmatic processes and plate tectonics. Geol. Soc. Lond. Special Publication 76:3–39.

Wallace ME, Green DH. 1988. An experimental determination of primary carbonatite magma composition. Nature 335:343–346.

Wallace PJ, Anderson AT, Jr., Davis AM. 1995. Quantification of pre-eruptive exsolved gas contents in silicic magmas. Nature 377:612–614.

Wark DA, Stimac JA. 1992. Origin of mantled (rapakivi) feldspars: Experimental evidence on dissolution- and diffusion-controlled mechanism. Contr. Mineral. Petrol. 111:345–361.

Warshaw CM, Smith RL. 1988. Pyroxenes and fayalites in the Bandelier Tuff, New Mexico: Temperatures and comparisons with other rhyolites. Am. Mineral. 73:1025–1037.

Waters AC, Fisher RV. 1971. Base surges and their deposits: Capelinhos and Taal volcanoes. J. Geophys. Res. 76:5596–5614.

Watson EB. 1994. Diffusion in volatile-bearing magmas. In: Carroll MR, Holloway JR, eds. Volatiles in magmas. Rev. Mineral. 30:371–411.

Watson EB, Brenan JM, Baker DR. 1990. Distribution of fluids in the continental mantle. In: Menzies MA, ed. Continental mantle. Oxford, Clarendon: 111–125.

Watson S. 1993. Rare earth element inversions and percolation models for Hawaii. J. Petrol. 34:763–783.

Watson S, McKenzie D. 1991. Melt generation by plumes: A study of Hawaiian volcanism. J. Petrol. 32:501–537.

Weaire D, Phelan R. 1994. A counter example to Kelvin's conjecture on minimal surfaces. Phil. Mag. Lett. 69:107–110.

Webb SL, Dingwell DB. 1990. Non-Newtonian rheology of igneous melts at high stresses and strain rates: Experimental results for rhyolite, andesite, basalt, and nephelinite. J. Geophys. Res. 95:15,695–15,701.

Weill DF, Hon R, Navrotsky A. 1980. The igneous system $CaMgSi_2O_6$–$CaAl_2Si_2O_8$–$NaAlSi_3O_8$. Variations on a classic theme by Bowen. In: Hargraves RB, ed. Physics of magmatic processes. Princeton, N.J., Princeton University Press: 49–92.

Weiss LE. 1972. Minor structures of deformed rocks. New York, Springer-Verlag.

Wendlandt RF. 1991. Oxygen diffusion in basalt and andesite melts: Experimental results and discussion of chemical versus tracer diffusion. Contr. Mineral. Petrol. 108:463–471.

Wentworth CK, Macdonald GA. 1953. Structures and forms of basaltic rocks in Hawaii. US Geol. Surv. Bull. 994.

White RS. 1993. Melt production rates in mantle plumes. Phil. Trans. R. Soc. London A 342:137–153.

White RS, McKenzie D. 1995. Mantle plumes and flood basalts. J. Geophys. Res. 100:17,543–17,585.

Whitham AG, Sparks RSJ. 1986. Pumice. Bull. Volc. 48:209–223.

Whitney JA. 1988. The origin of granite: The role and source of water in the evolution of granitic magmas. Geol. Soc. Am. Bull. 100:1886–1897.

Wiebe RA. 1996. Mafic-silicic layered intrusions: The role of basaltic injections on magmatic processes and the evolution of silicic magma chambers. Trans. Roy. Soc. Edinb. Earth Sci. 87:233–242.

Wilkinson JFG. 1965. Some feldspars, nephelines, and analcimes from the Square Top intrusion, Nundle, N.S.W. J. Petrol. 6:420–444.

Williams H. 1932. The history and character of volcanic domes. Univ. Calif. Publ. Geol. Sci. Bull. 21:51–146.

Williams H. 1932. Geology of Lassen Volcanic National Park, California. Univ. Calif. Publ. Geol. Sci. Bull. 21:195–385.

Williams H. 1941. Calderas and their origin. Univ. Calif. Publ. Geol. Sci. Bull. 25:239–346.

Williams H, McBirney AR. 1979. Volcanology. San Francisco, Freeman Cooper.

Williams H, Turner FJ, Gilbert CM. 1982. Petrography: An introduction to the study of rocks in thin section. New York, W.H. Freeman.

Wilshire HG, Meyer CE, Nakata JK, Calk LC, Shervais JW, Nielson JE, Schwarzman EC. 1988. Mafic and ultramafic xenoliths from volcanic rocks of the western United States. US Geol. Surv. Prof. Paper 1443.

Wilson CJN, Hildreth W. 1997. The Bishop tuff: New insights from eruptive stratigraphy. J. Geol. 105:407–439.

Wilson JT. 1963. Hypothesis of earth's behavior. Nature 198:925–929.

Wilson L, Sparks RSJ, Walker GPL. 1980. Explosive volcanic eruptions. IV. The control of magma properties and conduit geometry on eruption column behavior. Geophys. J. R. Astron. Soc. 63:117–148.

Wilson M. 1989. Igneous petrogenesis: A global tectonic approach. Boston, Unwin Hyman.

Wohletz KH. 1986. Explosive magma-water interactions: Thermodynamics, explosive mechanisms, and field studies. Bull. Volc. 48:245–264.

Wones DR, Eugster HP. 1965. Stability of biotite: Experiment, theory, and application. Am. Mineral. 50:1228–1272.

Woodhead J. 1989. Geochemistry of the Mariana arc (western Pacific): Source composition and processes. Chem. Geol. 76:1–24.

Woolsey TS, McCallum ME, Schumm SA. 1975. Modeling of diatreme emplacement by fluidization. Phys. Chem. Earth 9:29–42.

Worster MG, Huppert HE, Sparks RSJ. 1990. Convection and crystallization in magma cooled from above. Earth Planet. Sci. Lett. 101:78–89.

Wright TL, Okamura RT. 1977. Cooling and crystallization of tholeiitic basalt: Makaopuhi lava lake, Hawaii. US Geol. Surv. Prof. Paper 1004.

Yagi K. 1953. Petrochemical studies of the alkalic rocks of the Morotu district, Sakhalin. Geol. Soc. Am. Bull. 64:769–810.

Yoder HS, Jr. 1976. Generation of basaltic magma. Washington D.C. Natl. Acad. Sci.

Yoder HS, Jr., Stewart DB, Smith JV. 1957. Ternary feldspars. Annu. Rep. Dir. Geophys. Lab. Carn. Inst. Wash. Yrbk. 56:206–214.

Yoder HS, Jr., Tilley CE. 1962. Origin of basalt magmas: An experimental study of natural and synthetic rock systems. J. Petrol. 3:342–532.

Young DA. 1998. N. L. Bowen and crystallization-differentiation: The evolution of a theory. Min. Soc. Amer. Pub. 4.

Zen E. 1988. Phase relations of peraluminous granitic rocks and their petrogenetic implications. Annu. Rev. Earth Planet. Sci. 16:21–51.

Zhang Y, Walker D, Lesher CE. 1989. Diffusive crystal dissolution. Contr. Mineral. Petrol. 102:492–513.

GLOSSARY

A-type Refers to felsic rocks that contain abundant alkali feldspar, relatively high concentrations of incompatable and HFS ions, and are more anhydrous than arc rocks; occur in anorogenic regions.

Aa lava flow A thick basaltic to andesitic flow covered by an autoclastic mantle of broken vesicular pieces (generally <1 m across) of the crusted surface; treacherous to walk over.

Absarokite Very-high-K basalt.

Abyssal peridotite Seafloor peridotite; variably hydrated and has a high-T, solid-state strain fabric; variably depleted in basaltic components.

Accessory minerals Minerals that occur in modal proportions of no more than a small percentage and do not influence the naming of a rock, such as zircon, apatite, and magnetite.

Accidental clast A fragment of older rock unrelated to magma that forms the host lava or pyroclastic deposit; also called **xenolith** if rock or **xenocryst** if a crystal fragment.

Accreted Attached or fixed to; may refer to terranes of oceanic rock tectonically attached to the overriding plate in subduction zones as an accretionary prism.

Accretionary lapilli Concentrically layered, usually pea-sized and -shaped; formed in turbulent ash-and-steam clouds produced by explosive eruptions where fine moist ash adheres to some sort of nucleus.

Accuracy An indication of the closeness of a measurement to the "true" value.

Acid rock Contains >66 wt.% silica.

Activation energy The energy barrier between a metastable and a stable state.

Activity Effective concentration or availability of a component in a real liquid or solid solution.

Adakite Subduction-related andesite-dacite rock type that contains relatively high La/Yb ratios; magma is believed to be generated by partial melting of the hydrated basaltic oceanic crust under high pressure where garnet is a residual phase.

Adiabatic system An isolated system in which no heat is exchanged with its surroundings.

Advection One of four types of heat transfer, which involves flow of a liquid through openings in a rock of different T.

Aerosol Micrometer-sized droplets of a compound, such as sulfuric acid, dispersed in the atmosphere by a volcanic eruption.

AFC Assimilation and fractional crystallization acting concurrently.

Agglutinate Rock produced by welding of molten basaltic pyroclasts ("spatter") deposited at the base of a lava fountain.

Alkaline basalt Basalt rock type containing normative nepheline.

Alkaline rocks Rock suite that has a relative excess of alkalies over silica; rock types are usually silica-undersaturated.

Alpine peridotite Variably serpentinized rock in mountain belts; usually a dismembered segment of **ophiolite.**

Alteration Change in composition of a rock or individual mineral, usually involving replacement by a secondary mineral or minerals; generally of a more local nature than metamorphism or pertaining to changes related to hydrothermal solutions.

Alumina saturation index Molecular ratio $Al_2O_3/(K_2O + Na_2O + CaO)$.

Amygdaloidal Texture made of **amygdules.**

Amygdule Vesicle filled with low-T secondary minerals such as carbonate minerals, zeolites, and chalcedony that precipitate from fluid solutions percolating through rock.

Anatexis See **Partial melting.**

Andesite Glassy or aphanitic rock made essentially of plagioclase, Fe-Ti oxides, and some combination of pyroxene, hornblende, and perhaps biotite or olivine, contains 57–63 wt.% silica and less than about 7 wt.% total alkalies; see Figures 2.12 and 2.13.

Anhedral Irregularly shaped compact grains not bounded by any characteristic crystal faces.

Anisotropic fabric Directional fabric that looks different and has different properties in different directions.

Ankaramite Basaltic rock that contains abundant phenocrysts of pyroxene and olivine, usually by accumulation; if very abundant, rock may be ultramafic.

Anorthosite Phaneritic magmatic rock containing >90% plagioclase; typical mafic mineral is pyroxene and/or olivine.

Anorthosite suite Consists principally of anorthosite plus comagmatic gabbro, norite, and troctolite and their leucocratic varieties; includes minor rocks that contain abundant Fe-Ti oxides.

Aphanitic Texture that consists of a mosaic of crystals too small to be seen without magnification; may be **cryptocrystalline** or **microcrystalline.**

Aphyric Nonporphyritic aphanitic texture.

Aplite Textural variety (**aplitic**) of leucocratic granite; typically occurs in thin dikes within a coarser-grained, somewhat more mafic granitic pluton; also in layers with pegmatite.

Aplitic Fine-grained phaneritic, equigranular texture in which virtually all grains are equant and anhedral to subhedral.

Apophysis An offshoot, or branching, intrusion from an underlying larger pluton.

Apparent viscosity Refers to the property governing the permanent strain of a material whose behavior under applied stress is **non-Newtonian;** generally embodies a plastic yield strength.

Appinite A mafic, phaneritic, magmatic hornblende-rich rock that contains subordinate feldspar; usually associated with granitoids.

Arc magmatism Occurs where oceanic lithosphere subducts beneath overlying oceanic lithosphere in **island arcs** and beneath continental margins in **continental arcs.**

Arc signature Characteristic primitive mantle- or chondrite-normalized trace element pattern of arc magmatic rocks that is "spiky," unlike the smoothly varying pattern of mid–ocean ridge basalt; LIL elements are enriched relative to REE and HFS elements; see **Negative Nb-Ta-Ti anomaly.**

Ash Volcaniclast <2 mm; equivalent to sand, silt, and clay size for sedimentary fragments.

Ash flow A type of **pyroclastic flow** dominated by ash-size particles and possibly lesser lapilli formed by a collapsing volcanic column.

Ash-flow tuff Rock formed from an **ash flow;** also called **ignimbrite.**

Aspect ratio Ratio of thickness to length (diameter) of an extrusion or intrusion of magma.

Assimilation Physical incorporation and chemical dissolution of foreign material into a melt, modifying its composition; the magma or resulting rock is said to be contaminated or hybrid.

Autoclast Usually block-size rock fragments formed on the margins of active lava flows by breakup of the rigid crust.

Autolith See **Cognate inclusion.**

Ballistic clast Block or bomb ejected explosively from a volcanic vent; follows a trajectory like that of a cannon shell.

Ballooning Radially directed inflation of a magma chamber as additional magma is forcefully intruded.

Banakite Very-high-K andesite.

Basalt Glassy or aphanitic rock made mostly of plagioclase, pyroxene, and Fe-Ti oxides with or without olivine; contains 45–52 wt.% silica and less than 5 wt.% total alkalies; see Figure 2.12.

Basaltic andesite Glassy or aphanitic basaltic rock that contains 52–57 wt.% silica and less than about 5 wt.% total alkalies; see Figure 2.12.

Basanite Silica-poor, alkali-rich basaltic rock; see Figure 2.12.

Base surge See **Pyroclastic surge.**

Basic rock Contains 45–52 wt.% silica.

Batholith A usually **composite** pluton made of several separately intruded magmas that is exposed over hundreds to tens of thousands of square kilometers.

Bedding sag Caused by impacting **ballistic clasts** depressing soft, finer-grained underlying layers.

Benmoreite Intermediate-composition aphanitic or glassy rock more alkalic than basaltic andesite; see Figure 2.12.

Bimodal Refers to two distinct populations without intermediate members.

Binary Two components.

Block Angular **volcaniclast** >64 mm equivalent to cobble and boulder size for sedimentary fragments.

Block flow Resembles **Aa lava flow** but has a surface of more regularly shaped polyhedral chunks rather than jagged, highly vesicular blocks; can be as silicic as rhyolite.

Block-and-ash flow Small volume **pyroclastic flow** avalanches that are produced by disintegrative collapse of growing andesitic to rhyolitic domes or thick flows on composite volcanoes.

Boiling See **Exsolution.**

Bomb Streamlined **pyroclast** >64 mm; equivalent to cobble and boulder size for sedimentary fragments.

Boninite Basaltic andesitic or andesite in island arcs that has MgO > 8 wt.% and TiO_2 < 0.5 wt.%; contains abundant magnesian pyroxenes, usually in a glassy matrix.

Boundary layer A layer in which strongly varying T (**thermal boundary layer**) or composition (**compositional boundary layer**) lies between two regions of more uniform but contrasting T or composition; the lithosphere is an upper thermal boundary layer in the Earth between the cool hydrosphere and hot sublithospheric mantle; a compositional boundary layer lies between a growing crystal and an enclosing melt that has slow rates of chemical diffusion or lies at the margin of a magmatic intrusion.

Breakdown See **Replacement.**

Breccia pipe An intrusive magmatic **pipe** that contains abundant rock fragments, usually xenoliths.

Brecciated Refers to a fabric made of broken rock or mineral fragments.

Bridging oxygen In a melt, an O ion that is bonded with a network-forming Si or Al cation in a polymer.

Brittle behavior A real response in rocks subjected to high strain rates where P and T are low; characterized by elastic-like fracturing, breaking apart, and loss of cohesion when the applied stress reaches the **elastic strength;** highly crystalline magmas can also fracture in a brittle manner under high strain rates.

Brittle strength Essentially the **elastic strength,** or stress difference required to produce a fracture in a body.

Buffer reaction A chemical reaction in which the activity or fugacity of a component is maintained or fixed so long as all of the phases in the reaction coexist at equilibrium in a system.

Bulk chemical composition See **Whole-rock chemical composition.**

Bulk partition coefficient See **Partition coefficient.**

Buoyancy Gravity-related force acting on a body whose density differs from that of its immediate surroundings; **negative buoyancy** is a downward force on a body more dense than its surroundings; **positive buoyancy** is an upward force on a body less dense than its surroundings; **neutral buoyancy** is no net force on a body whose density is the same as that of its surroundings.

Calc-alkaline rocks Subalkaline rocks that have relative enrichment in silica and alkalies and little enrichment in Fe; see Figures 2.17 and 12.22.

Caldera A more or less circular or elliptical topographic depression created by collapse of roof rock into an underlying partially voided magma chamber; sometimes called a **cauldron.**

Carbonatite   A magmatic rock composed mostly of carbonate minerals.

Cataclasis   Closely spaced brittle fracturing, crushing, frictional sliding of broken fragments past one another, and rotation of grains, all of which increase the volume of the rock body because of included open spaces.

Cataclastic fabric   The product of **cataclasis.**

Cauldron   See **Caldera.**

Central eruption   Magma vented from a subvertical pipelike feeding conduit; builds a conical volcano.

Central magma intrusion   Usually upright, bottle-shaped or cylindrical intrusion from which smaller sheet intrusions may emanate.

Charnockitic rocks   A group of dark green, brown, or red felsic phaneritic rocks that contain various proportions of perthitic alkali feldspar, plagioclase, and quartz; mafic minerals are mostly orthopyroxene; have affinities with **A-type** granitic rocks; includes **charnockite** (pyroxene granite), **mangerite** (pyroxene monzonite), **jotunite** (pyroxene monzodiorite), **opdalite** (pyroxene granodiorite), and **enderbite** (pyroxene tonalite).

Chondrite   A type of meteorite believed to have accreted to form the inner planets in the solar system, including Earth; used as a reference standard to compare trace element concentrations.

Chondrite-normalized trace element diagram   Concentration ratios of trace elements in a rock to that in chondrite plotted on the logarithmic y axis and ordered along the x axis according to decreasing incompatibility.

Cinder cone   Deposit of basaltic lapilli and lesser blocks around a **strombolian eruption** vent; also called **scoria cone.**

Cinders   Lapilli-size **scoria** that are solid upon reaching the ground after discharge from a volcanic vent.

Clan   A group of rocks related in composition, mode of occurrence, or other attributes; see **Rock suite.**

Clapeyron equation   On a P-T diagram, gives the slope of a boundary line separating stability fields of any two phases, or two assemblages of phases, of identical composition in a closed system; the ratio of the entropy change to the volume change in a reaction between phases.

Closed system   Energy can flow across the system boundary but not matter.

Closure temperature T at which diffusion effectively ceases, arresting a chemical process.

Cogenetic   See **Comagmatic.**

Coignimbrite plume   See **Volcanic plume.**

Comagmatic   Kindred magmas having a common origin and usually related through differentiation processes; also called **cogenetic.**

Cognate inclusion (autolith)   Piece of rock that is genetically related to the host rock in which it is embedded.

Collapsing column   See **Volcanic plume.**

Columnar joints   Joints that form five-, six-, or seven-sided columns of rock; usually oriented perpendicular to the margins of tabular magma intrusions (dikes and sills) or extrusive deposits; formed by shrinkage during cooling.

Comb layering   Layered fabric expressed by long, branching skeletal to feathery, subparalled crystals that are oriented perpendicular to a planar boundary.

Comendite   A **peralkaline** rock in which $Al_2O_3 > 1.33 FeO + 4.4$ (wt.%).

Compaction foliation   See **Eutaxitic fabric.**

Compatible element   Preferentially included within a crystalline phase relative to coexisting liquid phase; **partition coefficient** >1.

Components   The smallest number of chemical entities required to describe the composition of every phase that exists in a system at equilibrium.

Composite intrusion   Has distinguishable parts of different composition and/or fabric reflecting emplacement of two or more contrasting magmas.

Composite volcano   Built mostly of andesitic and dacitic magmas extruded from a central vent; consists of innumerable alternating tongues of lava and volcaniclastic deposits, especially **lahars;** also called **stratovolcano.**

Compositional boundary layer   See **Boundary layer.**

Compositional layering   Planar contrasts in mineral or modal composition.

Compositionally zoned   Refers to a magmatic intrusion or pyroclastic deposit that has a systematic variation in chemical, modal, and/or mineralogical composition; also refers to systematic variation in chemical composition of a solid-solution mineral.

Compressive state of stress   Presses material together; prevails everywhere beneath the surface of the Earth.

Conduction   A type of heat transfer through a static body; accomplished by atomic motion.

Cone sheet   A **sheet intrusion** formed where magma invades a conical fracture system whose apex is an underlying **central intrusion.**

Confining pressure   Caused by weight of overlying material (rock or water); denoted by P.

Contact aureole   Country rock next to a magmatic intrusion that has been metamorphosed.

Contamination   See **Assimilation.**

Continental arc   Linear zone of **arc magmatism** along a continental margin overlying subducting oceanic lithosphere.

Continuous reaction relation   Chemical reaction in which solid-solution crystals and melt combine to yield solid-solution crystals and melt of different composition.

Continuous reaction series   The crystalline products of **continuous reaction relations** in a fractionally crystallizing magma.

Control line   In a variation diagram, a line that follows the compositional changes resulting from fractionation of a phase or phases, or mixing of two magmas, or assimilation of some contaminant into a magma.

Convection   Type of heat transfer accomplished by movement of different parts of a body at different temperature due to differences in their density.

Convective melt fractionation   Segregation of buoyant residual melt from partially crystallized magma, usually along margins of magma chamber by sidewall crystallization.

Cooling unit   A **pyroclastic flow** deposit that cooled as a single thermal entity with a unified history of compaction and welding; this **simple cooling unit** may be made up of two or more flows deposited rapidly enough so that there are no discontinuities in a normal compaction/welding profile between them; a **compound cooling unit** consists of a succession of flows emplaced closely in time so that only partial cooling breaks occur between them.

Country rock   Rock surrounding a magmatic intrusion; also called *roof, floor,* and *wall rock* as appropriate.

Critical point   *P* above which gaseous and liquid states are no longer distinguishable in a system and become a uniform fluid.

Cryptic metasomatism   See **Metasomatism.**

Cryptic layering   Systematic changes in the chemical composition of cumulus solid-solution minerals in a layered intrusion.

Cryptocrystalline   Texture that consists of a mosaic of minute crystals that cannot be resolved with an optical microscope.

Crystal-melt fractionation   Separation of crystals and melt; also called **fractional crystallization, fractionation,** or **crystallization differentiation.**

Crystalline residue   Crystals remaining after partial melting of a source rock.

Crystallinity   Proportion of crystals in a glassy rock or in a magma.

Crystallites   Minute crystals that do not react visibly to polarized light under the microscope.

Crystallization differentiation   See **Crystal-melt fractionation.**

Cumulate   Accumulation of crystals produced by crystal-melt fractionation.

Cumulophyric   A type of **porphyritic** texture in which several phenocrysts are aggregated together, the matrix is usually glassy or aphanitic.

Cumulus   Refers to mineral grains in a **cumulate** or the fabric formed by accumulation of crystals in a fractionating magma.

Cupola   Upper part of a pluton that extends higher into the roof rock than the main underlying mass.

Dacite   Glassy or aphanitic rock made essentially of plagioclase and lesser quartz and a combination of Fe-Ti oxides, pyroxene, hornblende, biotite, and minor sanidine; contains 63 wt.% to about 71 wt.% silica and less than about 7 wt.% total alkalies; see Figures 2.12 and 2.13.

Decomposition   See **Replacement.**

Decomposition (breakdown) reaction   What happens to a primary, usually higher-*T,* crystalline phase as it is replaced by one or more secondary crystalline phases as a result of changing intensive variables in the system.

Decomposition rim   Partial subsolidus replacement of an unstable mineral, usually biotite and amphibole, by an aggregate of secondary minerals, in response to declining water pressure (fugacity) or increasing oxygen in an extrusive rock.

Decompression   A reduction of pressure (*P*) in a system.

Deformation   Response of a body to applied stress; has three components—rotation, translation, and distortion (usually called **strain**); in most contexts, deformation is synonymous with strain, which is a change in shape and/or volume.

Degrees of freedom   The number of independent intensive variables that must be specified to characterize fully the state of equilibrium in a system; also called **variance.**

Dehydration melting   Partial melting of rocks in which the only water present is that bound structurally in hydrous minerals; no separate aqueous fluid phase exists; also called *water-deficient, water-undersaturated,* or *fluid-absent melting.*

Dendritic   Crystal shape resembling tree branches.

Density   Mass divided by volume; has units of grams per cubic centimeter or kilograms per cubic meter in SI units.

Density filter   Less dense crust blocks ascent of denser mafic mantle-derived magmas; see **Underplating.**

Depleted source   Magma-generating region, usually with reference to the peridotitic mantle, that has relatively low concentrations of the most incompatible elements, usually because of a previous episode of partial melting.

Deuteric alteration   A type of **alteration** that proceeds more or less automatically during cooling of a magmatic body in the presence of its own aqueous fluids.

Devitrification   Delayed crystallization of glass.

Diabase   Basaltic rock in which grain size is marginally phaneritic; also known as **dolerite;** typically occurs in dikes and sills.

Diapir   Body of buoyant magma that ascends slowly through surrounding ductile, viscous country rock in the lower continental crust or mantle.

Diatreme   A funnel-shaped **breccia pipe;** many are apparently emplaced at low *T* by volatile-rich, highly alkaline, mafic to ultramafic magmas, such as kimberlite.

Differentiation   In a magma, refers to processes that modify the composition of a primary magma after it leaves its source.

Diffusion   Movement of individual atoms through a group of atoms, usually driven by a concentration gradient.

Diffusion coefficient (diffusivity)   Proportionality constant between rate of diffusion and concentration gradient; related to the frequency at which atoms jump and their jumping distance.

Dihedral (wetting) angle   Angle between intersecting liquid/crystal interfaces.

Dike   A type of **sheet intrusion** that cuts discordantly across planar wall rock structure, such as bedding; also, a sheet intrusion in massive isotropic magmatic rock.

Dike swarm   Several to as many as hundreds of dikes emplaced more or less contemporaneously during an intrusive episode; in a **radial dike swarm** the dikes are arrayed like spokes of a wheel from a central point.

Diktytaxitic   Texture in some coarse-grained basalt lava flows; consists of small angular vugs interspersed pervasively among slightly larger plagioclase and pyroxene grains.

Dilatancy pumping   A deforming body may dilate (increase in volume) by developing internal openings, into which any nearby liquid may be drawn.

Diorite   Felsic, phaneritic, magmatic rock that contains, in terms of felsic minerals, <20 modal % quartz and has a plagioclase/alkali feldspar >9 and plagioclase less calcic than An_{50}; see Figure 2.8.

Discontinuous reaction series   A sequence of two or more **reaction pairs** that develop in a magma undergoing fractional crystallization; for example, clinopyroxene-hornblende-biotite.

Dissolution (resorption)   Crystal taken into solution in a liquid.

Divariant   A state of equilibrium that has two degrees of freedom.

Dolerite   See **Diabase.**

Ductile deformation (ductile flow)   Distributed, grain-scale permanent change in shape of a body without loss of cohesion.

Dunite   Ultramafic, phaneritic, magmatic rock that contains >90% olivine; see Figure 2.10b.

E-MORB   Derived from a mantle source enriched in incompatible trace elements relative to the source of normal mid-ocean ridge basalt **(N-MORB).**

Eclogite   Dense phaneritic rock of basaltic chemical composition formed at high P that is made essentially of red pyropic garnet and green omphacite clinopyroxene.

Effective normal stress   In a porous or cracked rock body, is the normal stress compressing the body minus the counteracting pressure of a liquid (volatile fluid or magma) in the openings.

Effusive eruption   Magma extruded from a volcanic vent as a coherent lava flow, as contrasted with an explosive eruption of **pyroclasts.**

Ejecta   See **Pyroclast.**

Elastic   A behavior or response to applied stress in which the strained body returns instantaneously to its initial undeformed state so long as the stress is less than the **elastic strength.**

Elastic strength   The stress difference in an elastic body at which it breaks or fractures, causing permanent deformation.

Elutriation   Entrainment of very fine ash particles in upward streaming gas in a **pyroclastic flow;** creates a **coignimbrite plume.**

En echelon dikes   A series of stepped separate dikes that are parallel to each other but oblique to the series as a whole.

Enclave   See **Mafic inclusion.**

Enderbite   Pyroxene tonalite; a type of **charnockitic rock.**

Endogenous   Grown from within, as a silicic lava dome.

Endothermic   A chemical reaction that absorbs heat.

Energy   Capacity for doing work. **Gravitational potential energy** of a mass is related to its position in the gravitational field of the Earth. **Kinetic energy** is possessed by a moving mass. **Thermal energy** resides in the motions and interactions of atoms internally within a body and is commonly referred to as **heat.**

Enthalpy   The amount of heat gained or lost in a system; roughly, the heat content of a system.

Enthalpy of melting   See **Latent heat of melting.**

Entropy   Uniformity in concentration of energy in a system or the degree of disorder in a system.

Equigranular   Phaneritic texture in which grains are of similar size; also called **granular.**

Equilibrium   State of a system in which the net result of forces acting on it is zero; there is no change in the system. **Metastable equilibrium** is a state that can be changed by some perturbation to a more stable, lower-energy state by surmounting an energy barrier. **Stable equilibrium** is a state that, if perturbed, readily returns to its original state of lowest energy. **Unstable equilibrium** is a state in which a slight disturbance causes a permanent change, usually to a more stable state of lower energy.

Equilibrium boundary line   Locus of values of intensive variables in a phase diagram where phases in adjacent stability fields coexist stably.

Equilibrium constant   Expresses the ratio of activities of components in a chemical reaction.

Eu anomaly   On a **chondrite-** or **primitive-mantle-normalized diagram,** Eu is either enriched (positive anomaly) or depleted (negative anomaly) relative to a smooth curve drawn through values of neighboring lighter and heavier rare earth elements.

Euhedral   Mineral grain completely bounded by its own rational crystal faces, forming a tabular, platy, columnar, or other habit; commonly, but not necessarily, formed by unrestricted growth in a liquid; also referred to as **idiomorphic.**

Eutaxitic fabric   Made of flattened, welded vitroclasts, defining a **compaction foliation** more or less parallel to the depositional surface of the pyroclastic-flow deposit in which the fabric is found; flattened pumice lapilli called **fiamme** (Italian, "flame") have ragged flamelike terminations; *not* a type of flow layering.

Eutectic   A system that consists of two or more crystalline phases plus a melt whose composition can be expressed in terms of proportions of the crystalline phases; all phases coexist in equilibrium at an isobaric invariant point (**eutectic point**) that is the minimum melting T (**eutectic T**) for the assemblage of crystalline phases; removal of heat causes a decrease in the proportion of melt to crystals but does not change the T of the system or the composition of any phase.

Exogenous   Grown by addition onto the surface.

Exothermic   A chemical reaction that releases heat.

Exsolution   Process of releasing excess volatiles from an oversaturated melt as their concentration exceeds their solubility.

Extension   Stretching or elongation of a body.

Extensional fractures   Cracks formed under brittle conditions that are more or less planar and oriented perpendicular to the least principal stress, σ_3; because they open in a direction perpendicular to the fracture they are readily filled with volatile fluid or magma that solidifies as veins or dikes.

Extensive variables   Parameters defining the nature of a system; depend on the amount of material in the system, such as volume and density.

Extrusion   The process and the product of magma emplaced onto the surface of the Earth, either as coherent lava or as pyroclastic material.

Fabric   Noncompositional properties of a rock, including textures and generally larger-scale structures. **Textures (microstructures)** are proportions of glass relative to mineral grains and their sizes, shapes, and mutual arrangements that are observable on a scale of hand sample or thin section under the microscope. **Structures** are generally seen in an outcrop and include, for example, bedding in a pyroclastic deposit or pillows in a submarine lava flow.

Facies   A chemically diverse group of rocks formed under similar conditions.

Fault (shear fracture)   A fracture along which parallel movement occurs when the shear stress on the fracture exceeds the frictional resistance; contrasts **extensional fracture.**

Feeder dike   Subvertical dike that supplies magma to an overlying intrusion or volcano.

Feldspar ternary   A system that comprises the three feldspar components, $NaAlSi_3O_8$-$KAlSi_3O_8$-$CaAl_2Si_2O_8$.

Felsic   Mnemonic adjective derived from *feldspar* and *silica*; used to describe a rock that contains abundant feldspar and either quartz or feldspathoids, such as granite; also the magma that contains abundant feldspar components.

Felsitic   Texture made of an aphanitic mosaic of mostly **felsic** minerals found commonly in rhyolite, dacite, and trachyte.

Felty   Holocrystalline, **microcrystalline** magmatic texture that consists of small interwoven plagioclase tablets; sometimes referred to as pilotaxitic texture.

Fenite   A metasomatic rock associated with **carbonatite,** ijolite, and other silica-poor rocks; composed of alkali feldspar and alkali amphibole and/or pyroxene.

Fertile (fertility)   See **Source-rock fertility.**

Fiamme   See **Eutaxitic fabric.**

Field relations   Properties and chronologic relations of a mass of rock discerned in exposures (outcrops) in hills, mountains, and human-made road cuts.

Filter pressing   Segregation of residual melt in partially crystallized magmas from the interlocking network of crystals caused by local gradients in pressure.

First law of thermodynamics   The total amount of energy remains constant in a system; also called the *law of conservation of energy.*

Fissure eruption   Magma vented from a subvertical crack or fracture.

Flow layering   Layered fabric in a magmatic rock expressed by alternating planar contrasts in composition, in sizes of crystals, or in concentration of vesicles; forms as a result of magma flow.

Flow markers   Suspended rigid crystals, volatile-fluid bubbles, or distinctive mineral aggregates embedded in magma that record attributes of the pattern of internal flow.

Flowage segregation   In flowing bodies of magma, grain-dispersive pressure pushes crystals and other solid particles away from the margins into the interior.

Fluid   See **Volatile fluid.**

Fluid-absent melting   See **Dehydration melting.**

Fluid pressure   The pressure exerted by volatile fluid in a system.

Fluidization   Mobilization of a **pyroclastic flow** due to upward movement of gas through the mass of cohesionless particles, which lifts them apart so that the flow behaves as if it were a low-viscosity fluid.

Foliation (foliated fabric)   Planar anisotropic fabric, such as various types of layering and preferred planar orientation of inequant mineral grains.

Forceful emplacement   Intrusion of magma that deforms country rock, resulting in anisotropic fabric that is more or less concordant to intrusive contact.

Fractional crystallization   See **Crystal-melt fractionation.**

Fractionation   See **Crystal-melt fractionation.**

Fugacity   Equivalent or effective partial pressure that is used to calculate free energy differences for real gases.

Fugacity coefficient   The ratio of **fugacity** to partial pressure; the measure of the departure of a real gas from a perfect one.

Fulgurite   Glassy rock that is formed by lightning strikes that locally melt rock or sediment; usually irregularly shaped crust or tubular structure that can be as much as 40 cm long and 6 cm in diameter.

Fumarole   A vent that releases volcanic gases.

Fused tuff   Usually ash-sized, initially cold and unconsolidated vitroclasts that are heated in some manner, such as by an overlying lava flow, so they become hot and stick together, forming a consolidated rock.

Ga   Giga annum = 10^9 years before present.

Gabbro   Mafic, phaneritic, magmatic rock type that contains 10–90 modal % plagioclase more calcic than An_{50} plus pyroxene, Fe-Ti oxides, with or without olivine; see Figure 2.10a; chemically equivalent to basalt.

Geobaric gradient   Rate at which P increases into the interior of the Earth.

Geobarometer   A mineral or assemblage of stably coexisting minerals whose chemical composition is a sensitive function of the P of crystallization.

Geothermal reservoir   Underground high-T water lodged in open spaces within rock; potential source of energy to create electric power.

Geothermal gradient (geotherm)   Rate at which T increases into the Earth.

Geothermometer   A mineral or assemblage of stably coexisting minerals whose chemical composition is sensitive to the *T* of crystallization.

Gibbs free energy   A thermodynamic energy function formulated in such a way that it is a minimum in a stable system of constant chemical composition.

Glass   An amorphous, usually silicate solid whose disordered atomic structure is that of a liquid; formed from a silicate melt that cooled too rapidly to crystallize.

Glassy (vitric)   A texture consisting of some proportion of glass.

Grain-boundary (intergranular) diffusion   Occurs along boundaries of mineral grains in an aggregate.

Grain dispersive pressure   In a flowing liquid, suspended solid particles in the velocity gradient near the fixed boundary tend to migrate into the flowing liquid.

Granite   Felsic phaneritic rock that contains, relative to felsic minerals, >20 modal % quartz and subequal amounts of alkali feldspar and plagioclase; see Figures 2.8 and 2.9.

Granite system   The $NaAlSi_3O_8$-$KAlSi_3O_8$-SiO_2-H_2O system.

Granitic rocks   Quartz-bearing phaneritic felsic rock; also called **granitoids**.

Granitoids   See **Granitic rocks**.

Granodiorite   Felsic phaneritic rock that contains, relative to felsic minerals, >20 modal % quartz and plagioclase > alkali feldspar; chemical composition similar to dacite; see Figures 2.8 and 2.9.

Granophyre   Felsic rock that has **granophyric** texture.

Granophyric   See **Graphic**.

Granular   See **Equigranular**.

Granulite   Metamorphic rock that consists of a mostly anhydrous mineral assemblage (such as plagioclase + pyroxene ± garnet); some may be crystalline residue in the deep continental crust left after extraction of granitic partial melts.

Graphic   Magmatic texture that consists of an intergrowth of alkali feldspar and quartz, the latter in triangular and hooklike forms resembling ancient writing; can be coarse where the quartz grains are several millimeters, or microcrystalline, visible only with a microscope, when it is called **micrographic** or **granophyric**.

Gravitational segregation   In a static magma, separation of melt and crystals because of contrasts in their density.

Greenstone belt   A usually Archean sequence of volcanic and lesser sedimentary rocks that has been variably deformed and metamorphosed generally at low *T* to green mineral assemblages that include epidote, actinolite, chlorite, and serpentine; dominant volcanic rock is subaqueous tholeiitic basalt.

Groundmass   See **Matrix**.

Harzburgite   Ultramafic phaneritic rock composed of orthopyroxene, 40%–90% olivine, and <10% clinopyroxene; a type of depleted mantle **peridotite**; see Figure 2.10b.

Hawaiian eruption   Extrusion of low-viscosity basaltic magma from fissures or central vents, creating lava fountains and sheetlike lava flows.

Hawaiite   Approximately, a sodic andesite; see Figure 2.12.

Heat   Transferred thermal energy.

Heat flux (flow)   Rate at which heat is transferred over time from surface of a body.

Heat conduction   See **Conduction**.

Heteromorphs (fabric)   Different fabrics formed by different kinetic paths in a rock or magma of the same composition; for example, aphanitic and vitrophyric rhyolite.

HFS   See **High-field strength element**.

High-field-strength (HFS) element   Element whose ion has a charge greater than or equal to 3 and radii = 0.7 − 1.1 Å, including Zr, Ti, Nb, and Ta.

High-K rocks   Calc-alkaline rocks that contain high K; see Figure 2.18.

Holocrystalline   Texture made wholly of crystals.

Horizon of neutral buoyancy   Level of density-stratified crust that is equal to density of magma.

Hyaloclastite   See **Hydroclasts**.

Hybrid magma (rock)   See **Assimilation** and **Mixed**.

Hydraulic fracturing   Occurs where volatile fluid or magma pressure is sufficiently large in rock openings that it can overcome compressive normal stresses and produce extensional fractures oriented perpendicular to the least principal stress, σ_3, into which the fluid or magma penetrates, forming veins or dikes.

Hydroclasts   Fragmental material produced by **hydromagmatic eruptions;** if composed of glassy fragments, the deposit is a **hyaloclastite**.

Hydroclastic deposit   See **Hydroclast**.

Hydromagmatic eruption   Interaction between magma and external water, as in the ground, a lake, or the ocean, that is explosively vaporized to steam.

Hydrostatic state of stress   Three **principal stresses** are of equal magnitude in all directions; also called **pressure**.

Hydrothermal alteration   Associated with advective percolation of hot hydrothermal solutions through a body, especially along fractures.

Hydrothermal solution   Hot aqueous fluid beneath the surface of the Earth; contains dissolved silica and many other chemical constituents.

Hypabyssal   Processes and products related to magma in the shallow continental crust.

Hypersolvus   Refers to granites and syenites that have crystallized from relatively dry magmas so that the feldspar is perthite and accompanying mafic minerals are commonly anhydrous; the perthite is a product of exsolution of an initially homogeneous high-*T* feldspar that crystallized from a melt and unmixed below the solvus.

Hypidiomorphic-granular   Magmatic phaneritic texture that consists of a mixture of euhedral, subhedral, and anhedral grains.

I-type   Refers to granitic rocks derived from magmas generated by partially melting igneous source rocks.

Idiomorphic   See **Euhedral**.

Igneous lamination   Texture in a magmatic phaneritic rock expressed by a planar preferred orientation of tabular feldspars.

Ignimbrite   See **Ash-flow tuff.**

Ijolite   Phaneritic nepheline-clinopyroxene rock; chemically equivalent to nephelinite.

Immiscible liquids   Two melts of different composition that coexist stably at some particular P and T.

Incompatible element   Element that is preferentially contained in loosely structured melt and excluded from more restrictive crystalline structures; **partition coefficient** <1.

Incongruent melting   A heated crystalline phase yields a melt plus a crystalline phase, neither of which has the same composition as that of the melted crystal.

Inequant   Refers to a mineral grain whose dimensions are unequal in different directions, for example, a platy or columnar grain.

Inequigranular   Texture in which grains are conspicuously variable in size.

Intensive variable   Parameter defining the nature of a system that is independent of the amount of material present and has a definite value at each point within the system; includes T, P, and concentration of chemical species.

Intercumulus   Melt or minerals in the space between **cumulus** mineral grains.

Interface   The boundary of any two neighboring phases.

Intergranular   Holocrystalline, **microcrystalline** texture in which randomly oriented crystals of plagioclase, pyroxene, and Fe-Ti oxides form a tight, interlocking mosaic.

Intermediate rock   Contains 52–66 wt.% silica.

Intersertal   Magmatic texture resembling **intergranular** texture except that brown glass is interspersed among the microcrystalline grains.

Intrusion   Process and product of magma insertion into preexisting rock beneath the surface of the Earth; compare **Pluton.**

Invariant   A state of equilibrium that has no degrees of freedom.

Ionic potential   Ratio of ionic charge to ionic radius.

Irreversible thermodynamic process   A unidirectional change in a system from an initial metastable state to a more stable, lower-energy final state brought about by a measurable change in an intensive variable.

Island arc   Arcuate string of volcanic islands built on oceanic crust overlying subducting oceanic lithosphere.

Isobaric   Constant pressure.

Isolated system   No matter or energy can be transferred across the system boundary with surroundings and no work can be done on or by the system.

Isopleth   A line of constant composition in a T-X or P-X phase diagram.

Isothermal line or **surface**   A line or surface of constant T; also called an **isotherm.**

Isothermal compressibility (coefficient of)   Isothermal change in volume of a body as P increases with depth in the Earth.

Isotopes   Atoms of an element whose nuclei contain the same number of protons but different numbers of neutrons. **Radiogenic** or **radioactive isotopes** are unstable and decay by nuclear processes into daughter isotopes that may be of the same or a different element. **Stable isotopes** do not decay. **Cosmogenic isotopes** are produced when high-energy cosmic rays interact with nuclei of atoms in the atmosphere.

Isotope exchange reaction T-dependent reaction involving isotopes of a particular element that change ratios in one or more phases; commonly occurs in the presence of some kind of liquid in which the isotopes can move about freely.

Isotope fractionation   Change in isotopic ratio of an element in a particular phase that occurs during a physical or chemical process because one isotope is preferentially incorporated at the expense of another isotope.

Isotropic fabric   Random fabric, appears the same in all directions.

IUGS   International Union of Geological Sciences; a Subcommission on the Systematics of Igneous Rocks created the classification used in this textbook.

Jotunite   Pyroxene monzodiorite; a type of **charnockitic rock.**

Juvenile   Refers to particles or fluid derived directly from magma.

Kamafugite   A clan of strongly silica-undersaturated volcanic rocks that contain combinations of olivine, clinopyroxene, Fe-Mg mica, perovskite, and (in katungite) melilite, (mafurite) leucite, or (ugandite) kalsilite ($KAlSiO_4$).

Kimberlite   See Section 13.12.2.

Kinetic path   Time-dependent history of a change in state of a rock-forming system; largely responsible for the fabric of the resulting rock.

Kinetics   Time-dependent dynamics of a system.

Komatiite   Glassy to aphanitic ultramafic rock formed from lava or shallow intrusions that contains >18 wt.% MgO and <1 wt.% ($Na_2O + K_2O$); composed essentially of olivine and pyroxene; chemically similar to peridotite.

Laccolith   Flat-floored intrusion with a domical upper surface; essentially concordant with the layered country rocks.

Lahar   Mass of intimately mixed water and poorly sorted rock material moving by gravity down the slopes of a volcano; also called a **volcanic debris flow;** also refers to the resulting deposit.

Laminar flow   Movement of a liquid in parallel sheets or "pencils," depending on the configuration of the liquid boundaries; **Reynolds number** <500.

Lamproite   See Section 13.12.2.

Lamprophyre   A generally mafic, alkaline, porphyritic dike rock that contains abundant mica, amphibole, clinopyroxene, and possibly olivine and melilite; feldspar may be present but never as phenocrysts.

Lapilli   Volcaniclasts 2–64 mm in size; equivalent to granule and pebble size for sedimentary fragments.

Lapilli tuff   Rock made of a mixture of ash- and lapilli-size volcaniclasts.

Large igneous province Large floods of basaltic and locally rhyolitic lava extruded during relatively brief episodes; occur as continental and submarine (basaltic only) oceanic plateaus.

Large-ion lithophile (LIL) element Element whose ions are mono- or divalent and have radii greater than 1.15 Å; includes Cs, Rb, K, Ba, U, and Th.

Latent heat (enthalpy) of melting The amount of heat absorbed at constant T during melting.

Lava Cohesive magma extruded from a volcanic vent.

Lava dome Lava extrusion having a relatively large **aspect ratio,** usually as a result of high apparent viscosity of a silicic lava; has a blocky autoclastic carapace and talus apron; a steep-sided **pelean dome** grows by expansion from within (**endogenously**), pushing slabs of rigid lava out of and away from the vent in fan fashion or along sled-runner-shaped ramps.

Lava fountain Blobs of relatively low-viscosity magma, usually of basaltic composition, ejected from an explosive volcanic vent.

Le Chatelier's principle If a change occurs in the state of a system, such as an increase in P, the system will respond in such a way as to minimize or moderate the effects of the change, such as by forming more compact, denser phases.

Leucitite Glassy to aphanitic rock made of leucite, clinopyroxene, Fe-Ti oxides, and possible olivine.

Leucocratic Refers to rocks that have <30 modal % mafic minerals.

Lever rule An equation (5.2) that expresses the weight or mole fractions of two stably coexisting phases in a phase diagram.

Lherzolite A type of peridotite composed of orthopyroxene and clinopyroxene and 40%–90% olivine; a widespread, fertile mantle source rock; see Figure 2.10b.

LIL See **Large-ion lithophile element.**

Lineation (lineated fabric) Linear fabric, such as preferred orientation of columnar mineral grains.

Liquid line of descent In a magma, the changing chemical composition of the residual melt during crystallization.

Liquids In this textbook includes volatile fluids and high-T silicate or carbonatite melts.

Liquidus line (surface) In a phase diagram, T above which any mixture of components in an equilibrium system is wholly melted; the locus of points representing the composition of melt coexisting stably with solid phases at some particular P and T.

Liquidus phase The first crystalline phase to precipitate in a melt as it cools below the liquidus T.

Lithic clast Rock fragment in volcaniclastic deposit or rock; typically lapilli or block size.

Lithophysae In silicic lava flows and compacted tuffs, hollow gas cavities on rock surfaces that have concentric, bubblelike microcrystalline shells.

Lithophysal Fabric made of **lithophysae.**

Littoral cone A type of **hydroclastic deposit** created where subaerially erupted lava contacts the ocean.

LOI See **Loss on ignition.**

Loss on ignition (LOI) Weight percentage (wt.%) of volatiles lost when rock powder is heated to 1000°C.

Low-K rocks Essentially synonymous with **tholeiitic rocks;** see Figure 2.18.

Ma Mega annum = 10^6 years before present.

Maar A broad, low-relief explosive volcanic crater whose floor lies below the general elevation of the preeruption land surface; is surrounded by a **tuff ring;** may be underlain by a **diatreme.**

Mafic Mnemonic adjective derived from *magnesium* and *ferrous*/**ferric;** used to describe a mineral or a rock that contains large concentrations of Mg and Fe.

Mafic inclusion (Enclave) Mafic aggregate embedded in a more leucocratic, felsic host rock; especially common in granodiorite, quartz diorite, and tonalite.

Magma Molten rock material; always consists of a melt that contains dissolved volatiles which may be exsolved into bubbles; suspended crystals generally present.

Magma overpressure Magma pressure that exceeds the confining (lithostatic) pressure, P, at a particular depth.

Major element A chemical constituent in a rock that, if expressed as an oxide, is >0.1 wt.%; usually includes Si, Al, Fe, Ca.

Mangerite Pyroxene monzonite; see **Charnockitic rocks.**

Mantle bedding Thin layers of ash or lapilli in pyroclastic-fall deposits created as tephra settles uniformly onto ground, like fallen snow.

Mantle plume Column of relatively hotter and less dense mantle that rises buoyantly toward the base of the lithosphere from the deep mantle.

Mantle reservoir Source of basaltic partial melts having distinct trace element and isotopic signatures.

Mass-balance equation Expresses the weight or mole fraction of phases and concentrations of constituents that are involved in a chemical reaction in a closed system.

Massif A massive, structurally distinct, commonly topographically elevated block of rock in a mountainous terrain.

Matrix Smaller grains (either **aphanitic** or **phaneritic**) surrounding larger ones (**phenocrysts**) that crystallized from magma, together defining **porphyritic** texture; matrix is also known as **groundmass.**

Medium-K rocks Calc-alkaline rocks that contain intermediate concentrations of K; see Figure 2.18.

Megacryst Exceptionally large crystal.

Melilitite Highly alkaline, strongly silica-undersaturated glassy or aphanitic rock composed essentially of melilite and clinopyroxene; see Figure 2.12.

Melt High-T liquid solution of ions; essential part of magma; also called silicate (carbonatite) liquid.

Melt fraction Proportion of partial melt in a source rock or of melt in a magma; can be expressed as a percentage.

Melt inclusion Minute volume of melt enclosed within a crystal; commonly forms by entrapment during skeletal growth of the crystal; quenches to an inclusion of glass.

Metaluminous rock   Alumina deficient rock in which **alumina saturation index** (mole basis) $Al_2O_3/(K_2O + Na_2O + CaO) < 1$; contains Al-poor biotite and hornblende and normative anorthite and diopside; usually refers to felsic rocks.

Metasomatism   Subsolidus modification of the chemical composition of a rock by invasive percolating liquids. In **cryptic metasomatism** the original solid-solution minerals remain but are changed in composition. In **modal metasomatism** original minerals are replaced by entirely new minerals.

Mg number   Atomic $100 \, Mg/(Mg + Fe)$ where Fe is total iron.

Miarolitic   In granitic rocks, texture defined by irregularly shaped cavities called **vugs** into which euhedral vapor-phase crystals have grown; cavities are generally many grain diameters apart.

Microcrystalline   Texture that consists of a mosaic of crystals that are visible only under a microscope.

Micrographic   See **Graphic.**

Microlites   Small crystals that display optical birefringence in polarized light; larger than **crystallites.**

Microstructure   See **Fabric.**

Migmatite   Layers, pods, and irregularly shaped masses of granite that are intimately mingled with mafic metamorphic rock; originate by partial melting and segregation of minimum-T-composition melts that solidify in situ before moving and collecting into a larger intrusive mass, by injection of magma from extraneous sources, and possibly by subsolidus metamorphic processes.

Mineral assemblage   Two or more minerals that coexist stably in a rock, such as quartz, feldspar, and biotite.

Mineralogical composition   Types and chemical compositions of minerals constituting a rock.

Minette   A common type of **lamprophyre** composed of abundant phenocrystic and groundmass biotite-phlogopite, alkali feldspar > plagioclase; other mafic minerals include clinopyroxene and possibly olivine.

Mingled   Refers to a **hybrid rock** or the magma from which it formed that has compositionally contrasting parts produced by physically merged dissimilar magmas.

Miscibility gap   Compositional range between stably coexisting phases across a solvus at a particular P and T.

Mixed   Refers to a **hybrid rock** or the magma from which it formed by homogenization of blended dissimilar magmas; phenocrysts may show disequilibrium textures and compositions.

Mobility   The ease with which an element can move about in a rock system; may depend on the solubility of an ion in aqueous solutions that is approximately proportional to ionic potential.

Modal metasomatism   See **Metasomatism.**

Modal composition (mode)   Volumetric proportions, usually expressed as a percentage, of the minerals constituting rock.

Mode   See **Modal composition.**

Model   A representation of a real system.

Molar heat capacity   The ratio of heat transferred into or out of a body to the incremental rise in its temperature; has units of joules per mole degree; compare **Specific heat.**

Molar volume   Inverse of density, or the volume of one mole of a material.

Mole fraction   The proportion of one component relative to all other components in a phase or system.

Monogenetic   Resulting from one process or having one source or originating at one time in one place, such as a volcano built in one extrusive episode.

Monomineralic   Consisting of one mineral.

Monzonite   A **granitoid** that contains subequal modal proportions of plagioclase and alkali feldspar and less quartz than granite

MORB   Mid–oceanic ridge basalt; constitutes the upper part of the oceanic crust; also called abyssal tholeiite, and seafloor, ocean-floor, and submarine basalt.

Mugearite   See Figure 2.12.

My 10^6 years.

Myrmekite   In granitic rocks, microcrystalline texture that consists of an intergrowth of vermicular ("wormy") quartz in a sodic plagioclase host; commonly in contact with K-rich alkali feldspar.

N-MORB   Normal **MORB;** derived from a **mantle reservoir** relatively depleted in incompatible elements.

Negative Nb-Ta-Ti anomaly   Depletion of these high-field-stress elements relative to adjacent more and less compatible elements in a normalized trace element pattern.

Nelsonite   Extremely rare rock made mostly of Fe-Ti oxides and apatite.

Nephelinite   Highly alkaline, strongly silica-undersaturated glassy or aphanitic rock composed essentially of nepheline and clinopyroxene and possible olivine; see Figure 2.12.

Network-forming cation   In a melt, Si and other small highly charged ions, such as Al and Ti, linked to O, forming a polymer.

Network-modifying cation   In a melt, a generally larger ion of lesser charge, such as Ca, Mg, or K, that is not part of a polymer.

Newtonian viscosity   See **Viscosity.**

Non-Newtonian   Compound rheologic behavior that has components of plasticity as well as viscosity; strain rate is not directly proportional to applied shear stress, as in **Newtonian viscosity.**

Nonbridging oxygen   In a melt, an O ion that is not part of a polymer and bonds with **network-modifying cations.**

Nonhydrostatic state of stress   Three principal stresses are unequal; can change the shape of a body.

Norite   A type of gabbro that has more orthopyroxene than clinopyroxene.

Norm   See **Normative composition.**

Normal stress   See **Stress.**

Normalized trace element diagram   A plot of concentration ratios of trace elements relative to chondrite or primitive mantle, usually arranged according to incompatibility.

Normative mineral   See **Normative composition.**

Normative composition (norm)   Assemblage of hypothetical water-free, standard **normative minerals** calculated from the chemical composition of a rock according to the rules outlined in Appendix B.

Nucleation   Process of formation of an embryonic cluster of ions, called a **nucleus,** during formation of a new phase in a changing system.

Nucleation density   Number of nuclei in a volume of melt.

Nucleus   See **Nucleation.**

Nuée ardente   See **Pyroclastic flow.**

Oceanic plateau   Broad topographic high made usually of basaltic rocks that rises 1 km or so above the surrounding seafloor and underlain by a crust as much as 40 km thick.

Ocelli   See **Variole.**

OIB   Oceanic island basalt.

Oikocryst   Literally, house crystal; surrounds many smaller grains in a magmatic rock; see **poikilitic** texture.

Olivine tholeiite   Basalt rock type that contains normative olivine and hypersthene.

Opdalite   Pyroxene granodiorite; a type of **charnockitic rock.**

Open system   Matter and energy can flow across the system boundary and work can be done on and by the system.

Ophiolite   A sequence of variably altered and usually tectonically fragmented oceanic rocks that includes, from the top down, marine sedimentary rock; volcanic rock (mostly basaltic pillow lava); **sheeted-dike complex** of basaltic rock; gabbro; cumulate ultramafic rocks; and metamorphic-textured mantle peridotite; is a segment of oceanic crust and underlying mantle tectonically emplaced onto overriding plate in a subduction zone.

Ophitic   Magmatic texture in which large clinopyroxenes partially to completely enclose smaller euhedral plagioclases; occurs in diabase and gabbro.

Orangeite   See Section 13.12.2.

Order of magnitude   Factor of 10.

Oscillatory zoning   Thin concentric zones in a solid-solution crystal of alternating enrichment and depletion in an end member component; commonly occurs in plagioclase where the zones are alternately enriched and depleted in the anorthite end member.

Ostwald ripening   Subsolidus process of increasing crystal size at the expense of less stable smaller crystals.

Ostwald's step rule   In a change of state, the kinetically most favored phase (or phases) forms before the more stable one of least possible free energy.

Overpressured system   State of disequilibrium in which the pressure of a magma system, or its volatile pressure, exceeds the confining pressure.

Oversaturated   The concentration of a component in a phase exceeds its solubility at some P and T so that a separate phase that contains the component forms; also refers to the concentration of silica and alumina in magmas and rocks; see **Silica saturation.**

Overstepping   A phenomenon caused by sluggish kinetic rates, especially of nucleation, in phase transitions where the change in the intensive variable must exceed the equilibrium value to induce growth of the new, more stable phase; for example, to freeze water the T must generally overstep or cool below 0°C.

P   See **Confining pressure.**

Pahoehoe lava flow   Usually formed of low-viscosity basalt lava that contains abundant subspherical volatile bubbles; consists of thin, glassy sheets, tongues and lobes, commonly overlapping one another, that are fed from distributary tubes.

Palagonite   Chemically altered basaltic glass; consists of a complex mixture of clay and zeolite minerals and hydrated ferric oxides; variably orange to brown; isotropic to weakly birefringent in thin section; commonly concentrically layered in colloform fashion.

Pantellerite   Peralkaline rock in which $Al_2O_3 < 1.33$ FeO $+ 4.4$ (wt.%).

Parasitic cone   Formed where magma is extruded from a flanking central vent on a **composite volcano** or **shield volcano.**

Parent magma   The beginning magma from which others are derived by differentiation.

Partial melting   Generation of a melt that is less than the whole rock; also called **anatexis. In equilibrium** or **batch partial melting** equilibrium is maintained between crystals and melt in the closed system. In **fractional partial melting** melt is separated from the residual crystals.

Partial molar volume   The change in molar volume of a phase that results when a small amount of one component is added to or taken away from the phase.

Partial pressure   The pressure of one gas component in a mixture of many.

Partially resorbed   Refers to an unstable, anhedral remnant of a crystal that was dissolving into the melt as magmatic conditions changed.

Partition coefficient   The ratio of the concentration of an element in a mineral to that in the stably coexisting liquid. **Bulk partition coefficient,** for a particular element, is the partition coefficient for each mineral multiplied by the mole fraction of the mineral in the rock, summed for all minerals.

Passive emplacement   Intrusion of magma that results in little deformation of country rock.

Pegmatite   Textural variety of leucocratic granite or, locally, of more mafic rock of highly variable grain size in which individual crystals can be as much as several meters. **Simple pegmatite** consists of albite, quartz, perthite, and possible minor muscovite, tourmaline, and Fe-Mn garnet; **zoned pegmatite** has a usually imperfect zonation of fabric and mineralogical composition concentric to margins of body; **complex pegmatite** has relatively high concentrations of incompatible elements that stabilize exotic minerals such as spodumene, pyrochlore, and cassiterite.

Pegmatitic   Fabric that consists of exceptionally large, but heterogeneously sized crystals, at least several centimeters to as much as several meters.

Pelean dome   See **Lava dome.**

Peralkaline rocks   Metaluminous rocks that have (mole basis) $Al_2O_3/(K_2O + Na_2O) < 1$; excess of alkalies relative to Al_2O_3 is manifest in alkali-rich mafic minerals and normative acmite or sodium metasilicate but no normative anorthite in the rocks.

Peraluminous rocks   Alumina-oversaturated felsic rock suite whose **alumina saturation index** (mole basis) $Al_2O_3/(K_2O + Na_2O + CaO) > 1$; excess Al_2O_3 is lodged in aluminous minerals such as muscovite, garnet, and cordierite.

Perfect equilibrium crystallization   Growing crystals react and reequilibrate completely with a melt as *P-T-X* conditions change; solid-solution crystals are homogeneous.

Perfect fractional crystallization   Crystals are immediately isolated, removed, or fractionated from the melt as soon as they form so that no crystal-melt reaction relations occur.

Peridotite   Ultramafic phaneritic rock composed of pyroxene and 40%–90% olivine; see Figure 2.10b.

Peritectic   An invariant reaction point in a cooling system at which a reaction relation ensues between crystals and melt to yield crystals of a different composition; same as an incongruent melting point in a heating system.

Perlite   Hydrated silicic glass that has **perlitic** texture; typically pearl-gray color, but may be green, red, or brown.

Perlitic   Texture made of concentric cracks in silicic glass; formed by hydration under subaerial conditions.

Permeability   Rate at which fluid moves through interconnected openings in rock.

Petrogenesis   Rock origin.

Petrogeny's residua system   In the $NaAlSi_3O_8$-$KAlSi_3O_8$-SiO_2-H_2O "granite" system, the minimum *T* melt and coexisting subequal proportions of alkali feldspar and quartz; corresponds to the composition of most naturally occurring granites.

Petrography   Observable properties of a rock, including its modal composition and fabric, from which a rock name is derived.

Petrology   The science of rocks, their nature and origin.

Petrotectonic association   Specific types of rocks and rock suites that occur together in a specific tectonic regime.

PGE   See **Platinum group element.**

Phaneritic   Texture in which grains of major rock-forming minerals are all large enough to be identifiable without magnification.

Phase   A part of a system that is chemically and physically homogeneous and bounded by a distinct interface with adjacent phases; may be solid, liquid, or gas.

Phase diagram   Shows which phase or phases in a system are more stable with respect to *T, P,* and concentration of chemical components; also called a **stability diagram.**

Phase layering   Abrupt appearance or disappearance of a particular mineral in a layered intrusion.

Phase rule   Indicates the number of intensive variables that may independently vary without changing the number of phases in a system at equilibrium.

Phenoclast   Irregularly shaped fragment of quartz or feldspar in pyroclastic flow deposits formed as crystals containing overpressured melt inclusions blown apart during eruption.

Phenocryst   Larger crystal precipitated from a melt embedded in finer-grained or glassy matrix.

Phonolite   Highly alkaline, leucocratic, intermediate-silica aphanitic or glassy rock composed of feldspathoids, alkali feldspar, and minor mafic minerals; see Figures 2.12 and 2.13.

Phreatic explosion   Vaporization of groundwater to steam due to contact with hot rock.

Phreatoplinian eruption   Highly explosive venting of volatile-rich, usually silicic magma enhanced by contact with external water that vaporizes to steam.

Picrite   An olivine-rich basalt having $MgO > 18$ wt.% and $(Na_2O + K_2O) = 1$–3 wt.%.

Pipe vesicle   See **Vesicle.**

Pipe (plug)   Slender, subvertical columnar or funnel-shaped body of intrusive magmatic rock.

Pitchstone   Massive, dark-colored glass that has a waxy luster in hand sample; contains 6–16 wt.% water absorbed at near-atmospheric conditions.

Plagiogranite   Phaneritic to granophyric rock composed mostly of sodic plagioclase, lesser quartz, and minor mafic minerals; originates by extreme differentiation of basaltic intrusions, such as in the oceanic crust; occurs in **ophiolite.**

Plastic   Behavior or response to applied stress in which strain is nonrecoverable (irreversible) at a stress above the **yield strength** of the material; below that stress no strain occurs.

Platinum group element (PGE)   Ru, Rh, Pd, Os, Ir, and Pt.

Plinian eruption   Highly explosive venting of volatile-rich, usually silicic magma that forms a steady turbulent **plinian plume.**

Plinian plume   See **Volcanic plume.**

Plug   See **Pipe.**

Pluton   An intrusion of magma that is not sheetlike.

Plutonic   Processes and products related to magma beneath the surface of the Earth.

Poikilitic   Magmatic texture in which larger crystals, called **oikocrysts,** enclose smaller, randomly oriented crystals.

Point defects   Imperfections in the atomic structure of a crystalline solid; allow for volume diffusion.

Polygenetic   Resulting from two or more processes, or having multiple sources, or originating at more than one time in more than one place.

Polymer   In a melt, chain of linked ions, usually Si and O.

Polymerization (degree of)   In a melt, the proportion of ions that form polymers; the ratio of nonbridging O to network-forming tetrahedrally coordinated cations, such as Si and Al; silica-rich melts are more polymerized than silica-poor melts.

Porosity   Relative proportion of openings in rock that can hold liquid.

Porous flow   Involves movement of liquid along grain boundaries or between connected pore spaces.

Porphyritic Inequigranular magmatic texture made of two grain sizes; larger crystals, commonly euhedral, called **phenocrysts,** embedded in a finer grained or glassy **matrix (groundmass).**

Porphyry Granitic rock that crystallizes from a crystal-bearing magma in a shallow intrusion as a result of reduction in water pressure; creates aphanitic porphyritic texture.

Ppb Parts per billion.

Ppm Parts per million.

Precision Reproducibility; a number that indicates how much statistical variation from the average or mean value occurs in replicate determinations.

Pressure Force acting over an area; has units of bars or pascals (Pa). **Confining pressure** (also referred to as *load* or *lithostatic pressure*) at some depth within the Earth denoted by *P* is produced by the weight of overlying rock or water; see also **Hydrostatic state of stress.**

Pressure ridge Elongate, commonly arcuate uplifts that are on the order of a meter in wavelength and amplitude formed on the surface of a moving lava flow.

Primary magma Unmodified magma leaving its source; has not experienced any overprinting differentiation.

Primitive mantle A hypothetical mantle composition before extraction of continental components but after segregation of the metallic core; used as a comparative reference in normalizing trace element data.

Primitive-mantle-normalized trace element diagram The concentration ratio of a trace element in a rock to that in primitive mantle plotted on the logarithmic *y* axis and ordered along the *x* axis according to decreasing incompatibility.

Primitive magma A magma that has been relatively unmodified by differentiation processes after leaving its source.

Bulk silicate Earth Nd and Sr isotopic ratios in chondritic Earth adjusted for the segregated metallic core; essentially mantle plus crust.

Principal planes Specially oriented orthogonal planes in a stressed body on which only **normal stresses** and no **shear stresses** prevail.

Principal stresses Three mutually perpendicular (orthogonal) **normal stresses** that act on **principal planes;** represent the total state of stress at a point in a body; in a general state of stress they are unequal and denoted by $\sigma_1 > \sigma_2 > \sigma_3$; all are compressive within the Earth.

Protoclastic Texture made of plagioclase crystals that were strained and locally broken during flow of partially crystallized magma.

Pseudotachylite Typically black glassy to aphanitic rock that is formed by frictional melting at high-strain rates along a fault.

Pumice Highly vesicular silicic glass.

Pumice flow A type of **pyroclastic flow** that consists of lapilli and/or blocks of pumice in a lesser matrix of ash.

Pumiceous Highly vesicular texture in silicic glass.

PV work See **Work.**

Pyroclast Fragment of volcanic material ejected explosively from a volcanic vent; are also called **ejecta** or **tephra.**

Pyroclastic Refers to processes and products related to explosive volcanic eruption.

Pyroclastic fabric Fragmental volcanic fabric; usually includes vitroclasts quenched from juvenile melt.

Pyroclastic fall (ash fall) A deposit of moderately to well-sorted bedded ash or lapilli settled out of a **volcanic plume.**

Pyroclastic flow Ground-hugging, fast-moving avalanche of hot ash, lapilli, and local blocks partly mobilized by gas; sometimes called an **ash flow** or a **nuée ardente** (French, "glowing cloud").

Pyroclastic surge Dilute concentration of solid particles in hot gas that moves across the ground at hurricanelike speed; a **base surge** is ring-shaped and moves radially outward from the base of a collapsing volcanic column.

Pyroxenite Ultramafic phaneritic rock composed of pyroxenes and <40% olivine; see Figure 2.10b.

Quartz diorite A **diorite** that contains, in terms of felsic minerals, >20 modal % quartz; see Figure 2.8.

Quartz tholeiite Basalt rock type that contains normative quartz and hypersthene.

Radiation A type of heat transfer by electromagnetic energy through a transparent medium; for example, solar heating.

Rapakivi Magmatic texture in which an alkali feldspar grain is rimmed by plagioclase.

Rapakivi granite **A-type** granite with **rapakivi** texture.

Rare earth element (REE) Coherent group of usually divalent ions comprising atomic numbers 57 to 71; can include Y; see Figure 2.20.

Rayleigh number A dimensionless ratio of buoyant driving force to viscous retarding force in thermal convection; larger ratios correspond to greater gravitational instability and convection.

Reaction pair Two minerals, one of which formed at the expense of the other, in a **reaction relation** with stably coexisting melt.

Reaction relation Chemical reaction in which crystals and melt combine to yield crystals of another composition and possibly a melt of a different composition.

Reaction rim Magmatic texture in which a more stable mineral or mineral assemblage surrounds an earlier formed, less stable mineral that was not completely resorbed in a **reaction relation;** the rim and rimmed mineral constitute a **reaction pair.**

Redox reaction Reduction-oxidation reaction.

REE See **Rare earth element.**

Relaxation "time" In a viscous melt, shear strain may accumulate faster than the time required for the atoms to relax into their equilibrium configuration, so that the melt breaks apart as if it were a solid.

Replacement A secondary mineral grain, or fine-grained aggregate of one or more minerals, that takes the place of a primary mineral grain formed under different conditions; also called **breakdown** or **decomposition.**

Replenishment   Refers to new draughts of commonly denser and usually hotter and more primitive magma introduced into a chamber of crystallizing magma.

Residual liquid (melt)   Silicate liquid (melt) that remains after some crystals have formed in a crystallizing magma.

Resorbed   Crystal dissolved into a melt.

Restite   Residual crystals or granular aggregates of the source rock remaining after partial melt generation that have been entrained into the ascending magma and preserved after its solidification.

Resurgent caldera   Created where a caldera floor is uplifted and domed some time after subsidence.

Resurgent boiling   See **Retrograde boiling.**

Retrograde boiling   Release of volatiles from an initially volatile-undersaturated melt caused by crystallization of volatile-free minerals in a cooling magma.

Reverse graded bedding   Particle size increases upward in a bed.

Reversible thermodynamic process   A change in a system that can be reversed by an infinitesimal change of an intensive variable.

Reynolds number   A dimensionless ratio of inertial to viscous forces used to characterize liquid flow as either laminar or turbulent; large ratios correspond to turbulent flow.

Rheology   The ways that rocks and magmas respond to applied stress by combinations of ideally elastic, plastic, and viscous behavior.

Rheomorphic   Refers to secondary flow of a mass of hot welded vitroclasts, such as happens where deposited on sloping ground.

Rhyolite   Leucocratic glassy or aphanitic rock made essentially of alkali feldspar, lesser quartz, and minor mafic minerals; contains generally >69 wt.% silica and >6 wt.% total alkalies; see Figures 2.12 and 2.13.

Rift   A linear zone of crustal extension; magma may ascend through subvertical extensional fractures striking parallel to the zone.

Ring dike   Subvertical cylindrical sheet intrusion; may be intruded along a ring fault.

Ring fault   Cylindrical subvertical fault bounding a subsided roof slab that forms a **caldera;** approximately follows the outline of the underlying evacuated magma chamber.

Ring-fracture stoping   Subsidence and engulfment of ring-fault-bounded roof block into underlying magma chamber.

Rock suite   Compositionally related kindred group of rock types, such as alkaline rocks.

Rock type   A rock that has a narrowly defined composition and a particular fabric, such as andesite.

Roof pendant   Erosional remnant of downward-projecting roof rock that is completely surrounded by plutonic rock.

Room problem   How was space in the crust provided for a magmatic intrusion?

Sanukitoid   Phaneritic Mg-rich dioritic rock; chemically similar to **boninite.**

Saturated   The state of a system in which the concentration of some component equals its solubility.

Schlieren   Anisotropic magmatic fabric defined by oriented, wispy, diffuse concentrations of mafic minerals in a more leucocratic matrix; common in granitic rocks.

Scoria   Highly vesicular mafic glass; there may be so many minute hematite grains in oxidized red-brown scoria that it does not appear glassy in hand sample; contrast **pumice.**

Scoria cone   See **Cinder cone.**

Scoriaceous   Texture of **scoria.**

Seamount   Submarine topographic prominence more than 100 m above seafloor; commonly formed by extrusions of lava.

Second boiling   See **Retrograde boiling.**

Second law of thermodynamics   Spontaneous natural processes even out the concentration of some form of energy; in an isolated system, spontaneous processes proceed in the direction of increasing entropy.

Seriate   Phaneritic, inequigranular magmatic texture in which grains range more or less continuously in size; contrast bimodal size distribution in **porphyritic** texture.

Shard   Usually ash-size fragment of glass; may be a broken piece of finely vesicular pumice or a vesicle wall.

Shear fracture   See **Fault.**

Shear stress   See **Stress.**

Sheet intrusion   Tabular magmatic body having very small thickness/length ratio (**aspect ratio**).

Sheeted dike complex   Subparallel dikes intruded into older dikes in an extending crust, such as at ocean spreading ridges.

Shield volcano   A low-aspect-ratio edifice having the shape of an overturned dinner plate or warrior's shield; built by innumerable extrusions of basaltic lava from a central vent complex and locally one or more fissure-rift systems.

Shoshonite   Andesitic rock that contains high concentrations of K; see Figures 2.12 and 2.18.

Sialic   Rock rich in Si and Al that contains abundant feldspar; used especially with reference to the continental crust.

Sidewall crystallization   Occurs in cooling bodies of magma at subvertical contacts as heat is transferred into the wall rock, creating a **thermal boundary layer.**

Silica saturation   Refers to the activity of SiO_2 or its concentration relative to other oxides in a rock. **Silica-oversaturated** rocks contain modal quartz or its polymorphs and normative quartz; corresponding melts have silica activity = 1. **Silica-saturated** rocks contain no modal quartz, feldspathoids, or olivine and no normative quartz, olivine, or nepheline. **Silica-undersaturated** rocks contain modal Mg-olivine and possibly feldspathoids, analcime, perovskite, melilite, normative olivine, and possibly normative nepheline; silica activity of melt <1.

Silicate liquid   A high-T solution of ions, mainly silicon and oxygen; also called **melt.**

Silicic   Refers to rocks, magmas, and melts that contain large concentrations of silica, manifested in rocks such as granite or rhyolite by an abundance of alkali feldspar, quartz, or its polymorphs, and glass rich in SiO_2.

Sill   A **sheet intrusion** that concordantly parallels planar structures in its host rocks; sometimes defined as a horizontal sheet intrusion.

Sill swarm   Many sills emplaced during an intrusive episode.

Solidus T below which any mixture of components in an equilibrium system consists only of crystalline solids.

Solubility   Maximum concentration of a particular component that can be uniformly dissolved in a melt or crystal under a particular P and T without the appearance of a separate stable phase made of that component.

Solution   A homogeneous mixture of two or more chemical components in which their concentrations may be freely varied within certain limits; may be solid, liquid, or gaseous.

Solvus   Locus of points in T- or P-space below which an initially homogeneous liquid or crystalline phase exsolves or unmixes into two or more different homogeneous liquids or crystalline phases.

Soret diffusion   Ions moving in a chemically homogeneous melt as a result of a thermal gradient.

Source rock   The rock that partially melts because of changes in P, T, or X; where magma is generated in the lower continental crust or upper mantle.

Source-rock fertility   Potential amount of components that are available in a source rock to yield a melt of a specific composition.

Spatter cone (mound)   Accumulation of welded, usually basaltic spatter around a lava fountain.

Specific heat   Ratio between the increment of heat leaving or entering a unit of mass to its incremental change in T, has units of joules per gram degrees; compare **Molar heat capacity**.

Specific volume   The reciprocal of density.

Spherulite   A spherical mass of fibrous or needlelike crystals in magmatic rocks that radiates from a central point; range in diameter from less than 1 mm to 1 m or more; most common in silicic aphanitic and glass rocks where radiating spray is composed of alkali feldspar and a polymorph of silica; compare **variole** and **ocelli** in mafic rocks.

Spherulitic   Texture that consists of radiating three dimensional sprays of needlelike minerals called **spherulites**; can form by **divitrification** or drastic undercooling of a melt.

Spinifex   Texture characterized by subparallel skeletal, platy, or bladed olivine and/or pyroxene grains; typically occurs in **komatiite**.

Stability diagram   See **Phase diagram.**

Stability field   Area in a stability (phase) diagram over which a phase or assemblage of phases is stable.

Standard state   A carefully defined reference state for a compound.

State   The nature of a system.

State properties (variables)   Parameters that define the nature of a system.

Stock   A pluton that is smaller than a **batholith**, commonly only a single intrusion; outcrop area generally <100 km².

Stokes's Law   Expresses the limiting or **Terminal velocity** of a solid particle moving in the gravitational field within a viscous medium.

Stoping   Process by which magma invades the brittle upper crust by engulfing pieces of country rock.

Strain   Distortion of a body, its change in shape and/or volume, under stress; one of three components of **deformation** in a stressed body.

Strain marker   Texturally or compositionally distinct body within a magma or rock that records the local strain.

Strain rate   The ratio of an increment of strain to the time during which it accumulates.

Stratovolcano   See **Composite volcano.**

Strength   Under specified conditions in a body, the **stress difference** required to produce permanent deformation.

Stress   Magnitude of a force divided by the area over which it is applied. **Normal stress** is perpendicular to the plane on which it acts. **Shear stress** is tangential (parallel) to the plane on which it acts.

Stress difference   Difference in magnitude between maximum and minimum applied stresses.

Strombolian eruption   Bursting of large gas bubbles in basaltic or andesitic magma near the top of a conduit; forms a **cinder** or **scoria cone.**

Structural order   In a melt, regularity in the array of atoms.

Structure   Features in a rock defined by aggregates of mineral grains seen at a scale of hand sample or larger, such as bedding; together with **texture** constitutes **fabric.**

S-type granite   Derived from metaluminous magmas generated by partially melting sedimentary source rocks that had high clay content.

Subalkaline rocks   A suite of rocks that have relatively low alkali-to-silica ratios; usually silica-saturated or silica-oversaturated; contain no normative nepheline.

Subhedral   Imperfectly formed compact grains only partly bounded by crystal faces.

Subsolvus   Refers to granitic rocks and syenites that have crystallized from relatively hydrous magmas so that a plagioclase and alkali feldspar crystallize directly from a melt; generally associated with biotite and/or hornblende.

Surface diffusion   Occurs essentially over a two-dimensional surface; for example, ions move about over the surface of a mineral grain through a static liquid in contact with it.

Surface free energy   Energy associated with a solid-solid or solid-liquid interface resulting from unbalanced ionic forces; is minimized at equilibrium.

Surface tension   Attractive force in a liquid body bounded by a gas where unbalanced atomic bonds on the surface of the liquid tend to pull it inward.

Surroundings   The part of the universe outside a system.

Syenite   Felsic phaneritic rock that contains, relative to felsic minerals, <20 modal % quartz and a high alkali-feldspar-to-plagioclase ratio; see Figure 2.8.

System   A part of the universe under study or discussion.

T   Temperature; measured in degrees Celsius (°C) or Kelvin (K); 0°C = 293K.

Tensile state of stress   Stresses acting in opposite directions to pull material apart; generally not realized beneath the surface of the Earth.

Tephra   See **Pyroclast.**

Terminal velocity   The limiting velocity of a solid particle sinking or rising in a liquid or gaseous medium after the buoyant driving force and retarding viscous drag force have been balanced.

Terrain   Refers to a topographic feature, such as mountainous terrain.

Terrane   A large, tectonically or compositionally distinct mass of rock, such as a terrane of accreted oceanic rocks along a continental margin overlying a subducting plate.

Ternary   Three components.

Texture   Size, shape, and mutual relations of mineral grains and proportions of crystals in a rock; together with **structure** constitute **fabric.**

Thermal conductivity   Proportionality constant between time rate of heat flow and the thermal gradient.

Thermal diffusion   Heat conduction.

Thermal diffusivity   Ability of a material to conduct heat relative to its accumulative capacity; ratio of thermal conductivity to the product of density and specific heat.

Thermal expansion (coefficient of)   Isobaric change in volume as T changes.

Thermal gradient   Difference in T between adjacent hotter and colder parts of a body divided by the distance between the parts.

Thermally activated   Refers to processes that are enhanced as a result of increased atomic mobility at elevated temperatures.

Thermochemical convection   Motion in a viscous body driven by contrasts in composition and T that create density contrasts sufficient to cause gravitational instability.

Thermodynamic potential energy   A thermodynamic energy function formulated in such a way that it is a minimum in a stable system of variable chemical composition.

Thermodynamic processes   Changes in a thermodynamic system from one equilibrium state to another.

Thermodynamics   A set of mathematical models and concepts that show interactions between forms of energy and how changes in T, P, and chemical composition affect the state of a system.

Thermomechanical feedback   As a viscous body is sheared, the work is transformed into heat, raising its T, which reduces the viscosity, localizing subsequent shear flow and heating, causing further reduction in viscosity, and so on.

Third law of thermodynamics   At absolute zero, where the Kelvin T is zero, crystals are perfectly ordered and all atoms are fixed in space so that there is only one possible distribution; **entropy** is zero.

Tholeiitic rocks   Subalkaline rocks that have relatively strong enrichment in Fe relative to Mg and little variation in silica; see Figures 2.17 and 12.22; compare **Calc-alkaline rocks.**

Tie line   Line in a phase diagram that connects points representing compositions of stably coexisting phases.

Tonalite   Intermediate composition, magmatic, phaneritic rock that contains, relative to felsic minerals, >20 modal % quartz and a plagioclase-to-alkali-feldspar ratio >9; see Figures 2.8 and 2.9.

Trace element   An element whose concentration is <1000 ppm (<0.1 wt.%).

Trachyandesite   Alkali-rich andesite; see Figure 2.12.

Trachybasalt   Alkali-rich basalt; see Figure 2.12.

Trachyte   Leucocratic glassy or aphanitic rock made essentially of alkali feldspar; contains about 60–69 wt.% silica and >6 wt.% total alkalies; see Figures 2.12 and 2.13.

Trachytic   Magmatic texture in an aphanitic rock expressed by a planar preferred orientation of tabular feldspars.

Transport phenomena   Movement of viscous material, atoms, or heat in a system.

Triple point   In a phase diagram, point of intersection of equilibrium boundary lines of three stability fields; all stable phases along these lines and in these fields coexist stably at the triple point.

Troctolite   A type of gabbro that has an olivine-to-pyroxene ratio >9; see Figure 2.10a.

Trondhjemite   Phaneritic, felsic, magmatic rock that consists of sodic plagioclase, quartz, and minor mafic minerals; essentially synonymous with **plagiogranite** and leucotonalite; see Figure 2.9.

TTG   Tonalite, trondhjemite, and granodiorite.

Tuff   Rock made of ash-size (<2-mm) volcaniclasts.

Tuff cone   Formed by **hydromagmatic** eruptions; intermediate in shape between a tuff ring and a cinder cone.

Tuff ring   Pyroclastic deposit in which the ratio of height to basal diameter <1:5; volume of crater space is larger than volume of ejecta; formed by hydromagmatic processes.

Turbulent flow   Rapid and chaotic motion of particles in small eddies superimposed on the overall flow of moving fluid; **Reynolds number** >4000.

Ultrabasic rock   Contains <45 wt.% silica.

Ultramafic rock   Especially rich in Mg and Fe; generally has little or no feldspar.

Undercooling   Decrease in T below some equilibrium point at which a phase becomes stable.

Underplating   Process of lodging mantle-derived magma at base of less dense continental crust; see **Density filter.**

Undersaturated   Concentration of a component in a phase less than its solubility at some P and T so that all of the component is dissolved homogeneously in the phase; also refers to the concentration of silica and alumina in magmas and rocks; see **silica saturation** and **alumina saturation index.**

Univariant   A state of equilibrium that has one degree of freedom.

Upper thermal stability limit   The maximum T at which a mineral or mineral assemblage is stable.

Vapor phase   Volatile fluid; usually refers to crystals precipitated from it (vapor-phase crystallization).

Variance   See **Degrees of freedom.**

Variation diagram   Graph that shows trends or patterns in compositional data.

Variole   Spherulitic aggregate of radially arrayed, needlelike crystals of plagioclase with or without clinopyroxene, also called **ocelli;** forms in mafic rock and magma, in contrast to a spherulite, which forms in silicic magma and rock.

Vesicle   A more or less spherical cavity in a magmatic rock formed by entrapment of a volatile fluid bubble during solidification of the magma; a **pipe vesicle** results from movement of the volatile bubble or the magma before solidification.

Vesicular   Texture made of **vesicles.**

Vesiculation   Creation of fluid bubbles in a volatile oversaturated melt by **exsolution.**

Viscosity   Resistance to flow; in a liquid, resistance results from immobility of atoms; in **Newtonian viscosity** the coefficient of viscosity, or the viscosity, is the ratio of applied shear stress to time rate of strain; has units of pascal seconds (Pa s).

Viscous   Behavior or response to applied stress that results in permanent strain no matter how small the stress; the viscous material has zero strength.

Vitric   See **Glassy.**

Vitroclast   Fragment of glass.

Vitrophyre   Texture in which large crystals (**phenocrysts**) lie in a glassy matrix.

Volatile fluid   Liquid phase that consists chiefly of **volatiles** and has a density generally <2 g/cm^3.

Volatiles   Water, carbon dioxide, sulfur, fluorine, chlorine, and so on, that are in gaseous form at near-atmospheric P but magmatic T.

Volcanic breccia   Rock made of predominantly block-size (>64 mm) volcaniclasts.

Volcanic chain   Line of volcanoes that have monotonically changing age, such as the Hawaiian chain formed as the oceanic lithosphere drifted over an essentially fixed mantle plume.

Volcanic debris flow   See **Lahar.**

Volcanic neck   A magma **pipe** that has fed a volcanic extrusion.

Volcanic plume   Mixture of pyroclasts and hot expanding gas, chiefly steam, that is discharged explosively from a vent into the atmosphere. **Plinian plume** is created by gas-rich magma that discharges turbulently from a relatively small-diameter vent and becomes buoyant as engulfed air is heated, causing convective ascent to tens of kilometers; neutrally buoyant umbrella broadcasts ejecta over large areas as ash fall. **Collapsing column** resembles a water fountain and is created by eruption of less volatile-rich magma from larger vents; discharge rate is so great that the plume contains more pyroclastic mass than can be lifted buoyantly and eruptive column collapses under its own weight, the resulting runout produces **pyroclastic flows. Coignimbrite plume** is a secondary plume produced as fine ash is flushed (elutriated) out of a pyroclastic flow by buoyantly rising gas.

Volcaniclast   Fragment of volcanic rock material produced by any process.

Volcaniclastic   Fabric made of **volcaniclasts;** also refers to any process by which they are produced.

Volume diffusion   Occurs within any single homogeneous phase.

Vug   Angular cavity in a rock; formed by collection of a volatile fluid between existing crystals.

Vuggy   Fabric made of **vugs.**

Vulcanian eruption   Complex explosive eruption of intermediate-composition magma that progresses from steamblast explosions ejecting ballistic blocks, to eruption of volcanic plumes that broadcast tephra over moderate distances, to extrusion of highly viscous lava from the vent.

Water-deficient (-undersaturated) melting   See **Dehydration melting.**

Welded tuff   Rock made mostly of vitric ash particles that are stuck together as a result of high T at emplacement; usually develops from an **Ash flow.**

Welding   Lithification process that consolidates loose vitroclasts hot enough to stick together when deposited.

Wetting angle   See **Dihedral angle.**

Whole-rock chemical composition   Weight concentrations of chemical species constituting a rock.

Work   Product of force times a displacement in the direction of the force. **PV work** is done in subvolcanic systems as high-P gas expands against its surroundings.

Xenocryst   Foreign crystal.

Xenolith   Foreign rock fragment.

Xenomorphic   See **Anhedral.**

Yield strength   The critical value of the **stress difference** at which a stressed plastic body permanently deforms.

Zoned ignimbrite   Systematic vertical and/or horizontal compositional variation through the sheetlike deposit.

Zoned intrusion (pluton)   Has more or less concentrically arrayed parts of contrasting composition and/or fabric.

Zoned crystal   A solid solution crystal that has a regular pattern of compositional variation (zoning).

INDEX